New Interpretations of Ape and Human Ancestry

ADVANCES IN PRIMATOLOGY

A Continuation Order Plan is available for this series. A continuation order will bring
delivery of each new volume immediately upon publication. Volumes are billed only upon
actual shipment. For further information please contact the publisher.

New Interpretations of Ape and Human Ancestry

Edited by
RUSSELL L. CIOCHON
University of California
Berkeley, California

and
ROBERT S. CORRUCCINI
Southern Illinois University
Carbondale, Illinois

PLENUM PRESS • NEW YORK AND LONDON

Library of Congress Cataloging in Publication Data

Main entry under title:

New interpretations of ape and human ancestry.

   (Advances in primatology)
   Includes bibliographical references and indexes.
   1. Primates, Fossil—Addresses, essays, lectures. 2. Human evolution—Addresses, essays, lectures. I. Ciochon, Russell L. II. Corruccini, Robert S. III. Series.
QE882.P7N48 1983   569'.8   83-2175
ISBN 0-306-41072-9

© 1983 Plenum Press, New York
A Division of Plenum Publishing Corporation
233 Spring Street, New York, N.Y. 10013

Printed in the United States of America

Dedicated To

SHERRY WASHBURN
Whose Ideas Are Embodied
In Many Chapters Of This Volume

and

CLARK HOWELL
In Recognition Of His Considerable
Contributions To Paleoanthropology

In Memory of
G. H. R. von KOENIGSWALD
A legendary figure in the field of paleoanthropology

G. H. R. von Koenigswald
1902-1982

Contributors

Peter J. Andrews
  Department of Paleontology
  British Museum (Natural History)
  Cromwell Road
  London SW7 5BD, England

Marietta L. Baba
  Department of Anthropology
  College of Liberal Arts
  Wayne State University
  Detroit, Michigan 48201

Raymond L. Bernor
  College of Medicine
  Department of Anatomy
  Laboratory of Paleobiology
  Howard University
  Washington, D.C. 20059

Noel T. Boaz
  Department of Anthropology
  New York University
  New York, New York 10003

S. R. K. Chopra
  Department of Anthropology
  Panjab University
  Chandigarh-160014, India

Russell L. Ciochon
  Department of Paleontology
  (temporary research appointment)
  University of California
  Berkeley, California 94720

Robert S. Corruccini
  Department of Anthropology
  Southern Illinois University
  Carbondale, Illinois 62901

John E. Cronin
  Departments of Anthropology and
  Biology
  Peabody Museum
  Harvard University
  Cambridge, Massachusetts 02138

Linda L. Darga
  Department of Anatomy
  School of Medicine
  Wayne State University
  Detroit, Michigan 48201

Louis de Bonis
  Laboratoire de Paléontologie des Ver-
  tébrés et Paléontologie Humaine
  Université de Poitiers
  Faculté des Sciences
  86022 Poitiers, France

Richard Dehm
  Institut für Paläontologie und historische Geologie
  Universität München
  8000 München 2, West Germany

Dean Falk
  Department of Anatomy and Caribbean
  Primate Research Center
  University of Puerto Rico
  Medical Science Campus
  San Juan, Puerto Rico 00936

John G. Fleagle
  Department of Anatomical Sciences
  Health Sciences Center
  State University of New York
  Stony Brook, New York 11794

David G. Gantt
  Institute of Dental Research
  University of Alabama School of
    Dentistry
  University Station
  Birmingham, Alabama 35294

Morris Goodman
  Department of Anatomy
  School of Medicine
  Wayne State University
  Detroit, Michigan 48201

Leonard O. Greenfield
  Department of Anthropology
  Temple University
  Philadelphia, Pennsylvania 19122

Donald C. Johanson
  Institute of Human Origins
  2700 Bancroft Way
  Berkeley, California 94704
  and Laboratory of Physical
    Anthropology
  Cleveland Museum of Natural History
  Cleveland, Ohio 44106

Richard F. Kay
  Department of Anatomy
  Duke University Medical Center
  Durham, North Carolina 27710

William H. Kimbel
  Laboratory of Physical Anthropology
  Cleveland Museum of Natural History
  Cleveland, Ohio 44106
  and Departments of Anthropology and
    Biology
  Kent State University
  Kent, Ohio 44240

Arnold G. Kluge
  Museum of Zoology and Department of
    Ecology and Evolutionary Biology
  The University of Michigan
  Ann Arbor, Michigan 48109

Adriaan Kortlandt
  Vakgroep Psychologie en Ethologie der
    Dieren
  Universiteit van Amsterdam
  Nieuwe Achtergracht 127
  1018 WS Amsterdam, The Netherlands

Jerold M. Lowenstein
  Department of Medicine (122 MR2)
  University of California
  San Francisco, California 94143

Larry L. Mai
  Division of Orthopaedic Surgery
  UCLA School of Medicine
  and Department of Anthropology
  University of California
  Los Angeles, California 90024

Henry M. McHenry
  Department of Anthropology
  University of California
  Davis, California 95616

Mary Ellen Morbeck
  Department of Anthropology
  University of Arizona
  Tucson, Arizona 85721

Martin Pickford
  National Museum of Kenya
  P.O. Box 40658
  Nairobi, Kenya

David R. Pilbeam
  Department of Anthropology
  Harvard University
  Cambridge, Massachusetts 02138

K. N. Prasad
  Geological Survey of India
  4-3-542, Bogulkunta
  Tilak Road, Alladin Building
  Hyderabad-500001, India

Michael D. Rose
  Department of Anatomy
  New Jersey Medical School
  University of Medicine and Dentis-
    try of New Jersey
  100 Bergen St.
  Newark, New Jersey 07103

Vincent M. Sarich
  Departments of Anthropology and
    Biochemistry
  University of California
  Berkeley, California 94720

Elwyn L. Simons
Duke University Center for the Study
of Primate Biology and History
Durham, North Carolina 27705

G. H. R. von Koenigswald
Section of Paleoanthropology
Senckenberg Museum
6000 Frankfurt 1, West Germany

Alan C. Walker
Department of Cell Biology and
Anatomy
The Johns Hopkins University
School of Medicine
Baltimore, Maryland 21205

Steven C. Ward
Department of Anthropology
Kent State University
Kent, Ohio 44242
and Department of Anatomy
Northeast Ohio Universities College of
Medicine
Rootstown, Ohio 44272

Tim D. White
Department of Anthropology
University of California
Berkeley, California 94720

Milford H. Wolpoff
Department of Anthropology
University of Michigan
Ann Arbor, Michigan 48109

Adrienne L. Zihlman
Department of Anthropology
University of California
Santa Cruz, California 95064

Preface

In the field of paleoanthropology there is perhaps no more compelling issue than the origin of the human lineage. From the 19th-Century proclamations of Darwin, Huxley, and Haeckel up to the present-day announcements of authorities such as Pilbeam, Johanson, and Leakey, controversy and debate have always surrounded the search for our earliest ancestors. Most authorities now agree that many important questions concerning ape and human ancestry will be solved by further investigation of the hominoid primates of the Miocene and Pliocene, and through exacting comparative anatomical and biomolecular analyses of their living representatives. Indeed, these studies will yield a better understanding of (1) the cladogenesis (branching order) of the hominoid primates, (2) the morphotype (structural components) of the last common ancestor of humans and the living apes, (3) the timing and geographical placement of the hominid-pongid (human-ape) divergence and (4) the adaptive nature and probable scenario for the initial differentiation of hominids from pongids.

In this context we organized a symposium entitled "Miocene Hominoids and New Interpretations of Ape and Human Ancestry" which met July 2–4, 1980, in Florence, Italy, just prior to the convening of the VIII Congress of the International Primatological Society. This Pre-Congress Symposium was attended by a small group of anthropologists, anatomists, biochemists, ecologists, and paleontologists, all dedicated to a reanalysis of the evolutionary position and a more parsimonious interpretation of the Miocene hominoids *vis à vis* modern apes and humans. The symposium was held in the 18th-Century Tribuna di Galileo of the Istituto di Zoologia. Under the arched marble columns of this magnificent chamber, symposium participants were able to address specifically some or all of the points mentioned above. From this symposium a consensus was reached concerning a rapprochement between the formerly irreconcilable biomolecular view of human origins and the paleontological view. This rapprochement was based on a more liberal presentation of the "molecular clock hypothesis" and on the view that *no* known Miocene hominoid including *Ramapithecus* could be considered a hominid, that is, an exclusively human ancestor.

xiii

The Pre-Congress Symposium held in Florence was a rather small gathering, representing less than half of the contributors to this volume. Yet we feel that events at this symposium in many ways foreshadowed developments and trends in the field of paleoanthropology still underway. With the symposium behind us we began in earnest to assemble contributions for the present volume. About this time a number of additional leading authorities, who had not been able to attend the symposium, agreed to contribute chapters to the volume. We also decided to enlarge the scope of the volume to include contributions on Oligocene hominoid ancestors and on Plio-Pleistocene hominoid and hominid descendants. The central theme of the volume, however, remained the Hominoidea of the Miocene.

The 30 chapters that follow approach the subject of ape and human ancestry from a wide variety of perspectives. Some of the topics covered include: biomolecular studies, craniofacial anatomy, cladistic methodologies, dental anatomy and histology, dental metrics, dietary adaptations, embryology, geochronology, karyology, paleoecology, paleomammalogy, paleoneurology, parasitology, postcranial anatomy and locomotor adaptations, postcranial morphometrics, Popperian historicoscientific analyses, traditional descriptive analyses, and zoogeography. Whether or not our understanding of ape and human ancestry is further elucidated by this broad treatment is left for the reader to decide. Certainly though, the sheer quantity of data presented in this volume should be of use to researchers for many years to come.

Chapters 1 and 2 provide an introductory background to the remaining chapters of the volume. As the authors of Chapter 1, we attempt to present an even-handed, slightly historical overview of hominoid phyletics without trying to anticipate the many new perspectives proposed by contributors to this volume. Chapter 2, on the other hand, provides an in-depth review and synthesis of the geochronology and zoogeographic relationships of Miocene hominoids, which serves as a useful backdrop to many of the succeeding chapters. Finally in Chapter 30, RCL presents a summary viewpoint of ape and human ancestry based primarily on the contributions to this volume. In the time since this volume went to press new specimens have been recognized from several Eurasian and East African localities, which may already force revision of some conclusions presented herein. Clearly, at the present time, the field of hominoid evolution is so rapidly changing that no volume can stand as a complete summary for very long. Our hope is that each new volume and each new major revision will take the field one step closer to a new paradigm of ape and human ancestry that will better resist attempts to falsify it.

R. L. Ciochon
R. S. Corruccini

Acknowledgments

Most cooperative scientific endeavors such as an edited volume owe their beginning in concept to a single idea or suggestion of one individual. In this regard, we express our sincere thanks to W. P. Luckett, who suggested the concept for this volume to RLC one cold December evening in 1978, while we waited at the Torino Airport. That suggestion grew into a 30-chapter volume with 37 contributors. In the process a number of individuals provided invaluable scientific stimuli. We acknowledge R. L. Bernor, E. Delson, J. G. Fleagle, R. F. Kay, and T. D. White for this role. Eric Delson deserves a special note of thanks for suggesting several contributions to this volume, which otherwise would not have been included, and for putting up with innumerable queries concerning bibliographic and other details. Reviewing our graduate years, we would like to thank our major professors, F. Clark Howell and S. L. Washburn, for introducing us in the early 1970s to the field of Miocene hominoid evolution, therefore providing the conceptual framework for this volume. We also wish to thank the following students: Paula Atchison, Al Foster, Enita Mullens, Joy Myers, and Donna Ryan, for invaluable technical and editorial assistance. Ms. Myers, now in the graduate program at Kent State University, deserves special mention for having assisted at *every* stage during the preparation of this volume and for single-handedly compiling all of the indexes. J. L. Simpson, G. M. Addington, and K. W. Culpepper of the Cartographic Laboratory, Department of Geography and Earth Sciences, at the University of North Carolina at Charlotte are gratefully acknowledged for preparation of much of the artwork and photography in this volume. We also acknowledge the assistance of other staff members at UNCC, such as G. Ballard, D. Carter, and A. Salvo, and similarly thank T. Thomas from the Department of Anthropology at Southern Illinois University. For providing photographs used in the jacket illustration we thank P. J. Andrews and the British Museum of Natural History, R. F. Kay and the Duke Primate Center, W. H. Kimbel and the Cleveland Museum of Natural History, D. R. Pilbeam, W. Sacco and Harvard University, and A. C. Walker and the National Museum of Kenya. The planning and production of the volume could not have taken place without the continued support and assistance of Kirk Jensen, Victoria Craven, and

Julliana Newell Ayoub at Plenum. On behalf of all the contributors we whole-heartedly thank them. With regard to the Pre-Congress Symposium held in Florence, Italy, we thank A. B. Chiarelli, Director of the VIII Congress of the International Primatological Society, for his organizational efforts, and B. Lanza, Director of the Istituto di Zoologia in Florence, for making the Tribuna di Galileo available for our symposium. Financial support for symposium participants was provided by an L. S. B. Leakey Foundation grant to RLC, a NATO Fellowship to RSC, and additional funds from the National Science Foundation International Travel Support Program and the UNCC International Studies Program. We also gratefully acknowledge the L. S. B. Leakey Foundation for granting additional research funds utilized throughout this volume's preparation. Finally we thank Francine Berkowitz of the Smithsonian Foreign Currency Program for providing both moral and financial support which greatly contributed to the completion of this work.

RLC
RSC

Contents

V.   EVIDENCE FROM PALEOENVIRONMENTAL STUDIES

VI.   DESCRIPTIVE ANALYSES OF SIWALIK MIOCENE HOMINOIDS

VIII.   CONCLUDING REMARKS AND SUMMARY COMMENTS

Guide to Specimen Abbreviations

The following is a list of alphabetic abbreviations of the institutional and/or locality designations of fossil specimens discussed in the chapters of this volume. Since no key to specimen abbreviations is systematically presented in each chapter, this master list has been drawn up to aid readers. These same abbreviations are used in the index to specimens which appears at the end of this volume.

A.L.	Afar Locality, Hadar, Afar Triangle, Northeastern Ethiopia
AMNH	American Museum of Natural History, New York, New York, U.S.A.
BMNH M	British Museum (Natural History), Mammal Division, London, United Kingdom
BP	Bursa, Paşalar Locality, Northwest Anatolia, Turkey
BSPhG	Bayerische Staatssammlung für Paläontologie und historische Geologie, Munich, West Germany
CGM	Cairo Geological Museum, Geological Surgey of Egypt, Cairo, Egypt
CYP	Chandigarh-Yale Project, Panjab, University, Chandigarh, India
DPC	Duke Primate Center, Duke University, Durham, North Carolina, U.S.A.
GSI	Geological Survey of India, Calcutta, India
GSI D-	Geological Survey of India, D Series, Calcutta, India
GSP	Geological Survey of Pakistan, Quetta, Pakistan
KA	Kromdraai, Northern Transvaal, Republic of South Africa
KNM-	Kenya National Museum, Nairobi, Kenya
BC	Baringo, Chemeron
BN	Baringo, Ngorora
CA	Chamtwara Member, Koru

CH	Chesowanja
ER	East Rudolf
FT	Fort Ternan
KA	Karungu
KO	Koru
KP	Kanapoi
LT	Lothagam
MB	Maboko Island
ME	Meswa Bridge
MJ	Majiwa
MW	Mfwangano Island
RU	Rusinga Island
SO	Songhor
X	Unknown
L.H.-	Laetoli Hominid -, Laetolil Beds, Northern Tanzania
MCZ	Museum of Comparative Zoology, Harvard University, Cambridge, Massachusetts, U.S.A.
MLD	Makapansgat Limeworks Dump, Northern Transvaal, Republic of South Africa
MTA	Maden Tetkik ve Arama Enstitüsü (Mineral Research and Exploration Institute), Ankara, Turkey
O.H.	Olduvai Hominid, Olduvai Gorge, Northern Tanzania
Omo	Omo Locality, Shungura Formation, Lower Omo Basin, Southern Ethiopia
ONGC	Oil and Natural Gas Commission, Dehra Dun, India
PA	Paleoanthropology Collection, Institute of Vertebrate Paleontology and Paleoanthropology, Academia Sinica, Beijing, China
PUA	Panjab University, Anthropology Department, Chandigarh, India
RPl	Ravin de la Pluie (Rain Ravine), Lower Macedonia, Greece
Rud	Rudabánya, Northeastern Hungary
SK	Swartkrans, Northern Transvaal, Republic of South Africa
SNM	Staatliches Museum für Naturkunde, Stuttgart, West Germany
STS	Sterkfontein (Type Site) Northern Transvaal, Republic of South Africa
STW	Sterkfontein (West Pit) Northern Transvaal, Republic of South Africa
TM	Transvaal Museum, Pretoria, Republic of South Africa
UMP	Uganda Museum, Primate Collection, Kampala, Uganda
USNM	United States National Museum, Smithsonian Institution, Washington, D.C., U.S.A.
YPM	Yale Peabody Museum, Yale University, New Haven, Connecticut, U.S.A.

Paleontological and Geological Background

I

Overview of Ape and Human Ancestry

1

Phyletic Relationships of Miocene and Later Hominoidea

R. S. CORRUCCINI AND
R. L. CIOCHON

Introduction

With increasing numbers of East African discoveries over the last 10 years, the hominid fossil record has become more complete for the upper Pliocene through Pleistocene time span. The greater part of the earlier Pliocene remains devoid of definite hominids, however, and one major reason for interest in descriptions of early hominids such as *Australopithecus afarensis* (Johanson and White, 1979) is the inference backward in time toward the possible appearance of the earliest hominid ancestor. In fact, there is no more challenging issue in anthropology than the origin of the hominid lineage. This particular ethnocentric emphasis began with the inception of the age of evolutionary reasoning. Darwin in his major writings inferred that an ancient anthropomorphous subgroup gave rise to humans, and he made it clear that the resemblance of humans to the anthropoid apes in so many respects could only be due to homologous characters shared with a common ancestor. He

R. S. CORRUCCINI • Department of Anthropology, Southern Illinois University, Carbondale, Illinois 62901.   R. L. CIOCHON • Department of Paleontology, University of California, Berkeley, California 94720.
Preparation of this chapter was supported in part by a grant to RLC from the L. S. B. Leakey Foundation and by a NATO Fellowship to RSC.

also emphasized that, although the early progenitors of humans would have been properly designated as catarrhines, we must not fall into the error of supposing that the last common ancestor of apes and humans necessarily closely resembled any existing ape or monkey:

> We are naturally led to enquire where was the birthplace of man at that stage of descent when our progenitors diverged from the Catarhine stock? . . . In each great region of the world the living mammals are closely related to the extinct species of the same region. It is therefore probable that Africa was formerly inhabited by extinct apes closely allied to the gorilla and chimpanzee; and as these two species are now man's nearest allies, it is somewhat more probable that our early progenitors lived on the African continent than elsewhere. But it is useless to speculate on this subject; for an ape nearly as large as a man, namely the Dryopithecus of Lartet, which was closely allied to the anthropomorphous Hylobates, existed in Europe during the Upper Miocene period; and since so remote a period the earth has certainly undergone many great revolutions, and there has been ample time for migration on the largest scale.
>
> At the period and place, whenever and wherever it may have been when man first lost his hairy covering, he probably inhabited a hot country; and this would have been favourable for a frugiferous diet on which, judging from analogy, he subsisted. We are far from knowing how long ago it was when man first diverged from the Catarhine stock; but this may have occurred at an epoch as remote as the Eocene period; for the higher apes had diverged from the lower apes as early as the Upper Miocene period, as shewn by the existence of the Dryopithecus. . . .
>
> The great break in the organic chain between man and his nearest allies, which cannot be bridged over by any extinct or living species, has often been advanced as a grave objection to the belief that man is descended from some lower form; but this objection will not appear as of much weight to those who, convinced by general reasons, believe in the general principle of evolution. Breaks incessantly occur in all parts of the series, some being wide, sharp and defined, others less so in varying degrees . . . but all these breaks depend merely on the number of related forms which have become extinct. . . .
>
> With respect to the absence of fossil remains, serving to connect man with his ape-like progenitors, no one will lay much stress on this fact, who will read Sir C. Lyell's discussion, in which he shews that in all the vertebrate classes the discovery of fossil remains has been an extremely slow and fortuitous process. Nor should it be forgotten that those regions which are the most likely to afford remains connecting man with some extinct ape-like creature, have not as yet been searched by geologists.
>
> (Darwin, 1871, pp. 199–201)

Subsequent to Darwin's time, fossil discoveries and advances in comparative morphology have largely narrowed the search for the progenitors of humans to the dryopithecines and related hominoids common in the Miocene epoch. It is widely, though not universally, believed that study of this group will yield better understanding of the nature of the last common ancestor of apes and humans, the time and place of hominid–pongid cladogenesis, and the adaptive nature and reason for initial differentiation of hominids from pongids.

Recent Fossil Discoveries and Interpretations

A great number of new fossil hominoid primate specimens have been discovered in recent years, expanding not only the sample size but the number of different anatomical areas known, and also expanding the geographic range of Miocene primates. The discoveries have been accompanied by various taxonomic schemes, all diverging in some respect from the last major revision of the dryopithecines (Simons and Pilbeam, 1965; see also Andrews, 1978a). While disagreement exists over the exact relationships of the new fossils, paleontologists and paleoanthropologists agree at least that the internal and external affinities of this group are fundamentally important to theories of the origin of humans and of extant apes species (see Figs. 1–4).

The first dryopithecine specimen, *Dryopithecus fontani*, was discovered 127 years ago in France. This species is known by specimens from the type locality, all of which are lower jaws. Little postcranial and no upper dental material has subsequently been recovered. According to Andrews (1978b), these shortcomings have had lasting repercussions on the diagnosis of other dryopithecine taxa because the African *Proconsul* and Eurasian *Sivapithecus* groups are best distinguished on the basis of upper teeth, while all these taxa differ but little in lower jaws. Nevertheless, it is becoming increasingly recognized that the consideration of these groups as mere subgenera of *Dryopithecus* does some injustice to their true taxonomic distinctness.

Many important new specimens of the African hominoids *Proconsul* and *Limnopithecus* have been described (Fleagle, 1975; Andrews and Walker, 1976; Andrews, 1978a; Martin, 1981; Andrews *et al.*, 1981a) and new genera, such as *Dendropithecus* (Andrews and Simons, 1976) and *Micropithecus* (Fleagle and Simons, 1978a) proposed. The smaller of these African genera are similar in many ways to gibbons (Simons and Fleagle, 1973), though these resemblences are probably due to the allometric affects of small size and the retention of symplesiomorphies (Ciochon and Corruccini, 1977; Harrison, 1982). The consideration of *Proconsul* as a subgenus of *Dryopithecus* (Simons and Pilbeam, 1965), common to our discipline for the last 15 years, is now increasingly rejected in favor of separate generic status based on postcranial differences and the pronounced molar cingula of *Proconsul*. The three early recognized species of *Proconsul* differ primarily in size (Le Gros Clark and Leakey, 1951; Pilbeam, 1969); nevertheless, historically there has been a long-standing attempt to assign these separate species to separate ancestral status for extant chimpanzees and gorillas (Simons and Pilbeam, 1965; Pilbeam, 1969; Simons, 1972) (see Fig. 1).

Many recent discoveries of Eurasian hominoids (e.g., Tekkaya, 1974; Chopra, 1978; Pilbeam *et al.*, 1977, 1980) can probably be assigned to *Sivapithecus*. Some taxa (larger in size than the previous species) had been recovered earlier but remained among collections now held in Europe which, until recently, were largely undescribed (see von Koenigswald, this volume, Chapter 19; Dehm, this volume, Chapter 20). Similarly, the taxa *Ankarapithecus meteai* and *Ouranopithecus macedoniensis* (Ozansoy, 1957; de Bonis *et al.*, 1974; de Bonis and Melentis, 1977) from Turkey and Greece should probably be considered as new species of *Sivapithecus* larger than the previously diagnosed Siwalik species (Andrews and Tekkaya, 1980). Another intriguing specimen is a crushed mandible from Greece named *Graecopithecus freybergi* (von Koenigswald, 1972), which has been considered a European representative of

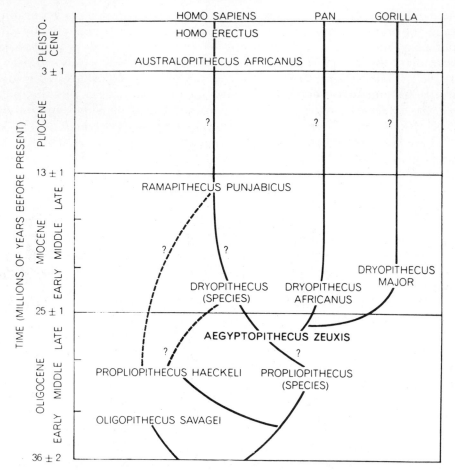

Fig. 1. Miocene hominoid evolutionary scheme according to Simons (1967). Note that *Pan, Gorilla,* and *Homo* are all derived independently from separate early Miocene hominoid taxa. This phylogenetic viewpoint was later reiterated by Pilbeam (1969). [Figure is reproduced from Simons (1967) and is courtesy of *Scientific American.*]

Ramapithecus (Simons, 1977). Additionally, new discoveries from Hungary have been referred to two new species, *Bodvapithecus altipalatus* and *Rudapithecus hungaricus* (Kretzoi, 1975). A previously known, but poorly represented *Sivapithecus* species, *S. darwini,* was recently assigned to new fossils from the Paşalar locality in Turkey (Andrews and Tobien, 1977), based on analogy with type specimens from central Europe; this may be the most phyletically primitive species of *Sivapithecus,* as it lies both temporally and morphologically intermediate between *Proconsul* and later Miocene Eurasian *Sivapithecus.* The new more complete Chinese discoveries (Xu and Lu, 1979, 1980; Wu, 1981) will also add considerably to the picture of Miocene hominoid evolution in Asia, as will the new skull described by Pilbeam (1982).

The enigmatic giant primate *Gigantopithecus* (von Koenigswald, 1935) has also been represented in the new collections from India (Simons and Chopra, 1969*a,b*), China (Jia, 1980), and Pakistan (Pilbeam *et al.*, 1977, 1980). While early theories that this primate represents a hominid are now largely discredited, *Gigantopithecus* continues to inspire great interest. Andrews believes he can document an evolutionary lineage encompassing *Proconsul major* → *Sivapithecus darwini* → *S. meteai* → *Gigantopithecus bilaspurensis* → *G. blacki* extending from the early Miocene to the Pleistocene (see Fig. 2); "this is the best documented fossil ape lineage" (Andrews, 1978*b*, p. 49). However, each step in this lineage will be disputed by one paleontologist or another.

Recent developments which have inspired more interest than all others concern the putative basal hominid, *Ramapithecus*. The sample of specimens of this controversial genus has rapidly expanded during the last two decades. In 1965, Simons and Pilbeam proposed that the Indian species *Ramapithecus punjabicus,* was actually the senior synonym of *R. brevirostris* also from the Siwaliks (Lewis, 1934), but this assignment of specimens is still disputed by G. E. Lewis and G. H. R. von Koenigswald (personal communication). The Turkish material from Paşalar and Çandir has been synonymized with the African *R. wickeri* sample by Andrews and Tobien (1977), who consider the Paşalar material to be the earliest and most primitive *Ramapithecus,* barely distinguishable from *Sivapithecus.* The Çandir mandible (Andrews and Tekkaya, 1976) is of some historical importance because it represents the first evidence for mandibular arcade shape that did not result from extensive reconstruction efforts (e.g., Walker and Andrews, 1973; Simons, 1961, 1964, 1968, 1977). New nearly complete mandibles assignable to *Ramapithecus punjabicus* from the Siwaliks of Pakistan (Pilbeam *et al.*, 1977; Pilbeam, 1978) resemble the Çandir mandible with reference to their shared nonhuman-like arcade shape.

Hominid-like traits of *Ramapithecus* affirmed by new discoveries include an abbreviated anterior portion of the arcade, robust and thickened mandible, buttressed symphysis, short face, and flat-wearing postcanine teeth with thick enamel. Less certain as hominid-like are the canine–premolar honing relationship, molar size progression, trigonid/talonid relationship, molar wear gradient, and buccal cingula (Corruccini, 1977; Frayer, 1978; Pilbeam, 1978; Greenfield, 1974, 1975, 1978, 1979). The microstructure of dental enamel was once thought hominid-like in *Ramapithecus* and *Sivapithecus* (Gantt *et al.*, 1977), but no longer (Gantt, this volume, Chapter 10). Altogether, the new material has not served to clarify the controversial affinities of *Ramapithecus* but rather has heightened disagreement (see Fig. 3). There is some consensus that *Ramapithecus* is the sister group of *Sivapithecus* and that it shares a functional complex (relating to heavy chewing in the mesial molar region) with early hominids. Other views proposed (for example, see Greenfield, 1979; Kay, 1982) suggest that *Ramapithecus* and *Sivapithecus* should no longer be considered generically distinct. Beyond this, the consensus position of the late 1960s that *Ramapithecus* was the basal hominid, ancestral to australopithecines, has greatly diminished (e.g., Fig. 4).

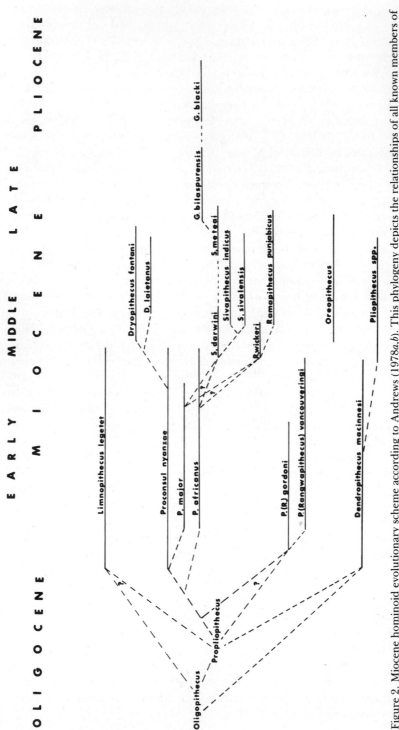

Figure 2. Miocene hominoid evolutionary scheme according to Andrews (1978a,b). This phylogeny depicts the relationships of all known members of the Dryopithecinae and indicates a possible origin to the *Ramapithecus* lineage. Note the absence of ancestor–descendant relationships linking the Miocene hominoids with extant hominids or pongids. [Figure is reproduced from Andrews (1978a) and is courtesy of the British Museum (Natural History).]

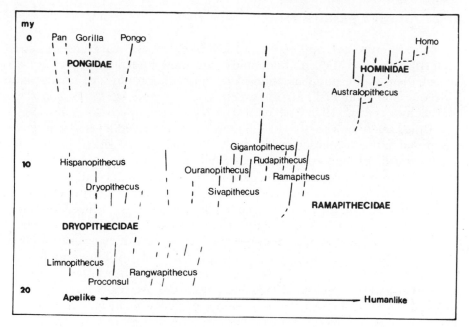

Fig. 3. Miocene hominoid evolutionary scheme according to Pilbeam (1979). Note that *four* distinct families of Miocene to Recent Hominoidea are depicted along an "ape–human continuum." This scheme recognizes a greater diversity at the familial level than had previously been proposed. The absence of hypothetical evolutionary connections between the four families is noteworthy. [Figure is reproduced from Pilbeam (1979) and is courtesy of Annual Reviews Inc.]

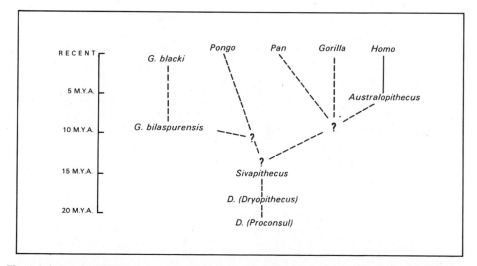

Fig. 4. Miocene hominoid evolutionary scheme according to Greenfield (1980). This phylogeny is based in part on biomolecular data indicating a relatively recent cladogenesis of the Hominoidea and on Greenfield's views regarding the synonomy of *Ramapithecus* with *Sivapithecus* (Greenfield, 1979; also this volume, Chapter 27). Note that *Sivapithecus* in this evolutionary scheme would be the last common ancestor of the extant Great Apes and humans. [Figure is redrawn from Greenfield (1980) and is courtesy of Alan R. Liss, Inc.]

Other Recent Considerations

In addition to description of new fossil finds, recent years have witnessed reinterpretation of Miocene hominoids from new interpretive perspectives. One of these is the popularization of cladistic philosophies of classification (Eldredge and Tattersall, 1975; Delson and Andrews, 1975; Delson *et al.*, 1977; Schwartz *et al.*, 1978; Wiley, 1981). Cladists attempt to assign taxonomic relatedness solely on the basis of uniquely shared derived (specialized) traits, as opposed to the "evolutionary" method of classification, which assigns value to character divergence and also occasionally recognizes the value of "wastebasket" horizontal taxonomic categories. These cladistic classifications have revised nomenclature, but not all workers are convinced the revisions entail equally far-reaching advances in understanding of relationships. Nevertheless, the cladistic process necessitates definition of character morphoclines and determination of ancestral conditions, and as such, if rigorously applied, it has much to offer to evolutionary interpretation of fossil hominoids.

Another rapidly increasing source of data involves functional analysis of postcranial fossils. Postcranial fragments lend themselves to functional, especially locomotor, interpretations, leading directly to conclusions about the basic adaptation of the concerned animals and thus their phylogeny. Indeed, it has been argued that the major hominoid specialization lies not in the teeth or diet but in postcranial adaptations to a "brachiating" or terminal branch hanging-and-feeding specialization that was central to diversification of extant large apes from generalized predecessors (Sarich, 1971; Washburn, 1972; Preuschoft, 1973; Morbeck, 1975, 1976; O'Connor, 1976; Corruccini *et al.*, 1976). According to this interpretation, Miocene hominoids were considered merely "dental apes." However, new material assigned to *Sivapithecus* from Pakistan and Hungary indicates that these earlier interpretations (based only on material representing *Proconsul*) may be incorrect. Detection of the beginning of increases in wrist, elbow, and shoulder mobility in *Sivapithecus* postcrania (Morbeck, 1979; McHenry *et al.*, 1980) is of central importance to theories about the emergence of hominoids that could be ancestral to the extant Great Apes. In this respect, the few fragments of Miocene postcranial fossils are receiving an inordinate amount of attention.

Ecological and paleogeographical considerations are also increasing in importance for interpretation of Miocene hominoid evolution (Kortlandt, 1974; Andrews and Van Couvering, 1975; Andrews *et al.*, 1981*b*; Pickford, 1981). Of interest are possible interpretations of niche specialization among the continental groups of dryopithecines, the latest time at which intercontinental migration could have occurred of forest-specialized hominoids across the spreading grasslands of the Pliocene, and possible dietary specialization of individual species based on dental features of occlusal significance (Kay, 1977*a*,*b*, 1981).

Considered from the other (earlier rather than later) temporal perspective, significant advances in our understanding of the origin of Miocene primates have resulted from current investigations in Oligocene deposits of

Egypt's Fayum (Simons *et al.*, 1978; Fleagle and Simons, 1978*b*). The dental and postcranial morphology of *Aegyptopithecus* and allied Fayum primates will, when fully understood, influence our interpretations of the origin and adaptive nature of the earliest and most primitive Miocene hominoids belonging to genus *Proconsul* (Fleagle *et al.*, 1975; Kay *et al.*, 1981).

No aspect of anthropological study directed toward the evolutionary relationships of Miocene hominoids (Fig. 5) has raised more controversy than the "molecular clock" (Uzzell and Pilbeam, 1971; Lovejoy and Meindl, 1972; Goodman and Lasker, 1975; Goodman, 1976; Sarich and Cronin, 1976; Cronin, 1977; Washburn, 1978; Corruccini *et al.*, 1979; Lowenstein, 1982). This concept is based on a hypothetical constancy of neutral mutation rates in

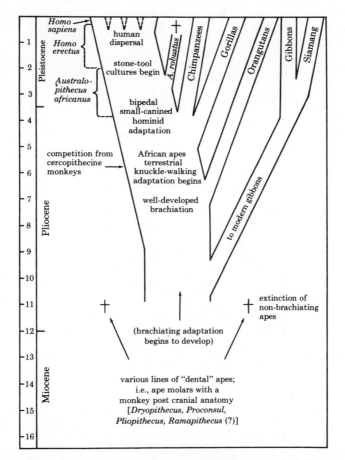

Fig. 5. Miocene hominoid evolutionary scheme according to Sarich (1971). This phylogeny is based on a strict interpretation of the "molecular clock" hypothesis of Sarich and Wilson (1967). Note that the biomolecular evidence indicates that the entire cladogenesis of the extant Hominoidea would have occurred in the late Miocene and Pliocene (modern view), all post-11 million years ago. This evolutionary scheme would rule out all but the latest Miocene hominoids from any ancestor–descendant relationships with extant hylobatids, pongids, and hominids [Figure is reproduced from Sarich (1971) and is courtesy of Little Brown and Co.]

primate macromolecules, leading to the use of immunological distance between living taxa as a direct measure of the amount of time they have been evolving separately. Although vigorously refuted at first by many hominoid paleontologists (Pilbeam, 1969; Simons, 1972; Simons and Pilbeam, 1972), the molecular clock has gained a slowly rising amount of respectability recently (Pilbeam, 1978; Gould, 1979; Johanson and White, 1979; Zihlman and Lowenstein, 1979). A recent contribution on the subject (Corruccini *et al.*, 1980) has, however, refined the notion of an "evolutionary slowdown" in molecular evolution, at least among anthropoid primates, through demonstration of nonlinear covariance obtained between different molecular matrices (see Fig. 6). The biochemical evidence of hominoid affinities remains important and controversial to determinations of relatedness among living species, and in particular the very close molecular relationship between human and chimpanzee continues to strengthen the position of those who

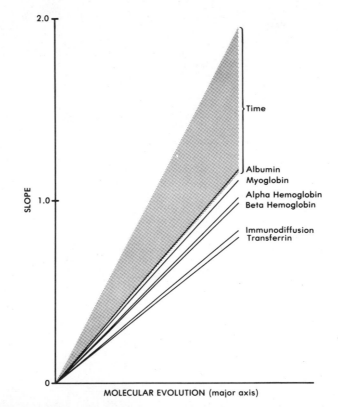

Fig. 6. This diagram briefly summarizes relative rates of change in different macromolecules; the slope of the lines indicate relative change. Note that the range of possible slope values for *geological time*, relative to molecules, is higher, meaning that perhaps all the molecules have slowed down in evolutionary rate over time. Such a modest slowdown in evolutionary rate common to all the macromolecular systems would help to reconcile the timing of the hominoid cladogenesis based on biomolecular data with the newly emerging views of that same cladogenesis based on paleontological data. [Figure is based on data from Corruccini *et al.* (1980).]

hold that diversification among ape and hominid species occurred later than the middle Miocene.

The molecular picture of close correspondence between human and chimpanzee received new impetus with the announcement and description of the geologically oldest species of *Australopithecus, A. afarensis* [Johanson *et al.,* 1978; Johanson and White, 1979, see also entire issue of *Am. J. Phys. Anthropol.,* volume 57, No. 4, (1982)]. The significance of *A. afarensis* for considerations regarding Miocene hominoid relationships lies in the following three points: (1) *A. afarensis* is demonstrably distinct from *A. africanus,* and is likely the common ancestor of both that species (plus its later descendants, the robust australopithecines) and of the lineage leading to genus *Homo;* (2) *A. afarensis* is the oldest demonstrable hominid; (3) the dental, facial, and temporal (but not postcranial) morphology of the species is perhaps more chimpanzee-like than *Dryopithecus*-like, leading Johanson and White to briefly question its descent from *Ramapithecus.* Detailed analysis of the morphological affinities of *A. afarensis* then may be seen as crucial to the question of the ancestry of hominids, especially by comparison with the newly recovered Miocene hominoid specimens and new studies on extant hominoids (e.g., Tuttle *et al.,* 1972; Tuttle, 1975; Oxnard, 1975; McHenry and Corruccini, 1976, 1980; Howell *et al.,* 1978; Corruccini, 1978; Susman, 1979; McHenry *et al.,* 1980; de Bonis *et al.,* 1981).

A recent development of comparable public acclaim to the announcement of *A. afarensis* has been the assertion (Zihlman *et al.,* 1978) that the "pygmy" species *Pan paniscus* represents the common ancestor of humans and chimpanzees, or at least a close analog of the last common ancestor of hominids and pongids. Although these claims are not without problems (Corruccini and McHenry, 1979; McHenry and Corruccini, 1980; Latimer *et al.,* 1981; Johnson, 1981), the work of Zihlman and colleagues is important for the new information provided about the least known and understood hominoid and for the theoretical issues raised concerning inference about the ancestral morphotype of the *Pan–Homo* clade. Zihlman and Lowenstein (1979) have further used morphological evidence comparing *P. paniscus* with *H. sapiens* together with the molecular evidence of their close affinity to argue against the possible hominid status of *Ramapithecus.*

Overview

Interest in Miocene hominoids and the origin of hominids has grown steadily since 1960. Probably the early 1980s will witness completion of a stage of reformulation of ideas about this group's taxonomic and evolutionary significance, a goal toward which this volume is directed. We shall say little more, as the following chapters speak for themselves. Several of them also provide the historical overview that would be appropriate at this point. We want only to reiterate several of the most worthwhile reflections.

It increasingly seems that hominoid evolution has been complex (Pilbeam, 1980), and many of the sacrosanct traits of the Hominidae have evolved more than once. Living hominoids are atypical relict forms. Their more varied ancestors may have not yet been sampled from more than a nonrandom part of their chronologic and paleoenvironmental range. Even what we have suggests that the 1960s trend toward simplifying all nomenclature and phylogenies must be completely reversed. The systematics of Pilgrim, Gregory, and Lewis were extravagant only at the species level; indeed, the phylogeny they conceived (Fig. 7) foreshadows the current idea that *Sivapithecus* is the sister group of hominids and living pongids.

The recovery of new and more complete fossils shows us inevitably how little we knew before. This is well illustrated when one compares what the concept of *Dryopithecus, Sivapithecus,* and *Australopithecus* represented during the time of Pilgrim, Gregory, and Lewis with what we know about these organisms today. Our view that these taxa have no modern analogues, yet at the same time represent vitally important steps in the evolution of the living apes and humans, is a significant advance. Accordingly, we may safely assume that what we know today about ape and human ancestry will compare little with what will be thought 50 years from now. What the fossil record really teaches us is the depth of our ignorance. It is well to recall the "F. Clark Howell Dictum": The more you know, the harder it is. New fossil finds will continue to be made, new advances in biochemistry and comparative anatomy

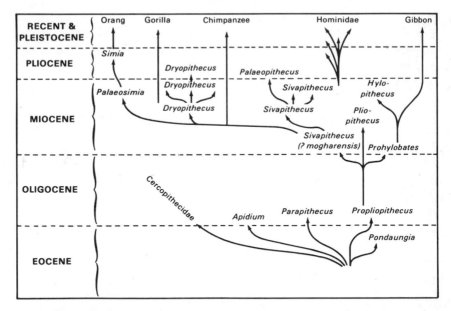

Fig. 7. Miocene hominoid evolutionary scheme according to Pilgrim (1927). Note the central position that *Sivapithecus* species play as the common ancestral stock of the Great Apes and humans. In Pilgrim's (1927, pp. 17–18) discussion of this phylogeny he acknowledges the impact of Gregory's (1922) views on the form of its construction. It is remarkable how similar Pilgrim's evolutionary scheme is to other, more recent ones discussed earlier in this chapter and to the variety of schemes put forth in the following chapters of this volume. [Figure is redrawn with no modification from Pilgrim (1927).]

will continue to flesh out our knowledge of the living hominoids and doubtlessly, new volumes such as this one will attempt to present this information in a coherent fashion. All of this will further our understanding of ape and human ancestry, just as certainly as it will highlight the great complexity of this issue.

ACKNOWLEDGMENTS

We would like to thank J. G. Fleagle for helpful comment throughout the development of this chapter and the UNCC Cartographic Laboratory Staff for preparation of the figures.

References

Andrews, P. J. 1978a. A revision of the Miocene Hominoidea of East Africa. *Bull. Br. Mus. (Nat. Hist.) Geol.* **30**:85–224.

Andrews, P. 1978b. Taxonomy and relationships of fossil apes, in: *Recent Advances in Primatology*, Volume 3, *Evolution* (D. J. Chivers and K. A. Joysey, eds.), pp. 43–56, Academic, London.

Andrews, P., and Simons, E. L. 1976. A new African gibbon-like genus *Dendropithecus* (Hominoidea, Primates) with distinctive postcranial adaptations: The significance to origin of Hylobatidae. *Folia Primatol.* **28**:161–168.

Andrews, P., and Tekkaya, I., 1976. *Ramapithecus* in Kenya and Turkey, in: *Les Plus Anciens Hominides* (P. V. Tobias and Y. Coppens, eds.), pp. 7–25, Colloque VI, IX Union Internationale des Sciences Prehistoriques et Protohistoriques, Nice. CNRS, Paris.

Andrews, P., and Tekkaya, I. 1980. A revision of the Turkish Miocene hominoid *Sivapithecus meteai*. *Palaeontology*. **23**:85–95.

Andrews, P., and Tobien, H. 1977. A new Miocene locality in Turkey with evidence on the origin of *Ramapithecus* and *Sivapithecus*. *Nature (Lond.)* **268**:699–701.

Andrews, P. J., and Van Couvering, J. A. 1975. Paleoenvironments in the East African Miocene, in: Approaches to Primate Paleobiology (F. S. Szalay, ed.). *Contrib. Primatol.* **5**:62–103.

Andrews, P., and Walker, A. 1976. The primate and other fauna from Fort Ternan, Kenya, in: *Human Origins: Louis Leakey and the East African Evidence* (G. L. Isaac and E. R. McCown, eds.), pp. 279–304, Benjamin, Menlo Park, California.

Andrews, P., Harrison, T., Martin, L., and Pickford, M. 1981a. Hominoid primates from a new Miocene locality named Meswa Bridge in Kenya. *J. Hum. Evol.* **10**:123–128.

Andrews, P., Meyer, G., Pilbeam, D., Van Couvering, J. A., and Van Couvering, J. A. H. 1981b. The Miocene of Maboko Island, Kenya: Geology, age, taphonomy and paleontology. *J. Hum. Evol.* **10**:35–48.

Chopra, S. R. K. 1978. New fossil evidence on the evolution of Hominoidea in the Siwaliks and its bearing on the problem of the evolution of early man in India. *J. Hum. Evol.* **7**:3–9.

Ciochon, R. L., and Corruccini, R. S. 1977. The phenetic position of *Pliopithecus* and its phylogenetic relationship to the Hominoidea. *Syst. Zool.* **26**:290–299.

Corruccini, R. S. 1977. Crown component variation in hominoid lower third molars. *Z. Morphol. Anthropol.* **68**:14–25.

Corruccini, R. S. 1978. Comparative osteometrics of the hominoid wrist joint, with special reference to knuckle-walking. *J. Hum. Evol.* **7**:307–321.

Corruccini, R. S., and McHenry, H. M. 1979. Morphological affinities of *Pan paniscus*. *Science* **204**:1341–1343.

Corruccini, R. S., Ciochon, R. L., and McHenry, H. M. 1976. The postcranium of Miocene hominoids: Were dryopithecines merely "dental apes"? *Primates* **17**:205–223.

Corruccini, R. S., Baba, M., Goodman, M., Ciochon, R. L., and Cronin, J. E. 1980. Nonlinear macromolecular evolution and the molecular clock. *Evolution* **34**:1216–1219.

Corruccini, R. S., Cronin, J. E., and Ciochon, R. L. 1979. Scaling analysis and congruence among anthropoid primate macromolecules. *Hum. Biol.* **51**:167–185.

Cronin, J. E. 1977. Anthropoid evolution: The molecular evidence. *Kroeber Anthropol. Soc. Pap.* **50**:75–84.

Darwin, C. 1871. *The Descent of Man and Selection in Relation to Sex*, Vol. 1. John Murray, London.

De Bonis, L., and Melentis, J. 1977. Un nouveau genre de Primate hominoïde dans le Vallésian (Miocène Supérieur) de Macedoine. *C. R. Acad. Sci. Paris D* **284**:1393–1395.

De Bonis, L., Bouvrain, G., Geraads, D., and Melentis, J. 1974. Première decouverte d'un Primate hominoïde dans le Miocène Supérieur de Macédoine. *C. R. Acad. Sci. Paris D* **278**:3063–3066.

De Bonis, L., Johanson, D. C., Melentis, J., and White, T. D. 1981. Variations métriques la denture chez les Hominidés primitifs: Comparaison entre *Australopithecus afarensis* et *Ouranopithecus macedoniensis. C. R. Acad. Sci. Paris D* **292**:373–376.

Delson, E., and Andrews, P. 1975. Evolution and interrelationships of the catarrhine primates, in: *Phylogeny of the Primates* (W. P. Luckett and F. S. Szalay, eds.), pp. 405–446, Plenum, New York.

Delson, E., Eldredge, N., and Tattersall, I. 1977. Reconstruction of hominid phylogeny: A testable framework based on cladistic analysis. *J. Hum. Evol.* **6**:263–278.

Eldredge, N., and Tattersall, I. 1975. Evolutionary models, phylogenetic reconstruction, and another look at hominid phylogeny, in: Approaches to Primate Paleobiology (F. S. Szalay, ed.). *Contrib. Primatol.* **5**:218–242.

Fleagle, J. G. 1975. A small gibbon-like hominoid from the Miocene of Uganda. *Folia Primatol.* **24**:1–15.

Fleagle, J. G., and Simons, E. L. 1978a. *Micropithecus clarki,* a small ape from the Miocene of Uganda. *Am. J. Phys. Anthropol.* **49**:427–440.

Fleagle, J. G., and Simons, E. L. 1978b. Humeral morphology of the earliest apes. *Nature (Lond.)* **276**:705–707.

Fleagle, J. G., Simons, E. L., and Conroy, G. C. 1975. Ape limb bone from Oligocene of Egypt. *Science* **189**:135–137.

Frayer, D. W. 1978. The taxonomic status of *Ramapithecus,* in: *Krapinski Pracovjek i Evolucija Hominida* (M. Malez, ed.), pp. 255–268, Jugoslavenska Akademija Znznosti i Umjetnosti, Zagreb, Yugoslavia.

Gantt, D. G., Pilbeam, D., and Steward, G. 1977. Hominoid enamel prism patterns. *Science* **198**:1155–1157.

Goodman, M. 1976. Toward a genealogical description of the Primates, in: *Molecular Anthropology* (M. Goodman and R. E. Tashian, eds.), pp. 321–353, Plenum, New York.

Goodman, M., and Lasker, G. W. 1975. Molecular evidence as to man's place in nature, in: *Primate Functional Morphology and Evolution* (R. H. Tuttle, ed.), pp. 71–101, Mouton, The Hague.

Gould, S. J. 1979. Our greatest evolutionary step. *Nat. Hist.* **88**(6):40–44.

Greenfield, L. O. 1974. Taxonomic reassessment of two *Ramapithecus* specimens. *Folia Primatol.* **22**:97–115.

Greenfield, L. O. 1975. A comment on relative molar breadth in *Ramapithecus. J. Hum. Evol.* **4**:267–273.

Greenfield, L. O. 1978. On the dental arcade reconstructions of *Ramapithecus. J. Hum. Evol.* **7**:345–359.

Greenfield, L. O. 1979. On the adaptive pattern of "*Ramapithecus.*" *Am. J. Phys. Anthropol.* **50**:527–548.

Greenfield, L. O. 1980. A late divergence hypothesis. *Am. J. Phys. Anthropol.* **52**:351–365.

Gregory, W. K. 1922. *The Origin and Evolution of the Human Dentition*, Williams and Wilkins, Baltimore. 548 pp.

Harrison, T. 1982. *Small Bodied Apes from the Miocene of East Africa,* Ph.D. Dissertation, University of London.

Howell, F. C., Washburn, S. L., and Ciochon, R. L. 1978. Relationship of *Australopithecus* and *Homo. J. Hum. Evol.* **7:**127–131.

Jia, L.-P. 1980. *Early Man in China,* Language Press, Beijing.

Johanson, D. C., and White, T. D. 1979. A systematic assessment of early African hominids. *Science* **203:**321–330.

Johanson, D. C., White, T. D., and Coppens, Y. 1978. A new species of the genus *Australopithecus* (Primates: Hominidae) from the Pliocene of Eastern Africa. *Kirtlandia* **28:**1–14.

Johnson, S. 1981. Bonobos: Generalized hominid prototypes or specialized insular dwarfs? *Curr. Anthropol.* **22:**363–375.

Kay, R. F. 1977a. The evolution of molar occlusion in the Cercopithecidae and early catarrhines. *Am. J. Phys. Anthropol.* **46:**327–352.

Kay, R. F. 1977b. Diets of early Miocene African hominoids. *Nature (Lond.)* **268:**628–630.

Kay, R. F. 1981. The nut-crackers—A new theory of the adaptations of the Ramapithecinae. *Am. J. Phys. Anthropol.* **55:**141–151.

Kay, R. F. 1982. *Sivapithecus simonsi,* a new species of Miocene hominoid with comments on the phylogenetic status of the Ramapithecinae. *Int. J. Primatol.* **3:**113–174.

Kay, R. F., Fleagle, J. G., and Simons, E. L. 1981. A revision of the Oligocene apes of the Fayum Province, Egypt. *Am. J. Phys. Anthropol.* **55:**293–322.

Kortlandt, A. 1974. New perspectives on ape and human evolution. *Curr. Anthropol.* **15:**427–448.

Kretzoi, M. 1975. New ramapithecines and *Pliopithecus* from the lower Pliocene of Rudabánya in north-eastern Hungary. *Nature (Lond.)* **257:**578–581.

Latimer, B. M., White, T. D., Kimbel, W. H., Lovejoy, C. O., and Johanson, D. C. 1981. The pygmy chimpanzee is not a living missing link in human evolution. *J. Hum. Evol.* **10:**475–488.

Le Gros Clark, W. E., and Leakey, L. S. B. 1951. The Miocene Hominoidea of East Africa. *Fossil Mammals of Africa* (Br. Mus. Nat. Hist.) **1:**1–117.

Lewis, G. E. 1934. Preliminary notice of manlike apes from India. *Am. J. Sci.* **27:**161–181.

Lovejoy, C. O., and Meindl, R. S. 1972. Eukaryote mutation and the protein clock. *Yearb. Phys. Anthropol.* **16:**18–30.

Lowenstein, J. M. 1982. Fossil proteins and evolutionary time, in: *Proceedings of the Pontifical Academy of Sciences,* The Vatican, Rome (in press).

Martin, L. 1981. New specimens of *Proconsul* from Koru, Kenya. *J. Hum. Evol.* **10:**139–150.

McHenry, H. M., and Corruccini, R. S. 1976. Affinities of Tertiary hominoid femora. *Folia Primatol.* **26:**139–150.

McHenry, H. M., and Corruccini, R. S. 1980. *Pan paniscus* and human evolution. *Am. J. Phys. Anthropol.* **54:**355–367.

McHenry, H. M., Andrews, P., and Corruccini, R. S. 1980. Miocene hominoid palatofacial morphology. *Folia Primatol.* **33:**241–252.

Morbeck, M. E. 1975. *Dryopithecus africanus* forelimb. *J. Hum Evol.* **4:**39–46.

Morbeck, M. E. 1976. Problems in reconstruction of fossil anatomy and locomotor behavior: The *Dryopithecus* elbow complex. *J. Hum. Evol.* **5:**223–232.

Morbeck, M. E. 1979. Hominoidea postcranial remains from Rudabánya, Hungary. *Am. J. Phys. Anthropol.* **50:**465–466 (Abstract).

O'Connor, B. L. 1976. *Dryopithecus (Proconsul) africanus:* Quadruped or non-quadruped? *J. Hum. Evol.* **5:**279–283.

Oxnard, C. E. 1975. The place of the australopithecines in human evolution: Grounds for doubt? *Nature (Lond.)* **258:**389–395.

Ozansoy, F. 1957. Faunes des mammifères du Tertiare de Turquie et leurs revisions stratigraphiques. *Bull. Min. Res. Explor. Inst. Turkey* **49:**29–48.

Pickford, M. 1981. Preliminary Miocene mammalian biostratigraphy for western Kenya. *J. Hum. Evol.* **10:**73–97.

Pilbeam, D. R. 1969. Tertiary Pongidae of East Africa: Evolutionary relationships and taxonomy. *Bull. Peabody Mus. Nat. Hist.* (Yale Univ.) **31:**1–185.

Pilbeam, D. 1978. Rearranging our family tree. *Hum. Nat.* **1**(6):38–45.

Pilbeam, D. R. 1979. Recent finds and interpretations of Miocene hominoids. *Annu. Rev. Anthropol.* **8**:333–352.

Pilbeam, D. R. 1980. Major trends in human evolution in: *Current Argument on Early Man* (L. K. Konigsson, ed.), pp. 261–285, Pergamon, Oxford.

Pilbeam, D. 1982. New hominoid skull material from the Miocene of Pakistan. *Nature (Lond.)* **295**:232–234.

Pilbeam, D. R., Meyer, G. E., Badgley, C., Rose, M. D., Pickford, M. H. L., Behrensmeyer, A. K., and Shah, S. M. I. 1977. New hominoid primates from the Siwaliks of Pakistan and their bearing on hominoid evolution. *Nature (Lond.)* **270**:689–695.

Pilbeam, D. R., Rose, M. D., Badgley, C., and Lipschutz, B. 1980. Miocene hominoids from Pakistan. *Postilla* (Peabody Mus. Nat. Hist., Yale Univ.) **181**:1–94.

Pilgrim, G. 1927. A *Sivapithecus* palate and other primate fossils from India. *Mem. Geol. Surv. India (Palaeontol. Ind.)* **14**:1–26.

Preuschoft, H. 1973. Body posture and locomotion in some East African Miocene Dryopithecinae, in: *Human Evolution* (M. H. Day, ed.), pp. 13–46, Symposium of the Society for the Study of Human Biology, Volume 11, Barnes and Noble, New York.

Sarich, V. M. 1971. A molecular approach to the question of human origins, in: *Background for Man* (V. M. Sarich and P. Dolhinow, eds.), pp. 60–81, Little Brown, Boston.

Sarich, V. M., and Cronin, J. E. 1976. Molecular systematics of the Primates, in: *Molecular Anthropology* (M. Goodman and R. E. Tashian, eds.), pp. 141–170, Plenum, New York.

Sarich, V. M. and Wilson, A. C. 1967. Immunological time scale for hominid evolution. *Science* **158**:1200–1203.

Schwartz, J. H., Tattersall, I., and Eldredge, N. 1978. Phylogeny and classification of the primates revisited. *Yearb. Phys. Anthropol.* **21**:95–133.

Simons, E. L. 1961. The phyletic position of *Ramapithecus. Postilla* (Peabody Mus. Nat. Hist., Yale Univ.) **57**:1–9.

Simons, E. L. 1964. On the mandible of *Ramapithecus. Proc. Natl. Acad. Sci. USA* **51**:528–535.

Simons, E. L. 1967. The earliest apes. *Sci. Am.* **217**(6):28–35.

Simons, E. L. 1968. A source for dental comparison of *Ramapithecus* with *Australopithecus* and *Homo. S. Afr. J. Sci.* **64**:92–112.

Simons, E. L. 1972. *Primate Evolution,* Macmillan, New York. 322 pp.

Simons, E. L. 1977. *Ramapithecus. Sci. Am.* **236**(5):28–35.

Simons, E. L., and Chopra, S. R. K. 1969*a.* A preliminary announcement of a new *Gigantopithecus* species from India, in: *Proceedings Second International Congress Primatology, Atlanta, GA. 1968,* Vol. 2, pp. 135–142, Karger, Basel.

Simons, E. L., and Chopra, S. R. K. 1969*b. Gigantopithecus* (Pongidae, Hominoidea) a new species from North India. *Postilla* (Peabody Mus. Nat. Hist., Yale Univ.) **138**:1–18.

Simons, E. L., and Fleagle, J. G. 1973. The history of extinct gibbon-like primates. *Gibbon and Siamang* **2**:121–148.

Simons, E. L., and Pilbeam, D. R. 1965. Preliminary revision of the Dryopithecinae (Pongidae, Anthropoidea). *Folia Primatol.* **3**:81–152.

Simons, E. L., and Pilbeam, D. R. 1972. Hominoid paleoprimatology, in: *The Functional and Evolutionary Biology of Primates* (R. Tuttle, ed.), pp. 36–62, Aldine Atherton, Chicago.

Simons, E. L., Andrews, P., and Pilbeam, D. R. 1978. Cenozoic apes, in: *Evolution of African Mammals* (V. J. Maglio and H. B. S. Cooke, eds.), pp. 120–146, Harvard University Press, Cambridge.

Susman, R. L. 1979. The comparative and functional morphology of hominoid fingers. *Am. J. Phys. Anthropol.* **50**:215–236.

Tekkaya, I. 1974. A new species of Tortonian anthropoid (Primates, Mammalia) from Anatolia. *Bull. Min. Res. Explor. Inst. Turkey* **83**:148–165.

Tuttle, R. H. 1975. Parallelism, brachiation, and hominoid phylogeny, in: *Phylogeny of the Primates* (W. P. Luckett and F. S. Szalay, eds.), pp. 447–480, Plenum, New York.

Tuttle, R. H., Basmajian, J. V., Regenos, E., and Shine, G. 1972. Electromyography of knuckle-walking: Results of four experiments on the forearm of *Pan gorilla. Am. J. Phys. Anthropol.* **37**:255–266.

Uzzell, T., and Pilbeam, D. R. 1971. Phyletic divergence dates of hominoid primates: A comparison of fossil and molecular data. *Evolution* **25:**615–635.

Von Koenigswald, G. H. R. 1935. Eine fossile Saugetierfauna mit Simia aus Sudchina. *Proc. K. Ned. Akad. Wet. Amsterdam* **38:**872–879.

Von Koenigswald, G. H. R. 1972. Ein Unterkiefer eines fossilen Hominoiden aus dem unterpliosan Greichenlands. *Proc. K. Ned. Akad. Wet. Amsterdam B* **75:**385–394.

Walker, A. C., and Andrews, P. 1973. Reconstruction of the dental arcades of *Ramapithecus wickeri. Nature (Lond.)* **244:**313–314.

Washburn, S. L. 1972. Human evolution, in: *Evolutionary Biology,* Vol. 6 (Th. Dobzhansky, M. K. Hecht and W. C. Steere, eds.), pp. 349–361, Appleton-Century-Crofts, New York.

Washburn, S. L. 1978. The evolution of man. *Sci. Am.* **239**(3):194–208.

Wiley, E. O. 1981. *Phylogenetics: The Theory and Practice of Phylogenetic Systematics,* John Wiley, New York. 439 pp.

Wu, R. 1981. First skull of *Ramapithecus* found. *China Reconstructs* **30**(4):68–69.

Xu, Q., and Lu, Q. 1979. The mandibles of *Ramapithecus* and *Sivapithecus* from Lufeng, Yunnan. *Vertebr. Palasiat.* **17:**1–13.

Xu, Q., and Lu, Q. 1980. The Lufeng ape skull and its significance. *China Reconstructs* **29**(1):56–57.

Zihlman, A. L., and Lowenstein, J. 1979. False start in the human parade. *Nat. Hist.* **88:**85–91.

Zihlman, A. L., Cronin, J. E., Cramer, D. L., and Sarich, V. M. 1978. Pygmy chimpanzee as a possible prototype for the common ancestor of humans, chimpanzees and gorillas. *Nature (Lond.)* **275:**744–746.

Geochronology and Zoogeographic Relationships of Miocene Hominoidea

2

R. L. BERNOR

Introduction

The Miocene epoch, spanning 23.5–5 m.y.a., documents important changes in Earth history. The Eurasian and African chapters of the Miocene record have been particularly well developed during the last 10 years because of heightened interests in several subdisciplines of geology and paleontology. From these studies it has become increasingly apparent that there are complex interrelationships among Miocene global tectonics, paleogeographic factors, climatic change, shifting environments, and community evolution. Whereas patterns of evolutionary change during the Miocene were once generalized as marking a "modernization" of mammalian families (Osborn, 1910), the recent efforts of paleontologists, geophysicists, and geologists offer a more detailed unraveling of the natural history of this period.

Some very fundamental changes occurred in Eurasian geography, environments, and mammalian communities during the Miocene. Geographically, Eurasia and Africa alike changed dramatically from the earlier Cenozoic configuration of a relatively uniform, low topography to a highly variegated topographic profile resulting from mountain-building of the Alpine and Hi-

R. L. BERNOR • College of Medicine, Department of Anatomy, Laboratory of Paleobiology, Howard University, Washington, D.C. 20059.

malayan Systems in Eurasia and development of the Rift Valley System in East Africa. The Tethys epicontinental seaway, which once had spanned much of Europe and western Asia and nourished higher latitudes with subtropical currents from the Indian Ocean, became severed and eventually was restricted to the Mediterranean, Black Sea, and Caspian Sea Basins. In East Africa the uplift which accompanied extensive rifting severed the continuity of moist tropical currents from the Atlantic and Indian Oceans which formerly had supplied year-round rainfall across equatorial Africa. These paleogeographic shifts influenced major changes in climatic patterns, which in turn shaped "modernized" environments with greater seasonality and latitudinal zonation. Mammalian communities underwent extensive restructuring during this interval also. However, the long-standing view that Paleogene and early Neogene mammalian forest communities were nearly synchronously replaced during the late Miocene by savanna-like ("Pontian") mammal communities is certainly incorrect. Instead, Miocene continental mammal communities appear to have changed diachronously across Eurasia and Africa, and at times were isolated from one another by ecological and/or physical barriers, whereas at other times they became more closely allied.

Within this broad environmental context we see the African radiation of hominoid primates during the late Oligocene–early Miocene, their dispersal throughout Eurasia at the beginning of the middle Miocene, and eventual isolation to the tropical areas of southeast Asia and sub-Saharan Africa by the latest Miocene. Subjects such as dispersal times, likely migration routes, association of different hominoid groups and biotic communities, patterns of evolutionary replacements and extinctions, and of course the phylogenetic relationships of extant hominoids all are issues that are being interpreted from the geological and paleontological record. The first requirement for interrelating these various paleontological parameters is a sound geochronologic framework that temporally ranks hominoid fossils and their associated faunas. The second step unites continental faunas into regions of common zoogeographic affinity, or zoogeographic provinces, and compares these provinces at successive time steps. Detailed studies of taphonomy and paleoecology allow a more refined picture of local environments which potentially can be extrapolated, to a limited degree, over a single province. Finally, the study of morphological character states allows interpretations of phylogenetic relationships of known fossils among themselves and with regard to extant species. The focal points of this presentation are twofold. First, an up-dated geochronologic ranking of major Miocene Eurasian–African mammalian fossil localities will be given largely excluding sub-Saharan Africa [see Van Couvering and Van Couvering (in preparation) and the series of articles compiled by Andrews (1981) for reviews of sub-Saharan faunas]. Second, patterns of hominoid evolution, dispersal, and extinction will be integrated into a comprehensive mammalian geochronologic and zoogeographic framework. The intention of this presentation is to provide a background for phylogenetic interpretations of hominoid evolution developed in this volume and elsewhere.

Geochronologic Framework

During the last 25 years substantial advances have been made in the methods for improving geochronologic resolution of continental vertebrate-bearing deposits. Since their introduction, radiometric techniques have proliferated in variety and have been refined to a considerable degree. Despite the broad popularity of producing a numerical date, many who use this kind of information overlook some very fundamental aspects of a radiometric determination. A question which needs to be asked, and is most often disregarded, is what other bases of age evaluation are available, besides technical ones that usually only the geophysicist is capable of understanding? When a series of consecutively superposed dates from a fossiliferous basin is available, then a succession of dates from oldest at the bottom to youngest at the top often is assumed to be an adequate verification of a basin's chronologic range [but is not always accurate, because of geochemical problems (Drake, personal communication)]. Use of more than one radiometric technique for a given source is yet another means of comparison which is relied upon. While these methods are coupled with an increasingly sophisticated radiometric technology, so-called "absolute" dates are not the panacea that many investigators have made them out to be, for a number of reasons.

Specifically, a majority of the Miocene mammalian faunas of Eurasia and Africa do not have the volcanic materials associated with them necessary for radiometric dating. East African early and middle Miocene faunas, central European early to late Miocene faunas, and a few localities in Spain, Turkey, Greece, Iran, and Pakistan have yielded radiometric determinations. Of these, only a selected number of localities in East Africa, central Europe, and Iran have yielded a succession of dates from stratigraphically superposed fossil localities. As an example, the late Miocene age Maragheh fauna from northwest Iran has illustrated that despite the quantity of volcanic materials, their stratigraphic superposition associated with an abundant mammalian fauna, and dating by two independent radiometric techniques (potassium–argon and fission track), contradictions concerning the age of all dated levels occur (Bernor *et al.*, 1980; Campbell *et al.*, 1980). Evaluation of these dates, and indeed the refinement of the age of the Maragheh succession, has been made by the use of an extensive suite of mammalian species which can biochronologically rank the different biostratigraphic intervals (Bernor, 1978; Bernor *et al.*, 1979*b*, 1980; Campbell et al., 1980). The radiometric determinations clearly were inadequate in themselves because of geochemical problems with the individual tuffs. Use of biochronologic methods allowed important evaluation of the fauna's ages which otherwise would have yielded a less certain age referral.

Because of the relative paucity of radiometric dates that have been determined and thoroughly evaluated, most of those who work with Eurasian mammal faunas prefer to use a geochronologic network which ranks faunas biochronologically by the stage of evolution of temporally sensitive taxa and incorporates migration events, radiometric dates, and marine microfossil correlations (which fortuitously interdigitate with many continental deposits around the Mediterranean Basin). Within this framework of multiple overlapping geochronologic systems, a single fauna's age with respect to several other faunas can be constantly evaluated and shifted upward or downward in the time scale as new information of these various kinds becomes available. Effectively, this methodology is removed from the fixity of "absolute" dates and its robustness can be evaluated by the diversity and quality of geochronologic tools available for its age ranking.

The development of a European Neogene mammalian biochronology has undergone a major philosophical revolution during the last several years. Traditionally, European stratigraphers biochronologically ranked continental mammal faunas solely by their interpretation of the "stage of evolution" of the species present, and then referred the fauna to a marine stage. Because there was usually no direct lithostratigraphic tie-in with the marine record, these age referrals lacked real geochronologic integrity [see Berggren and Van Couvering (1974) for a comprehensive review]. More recent studies of European mammalian biochronology have proceeded by developing some local biostratigraphic evolutionary sequences (Van de Weerd, 1976; Daams *et al.*, 1977), mostly using rodent lineages, within closely controlled lithostratigraphic sequences. As a result, correlations are now made between faunas with similar species evolutionary steps, not with marine units. A number of Spanish basins are becoming particularly well developed in this regard (including paleomagnetic investigations), and currently stand as the biochronologic referral standard for Eurasian and North African Neogene mammalian correlations. Mein (1975, 1979) has assembled a comprehensive Neogene mammalian "zonation" scheme for Europe, North Africa, and western Asia which biochronologically ranks faunas based on the occurrence of characteristic forms of evolving lineages, common species associations, and first appearances of species. While it must be emphasized that Mein's "zones" are commonly idealized, and do not benefit from proper characterization and definition as outlined by Woodburne (1977), they do identify and incorporate radiometric and marine correlations (only when they interdigitate continental mammal-bearing deposits) into the mammalian biochronologic framework to render a comprehensive geochronologic scheme.

A fundamental problem with trans-Eurasian–African Miocene mammalian correlations is the sharp contrast in mammalian communities across these continents. Crusafont (1958) and Tobien (1970) recognized a significant difference in late Miocene mammalian communities between southwest, central, and eastern Europe. Recent studies of vertebrate assemblages in Turkey (Sickenberg *et al.*, 1975), Iran (Bernor, 1978; Campbell *et al.*, 1980), and Greece (Solounias, 1979, 1981) have shown even greater differences in middle and late Miocene mammalian communities than previously realized. In an effort to refine our understanding of these biofacies differences, Bernor (1978) studied the zoogeographic affinities of 38 middle and late Miocene Eurasian and African large-mammal localities. He applied Simpson's (1947,

1960) faunal resemblance index and a cluster analysis of these indices to test whether these mammal-bearing localities grouped by temporal intervals or geographic proximity. While there was a definite temporal component to locality groupings, geographic proximity appeared to be a strongly influential variable also. These results led Bernor (1978) to initially propose four zoogeographic provinces: (1) an Eastern European–Southwest Asian Province; (2) an East African Province; (3) a Siwalik Province; and (4) a Western and Central European Province.

Subsequent to these preliminary zoogeographic investigations, some attempts to further delineate provincial mammalian communities and environments have been made. Bernor (1978) initially suggested a diachroneity in the appearance of "savanna-like" large-mammal community dispersal across Eurasia and Africa. Bernor *et al.* (1979a) investigated the evolution of this "chronofauna" (*sensu* Olson, 1966; J. H. Van Couvering, 1980) and outlined its geographic spread with environmental change between 17 and 5 m.y.a. De Bonis *et al.* (1979) and Sen (1980) have shown further that the paleogeographic distributions of several Miocene ruminant and rodent groups are similarly aligned into at least some of these paleozoogeographic provinces. These lines of evidence led Bernor (in preparation) to reorganize Mein's (1975, 1979) "biozonation" into a number of potential zoogeographic provinces, in which the known Neogene faunas are biochronologically ranked by the same criteria as used by Mein: characteristic forms of evolutionary lineages, characteristic associations, and first appearances. Instead of using the western European mammalian biozonations as a standard to extrapolate ages across all of Eurasia and Africa, the provincial format ranks geographically restricted faunas into a chronologic sequence and seeks mammalian biochronologic correlations between adjacent and, whenever possible, more distantly removed provinces. In a few instances, radiometric dates and/or marine correlation information for mammalian localities can be used to establish provincial chronologic "anchor points" by assigning "absolute" ages to particular faunas.

The ultimate goal of this methodology is to establish more precise interprovincial continental correlations by first developing a number of intraprovincial, geochronologically ranked sequences. It is hoped that this will promote a more accurate biochronologic referral of faunas which were too far removed geographically from the western European standard sequence to be incorporated initially, and allow more extensive documentation about zoogeographic features such as immigration events and changes in faunal isolation and affiliation between adjacent provinces. As a result of these preliminary investigations, six zoogeographic provinces are proposed here: (1) a Western and Southern European Province; (2) an Eastern and Central European Province; (3) a Rumanian and Western USSR Province; (4) a Sub-Paratethyan Province; (5) a North African Province; and (6) a Siwalik Province (see Fig. 1). Three other Miocene provinces also can be suggested from their mammalian assemblages: an East African Province, a North Asian Province, and a Southeast Asian Province. The first is currently undergoing a major review (see Andrews, 1981) while the last two are just beginning to be devel-

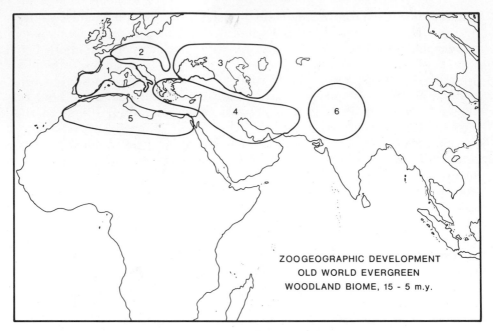

Fig. 1. Proposed zoogeographic provinces of Eurasia and North Africa: (1) Western and Southern European Province; (2) Eastern and Central European Province; (3) Rumanian and Western USSR Province; (4) Sub-Paratethyan Province; (5) North African Province; (6) Siwalik Province.

oped in a modern geochronologic sense (Chiu *et al.,* 1979). Of the six provinces referred to here, mammal faunas which include radiometric and/or marine correlation information are noted, and hominoid bearing localities are signaled (Table 1). The delineation of provinces given here stands as a first-step hypothesis which must bear future testing by detailed systematic studies, paleogeographic reconstructions, and geochronologic refinements.

Paleogeographic and Paleoenvironmental Overview

The Miocene epoch has long been characterized as recording substantial changes in continental environments and mammalian communities. What has not been recognized clearly enough by students of these fields is that such changes are directly related to longer term tectonic and climatic changes. Dewey *et al.* (1973) developed a comprehensive plate tectonic model for explaining the evolution of the Mediterranean region during the past 200 million years. This region's tectonic evolution had far-reaching environmental consequences, which affected Eurasian and African continental climates, floras, and faunas. The authors presented a preliminary tectonic scenario of the Mediterranean area that depicts a complex of ancient microplates that col-

Table 1. Miocene Biochronologic Rankings

Age, m.y.a.	European Land Mammal Age	European Mammal Neogene (MN) Zone	Western and Southern European Province	East and Central European Province	Rumanian–Western USSR Province	Sub–Paratethyan Province	North African Province	Siwalik Province	Stage[a]	Epoch[b]
5		13	Baccinello V3 Arenas del Rey La Alberca[d] Librilla[c] La Fontana Baccinello V2 Casino	Baltavar Polgardi	Mamay Tudorovo[d]	Amasya Ano Metochi Ditiko Kinik	Natrun Sahabi[c] Khendek-el-Ouaich Marceau		Me	mi
6										
7	Turolian	12b	Mt. Luberon Ratavoux Concud Los Mansuetos		Taraklia N. Elisavet. Grebeniki[d]	U. Maragheh[c] Samos[c] Garkin		Dhok Pathan Fm.[c]	To	
8		12a	Baccinello V1 Crevillente 4 and 5[d]		Cimislia[d]	Kayadibi Saloniki M. Maragheh[c] Pikermi	Amama 2			
9		11	Crevillente 1 Vivero Mollon Lobrieu	Eichkogel D. Durkheim Kohfidisch	Grossulovo[d] Berislave[e]	L. Maragheh[c]	Sidi Salem	Nagri Fm.[c]		
10		10	Masia d Barbo	Vosendorf	Sebastopol[d] Ialovena[d] Varnitsa[d]	Kastellios 2–3 Pyrgos[c] Ravin Pluie[c] Esme Akcakoy	Amama 1 Oued Zra Jeb. Semmene	Chinji Fm.[c]	Se	
11	Vallesian	9	Soblay Can Ponsich[c] Llobateres[c] Hostalets Sup. Nombrevilla	Csakvar Eppelsheim[c] Rudabánya[c] Howenegg[d,e]	Lapouchna[d] Braila[d] Kalfa[d]	Kastellios 1	Bou Hanifia[d,e] Beglia Sup.			
12										
13	Astaracian	8	La Barbera[c] Hostalets Inf.	Anwil Trimmelkham[c] Ohningen	Eldar[c,d] Udabno[c,d]	Y. Eskihisar Chios	Beglia Inf. Pataniak 6			
14		7	La Grive L3[c] St. Gaudens[c] La Grenatier La Grive M Simorre	Steinheim Opole[c]	Korethi	Sofca Çandir[c] Hofuf Fm.[d]	Beni Mellal			

(continued)

Table 1 (*Continued*)

Age, m.y.a.	European Land Mammal Age	European Mammal Neogene (MN) Zone	Western and Southern European Province	East and Central European Province	Rumanian–Western USSR Province	Sub–Paratethyan Province	North African Province	Siwalik Province	Stage[a]	Epoch[b]
		6	P. Santerem	Goriach[c]	Byelometchet-skaya	Dam Fm.[c,d]				
			Liet[d]			Prebreza				
15			Sansan[c]			Paşalar[c]				
			La Grive L							
		5	Lisboa 5b	Neudorf Sand.[c]		Sibnica		Kamlial	La	
			P. L. Thenay[c,d]	Langenmoosen		Mala Miliva		Fm.[?c]		
16			Sos[d]	Neudorf Splt.[c]		Lozovik				
		4	V. Collonges	Eibiswald		Banja Luka	Wadi Moghara		Bu	
			La Romieu	Orechov						
17			Búnol	Eckertshofen		Cucale	Gebel Zelten			
	Orleanian		Artenay	Ipolytarno						
		3b	Moli Calopa							
			Rubielos de Mora	Wintershof W.						
18					Nakhicevan					
		3a	Lisboa 1[d]	Bissingen				Dera Bugti		
19			Ateca 1,3							
			Chilleurs							
			Estrepouy	Eggenburg[d]						
20			Chitenay							
		2	Bouziques	Haslach					Aq	
			Langnac	La Chaux						
21			Cetina Aragon	Ulm						
			Caunelles[d]	Bundenheim						
22	Agenian	1	Saulcet	Eger[d]						
23				Weissenau	Agyspe					
				Tomerdingen	Benara					
24			Paulhiac							o
25										

[a]Me, Messinian; To, Tortonian; Se, Serravallian; La, Langhian; Bu, Burdigalian; Aq, Aquitanian.
[b]mi, Miocene; o, Oligocene.
[c]Hominoid-bearing locality.
[d]Marine correlated locality.
[e]Radiometrically dated locality.

lided with the Eurasian Plate and consolidated during the early Cenozoic, and subsequently was closely followed in the early Miocene by the docking of the African continent with southwest Asia (Zagros Crush Zone of Iran).

At the end of the Oligocene and beginning of the Miocene, a distinct marine bioprovinciality existed in southern Europe, southwest Asia, and North Africa (Fig. 2; in part from Steininger, 1977). Because of the accelerated orogeny of the Alpine System, these marine biogeographical provinces became progressively more distinct throughout the Miocene, and by the end of this interval, the Mediterranean, Caspian, and Black Sea Basins approached their current configurations. Steininger (1977) has shown that by the end of the Oligocene the Paratethys, separated from the Mediterranean Basin by the rising Alps, no longer had direct continuity with the Atlantic Ocean. The only connection between the European seaway complex and the Atlantic Ocean was through the Gibraltar portal. However, during the earliest Miocene, the European seaways continued to maintain a shallow marine continuity with the Indo-Pacific marine provinces, and tropical invertebrate faunas appeared in Europe (Steininger, 1977, p. 239). To the north, the Paratethys Sea, which extended from the southwestern limit of the Alps, eastward across the northern flank of the Alps through central Europe, to the present location of the Aral Sea in the USSR, shared two distinct waterway connections with the Mediterranean marine basin. The westernmost was via the Rhone River Basin in southeastern France, the other, from the northwestern limit of the Alps through the Vienna Basin across Italy (Fig. 2). The connection of these seaways with the Indo-Pacific realm was severed during the early Miocene, and the interconnections of the Paratethys and Mediterranean Basins occurred in several stages during the remainder of the Miocene. These eventually culminated with a complete desiccation of the Mediterranean Basin *ca.* 5.5 m.y.a. (Hsü *et al.*, 1973, 1977). The marine paleogeographic events cited here are summarized in Figs. 2–6. Steininger and Papp (in preparation) currently are developing more detailed paleogeographic maps for time intervals on the order of 1 million years each, more closely depicting these changes.

The effects of transforming the southern European and northern African continental region from a subtropical marine-nourished realm to one of progressively cooler and more seasonal environments is reflected in the climatic and floral records. Axelrod and Raven (1978) have suggested that the progressive northward motion of the African Plate during the Paleogene restricted Indo-Pacific marine currents and shifted the slight contrasts between sea-surface and land-surface temperatures to progressively greater ones. This tectonically controlled temperature contrast resulted in an intensified stationary high pressure center that blocked the passage of Atlantic moisture across the African west coast. This shift was accompanied by progressive uplift of the Himalayan and Alpine Mountain Systems and formation of the Rift Valley System in East Africa. Andrews and Van Couvering (1975) have suggested that the increased local orographic relief of the East African Rift System blocked the tropical moist air ingress into the interrift region, causing greater regional climatic seasonality.

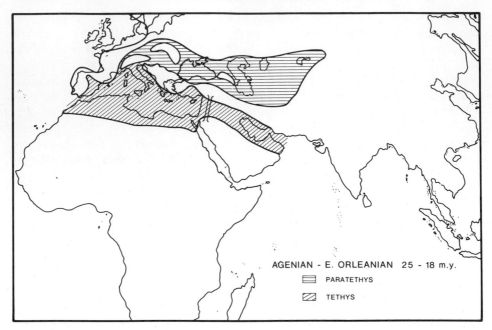

Fig. 2. Agenian–early Orleanian Mammal Ages, 25–18 m.y.a. This interval records the docking of the African Plate with the Eurasian Plate. The extensive Tethys Epicontinental Seaway which once spanned much of Europe, western Asia, the northern fringe of Africa, and part of the Saudi Arabian peninsula underwent a regional regression. The establishment of a land bridge depicted here (][) allowed the initial Neogene faunal interchange of Eurasian and African mammals, and at the same time discontinued the nourishing of European, west Asian, and North African environments with tropical/subtropical Indo-Malaysian marine currents. The change in this ocean circulation later catalyzed significant climatic shifts, which caused increased continental seasonality in Europe and North Africa during the Miocene.

Axelrod (1975) notes that seasonal, summer-drought-adapted sclerophyllous vegetation progressively evolved and spread geographically during the Miocene, replacing laurophyllous evergreen forests which were adapted to moist, subtropical and tropical conditions with temperate winters and abundant summer rainfall. The ecological shift from forested to open woodland conditions is related directly to climatic trends of cooler winters and decreased summer rainfall. However, this process was not uniform across Eurasia and Africa. Whereas some areas became more seasonal in the early Miocene, other areas retained less seasonal subtropical forests and closed woodlands quite late in the Miocene (Bernor *et al.*, 1979a). The more open-habitat woodland communities eventually occupied a distinct belt across southern Eurasia and formed a broad ecotonal interface with the Turgyan forests, which contained moist-temperate-adapted mixed deciduous hardwoods and conifers north of the Paratethys Sea, and the Poltavian tropical evergreen forests to the south (Axelrod, 1975).

The increased seasonality of northern Africa accelerated during the Miocene. Axelrod and Raven (1978) cite the occurrence of an extant montane rain forest in the Canary Islands, a volcanic archipelago no older than middle Miocene, as a remnant of similar montane forests on the adjacent coast of northwestern and northern Africa during the early Miocene. According to Axelrod and Raven (1978), the Sahara Desert is a relatively recent feature, with a rise of 200 mm in annual rainfall being adequate to support a sclerophyll vegetation. Bernor (1978), Bernor *et al.* (1979*a*), and Thomas (1979) have argued from the mammalian faunal evidence that more seasonally adapted open-country mammalian communities existed in northern Africa by late–early and early–middle Miocene times. Axelrod and Raven (1978) believe that a sclerophyll vegetation probably occupied all of the present semi-desert and desert areas of North Africa in the Miocene, with relict patches of forest possible in favorable areas. Hence, mammalian faunal community and paleobotanical evidence jointly suggest that North Africa was substantially drier than the northern Mediterranean shores, as currently is the case, presumably because of a longer and more severe drought season.

East Africa was undergoing analogous changes during the early and middle Miocene because of the tectonic rifting activity referred to above. Axelrod and Raven (1978) have cited a significant shift in East African floral communities directly associated with tectonic activity and the development of a "double" rain shadow which existed between the Western Rift and Eastern Rift Highlands (Andrews and Van Couvering, 1975). The decrease in annual moisture to this interrift area favored the spread of seasonally adapted open woodlands and their accompanying fauna, beginning in the middle Miocene (J. H. Van Couvering, 1980). Axelrod and Raven (1978) also indicate that development of the Ethiopian highlands occurred during the Miocene, as a result of rifting and uplift of northeastern plateaus bordering the rift zone. These authors suggest that this triggered a regional climatic change which promoted the displacement of lowland forests by more seasonally adapted woodlands along the east coast of Africa.

Miocene mammalian communities certainly responded to the tectonic, climatic, and floral evolutionary trends outlined above. A close interrelationship between successional floral shifts and observable mammalian community responses is not presently possible because of discrepancies in geochronologic referrals of floras on the one hand and faunas on the other. It is not often that fossil floras are accurately calibrated or properly referred to the geochronologic time scale. Too often, Eurasian and African floras are referred to "early," "middle," or "late" Miocene of old usage, confounding the integration of faunal and floral information. Moreover, there is often little sound geochronologic basis for establishing the age of a given flora. Recently there has been a concerted effort by the participants of the International Congress on Mediterranean Neogene Stratigraphy (1971, 1975, 1979) to geochronologically interdigitate fossil floras and faunas. The excellent stratigraphic and radiometric framework for East Africa is also promoting refined resolution for associating floras and faunas. However, a major proportion of the Eurasian and African land masses presently have from moderate to poor geochronologic control for interrelating floral and mammalian communities.

Therefore, while the integration of mammalian zoogeographic community evolution with floral community shifts (Figs. 2–6) presently appears to be generally accurate, much more information is needed to adequately understand these relationships.

Recently, paleoecologists have had some measure of success in determining habitat types from fossil mammal assemblages alone. Andrews *et al.* (1979), Andrews and Evans (1979), and J. H. Van Couvering (1980) have implemented studies of ecological diversity and habitat spectra of extant mammalian communities to determine likely ecotypes of fossil communities. Andrews *et al.*'s (1979) and Andrews and Evans' (1979) use of ecological diversity studies entails characterization of mammalian assemblages by systematic division, size (by weight), locomotion categories, and feeding classes. A variety of extant habitats have been typified by these methods and used as standards for comparison with fossil mammalian communities. Discrimination of likely fossil habitats has proved remarkably good, and forms the basis for future ecological interpretations of fossil mammal communities. J. H. Van Couvering's (1980) approach is similar, but more specifically addresses ecological reconstruction in terms of successional community evolution from tropical forest to savanna mosaic habitats. While this approach is in its earliest phases of development, it already has successfully integrated climatic and environmental changes, plant community succession, and mammalian community responses to aid interpretations of the evolution of African savanna-mosaic mammalian communities. These studies of fossil communities have yielded the very important observation that ancient environments do not usually have direct parallels with extant ones. The ecological interpretations given in the following section attempt to adapt the ecological categories of these authors in a broader sense, but cannot as yet be as succinct, because of the need for similar methodological analysis.

Geochronology, Mammalian Zoogeography, and the Hominoid Sequence

The Oligocene–Miocene boundary has recently been marked as a specific stratigraphic point at the base of the stratotype of the Aquitanian Marine Stage, located at Moulin de Bernachon, France (George *et al.*, 1969). However, J. A. Van Couvering and Berggren (1977) have shown that this lithostratigraphic point excludes a lowermost "piece" of the fossil record traditionally included within the Miocene. They argue that this problem surfaced when Blow (1969) considered that the presence of the planktonic foraminiferan *Globigerinoides primordia* with an N.4 microfauna in the stratotype of the Aquitanian Stage in France justified equating the base of Zone N.4 with the base of the Aquitanian interval. However, according to Eames (1970), the Aquitanian stratotype rests on a transgressive unconformity. J. A. Van Couvering and Berggren (1977) cite a number of sequences in southern France which rest stratigraphically below the Type Aquitanian and contain *G. primordia*. Their review of the calibrated latest Chattian (latest Oligocene) Marine Stage of the North Sea Bioprovince (*sensu* Steininger, 1977; here Fig. 2), central Para- tethyan (of Central Europe) Egerian–Eggenburgian marine microfossil faunas, western North American late Oligocene–early Miocene marine faunas, and Mediterranean marine invertebrate faunas has enabled them to estimate that the base of the *Globigerinoides primordia* first appearance is *ca.* 25 m.y.a. As a result of this estimation, J. A. Van Couvering and Berggren (1977) suggest that the base of the Aquitanian stratotype, and therefore the base of the Miocene, is best placed at 23.5 m.y.a.

The continental Miocene of Europe, as characterized by mammalian bio- chronology, includes five mammal "stage"/ages, including two for the early Miocene (the Agenian and Orleanian); one for the middle Miocene (the As- taracian); and two for the late Miocene (the Vallesian and Turolian). The current geochronologic ranges for these intervals is given in Table 1, and a detailed description of several of the zoogeographic provincial faunas re- ferred to here, as well as an evaluation of the bases for their age referral, has been given by Bernor (in preparation). This geochronologic framework will be developed here in a zoogeographic sense, highlighting significant aspects of provincial evolution, moments of faunal exchange, geographic location of hominoid primate occurrences, and likely corridors for their dispersal.

Agenian

The Agenian Mammal "Stage"/Age, formerly termed the Aquitanian "Mammal Stage," has been characterized by Ginsburg (1971) after a mam- malian assemblage from the Sables de l'Agenais in the southern marginal portion of the Aquitaine Basin. Continental deposits from the Aquitaine Basin bear fossil vertebrate assemblages ranging from late Eocene to late Miocene in age and owe their occurrences there to the formation of a north- ern piedmont structure which flanks the Pyrenees Mountains (Richard, 1948). The name "Aquitanian" was used for the early Miocene assemblage because it was originally believed to be bounded by the same convention as the marine Aquitanian Stage. Actually, the mammalian age has only to do with the supposed "stage of evolution" of its contained fauna and has no litho- stratigraphic tie-in with the marine stratotype Aquitanian (J. A. Van Cou- vering and Berggren, 1977). Mein (1975, 1979) reports that the earliest Age- nian-aged faunas are latest Oligocene in age [i.e., the famous Paulhiac fauna of Richard (1948, pp. 209–210)].

The Agenian Mammal Age includes the first two MN zones of Mein (1975, 1979) and ranges in age from approximately 25 to 20 million years. As Table 1 indicates, the lower half of MN Zone 1 is referred to the Oligocene, with the beginning of the earliest Miocene corresponding roughly with upper MN 1. Mammalian faunas equivalent in age to European MN Zones 1 and 2 are known to occur in only three provinces of Eurasia and North Africa: the

Western and Southern European Province, the Eastern and Central European Province, and the Rumanian and Western USSR Province. Mein (1979) had commented that the beginning of the Neogene does not document any significant changes in the European fauna, lacking immigrations, but records some autochthonous first appearances. Indeed, he refers to this interval as harboring an impoverished Oligocene fauna. Paleogeographically, Europe was still broadly inundated by the extensive Tethyan epicontinental sea complex, which acted as an absolute geographic barrier between the Eurasian and African land masses. Chiu *et al.* (1979) have recently reported a series of Miocene faunas from northern China, the oldest of which appears to just postdate the Agenian. Therefore, there is no basis for suggesting trans-Eurasian provinciality during the Agenian interval. However, contrasting with this uncertainty about Eurasian provinciality, we note that the greatest Miocene provincial divergence existed between Eurasia and Africa due to their separation by the Tethys Seaway (see Fig. 2).

Hominoid primates certainly originated in the African continent (Campbell and Bernor, 1976). The most recent systematic revision of Miocene Hominoidea from East Africa (Andrews, 1978a) reports some seven species of dryopithecines, including three genera: *Proconsul*, with two subgenera, *P.* (*Proconsul*) and *P.* (*Rangwapithecus*); *Limnopithecus;* and *Dendropithecus*. Andrews has grouped the first two genera with the Pongidae, and the last, *Dendropithecus*, with the Hylobatidae. Andrews (1978a,b) has further proposed a close phylogenetic relationship between these and the Oligocene Fayum primates *Aegyptopithecus* (to *Proconsul*) and *Propliopithecus* (to *Dendropithecus*).* The early Miocene African dryopithecines appear to have been largely restricted to forest environments, with more open nonforest habitats absent until the latest phases of the early Miocene and middle Miocene (Andrews and Van Couvering, 1975; Andrews, 1978a).

Orleanian

The Orleanian Land-Mammal Age is named after a suite of late–early Miocene vertebrate localities from the Sables de l'Orleanais of the Loire Valley, near Orleans, France (Ginsburg, 1971). This interval includes MN Zones 3–5 of Mein (1975, 1979) and has been estimated as spanning 20–15 m.y.a. (J. A. Van Couvering, personal communication). Recently, Daams *et al.* (1977) have grouped the Orleanian and succeeding Astaracian Land-Mammal Ages into the Aragonian "Superstage," including MN "zones" 3–8, with type sections designated in the Calatayud–Teruel Basins of east central Spain.

Chronostratigraphic application of the Type Aragonian mammal faunas

---

*Note that while other authors (Szalay and Delson, 1979; Greenfield, 1979, 1980) have suggested alternative systematic schemes for hominoid primates, I prefer to adopt Andrews here on general philosophical grounds and because his evolutionary interpretations integrate well with the zoogeographic observations developed here.

is problematic because of the lack of precise lithostratigraphic/species appearance definition for both the base and top of the interval. Moreover, time-stratigraphic definition applying mammalian species is made difficult because of their limited geographic range and the resulting problem of recognizing "stage" boundaries between provinces. Yet Daams *et al.* (1977) have given detailed documentation of species ranges, evolutionary appearances, characteristic associations, overlap range zones, and extinctions, which provides an important litho-biostratigraphic framework for early Miocene European mammalian geochronology. For the purpose of this presentation, however, the Orleanian and Astaracian intervals will be discussed separately because of the great changes in faunal communities and paleogeographic and zoo-geographic characteristics noted at these times.

Mein (1979) and Bernor (in preparation) have collectively reported mammalian faunas from all six of the Eurasian and North African provinces cited here. Of these, the Western and Southern European Province is best known and has been most thoroughly studied. The biochronologic ranking by Mein (1979) of the Central European Province has not been similarly evaluated since the work of Cicha *et al.* (1972). Cicha *et al.* (1972) developed correlations by the use of small mammal homotaxis with southwestern Europe and of local marine tie-ins to the Paratethys biochronologic sequence. Large mammals, which have proved particularly useful for geographically extended correlations and depiction of provinciality, have never been substantively reviewed for the purpose of continental biochronology. Efforts are currently being initiated by a number of investigators to correct this problem. The Rumanian and Western USSR Province contains only one fauna, Nakhicevan (Mein, 1979). This last province is poorly known because of the general lack of recent literature on its faunas; the faunas cited in Table 1 are taken from Mein (1979) and cannot at this time be evaluated. The Sub-Paratethyan Province includes some faunas, but no publications on these are available other than Mein's (1979). The North African Province contains two well-known faunas, Gebel Zelten and Wadi Moghara [see Hamilton (1973) for a review]. The Siwalik Province is represented by a medial Orleanian age locality, Dera Bugti. Madden and Van Couvering (1976) have reported that this fauna is dominated by African taxa (15/20) which record their first major Cenozoic emigration. These authors further report that this emigration led to a rapid provincial extinction of several autochthonous taxa. Finally, recent work in the north China Neogene (Chiu *et al.,* 1979) has yielded a mammal fauna which tentatively appears referable to the Orleanian (Bernor, in preparation).

The Western and Southern European Province contains a number of Orleanian faunas from Spanish and French continental basins. Spanish sequences of this age are included in the Calatayud and the Valles-Penedes Basins. French sequences include ones in the Aquitaine and Loire Basins. The litho- and biostratigraphic ordering of these faunas allows a biochronologic ranking and stands as the principal area for the referral of other European faunas which usually are stratigraphically isolated. The earliest interval of the Orleanian, the lower portion of MN 3a, still witnesses Africa and Eurasia separated by the Tethys Sea. At this time, southwestern and central Europe were zoogeographically contiguous and largely harbored more highly forested habitats. These European faunas were geographically contiguous with northern China (i.e., no physical barriers existed) and received a number of immigrants at that time by a high-latitude (north of the Paratethys Sea) exchange route. The most notable immigrant is the North American equid *Anchitherium*, whose first appearance marks the base of the Orleanian Land-Mammal Age.

The first emigration of African mammals is recorded in MN 3. Berggren and Van Couvering (1974) cite this "event" as occurring *ca.* 17.5 m.y.a., and initially involving only proboscideans, hence their reference to the "Proboscidean Datum" (MN 3b here). However, Mein (1979) cites the first occurrence of the African anthracothere *Brachyodus* from the beginning of MN 3a (20 m.y.a.), and Ginsburg (1971) cites the possible occurrence of a proboscidean tooth from the MN 3a locality of Condom. These lines of evidence suggest an earlier beginning of Eurasian–African faunal exchange, *ca.* 20–18.5 m.y.a. Throughout the remainder of the Orleanian, a number of African and Asian immigrations are recorded. A migration route and its ecological typification have been proposed by Van Couvering and Van Couvering (1976). They suggest that both open habitat and forest environments were supported (their "high" and "low" roads, respectively) when the African Rift System, supporting elongated basins with an interconnected series of rivers and lakes, completed its formation in the early Miocene. The diverse African fauna presumably used this corridor for its emigration, dispersing into the European and Asian continents several times during the Miocene interval.

The Southern and Western European Province was periodically isolated from the extensive African emigration because of a marine barrier posed by the Paratethys and remnant Tethys Seas (see Figs. 2 and 3). Antunes (1979) has demonstrated the existence of a late Orleanian (MN 5) land-mammal migration route between Spain, Turkey, and Asia via the Sub-Alpine Arch. According to his interpretations, periodic marine regressions in this area opened a southern land route during the late Orleanian. While a more northerly route around the Paratethys was geographically feasible, the existence of deciduous hardwood forests there (Axelrod, 1975) could have posed an ecological barrier for tropical mammals exiting from Africa.

The recent report by Chiu *et al.* (1979) of northern Chinese mammal faunas include an Orleanian age locality from the Xiejia Formation. They state that most of the genera in this fauna are endemic Oligocene forms. While the absence of *Anchitherium* here would suggest a pre-Orleanian age, the occurrence of an advanced *Eucricetodon* supports an Orleanian referral. J. H. Van Couvering (1980; and in Bernor *et al.*, 1979*a*) has suggested that northern China harbored open habitats in the late Oligocene, based largely on observations of the community structure of the Hasanda Gol mammal fauna (Mellett, 1968). Her argument finds support from the paleogeographic–climatic interpretation that the progressive uplift of the Himalayas, first initiated by the late Eocene docking of the Indian continent with Eurasia

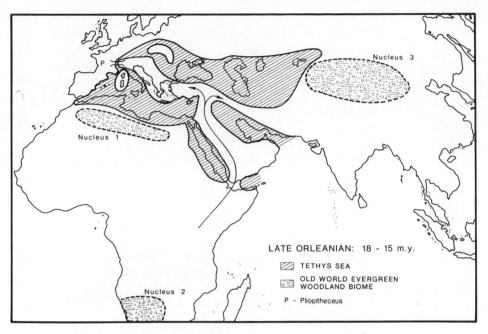

Fig. 3. Late Orleanian mammal Age, 18–15 m.y.a. Following the docking of the African Plate with Eurasia and the establishment of a land corridor across the Arabian Peninsula, there was a substantial immigration of African mammals into Eurasia. Sometime during this interval, Neogene catarrhine primates make their first appearance in Eurasia. This interval also records the development of more seasonal environments and the early evolution of open-country woodland habitats and mammalian communities.

(Dewey *et al.*, 1973), placed Mongolia in a rain shadow, possibly initiating the earliest occurrence of seasonal conditions in the Old World. The very early occurrence of an advanced bovid genus (*Oioceros*) there during the early Miocene renders some further support to this argument.

The North African Province appears to be yet another early theater for open habitat faunas (Bernor *et al.*, 1979a; Thomas, 1979). Here giraffids (two genera) and bovids (three genera) make their first appearances. These two groups are significant ecologically because they are the first of six principal mammalian families (Felidae, Hyaenidae, Equidae, Rhinoceratidae, Giraffidae, and Bovidae) that underwent extensive adaptive radiations and formed a chronofaunal woodland community spanning most of Eurasia and North Africa in the latest Miocene (Bernor *et al.*, 1979a). The occurrence of these two ruminant groups at Gebel Zelten has led J. H. Van Couvering (in Bernor *et al.*, 1979a) to suggest the existence of a second open-country nucleus by the early Miocene (Fig. 3).

The first occurrence of hominoid primates in Eurasia is not recorded until MN 5. Ginsburg (1971) has reported a primitive *Pliopithecus, P. piveteaui,* from the Pontlevoy fauna in the Sables de l'Orleanais. Ginsburg (1971, 1975;

also see Szalay and Delson, 1979) has referred a number of other small French localities in and around the Loire Basin with *Pliopithecus* to this zone (*ca.* 16–15 m.y.a.), including Manthelan, Lassé, Pontigné, Rillé, Savigné, Dénèze, and Noyan Sous-le-Lude. Delson (personal communication) [also suggested in Szalay and Delson (1979)] has recently suggested that *Pliopithecus* gained entry into western Europe via the MN 5 Sub-Alpine Arch migration route cited above (Antunes, 1979). *Pliopithecus* has long been referred to the Hylobatidae because of the gibbon-like appearance of its skull and dentition. However, Delson and Andrews' (1975) review of *Pliopithecus* has demonstrated that many of the cranial similarities are largely primitive and common to the Catarrhini. Ciochon and Corruccini (1977) came to a similar conclusion regarding the postcranial skeleton and proposed that *Pliopithecus* no longer be considered ancestral to the Hylobatidae. Delson and Andrews (1975) and Szalay and Delson (1979) have now formally referred *Pliopithecus* to a distinct family, the Pliopithecidae, which unites it and the East African Miocene genus *Dendropithecus* with the Fayum Oligocene forms "*Aegyptopithecus*" [*Propliopithecus* of Szalay and Delson (1979)], *Propliopithecus,* and "*Aeolopithecus*" [*Propliopithecus* of Szalay and Delson (1979)]. Ginsburg and Mein (1980) also derive *Pliopithecus* from a grade of evolution similar to that of the Fayum genera "*Aegyptopithecus*" and *Propliopithecus.**

Ginsburg (1975) has recently studied *Pliopithecus* and recognized a morphocline polarity between several species. He cites *P. piveteaui* as being the most primitive form, especially in the retention of a well-developed P^4, constricted P_4 crown, external cingulum on M_{1-2}, and distal constriction with lack of cingulum on M_3. From *P. piveteaui*, Ginsburg posits a temporal succession: *P. piveteaui, P. antiquus* (Sansan and La Grive, MN 6 faunas), pliopithecid from Goriach (MN 6), *P. vindobonensis* (from Neudorf-Spalte, MN 6), *P. lockeri* (from Trimmelkan, MN ?8), citing a trend of progressive elongation of the molars. Szalay and Delson (1979), however, cite extreme variability in all *Pliopithecus* materials, both within a single locality and between localities, and caution against the recognition of several species. With some reservations, they characterize six European and one Asian species of *Pliopithecus*. According to their inventory, *Pliopithecus* occurs throughout Europe, except for Italy and Greece, and ranges in age from MN 5 to MN 9 (early Vallesian). *Pliopithecus* was evidently adapted to folivory rather than frugivory, was an active arborealist, and may have engaged in some suspensory postures as

---

*While this manuscript was in review, Ginsburg and Mein (1980) published a description of a new species of Pliopithecidae, *Crouzelia rhodanica*, and took the opportunity to make a revision of the family Pliopithecidae Remane, 1965. In their systematic revision of the family, the authors recognize two subdivisions of the Catarrhini: the Eocatarrhini, to which the Pliopithecidae belong, and Eucatarrhini, to which the Cercopithecoidea and Hominoidea belong. Within the Pliopithecidae are included two subfamilies: (1) Pliopithecinae, including the species *Pl. antiquus, Pl. pivteaui,* and *Pl. vindobonensis;* (2) Crouzelinae, including the species *C. auscitanensis, Plesiopithecus lockeri,* and *Anapithecus hernyaki.* I refrain from adopting this systematic framework here only because of its very recent publication and the very short time to thoroughly evaluate the revision. Ginsburg and Mein's (1980) paper does not significantly alter the geochronologic and zoogeographic presentation given here.

well as running, climbing, and leaping (Szalay and Delson, 1979). These lines of evidence suggest a morphological and behavioral (feeding–locomotion) convergence with arboreal monkeys. *Pliopithecus'* extinction in Europe during the late Miocene has been linked with competition from monkeys (Delson, personal communication); however, greater seasonality, and the resulting changes in temperature and floral communities, may very well have contributed to its demise.

Astaracian

The Astaracian Land-Mammal Age is named after a stratigraphic sequence of mammal-bearing localities from the Calcaire d'Astarac in the Aquitaine Basin, southern France (Ginsburg, 1971). The most celebrated locality here is the Sansan fauna, which documents a number of mammalian "evolutionary steps" and immigration events used to characterize MN 6. The Astaracian interval includes MN Zones 6 to 8, approximately ranging from 15 to 12 m.y.a. Mammalian faunas have been reported from all six provinces discussed here, as well as northern China and eastern Africa. The fossil records of the Western and Southern European, Sub-Paratethyan, North African, and Siwalik Provinces have recently undergone extensive investigations by a number of paleontologists and will form the basis of discussion developed here (Bernor, in preparation). The Eastern and Central European Province has most recently been addressed by Cicha *et al.* (1972), who integrated small-mammal evolutionary sequences with marine tie-ins to correlate these faunas with western Europe. The Rumanian and Western USSR Province continues to be a problem for biochronologic and zoogeographic characterizations because of the lack of recent definitive studies. The Astaracian interval is particularly important for mammalian geochronologic and zoogeographic studies because it is a pivotal period for large-mammal community evolution. At the beginning of this interval, geographic expansion of the open-country woodland chronofauna began and diachronically reached its successional climax as well as its maximum geographic extent in the late Turolian (Bernor, 1978; Bernor *et al.,* 1979a).

The Western and Southern European Province has the most extensively developed sequence for Eurasia and North Africa. The base of MN 6 here records a major immigration of African and Sub-Paratethyan mammals into western Europe. Possibly involved in this immigration were hominoid primates (*Dryopithecus, Crouzelia*), proboscideans (*Deinotherium* and *Platybelodon*), bovids (*Protragocerus*), and pigs (*Kubanochoerus* and *Listriodon*) (Mein, 1975, 1979). A number of autochthonous evolutionary steps are also recorded here (Mein, 1979). Crusafont (1958) suggested that an Iberian–North African migration route opened at the Gibraltar portal during the middle "Vindobonian" (Astaracian). However, Berggren and Phillips (1971) have refuted this on the basis of tectonic arguments. The marked difference in the characteristic mammal communities of the Iberian and North African regions further argues against this connection (Bernor *et al.,* 1979a; Thomas, 1979). The most likely exchange route again would have been the Sub-Alpine Arch. The sharpened divergence of western Asian mammalian communities at this time, harboring more open-country forms, and the existence of more temperate conditions with deciduous forests in higher latitudes may have posed an ecological filter barring access by tropical forms.

Recent excavations by Ginsburg and Tassy (1977) at Simorre (MN 7) have illuminated the community ecology of the Astaracian in the Aquitaine Basin. The authors cite two distinct ecological associations there. The first is a "closed forest" with the MN 7 local characteristic association of *Zygolophodon turicensis, Dicrocerus elegans, Dorcatherium crassum,* and *Brachypotherium brachypus.* The second is a more open (yet probably still fairly closed) subtropical woodland with the characteristically associated mammals *Gomphotherium angustidens, Eotragus sansanensis,* and *Heteroprox larteti.* Provincially, these associations suggest a subtropical forest–woodland ecotone probably with several vegetation strata. The rest of this province's characteristic species includes a diverse rodent, insectivore, primate, ursid, and cervid fauna as well as some pigs. The underrepresentation of species characteristic of the open-country chronofauna typical of other districts supports this interpretation. Thenius' (1959) review of central European faunas shows a similar mammalian community composition. The myriad of localities rimming the northern flanks of the Alps in Germany, Switzerland, and Austria shows that the Astaracian mammal faunas there (his "Vindobonium") has a high diversity of small mammals, including insectivores, primates, rodents, and lagomorphs, small- to medium-sized carnivores, perissodactyls (*Anchitherium, Chalicotherium,* and *Brachypotherium*), and artiodactyls (especially diverse are cervids and suids) jointly characteristic of this same subtropical forested–woodland environment. Cicha *et al.* (1972) have more recently characterized central European small-mammal faunas, and suggested that it was zoogeographically contiguous with western Europe at this time.

The Sub-Paratethyan Province sharply contrasts with the western and central European zoogeographic theaters at this time. Recent work in Turkey (Sickenberg *et al.,* 1975) and Saudi Arabia (Hamilton *et al.,* 1978; Thomas *et al.,* in preparation; Whybrow *et al.,* in preparation) has added a great deal of information about their provincial faunas and geochronology. While there is some homotaxis between this and the Western and Central European Provinces, particularly the carnivores and rodents, the Sub-Paratethyan Province further records the evolution of progressive open-country large-mammal families: the Hyaenidae, Giraffidae, and Bovidae. These families include a number of genera common to the late Miocene open-country chronofauna which dominated the Old World. Faunal exchanges here establish the Sub-Paratethyan Province as the woodland environmental "hub" for a corridor of open habitats which extended from western North Africa eastward across Arabia into Afghanistan, northwest into the eastern Mediterranean area, and northeast into north China [Shanwang and Tung Gur faunal groups (Chiu *et al.,* 1979)]. East Africa also records the extension of open-habitat woodlands, as recorded by the Fort Ternan fauna, *ca.* 14 m.y.a. (Andrews and Walker, 1976). At the beginning of the Astaracian, dispersal barriers were relaxed and mammalian exchanges between provinces of these corridors were active. Advanced hominoid primates, *Sivapithecus* and *Ramapithecus,* bovids, and giraffids appear to have diversified within this province and dispersed into East Africa and the Siwalik Provinces. Species of the Giraffidae and Bovidae also appear to have extended their range into central Asia. Sub-Saharan rodents, including pedetids and phiomyids, exited Africa (Sen and Thomas, 1979), as did the suid *Listriodon* and tubulidentate *Oryceteropus.* Bernor *et al.* (1979*a*) have concluded that the development of a diverse open-country chronofauna is closely tied to the development of evergreen woodland vegetation type in which three vegetation strata—tree, chaparral, and herbaceous undergrowth—existed. This ecological setting offered a multistoried herbaceous growth for a variety of browsers, which in turn supported the diversity of small to large carnivores characteristic of the faunas. These ecological parameters, coupled with the free exchange permitted by the open-country zoogeographic corridor, promoted the rapid evolution of this "savanna-mosaic" chronofauna (*sensu* J. H. Van Couvering, 1980).

The North African Province has a number of faunas referred to MN Zones 7 and 8. This province has an interesting mixture of endemic small mammals and nonendemic large mammals. Jaeger (1977), who has developed the small-mammal biochronologic framework for this province, reports that an endemic rodent subfamily, the Myocricetodontinae, underwent an isolated adaptive radiation in North Africa. Ctenodactylid rodents appear to have been derived from a Paleogene central Asian stock which subsequently also underwent a local radiation in North Africa. The sub-Saharan rodent families Thryonomyidae and Pedetidae evidently entered North Africa prior to the middle Miocene, where they are last recorded from MN 7. Contrary to this pattern of rodent endemicity, the large carnivore groups Hyaenidae, Felidae, and Ursidae (Ginsburg, 1977) and ungulates Rhinoceratidae (Guerin, 1976), Giraffidae (Heintz, 1976), and Bovidae (Heintz, 1973) are all open-country forms that shared affinities with Sub-Paratethyan, East African, and Central Asian species. North African faunas are often coastal, and as such formed a lagoonal marine/continental interface that harbored environments which graded from marshy, to riverine fringe woodland near the coast to a "savanna-mosaic" woodland distally [for example, Sahabi (Boaz *et al.,* 1979)].

The Siwalik Province continues its record during the Astaracian. Recent work in the Potwar Plateau by the joint Yale–Geological Survey of Pakistan (GSP) project has done much to clarify the geochronology and mammalian faunal characteristics of this province (Pilbeam *et al.,* 1979). The Yale–GSP groups have initiated an extensive paleomagnetic–mammalian biostratigraphic–paleoecologic program for the Miocene-aged faunas. The authors have defined a sequence of 11 biostratigraphic and in part paleomagnetically calibrated "chronozones," which range through the Kamlial, Chinji, Nagri, and Dhok Pathan Formations. The first three zones include the faunas from the Kamlial and Chinji Formations, Zones 4–7, the Nagri Formation faunas;

and Zones 8–11, the Dhok Pathan faunas. Pilbeam *et al.* (1979) have correlated the Kamlial and Chinji faunas (Biozones 1–3) with the Astaracian faunas
of Europe and the Fort Ternan–Ngorora faunas of East Africa. Similarly,
Pilbeam *et al.* correlate the Nagri and Dhok Pathan faunas with Vallesian and
Turolian ones of Europe. Their recent paleomagnetic calibration of the Chin-
ji/Nagri boundary at 10.2 m.y.a. has led them to suggest that the Astaracian/
Vallesian boundary is the same.

Bernor (in preparation) has contested Pilbeam *et al.*'s (1979) use of Siwalik biostratigraphy/paleomagnetic parameters for defining the Astara-
cian/Vallesian boundary, primarily on the basis of calibration of the well-
documented first occurrence of the three-toed horse *Hipparion* at *ca.* 12.0
m.y.a. in Europe and North Africa (Bernor *et al.*, 1980; Bernor, 1978, in
preparation; Ameur *et al.*, 1979). Bernor *et al.* (1979*a*) and Bernor (in preparation) have noted that the Miocene Siwalik faunas are provincially distinct
for most of the Miocene, and do not assimilate the savanna-mosaic open-
country chronofauna until the late Turolian (MN 13, *ca.* 7 m.y.a.).

In evaluating the Astaracian/Vallesian chronologic boundary for the
Siwaliks, it would appear that the base of Pilbeam *et al.*'s (1979) Biozone 3,
within the Chinji, is closer to the Vallesian/Turolian boundary. The first reported occurrence of the murid, cf. *Progonomys*, at the base of Chinji Biozone
3 suggests a chronologically post-*Hipparion* first appearance (defined locally in
Europe as basal Vallesian) age for this boundary. Therefore, the chronologic
boundary between the Astaracian and Vallesian is probably best referred to
an unspecified point within Chinji Biozone 2, below the base of Biozone 3.

The Kamlial and Chinji faunas reported by Pilbeam *et al.* (1979) have
their own distinctive provincial characteristics. A number of the carnivores,
including *Hyainailouros*, *Dissopsalis*, and *Metapterodon*, were probably archaic
holdovers from the Orleanian emigration of tropical African forms. The
bovids are largely Astaracian in aspect, but their diversity is low compared
with the Sub-Paratethyan and North African districts. The giraffids also are
not as diverse, with only one genus, the curious *Giraffokeryx*, occurring. The
rodents, suids, and hominoid primates underwent a major evolutionary radiation in the Siwaliks during the Chinji and Nagri intervals. This mixed assemblage of archaic carnivores, a low diversity of browsing ruminants, and
woodland/bushland omnivores (pigs, ground-dwelling hominoids) sharply
contrasts with Sub-Paratethyan, North African, and Chinese faunas, which
during the same 15–10 m.y.a. interval harbored a diverse "savanna-mosaic"
fauna. The retarded appearance of this chronofauna in the medial Dhok
Pathan appears to be closely associated with the delayed late Miocene appearance of a Mediterranean-type sclerophyllous vegetation in northwestern
India and Pakistan (Axelrod, 1975, p. 297).

Astaracian hominoid primates appear to segregate into distinct provinces, as do other mammals. Reference to Fig. 4 reveals that *Dryopithecus* and
Pliopithecus are nearly limited in the Astaracian to the Western and Southern
European and Central European Provinces. The occurrences of *Dryopithecus*
at the late–middle Miocene localities of Udabno and Eldar on the western

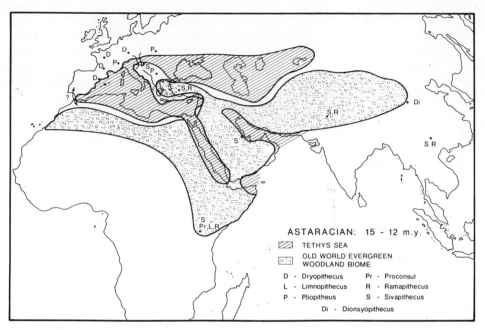

ASTARACIAN: 15 - 12 m.y.

TETHYS SEA

OLD WORLD EVERGREEN WOODLAND BIOME

D - Dryopithecus   Pr - Proconsul
L - Limnopithecus   R - Ramapithecus
P - Pliopitheus   S - Sivapithecus
Di - Dionsyopithecus

Fig. 4. Astaracian Mammal Age, 15–12 m.y.a. The more seasonally adapted Old World Evergreen Woodland Biome and its more open country environments become geographically extended during this interval. Astaracian mammal faunas show an increased provinciality, with more open habitats in the area spanning North and East Africa, Saudi Arabia, the eastern Mediterranean, and west and central Asia. More wooded areas were present in central and western Europe, the Siwaliks, southeast Asia, and sub-Saharan Africa. Hominoid primates show relatively strong provinciality during this interval also.

coast of the Caspian Sea are the only reported exceptions (Szalay and Delson, 1979). The several species of *Proconsul,* and *Limnopithecus legetet,* are only known to occur in sub-Saharan Africa at this time. *Sivapithecus* and *Ramapithecus* occur predominantly in the Sub-Paratethyan and Siwalik Provinces, as well as in China (Chiu *et al.,* 1979). *Sivapithecus* also made a limited incursion into the eastern part of the Central European Province, as evidenced by the limited material from Neudorf Sandberg (*S. darwini*) and East Africa ["*Sivapithecus africanus,*" Maboko Island (Andrews and Molleson, 1979)]. *Ramapithecus* also was largely a Sub-Paratethyan and Siwalik provincial form, but similarly extended its geographic range into East Africa (Maboko and Fort Ternan), eastern Europe [early Vallesian form, ?*Ramapithecus hungaricus* (Kretzoi, 1975; Andrews, 1978*b*)], and possibly southern Asia. The chronological ranges of middle and late Miocene hominoids in the Western and Southern European, Central European, Sub-Paratethyan, Siwalik, and East African Provinces are arrayed in Table II, and will be further elaborated upon in the sections which follow, especially in the final section.

    The genus *Dryopithecus* first appears in the European record during MN 8. While Mein (1979) cites the first appearance of this species as being MN 6, his reference is to large hominoids from the Neudorf Sandberg locality now recognized as being *Sivapithecus* (Delson, personal communication). Szalay and Delson (1979) recognize two species of *Dryopithecus* in Europe, *D. fontani* and *D. brancoi* (formerly *D. laietanus*). The known chronological range for *D. fontani* in the Western and Southern European Province, as well as the Central European Province, is MN 8–10. The smaller species, *D. brancoi,* ranges from MN 8 to 10 in the Western and Southern European Province, and from MN 9 to 10 in the Central European Province.

Andrews (1978a) has most recently studied the morphological attributes of *D. fontani* and concluded that it is very similar to the East African species *P. nyanzae* in its lack of a maxillary molar cingulum and in having a lower instead of an upper mandibular torus. Szalay and Delson (1979) recognize only a subgeneric distinction between the European and East African species: *D. (Dryopithecus)* in Europe, *D. (Proconsul)*, *D. (Rangwapithecus)*, and *D. (Limnopithecus)* in Africa. They have also closely united the larger *D. fontani* and smaller *D. brancoi*, stating that they essentially differ only in size. Andrews (1978a) has noted that for the most part, species of East African *Proconsul* were forest animals. Their morphology, particularly in the molar dentition (i.e., thin enamel, more constricted and pointed cusps compared to the sivapithecines*), supports Andrews' ecological observations. The faunal associations of Astaracian and Vallesian (next section) hominoids similarly suggest forested environments, which sharply contrasted with North African and Sub-Paratethyan open woodlands at this time. The fossil record includes no *Proconsul* or *Dryopithecus* species from middle and late Miocene woodland provinces which lie between the European and sub-Saharan areas (i.e., North African and Sub-Paratethyan Provinces), or, for that matter, anywhere outside of East Africa, Europe, and the western USSR. These various lines of evidence suggest that European dryopithecines probably exited East Africa prior to the early Astaracian (probably MN 5 with *Pliopithecus*) spread of woodland environments across the North African and Sub-Paratethyan and into the Central Asian Provinces. However, a later, Astaracian dispersal via a forest corridor [Rift Valley gallery forest of Van Couvering and Van Couvering (1976)] remains a possibility.

Sivapithecine primates make their first appearance in a number of geographic locations at the very beginning of the Astaracian Mammal Age, *ca.* 15 m.y.a. The earliest forms include the Czechoslovakian and Turkish species *Sivapithecus darwini,* the Saudi Arabian form "?*Sivapithecus*" sp. (Andrews *et al.*,

---

*I use the term sivapithecines in a generalized sense to denote the several genera of thick-enameled middle and late Miocene hominoids which occur in eastern Europe, southwest Asia, and east Africa. Szalay and Delson (1979) refer these to the tribe and subfamily Sugrivapithecini and Homininae, while Pilbeam (Pilbeam *et al.*, 1977b) has advocated a family distinction Ramapithecidae for these with two subfamilies or tribes: Sivapithecinae or Sivapithecini and Ramapithecinae or Ramapithecini. I leave the discussion of these systematic issues to others in this volume, but choose to use the term "sivapithecine complex" (no "*Sivapithecus* complex") as an ill-defined taxonomic rank to underline the uncertainty surrounding the correct systematic ordering and evolutionary relationships of these species of Hominoidea.

1978), and the East African species "*Sivapithecus africanus.*" While Andrews (1978*a,b*) and Andrews and Tobien (1977) have claimed that the Paşalar fauna, which includes *Sivapithecus darwini* and *Ramapithecus* cf. *wickeri,* is the oldest sivapithecine-bearing fauna (MN 5), Bernor (1978) and J. A. Van Couvering (personal communication, 1979) suggest that the Paşalar fauna is better referred to the earliest Astaracian (MN 6). Andrews *et al.* (1978) have claimed that the Saudi Arabian species from the Dam Formation (locality of Ad Dabtiyah), as well as the East African species from Maboko Island, "*S. africanus,*" share what is probably a derived dental complex (enlarged premolars relative to molars and thickened enamel) but differ from other sivapithecines in having a narrow palate, greater height of buccal cusps, and greater cingulum development. While Andrews prefers to separate these species from *Sivapithecus,* their derived characters may suggest the earliest morphological phases in the evolution of a mosaic of dental features characteristic of later sivapithecine species.

Andrews (1978*a,b*) derives *Sivapithecus darwini* from *Proconsul major,* stating that *S. darwini*'s lower molars retain a buccal cingulum and narrow trigonids as in *P. major.* However, the enlargement of the premolars, thick enamel, and expanded maxillary molar cusps presage *S. indicus'* morphology. Andrews' observations have led him to suggest a potential lineage: *P. major, S. darwini, S. meteai, Gigantopithecus bilaspurensis* [=*G. giganteus* of Szalay and Delson (1979)], *G. blacki.* He suggests two possibilities for the origin of *Ramapithecus:* (1) from the East African subgenus *Rangwapithecus,* and (2) from *Sivapithecus darwini. Ramapithecus* appears in the fossil record concurrently with early Astaracian *Sivapithecus* species. It is first reported from the early Astaracian (and equivalent age) localities of Paşalar, Turkey, and Maboko Island, East Africa. Its chronologic range continues in the Astaracian, where it is reported from the Sub-Paratethyan provincial locality of Çandir [MN 7; reported by Andrews and Tekkaya (1976)], the East African locality of Fort Ternan [*ca.* 14 m.y.a.; reported by Andrews and Walker (1976)], and the Siwalik Hills (Szalay and Delson, 1979). *Ramapithecus* probably extended its geographic range into eastern Europe during the Astaracian, but is not reported there until the earliest Vallesian [Rudabanya, early MN 9 (Kretzoi, 1975; Andrews, 1978*b*)].

Andrews and Evans (1979) have studied the local environmental context of *Ramapithecus* in East Africa and concluded that it inhabited nonforest woodlands. Preliminary observations of the Paşalar and Çandir faunas by Andrews (personal communication) have suggested a very similar environment for both the *Sivapithecus* and *Ramapithecus* species there. The first occurrence of sivapithecines in woodland environments appears to be closely associated with the earliest Astaracian spread of woodland habitats across the North African and Sub-Paratethyan Provinces (Bernor *et al.,* 1979*a*). Furthermore, the evolution, dispersal, and extinction of this so-called "sivapithecine-complex" would appear to be closely tied to the appearance, expansion, and replacement of subtropical woodlands in these provincial districts. A principal focus of paleoecologists working with this problem is to gain a better understanding about the mammalian community structure and range of habitats represented in these woodland environments. Andrews and Evans' (1979) characterization of the Fort Ternan mammalian fauna as containing predominantly browsing large ground mammals mixed with higher proportions of insectivores and frugivores than known from contemporary woodlands suggests an ecosystem very similar to the kind that Bernor *et al.* (1979*a*) and J. H. Van Couvering (1980) suggested as being characteristic of Eurasian and African Astaracian woodlands (Fig. 4). These woodland environments differed from extant African ones in having less strongly seasonal conditions, with rainfall being more abundant and evenly distributed during the summer season. During the late Miocene, these woodlands probably shifted to a more seasonal pattern. This shift promoted the further evolution of the large-mammal chronofauna, but selected against the diversity of smaller mammals, including insectivores, rodents, hominoid primates, and suids, which coexisted with a diverse assemblage of medium to large ungulates and small to large carnivores typical of the early woodland chronofauna.

Vallesian

The late Miocene corresponds to Daams *et al.*'s (1977) "Catalonian" Superstage, *ca.* 12–5 m.y.a. This interval includes two mammal ages: the Vallesian, MN 9–10 (*ca.* 12–10 m.y.a.), and the Turolian, MN 11–13 (*ca.* 10–5 m.y.a.). The Vallesian Land-Mammal "Stage" was named by Crusafont (1950) after a suite of localities from the Valles-Penedes Basins, near Barcelona, Spain. Crusafont and co-workers have on a number of occasions (Crusafont 1950, 1965; Crusafont and Golpe 1971, 1972; Crusafont and Santonja, 1964) commented on these particular faunas. His original intent in erecting the Vallesian "Stage" was to announce the intermediate ecological character and chronologic position of these Spanish faunas between the local "forested" biotopes of the Astaracian and the "savanna" biotopes of the Turolian interval. The Vallesian localities in the Barcelona district have the additional advantage of directly overlying the latest Astaracian fossiliferous conglomerates, sands, and marls, giving a stratigraphically continuous sequence across this boundary.

Vallesian-aged faunas are represented in all six of the provinces listed in Table 1, and have been additionally identified in the north and south China districts (Chiu *et al.*, 1979) as well as in East Africa (Bishop and Pickford, 1975). Zoogeographically, the Vallesian includes the immigration of some elements of the large mammal chronofauna into western and southern Europe, and fewer of these into the Central and Eastern European Province. The Siwalik Province apparently shows no significant community restructuring, retaining its archaic early and middle Miocene forest character. However, in direct contrast to this relict pattern, the North African, Sub-Paratethyan, and northern Chinese faunas are relatively progressive in their evolutionary succession of the large-mammal chronofauna. East Africa appears to have been at least partially isolated from the northern provinces, presumably by an ecological filter (Bernor, 1978; Thomas, 1979), and retained several older Astaracian taxa. The paleogeographic configuration represented in Fig. 5 shows that by the early Vallesian, the Tethys Sea had become disconnected in eastern Europe, isolating the Paratethys and Paleo-Mediterranean Seas. The Paratethys became progressively brackish, terminating any marine correlations between it and the Mediterranean Basin. During this interval, free faunal exchanges between central Europe and open habitat provinces to the east, southeast, and southwest would have been limited only by ecological filters (Bernor, 1978).

The Vallesian record of the Western and Southern European Province begins with a major immigration of the large-mammal chronofauna from the Sub-Paratethyan and possibly the western and central Asian areas. The North American equid *Hipparion* is the most celebrated immigrant here, as it was throughout almost all of Eurasia and Africa. Also occurring at this time are the south Asian murid *Progonomys;* several carnivores, *Eomellivora, Ictitherium, Lycyaena, Percrocuta, Machairodus,* and *Indarctos;* the rhinoceratid *Diceros;* and the giraffids *Decennatherium, Birgerbohlinia,* and *Helladotherium.* Mein (1979) also cites some autochthonous evolutionary steps in a few rodent lineages. This series of immigrations does not appear to have caused a mass extinction

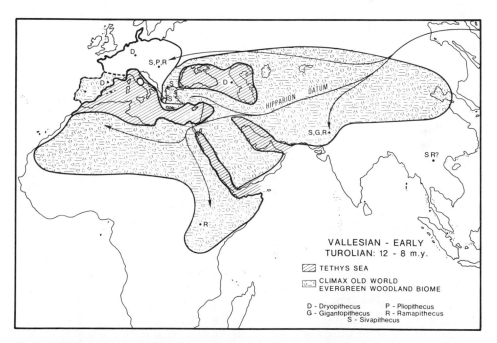

Fig. 5. Vallesian–Early Turolian Mammal Ages, 12–8 m.y.a. Regression of the Tethys and Paratethys Seas continued, as did the spread of the open-country large-mammal chronofauna. As environments became increasingly seasonal, geographic ranges of hominoid primates regressed. By 8 m.y.a., Eurasian hominoid primates appear to have become extinct except for the subtropical and tropical environments of southeast Asia.

of Astaracian species, but rather a provincial enrichment of the mammalian community. This province retained a diverse insectivore, rodent, and hominoid primate fauna. Some archaic carnivores also persisted, including amphycyonines, ursids (*Ursavus*), and mustelids (*Ischyrictis, Taxodon,* and *Trocharion*), and the primitive ictitherine hyaenid "*Progenetta.*" Proboscideans also include Astaracian genera (*Gomphotherium* and *Tetralophodon*), but the artiodactyls are most notable for retaining relicts, especially among the cervids (*Dorcatherium, Euprox, Micromeryx, Capreolus*) and suids (*Listriodon* and *Conohyus*). The second half of the Vallesian, MN 10, witnesses little provincial change in this pattern, only adding another carnivore (*Simocyon*), giraffid (*Samotherium*), and bovid (*Gazella ?deperdita*) (Mein, 1979) from the Sub-Paratethyan Province [see Alberdi (1974) and Bernor (in preparation) for a review of these faunas]. Ecologically, this pattern of immigration of carnivores and browsing ungulates superimposed upon the retention of several Astaracian forest denizens suggests the possibility that provincial environments were enriched by the shift toward more open woodlands.

The Central European Province appears to have received a more limited immigration of western Asian open-country species. The late Miocene large-mammal faunas from this province have not been studied since Thenius' (1959) overview, making provincial biochronologic rankings and zoogeographic interpretations less certain than for other districts. However, Thenius' (1959) listing of central European faunas suggests that while some large carnivores (*Machairodus,* "*Paramachairodus,*" *Ictitherium, Indarctos,* and *Simocyon*) and ungulates (*Hipparion, Microstonyx,* and *Miotragocerus*) entered here in the Vallesian, the faunas otherwise retained an Astaracian community character. Bernor (1978) and Bernor *et al.* (1979*a*) have interpreted this as meaning that the Central European Province largely retained a forest character, at least along the northern and eastern periphery of the Alpine System.

In contrast to the European Provinces, the Sub-Paratethyan Province appears to have progressively evolved its open woodland character during the Vallesian. While there are only a few large-mammal faunas from this province, the localities closely agree in their mammalian faunal composition. Advanced members of the Hyaenidae (*Percrocuta*), Felidae (*Machairodus*), Proboscidea (*Choerolophodon pentelici*), Rhinoceratidae (*Diceros neumayri*), Giraffidae (two species of *Palaeotragus, Bohlinia, Decennatherium*), and Bovidae (*Mesembriacus, Oioceros,* and *Prostrepsiceros*) occur here (Sickenberg *et al.,* 1975), while only a very few Astaracian taxa are present (Bernor, in preparation). This line of evidence suggests that seasonality continued to increase in the Sub-Paratethyan Province, and that more open habitats were in existence here than at contemporaneous localities of the European provinces.

The North African Province also has a relatively limited Vallesian fossil record. However, the pattern of mammalian community evolution here is very similar to the one seen in the Sub-Paratethyan Province. Advanced large mammal genera of the Hyaenidae, Felidae, Rhinoceratidae, Giraffidae, and Bovidae appear in the provincial record, with no record of small to large forest mammals typical of the higher latitudes of Europe. Chiu *et al.*'s (1979,

p. 267) report of Vallesian-equivalent north China localities of the Baho Formation similarly contain carnivore and ungulate members of the open country woodland chronofauna. An interesting feature of the North China
Province is that it shows a distinct provincial divergence from southern Chinese forested faunas (which contain species of *Sivapithecus*), as exemplified by
the Lufeng fauna (Chiu *et al.*, 1979). South of the Sub-Paratethyan Province,
the East African Province appears to have been partially isolated from the
events in the north. The Ngorora fauna (Bishop and Pickford, 1975) includes
some MN 9 immigrants (including *Percrocuta* and *Hipparion*), but the remainder of the identified ungulates (*Palaeotragus, ?Samotherium, Protragocerus,
?Pseudotragus,* and *Gazella*) appear to be holdovers from provincial middle
Miocene faunas. Delson (1975), Bernor (1978), and Thomas (1979) have suggested that the Sahara climate became increasingly seasonal during this period, and acted as an ecological barrier to free faunal exchanges with Eurasia.

The Siwalik Province continues to retain its own mammalian community
characteristics during the Vallesian. Pilbeam *et al.*'s (1979) calibration of the
Chinji/Nagri boundary at 10.2 m.y.a. corresponds closely with estimates of the
Vallesian/Turolian boundary as being *ca.* 10–9.5 m.y.a. (Berggren and Van
Couvering, 1974; Jaeger, 1976; Jaeger *et al.*, 1977). As discussed in the section
on the Astaracian, the Chinji Formation retained a number of relict genera of
the Orleanian African emigration, some Astaracian immigrants, as well as a
number of autochthonous lineages. This province appears not to have
changed its mammalian community structure in any significant way from
earlier Miocene times.

During the Vallesian, hominoid primates largely continued to persist in
their local provinces. In the Western and Southern European Province,
Pliopithecus became extinct, but *Dryopithecus fontani* and *D. brancoi* made their
final appearances during the late Vallesian. In the more forested environments of east and central Europe, hominoid primates enjoyed their greatest
diversity, including *Pliopithecus ?hernyaki, Dryopithecus fontani, D. brancoi,
Sivapithecus ?darwini* (or *?meteai*), and *?Ramapithecus hungaricus*. The Sub-Paratethyan Province has only one species of hominoid present, *S. meteai,* from its
northwest limit [see Andrews and Tekkaya (1980) for a review of this species].
Three localities have yielded this species: Middle Sinap, Turkey (Ozansoy,
1965; formerly "*Ankarapithecus*" *meteai*); Pyrgos, Greece (von Koenigswald,
1972; formerly "*Graecopithecus freybergi*"); and Ravin de la Pluie (de Bonis *et
al.,* 1973, 1975; de Bonis and Melentis, 1977, 1978, 1979; formerly "*Ouranopithecus macedoniensis*"). Hominoid primates continued to thrive in the Siwalik
Province, with reported occurrences of *Sivapithecus indicus, S. sivalensis,* and
Ramapithecus punjabicus (Szalay and Delson, 1979). It should be mentioned
that Pilbeam *et al.*'s (1979) recent report on Siwalik faunas does not document
these specific taxa within the Chinji Formation, noting only the existence of
"hominoids" there. As mentioned earlier, the south China Lufeng fauna contains a *Sivapithecus* species. This species is markedly sexually dimorphic, as is
S. meteai (de Bonis *et al.,* 1975; de Bonis and Melentis, 1979; Andrews and
Tekkaya, 1980), with the female possibly having been confused with the

smaller sivapithecine *Ramapithecus*. Finally, the East African hominoid record is very poor. Bishop and Pickford (1975) reported, and Howell (1978) has reviewed, a single M^2 from the Ngorora Formation Member C, which these authors tentatively have referred to Hominidae gen. et sp. indet.

Turolian

The Turolian Land-Mammal "Stage" was named by Crusafont (1965), and intended to replace his previous nonprovincial term the "Pikermian" (Crusafont 1950). The Turolian is a long interval [MN 11–13, *ca.* 10–5 m.y.a. (Berggren and Van Couvering, 1974; Bernor, in preparation)], and marks the maximum geographic extension of the large-mammal chronofauna. The transgression of this climax chronofauna is diachronic through Eurasia and Africa, and despite traditional views of pan-Eurasian zoogeographic homogeneity, some degree of provinciality is retained. The Western and Southern European Province continued to receive immigrants from Asia and Africa via the Sub-Alpine Arch throughout this interval, but still harbored a remarkable number of Astaracian relicts. Most of the Central European Province retained its forest character throughout the Turolian, with low diversities of typical open-country carnivores and ungulates (Bernor *et al.*, 1979a). Recent accounts of the Rumanian and Western USSR Province (Gabunia, 1979) suggest a close zoogeographic affinity with the Sub-Paratethyan Province, and paleogeographic reconstructuions (Figs. 5 and 6) reveal the lack of a prohibitive physical barrier (i.e., the Paratethys) between these. The Sub-Paratethyan, North African, and North China Provinces were the most progressive evolutionary theaters for the large-mammal chronofauna. The endemicity of the Siwalik and East African Provinces relaxed during the later Turolian, as evidenced by major faunal migrations between these and the North African and Sub-Paratethyan Provinces (Bernor, 1978; Bernor, in preparation).

The fossil record and its biochronologic subdivision is again best developed in the Western and Southern European Province. The first zone, MN 11, is a short interval (*ca.* 10–9 m.y.a.), and is marked by the first appearances of several rodent genera. Most notable is the "punctuated" diversification of the local murid fauna [first appearances of *Parapodemus*, *Valerymys*, and *Occitanomys* (Van de Weerd, 1976; Mein, 1979)], which continued through the Turolian. The ensuing intervals (MN 12 and 13) are distinguished by increased faunal resemblance of these provincial faunas with Sub-Paratethyan ones, marking more limited ecological filtering of its large-mammal chronofauna (Bernor, 1978). Here MN 12 includes the immigration of several medium to large mammals, including carnivores (*Enhydriodon*, *Baranogale*, ?*Canis*), advanced bovids (?*Prostrepsiceros*, *Protoryx*), proboscideans (*Choerolophodon*), and the African hippo, *Hexaprotodon primaevus*. Refined correlations between this and the Sub-Paratethyan Province have been difficult because of the lack of identification of evolutionary stages in the large-mammal groups which are

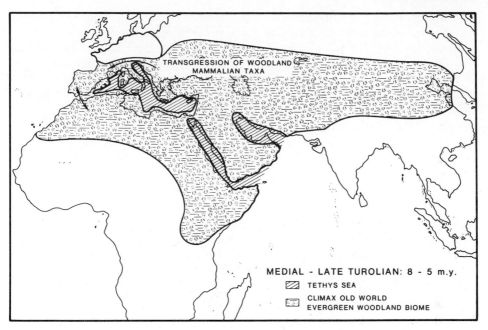

Fig. 6. Medial to Late Turolian Land-Mammal age, 8–5 m.y.a. This interval records the maximum transgression of the large-mammal open-country chronofauna. The transgression of this mammalian community type is diachronic, and provinciality reduces somewhat by the number of large- and small-mammal faunal exchanges. At the end of this interval the Mediterranean desiccates, permiting a faunal exchange across Gibraltar and perhaps other North African–southern European land routes. No hominoid primates are recorded at this time from Eurasia. The African hominoid record, though sparse, indicates the potential for being the nexus of early australopithecines.

common between these two areas. Recently, some success in correlating these provinces by comparable stages of evolution in hipparionine horses has been realized (Bernor *et al.*, 1980) and has allowed direct age rankings within and between these provinces. While this interval marks an even greater transgression of the large-mammal chronofauna into western Europe, the diversity of its characteristic taxa is not nearly as great, and a number of middle Miocene endemic species continue to be harbored there [*Proscampanus sansaniensis, Prolagus oeningensis, Steneofiber jaegeri, Palaeomeryx meyeri,* and *Protragocerus* (Alberdi, 1974)].

MN 13 is the terminal Miocene zone. The Turolian trend of transgression of the Sub-Paratethyan chronofauna into western Europe continues with the immigration of a full range of small to large mammals. Western Asian genera (*Protatera, Diceros, Iranotherium, Sus*), North African genera (*Paraethomys, ?Parabos, Myocricetodon*), and the East African genus *Anancus* occur for the first time in western Europe (Mein, 1979; Howell, 1980). Although the Sub-Alpine Arch remained the major migration route at this time, Jaeger *et al.* (1977) and Howell (1980) cite a brief interval of exchange across the Gibraltar portal during the Messinian desiccation of the Mediterranean Basin, *ca.* 5.5–5 m.y.a. Remarkably, Spain still retained a number of early and middle Miocene relicts, including a carnivore (*Hemicyon*), proboscidean (*Zygolophodon*), cervid (*Euprox*), and a very primitive bovid (*Eotragus*) (Alberdi, 1974). The Italian faunal sequence at Baccinello, famed for its enigmatic cercopithecoid primate *Oreopithecus* (Szalay and Delson, 1979), ranges through MN Zones 12 and 13 and shows an insular island aspect until the uppermost V3 fauna. Although Italy consisted of a number of isolated islands during most of the Miocene and its fauna was highly endemic, some western European mammals (murids) immigrated there (Hürzeler and Engesser, 1976). This suggests that the Baccinello fauna is better aligned with the Western and Southern European instead of North African Province.

The Central European Province shows an ecological gradient between the forests of the northern Alpine flanks, the Vienna Basin, Hungarian Plain, and Yugoslavian area. Franzen and Storch (1975) have reported an early Turolian locality from Germany, Dorn Durkheim, which contains a low diversity of the large-mammal open-country chronofauna, including one species of hyaenid, one species of equid, no giraffids, and only one species of bovid. Thenius' (1959) report of the Vienna Basin locality of Eichkogel indicates that open-country forms including an ostrich (*Struthio*), aardvark (*Orycteropus*), rhinoceratid (*Diceros pachygnathus*), a giraffid, and bovids (*Gazella* and *Miotragocerus*) immigrated this far north. Further south in the Hungarian Plain, a cluster of three Turolian localities, Kohfidisch, Baltavar, and Polgardi (as well as the late Vallesian locality Csakvar), rim Lake Balaton. Thenius (1959) reports a similar assimilation of the open-country large-mammal chronofauna with the occurrence of carnivores (*Simocyon, Ictitherium, Percrocuta, Machairodus*), the terrestrial colobine monkey *Mesopithecus pentelici,* a suid (*Microstonyx*), and bovids (*Gazella* and *Miotragocerus*). A preliminary analysis of the Yugoslavian locality of Titov Veles (Solounias, personal communication) suggests that it was the northernmost eastern European extension of true Sub-Parathethyan large-mammal faunas. As cited earlier, Gabunia's (1979) brief summary of the western USSR Turolian faunas suggests that Sub-Parathethyan faunas also fully extended into this province during the Turolian.

The Sub-Parathethyan Province retains its distinction as the zoogeographic "hub" for the large-mammal woodland chronofauna and interprovincial mammalian exchanges. Recent work on provincial large-mammal faunas in Greece (Solounias, 1979, 1981; de Bonis *et al.,* 1973, 1975, 1979), Turkey (Sickenberg *et al.,* 1975; Gaziry, 1976; Schmidt-Kittler, 1976), Iran (Bernor, 1978; Bernor *et al.,* 1979*b*; Campbell *et al.* 1980), and Afghanistan (Heintz *et al.,* 1978) has done much to enhance our knowledge of the diversity of open-country mammals and the biochronology and zoogeographic characteristics of these faunas. The hyaenids, felids, proboscideans, equids, rhinoceratids, giraffids, and bovids reach their greatest known diversity in this province. Geochronologic ranking of these faunas has been difficult because of the long chronologic stability of this mammalian community (Bernor *et al.,* 1979*a*). However, recent work in the Maragheh Basin of northwestern Iran by the Lake Rezaiyeh Project has preliminarily subdivided the 2.5 million-year-long early and medial Turolian sequence into three biostratigraphic intervals by the stages of evolution of a single potential *Hipparion* lineage (Bernor, 1978; Bernor *et al.*, 1979*b*, 1980). Biochronologic correlations with the peri-Mediterranean and East African localities (Bernor *et al.*, 1980), and recent radiometric calibrations of the Maragheh sequence (Campbell *et al.*, 1980) have established this locality as a principal "anchor point" for provincial geochronology. Mammalian systematic studies of the classic Greek Turolian localities of Pikermi and Samos by Solounias (1979, 1981) have revealed the previously unrealized diversity of the large-mammal chronofauna in this province. The mammalian communities here were far more diversified in open-country large mammals than are extant savanna communities, a phenomenon which Bernor *et al.* (1979*a*) have attributed directly to the presence of a sclerophyllous evergreen woodland. Climatic conditions at this time were slightly drier and cooler than in the Vallesian, but winters were frost-free, providing year-round growth (and thus food for the diversity of browsers present), and summers were drier but still relatively wet. Bernor *et al.* (1979*a*) have suggested that the presence of three vegetation strata (tree, chaparral, and herbaceous layers) was probably the single most important factor in the evolution of such a species-rich biome.

A preliminary study of the Chinese Turolian faunas (Chiu *et al.*, 1979; Solounias, 1981) reveals a close zoogeographic affinity between this and the Sub-Paratethyan Province. Once again, a diverse fauna of open-country forms inhabited this province. Solounias (personal communication) has suggested that generic faunal resemblances between Chinese and Sub-Paratethyan Turolian faunas may be as high as 90%. On the other hand, North African large-mammal faunas are very poorly known for this interval. However, recent work at the Libyan locality of Sahabi by the International Sahabi Project (Boaz *et al.*, 1979) has yielded an extensive suite of MN 13–14 Boundary large mammals which shows a mixture of western European, Sub-Paratethyan, and East African large-mammal affinities (Thomas *et al.*, 1982).

The Siwalik Province seemingly retains its endemic character until the medial Dhok Pathan, *ca.* 8–7 m.y.a. Pilbeam *et al.* (1979) have reported a major faunal turnover during local Siwalik Biozones 9–11, *ca.* 8–6 m.y.a. Small mammals are not well represented at this time (which is typical for Turolian large-mammal faunas), and hominoid primates appear to have become extinct by Biozone 9. The hyaenids *Lycyaena* and *Palhyaena indicum* first appear in Biozones 9 and 10, respectively. The smaller carnivores common to Sub-Paratethyan faunas, *Felis* and *Plesiogulo*, appear in Biozone 10, and *Enhydriodon* in Biozone 11. The suids undergo a major replacement of several genera at the base of Biozone 10. The hippo genus common in medial and late Turolian peri-Mediterranean localities, *Hexaprotodon*, first occurs in Biozone 10. Giraffids diversify in Biozones 10 and 11, and East African (*Kobus*) and Sub-Paratethyan (*Prostrepsiceros*) bovids first occur in Biozone 9. The extensive transgression of open-country large mammals at this time suggests a close correlation with Axelrod's (1975) citation of a late Miocene transgression of a sclerophyllous woodland community into India and Pakistan.

Turolian transgression of the open-country chronofauna across Eurasia is coincident with hominoid extinctions (Figs. 5 and 6). By the earliest Turolian, middle Miocene hominoids became extinct in the Western and Southern European and Central European Provinces. The occurrence of *Sivapithecus meteai* at Pyrgos, Greece has been reported by von Koenigswald (1972; "*Graecopithecus freybergi*") as early Turolian, but the faunal information is insufficient for distinguishing the age of this localities from late Vallesian localities. In the Siwalik Province, the last reported occurrence of hominoids in Siwalik Biozone 8 is followed directly by the transgression in local Biozones 9–11 of the large-mammal chronofauna, and evidently the late extension of typical sclerophyllous woodlands (Axelrod, 1975). In East Africa, the record is unclear. Howell (1978) has most recently reviewed MN 13 equivalent (7–5 m.y.a.) hominoid material from Lukeino and Lothagam I. A current issue among students of hominid evolution is not only at what time did the earliest known bipedal hominid *Australopithecus* diverge from an earlier hominoid stock, but whether or not a late Miocene Siwalik hominoid was ancestral to this African hominid.

Summary and Conclusions

This presentation has attempted to assess the zoogeographic relationships of Eurasian and African hominoids within a network of geochronologic, paleoenvironmental, and paleogeographic information. The geochronologic framework has been strongly emphasized here because interpretations of the timing of hominoid evolutionary events, dispersals, and extinctions is dependent upon as accurate a chronologic referral as possible. From the geochronologic framework, mammalian zoogeographic relationships are contrasted across provinces, and paleoenvironmental and paleogeographic interpretations are enlisted to help explain the timing of faunal exchanges and likely migration corridors. Similarly, divergence of the various zoogeographic provinces is noted, and explanation by paleogeographic and/or paleoenvironmental features is given. The resultant interpretive framework is dynamic and requires constant testing by the empirical "building blocks" of floral and faunal systematics and physical stratigraphy.

The chronologic ranges of late–early Miocene through late Miocene Eurasian and African hominoids is given in Table 2. Miocene hominoids most certainly draw their origins from an early African group (Campbell and Bernor, 1976). The Hominoidea flourished exclusively on that continent until the latest Orleanian (MN 5, *ca.* 16 m.y.a.), when an African–Arabian–Eurasian land bridge allowed their prochoresis out of Africa. The European hominoid *Pliopithecus* first appeared during MN 5, probably reaching the Western and Southern European Province first by the Sub-Alpine migration corridor of Antunes (1979). *Pliopithecus* flourished in the Western and Southern European Province and the Eastern and Central European Province from MN 5 to

Table 2. Provincial Chronologic Ranges of Middle and Late Miocene Hominoidea[a]

Age, m.y.a.	15	14	13	12	11	10	9	8	7	6	5
European Land Mammal Age	Orleanian	Astaracian			Vallesian			Turolian			
European Mammal Neogene Zone	5	6	7	8	9	10	11	12a	12b	13	

Western and Southern European Province:
- Pliopithecus piveteaui
- Pliopithecus antiquus
- Crouzelia auscitanensis
- Pliopithecus sp.
- Dryopithecus fontani
- Dryopithecus brancoi

East and Central European Province:
- Pliopithecus vindobonensis
- Pliopithecus antiquus
- Pliopithecus lockeri
- Pliopithecus ?hernyaki
- Dryopithecus fontani
- Dryopithecus brancoi
- Sivapithecus darwini (or meteai here?)
- ?Ramapithecus hungaricus

Sub-Parathethyan Province:
- Sivapithecus darwini
- cf. Sivapithecus sp.
- Ramapithecus wickeri
- Sivapithecus meteai

(continued)

Table 2 (*Continued*)

Age, m.y.a.	15	14	13	12	11	10	9	8	7	6	5
Siwalik Province					?	*Sivapithecus indicus*	*Sivapithecus sivalensis*	*Ramapithecus punjabicus* *Gigantopithecus bilaspurensis*			
East African Province	*Dendropithecus macinnesi* *Proconsul africanus* *Proconsul nyanzae* *Proconsul major* *Limnopithecus legetet* *Rangwapithecus vancouveringi* "*Sivapithecus africanus*" *Ramapithecus wickeri*					Hominidae gen. et sp. indet.			*Australopithecus* sp.		
Marine stage	Langhian		Serravallian			Tortonian				Messinian	
Epoch	Middle Miocene					Late Miocene					

ᵃ Solid horizontal time lines indicate chronological ranges; dashed lines indicate a more general estimate of chronological ranges.

MN 9 (*ca.* 16–11 m.y.a.), when it became extinct. A close relative of *Pliopithecus, Crouzelia auscitanensis,* has been reported from a brief interval in France [MN 6 (Ginsburg, 1975)]. The European dryopithecine *Dryopithecus* first appears in the fossil record during MN 8. Andrews (1978*a,b*) has suggested that *Dryopithecus fontani* is morphologically similar to *Proconsul nyanzae* and Szalay and Delson (1979) claim a close relationship between *D. fontani* and a smaller form *D. brancoi.* It is argued here that because of the close morphological distance between *Dryopithecus* and the African tropical forest *Proconsul* species, and because of the apparent subtropical forest/closed woodland communities inhabited by *Dryopithecus,* it entered Europe by a forest corridor at the end of the Orleanian. Southern, western, central, and eastern European hominoids appear to have become extinct during the later Vallesian. Major environmental change to more seasonal environments and competition from monkeys (Delson, personal communication) are cited as plausible causes for their demise.

A group of hominoids with thick cheek tooth enamel, referred to here as the "sivapithecine complex," is most interesting to students of hominoid evolution because of their potential ancestry to the latest Miocene and Pliocene African genus *Australopithecus.* The first appearance of these nonforest apes is at the base of the Astaracian, *ca.* 15 m.y.a. Their earliest record includes a broad distribution centered in the Sub-Paratethyan Province (*Sivapithecus darwini,* cf. "*Sivapithecus*" sp., *Ramapithecus wickeri*), and extending to the northeast in the Vienna Basin of the Eastern and Central European Province (*S. darwini*), into the East African Province ("*S. africanus*" and *R. wickeri*), and probably eastward into the Siwalik and South Asian Provinces ["hominoids" in the Kamlial and Chinji Formations (Pilbeam *et al.,* 1979)]. The appearance of these nonforest denizens coincides closely with the appearance of a seasonally adapted sclerophyllous evergreen woodland biome, the early adaptive radiation of woodland ruminants (Bovidae and Giraffidae), and the subsequent diversification of their principal predators, the Hyaenidae and Felidae. The geographic ranges of these nonforest apes and the European ape *Pliopithecus* overlap only in eastern Europe (MN 6, Neudorf Sandberg, Vienna Basin; MN 9, Rudabánya, Hungary).

Species of the "sivapithecine complex" became extinct in eastern Europe and the Sub-Paratethyan Province by the early Turolian (*ca.* 10–9 m.y.a.), while they apparently survived in the Siwalik Province until *ca.* 8–7 m.y.a. [last appearance, Biozone 8 of Pilbeam *et al.* (1979)]. In East Africa the record is disappointingly poor, and the time of extinction of this group is not known. In Eurasia, extinction of species belonging to the "sivapithecine complex" is heralded by the evolution of more open-country faunas which include diversified ungulate (Equidae, Rhinoceratidae, Giraffidae, and Bovidae) and carnivore (Hyaenidae and Felidae) families. These Miocene hominoid extinctions also appear to correspond with broad interprovincial environmental change to increasingly seasonal environments harboring more highly open woodlands. Bernor *et al.* (1979*a*) and J. H. Van Couvering (1980) have emphasized that the Turolian Biome was a woodland that cannot be directly compared to extant savannas, because of the existence of three distinctive vegetation strata: tree, bush, and herbaceous levels. This habitat was capable of supporting a much richer diversity of herbivores and their predators than are extant savannas.

A compelling issue is the immediate ancestry of the first demonstrably bipedal hominid, *Australopithecus*. The hominoid-rich faunas of the Lower and Middle Siwaliks have focused attention on this region as a likely spot for hominid origins (Pilbeam *et al.*, 1977*a,b;* 1979). Indeed, species of the "sivapithecine complex" from the Potwar Plateau are morphologically quite close to *Australopithecus*. However, the zoogeographic patterns depicted here suggest that the Siwalik hominoids may not include the direct ancestor of *Australopithecus*.

The faunal and environmental information presented here depicts the Siwalik Province as being quite distinct from at least the early Astaracian. While the Sub-Paratethyan, North African, and North Asian zoogeographic provinces were rapidly evolving an open-woodland chronofauna during the Astaracian and Vallesian, faunal and geochronologic information (Pilbeam *et al.*, 1979) suggests that the Siwalik Province retained more closed environmental conditions which harbored early Miocene forest closed-woodland relics and endemically evolved forms. This interpretation gains further support from faunal migration information, which shows a predominant dispersal of mammals *into* and not *out of* the Siwalik Province during the Miocene [the most notable exceptions being some murid rodents (Jacobs, 1979)]. Major pulses of immigration across the Sub-Paratethyan Province into the Siwalik Province were during the medial Orleanian, early Astaracian, and medial/late Turolian. The reason for strong zoogeographic provinciality in the Siwaliks may be found in a paleogeographic–environmental explanation. Today, the Potwar Plateau is flanked to the north by the northwest - southeast-trending Himalayas and to the west by the north-south-trending Sulaiman and Kirthar ranges of Baluchistan and Sind. Sahni and Mitra (1980) have reported that the Himalayan System underwent one of its strongest orographic pulses in the middle Miocene. The uplift of these encircling mountain ranges would have trapped moist Indo-Pacific monsoons on the slopes facing the Siwalik Province and nourished the less seasonal environments there. While faunal migrations in and out of the Siwalik Province may have been possible by a lowland corridor through southern Baluchistan, the ecological differences caused by the local geography and climatology probably supported this distinct biome.

While species of the "sivapithecine complex" appear to have been adapted to more open habitats than were their sub-Saharan African or European provincial relatives, they have not been found in any Turolian locality (with the only possible exception being Pyrgos). The evolution of the Turolian open-country chronofauna was most diversified in the more seasonal environments of the North Asian, Sub-Paratethyan, and North African Provinces. The presence of biotopes that were more seasonal than those that species of the "sivapithecine complex" appear to have been adapted to, intermediately positioned between the Siwalik and east African provinces, could have posed an effective ecological barrier between the two areas by the earliest Turolian. If this were the case, the evolution of *Australopithecus* would most likely have been an African event. Bernor (1978) and Thomas (1979) have suggested a seasonal environmental barrier between North Africa and sub-Saharan Africa which may also have strongly influenced the apparent restriction of *Australopithecus* in sub-Saharan Africa.

This interpretation of *Australopithecus* origin must stand strictly as a hypothesis. The fact remains that faunal migrations took place between the East African and Siwalik Provinces for most of the Miocene. The notion that late Miocene desertic environments existed between these areas has recently been rejected by Axelrod (1975, and personal communication) and Axelrod and Raven (1978). However, Burckle (personal communication) has reported the presence of sand in deep sea cores taken off the west coast of Africa, and suggested the possibility of some desertic conditions in North Africa. Although some highly seasonal (desertic?) environments may have existed in North Africa, the very fact that several mammal taxa are reported as having migrated across the North African and Sub-Paratethyan Provinces in itself suggests that at least some nondesertic habitats existed during the late Miocene. As an example, the emigration of the hippopotamid *Hexaprotodon primaevus* from East Africa into the Mediterranean (Mein, 1975, 1979) and Siwalik regions (Pilbeam *et al.*, 1979) would have required a network of watercourses for this animal's passage. These same watercourses could have harbored the potential combination of habitats for the passage of hominoid primates. The eventual acceptance or rejection of hypotheses about *Australopithecus* origins must be verified by well-calibrated fossil evidence. The enlistment of zoogeographic, paleogeographic, and paleoenvironmental information requires more data and refined methods for their evaluation.

ACKNOWLEDGMENTS

This presentation is a direct outgrowth of research made under the auspices of three paleontological research projects: the Lake Rezaiyeh Project, 1973–1978; the International Sahabi Research Project, 1977–current; the Spanish Neogene Mammalian Evolution and Geochronology Project, 1979–current. I am grateful to the Director of the Lake Rezaiyeh Project, Dr. Bernard G. Campbell, the Director of the International Sahabi Research Project, Dr. Noel T. Boaz, and the National Museum of Natural History, Madrid, for invitations to make extensive faunal investigations during the last several years. Virtually all of my travel funds for field collection and museum study were graciously provided by Mr. Gordon Getty and the L. S. B. Leakey Foundation, to whom I am inestimably thankful. I would also like to thank P. Andrews, E. Delson, C. Hansen, P. Whybrow, and M. O. Woodburne for reviewing the text and providing many useful comments and suggestions.

The illustrations were drawn by G. Thomas and the photographs taken by B. Hicks, Department of Earth Sciences, University of California, Riverside. Finally, I would like to thank the numerous individuals of the European and American museums and universities who made my studies so enjoyable.

References

Alberdi, M. T. 1974. Las faunas de *Hipparion* de Los Yacimientos Espanoles. *Estud. Geol.* **30:**189–212.

Ameur, R., Bizon, G., Jaeger, J.-J., Michaux, J., and Muller, C. 1979. A propos de l'immigration des hipparions en Afrique du Nord, in: *7e Reun. Sci. de Terre, Lyon*, p. 8.

Andrews, P. J. 1978a. A revision of the Miocene Hominoidea of East Africa. *Bull. Br. Mus. (Nat. Hist.) Geol.* **30:**85–224.

Andrews, P. J. 1978b. Taxonomy and relationships of fossil apes, in: *Recent Advances in Primatology,* Volume 3, *Evolution* (D. J. Chivers and K. A. Joysey, eds.), pp. 43–56, Academic, London.

Andrews, P. J. (ed.). 1981. The Miocene of East Africa. *J. Hum. Evol.* **10:**1–158.

Andrews, P. J., and Evans, E. N. 1979. The environment of *Ramapithecus* in Africa. *Paleobiology* **5**(1):22–30.

Andrews, P. J., and Molleson, T. I. 1979. The provenance of *Sivapithecus africanus. Bull. Br. Mus. (Nat. Hist.) Geol.* **32:**19–23.

Andrews, P. J., and Tekkaya, I. 1976. *Ramapithecus* in Kenya and Turkey, in: *Les Plus Anciens Hominides* (P. V. Tobias and Y. Coppens, eds.), pp. 7–25, Colloque VI, XI Union Internationale des Sciences Prehistoriques et Protohistoriques, Nice. CNRS, Paris.

Andrews, P. J., and Tekkaya, I. 1980. A revision of the Turkish Miocene hominoid *Sivapithecus meteai. Palaeontology* **23:**85–95.

Andrews, P. J., and Tobien, H. 1977. New Miocene locality in Turkey with evidence of the origin of *Ramapithecus* and *Sivapithecus. Nature* **268:**699–701.

Andrews, P. J., and Van Couvering, J. H. 1975. Palaeoenvironments in the East African Miocene, in: Approaches to Primate Paleobiology (F. S. Szalay, ed.). *Contrib. Primatol.* **5:**62–103.

Andrews, P. J., and Walker, A. 1976. The primate and other fauna from Fort Ternan, Kenya, in: *Human Origins: Louis Leakey and the East African Evidence* (G. L. Isaac and E. R. McCown, eds.), pp. 279–304, Benjamin, Menlo Park, California.

Andrews, P., Hamilton, W. R., and Whybrow, P. J. 1978. Dryopithecines from the Miocene of Saudi Arabia. *Nature (Lond.)* **274:**249–250.

Andrews, P. J., Lord, J. M., and Evans, E. M. N. 1979. Patterns of ecological diversity in fossil and modern mammalian communities. *Biol. J. Linn. Soc.* **11:**177–205.

Antunes, M. T. 1979. *Hispanotherium* fauna in Iberian Middle Miocene, its importance and paleogeographic meaning. *Ann. Geol. Pays Hellen.* **1979**(1):19–26.

Axelrod, D. I. 1975. Evolution and biogeography of Madrean-Tethyan sclerophyll vegetation. *Ann. Mo. Bot. Gard.* **62**(2):280–334.

Axelrod, D. I., and Raven, P. H. 1978. Late Cretaceous and Tertiary vegetation history of Africa, in: Biogeography and Ecology of Southern Africa (M. J. A. Werger, ed.). *Monogr. Biol.* **31:**77–130.

Berggren, W. A., and Phillips, J. D. 1971. Influence of continental drift on the distribution of the Tertiary benthonic foraminifera in the Caribbean and Mediterranean regions, in: *Symp. Geo. Libya Fac. Sci. Univ. Libya*, pp. 263–297, Catholic University Press, Beirut.

Berggren, W. A., and Van Couvering, J. A. 1974. The late Neogene biostratigraphy, geochronology, and paleoclimatology of the last 15 million years in marine and continental sequences. *Palaeogeogr. Palaeoclimatol. Palaeoecol.* **16**(1/2):1–216.

Bernor, R. L. 1978. *The Mammalian Systematics, Biostratigraphy and Biochronology of Maragheh and Its Importance for Understanding Late Miocene Hominoid Zoogeography and Evolution*, Ph.D. Dissertation, University of California, Los Angeles, 314 pp.

Bernor, R. L., Andrews, P. J., Solounias, N., and Van Couvering, J. H. 1979a. The evolution of "Pontian" mammal faunas: Some zoogeographic, paleoecologic, and chronostratigraphic considerations. *Ann. Geol. Pays Hellen.* **1979**(1):81–90.

Bernor, R. L., Tobien, H., and Van Couvering, J. A. 1979b. The mammalian biostratigraphy of Maragheh. *Ann. Geol. Pays Hellen.* **1979**(1):91–100.

Bernor, R. L., Woodburne, M. O., and Van Couvering, J. A. 1980. A contribution to the chronology of some Old World Miocene faunas based on hipparionine horses. *Géobios* **13**(5):25–59.

Bishop, W. W., and Pickford, M. H. L. 1975. Geology, fauna and palaeoenvironments of the Ngorora Fm., Kenya Rift Valley. *Nature (Lond.)* **254**:185–192.

Blow, W. H. 1969. Late Middle Eocene to Recent planktonic foraminiferal biostratigraphy, in: *International Conference on Planktonic Microfossils, Geneva 1967* (P. Bronnimann and H. H. Renz, eds.), pp. 199–421, Brill, Leiden.

Boaz, N. T., Gaziry, A. W., and El-Arnuati, A. 1979. New fossil finds from the Libyan upper Neogene site of Sahabi. *Nature (Lond.)* **280**:137–140.

Campbell, B. G., and Bernor, R. L. 1976. The origin of the Hominidae: Africa or Asia? *J. Hum. Evol.* **5**:441–454.

Campbell, B. G., Amini, M. H., Bernor, R. L., Dickenson, W., Drake, R., Morris, R., Van Couvering, J. A., and Van Couvering, J. H. 1980. Maragheh: A classical late Miocene vertebrate locality in northwestern Iran. *Nature (Lond.)* **287**:837–841.

Chiu, C. S., Li, C. K., and Chiu, C. T. 1979. The Chinese Neogene. A preliminary review of the mammalian localities and faunas. *Ann. Geol. Pays Hellen.* **1979**(1):263–272.

Cicha, I., Fahlbusch, V., and Fejfar, O. 1972. Biostratigraphic correlation of some late Tertiary vertebrate faunas in Central Europe. *Neues Jahrb. Palaeontol. Abh.* **140**:129–145.

Ciochon, R. L., and Corruccini, R. S. 1977. The phenetic position of *Pliopithecus* and its phylogenetic relationship to the Hominoidea. *Syst. Zool.* **26**:290–299.

Crusafont, M. 1950. El sistema Miocenico en la depression Española del Valles-Penedes, in: *Proc. Intern. Geol. Congr. 18th Sess., London 1948*, Part 11, pp. 33–42.

Crusafont, M. 1958. Endemism and paneuropeism in Spanish fossil mammalian faunas, with special regard to the Miocene. *Soc. Sci. Fenn. Comm. Biol.* **18**(1):3–31.

Crusafont, M. 1965. Observations á un travail de M. Freudenthal et P. Y. Sondaar sur des nouveaux gisements á *Hipparion* d'Espagne *Proc. K. Ned. Akad. Wet. Amsterdam B* **68**:121–126.

Crusafont, M., and Golpe, J. M. 1971. Biozonation des mammiferes Neogenes d'Espagne. *Bur. Recher. Geol. Min. Lyon.* **78**:121–129.

Crusafont, M., and Golpe, J. M. 1972. Dos nuevos yacimientos del Vindoboniense en el Valles. *Acta Geol. Hisp.* **VII**(2):71–72.

Crusafont, M., and Santonja, J. T. 1964. Apercu chronostratigraphique des bassins de Calatayud-Teruel. *Inst. Luc. Mallad.* **IX**:89–92.

Daams, R., Freudenthal, M., and Van de Weerd, A. 1977. Aragonian, a new stage for continental deposits of Miocene age. *Newsl. Stratigr.* **6**(1):42–55.

De Bonis, L., and Melentis, J. 1977. Un nouveau genre de primate hominoide dans le Vallesian (Miocene Superieur) de Macedoine. *C. R. Acad. Sci. Paris D* **284**:1393–1396.

De Bonis, L., and Melentis, J. 1978. Les primates du Miocene Superieur de Macedoine, étude de la Machoire superieure. *Ann. Paleontol. Vertebr.* **64**(2):185–202.

De Bonis, L., and Melentis, J. 1979. Les hominoides du Miocene de Grece. Leur place dans l'histoire biogeographique des primates. *Ann. Geol. Pays Hellen.* **1979**(1):177–182.

De Bonis, L., Bouvrain, G., Keraudren, B., and Melentis, J. 1973. Premier resultats des fouilles recentes en Grece Septentrionale (Macedoine). *C. R. Acad. Sci. Paris D* **277**:1431–1434.

De Bonis, L., Bouvrain, G., and Melentis, J. 1975. Nouveaux restes de primates hominoides dans le Vallesian de Macedoine (Gréce). *C. R. Acad. Sci. Paris D* **281**:379–382.

De Bonis, L., Bouvrain, G., and Geraads, D. 1979. Artiodactyles du Miocene Superieur de Macedoine. *Ann. Geol. Pays Hellen.* **1979**(1):167–176.

Delson, E. 1975. Paleoecology and zoogeography of the Old World monkeys, in: *Primate Functional Morphology and Evolution* (R. H. Tuttle, ed.), pp. 37–64, Mouton, The Hague.

Delson, E., and Andrews, P. J. 1975. Evolution and interrelationships of the catarrhine primates, in: *Phylogeny of the Primates* (W. P. Luckett and F. S. Szalay, eds.), pp. 405–446, Plenum, New York.

Dewey, J. F., Pitman, W. C., Ryan, W. B. F., and Bonnin, J. 1973. Plate tectonics and the evolution of the Alpine System. *Bull. Geol. Soc. Am.* **84**:3137–3180.

Eames, F. E. 1970. Some thoughts on the Neogene/Paleogene boundary. *Palaeogeogr. Palaeoclimatol. Palaeoecol.* **8**:37–48.

Franzen, J. L., and Storch, G. 1975. Die unterpliozäne (Turolisch) Wirbeltierfauna von Dorn-Durkheim, Rheinhessen (S. W. Deutschland). Geology, Mammalia: Carnivora, Proboscidea, Rodentia. *Senckenbr. Lethaea* **56**(4/5):233–303.

Gabunia, L. K. 1979. Biostratigraphic correlations between the Neogene land mammal faunas of the East and Central Paratethys. *Ann. Geol. Pays Hellen.* **1979**(1):421–423.

Gaziry, A. W. 1976. Jungtertiäre Mastodonten aus Anatolien (Turkei). *Geol. Jahrb. Beih.* **22**:1–143.

George, T. N., Chairman, and Members of the Stratigraphy Committee of the Geological Society of London 1969. Recommendations on stratigraphic usage. *Proc. Geol. Soc. Lond.* **1656**:139–166.

Ginsburg, L. 1971. Les faunes de mammiferes Burdigaliens et Vindoboniens des bassins de la Loire et de la Garonne, in: *Ve Cong. Neog. Medit. Lyon.*, Volume 1, pp. 153–167, Bureau de Recherches, Orléans.

Ginsburg, L. 1975. Le Pliopithèque des faluns Helvetiens de la Touraine et de l'Anjou, in: Évolution des Vértebrés—Problèmes Actuels de Paléontologie. *Colloq. Int. Cent. Nat. Rech. Sci.* **218**:877–885.

Ginsburg, L. 1977. Les carnivores du Miocene de Beni Mellal (Maroc). *Geol. Medit.* **IV**(3):225–240.

Ginsburg, L., and Mein, P. 1980. *Crouzelia rhodanica*, nouvelle espece de primate catarhinien, et essai sur la position systematique des Pliopithecidae. *Bull. Mus. Natn. Hist. Nat. Paris*, Sér. 4 **2**(c):57–85.

Ginsburg, L., and Tassy, P. 1977. Les fouilles paleontologiques dans la region de Simorre. *Bull. Soc. Arch.* **1977**:1–19.

Greenfield, L. O. 1979. On the adaptive pattern of *Ramapithecus*. *Am. J. Phys. Antropol.* **50**:527–548.

Greenfield, L. O. 1980. A late divergence hypothesis. *Am. J. Phys. Anthropol.* **52**:351–365.

Guerin, C. 1976. Les restes de rhinoceros du gisement Miocene de Beni Mellal, Maroc. *Geol. Medit.* **III**(2):105–108.

Hamilton, W. R. 1973. North African Lower Miocene rhinoceroses. *Bull. Br. Mus. (Nat. Hist.)* **24**(6):351–395.

Hamilton, W. R., Whybrow, P. J., and McClure, H. A. 1978. Fauna of fossil mammals from the Miocene of Saudi Arabia. *Nature (Lond.)* **274**:248.

Heintz, E. 1973. Un nouveau bovide du Miocene de Beni Mellal, Maroc: *Benicerus theobladi* n. gen., n. sp. (Bovidae, Artiodactyla, Mammalia). *Geol., 3e Ser.* **18**:245–248.

Heintz, E. 1976. Les Giraffidae (Artiodactyla, Mammalia) du Miocene de Beni Mellal, Maroc. *Geol. Medit.* **III**(2):91–104.

Heintz, E., Ginsburg, L., and Hartenberger, J.-L. 1978. Mammiferes fossiles en Afghanistan: Etat des connaissances et resultats d'une prospection. *Bull. Mus. Natn. Hist. Nat. Paris, Ser. 3* **69**:101–119.

Howell, F. C. 1978. Hominidae, in: *Evolution of African Mammals* (V. J. Maglio and H. B. S. Cooke, eds.), pp. 154–248, Harvard University Press, Cambridge.

Howell, F. C. 1980. Zonation of late Miocene and early Pliocene circum-Mediterranean faunas. *Géobios* **13**(4):653–657.

Hsü, K. J., Cita, M. B., and Ryan, W. B. F. 1973. The origin of the Mediterranean Evaporite, in: *Initial Reports of the Deep Sea Drilling Project*, Volume XIII(2), pp. 1203–1231, U.S. Government Printing Office, Washington, D.C.

Hsü, K. J., Mantadert, L., Bernoulli, D., Cita, M. B., Erickson, A., Garrison, R. E., Kidd, R. B., Melieres, F., Muller, C., and Wright, R. 1977. History of the Mediterranean Salinity Crisis. *Nature (Lond.)* **267**:399–403.

Hürzeler, J., and Engesser, B. 1976. Les faunas de mammiferes Neogenes du Bassin de Baccinello (Grosseto, Italie). *C. R. Acad. Sci. Paris D* **283**:333–336.

Jacobs, L. L. 1979. Fossil rodents (Rhizomyidae and Muridae) from Neogene Siwalik deposits, Pakistan. *Bull. Mus. Ariz.* **52**:1–103.

Jaeger, J.-J. 1977. *Les Rongeurs du Miocene Moyen et Superieur du Maghreb, Palaeovertebrata* **8**:(1)1–166.

Jaeger, J.-J., Martinez, N. P., Michaux, J., and Thaler, L. 1977. Les faunes de micromammiferes du Neogene Superieur de la Mediterranee Occidentale. Biochronologie, correlations avec les formations marines et echanges intercontinentaux. *Bull. Soc. Geol. Fr.* **19**(3):501–506.

Kretzoi, M. 1975. New ramapithecines and *Pliopithecus* from the lower Pliocene in northeastern Hungary. *Nature (Lond.)* **257**:578–581.

Madden, C. T., and Van Couvering, J. A. 1976. The proboscidean event: Early Miocene migration from Africa, in: *Abstr. Geol. Soc. Am., Annual Meeting, 1976.*

Mein, P. 1975. *Report on Activity of the R.C.M.N.S. Working Groups (1971–1975) Brataslava: Vertebrata,* pp. 78–81, I.U.G.S. Commission on Stratigraphy, Subcommission on Neogene Stratigraphy, Brataslava.

Mein, P. 1979. Rapport d'activite du group de travail vertebres mise a jour de la biostratigraphie du Neogene bassé sur les mammiferes. *Ann. Geol. Pays Hellen.* **1979**(III):1367–1372.

Mellett, J. 1968. The Oligocene Hsanda Gol Formation, Mongolia: A revised faunal list. *Am. Mus. Novit.* **2318**:1–16.

Olson, E. C. 1966. Community evolution and the origin of mammals. *Ecol.* **47**:291–302.

Osborn, H. F. 1910. *The Age of Mammals,* Macmillan, New York. 635 pp.

Ozansoy, F. 1965. Etude des gisements continentaux et des mammiferes du Cenozoique de Turquie. *Mem. Soc. Geol. Fr.* **44**:5–89.

Pilbeam, D., Barry, J., Meyer, G. E., Shah, S. M. I., Pickford, M. H. L., Bishop, W. W., Thomas, H., and Jacobs, L. L. 1977a. Geology and paleontology of Neogene strata of Pakistan. *Nature (Lond.)* **270**:684–689.

Pilbeam, D., Meyer, G. E., Badgley, C., Rose, M. D., Pickford, M. H. L., Behrensmeyer, A. K., and Shah, S. M. I. 1977b. New hominoid primates and their bearing on hominoid evolution. *Nature (Lond.)* **270**:689–695.

Pilbeam, D. R., Behrensmeyer, A. K., Barry, J. C., and Shah, S. M. I. 1979. Miocene sediments and faunas of Pakistan. *Postilla* (Peabody Mus. Nat. Hist., Yale Univ.) **179**:1–45.

Richard, M. 1948. Contribution á l'etude du Bassin d'Aquitaine. Les gisements de mammiferes Tertiaires. *Mem. Soc. Geol. Fr. Ser. 5* **52**:1–380.

Sahni, A., and Mitra, H. C. 1980. Neogene palaeobiogeography of the Indian Subcontinent with special reference to fossil vertebrates. *Palaeogeogr. Palaeoclimatol. Palaeoecol.* **31**:39–62.

Schmidt-Kittler, N. 1976. Carnivores from the Neogene of Asia Minor. *Palaeontographica A* **151**:1–131.

Sen, S. 1980. Cricetodontini (Rodentia, Mammalia) du Miocene de Grece et de Turquie. Implications paleobiogeographiques et paleogeographiques, in: *8e Reun. Ann. Sci. Terr.,* p. 323, Marseille.

Sen, S., and Thomas, H. 1979. Decouverte de Rongeurs dans le Miocene moyen de la Formation Hofuf (Province du Hasa, Arabie Saoudite). *C. R. Somm. Soc. Geol. Fr.* **1979**(1):34–37.

Sickenberg, O., Becker-Platen, J. D., Benda, L., Berg, D., Engesser, B., Gaziry, W., Heissig, K., Hunermann, K. A., Sondaar, P. Y., Schmidt-Kittler, N., Staeche, K., Staesche, U., Steffens, P., and Tobien, H. 1975. Die Gliederung des hoheren Jungtertiärs und Altquartärs in der Turkei nach Vertebraten und ihre Bedeutung für die internationale Neogen-Stratigraphie. *Geol. Jahrb.* **15**:1–167.

Simpson, G. G. 1947. Holarctic mammalian faunas and continental relationships during the Cenozoic. *Bull. Geol. Soc. Am.* **58**:613–688.

Simpson, G. G. 1960. Notes on the measurement of faunal resemblance. *Am. J. Sci.* **258A**:300–311.

Solounias, N. 1979. *The Turolian Fauna from the Island of Samos, Greece, with Special Emphasis on the Hyaenids and Bovids,* Ph.D. Dissertation, University of Colorado, Boulder. 422 pp.

Solounias, N. 1981. The Samos fauna. *Contrib. Vert. Paleontol.* **6:**1–232. Karger, Basel.

Steininger, F. F. 1977. Integrated assemblage-zone biostratigraphy at marine–non-marine boundaries: Examples from the Neogene of central Europe, in: *Concepts and Methods of Biostratigraphy* (E. G. Kauffman and J. E. Hazel, eds.), pp. 235–256, Dowden, Hutchinson and Ross, Stroudsbourg.

Szalay, F. S., and Delson, E. 1979. *Evolutionary History of the Primates,* Academic, New York. 580 pp.

Thenius, E. 1959. *Tertiär,* Ferdinand Enke Verlag, Stuttgart. 328 pp.

Thomas, H. 1979. Le role de barriere ecologique de la ceinture Saharo-Arabique au Miocene: Arguments paleontologiques. *Bull. Mus. Natn. Hist. Nat. Paris Ser. 4* **1:**127–135.

Thomas, H., Bernor, R. L., and Jaeger, J.-J. 1982. Origines du peuplemant Mammalien Afrique du nord durant le Miocene terminal. *Géobios.* (in press).

Tobien, H. 1970. Biostratigraphy in the mammalian faunas at the Pliocene–Pleistocene boundary in middle and western Europe. *Palaeogeogr. Palaeoclimatol. Palaeoecol.* **8:**77–93.

Van Couvering, J. A., and Berggren, W. A. 1977. Biostratigraphical basis of the Neogene time scale, in: *Concepts and Methods of Biostratigraphy* (E. G. Kauffman and J. E. Hazel, eds.), pp. 283–306, Dowden, Hutchinson and Ross, Stroudsbourg.

Van Couvering, J. H. 1980. Community evolution in Africa during the late Cenozoic, in: *Fossils in the Making* (A. K. Behrensmeyer and A. P. Hill, eds.), pp. 272–298, University of Chicago Press, Chicago.

Van Couvering, J. H., and Van Couvering, J. A. 1976. Early Miocene mammal fossils from East Africa: Aspects of the geology, faunistics and paleoecology, in: *Human Origins: Louis Leakey and the East African Evidence* (G. L. Isaac and E. R. McCown, eds.), pp. 155–207, Benjamin, Menlo Park, California.

Van de Weerd, A. 1976. Rodent faunas of the Mio-Pliocene continental sediments of the Teruel–Alfambra region, Spain. *Utrecht Micropaleontol. Bull. Spec. Publ.* **2:**1–218.

Von Koenigswald, G. H. R. 1972. Ein Unterkiefer eines fossilen Hominoiden aus dem Unterpliozän Griechenlands. *Proc. K. Ned. Akad. Wet. Amsterdam B* **75:**386–394.

Woodburne, M. D. 1977. Definition and characterization in mammalian chronostratigraphy. *J. Paleo.* **50:**220–234.

Evidence from Molecular Biology and Comparative Anatomy

II

The Bearing of Molecular Data on the Cladogenesis and Times of Divergence of Hominoid Lineages

3

M. GOODMAN, M. L. BABA, AND
L. L. DARGA

Introduction

The analysis of amino acid sequence data by the maximum parsimony method can provide valuable insights into the evolutionary processes which shaped the hominoid radiation and the emergence of *Homo*. Recent discoveries of hominine and ramapithecine fossils from East Africa and Asia have stimulated reevaluation of the evolutionary history of Anthropoidea, especially of that branch leading to *Homo sapiens* (Johanson and White, 1979; Greenfield, 1980).

These new finds, examined in light of the rapidly increasing knowledge of evolutionary change at the molecular level, have yielded exciting insights into the processes leading to the divergence of African apes and humans, and have sparked increased dialogue among specialists in a number of fields, from molecular genetics to paleontology, archeology, and geology. The topic of this volume reflects the intense interest that higher primate evolution enjoys today.

M. GOODMAN • Department of Anatomy, School of Medicine, Wayne State University, Detroit, Michigan 48201.   M. L. BABA • Department of Anthropology, College of Liberal Arts, Wayne State University, Detroit, Michigan 48201.   L. L. DARGA • Department of Anatomy, School of Medicine, Wayne State University, Detroit, Michigan 48201.

Genealogical reconstruction carried out on amino acid sequence data from corresponding (*i.e.,* evolutionarily related) proteins of extant species can elucidate phylogenetic relationships within and among higher taxonomic units. The reconstructions also permit investigation of the tempo of molecular change and the forces underlying that process of change. A pattern of marked acceleration in rates of change followed by increasingly sharp decelerations during the past one-half billion years has characterized the tempo of molecular evolution for several proteins (Goodman *et al.,* 1975; Goodman and Czelusniak, 1980; Goodman, 1981; Baba *et al.,* 1981). Such a pattern of molecular evolution could result from alternating forces of positive and stabilizing selection.

For proteins demonstrating striking shifts in rates of amino acid substitutions over time, it is not possible to calculate accurate divergence dates within Anthropoidea using the molecular clock approach. Our analysis of amino acid sequence data for several proteins by the clock model yields divergence dates, particularly within Hominoidea, that are far too recent in view of well-established fossil evidence. Such errors are to be expected if molecules have experienced a slowdown in evolution between the time of emergence of the anthropoid ancestor and the present. In this chapter, we will focus upon the effects of positive and stabilizing selective forces in the evolution of molecules during the hominoid radiation.

Maximum Parsimony Analysis of Amino Acid Sequence Data

The maximum parsimony method is a powerful tool in the analysis of molecular evolution. The theoretical foundations, mathematical proof, and detailed description of computer programs behind this approach are described elsewhere (Moore, 1976; Goodman, 1976; Goodman *et al.,* 1979; Goodman, 1981). The object of the parsimony algorithm is to construct ancestral messenger RNA (mRNA) sequences and a branching network which, in total, requires the fewest number of nucleotide replacements to account for observed sequence differences among lineages. An augmentation procedure corrects for the underestimation of nucleotide replacements in regions of the tree which are represented by only a few sequences (Baba *et al.,* 1981).

The parsimony method conforms to the Hennigian principle that primitive features and similarities due to convergence be distinguished from shared, derived features and the latter determine the branching order of the tree. The accuracy of the method is relatively unaffected by varying rates of molecular change within and among lineages. Sometimes, however, because of an erratic distribution of convergent substitutions among sequences, the parsimony tree with lowest nucleotide replacement (NR) score depicts some incorrect genealogical relationships. However, one can maximize the fit of the gene phylogeny to well established features of the species phylogeny by finding the tree with lowest NR plus gene duplication (GD) plus gene expression event (GE) score (Goodman *et al.*, 1979) and thereby produce a more accurate genealogical tree.

When our objective is to find the correct species phylogeny we combine amino acid sequence data in a tandem alignment, as if each contemporary species or operational taxonomic unit (OTU) was represented by an extended polypeptide chain encoded by a single giant gene. The parsimony trees constructed for the OTUs with such a tandem alignment should more closely

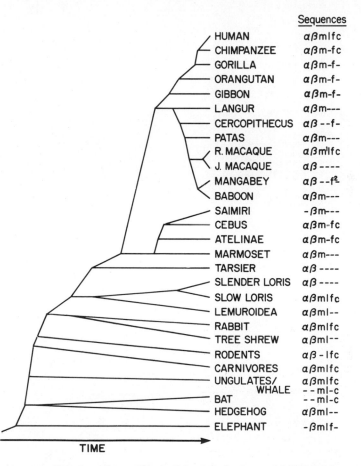

Fig. 1. Eutherian genealogy based on tandem alignment of α-hemoglobin (α), β-hemoglobin (β), myoglobin (m), lens α-crystallin A (l), fibrinopeptides A and B (f), and cytochrome *c* (c). Scientific names and taxonomic units of lineages depicted in this dendrogram are as follows: human (*Homo sapiens*); chimpanzee (*Pan troglodytes*); gorilla (*Gorilla gorilla*); orangutan (*Pongo pygmaeus*); gibbon (*Hylobates* sp.); langur (*Presbytis*); cercopithecus (*Cercopithecus*); patas (*Erythrocebus patas*); R. macaque (*Macaca mulatta*); J. macaque (*Macaca fuscata*); mangabey (*Cercocebus*); baboon (*Papio* sp.); saimiri (*Saimiri* sp.); cebus (*Cebus* sp.); atelinae (*Ateles* sp.); marmoset (*Saguinus* sp.); tarsier (*Tarsius*); slender loris (*Loris tardigradus*); slow loris (*Nycticebus coucang*); Lemuroidea (*Lepilemur, Lemur fulvus*); rabbit (Lagomorpha); tree shrew (*Tupaia* sp.); rodents (Rodentia); carnivores (Carnivora); ungulates–whale (Ungulata–Cetacea); bat (Chiroptera); hedgehog (Insectivora); elephant (Proboscidea). See Goodman (1981) for detailed identification of the species represented by the polypeptides in this genealogical reconstruction.

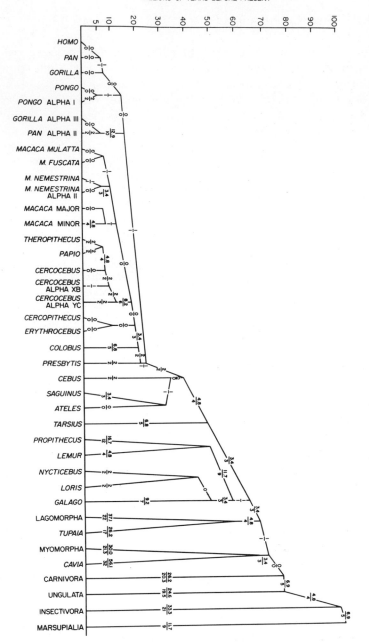

Fig. 2. Mammalian portion of tree constructed for α-hemoglobin sequences by our maximum parsimony approach. In this reconstruction the branching order of major mammalian lineages was made to follow that obtained for the tandem alignment of seven polypeptide chains. Mammalian species represented by α-hemoglobin sequences are as follows: *Homo* (*H. sapiens*); *Pan* (*P. troglodytes*); *Gorilla* (*G. gorilla*); *Pongo* (*P. pygmaeus*); *Pongo* α I; *Gorilla* α III; *Pan* α II; *Macaca mulatta; M. fuscata; M. nemestrina; M. nemestrina* α II; *Macaca* major (*M. irus* major α); *Macaca* minor (*M. irus* minor α); *Theropithecus* (*Theropithecus gelada*); *Papio* (*P. anubis*); *Cercocebus* (*C. atys*); *Cercocebus* α XB (*C. atys* α XB); *Cercocebus* α YC (*C. atys* α YC); *Cercopithecus* (*C. aethiops*);

70

approximate the true species phylogeny since the effects of convergence are reduced when our sample of the genome is increased.

Combined-Protein Analysis

Figure 1 is the parsimony cladogram for a tandem alignment of α- and β-hemoglobin, myoglobin, fibrinopeptides A and B, cytochrome *c*, and lens α-crystallin A. This divergence tree depicts Primates as a monophyletic assemblage which is most closely related to a branch containing the orders Scandentia and Lagomorpha.

Living Primates divide into Strepsirhini (lorises and lemurs) and Haplorhini (tarsiers and Anthropoidea): Within Haplorhini, Anthropoidea is a monophyletic unit which bifurcates to form the sister groups Platyrrhini and Catarrhini. Platyrrhini bifurcates to form Callitrichidae and Cebidae; Cebidae contains the traditional *Cebus–Saimiri* cluster. Catarrhini divides into Cercopithecoidea and Hominoidea, with Cercopithecoidea including a monophyletic Cercopithecinae and, within this subfamily, the papionine cluster of *Papio* and *Cercocebus,* most clearly related to *Macaca.*

Although Hominoidea always appears as a monophyletic assemblage, three equally parsimonious arrangements for subbranching within Hominoidea have been found. One of these arrangements joins Hylobatidae and *Pongo* as a single cluster forming the most ancient splitting within Hominoidea. Another arrangement depicts *Pongo* as the most ancient lineage, fol-

---

Erythrocebus (*E. patas*); *Colobus* (*C. badius*); *Presbytis* (*P. entellus*); *Cebus* (*C. apella*); *Saguinus* (*S. fuscicollis*); *Ateles* (*A. geoffroyi*); *Tarsius* (*Tarsius syrichta*); *Propithecus* (*Propithecus* sp.); *Lemur* (*L. fulvus*); *Nycticebus* (*N. coucang*); *Loris* (*L. tardigradus*); *Galago* (*G. senegalensis*); Lagomorpha (*Oryctolagus cuniculus*); Scandentia (*Tupaia glis*); Myomorpha (*Mus musculus* C57 B1 mouse, *M. musculus* NB mouse, *Ondatra zibethecus, Lemmus sibiricus, Clethrionomys rutitus, Microtus xanthognathus, Rattus* sp.); *Cavia* sp.; Carnivora (*Canis familiaris* duplicated α, *C. latrans, C. familiaris, Urocyon cineroargenteus, Thalarctos maritimus, Meles meles, Procyon lotor, Nasua narica, Felis felis, Panthera leo*); Ungulata (*Ovis aries, Capra* sp., *Capra* 2-α duplicated, *Bos taurus, Llama* sp., *Camelus dromedarius, Sus scrofa, Equus asinus, Equus equus, E. equus* fast α 24 Y); Insectivora (*Erinaceus europueus*); Marsupialia (*Macropus cangaru, Didelphis* sp. fast α). α I, α II, α III, α XB, and α YC are terms designating variant α-hemoglobins found in primate species which may be the products of allelic variation or gene duplication. The terms for these variant α chains were assigned by the investigators who discovered them. Unaugmented and augmented link length values are given for each link between internal nodes or between an internal node and a terminal point. For each link the unaugmented value is the number shown below the line, and the augmented value above the line. Augmented link lengths are the number of nucleotide replacements (fixed mutations) between adjacent ancestor and descendant sequences corrected for superimposed fixations by an augmentation algorithm. The ordinate is a time scale in millions of years before present with major ancestral nodes fixed by paleontological views concerning ancestral separations of the organisms from which the α-hemoglobin came. Where fossil evidence was insufficient to establish ancestral branch points, a heuristic procedure was used in which times were estimated from the magnitude of the link lengths in these areas of the tree and by interpolation between the points that were placed on the basis of paleontological evidence.

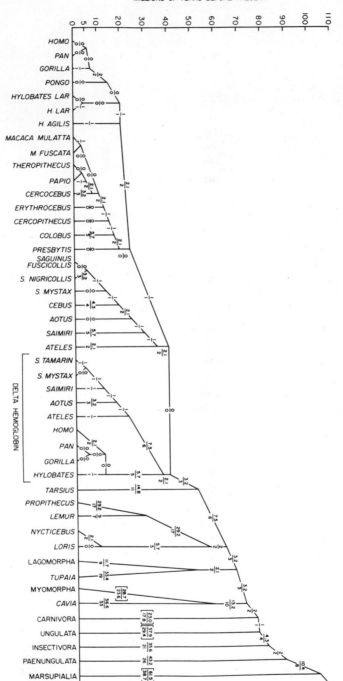

Fig. 3. Mammalian portion of tree constructed for β-type hemoglobin sequences by our maximum parsimony approach. In this reconstruction the branching order of major mammalian lineages was made to follow that obtained for the tandem alignment of seven polypeptide chains. Mammalian species represented by β-type hemoglobin sequences and/or groups of species which

lowed by Hylobatinae. The third arrangement, seen in Fig. 1, supports immunological (Goodman and Moore, 1971; Dene *et al.*, 1976; Baba *et al.*, 1979) and DNA hybridization (Benveniste and Todaro, 1976) findings by placing *Pongo* closer to *Homo* and the African apes.

Members of the Homininae (*Homo, Pan,* and *Gorilla*) are always more closely related to one another than to the Asiatic apes *Pongo, Symphalangus,* and *Hylobates.* The close genetic relationship of the African apes and humans suggested by all the molecular evidence gathered to date has encouraged reevaluation of hominoid fossil evidence, especially ramapithecine material. Recent finds from Afar and Laetolil (Johanson and White, 1979) are consistent with a more recent date of human–ape divergence than once thought.

Single-Protein Analysis

Figures 2–7 depict phylogenetic relationships within the Hominoidea from the perspective of five individual proteins analyzed by the maximum parsimony method. Link lengths representing the number of nucleotide replacements (NRs) between internal nodes and between terminal points and internal nodes are given for each connecting branch. Unaugmented and augmented NR values are shown. Unaugmented NR values represent the raw number of nucleotide replacements (fixed mutations) occurring between two nodes in a maximum parsimony tree. Augmented NR values represent the number of nucleotide replacements between ancestor and descendant sequences corrected for superimposed fixations by an augmentation algorithm (Baba *et al.*, 1981).

Each of the trees shown in Figs. 2–7 follows the species phylogeny depicted in Fig. 1 and therefore represents a most parsimonious arrangement of lowest NR + GD + GE length, rather than of lowest NR length alone. Thus, e.g., 1 NR would be saved in the myoglobin tree (Fig. 4) if the position of

---

differ from those listed in the legend of Fig. 2 are as follows: *Aotus* (*Aotus* sp.); *Saimiri* (*Saimiri* sp.); *Suguinus* (*S. mystax* δ, *S. tamarin* δ); Myomorpha (*Mus musculus* C57 B1 mouse, *M. musculus* AKR mouse, *M. cervicolor* major D-like, *M cervicolor* major S-like, *M. musculus* minor, *Rattus* sp., *Lemmus sibiricus, Ondatra zibethecus, Clethrionomys rutitus, Microtus xanthognathus*); Carnivora (*Canis familiaris, C. latrans, Urocyon cineroargenteus, Thalarctos maritimus, Meles meles, Procyon lotor, Nasua narica, N. narica* slow β, *N. nasua, Felis felis, Panthera leo, Felix catus* slow β); Ungulata (*Ovis aries, Capra* sp., *Ovis aries* C, *Capra* sp. C, Barbary sheep C, *Bos taurus* A, *B. taurus, B. taurus* fetal, *Ovis aries* fetal, *Llama* sp., *Camelus dromedarius, Sus scrofu, Equus equus*); Paenungulata (*Elephas maximus*); Marsupialia (*Macropus cangaru, M. cangaru* two, *Didelphis* sp.). This tree depicts the evolution of both β and δ forms of β-type hemoglobin. The δ polypeptide is a variant form of β-type hemoglobin which resulted from a gene duplication at the β locus occurring prior to the platyrrhine–catarrhine divergence. The δ polypeptide is present as a minor component in the adult hemoglobin of most anthropoid primates. The δ gene occurs in all members of Anthropoidea, but is expressed only among ceboid and hominoid species; it is silent in cercopithecoids. Unaugmented and augmented link length values are displayed as in Fig. 2. Divergence dates for ancestral nodes were established by the procedure described in the legend of Fig. 2.

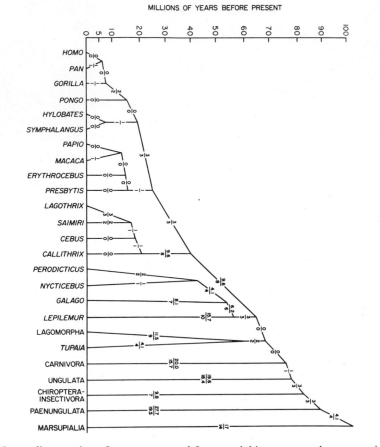

Fig. 4. Mammalian portion of tree constructed for myoglobin sequences by our maximum parsimony approach. In this reconstruction the branching order of major mammalian lineages was
made to follow that obtained for the tandem alignment of seven polypeptide chains. Mammalian
species or groups of species represented by myoglobin sequences which differ from these listed in
the legend of Fig. 2 are as follows: *Symphalangus* (*S. syndactylus*); *Lagothrix* (*L. lagothrix*); *Saimiri* (*S.
sciureus*); *Callithrix* (*C. jacchus*); *Perodicticus* (*P. potto*); *Galago* (*G. crassicaudatus*); *Lepilemur* (*L. mus-
telinus*); Carnivora (*Phoca* sp., *Eumetopias* sp., *Canis familiaris, Urocyon cineroargenteus, Lycaon pictus,
Meles meles*); Ungulata (*Sus scrofa, Cervus elephus, Bos taurus, Ovis aries, Phocoenoides dalli, Delphinus
delphis, Orcinus orca, Globicephala melaena, Balaenoptera acutorostrata, B. physalus, Megaptera
nouaengliae, Eschrichtius gibbosus, Kogia simus*); Ungulata (*Equus equus, E. zebra*); Chiroptera–Insectivora (*Rousettus aegyptiacus, Erinaceus* sp.); Paenungulata (*Elephas maximus, Loxodonta
africana*); Marsupialia (*Macropus cangaru, Didelphis* sp.). Unaugmented and augmented link length
values are displayed as in Fig. 2. Divergence dates for ancestral nodes were established by the
procedure described in the legend of Fig. 2.

Pongo and *Hylobates* were reversed. However, this would cost 1 GD and 3 GEs
on the assumption that the overall molecular evidence is correct in having
Pongo group first with the branch leading to *Homo, Pan,* and *Gorilla.* Consequently, it is more parsimonious to have the arrangement for myoglobin the
same as that for the other proteins.

    In these trees, the African apes and *Homo* always cluster in a monophyle-

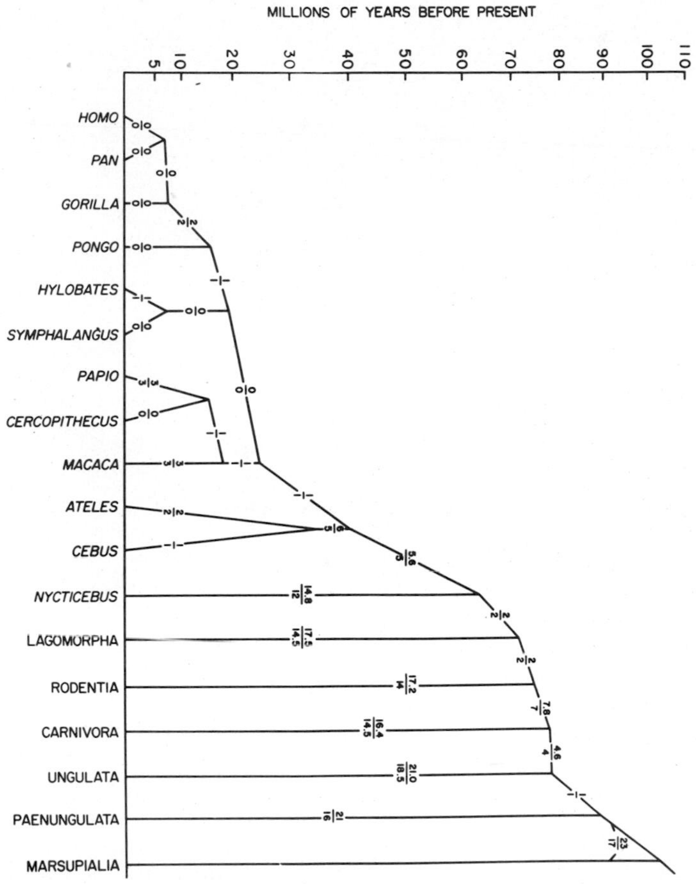

Fig. 5. Tree constructed for fibrinopeptide A and B sequences by our maximum parsimony approach. In this reconstruction the branching order of major mammalian lineages was made to follow that obtained for the tandem alignment of seven polypeptide chains. Mammalian species or groups of species represented by fibrinopeptide A and B sequences which differ from those listed in the legend of Fig. 2 are as follows: *Symphalangus* (*S. syndactylus*); *Papio* (*P. leucophaeus*); Rodentia (*Rattus* sp.); Carnivora (*Urocyon* sp., *Canis familiaris*, *Felis felis*, *Panthera leo*); Ungulata (*Cervas elephas*, *C. canadensis*, *C. nippon*, *Muntiacus muntjak*, *Odocoileus hemionus*, *Rangifer tarandus*, *Bison bonasus*, *Bos taurus*, *Bubalus bubalus*, *Syncerus caffer*, *Capra* sp., *Ovis aries*, *Gazella* sp., *Giraffa camelopardalis*, *Antilocapra americana*, *Llama* sp., *Vicugna vicugna*, *Camelus dromedarius*, *Sus scrofa*, *Equus asinus*, *E. caballus* × *asinus*, *E.* sp. *zebra*, *E. equus*, *Tapirus terrestris*, *Dicerus simus*); Paenungulata (*Elephas maximus*); Marsupialia (*Macropus cangaru*). Unaugmented and augmented link length values are displayed as in Fig. 2. Divergence dates for ancestral nodes were established by the procedure described in the legend of Fig. 2.

tic assemblage. In Figs. 3–5, the trees for β-hemoglobin, myoglobin, and fibrinopeptide A and B sequences, and the shared, derived substitutions on the common stem of *Pan, Homo,* and *Gorilla* separate these genera from the Asiatic apes. This arrangement lends support to the grouping of *Gorilla, Pan,* and *Homo* in a single subfamily Homininae (Goodman and Moore, 1971). In these three trees, after the *Hylobates* and then *Pongo* divergences, *Homo, Pan,* and

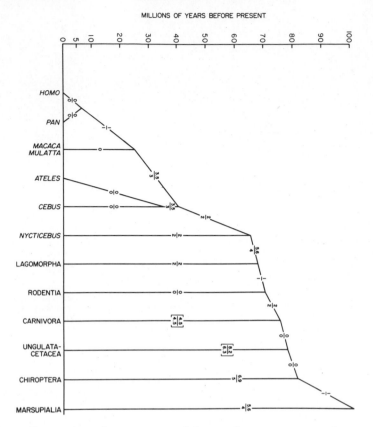

Fig. 6. Mammalian portion of tree constructed for cytochrome *c* sequences by our maximum parsimony approach. In this reconstruction the branching order of mammalian lineages was made to follow that obtained for the tandem alignment of seven polypeptide chains. Mammalian species or groups of species represented by cytochrome *c* sequences which differ from those listed in the legend of Fig. 2 are as follows: Rodentia (*Mus musculus*); Carnivora (*Canis familiaris*, *Phoca* sp.); Ungulata–Cetacea (*Equus asinus*, *E. equus*, *Ovis aries*, *Camelus dromedarius*, *Hippopotamus* sp., *Sus scrofa*, *Eschrichtus glaucus*); Chiroptera (*Rousettus* sp.); Marsupialia (*Macropus cangaru*). Unaugmented and augmented link length values are displayed as in Fig. 2. Divergence dates for ancestral nodes were established by the procedure described in the legend of Fig. 2.

Gorilla form a trichotomy from the node for their common ancestor. However, in the α-hemoglobin tree (Fig. 2), *Homo* and *Pan* group together first, sharing a common substitution not shared by *Gorilla*. Furthermore, in the β-hemoglobin tree, *Homo* and *Pan* sequences are identical to the putative (*Homo, Pan, Gorilla*) ancestral sequence, while *Gorilla* has diverged from the ancestral state by 1 NR. In the carbonic anhydrase tree (Tashian *et al.*, 1980), where *Pongo, Homo,* and *Pan* are represented in the carbonic anhydrase I region, *Homo* and *Pan* join together first, again supporting the above taxon Homininae. In Figs. 6 and 7, the trees for cytochrome *c* and lens α-crystallin A sequences, the hominoid region of the tree is too sparsely represented to provide detailed insights on relationships within Hominoidea. *Homo* and *Pan* share identical cytochrome *c* sequences (Fig. 6).

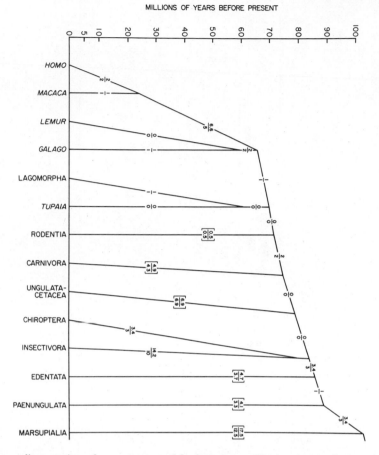

MILLIONS OF YEARS BEFORE PRESENT

Fig. 7. Mammalian portion of tree constructed for lens α-crystallin A sequences by our maximum parsimony approach. In this reconstruction the branching order of major mammalian lineages was made to follow that obtained for the tandem alignment of seven polypeptide chains. Mammalian species or groups of species represented by lens α-crystallin A sequences which differ from those listed in the legend of Fig. 2 are as follows: Lagomorpha (*Oryctolagus cuniculus, Ochotona princeps*); Rodentia (*Cavia* sp., *Rattus* sp.); Carnivora (*Canis familiaris, Mustela* sp., *Phoca* sp., *Felis felis, Manis pentadactyla*); Ungulata–Cetacea (*Tapirus terrestris, Dicerus simus, Equus equus, Giraffa camelopardalis, Bos taurus, Camelus dromedarius, Sus scrofa, Balenoptera aeutorostrata, Phocoenoides dalli*); Chiroptera (*Artibeus* sp.); Insectivora (*Erinaceus europaeus*); Edentata (*Bradypus tridactylus, Tamandua* sp., *Choloepus* sp.); Paenungulata (*Procavia capensis, Orycteropus afer, Trichechus manatus, Loxodonta africana*); Marsupialia (*Macropus rufus, Didelphis marsupialis*). Unaugmented and augmented link length values are displayed as in Fig. 2. Divergence dates for ancestral nodes were established by the procedure described in the legend of Fig. 2.

Rates of Molecular Evolution Investigated by the Maximum Parsimony Method

From the NR values between ancestral and descendent mRNA sequences in a maximum parsimony reconstruction it is possible to calculate rates of molecular change for individual lineages during particular periods of time in

Table 1. Rates of Evolution in α- and β-Hemoglobin, Myoglobin, and Cytochrome c[a]

Evolutionary period	Age, m.y.a.	α-Hemoglobin NR%	β-Hemoglobin NR%	Myoglobin NR%	Cytochrome c NR%
Vertebrate to gnathostome ancestor	500–425	96.7	97.4	88.0	4.4
Gnathostome ancestor to present	425–0	28.5	28.1	27.4	5.0
		(20.3–36.1)	(20.4–34.3)	(19.7–30.3)	(2.9–13.0)
Gnathostome to amniote ancestor	425–300	62.4	54.8	65.9	8.5
Amniote ancestor to present	300–0	15.0	17.2	12.3	3.9
		(8.9–25.1)	(8.7–24.8)	(4.8–15.8)	(1.3–15.7)
Amniote to eutherian ancestor	300–90	10.5	13.0	10.6	1.4
Eutherian ancestor to present	90–0	25.7	31.2	17.8	6.0
		(14.0–37.0)	(16.8–49.9)	(6.0–27.9)	(1.1–17.2)
Eutherian to Anthropoidea ancestor	90–40	36.9	43.8	22.2	19.0
Anthropoidea ancestor to humans	40–0	5.3	8.6	14.7	7.0

[a]m.y.a., millions of years before the present. NR%, nucleotide replacements per 100 codons per 100 million years.

the past. These rates of change are calculated as nucleotide replacements per 100 codons per 100 million years.

Table 1, from data previously presented in Goodman and Czelusniak (1980), summarizes and compares rates of change in four molecules during the past one-half billion years. The evolution of α- and β-hemoglobin, myoglobin, and cytochrome *c* depicted in this table reflects not only nonuniform rates of change over time, but also a remarkable synchrony in the tempo of change for these four protein chains.

Rates of nucleotide replacements accelerated for all of the globin genes during the early vertebrate period, and continued to be elevated for all four proteins between the emergence of the gnathostome ancestor and the emergence of the amniote ancestor. A second acceleration for all of these proteins occurred with the emergence of the eutherian ancestor and continued through the early primate stem until the emergence of the anthropoid ancestor. Finally, change in all of the proteins decelerated in the Anthropoidea, with the slowdown continuing in the line to *Homo* until the present.

The genealogical tree results shown in Fig. 8 depict the synchronized

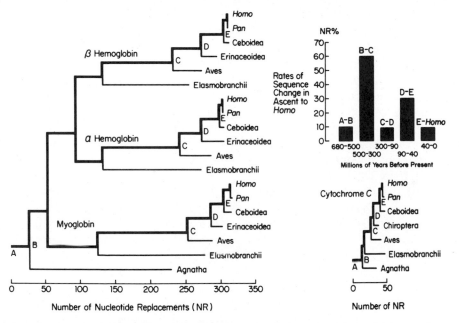

Fig. 8. Acceleration–deceleration pattern in globin and cytochrome *c* evolution in the vertebrates. Here NR% is the number of nucleotide replacements per 100 codons per 100 million years. Divergence dates for major nodes in this dendrogram reflect palaeontological views such as those of McKenna, Romer, Young, Schopf, Cloud, and Valentine. The mollusc vs. vertebrate divergence point (node A) is placed at 680 m.y.a.; the earliest divergence within Vertebrata (node B) is dated at 500 m.y.a.; the bird–mammal split (node C) is placed at 300 m.y.a.; 90 m.y.a. is set as the time of earliest splitting within Eutheria (node D); platyrrhine–catarrhine divergence (node E) is placed at 40 m.y.a. Within the range of plausible divergence dates, the most recent was chosen for the platyrrhine–catarrhine split; conversely, the most ancient splitting time was chosen for the vertebrate ancestor. Nevertheless, the acceleration–deceleration pattern of change is still found in this figure.

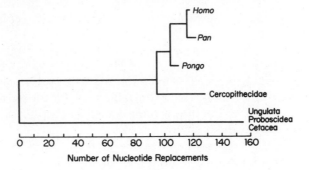

Fig. 9. Cladogram based on sequences of α- and β-hemoglobin, myoglobin, fibrinopeptides A and B, and carbonic anhydrase I. The data set used to determine the branching order depicted in this figure groups Ungulata closer to Proboscidea than to Cetacea. An expanded data set used to produce the tree in Fig. 1 groups the Ungulata–Cetacea branch closer to Primates than to Proboscidea; the latter emerges as the most ancient lineage diverging from the basal eutherian stem in Fig. 1.

tempos of molecular change in the descent to *Homo* seen in α- and β-hemoglobin, myoglobin, and cytochrome *c*. The number of nucleotide replacements recorded are those that accumulated in these four protein chains, and only those branches are shown that are represented by all four chains. The length of each branch equals the number of nucleotide replacements calculated to have occurred along that line of evolution. Figure 8 illustrates that a highly accelerated rate of molecular evolution (67 per 100 codons per 100 million years) occurred between the emergence of the vertebrate ancestor and the emergence of the bird–mammal ancestors. The rate then falls steeply to 9 NR% between the bird–mammal separation and the emergence of the eutherian ancestor. An accelerated rate of 26 NR% occurs between the emergence of the eutherian and anthropoid ancestors, and then the rate falls back to 10 NR% for the last 40 or so million years of evolution to *Homo*. Indeed from the *Pan–Homo* most recent common ancestor to present-day humans the rate is 0 NR% for these four protein chains.

More extensive hominoid sequence comparisons, permitting inclusion of *Pongo* as well as *Homo* and *Pan,* were obtained by substituting fibrinopeptides A and B and carbonic anhydrase I for cytochrome *c* (Fig. 9). As can be noted, although ten nucleotide replacements occurred on the common *Homo–Pan* stem after the separation of the *Pongo* branch, only five fixed mutations separate *Homo* and *Pan.* Of these, just 1 NR is on the *Homo* branch, while 4 NRs occur on the *Pan* branch. This relatively low amount of change contrasts with the 140 NRs that accumulated along the average mammalian lineage from the beginning of the eutherian radiation to the present (representing a replacement rate of 22% NR for these six proteins).

In view of *Australopithecus* fossils the ancestral divergence between *Homo* and *Pan* must have occurred before 5 million years ago (m.y.a.), yielding a maximum evolutionary rate of 2.7% NR% for the hominine lineage to modern humans. This hominine rate, which is eight times slower than the average mammalian rate, would be even more greatly reduced if *Ramapithecus* was closer to *Homo* than to any other extant genus.

Application of a Clock Model to Amino Acid Sequence Data

Clearly, marked fluctuations in rates of molecular change over time would prevent calculation of accurate divergence dates by use of the molecular clock model. This is apparent on calculating putative branch times for the major nodes shown in Figs. 2–6 by the clock model on the assumption that molecular evolution proceeds at approximately equal rates. The "clock" is calibrated by setting the time of earliest divergence within the ancestral stock of eutherian mammals to 90 m.y.a. Clock dates calculated from unaugmented NR values are shown in Table 2 and the corresponding clock dates from augmented values in Table 3. A set of average clock dates from the eight proteins is given in Table 4. The dates were obtained by averaging the individual protein dates shown in Tables 2 and 3. Another set of dates averaged by weighting the value of each nucleotide replacement equally did not differ significantly from those shown in Table 4. The results show that the clock dates for the divergence points within Primates, especially within Anthropoidea, are far too recent, given well-established fossil evidence.

The splitting times for (*Homo, Pan, Gorilla*) and *Homo* vs. *Pan* are particularly glaring since recent discoveries in East Africa suggest that *Australopithecus*

Table 2. Clock Dates for Major Divergence Events Calculated by Maximum Parsimony Analysis of Amino Acid Sequence Data for Individual Proteins, Using *Unaugmented* Nucleotide Replacement Values[a]

Divergence event	Clock date, m.y.a.							
	β-Hb	Myo	Fib A&B	lens α-cryst.	α-Hb	Cyt *c*	CA I	CA II
Paenungulata vs. remaining Eutheria	90	90	90	90	—	—	—	—
Chiroptera–Insectivora vs. remaining Eutheria	84.7	78.4	—	50.9	80.5	69.9	—	—
Ungulata—Cetacea vs. remaining Eutheria	74.1	68.2	86.3	49.9	69.9	69.9	69.9	69.9
Strepsirhini vs. Haplorhini	45.9	61.0	45.2	43.8	41.6	58.9	—	—
Playrrhini vs. Catarrhini	17.8	29.5	19.5	—	25.2	40.3	—	14.1
Cercopithecoidea vs. Hominoidea	13.9	13.0	13.3	14.6	23.3	7.9	18.0	2.9
Hylobatidae vs. Hominidae	4.9	6.7	8.7	—	—	—	—	—
Pongo vs. Homininae	4.8	8.1	7.5	—	3.3	—	11.0	—
(*Homo, Pan, Gorilla*)	0.9	2.7	0	—	1.8	—	—	—
Homo vs. *Pan*	0	2.0	0	—	0	0	4.0	—

[a]The clock dates shown in this table were calculated on the basis of a calibration date of 90 m.y.a. for the earliest splitting within the Eutheria. Maximum parsimony analysis of data sets including Proboscidea sequences depict Proboscidea as the most ancient lineage within Eutheria; thus the point of divergence for Paenungulata (a superorder including the order Proboscidea) vs. remaining Eutheria was set at 90 m.y.a. Clock dates for other early divergence points within Eutheria (e.g., Chiroptera–Insectivora vs. remaining Eutheria and Ungulata–Cetacea vs. remaining Eutheria) were initially calculated from data sets calibrated at 90 m.y.a. for the Paenungulata vs. remaining Eutheria ancestor. Such clock dates were then used to calibrate the clock for those data sets that did not include Proboscidea.

Table 3. Clock Dates for Major Divergence Events Calculated by Maximum Parsimony Analysis of Amino Acid Sequence Data for Individual Proteins, Using *Augmented* Nucleotide Replacement Values[a]

Divergence event	Clock date, m.y.a.							
	β-Hb	Myo	Fib A&B	lens α-cryst.	α-Hb	Cyt c	CA I	CA II
Paenungulata vs. remaining Eutheria	90	90	90	90	—	—	—	—
Chiroptera–Insectivora vs. remaining Eutheria	85.3	78.4	—	62.2	83.5	72.7	—	—
Ungulata–Cetacea vs. remaining Eutheria	75.4	69.2	86.4	59.7	72.7	72.7	72.7	72.7
Strepsirhini vs. Haplorhini	45.6	62.8	44.6	45.2	43.8	58.7	—	—
Platyrrhini vs. Catarrhini	15.4	28.2	18.5	—	27.6	43.6	—	12.7
Cercopithecoidea vs. Hominoidea	12.1	11.9	12.1	12.8	24.9	7.5	17.3	2.6
Hylobatidae vs. Hominidae	4.7	6.2	7.9	—	—	—	—	—
Pongo vs. Homininae	4.2	7.4	6.8	—	2.7	—	10.2	—
(*Homo, Pan, Gorilla*)	0.8	2.5	0	—	1.5	—	—	—
Homo vs. *Pan*	0	1.8	0	—	0	0	3.9	—

[a]Calibration dates for data sets represented in this table were determined according to the procedure described in footnote *a* of Table 2.

Table 4. Clock Date Averages for Major Divergence Events

Divergence event	Clock date average, m.y.a.		
	Unaugmented NR value	Augmented NR value	Protein data used[a]
Paenungulata vs. remaining Eutheria	90	90	β, m, F, l
Chiroptera–Insectivora vs. remaining Eutheria	72.8	76.4	β, m, l, α, c
Ungulata–Cetacea vs. remaining Eutheria	69.9	72.7	β, m, F, l, α, c, I, II
Strepsirhini vs. Haplorhini	49.4	50.1	β, m, F, l, α, c
Platyrrhini vs. Catarrhini	24.4	24.3	β, m, F, α, c
Cercopithecoidea vs. Hominoidea	13.4	12.7	β, m, F, l, α, c, I, II
Hylobatidae vs. Hominidae	6.8	6.3	β, m, F
Pongo vs. Homininae	7.0	6.3	β, m, F, α, I
(*Homo, Pan, Gorilla*)	1.4	1.2	β, m, F, α
Homo vs. *Pan*	1.0	1.0	β, m, F,α, c, I

[a]β, β-hemoglobin; m, myoglobin; F, Fibrinopeptides A & B; l, lens α-crystallin A; α, α-hemoglobin; c, cytochrome *c*; I, carbonic anhydrase I; II, carbonic anhydrase II.

existed at least 4 or 5 million years ago. Clearly, these clock dates will become more erroneous if *Ramapithecus* is included in the ancestral lineage to *Homo* after this lineage separated from that to *Pan.*

Natural Selection and Rates of Evolution in Single Proteins

As can be noted in Table 1 and Fig. 8, a synchronized acceleration–deceleration pattern of evolution for molecules involved in energy metabolism occurred during the past one-half billion years. Such a pattern is to be expected if certain adaptive radiations placed greater selective pressure on energy-associated proteins to develop new or altered functions. If the selectionist model is correct, decelerations represent the action of stabilizing selection holding newly perfected functions constant. During these periods of relatively slow evolution the small amount of actual change which occurs in the protein may result more from random drift of neutral mutations than from positive selection for change in function. However, even where neutral mutations occur, only a limited number of the 20 amino acids are likely to be considered equivalent to one another at each of the positions accepting such substitutions. Therefore, stabilizing selection remains a strong force excluding most nucleotide replacements except those that are silent at the amino acid level because they go from one synonymous codon to another.

The Darwinian hypothesis that positive selection was a dominant force directing change during periods of accelerated molecular evolution may be tested by examining relative amounts of change in residue positions grouped according to their function. If selection does drive the process of nucleotide replacement during periods of accelerated change, substitutions occurring during those periods should concentrate in residue groupings associated with critical functions that may be altered as organisms radiate into new ecological zones. This hypothesis has received support from structural-functional analysis of evolutionary rates conducted for α- and β-hemoglobin (Goodman *et al.*, 1975; Goodman and Czelusniak, 1980; Goodman, 1981) and for cytochrome *c* (Baba *et al.*, 1981).

In α- and β-hemoglobin, the residue positions with the most crucial functions, namely those involved in heme and $\alpha_1\beta_2$ contacts and the Bohr effect, evolved the most slowly and show few normal human variants. The most rapidly evolving positions in the eutherian lineage to *Homo* are the $\alpha_1\beta_2$ contact sites. Although these sites evolved about twice as fast as did the surface positions without sharply defined functions, they have very few normal variants in the human species. The pattern is that of positive selection for adaptive substitutions, followed by stabilizing selection to preserve an improved set of $\alpha_1\beta_2$ contact sites. Indeed, selection is now apparently acting strongly against any further evolutionary change in the amino acid sequence of human hemoglobin variants; almost none are at polymorphic frequencies, indicating that such variants harm the subtler adaptations of the molecule.

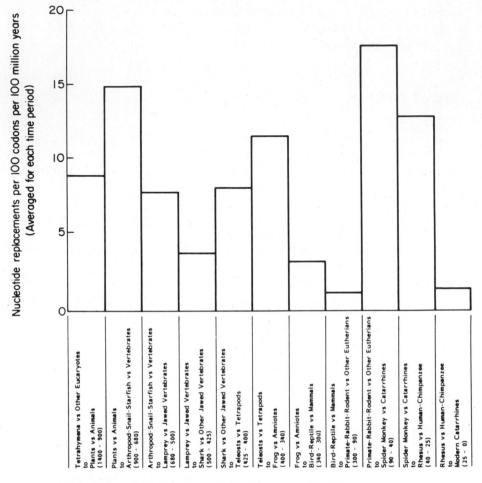

Fig. 10. Bar graph depicting fluctuations in the tempo of evolutionary change for major evolutionary periods in the phylogeny of cytochrome *c*. Bars represent averaged rates of change for each evolutionary period; absolute rates are given in Baba *et al.* (1981).

Cytochrome *c*, like the globins, contributes to the pattern of accelerations and decelerations in rates of protein evolution demonstrated in Fig. 7, and provides evidence that natural selection caused this pattern. The histogram in Fig. 10 depicts the fluctuating tempo of change in the cytochrome *c* molecule during the past one-half billion years. The dramatic speedup in evolution which occurred between the time of the emergence of the eutherian ancestor and the Anthropoidea ancestor is particularly significant. It correlates with the fact that there is a marked difference between the cytochromes *c* of higher primates and those of other mammals in ability to interact with cytochrome oxidase (Borden *et al.*, 1978). Old World monkey and ape cytochromes *c* bind more tightly and incorrectly than does beef cytochrome *c* to beef oxidase and thus inhibit electron transport activity. Lower primate cytochromes *c* are as active as beef cytochromes *c*. However, if the heart mitochondria are from primates, primate cytochromes *c* are highly efficient. In view of these findings, a credible explanation for the accelerated rate of cytochrome *c* evolution in the early primate lineage is that positive natural selection was the cause. This explanation is supported by the finding that almost all of the amino acid replacements in this early primate lineage occurred at sites involved in the interaction between cytochrome *c* and cytochrome oxidase (Baba *et al.*, 1981).

The genealogical trees constructed from amino acid sequence data reveal a pattern of accelerations and decelerations in rates of molecular evolution, where the trend is toward decelerations after major phylogenetic advances. Particularly important is the mounting evidence for a synchronized molecular slowdown within hominoid lineages. Such a pattern is indicative of anagenesis, the process of phyletic change which increases the level of molecular organization within organisms in such a way that organisms have greater independence from, and control over, their external environment. As a molecular process, anagenesis increased the number of stereochemical recognition sites among macromolecules. Not only did levels of internal hemeostasis increase and, therefore, the force of stabilizing natural selection, but in the human lineage the superstructure of culture erected a further buffer against external perturbations. Thus hominines in general, and humans in particular, became increasingly less subject to external pressure for the selection of new mutations in their genes.

Conclusions

The nature and timing of evolutionary events that have led to the extant primates, especially to humans and the African apes, can only be partially clarified by the molecular clock hypothesis. In general terms, the longer ago two species last shared a common ancestor, the greater is the genetic divergence between them. Consequently, the amounts of genetic change can be related to dates of divergence. However, the accelerations and decelerations in the evolutionary history of several sequenced proteins tend to invalidate certain of the dates obtained by molecular clock calculations, such as those determining the time course of hominoid phylogeny.

Since protein sequence evolution has tended to decelerate over time, with the magnitude of deceleration increasing toward the present, divergence dates calculated by the clock model become ever more unreasonable with the recency of the evolutionary event. However, it may be noted that the vast majority of silent mutations are not revealed in amino acid sequence data. Detection of these silent nucleotide replacements, which is now possible by nucleotide sequencing, should eventually produce sufficient data for a more thorough test of the clock concept. Theoretically, if silent mutations constitute a large proportion of change in a molecule, and if generation time is not proven to be a significant factor affecting rates of molecular evolution, then clock dates based on nucleotide sequence data may fall more closely in line with divergence dates suggested by the fossil record.

Given the above qualifications, however, the theoretical implications of the clock may still be misleading. A generally accurate clock would require that neutral mutations and random walk have been the most important processes responsible for evolution at the molecular level (Fitch and Langley, 1976). Functional analysis of α- and β-hemoglobin and cytochrome c have demonstrated that selection has been crucial in orchestrating change in the history of molecules.

References

Baba, M. L., Darga, L. L., and Goodman, M. 1979. Immunodiffusion systematics of the primates. Part V. The Platyrrhini. *Folia Primatol.* **32**(3):207–238.

Baba, M. L., Darga, L. L., Goodman, M., and Czelusniak, J. 1981. Evolution of cytochrome c investigated by the maximum parsimony method. *J. Mol. Evol.* **17**:197–213.

Benveniste, R. E., and Todaro, G. J. 1976. Evolution of type C viral genes: Evidence for an Asian origin of man. *Nature (Lond.)* **261**:101–108.

Borden, D., Ferguson-Miller, S., Tarr, G., and Rodriguez, D. 1978. Sequence, function, evolution of spider monkey cytochrome *c*. *Fed. Proc.* **36**(6):1517.

Dene, H. T., Goodman, M., and Prychodko, W. 1976. Immunodiffusion evidence on the phylogeny of the Primates, in: *Molecular Anthropology* (M. Goodman and R. E. Tashian, eds.), pp. 171–195, Plenum, New York.

Fitch, W. M., and Langley, C. H. 1976. Protein evolution and the molecular clock. *Fed. Proc.* **35**(10): 2092–2097.

Goodman, M. 1976. Toward a genealogical description of the primates, in: *Molecular Anthropology* (M. Goodman and R. E. Tashian, eds.), pp. 321–353, Plenum, New York.

Goodman, M. 1981. Decoding the pattern of protein evolution. *Prog. Biophys. Mol. Biol.* **38**:105–164.

Goodman, M., and Czelusniak, J. 1980. Mode, tempo, and role of natural selection in the evolution of heme proteins, in: *Proteins of the Biological Fluids* (H. Peeters, ed.), pp. 57–60, Pergamon, Oxford.

Goodman, M., and Moore, G. W. 1971. Immunodiffusion systematics of the primates. I. The Catarrhini. *Syst. Zool.* **20**:19–62.

Goodman, M., Moore, G. W., and Matsuda, G. 1975. Darwinian evolution in the genealogy of hemoglobin. *Nature (Lond.)* **253**:603–608.

Goodman, M., Czelusniak, J., Moore, G. W., Romero-Herrera, A. E., and Matsuda, G. 1979. Fitting the gene lineage into its species lineage: A parsimony strategy illustrated by cladograms constructed from globin sequences. *Syst. Zool.* **28**:132–163.

Greenfield, L. O. 1980. A late divergence hypothesis. *Am. J. Phys. Anthropol.* **52**:351–366.

Johanson, D. C., and White, T. D. 1979. A systematic assessment of early African hominids. *Science* **203**:321–330.

Moore, G. W. 1976. Proof for the Maximum Parsimony ("Red King") Algorithm, in: *Molecular Anthropology* (M. Goodman and R. E. Tashian, eds.), pp. 117–137, Plenum, New York.

Tashian, R. E., Hewett-Emmett, D., and Goodman, M. 1980. Evolutionary diversity in the structure of carbonic anhydrase, in: *Proteins of the Biological Fluids* (H. Peeters, ed.), pp. 153–156, Pergamon, Oxford.

A Model of Chromosome Evolution and Its Bearing on Cladogenesis in the Hominoidea

4

L. L. MAI

Introduction

Chromosomes may be thought of as the vehicles by which encapsulated genetic material is transmitted between organisms. Chromosomes, like other traits or characters, present a morphology which can be used to distinguish between species, and which changes discretely over time. As one is able to speak of the evolution of the canine tooth or of brain size, one may also speak with equal acuity of the evolution of chromosomes.

Human chromosomes are becoming so well described in terms of their morphology that human cytogeneticists now think of individual banding patterns on chromosomes almost as fingerprints; a numerical convention has been established to delimit dark from light bands, and many genetic loci have been assigned to individual chromosome bands (Fig. 1).

Nonhuman primate chromosomes, as one might expect, resemble human chromosomes both in morphology and in the technical characteristics by which they can be cultured and harvested. Consequently, it is no surprise that both qualitative and quantitative differences in primate chromosomes should be expected to reflect their taxonomic relationships, and therefore evolutionary histories (Dutrillaux, 1975; Seuanez, 1979). Also, since chromosome fea-

---

L. L. MAI • Division of Orthopaedic Surgery, UCLA School of Medicine, and Department of Anthropology, University of California, Los Angeles, California 90024.

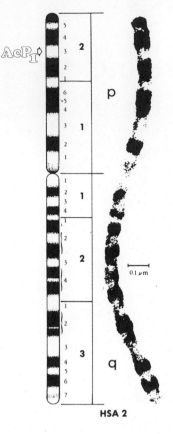

Fig. 1. Human chromosome 2 (HSA2) alongside a schematic representation of banding patterns and standardized labeling nomenclature. The dark bands are constituted by heterochromatin, the light by euchromatin. Well over 100 loci have been assigned to specific bands in the human genome. The locus for red cell acid phosphatase (AcP_1) has been assigned to band p2.3, for example.

tures are derived characters, it becomes axiomatic that chromosomes may be used to reconstruct phylogenies, because "a phylogeny does not need to be based on fossils but can be inferred from a careful comparative analysis of morphological characters" (Mayr, 1974, p. 98).

This chapter is concerned primarily with the presentation of a model of primate evolutionary relationships, specifically among hominoid primates, using only a single class of neontological data: chromosomes. Secondarily, it deals with the cladistic value of cytogenetic data.

A Cartesian Model of Chromosome Evolution

The mechanisms of chromosome repatterning are an inferred class of phenomena whose consequences may be observed under the light microscope, and thus are in a practical sense distinct from point mutations of nucleotide triplet sequences, as these latter occur at the submicroscopic level.

Known chromosomal repatterning mechanisms resulting in anomalies may be grouped into two major classes: numerical and structural. The first class (numerical) includes the *aneuploidies*—polyploidies, endoreduplications, and the like. The second class (structural) includes the many kinds of *translocations,* the two types of *inversions,* and the *intrachanges.* The two inversion types are (1) *para*centric, in which the banding pattern of a portion of a chromosome not involving the centromere is inverted, and (2) *peri*centric, in which the centromere is involved.

The behavior and consequences of these repatterning mechanisms is well known (e.g., White, 1973, Chapter 7). Many of these yield meiotic products that are lethal or sublethal (in the chromosomal heterozygote, or hetero-karyote), and are of interest to others because of the consequences of fetal wastage, birth spacing, and differential fecundity. But the systematicist is concerned only with the viable chromosome repatterning event and its evolutionary consequences. It is this event that seems to contribute significantly to the reproductive isolation of demes and species over time.

The mechanisms which seem to have predominated in the better known evolutionary series (at least in the vertebrates) are Robertsonian translocations (fissions and fusions), and inversions. The consequences of a single repatterning event by a mechanism of Robertsonian translocation is diagrammed in Fig. 2. Given a fusion event (left to right in Fig. 2), a hypothetical ancestral population *a* is changed into a daughter population *b* and reduced in diploid number by two for each such event. The small chromosomes in *a* are usually termed telocentric chromosomes,* while the X-shaped chromosomes in *b* are usually termed metacentric chromosomes. In the case of a fusion event, then, the daughter population is reduced in the number of telocentric, and increased in the number of metacentric, chromosomes relative to the ancestral population in a very systematic way. The opposite would be true of a fission event (*b* to *a*). If parent individuals representing the two karyotypes produce a hybrid offspring, the karyotype of the offspring would be intermediate, and the offspring may often be sterile, as is the case in mules, or of lowered fecundity, as in humans (Boué *et al.,* 1975).

The consequences of a single repatterning event by the mechanism of pericentric inversion are diagrammed in Fig. 3. Given an ancestral population consisting again entirely of telocentric chromosomes (*a*), a pericentric inversion event can result only in a daughter population (*b*) with the same diploid number but reduced in the number of telocentrics, and therefore also in the number of chromosome "arms" (the fundamental number), by two, for every inversion event. If, on the other hand, the inversion were to occur in a metacentric chromosome, the daughter population could, if the breakpoints were as in the example (*b* to *a*), result in a karyotype with the same diploid number but increased by two in telocentrics and fundamental number. If the breakpoints were equidistant from the centromere, it follows that, numerically,

---

*There are in fact five chromosome morphotypes in an idiogram (Levan *et al.,* 1964). In this chapter, however, metacentrics and submetacentrics will be collapsed into a single class called metacentrics, while submetacentrics, acrocentrics, and telocentrics will be called telocentrics.

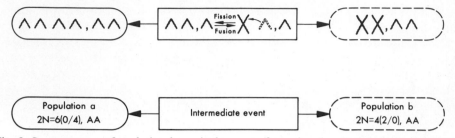

Fig. 2. Consequences of evolution by a single event of translocation (Robertsonian fusion or fission).

nothing would be changed, but, qualitatively, the banding pattern of the particular chromosome involved would be repatterned.

These two examples show clearly that quantitative changes are very systematic, and can be detected by simply measuring the number and types of chromosomes in a karyotype. This is fortunate for the systematist, as karyologists have been measuring just these parameters for decades, and abundant data are available to construct models and describe evolutionary lineages (Benirschke and Hsu, 1967).

The number and types of chromosomes in a standard karyotype are presented by various conventions in some format such as the following:

$$\theta = (2n/X.V.nS/F_n)$$

where θ (theta) is the numerical "formula" for a karyotype; $2n$ is the diploid complement of chromosomes; V is the number of telocentric chromosomes in the complement; X is the number of nontelocentric chromosomes (that is, the metacentrics plus submetacentrics); nS is the number of sex chromosomes in the complement; and F_n is the fundamental number of chromatid arms (usu-

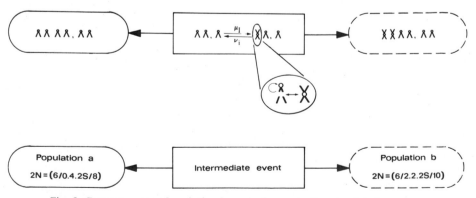

Fig. 3. Consequences of evolution by a single event of pericentric inversion.

ally $2n + X$, or $2n + X + nS$). The karyotype for *Homo sapiens*, using this convention, is

$$\theta = (46/34.10.2S/80)$$

Figure 4 illustrates a two-dimensional surface on which, for any given karyotype, the number of telocentric autosomes V may be plotted vs. the fundamental number of arms F_n present in a karyotype, resulting in a point representing θ. The space in the upper left corner represents impossible outcomes, as the number of telocentrics can never exceed the fundamental number of chromosome arms. Since the diploid number $2n$ of a species is also systematically related to the fundamental number, the point θ can also be plotted if only $2n$ and V are known (which is true of much of the published data).

Since only part of the available plotting space is ever used, and since $2n$ and V are the most common parameters used to locate θ, the rectangular plot

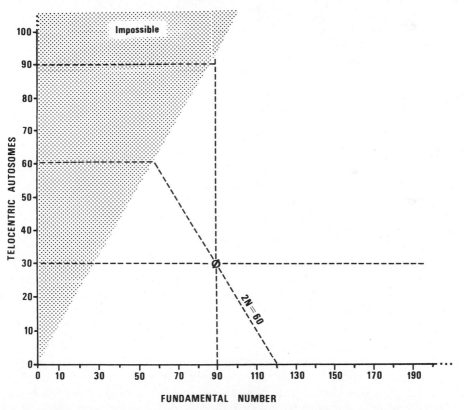

Fig. 4. Two-dimensional surface on which any karyotype may be located by plotting V (or telocentric autosomes) vs. either F_n (fundamental number) or $2n$ (diploid number), yielding a point θ. Using $F_n = 2n + X$, the example illustrates θ for $\theta = (60/30.30.2S/90)$.

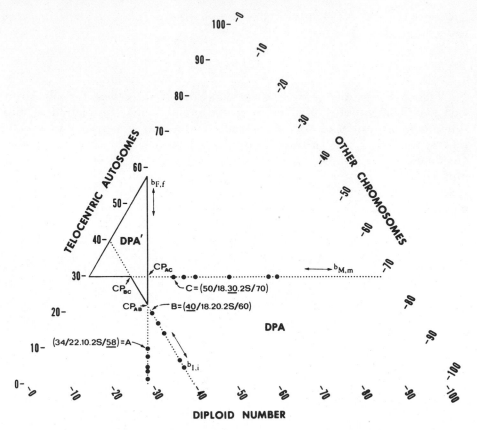

Fig. 5. Plotting surface abridged from that illustrated in Fig. 4. Several exemplary θ's (A, B, C) are plotted as members of arrays assuming fusion/fission ($b_{F,f}$), inversion ($b_{I,i}$), and metacentric aneuploidy ($b_{M,m}$). The common progenitors (CP_i) and the domain of potential ancestors (DPA') for these arrays are also illustrated. These latter terms are explained later in the chapter.

in Fig. 4 can be conveniently collapsed into a closed triangular plot with two degrees of freedom (Fig. 5).

The utility of this Cartesian scheme of plotting is at least threefold: (1) it may be generalized into an *n*-dimensional model as more data become incorporated; (2) karyotypes are located spatially in an approximate reflection of taxonomic relationships; (3) the *mechanisms* contributing to the evolution of a lineage or of related lineages may be inferred by simple inspection of the plot. This latter point is more important and is illustrated in Fig. 6. The known karyotypes of the mouse genus *Peromyscus* are plotted and can be seen to form a sloping line of approximately 60° from the upper left to the lower right. Interestingly, those species judged on traditional taxonomic and paleontological grounds to be the most primitive fall near the origin of the pattern at the upper left, while in general the morphologically more derived species fall to the right and below this "origin" (Hsu and Arrighi, 1968). Thus the assumption that relatively "primitive" karyotypic states are characterized predominately by telocentric autosomes (the ratio of telocentrics to autosomes approaches unity, V/aut $\rightarrow$ 1), while relatively "specialized" karyotypic states are characterized predominately by nontelocentric autosomes (X/aut $\rightarrow$ 1), seems justified.

White (1973) maintains that derived or specialized karyotypes, which contain two or more classes of autosomes ["asymmetric" karyotypes by Stebbins' definition (1950)], are evidence for canalization of chromosome repatterning within anagenetic lineages, an effect he labels "karyotypic orthoselection." According to White, many examples may be cited in which only one or two repatterning mechanisms seem to have predominated in the karyotypic history of these lineages.

In the case of *Peromyscus* the primary mechanism of chromosome evolution, pericentric inversion, has been well documented (Caire and Zimmerman, 1975). It is then possible to generalize and state that any set of taxa

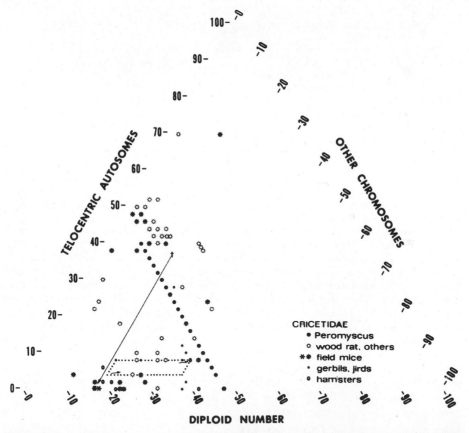

Fig. 6. Preliminary karyotypic domain of the extant progeny (DEP) of the members of the long-tailed rodent family Cricetidae. The deer mouse genus *Peromyscus*, indigenous to the arid southwestern United States, has been extensively studied, and is characterized by a DEP slope congruent with a Sturtevant line of $2n = 48$. The tetragram illustrates a documented case of tetraploidy in the golden hamster (Ohno *et al.*, 1966), and the arrow indicates a possible case of chromosomal macroevolution in the genus *Microtus*, possibly explained by tandem fusion(s).

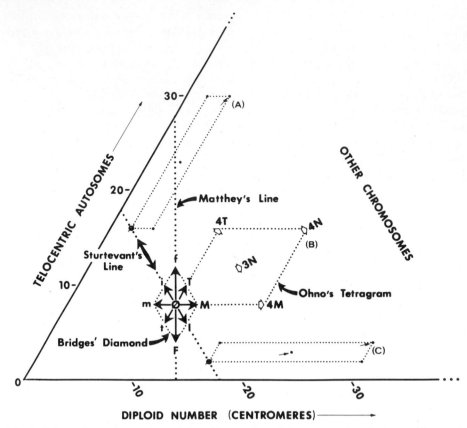

Fig. 7. Schematic representation of the major modes of chromosomal evolution. Bridges' "diamond" represents the potential area of one-event displacements from the parental karyotype θ, where the events are fusion and fission (F,f), inversions (I,i), and aneuploidy involving telocentric (T,t) or nontelocentric (M,m) autosomes. Sturtevant's and Matthey's lines are slopes along which evolutionary products are to be found when repatternings are exclusively by modes (I,i) or (F,f), respectively (with the latter often referred to as a "Robertsonian series"). Ohno's "tetragram" represents the potential area of one-event displacements from the original karyotype θ where the repatterning event is partial or full polyploidy (Ohno, 1970). The configuration of this last space is a function of the number of telocentric autosomes in the parental karyotype, with case A representing many, B representing half, and C representing only a few telocentric autosomes in the parental karyotype.

assumed to be related and forming a pattern such as that observed in *Peromyscus* must have evolved primarily by the repatterning mechanism of pericentric inversion. Any variability about such a line, of course, suggests that other mechanisms have also been active.

Other known repatterning mechanisms can be similarly visualized using this Cartesian plotting scheme. Fission–fusion events, for example, fall only on the vertical. Metacentric trisomies fall on a line with a slope of 120°, and so forth. A few exemplary points are plotted in Fig. 5. I have elsewhere plotted most of the data available as of 1978, using hundreds of taxa (Mai 1979), and the result of this exercise is presented in Fig. 7.

Here it can be seen that each of the major mechanisms occupies a unique axial relationship relative to the other mechanisms of chromosome repatterning. In the example, a hypothetical karyotype θ = (18/8.8.2S/28) has been located by plotting V vs. $2n$; hence this karyotype may also be specified as θ = (8,18). If a daughter population were to "bud" from this parental karyotype and were to be plotted in the same illustration, the direction of the "bud" would allow the mechanism to be inferred, and the displacement would allow the number of events to be calculated.*

In Fig. 7, the arrows leading to F, I, . . . , t represent single-displacement events which, if extrapolated, become regression slopes describing arrays of karyotypes derived from the parental karyotype and related in proportion to the number of discriminating events. The upper case letters F, I, M, and T signify "forward" mutations by the mechanisms of Robertsonian fusion, inversion (in favor of metacentricity), metacentric trisomy, and telocentric trisomy, respectively. Lower case letters signify "back" mutations: Robertsonian fission, inversion (in favor of telocentricity), metacentric monosomy, and telocentric monosomy. The tetragrams represent areas in which daughter karyotypes derived by triploidy or tetraploidy would be expected to fall. For historical perspective I have named the major axes and surface areas of interest after geneticists associated with the mechanisms: Matthey (translocations), Sturtevant (inversions), Bridges (many mechanisms), and Ohno (polyploidy).

From θ = (18/8.8.2S/28), the following single-event displacements—that is, displacements to the periphery of what I have termed "Bridge's diamond"—are possible:

$$(20/6.12.2S/28) \xleftarrow{f} θ \xrightarrow{F} (16/10.4.2S/28)$$
$$(18/6.10.2S/20) \xleftarrow{i} θ \xrightarrow{I} (18/10.6.2S/30)$$
$$(16/6.8.2S/24) \xleftarrow{m} θ \xrightarrow{M} (20/10.8.2S/32)$$
$$(16/8.6.2S/26) \xleftarrow{t} θ \xrightarrow{T} (20/8.10.2S/30)$$

These displacements represent the single-event dimensions potentially available to a population of organisms with a modal karyotype θ. Note again that the arrows leading from θ along F or f, for example, represent not only single-event displacement dimensions but, when extrapolated or documented over several events, define regression slopes representing the inversion repatterning mode, as each of the unique slopes represents one of the major modes of repatterning mechanism radiating from θ. This pattern recalls the plotted array of *Peromyscus* in Fig. 6.

It can be seen that this Cartesian representation of karyotype evolution presents data in an easily digestible format. It is also amenable to formal analysis of ancestry derivation and proposition testing. Brief examples of such tests in the primates follow.

---

*A third dimension is required to visualize some pericentric inversions.

Implications of the Model for Cladogenesis in the Hominoidea

Catarrhine Chromosome Evolution and Cytotaxonomy

If the plotted array of related karyotypes is called the domain of the extant progeny (DEP), then the relationship of the predominant repatterning mechanism to the DEP of interest may be defined as:

Definition DEP. Within a monophyletic group, the slope of that line described by points characterizing those populations thought on independent grounds to belong to taxa different at the level of the genus (or above) shall be considered the slope or line representing the modal mechanism(s) of that group of taxa (b_{DEP}).

In the case of the Catarrhini, that group of primates indigenous to the Old World, the pattern of cytogenetic evolution and the mechanisms leading to their DEP(s) are unfortunately not as dramatic as in some rodent lineages, and the least obvious of any of the primate suborders. By Definition DEP, however, several "modal" slopes within the Catarrhini are obvious. Figures 8 and 9 illustrate the DEPs of the Catarrhini. The most dominant slope is described by a Sturtevant line of $2n = 48$. This first slope S_I is indicated by a dotted line in Fig. 9. At least ten genera fall "on or near" this slope, including *Pongo,** *Pan, Homo, Gorilla, Presbytis, Nasalis, Allenopithecus* (=*Cercopithecus nigroviridis*), and possibly *Symphalangus* and *Hylobates* (*klossi*) (the latter pair are both $2n = 50$).

A second possible slope S_{II} is described by a Matthey's line of $F_n = 98$. At least eight genera fall "on or near" this slope (Fig. 9), including *Nasalis, Presbytis, Allenopithecus, Symphalangus, Hylobates* (*klossi, concolor*), *Erythrocebus, Miopithecus,* and *Cercopithecus aethiops.*

A third possible slope (S_{III} in Fig. 9) is described by a horizontal cluster of populations having in common the property that all have a "constant" number of telocentrics: 0–2. Thirteen genera fall "on or near" this slope, including *Symphalangus, Hylobates, Allenopithecus, Nasalis, Presbytis, Rhinopithecus,* and the six genera of the 42-chromosome tribe Papionini (*Macaca, Papio, Theropithecus, Mandrillus, Cercocebus,* and *Cynopithecus*).

A fourth slope of 120° (unlabeled) represents the DEP of many species of Cercopithecoidea. This highly heteromorphic genus seems to be characterized at present by a slope predicted by the mechanism(s) of telocentric aneuploidy (T, t), possibly fission (f), or possibly even partial tetraploidy, and therefore appears to be an exception to the general patterns of chromosomal evolution observed within Catarrhini as a group, the Primates as an order, and even those vertebrates, in general, surveyed earlier in a larger work (Mai 1979).

---

*The Bornean and Sumatran varieties of orangutans differ by a pericentric inversion (Seuanez *et al.,* 1979).

There must be a limit to the number of events of a certain type that an evolutionary series of karyotypes can sustain:

Proposition A.   In a case of strict karyotypic orthoselection progeny derived from P will radiate from P toward those points of maximum deviation (R_{min} and R_{max}) along slope b_{DEP}.

By Definition DEP, slope S_{III} qualifies as the literally "modal" slope within the Catarrhini. This slope, however, constituted by at least nine genera all described by having "no" telocentrics (that is, they are described by the "line" here called R_{max}), probably represents a case of independent convergence on the "orthoselectional limits" of chromosome evolution R_{max} when one or two mechanisms (such as Robertsonian fusion and pericentric inversion) are available to or selected for in these lineages.

Figure 6 schematically illustrates this suggestion that populations "orthoselected" for one or two cytogenetic mechanisms must eventually reach the maximum range of orthoselectional possibilities R_{max}. In the example, it oc-

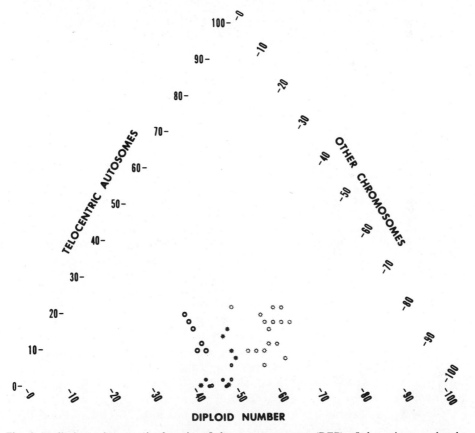

Fig. 8. Preliminary karyotypic domain of the extant progeny (DEP) of the primate suborder Catarrhini (order Primates), with points left unlabeled to illustrate tentative modal slopes, which are more fully indicated in Fig. 9. Data are from Mai (1979) and references cited therein.

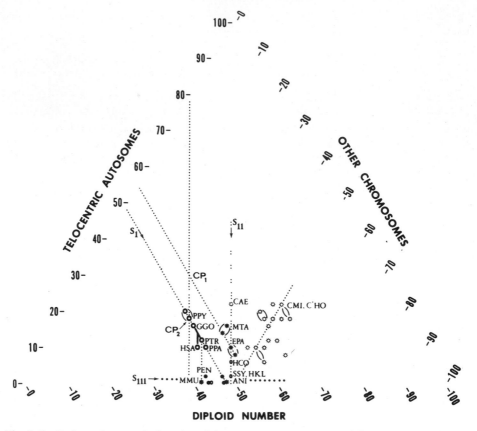

Fig. 9. Preliminary karyotypic domains of the extant progeny (DEPs) of the catarrhine primates, with points labeled and modal slopes S_I, S_{II}, and S_{III} indicated by dotted lines. Abbreviations for the taxa illustrated are as follows: ANI, *Allenopithecus* (=*Cercopithecus?*) *nigroviridis;* CAE, *Cercopithecus aethiops;* CMI, *Cercopithecus mitis;* CHO, *Cercopithecus l'hoesti;* EPA, *Erythrocebus* (=*Cercopithecus?*) *patas;* GGO, *Gorilla gorilla;* HCO, *Hylobates concolor;* HKL, *Hylobates klossi;* HSA, *Homo sapiens;* MMU, *Macaca mulatta;* MTA, *Miopithecus* (=*Cercopithecus?*) *talapoin;* PEN, *Presbytis entellus;* PPA, *Pan paniscus;* PPY, *Pongo pygmaeus;* PTR, *Pan troglodytes;* SSY, *Symphalangus syndactylus.*

curs when no chromosomes remain telocentric, in which case certain pericentric inversions and fusions are no longer possible. When several taxa independently converge on R_{max}, as appears to have been the case here, an artifactual conglomeration of points forming a "slope" is the outcome. That this must have been so is testable by comparing the banding patterns of the converged taxa (which should be considerably different under the hypothesis of convergence), but such a test is beyond the scope of this chapter.

Slope S_{II}, a Matthey's line described by a constant $F_n = 98$ and occupied by eight genera, intercepts the Sturtevant's line of $2n = 62$ (thought to be modal for the Atelinae and Lorisidae) at (62/34.26.2S/98), near the present domain of *Lagothrix* (62/20.40.2S/84) (Fig. 9). Slope S_I, a Sturtevant's line described by a constant $F_n = 48$ (or 46 to 50), and occupied by at least ten genera (including apes and humans), intercepts the modal slope of the Platyrrhini at (48/24.22.2S/74) (a Matthey's line of $F_n = 74-76$), near the present domain of the heteromorphic platyrrhine genus *Callimico* (48/30.16.2S/80) (Fig. 9).

If derived karyotypes do indeed form such arrays, it follows that the most primitive karyotype should at one time have been located somewhere on that array:

Proposition B.　In a case of strict karyotypic orthoselection the progenitor P of any array of extant progeny will be found on or near b_{DEP}, the line defining both the empirical array or "domain of extant progeny" and the slope expected under the assumption of the exclusive operation of mechanism (b_i).

Further, when two karyotypic arrays, known on independent evidence to be genetically related, intersect, we have the following:

Proposition C.　In a case of karotypic orthoselection involving genetically related lineages but with two different repatterning mechanisms and therefore two intersecting arrays, the point of intersection represents the "logical" location of the last progenitor common to the two lineages (CP).

But the "logical" common progenitor (CP) may not have been the actual CP. After a direction of chromosomal evolution from many to few telocentrics is assumed, it is possible at least to delimit the surface domain in which the actual CP must have resided:

Definition DPA.　In the general case of karyotypic orthoselection for intersecting arrays of monophyletic populations described by DEP_1, DEP_2,..., DEP_n, the inferred domain of common potential ancestors shall be defined as that area or domain circumscribed by DEP_1, DEP_2,..., and R_{max}.

This is schematically illustrated for the primates in Fig. 5. The DPAs of three taxa, the "Prosimii," the Platyrrhini, and the Catarrhini, seem to be related in just such a systematic way. This relationship is illustrated in Fig. 10, in which it may be seen that the estimated DPAs of these three taxa form a nested set in which the DPA‴ of the "Prosimii" is circumscribed by the DPA″ of the Platyrrhini, and that the DPA′ of the Catarrhini circumscribes both of these former DPAs. In other words,*

$$DPA''' \subset DPA'' \subset DPA'$$

From this case and other good examples found in the rodents, the following rule applies:

---

*The symbol $\subset$ signifies the logical concept "is contained in." The relationship $A \subset B$ means both A is contained in B, and B contains A.

Rule: If, in the case of karyotypic orthoselection for intersecting arrays of monophyletic populations with inferred domains of potential ancestors described by DPA_1, DPA_2, . . . DPA_n, the following relationship is true:

$$DPA_1 \subset DPA_2 \subset DPA_n$$

then DPA_1 is ancestral to DPA_2 is ancestral to ... is ancestral to DPA_n.

By this rule, which I have called the rule of inferred domains, it may be concluded that primate DPA‴ is ancestral to primate DPA″ is ancestral to primate DPA′. Several implications follow from this conclusion. First, this conclusion implies that the common progenitor of the Old World primates (CP_3 in Fig. 10) arose from a primate population described by a karyotype found within the area of DPA′. In other words, either a karyotypically primitive "prosimian" or a protoplatyrrhine could have given rise to the catarrhine primates. Second, this conclusion implies that the common progenitor of the New World primates (CP_2 in Fig. 10) arose from a primate population described by a karyotype found within the area of DPA″. That is, the Platyrrhini probably arose from a protoprosimian population described at some point in the past by a karyotype found within DPA‴, an area at present occupied by karyotypically "primitive prosimians," such as the microcebids (Dutrillaux, 1979). Third, the above implications one and two, when conversely stated, are very probably *not* true. It seems highly unlikely, in other words, that a population described by a karyotype lying within the catarrhine DEP could have given rise to either the platyrrhine or the "prosimian" karyotypic arrays as seen in present populations, or that a population described by a karyotype lying within the platyrrhine DEP could have given rise to the present "prosimian" array.

While these implications may seem trivial at a first reading, i.e., a protoprosimian gave rise to a protoplatyrrhine and a protoplatyrrhine gave rise to a protocatarrhine, they are important for two reasons. First, in terms of veracity, support for one primate origin hypothesis, based on morphological and molecular evidence, is supported: a protoprosimian gave rise to a protoplatyrrhine. Second, this evidence bears directly on the controversy of the origin of the platyrrhines (Chiarelli, 1980). The chromosome data do not support the hypothesis of independent and parallel origins for the platyrrhines and the catarrhines from within the protoprosimians. As has been recently suggested, an African origin for the platyrrhines is possible (Ciochon and Chiarelli, 1980). The chromosomal data support this latter possibility.

Within the Hominoidea it is not surprising to note that the Lesser Apes should differ from the "logical" hominoid CP by a minimum of ten pericentric inversions plus two other major fission and/or aneuploidy events. Gibbon chromosomes have been considered generally nonhomologous when compared directly to other hominoid taxa (de Grouchy *et al.*, 1978). There is little doubt that one of the major repatterning mechanisms involved in gibbon

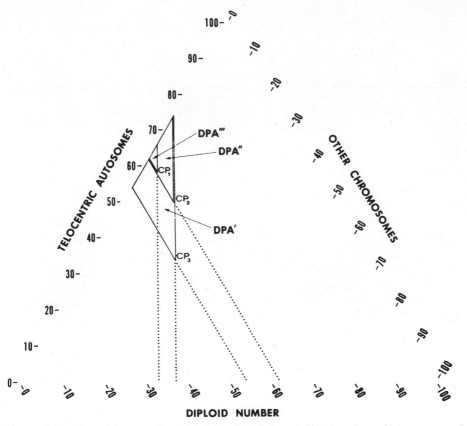

Fig. 10. Illustration of the nested relationship of the inferred domains of potential ancestors of the "Prosimii" (DPA‴), the Platyrrhini (DPA″), and the Catarrhini (DPA′). Such a relationship may indicate that these groups represent an evolutionary sequence in which populations chromosomally similar but ancestral to DPA‴ gave rise to populations ancestral to either or both of the larger DPAs (DPA″ and DPA′).

evolution has been pericentric inversion. If the gibbons are considered an extrapolation of the Sturtevant line that passes through the Great Apes, then it follows that the karyotypes of the Lesser Apes reflect a dramatic deviation from the more conservative karyotypic evolution seen in the Great Apes. Further, their gross chromosome morphology is convergent with certain cercopithecoid taxa. Both the Lesser Apes and these cercopithecoid taxa have independently converged on the X axis principally by pericentric inversion. Because of the large differences seen between the Lesser and Great Ape karyotypes, however, it is not presently possible to suggest a point of reticulation of the Lesser Apes with a cladogram of the Great Apes. Excluding the gibbons, then, it is of interest next to infer the cladogram for the rest of the Hominoidea at a higher level of resolution.

The Cladistic Value of Cytogenetic Data: An Example from the Hominoidea

Phenetic vs. Cladistic Analyses

Dorothy Miller recently suggested that "man is more closely related to the gorilla than to the chimpanzee" because human and gorilla share some morphological characters on some homologous chromosomes which she feels may be important (Miller, 1977). Miller's argument in support of this hypothesis is tenuous for several reasons which have been addressed in detail elsewhere (Mai, 1979), but these exceptions can be fairly collapsed into the statements that (1) the inference of a branching sequence, on the basis of the type of data employed by Miller, is inappropriate, (2) the use of weighted "homologies" is questionable, and (3) her central hypothesis of cladistic relationship can be falsified when subjected to a test utilizing a different type of cytogenetic data: structural rearrangements.

A potential source of confusion seems to exist over the meaning of the term relationship. According to Cain and Harrison (1960), there are two kinds of relationship: phenetic and phyletic. Phenetic relatedness is simple affinity, or overall similarity as suggested by the characters of organisms (or objects) without any implication as to their ancestral connections. A phyletic relationship, on the other hand, "aims to show the course of evolution" (p. 3), and may be either (a) patristic, which is affinity due to sharing of many characters derived from a common ancestor, or (b) cladistic, which is relatedness due to recency of common ancestry. In the latter case, only the branching sequence of ancestral lineages can be deduced; in the former case, branching sequences may be inferred, and patristic relationships can be estimated. The fundamental difference between the two approaches lies in the nature of the data being employed. Phenetic analyses are appropriate when the characters have nonzero reverse mutation rates, exhibit continuous variation or polymorphism, etc.; such analyses are therefore probabilistic (Holmquist, 1976). Cladistic analyses are appropriate when the data consist of more complex characters which have evolved in exactly the same way only once (LeQuesne, 1974). Thus there is no *a priori* expectation that phenetic and cladistic analyses performed on the same taxa should yield congruent results in terms of branching sequences (Fitch and Margoliash, 1968), and "phylogenetic enigmas" may thus sometimes arise (Romero-Herrera *et al.*, 1976). Well-known examples of the applications of phenetic analyses (with patristic generalizations) employ data derived from techniques such as electrophoresis (Avise, 1974), immunology (Goodman, 1976; Sarich and Cronin, 1976), amino acid sequencing (Goodman, 1976), and DNA–DNA hybridization (Benveniste and Todaro, 1976).

The data employed by Miller (1977) in support of her hypothesis are appropriate *only* for the demonstration of phenetic relationships. These data are *not* appropriate for the demonstration of phyletic relationships. This statement follows from the observation that fluorescence, C-band size, and (possibly) 5-MeC concentration are heteromorphic character states in human and gorilla. In such cases, apparent structural similarities may understandably be misclassified as homologies, when similarity is due actually to concurrently achieved heteromorphic states in the different species. For example, Miller's observation of "intense fluorescence" of the pericentromeric regions of the homologous chromosomes 13, 14, and 15 and the distal Y, and in the satellites of chromosomes 13, 14, and 15, is a variable phenomenon in both human (Buckton *et al.*, 1976; Bogart and Benirschke, 1977; Pearson, 1977), and gorilla (Miller *et al.*, 1976). A C-band-size heteromorphism has been documented for chromosomes 1, 9, 16, and the Y in humans (Crossen, 1975; Halbrecht and Shabtay, 1976; Craig-Holmes, 1977). Further, structural rearrangements which are likely to have affected the pattern of distributions of these heteromorphisms among these species are thought to have occurred on each of these chromosomes.

Structural rearrangements have also affected the Y chromosomes of these species, and the nature of the heteromorphism is perhaps clearer in this specific example. Human Y-specific reiterated sequences (It-Y DNA), comprising an estimated 10% of Y-chromosome DNA, have been isolated from male DNA (Kunkel *et al.*, 1976). The quantity of It-Y present in individuals with Y-chromosome anomalies permitted localization of It-Y on the long arm of the Y in the region of intense fluorescence. Absence of It-Y and commensurate fluorescence in phenotypically normal Yq males established that It-Y has no evident role in male determination (Kunkel *et al.*, 1977). A larger sequence comprising an estimated 50% of Y-chromosome DNA (and possibly also including It-Y) may function either as genetically neutral DNA added to small amounts of male-determining DNA to facilitate segregation during spermatogenesis, or to suppress crossing-over with other chromosomes (Cooke, 1976). These studies suggest that the presence of "intense fluorescence" on the Y chromosomes of human and gorilla (and "absence" in chimpanzee and orangutan) may be a pattern of similarity due more to independent heteromorphic states than to an exclusive human–gorilla homology of structural genes. As in the case of 1qh, 9qh, and 16qh, there is also the possibility that observed changes in the centromeric indices of these chromosomes (and usually attributed to pericentric inversions) may be in fact due to changes in heteromorphic states of the constitutive heterochromatin located in these regions (Angell and Jacobs, 1975).

Other evidence supporting this point is the lack of congruity expected for a pattern of homologies that are also "true" synapomorphies.* The chimpanzee and gorilla, for example, exhibit homologous fluorescent zones where neither human nor orangutan exhibit similar fluorescence (Miller, 1977, Table 3). In cases where such incongruent homologies exist, it can only be

---

*Characters or character states which are derived from an ancestral state by transformation in a monophyletic group are *apomorphous;* the presence of an apomorphic character state in different species is a *synapomorphy* (Hennig 1966).

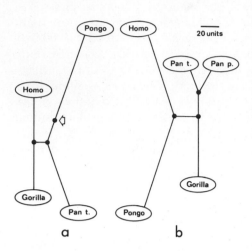

Fig. 11. Phenograms representing (a) Miller's (1977) hypothesis and (b) a maximum parsimony phenogram generated from the unweighted data matrix illustrated in Table 1. Branch lengths of (a) were grossly estimated from Fig. 2 of Miller (1977). An arrow indicates the node at which the gibbon articulates with this arrangement. Branch lengths of (b) were computed from the same data but according to the method of Sokal and Sneath (1973). Note that phenogram (b) implies both phenetic and cladistic relationships different from those arrived at by Miller.

assumed that the largest set of congruent homologies are actually "true" synapomorphies (i.e., reflecting the true evolutionary history of the organisms), and that the smaller (incongruent) sets mimic the "true" synapomorphies (Platnick, 1977). Since the data employed by Miller are heteromorphic and display incongruent patterns of "homologies," the relationship suggested must be phenetic. In this case, Miller's attempt to infer a *branching* pattern using these data alone seems inappropriate.

Finally, Miller seems to have weighted (or biased) those heteromorphic states in which human and gorilla are "homologous" such that the hypothesis that human and gorilla shared a common ancestor more recently than human and chimpanzee is favored (cf. Fig. 11a). This becomes more apparent when the entire data set accumulated by Miller (Miller, 1977, Table 3), consisting of a minimum of 886 pairwise "homologies," is retabulated and collapsed into a 5 × 5 matrix (Table 1). When, by a conventional method (Sokal and Sneath, 1973; Farris, 1973), these data are "unweighted," this matrix of unweighted similarities can be transformed into a maximum parsimony phenogram (Fig.

Table 1. Heteromorphic Similarities among the Hominoidea[a]

	HSA	PTR	PPA	GGO	PPY
Homo sapiens (HSA)	—	171	173	186	200
Pan troglodytes (PTR)	82	—	55	120	185
Pan paniscus (PPA)	81	150	—	121	186
Gorilla gorilla (GGO)	83	112	111	—	200
Pongo pygmaeus (PPY)	65	74	73	65	—

[a]The data matrix consisting of 886 pairwise "homologies" or heteromorphic similarities reduced from Table 3 of Miller (1977) is presented in the lower left portion, while the upper right portion of the matrix consists of a simple transformation of Miller's raw data such that the phenetic distance between human and orangutan equals 200 arbitrary units.

11b) with implied phyletic relationships different from that arrived at by Miller. The difference seems to be due to Miller's biased weighting of a selected subset of homologies favoring her hypothesis. Ernst Mayr's warning against just such an error was: "Only derived [apomorphous] characters are considered legitimate for relationship, and taxa are therefore based on the joint distribution of derived characters [=synapomorphies]" (Mayr, 1974, p. 98).

Cladogenesis in the Hominoidea

Structural rearrangements of chromosomes, in contrast, appear on available evidence to represent a class of cytogenetic differences appropriate to a cladistic analysis of relationships between taxa. Structural rearrangements (such as pericentric inversions, for example) are commonly invoked to explain linear differences in the banding patterns observed between homologous chromosomes of primate species (Table 2). Structural rearrangements are apparently not subject to quantitative variation; they are complex and often possibly unique mutations (Hansmann, 1976), with associated fecundability differentials in heterozgotes, and exhibit unambiguous segregation patterns with a highly predictable pattern of inheritance.

Because of these characteristics, when two related taxa share the occurrence of an inferred chromosome structural rearrangement, the probability is high that the observed joint distribution is due to synapomorphy, i.e., that each is derived from the same deme or common ancestral population in which the rearrangement occurred *de novo* (Bobrow and Madan, 1973). This latter high probability allows one to exploit the proposition that structural rearrangements represent a class of data more robust, in terms of testing hypotheses regarding branching sequences, than a class of data characterized by a relatively lower heritability, a known nonzero reverse mutation rate, and segregation patterns that can be ambiguous, such as the data employed by Miller (1977). Hypotheses generated by the former (cladistically robust data) may therefore be utilized to exclude hypotheses suggested by the latter (phenetically robust data). Specifically, cladograms based on structural rearrangements may be used to test phenograms based, in this case, on heteromorphic cytogenetic data. In this light, Miller's phenogram (Fig. 11a) and Fig. 11b thus represent two competing phenetic hypotheses generated from the same data set by weighted and unweighted algorithms, respectively, and testable by comparison with a cladogram generated by structural differences.

More than 30 structural rearrangements have become fixed in the separate lineages leading to the living hominoid species (Table 2 and Fig. 12). Of these events, at least 23 are inversions. Figure 12 represents one of three possible unrooted Wagner networks appropriate for these taxa (Wagner, 1969; Farris, 1973). Two networks have been previously excluded and are not illustrated here since each of these requires the improbable redundancy of the entire V set of four pericentric inversion events (Mai 1979).

Table 2. The Five Sets of Homologies (or Cytogenetic Apomorphies and Synapomorphies) Corresponding to Events That Must Be Assigned to Lineages V, W, X, Y, and Z in Fig. 12

Set[a]	Event	Homologies	Breakpoints[b]
	5a	PPY, HSA↔GGO, PTR, PPA	*inv(5)(p14q14)
	12a	PPY, HSA↔GGO, PTR, PPA	*inv(12)(p12q14)
V	16a	PPY, HSA↔GGO, PTR, PPA	*inv(16)(p12q12)
	17a	PPY, HSA↔GGO, PTR, PPA	*inv(17)(p12q21)
	2a	PPY↔HSA, GGO, PTR, PPA	*inv(2)(p14q11)
	3a	PPY↔HSA, GGO, PTR, PPA	*inv(3)(p21q21)
	3b	PPY↔HSA, GGO, PTR, PPA	t(3;3)
	7a	PPY↔HSA, GGO, PTR, PPA	*inv(7)(p21q11)
	10a	PPY↔HSA, GGO, PTR, PPA	inv(10)(q11q22)
W	11a	PPY↔HSA, GGO, PTR, PPA	t(11)(p14q13)+
	11b	PPY↔HSA, GGO, PTR, PPA	*inv(11)(p13q13)
	12b	PPY↔HSA, GGO, PTR, PPA	*inv(12)(p13q21)
	12c	PPY↔HSA, GGO, PTR, PPA	*inv ins(12)(p13q13q21)
	17b	PPY↔HSA, GGO, PTR, PPA	*inv(17)(p11q12)
	20	PPY↔HSA, GGO, PTR, PPA	*inv(20)(p12q23)
	1a	HSA↔PPY, GGO, PTR, PPA	ins(1)(q11+)
	2b	HSA↔PPY, GGO, PTR, PPA	t(2)(pq)
X	4b	HSA↔PPY, GGO, PTR, PPA	*inv(4)(p12q21.2)
	9b	HSA↔PPY, GGO, PTR, PPA	t(9)(q12q14)
	18	HSA↔PPY, GGO, PTR, PPA	*inv(18)(p11q11)
	2c	PTR, PPA↔HSA, PPY, GGO	inv(2)(centromere)
Y	4a	PTR, PPA↔HSA, PPY, GGO	*inv(4)(pq)+del(4)(p-)
	15	PTR, PPA↔HSA, PPY, GGO	*inv(15)(p11q12)
	9a	PTR↔PPA, HSA, PPY, GGO	*inv(9)(p21q31)
Y'	13	PTR↔PPA, HSA, PPY, GGO	inv(13)(q13+)
	1b	PPA↔PTR, HSA, PPY, GGO	*inv(1)(p1200q12)
	2d	PPA↔PTR, HSA, PPY, GGO	inv(2)(q11q13)
Y''	7c	PPA↔PTR, HSA, PPY, GGO	t(7;21)
	10c	PPA↔PTR, HSA, PPY, GGO	*inv(10)(p12q21)
	22	PPA↔PTR, HSA, PPY, GGO	t(22+)
	5b	GGO↔PPA, PTR, HSA, PPY	t(5;17)
	7b	GGO↔PPA, PTR, HSA, PPY	inv(7)(q11q12)
	8	GGO↔PPA, PTR, HSA, PPY	*inv(8)(p12q21)
Z	10b	GGO↔PPA, PTR, HSA, PPY	*inv(10)(p15q24)
	14	GGO↔PPA, PTR, HSA, PPY	*inv(14)(p13q13)
	16b	GGO↔PPA, PTR, HSA, PPY	*inv(16)(p12q11)

[a]The orangutan (PPY) is synapomorphic (or symplesiomorphic, depending upon which character states are assumed to be derived or ancestral, respectively) with humans (HSA) for the V set of events, while the orangutan is apomorphic (or plesiomorphic) for the W set, and so forth. [Adapted from Dutrillaux (1975).]
[b]Asterisk indicates pericentric inversions.

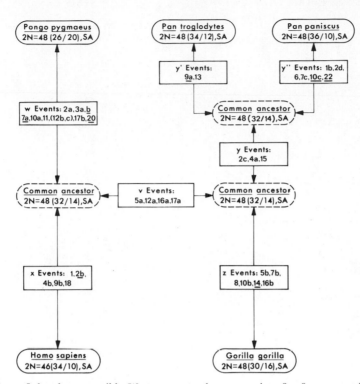

Fig. 12. One of the three possible Wagner networks appropriate for four neontological taxa. Genera of the Hominoidea (excluding the Hylobatidae) are illustrated. Two additional networks (not pictured) are possible, but have been excluded since they require that the V set of events must have occurred independently in each of two different lineages. One of the networks excluded by this parsimony method is compatible with Miller's hypothesis that the gorilla is more closely related to humans than is the chimpanzee. Ovals contain the binomial of the extant species (e.g., *Pongo pygmaeus* = orangutan), the diploid number characteristic of that species ($2n = 48$), the number of nontelocentrics and telocentrics in a modal karyotype (26/20), and a description of the X and Y chromosomes, respectively (S = submetacentric, A = acrocentric). Rectangles contain the sets of events distributed to the several lineages on the basis of observed synapomorphy: the W set, for example, consists of ten events for which the orangutan appears to be unique, while the V set consists of four events for which humans and the orangutan are synapomorphic (alternatively, the gorilla and the two chimpanzees are also synapomorphic), etc. A rearrangement which altered the centromeric index of a chromosome, from meta- to telocentric or *vice versa*, for example, is underlined. Thus the human karyotype, *Homo sapiens*, $2n = 46(34/10)$,SA, can be derived from the human–orangutan common ancestor, $2n = 48(32/14)$,SA, via event 2b, which changed the number of acrocentrics in the diploid karyotype from 14 to 10 (or *vice versa*), the number of metacentrics by two, and the diploid number by two.

Following the testing procedure mentioned above, it may be seen that Miller's hypothesis (that human and gorilla shared a more recent common ancestor than human and chimpanzee, Fig. 11a) can be excluded (falsified) since her hypothesis is incompatible with the branching pattern produced by these cladistically more robust repatterning data (Fig. 12). Note that the reticulation of the branch leading to *Gorilla* is different in the two cladograms.

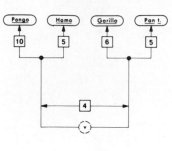

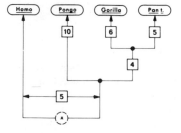

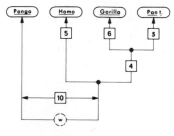

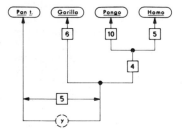

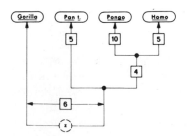

Fig. 13. The minimum set of five Wagner trees representative of four neontological taxa, all compatible with the Wagner network illustrated in Fig. 12. These "trees" differ from each other by the preference of root placement at V, W, X, Y, or Z. Squares contain the number of cytogenetic events assigned to each branch. Note the compatibility of the maximum parsimony phenogram generated from the matrix of unweighted "homologies" (Fig. 11b) with the tree rooted at W. The tree suggested by Miller (in which gorilla and human would share a most recent common ancestor, Fig. 11a), is not compatible with these data on chromosomal repatterning in the Hominoidea.

Fig. 14. Alternative arrangement of the tree rooted at W in Fig. 13, illustrating the possibility of reticulation at the node or trifurcation of the lineages leading to *Homo, Gorilla,* and *Pan.* This hypothesis is a possibility that can be excluded by current cytogenetic data, following the assumption that the V set of four pericentric inversions occurred only once in the Hominoidea.

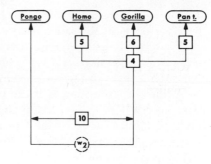

The competing phenogram generated from the matrix of unweighted "homologies" (Fig. 11b), however, is entirely compatible with the cytogenetic data available.

Similar tests may be made regarding other phenograms which have been proposed for these same taxa but which have been generated from other data. For each unrooted Wagner network such as Fig. 12 there exists a derivative set of rooted trees, or evolutionary hypotheses. For four terminal genera there is a minimum set which consists of five such trees (Fig. 13).* These individual trees are equally probable events given the current state of the chromosome data. Only the further high-resolution comparison of Lesser with Great Ape karyotypes will permit rejection of four of these trees. However, it is possible even at this point to assess the reality of the (*Homo,Gorilla,Pan*) trifurcation hypothesis suggested by the molecular data. This hypothesis, illustrated as the cladogram in Fig. 14, is simply a modification of the tree rooted at W in Fig. 13. A little reflection produces the conclusion that the only explanation compatible with the trifurcation hypothesis is that four inversion events, with exactly the same breakpoints, have occurred independently (twice each), once each in the *Gorilla* and the *Pan* lineages (or once each in the *Pongo* and the *Homo* lineages). But, as discussed earlier, each inversion event results in a daughter population in which genetic introgression with the parent population is not possible, due to a breeding barrier imposed by the inversion itself. Because of this it is therefore unlikely that the V set of four events has occurred exactly the same way twice each within the hominoid clade. Because this requirement of the trifurcation hypothesis is not met, the chromosome data therefore suggest that the trifurcation hypothesis is untenable.

Conclusions

Miller's hypothesis can be falsified by the cladistically more robust structural rearrangement chromosome data illustrated in Fig. 12. As both of the

---

*A rather small set ($2n \times 3$) exists when neotological data are employed and branching patterns are assumed to be dichotomous. If nodal reticulation is assumed (Sneath, 1975), additional

alternative networks assume that the V set of four events occurred at least twice with exactly the same breakpoints, which seems highly unlikely, it would be of interest to estimate elsewhere the likelihood that such an inferred set of pericentric inversions is a synapormorphic set, as opposed to the alternative that such a set is in fact two independent apomorphic sets.

Miller's assumption that the lineage leading to the gibbon reticulates as illustrated in Fig. 11a (noted by the arrow) represents another test of interest, but this cannot be confirmed or denied on the basis of currently available cytogenetic evidence. If one prefers to think in terms of any one of the five available evolutionary hypotheses illustrated in Fig. 13, however, the one rooted in W is probably the most conservative. The cytogenetic events depicted on this tree may be thought of as dynamic population changes, as pictured in Fig. 15. Given this scenario, it must be concluded that during the course of hominoid evolution, populations homokaryotic for newly arisen morphs budded from parental populations, and that several of these chromosomal morphs have subsequently become "extinct"—that is, do not appear as extant populations without additional changes—and have been replaced by chromatids with the repatterned linkage groups observed in existing hominoids.

The probable requirements for fixation of a pericentric inversion in a population are: (1) a small effective breeding population (the gibbon mating system is a good model), or (2) presence of alleles within the limits of the fixed inversion which confer superior fitness to the bearer relative to the ancestor such that the expected fecundity differential in the heterokaryote is overcome (Boué et al., 1975; Mai, 1979). These requirements bring to mind the classic model of fission described by disruptive selection in which homokaryotes are favored over the heterokaryote. This is illustrated in Fig. 16, which is a cladistic representation of the consequences of disruptive selection with unequal fitness of homokaryotes. When the new morph is a repatterned chromosome, and parental populations homokaryotic for the ancestral morph become extinct, the daughter populations homokaryotic for the new morph replace the ancestral populations. This mode of fission has been termed "stasipatric" (White, 1978). Further, since there can be no genetic introgression once the daughter population is established, because the inversion itself precludes this possibility, each inversion fixation represents a very real cladistic event. And since the products of the intermediate ancestral cytogenetic events no longer appear as living taxa, one must conclude that the demes represented by these morphs became extinct.

It would be inappropriate to propose a single classification of the Hominoidea until the nodal reticulation of the gibbon is finalized. At this point the cytogenetic evidence cannot exclude any of the five hypotheses (and therefore any of five classificatory schemes) illustrated in Fig. 13. It has been possible,

---

possibilities arise (Cracraft, 1974). If paleontological taxa are also employed, even more possible arrangements exist, as indicated by Harper (1976) in an excellent review of these problems.

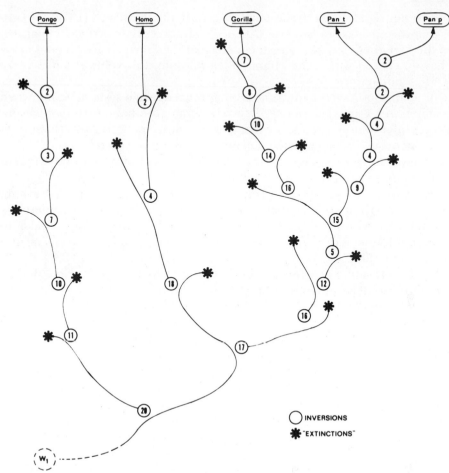

Fig. 15. Cladistic representation of the hominoid Wagner network (Fig. 12) as rooted at W (Fig. 13). In this representation it may be seen that several chromosomal morphs have become "extinct"—that is, have been replaced by chromatids with repatterned linkage groups. Numbers within the circles represent the chromatid in which the repatterning occurred in terms of its homology with the human chromatid of the same number.

Fig. 16. Cladistic representation of the consequences of disruptive selection with unequal fitness of homo-karyotes. If the new morph is a repatterned chromo-some, parental populations homokaryotic for the an-cestral morph will eventually become extinct, while daughter populations homokaryotic for the new mor-ph will replace the ancestral populations.

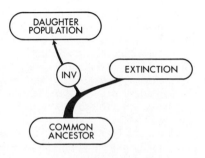

on the other hand, to render untenable or falsify both Miller's (1977) hypothesis that the gorilla is more closely related to humans than is the chimpanzee, and the hypothesis suggested by the molecular data that a trifurcation is likely to have occurred within the Hominoidea (Cronin, this volume, Chapter 5).

In summary:

1. Chromosomal repatterning among related taxa is systematic and when presented in a Cartesian model of chromosome evolution accommodates inferences of evolutionary mechanisms, directions, and histories. In this respect, the order Primates is not exceptional among the vertebrates.

2. Excluding *Hylobates,* all other catarrhine taxa appear to be cytogenetically more derived than the Hominoidea.

3. Intersecting karyotypic arrays within the "Prosimii," the Platyrrhini, and the Catarrhini suggest that a proto-Platyrrhine/proto-Catarrhine common ancestor was, in terms of gross chromosome morphology, similar to karyotypes found at present located within the domain of the Platyrrhini, and not the Catarrhini.

4. The Hominoidea form a karyotypic orthoselectional series (that is, generalized to derived types of karyotypes) as follows:

Pongo, (Gorilla, Homo, Pan), Hylobates

with the Lesser Apes clearly more derived than the other taxa.

5. Structural repatterning data clearly do not support the hypothesis offered by Miller (1977) that *Gorilla* is more closely related to *Homo* than is *Pan;* these two pongids are approximately equidistantly removed from *Homo.*

6. The proto-*Pan*/proto-*Gorilla*/proto-*Homo* trifurcation hypothesis seems untenable if not falsifiable.

7. It is premature to propose a classification of the Hominoidea based on cytogenetic data until the node of reticulation of the gibbons has been resolved.

References

Angell, R. R., and Jacobs, P. A. 1975. Lateral asymmetry in human constitutive heterochromatin: Frequency and inheritance. *Am. J. Hum. Genet.* **30:**144–152.

Avise, J. C. 1974. Systematic value of electrophoretic data. *Syst. Zool.* **23:**465–481.

Benirschke, K., and Hsu, T. C. 1967. *An Atlas of Mammalian Chromosomes,* Springer-Verlag, New York.

Benveniste, R. E., and Todaro, G. J. 1976. Evolution of type C viral genes: Evidence for an Asian origin of man. *Nature (Lond.)* **261:**101–108.

Bobrow, M., and Madan K. 1973. The effects of various banding procedures on human chromosomes, studied with acridine orange. *Cytogenet. Cell Genet.* **12:**145–156.

Bogart, M. H., and Benirschke, K. 1977. Q-banded polymorphism in a family of pygmy chimpanzees *(Pan paniscus). J. Med. Primatol.* **6:**172–175.

Boué, J., Taillemite, J. L., Hazael-Massieux, P., Leonard, C., and Boué, A. 1975. Association of pericentric inversion of chromosome 9 and reproductive failure in ten families. *Humangenetik* **30:**217–224.

Buckton, K. E., O'Riordan, M. L., Jacobs, P. A., Robinson, J. A., Hill, R., and Evans, H. J. 1976. C- and Q-banded polymorphisms in the chromosomes of three human populations. *Ann. Hum. Genet.* **40:**99–112.

Cain, A. J., and Harrison, G. A. 1960. Phyletic weighting. *Proc. Zool. Soc. Lond.* **135**:1–31.

Caire, W., and Zimmerman, E. G. 1975. Chromosomal and morphological variation and circular overlap in the deer mouse, *Peromyscus maniculatus,* in Texas and Oklahoma. *Syst. Zool.* **24**:89–95.

Chiarelli, A. B. 1980. The karyology of South American primates and their relationship to African and Asian species, in: *Evolutionary Biology of the New World Monkeys and Continental Drift* (R. L. Ciochon and A. B. Chiarelli, eds.), pp. 387–398, Plenum, New York.

Ciochon, R. L., and Chiarelli, A. B. 1980. Paleobiogeographic perspectives on the origin of the Platyrrhini, in: *Evolutionary Biology of the New World Monkeys and Continental Drift* (R. L. Ciochon and A. B. Chiarelli, eds.), pp. 459–494, Plenum, New York.

Cooke, H. 1976. Repeated sequence specific to human males. *Nature (Lond.)* **262**:182–186.

Cracraft, J. 1974. Phylogenetic models and classification. *Syst. Zool.* **23**:71–90.

Craig-Holmes, A. P. 1977. C-band polymorphism in human populations, in: *Population Cytogenetics* (E. B. Hook and I. H. Porter, eds.), pp. 161–178, Academic, New York.

Crossen, P. E. 1975. Variation in the centromeric banding of chromosome 19. *Clin. Genet.* **8**:218–222.

De Grouchy, J., Turleau, C., and Finaz, C. 1978. Chromosomal phylogeny of the primates. *Annu. Rev. Genet.* **12**:289–328.

Dutrillaux, B. 1975. Sur la nature et l'origine des chromosomes humains. *Monogr. Ann. Genet.* **3**:1–104.

Dutrillaux, B. 1979. Chromosomal evolution in primates: Tentative phylogeny from *Microcebus murinus* to man. *Hum. Genet.* **48**:251–314.

Farris, J. S. 1973. Probability model for inferring evolutionary trees. *Syst. Zool.* **22**:250–256.

Fitch, W. M., and Margoliash, E. 1968. The construction of phylogenetic trees: How well do they reflect past history? *Brookhaven Symp. Biol.* **21**:217–242.

Goodman, M. 1976. Toward a genealological description of the primates, in: *Molecular Anthropology* (M. Goodman and R. E. Tashian, eds.), pp. 321–353, Plenum, New York.

Halbrecht, I., and Shabtay, F. 1976. Human chromosome polymorphism and congenital malformations. *Clin. Genet.* **10**:113–122.

Hansmann, I. 1976. Structural variability of human chromosome 9 in relation to its evolution. *Hum. Genet.* **31**:247–262.

Harper, C. W. 1976. Phylogenetic inference in paleontology. *J. Paleontol.* **50**:180–193.

Hennig, W. 1966. *Phylogenetic Systematics,* Illinois University Press, Urbana, Illinois. 320 pp.

Holmquist, R. 1976. Random and nonrandom processes in the molecular evolution of higher organisms, in: *Molecular Anthropology* (M. Goodman and R. E. Tashian, eds.), pp. 89–116, Plenum, New York.

Hsu, T. C., and Arrighi, F. E. 1968. Chromosomes of *Peromyscus* (Rodentia, Cricetidae). I. Evolutionary trends in 20 species. *Cytogenetics* **7**:417–446.

Kunkel, L. M., Smith, K. D., and Boyer, S. H. 1976. Human Y-chromosome-specific reiterated DNA. *Science* **191**:1189–1190.

Kunkel, L. M., Smith, K. D., Boyer, S. H., Borgaonkar, D. S., Wachtel, S. S., Miller, O. J., Berg, W. R., Jones, H. W., and Ray, J. M. 1977. Analysis of Y-chromosome-specific reiterated DNA in chromosome variants. *Proc. Natl. Acad. Sci. USA* **74**:1245–1249.

LeQuesne, W. J. 1974. The uniquely derived character concept and its cladistic application. *Syst. Zool.* **23**:513–517.

Levan, A., Fredga, K., and Sandberg, A. A. 1964. Nomenclature for centromeric position on chromosomes. *Hereditas* **53**:201–207.

Mai, L. L. 1979. *A General Model of Cytogenetic Evolution with Preliminary Tests of Some Primate "Origins" and "Molecular Clock" Hypotheses,* Ph.D. Dissertation, Department of Anthropology, University of California, Los Angeles.

Mayr, E. 1974. Cladistic analysis or cladistic classification? *Zool. Syst. Evol.-Forsch.* **12**:94–128.

Miller, D. A. 1977. Evolution of primate chromosomes. *Science* **198**:1116–1124.

Miller, O. J., Schnedl, W., and Erlanger, B. F. 1976. Chromosomal localization of 5-methylcytocine. *Chromosomes Today* **5**:457.

Ohno, S. 1970. *Evolution by Gene Duplication,* Springer-Verlag, Berlin. 160 pp.

Ohno, S., Weiler, C., Poole, J., Christian, L., and Stenius, C. 1966. Autosomal polymorphism due to pericentric inversions in the deer mouse (*Peromyscus maniculatus*) and some evidence of somatic segregation. *Chromosoma* **18**:177–187.

Pearson, P. L. 1977. Banding patterns, chromosomal polymorphism, and primate evolution, in: *Molecular Structure of Human Chromosomes* (J. J. Yunis, ed.), pp. 267–282, Academic, New York.

Platnick, N. I. 1977. Cladograms, phylogenetic trees, and hypothesis testing. *Syst. Zool.* **26**:438–442.

Romero-Herrera, A. E., Lehmann, H., Joysey, K. A., and Friday, A. E. 1976. Evolution of myoglobin amino acid sequences in primates and other vertebrates, in: *Molecular Anthropology* (M. Goodman and R. E. Tashian, eds.), pp. 289–300, Plenum, New York.

Sarich, V. M., and Cronin, J. E. 1976. Molecular systematics of the primates, in: *Molecular Anthropology* (M. Goodman and R. E. Tashian, eds.), pp. 141–170, Plenum, New York.

Seuanez, H. N. 1979. *The Phylogeny of Human Chromosomes,* Springer-Verlag, Berlin. 189 pp.

Seuanez, H., Evans, H. J., Martin, D. E., and Fletcher, J. 1979. An inversion of chromosome 2 that distinguishes between Bornean and Sumatran orangutans. *Cytogenet. Cell Genet.* **23**:137–140.

Sneath, P. H. A. 1975. Cladistic representation of reticulate evolution. *Syst. Zool.* **24**:360–367.

Sokal, R. R., and Sneath, P. H. A. 1973. *Principles of Numerical Taxonomy,* 2nd ed., Freeman, San Francisco. 573 pp.

Stebbins, G. L. 1950. *Variation and Evolution in Plants,* Columbia University Press, New York. 643 pp.

Wagner, W. H. 1969. The construction of a classification, in: *Systematic Biology,* pp. 67–99, National Academy of Sciences Publication No. 1692, Washington, D.C.

White, M. J. D. 1973. *Animal Cytology and Evolution,* 3rd ed., Cambridge University Press, Cambridge. 461 pp.

White, M. J. D. 1978. *Modes of Speciation,* Freeman, San Francisco. 455 pp.

Apes, Humans, and Molecular Clocks

A Reappraisal

5

J. E. CRONIN

Overview

> But in enunciating this important truth I must guard myself against a form of misunderstanding, which is very prevalent. I find, in fact, that those who endeavour to teach what nature so clearly shows us in this matter, are liable to have their opinions mispresented and their phraseology garbled, until they seem to say that the structural differences between man and even the highest apes are small and insignificant. Let me take this opportunity then of distinctly asserting, on the contrary, that they are great and significant; that every bone of a Gorilla bears marks by which it might be distinguished from the corresponding bone of a Man; and that, in the present creation, at any rate, no intermediate link bridges over the gap between *Homo* and *Troglodytes*.
>
> It would be no less wrong than absurd to deny the existence of this chasm; but it is at least equally wrong and absurd to exaggerate its magnitude, and, resting on the admitted fact of its existence, to refuse to inquire whether it is wide or narrow.
>
> Huxley (1863, p. 104).

It has been over 100 years since T. H. Huxley wrote these words. Yet debate still centers on the exact genealogical relationships of apes and humans. The magnitude of this gap is of great concern. Fifteen years ago Sarich (1967) noted that the gap was "narrow yet deep." Narrow in the sense that at the structural gene level, humans and the African apes are almost identical; deep because there are substantial morphological and profound behavioral differences between these close relatives. These differences may result from some few small genetic changes each with large phenotypic effects. An examination of the evidence of the genetic differences among the living species can place into perspective the nature of the gap separating hominids from other hominoids.

J. E. CRONIN • Departments of Anthropology and Biology, Peabody Museum, Harvard University, Cambridge, Massachusetts 02138.

Introduction

Since the pioneering work of Nuttall (1904) in comparing differences among various species by means of immunological precipitin work, genetic data have become powerful tools in elucidating phylogenetic history. Molecular studies involving humans and their nearest relatives, the pongids, have greatly expanded in the last 20 years. Except for the genus *Drosophila* and certain microorganisms, there are substantially more genetic and evolutionary data available from hominoids than for any other comparable group of organisms.

While not absolutely definitive, the major relationships in the Hominoidea have been fairly clearly defined. The cladistic conclusions drawn from such data are not readily accepted by paleontologists or neontologists (Kluge, this volume, Chapter 6; Kay and Simons, this volume, Chapter 23). And yet, as I will discuss, interpretations of the fossil record of primates are more and more compatible with observations initially made over the last 15 years by V. Sarich, J. Cronin, and A. Wilson. Moreover, beyond the problem of molecular cladistics, the use of proteins or nucleic acids as molecular clocks in order to date divergence events is far from unanimously accepted, and is not even seriously considered by most evolutionary biologists. Yet, the data do not allow an open-minded scholar to lightly dismiss the concept of a clock. Proteins do have both varying and regular rates of evolution. Furthermore, it is clear that the clock is one of an approximate, not metronomically perfect, nature (Cronin and Meikle, 1979, 1982).

The specific controversy at hand is centered around a proposed recent divergence date for hominids and pongids on the order of 5 m.y.a. This was suggested by Sarich and Wilson (1967a,b). This date is in contrast to dates estimated by certain scholars to be in the early or mid-Miocene. Differences in interpretations of events force quite substantial changes in the "scenarios" of this history (Zihlman *et al.*, 1978; Lovejoy, 1981; Cronin *et al.*, 1981; Kay and Simons, this volume, Chapter 23). Different modes of human evolution are delineated when different assumptions that are either implicity or explicitly stated are entailed in the construction of the various models. However much the views of paleoprimatologists have changed, little credit is usually given to these students of molecular evolution for important suggestions. It is as if molecular interpretations have imperceptibly pervaded and influenced thought processes without some explicit cognizance being taken of their existence. Perhaps acceptance of the molecular data can be analogized to an infection by a slow virus ("Molecular Kuru") that grows imperceptibly until signs of its presence are manifest in the aberrant behavior of the victim. Even treatment early in the course of the infection by *Ramapithaseptics* to counter the fever of "recent divergence" or with anti-African bodies to interfere with raft development have not eradicated the disease. In fact, the molecular disease is now seen to be a beneficial symbiote by some anthropological witch doctors. New molecular ideas have become incorporated into the genome and like middle repetitive (selfish) DNA have self-replicated and have passed the filter of natural selection.

This chapter presents the various data that have been assembled to resolve the question of the nature and timing of human origins and offers a reappraisal of the molecular clock in light of almost 20 years of controversy. A "most likely" scenario of human origins is then offered, given the molecular determination of the sequence and timing of the various events in the evolution of the hominoids culminating in modern humans. Finally, the history and conceptual framework surrounding the attempt to resolve the question of human origins from a biochemical perspective are reviewed. In the Appendix an evaluation of objections to the molecular clock is given by V. M. Sarich as a personal retrospective.

Macromolecular Data Bearing on Hominoid Affinities

Where has the last decade brought us in terms of understanding hominoid evolution? In 1970, there was available a substantial amount of immunodiffusion data, some albumin microcomplement fixation data, and only a few data points generated by comparisons of hominoid proteins through sequencing. The basic cladistic conclusions drawn were: (1) the hominoids are undoubtedly a monophyletic group; (2) the *Hylobates* lineage represents the most ancient speciation event; and (3) *Homo, Pan,* and *Gorilla* then share a common ancestry subsequent to the separation of the other apes (see Appendix). The suggestion was made by Sarich and Wilson (1967) that these latter three forms separated around 5.0 m.y.a.

At the beginning of the decade, the three major outstanding problems were: (1) the exact provenience of the orangutan relative to *Hylobates,* the African apes, and humans; (2) the unresolved sequence of speciations among *Homo, Pan,* and *Gorilla;* and (3) the degree of precision and universality of the molecular clock when applied to other proteins or segments of nucleic acids. Specific experiments were either underway or being designed in the early part of the last decade in order to collect the appropriate data to answer these specific questions, as well as to contribute to our understanding of molecular evolution in general. For an overview of this period and the development of the conceptual framework, see the Appendix.

The last 5 years have brought profound changes in molecular techniques. The ability to directly sequence genes is truly a revolution. However, the extent of available data is limited. Our primary information has been gathered through techniques developed during the decade of the 1960s, which have provided the macromolecular data concerning hominoid phylogeny that will be presented, from (1) immunology, (2) electrophoresis, (3) amino acid sequencing, and (4) nucleic acid techniques, including endonuclease restriction mapping, sequencing, and hybridization.

The major issues are the confirmation and documentation of the existence of specific lineages. The conclusions drawn from the data relate to: (1) the clear indication of monophyly for the hominoids *vis à vis* the Old

World monkeys; (2) the initial cladogenic event leading to the lineage of modern *Hylobates;* (3) the subsequent, relatively short-lived common stock of *Pongo* and the African apes and humans; (4) the divergence of the orangutan lineage; (5) the common clade uniting hominids, chimpanzee, and gorilla; and (6) the inability of the molecular data even after 20 years of intensive data collection to confidently resolve a sequence of divergence among a (*Homo, Pan, Gorilla*) trichotomy.

Each technique will be critically examined to determine what its resultant data have and have not added to our knowledge, what they confirmed, and, just as importantly, what their relative importance is both in contribution and resolving power for understanding human origins. No law gives equal importance to all molecular data, nor should it.

Immunology

The immunological data basically stem from two sources: immunodiffusion techniques with antisera directed against whole serum (Goodman, 1961; Goodman *et al.,* this volume, Chapter 3), and microcomplement fixation (MC'F) technique, using antisera directed against specific purified serum proteins (Sarich and Cronin, 1976). The raw data are converted into antigenic distance units in the former case and to immunological distance (I.D.) units in the latter case (Table 1). Procedures are available for converting the raw distance data into phylogenetically informative hypotheses (Goodman and Moore, 1971; Cronin and Sarich, 1975; Sarich and Cronin, 1976).

The conclusions drawn from the two types of data are remarkably similar. The Hominoidea are clearly a monophyletic clade relative to the Old World monkeys. No immunological data show *any* overlap in genetic distances between these two major subdivisions of the Catarrhini. Within the ape and human clade both the immunological (including the initial albumin results and the subsequent transferrin study) as well as the immunodiffusion data are concordant in recognizing that the initial branching is of the lineage leading to the modern *Hylobates* species (Fig. 1.) (Cronin *et al.,* 1983).

Table 1.   Immunological Distances among Hominoid Genera

	Homo	*Pan*	*Gorilla*	*Pongo*	*Hylobates*
Homo	—	6[a]	4	12	12
Pan	10[b](16)[c]	—	8	10	12
Gorilla	6(10)	8(16)	—	11	11
Pongo	18(30)	20(30)	23(34)	—	11
Hylobates	24(36)	24(36)	24(35)	24[d](35)	—

[a]Data above diagonal represents albumin distances.
[b]Data below diagonal represents transferrin immunological distances.
[c]Data represents the sum of albumin and transferrin distances.
[d]The anti-*Pongo* transferrin distances are corrected as detailed in Cronin and Sarich (1975).

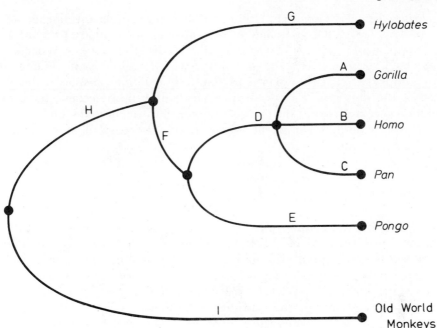

Fig. 1. A cladistic analysis of macromolecular evolution in the Hominoidea. Lineage: A. *Gorilla:* NADH 100 - 85, 95; PGM2 100 - 96; 6PGD 100 - 97, 105; Cer-1 100 - 98; Est A ? - 101; Hpt 105 - 107. B. *Homo:* LDH-A 96 - 100; PGI 102 - 100; Hpt 105 - 100. C. *Pan:* IDH-S 100 - 96. *Pan troglodytes:* AK 100 - 96; PGM 100 - 96; 6PGD 100 - 97. *Pan paniscus:* CAT 100 - 98; fum 100 - 98. D. African Ape & *Homo:* GOT-S 96 - 100; Est A ? - 100. E. *Pongo:* IDH-S 100 - 96; Est A ? - 98; Sumatra 100 - 97. F. Large Ape: G6PD 102 - 100. G. *Hylobates:* 1) Common *Hylobates* lineage - Cer-1 100 - 98; Est A. ? - 102; Est B ? - 102; Hpt 105 - 107; 2) *H. lar* & *H. concolor* appear to be derived, ADA ? - 92 (*H. syndactylus* is 96); 6PGD-1 100 - 94 (*H. syndactylus* 100 - 107); 3) *H. concolor* & *H. syndactylus* appear to be derived, HbαB 100 - 96 (*H. lar* appears to have the primitive allele 100); ALB 100 - 102 (*H. lar* appears to have the primitive allele 100). H. Apes: There are no clearly *derived* changes uniting the apes along this lineage. The polarity of change is difficult to determine.

The exact positioning of the next bifurcation remains somewhat more problematic. The nonadditive and un-rate-tested antigenic distance data seem to indicate that *Pongo* is closest to the African apes and *Homo*. The MC′F data showed at first that *Pongo* was as distantly removed from this group as is *Hylobates*. However, recent analysis of the same data suggests marginally definable lineage. The analysis of the albumin and transferrin data was presented in 1976 as a possible solution. Yet at that period we felt it would be a more conservative analysis to not place tremendous reliance upon a lineage that was definable by only 2–3 ID units (Sarich and Cronin, 1976; see this chapter, Appendix). As will be shown, other molecular data have been just as ambiguous in the placement of *Pongo* (Andrews and Cronin, 1982).

Lastly, and most conclusively, the immunological data clearly demonstrated a common lineage uniting human, chimpanzee, and gorilla. This was apparent in the earliest molecular publications and remains so to this day (see the Appendix; also Goodman, 1961; Sarich 1967). Along this lineage are a number of shared derived changes in both albumin and transferrin. The immunodiffusion data strongly support this cladistic hypothesis. What is unresolvable in this data set is the question of whether there is any sequence of bifurcations. Due to either (1) the probably short temporal span of the existence of this common lineage, during which time no detectable molecular evolution occurred; or (2) the margin of error of the techniques; or (3) homoplasy, this trichotomy cannot be resolved with the current immunological techniques at hand.

Electrophoresis

Electrophoretic comparisons of proteins among species are useful for systematic purposes (Avise, 1974; Sarich and Cronin, 1976; Sarich, 1977). Three sets of electrophoretic data have utility in sorting out hominoid phylogeny. The first study was that by King and Wilson (1975). This study examined 44 loci coding for red cell and plasma proteins and enzymes. They found that humans and chimpanzees differed by a genetic distance value of $D = 0.62$. This is an amount characteristic of other taxa that are congeneric species. While truly a historic article, there were no other hominoids compared in this study. Thus, the data must be used in conjunction with the other molecular data in order to generate a scale. By itself, the study clearly demonstrated a close genic similarity of humans and chimpanzees, the average gene being 99% identical in the two species.

What is intriguing about the King and Wilson data is that there are different subsets of information hidden within the overall D value. There are 21 loci that are identical between humans and chimpanzees and 20 that are completely different, with alternative alleles being fixed. Only three loci have common polymorphisms. The first set of electrophoretically identical proteins are quite slowly evolving, on the order of one substitution per 10–20 million years. These loci are mainly intracellular. The set of proteins that are almost all different between the two taxa are the rapidly evolving loci. This latter set has a rate of substitution of one detectable electrophoretic substitution per 1–2 million years (Sarich, 1977; Sarich and Cronin, 1976; Cronin et al., 1980). The conclusion to be drawn is that different loci are informative phylogenetically at different temporal levels subsequent to speciation. Thus one must choose the appropriate level of analysis to answer different questions.

Comparisons of hominoid taxa using strictly the evidence of the rapid set of loci, predominantly the plasma proteins, suggest that human, chimpanzee, and gorilla differ by a plasma protein electrophoretic (PPE) D value of approximately 1.6 ± 0.2 (Table 2). The orangutan and the gibbon are too distant, with all or substantially all loci having changed, to provide useful cladistic information. Thus, the technique of using the rapid loci allows us to place *Homo, Pan,* and *Gorilla* into a cluster relative to the Asiatic species due to

Table 2.   Plasma Protein Electrophoretic (PPE) Genetic Distances (D)

Taxa Compared	D
Homo sapiens vs. *Pan troglodytes*	~ 1.6
vs. *Pan paniscus*	~ 1.7
vs. *Gorilla gorilla*	~ 1.7
Pan troglodytes vs. *Pan paniscus*	~ 1.0
vs. *Gorilla gorilla*	~ 1.6
Pan paniscus (within)	~ 0.8
Gorilla gorilla gorilla vs. *G. g. beringei*	~ 0.8
Pongo pygmaeus pygmaeus vs. *P. p. abelli*	~ 0.8
Hylobates lar vs. *Hylobates syndactylus*	~ 1.6
vs. *Hylobates concolor*	~ 1.6

the close genic similarity of the former. However, the distances within the subset are unapportionable, due to the lack of an appropriate outside reference point.

The loss of resolving power among the major lineages is a gain of resolving power at the lower taxon levels. Within *Pan*, *P. troglodytes* and *P. paniscus* differ by a value of $D = 1.0$. The two subspecies of *Gorilla* differ by a somewhat smaller value. Detectable differences of this level or less are also found between the Bornean and Sumatran orangutan. Work within the *Hylobates* cluster reveals a similar magnitude of difference as that found between humans and the African apes. The three major lineages within this cluster are *H. concolor*, *H. lar*, and *H. syndactylus*. Species within the *H. lar* complex are considerably closer at the genetic level (Cronin *et al.*, 1983).

A third study of note is that by Bruce and Ayala (1979). Twenty-three loci were studied in the extant hominoids (Table 3). The data suggest that the lineage containing all of the *Hylobates* species is *definitely* the sister taxon of the large apes and humans. However, the resultant tree, did not resolve a tetrachotomy among *Homo*, *Pan*, *Gorilla*, and *Pongo*. *Pongo* did cluster with the

Table 3.   Electrophoretic Distances among Hominoid Species

	Homo	*Pan t.*	*Pan p.*	*Gorilla*	*Pongo*[a]	*Hylobates*[b]
Homo	—	0.386[c]	0.312	0.373	0.349	0.887
Pan troglodytes	0.383[d]	—	0.103	0.373	0.264	0.758
Pan paniscus	0.302	0.095	—	0.385	0.177	0.887
Gorilla	0.452	0.405	0.452	—	0.461	0.637
Pongo[a]	0.427	0.258	0.427	0.526	—	0.662
Hylobates[b]	0.916	0.799	0.799	0.693	0.799	—

[a] Values are averaged to two *Pongo* subspecies.
[b] Values are averaged to three *Hylobates* species.
[c] D values above the diagonal taken from Bruce and Ayala (1979).
[d] Values below the diagonal are D values based on a locus by locus, most common allele comparison at 22-23 loci (Cronin *et al.*, 1980).

other three, which is by itself an important observation. But due to the relatively few loci surveyed, the ability to resolve lineages is limited. Resolution increases as a function of the square root of the number of loci compared. On the other hand, the number of individuals necessary to examine in order to gain an estimate of the "true" distance involved is quite small (Sarich, 1977; Cronin *et al.*, 1980). After 5–6 individuals are compared, the gain in information at the cladistic level greatly diminishes relative to time and expense. The results within *Pongo* and *Pan* subspecies are in close agreement with the extent of divergence initially detected by immunological and plasma protein comparisons (Sarich and Cronin, 1976).

In a cladistic analysis of the electrophoretic mobility of 23 hominoid proteins and enzymes as published by Bruce and Ayala (1979) specific lineages are tentatively resolvable (Table 3). The ancestral ape has 17 of 23 loci the same as the ancestral catarrhine condition. However, the remaining 6 loci's primitive states cannot be easily determined. Thus, there do not appear to be any derived hominoid substitutions. *Hylobates* is the initial clade to diverge. Subsequent to this, there is one change at the G6PD locus that unites *Pongo* with the African apes and humans. The lineage leading to *Pongo* has some 4–6 derived changes. *Homo, Pan,* and *Gorilla* appear to share only one derived electrophoretic change at the GOT-S locus along a common clade. Quite possibly, *Homo* and *Pan* may share one derived change relative to the *Gorilla*, thereby indicating a common lineage. Or, alternatively, *Gorilla* may be the derived lineage. Unfortunately, the polarity of change at the Esterase A locus is equivocal. This type of analysis on a limited body of data does leave one feeling unsatisfied. Given the possibilities of parallel mutations to a similar phenotype, that is, net electric charge on the molecule, caution must be exercised in assessing the importance of one or two apparent derived similarities. The larger the number of apparent shared character states (electrophoretic identity), the greater the confidence in defining specific lineages.

Unfortunately, the electrophoretic data are only marginally helpful in sorting out either the position of the orangutan or in defining the sequence of events in the African ape–human divergence. However, the data are quite strong in two areas: (1) in positioning *Hylobates;* and (2) in clustering *Homo, Pan,* and *Gorilla* in a common clade.

Amino Acid Sequence

Protein sequence data have added to our knowledge of the evolutionary relationships of many organisms. This information has been most informative at taxonomic levels usually much higher than among members of the superfamily Hominoidea, such as at the level of between superfamilies, orders, or classes. Furthermore, due to the relatively time-consuming nature of protein sequencing, small proteins or polypeptides tend to be studied, therefore yielding quantitatively few data points. So, to date, only completed sequences of the globin polypeptides [α, β, γ, δ, and muscle (myo)] and the fibrinopeptides have been of much utility in deriving evolutionary information within the hominoid cluster. These data have been extensively reviewed by Goodman (1976; Goodman *et al.,* this volume, Chapter 3), and I will only touch upon the major conclusions. Furthermore, the globin data magnificently demonstrate the difficulties in determining homology among loci. Specifically the globin genes have demonstrated gene duplications, suppressions, and concerted changes occurring in their evolution (Liebhaber *et al.,* 1981; Martin *et al.,* 1980; Slighton *et al.,* 1980; Zimmer *et al.,* 1980). All molecular data are inherently susceptible to these problems; but the globin data provide specific concrete examples.

The fibrinopeptides A and B, totaling some 30 amino acids, have been sequenced in a number of mammals (Wooding and Doolittle, 1972). All of the fibrinopeptides of major species of hominoids are available. *Homo, Pan,* and *Gorilla* are identical in sequence in both the A and B peptides. These three species share two changes that unite them in a common clade. *Pongo* shares one change with these three, indicating a common ancestral stock for these four forms. The most ancient divergence within the apes is the lineage leading to *Hylobates.* One single amino acid change shared among *Pongo* and the African apes and humans, along with the nucleic acid data (see Appendix), are the most *definitive* of the molecular data in positioning the orangutan lineage *vis à vis Hylobates* and the African apes and humans. Perhaps the paucity of detectable change along this clade argues for temporal brevity.

The globin data (Goodman *et al.,* this volume, Chapter 3) clearly cluster all apes relative to the Old World monkeys, and on the other hand cluster human, chimpanzee, and gorilla. However, as interpreted by maximum parsimony analysis, the most parsimonious tree links *Pongo* and *Hylobates* as a clade and sister taxon to the African apes and humans. Analysis of the myoglobin sequences most parsimoniously places *Pongo* as the most ancient divergence lineage, even prior to the branching of the gibbon lineage. Within the human–chimpanzee–gorilla group, the α-globin sequences link *Pan* and *Homo* by one shared derived change. All other substitutions are autapomorphic along one lineage, with similarities in the sequences of the other two genera being primitive hominoid retentions. The data thus again yield an unresolved trichotomy. Of the relatively few hominoid protein sequences available, the fibrinopeptide sequences have been the most informative. However, of 180 amino acids (640 bases) and six species, five amino acid changes in these polypeptides are our data base. In the myoglobin sequences, comparing 918 amino acids (2754 bases), four amino acid changes, although maximally parsimonious, yield an unlikely tree. Five hits result in a likely tree, but entail homoplasy. The hemoglobin data lack completeness for hominoid comparisons, but what exist confirm most of the results obtained from other methodologies.

Nucleic Acid Data

The nucleic acid hybridization data and the immunological data are the most informative results for interpreting hominoid relationships. DNA hybridization involves measuring the degree of homology among the cellular DNA of different species. Differences are expressed as percent nucleotide sequence divergence. The most important nonimmunological additions to the knowledge of hominoid cladistics were the results of Hoyer *et al.* (1972) and Benveniste and Todaro (1976). The latter data revealed that a single-copy DNA of *Pongo* was more similar to that of *Homo, Pan,* and *Gorilla* than to that of *Hylobates.* One note of caution is that there is only one *Pongo* data point and there is no reciprocal measurement in the latter comparison. The results of Hoyer were equivocal as to this placement (see Appendix). Again, the trichotomy among *Homo, Pan,* and *Gorilla* is unresolvable.

Recently, mitochrondrial DNA of hominoid species has been studied by endonuclease restriction enzymes (Ferris *et al.,* 1981). Because of the fast rate of evolution of this 16.5-kilobase-long genome, about ten times as fast as that of nuclear DNA, there is no suitable outside reference point to align the branching sequences of the hominoid lineages. The Old World monkeys are too distant an out-group, as their DNA has been substantially altered. The study assumes that *Hylobates* is the out-group relative to the other ape lineages. Given this, then, *Pongo* is the second major lineage diverging as the sister lineage to a common (*Homo–Pan–Gorilla*) clade. Within this subset, *Pan* and *Gorilla* are briefly joined, with *Homo* being the out-group. However, when alternative trees are constructed, only a few more mutations unite *Homo* and *Pan,* with *Gorilla* being the out-group. A few more mutations yield a true trichotomy. The importance of the data is in indicating a possible *Homo–Pan* or *Pan–Gorilla* clade and in possibly excluding a true trichotomy. Unfortunately, even with the rapid rate of evolution of this molecule and our great expectations, this trichotomy has not been resolved through the use of restriction site analysis. All of us await the results of a nucleotide sequence study of an m+DNA fragment currently underway (Wilson, personal communication).

Implications of the Molecular Results: Models of Human Origins and the Most Likely Scenario of Hominoid Evolution

The Molecular Framework

The macromolecular data provide a basic framework within which to interpret the course of hominoid evolution. The range of possible cladistic and temporal interpretation of hominoid phylogeny is severely constrained by the molecular evidence. The biomolecular data presented as a composite molecular tree in Fig. 2 clearly argue for three major cladogenic events in hominoid evolution since the separation from the lineage leading to the Old World monkeys. Given the approximately regular nature of the evolution of certain molecules, the time set for that event is about 20 ± 2 m.y.a. (Sarich and Cronin, 1976).

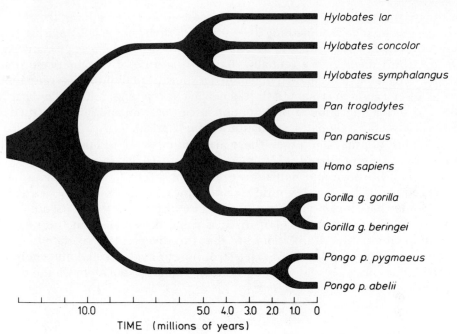

Fig. 2. A phylogeny of the Hominoidea using all molecular data in order to position each specific lineage. Times of divergence are estimated from calculations based on the molecular clock model as calibrated in Sarich and Cronin (1976) and Cronin and Meikle (1982) and as discussed in the text. The nodes are drawn wider than an exact point in order to emphasize either ambiguity in the cladistics or the margin of error of the divergence estimates.

The first cladistic event within the hominoids is the separation of the gibbon lineage from that leading to the large apes and humans. That event is best dated at about 12 ± 3 m.y.a., during the middle Miocene. The second event, the divergence of the orangutan lineage from that linking the African apes and humans, occurred approximately 10 ± 3 m.y.a. After this, the African apes and humans share a common lineage. The duration of this clade is at a minimum 3 million years and at a maximum 9 million years. The divergence of the African apes and humans, then, is 5 ± 1.5 m.y.a. (Cronin and Meikle, 1982).

None of the data is conclusive in clearly defining a sequence of two bifurcations in the African apes–human clade. The most likely linkage is a *Pan–Gorilla* alignment. This pairing is favored by the mitochondrial DNA data (Ferris *et al.*, 1981). The second most likely pairing is that of *Homo* and *Pan* excluding *Gorilla*. Sequence data partially support this interpretation. The third alternative, a *Gorilla–Homo* pairing, is the least likely from the molecular data. Miller's (1977) analysis of the chromosomal data does support this association [but, see Mai (this volume, Chapter 4) for an alternative interpretation]. Although certainly possible, a real trichotomy—three lineages simultaneously originating from one ancestral population—seems extremely unlikely, that possibility being probably no more than 1%. Whereas I would estimate the possibility of a *Pan–Gorilla* association at something like 50%, a *Homo–Pan* linkage at something like 40%, with the *Homo–Gorilla* pairing having a probability of less than 10%. The proposed common lineage for any pair vs. the third member is essentially "hidden" because of the absence of currently detectable molecular change. Such a short period of time is compatible with either a sudden ecological shift (a new niche opening) followed by speciation, or with a migration of the common ancestor from another area (such as Asia) followed by allopatric speciation *in situ* (Africa).

Within the major lineages, there are a number of additional detectable radiations. *Hylabates concolor, H. syndactylus,* and *H. lar* radiate around 5–6 m.y.a. The Bornean and Sumatran orangutan differ. They appear to have diverged about 1 m.y.a. Within the *Gorilla* lineage, the two subspecies differ by a somewhat similar time span, and, as reported previously (Cronin and Sarich, 1977), the pygmy chimpanzee is quite different from the western common chimpanzee. Indications are that this divergence may date to some 2.0 m.y.a. Within the genus *Homo,* electrophoretic evidence and mitochondrial DNA data place the radiation of modern human populations on the order of 100,000–500,000 years ago.

Implications of the Molecular Data for the Hominoid Fossil Record

The Miocene epoch was the period that heralded mammalian faunas of modern appearance. It was also the period that saw the primary radiation of the hominoid primates, culminating in the appearance of forms that share many of the characteristics of living apes and humans (Pilbeam, 1979; Andrews, 1981). The specific affinities within this complex of fossils, however, and their relationships with living hominoids are still the subject of intense controversy (Pilbeam, 1979; Andrews, 1981). There are recent Pleistocene forms which are obviously related to either the living orangutan (Delson, 1977) or the gibbon (Szalay and Delson, 1979). Undisputed Plio-Pleistocene forms of early hominids from East Africa also exist (Johanson and White, 1979; Howell 1978; Boaz, this volume, Chapter 28). However, no Miocene fossil hominoid has been convincingly shown to be uniquely related to any of the living apes or hominids.

The molecular evidence suggests an origin of the hominoid lineage some 20 ± 2 m.y.a., in the early Miocene. This is definitely compatible with the fossil record. It was suggested some years ago that likely candidates for the catarrhine ancestor to both hominoids and cercopithecoids might be found in forms from the Fayum deposits in Egypt, such as *Aegyptopithecus* (Sarich, 1970; Cronin and Sarich, 1975). It is heartening to see similar interpretations now emanating from paleoprimatologists (Kay *et al.,* 1981; Fleagle and Kay, this volume, Chapter 7). However, this view is directly counter to earlier and some continuing interpretations of *Aegyptopithecus* as a hominoid (Simons, 1970, 1976, 1981). There is no cercopithecoid or hominoid fossil that directly contradicts the 20 m.y.a. time of divergence (Cronin and Meikle, 1982). Undoubted hominoid fossils, *Proconsul,* are found very close to the 20 m.y.a. date and specifically derived cercopithecoid fossils date to no older than this period and perhaps a few million years younger (Szalay and Delson, 1979).

Table 4 presents specific dates of divergence as generated by molecular data compared to the first appearance of those *specific* lineages in the fossil record. None of the estimates is directly contradicted by known fossil specimens. The first appearance of a lineage in the fossil record may be close to its actual time of origin. Certain models of speciation suggest that rapid divergence of peripheral populations may lead to new taxa in the fossil record (Gould and Eldredge, 1977). One implication of these models is that one would not expect to find a long chain of gradually changing species as fossils; a large amount of morphological change may take place in a geologically short period of time. Therefore, we do not believe that it is necessary to add a large proportion of the known time of existence of a lineage to estimate its age of origin (Cronin and Meikle, 1982).

A recent analysis of a specific case of contention in primate evolution, the time of separation of the *Papio* and *Theropithecus* lineages, has been analyzed using both molecular and paleontological data (see Fig. 3). A calculation of the biochemical data gives a time of separation of 3.2 ± 1.3 m.y.a. The minimum age established by the fossil record is about 4.0 m.y.a. The data drawn from differing lines of evidence are quite concordant (Cronin and Meikle, 1979, and 1982). Molecular and fossil data are not and need not be at odds.

Table 4. Predicted Dates of Divergence as Generated by the Molecular Data Compared with the First Appearance of the Lineage in the Fossil Record

Lineages	Molecular Estimate of Time of Origin	Oldest Fossil Evidence
Hominoidea	20 ± 3 million years[a,b]	ca. 23 million years[g]
Pongo	10 ± 3 million years[a]	ca. 1–2 million years[g]
Hylobates	12 ± 3 million years[a]	ca. 1–2 million years[g]
Hominidae	5.0 + 1.5 million years[a,c]	ca. 5.5 million years[g]
Pan	5.0 ± 1.5 million years[a,c]	none
Gorilla	5.0 ± 1.5 million years[a,c]	none
Cercopithecidae	20 ± 3 million years[a,b]	ca. 20 million years[g]
Colobinae and Cercopithecinae	12 ± 2 million years[b]	ca. 15 million years[g]
Cercopithecus	7 ± 2 million years[a,b,d]	ca. 3.5 million years[g]
Macaca	6–8 million years[e]	ca. 7–8 million years[g]
Theropithecus	2–4.5 million years[f]	ca. 4 million years[g]

[a] Sarich and Cronin (1976)
[b] Cronin (1975)
[c] Cronin and Meikle (1982)
[d] Cronin and Sarich (1976)
[e] Cronin, *et al.* (1980)
[f] Cronin and Meikle (1979)
[g] Szalay and Delson (1979)

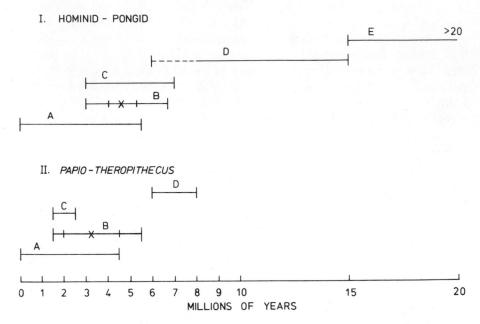

Fig. 3. A comparison of suggested time scales for (I) the hominid–pongid divergence (II) the *Papio–Theropithecus* divergence. Part I: (A) Range of undoubted fossil hominids; (B) range of molecular estimates of the time of separation of the hominid and African ape lineages. The cross is the mean estimate using data in Cronin and Meikle (1979) and calibration of rates in Sarich and Cronin (1976). Inside bars are ±1 standard deviation, calculated from the plus and minus error of each measurement; (C) previous molecular estimates of divergence time (Sarich and Cronin, 1976; Sarich, 1971); (D) range of *Ramapithecus* (Pilbeam, 1979); (E) range of paleontologically based estimates of time of separation of hominids and pongids (Simons, 1976; Walker, 1976; Simons and Fleagle, 1973). Part II: (A) Range of undoubted fossil *Theropithecus;* (B) range of molecular estimates of the time of separation of the *Theropithecus* and *Papio* lineages (symbols as in I); (C) previous molecular estimates of divergence time (Sarich, 1970); (D) range of paleontologically based estimates of the time of separation of *Theropithecus* and other papionine lineages (Delson, 1975; Simons, 1970). [This figure is taken from Cronin and Meikle (1982).]

Origin of the Hominid Lineage

Over the years, the recent-divergence hypothesis was challanged by the belief that a specific fossil form, *Ramapithecus,* was directly and solely ancestral to hominids. If *Ramapithecus* were a hominid, the middle to late Miocene date for the form invalidated the molecular claims (Simons, 1976). Yet recently the status of *Ramapithecus* as a hominid has been questioned on morphological as well as molecular grounds (Greenfield, 1979; Sarich, 1969; Sarich and Cronin, 1976; Zihlman *et al.,* 1978; Cronin and Meikle, 1982). Moreover a fossil ape, *Sivapithecus meteai,* has been shown to share a number of characteristics with the living orangutan. *Sivapithecus meteai* is part of the middle Miocene *Sivapithecus–Ramapithecus* complex. If this group is a valid clade, then

Ramapithecus must also be considered as being more closely related to the orangutan than to humans (Andrews and Cronin, 1982). Specifically, there are a number of morphological traits, such as the size and relative proportion of the incisors, the interorbital distance, the positioning of the zygomatic, and the formation of the nasal area, that appear to be shared derived characters between *S. meteai* and the orangutan (Andrews and Cronin, 1982). *Ramapithecus* has clearly been linked with the forms in the genus *Sivapithecus* (Pilbeam, 1979; Kay and Simons, this volume, Chapter 23). There are two consequences of this linkage. If one member of a clade is closely related to the orangutan, then the other members of the clade are linked as well. *Ramapithecus*, then, cannot uniquely be linked with hominids (see Fig. 4). The

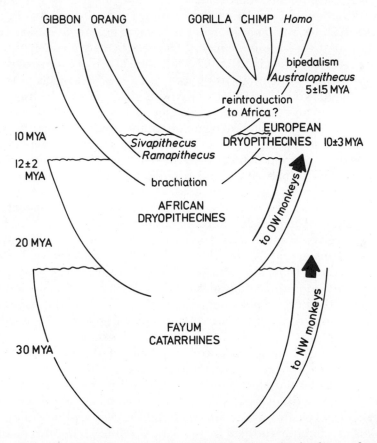

Fig. 4. A scenario of hominoid evolution as developed using the molecular data as a framework in order to interpret the cause of morphological evolution and potential fossil relationships. Hominids are seen to evolve *ca.* 5 ± 1.5 m.y.a. In this scheme, *Sivapithecus–Ramapithecus* are best related to the orangutan clade (Andrews and Cronin, 1982). An alternative possibility is for some African form of late Miocene–Pliocene *Proconsul* sp. to give rise in Africa to the common ancestor of *Homo, Pan,* and *Gorilla.*

hominid affinity of *Ramapithecus* can no longer be quoted as invalidating the molecular prediction of the recency of hominid origins.

The direct comparison of hominids to *Ramapithecus* also fails to strongly associate the two forms (Andrews and Cronin, 1982). The characters cited most recently by Kay and Simons (this volume, Chapter 23) such as reduced canines, reduced sexual dimorphism in canines, and low-crowned molars with thick enamel, are equivocal in their phylogenetic importance. Sexual dimorphism is a labile characteristic, and low sexual dimorphism in canine –premolar areas may be primitive for hominoids. Other characters, such as robust mandibles, may have evolved in parallel. Little phylogenetic weight can be given to a quantitative character such as thick enamel; *Pongo* has thick enamel, as do hominids. Perhaps thick enamel is the primitive condition for the Great Apes, with *Pan* and *Gorilla* having a secondary reduction. Thus, even on morphological arguments, *Ramapithecus* cannot be unequivocally uniquely associated with hominids (Andrews and Cronin, 1982; also see Fig 4).

As stated above, the separation of the lineages leading to humans and the African apes occurred about 5.0 m.y.a. The specific date given by three separate sets of molecular data is 4.6 ± 0.7 m.y.a. The range of estimates is 3.0–6.7 m.y.a. The plasma protein electrophoretic data yield estimates at the lower end of this range, while the DNA comparisons give values at the upper end of the range (Cronin and Meikle, 1982; also see Fig. 4). Since no Miocene hominoids have been demonstrated to be uniquely ancestral to hominids, what can one say of the first appearance of ancestral humans in the fossil record? And how well do the molecular dates fit with the fossil record? The fossil record of hominids extends to between 5 and 6 million years at Lothagam (Howell, 1978); older specimens sometimes considered to be hominids are either quite fragmentary or otherwise of uncertain status (Howell, 1978; Pilbeam, 1979; Boaz, this volume, Chapter 28). Recent additions to the fossil record have for the first time given a more than fragmentary view of 3–4 million-year-old hominids (Johanson and White, 1979). These specimens are said to display many dental/cranial traits which make them more similar to an ape-like ancestor than are any undoubted hominids previously known. The *estimated* time of appearance of the first possible hominid in the fossil record (Lothagam) is less than two standard deviations from the mean time of divergence as calculated from the molecular data (Figs. 3, 4). The molecular dates are also within two standard deviation of the appearance of the first well-dated and true hominid *populations* in the fossil record (Laetoli).

Since the molecular clock model states that some time around 5 m.y.a. hominids and pongids diverged, one of the consequences of this prediction is that we should expect to find primitive hominids at a slightly younger date. This seems to be amply fulfilled by the nature of the hominid remains from Laetoli and Afar. Whatever their specific designations, these forms of *Australopithecus* are primitive in dental and cranial features. On the opposite side of this cladistic event, at perhaps 5–7 m.y.a. one should expect to find a suitable candidate for a common ancestor of the African apes and humans. No fossil forms, especially with the demise of *Ramapithecus* as a candidate, fill this role. It has recently been suggested how a potential ancestral form might have appeared given the constraints of the molecular data as a framework. With the cladistics determined, one can build a valid model (Zihlman *et al.*, 1978; Zihlman and Lowenstein, this volume, Chapter 26).

Molecular data are thus not at variance with the time of appearance of the first undisputed hominids in the geological record. The models of either punctuated equilibrium or rapid gradualism after speciation could account for the hominid morphological transition (Cronin *et al.*, 1981). This interpretation is consistent with a relatively short period of time, perhaps 1–2 million years, during which hominid morphological changes could occur (one million years is approximately 50,000 human generations; to a geneticist this is more than sufficient time for many major changes to occur). A rapid shift in locomotor adaptations was the first major derived change, followed by unique changes in the brain and dentition, etc. One need not posit millions of years for the initial hominid adaptations to evolve from the ape ancestral conditions. Rates of evolution vary dramatically and traits may be responding to either intense selection or random fixation. Small population size, coupled with inbreeding, may facilitate rapid change during the process of speciation. This type of small population bottleneck, or small population mode of allopatric speciation, may have predominated in ancestral hominid populations. In any case, whatever the model of evolution, one need not postulate long periods of time for evolution to occur.

It is a mandatory conclusion, drawn from the molecular comparisons, that during the period of late Miocene–earliest Pliocene a common ancestor of *Homo*, *Pan*, and *Gorilla* existed. One species, one clade, gave rise ultimately to these three living genera. All three resultant forms must trace back to *one* and only *one* ancestor. It is possible, and indeed probable, that two of the three lineages were united for a brief period, during which time common features evolved, but it is still mandated that all three lineages stem from one common species. Lacking good fossil evidence one can turn to the comparative anatomy of the living forms to provide data from which to reconstruct an ancestral morphotype (Kluge, this volume, Chapter 6). A reconstructed potential ancestral form must be compatible with the constraints of the molecular framework. Only after the cladistics are determined can one build a valid model.

What is such a common ancestral form likely to resemble? Sarich (1968), over a decade ago, suggested that the common ancestor of the African apes and humans would appear to be not unlike a small chimpanzee. Similar observations were made by Washburn (1963). This notion culminated in a more formal proposal by Zihlman *et al.*, (1978) suggesting that *Pan paniscus* represented a form that was morphologically close to what one might expect a common ancestral African hominoid to look like (also see Zihlman and Lowenstein, this volume, Chapter 26). That is, the pygmy chimpanzee has retained the primitive ancestral morphotype in many features, or at least is a good analog. The conclusion was drawn from three lines of evidence: (1) biochemistry, (2) morphology, and (3) the fossil record.

This paper quickly drew a number of replies attempting to refute this hypothesis (Johnson, 1981; Latimer *et al.*, 1981). The objections were centered around all three lines of evidence cited to support the original hypothesis. The one that concerns us most here is the disagreement with the biochemical evidence. We stated (Cronin, 1977; Zihlman *et al.*, 1978) that *Pan paniscus* and *P. troglodytes* are closely related to each other. We never implied that *Homo* and *P. paniscus* had any special affiliation, to the exclusion of *P. troglodytes* or the *Gorilla* lineage. Numerous biochemical studies confirm this on immunological and electrophoretic critera (Cronin, 1975, 1977; Sarich and Cronin, 1976; Zihlman *et al.*, 1978; Bruce and Ayala, 1979), as do studies of mitochondrial DNA (Ferris *et al.*, 1981). To assert that we said otherwise is not true and to generate a smokescreen. If *Pan* and *Gorilla* share a common lineage, it is brief, and all *Pan* and *Gorilla* species still *must* share a common ancestor with hominids. We consulted the molecular framework in order to secondarily delineate the nature of morphological change. Kluge's attempt (this volume, Chapter 6) at reconstruction fails precisely because he does not begin with the proper phylogenetic framework.

Second, the morphological comparisons involving limb morphology, proportions, and dental area relative to body weight have been discussed at length by Johnson (1981), Jungers and Susman (1981), Shea (1981), and Zihlman (1981). Finally, objections to our model center around the observation that *Australopithecus* sp. from Laetoli and Afar dated at about 2.8–3.8 m.y.a. do not look like *P. paniscus*. This constitutes an absurd mastery of the obvious. This is like saying a hominid does not look like a pongid. We did not say that *P. paniscus* would be identical to the *Australopithecus* sp., but instead that the earliest hominids would resemble *P. paniscus* more than the other living African hominoids. We do not yet have undoubted fossil evidence of the common ancestral node. The speciation event could lie 0–3.0 million years before the Laetoli and Afar dates and still be compatible with both the molecular clock results and the recent-divergence hypothesis. Aspects of *Australopithecus* sp. that are not identical with extant pygmy chimpanzees may have been acquired subsequent to the separation from the ancestral form along one or both of the lineages. The patently obvious specialization on the *Australopithecus* lineage relates to the acquisition of bipedalism. Although the oldest specimens do demonstrate some derived dental features, they are still quite primitive in others and in overall cranial anatomy (Johanson and White, 1979).

For the *P. paniscus* lineage, postcanine tooth size seems to be smaller than expected for body size and may represent a recent specialization (Jungers and Susman, 1981). The tooth size/body size ratios are especially disparate when comparing living pygmy chimpanzees and the *Australopithecus* fossils from Hadar and Laetoli. But in all probability, these *Australopithecus* forms reflect the condition closer to the common ancestor than does *P. paniscus* for this feature. Thickly enameled large molars relative to body size may be the primitive condition for the Great Apes and hominids, as mentioned above. Yet the case for parallelisms between *Sivapithecus* and *Australopithecus* cannot be ignored (Andrews and Cronin, 1982). Thus, all we can say with a great deal of confidence is that the pygmy chimpanzee appears to have reduced dental dimensions and *Australopithecus* has large molars relative to body size. The fact that *Australopithecus* has or retains a large tooth size/body size rate at about 3.5 m.y.a. is not direct proof against the pygmy chimpanzee model as a whole.

The extent of sexual dimorphism in the early hominids has also been used as a source of refutation of the pygmy chimpanzee model. However, the extent of dimorphism in the early *Australopithecus* sample remains to be proven. Even if it is shown to be large, this does not mandate that 1–2 million years before this period the common ancestor had high sexual dimorphism. In fact, low sexual dimorphism in canine-premolar size may be the ancestral condition for hominoids (Andrews and Cronin, 1982). As should be well known, sexual dimorphism is an extremely labile trait, which can change rapidly. Note the differences within *Homo* populations as well as the different amounts of dimorphism between closely related species (Brunker, 1980). The extent of sexual dimorphism in early *Australopithecus* or the common homi-nid–pongid ancestor remains an open question, as far as testing the model is concerned. At the least, the data are inconclusive.

Furthermore, the tendency to link *A. afarensis* with the *Sivapithecus* group (White *et al.*, 1981) as broadly defined (Andrews and Cronin, 1982) is clearly based on shared primitive traits. If not, then the traits in common are parallel, or then *A. afarensis* is linked to the *Pongo* clade, which in effect, removes it from human ancestry. More seriously, this tendency to link early hominids with specific "known" or "true" ancestors on the basis of little evidence may satisfy the soul or the Western mind that there is a truth; but, it points up the danger of "inherent desires" not being a part of the scientific methodology of hypothesis testing. And in the latter case what it suggests is that for our purposes the pygmy chimp and early African australopithecine forms may share primitive traits that are useful for ancestral morphotype reconstruction.

Thus, I feel that the suggestion that the pygmy chimpanzee serves as a model for a common African ape–human ancestor remains a viable hypothesis. If nothing else, it serves a heuristic purpose. Not all traits of the pygmy chimpanzee must be identical to those of the common ancestor. Nor should one expect them to be, in order for the model to be generally valid. Traits change at different rates in different lineages. I would not expect the pygmy chimpanzee lineage to escape the mosaic nature of evolution. It is obvious that *Australopithecus* did not. Tracking the lineage back in time, one *must* find these forms converging morphologically in a common ancestor. The fossil record of about 5.0 m.y.a. (as initially suggested by the molecular workers) may contribute to a resolution of this issue. Yet one should not expect the fossil record to be the only arbiter of phylogeny or determinant of the course of evolution. As Patterson (1981) has said "instances of fossils overturning theories of relationship based on recent organisms are very rare, and may be nonexistent. It follows that the wide spread belief that fossils are the only, or best, means of determining evolutionary relationships is a myth . . . it is rare, perhaps unknown, for fossils to overthrow theories of relationships based on recent forms. If fossils do not contradict such theories, they must corroborate, or support them" (p. 219). Patterson (1981) also quotes Mayr on recent mammals: "Our ideas of their relationships, based on a study of their comparative anatomy, have in no case been refuted by subsequent discoveries in the fossil record" (p. 219). Patterson goes on to note, "We should not forget that the textbooks of twenty years ago placed the human–ape split in the early Miocene or Oligocene at about 25 million years ago; subsequent molecular evidence, not new fossils, is behind the more modest estimates now in vogue" (p. 218). Fossils certainly provide important data, but are not the only source of information in determining phylogeny.

I would suggest that, of the living forms, after analysis of the biochemistry, morphology, and the fossil record, *P. paniscus* remains a close, though not perfect, approximation of our common ancestors with the African apes. The model may need to be refined, but nothing has been suggested that takes its place.

Conclusion

The biochemical approach has had a large effect on the various branches of biology and biological anthropology. The various molecular techniques as discussed in this chapter have each contributed data toward resolving hominoid phylogeny. The broad outline of human evolution has now been painted. What remains is to continue to resolve the uncertainties of the exact nature and timing of the origins of the hominid lineage. In the future, newer techniques, such as nucleotide sequencing, will make even more of a contribution in this area. The older technologies will also continue to yield valuable data in the areas of population genetics, molecular evolution, and systematics.

As a result of continually changing interpretations by paleontologists, we feel vindicated in both our approach to problems in human evolution and in our interpretations of the data. More and more of the interpretation of primate and human phylogeny is explicitly or implicitly influenced by the molecular data. Furthermore, I am increasingly confident that the hominid–pongid divergence is in the range of 4.0–8.0 m.y.a. We are not shaken in our belief in the strength of the molecular clock model when used appropriately and not naively. We have not changed in substantial matters over the years. As a result we take a measure of satisfaction in the coming confirmation of a recent divergence of hominids and pongids. Yet, at the same time, one cannot help but wonder how far the field might have advanced if Huxley's remarks had been taken more seriously, or if Nuttall's work had been followed earlier, or if most paleontologists had not resisted the acceptance of molecular work.

Finally, the magnitude of the gap between humans and chimpanzees is narrow at the genetic level. It is deep at least in terms of behavioral differences, but it becomes increasingly shallow at the morphological level as more fossil evidence is recovered. I fervently hope that not another 100 years will pass before the gap narrows between the students of molecular evolution and those of neontology or paleontology. This would indeed be unfortunate.

As far as I am concerned, we shall not wait idly for such a reconciliation nor shall we quietly fade away. The new revolution in genetics will continue to yield data that impinge on the field of evolution and these data will continue to have implications for the field of human evolution.

ACKNOWLEDGMENTS

I would like to thank E. Meikle and L. Brunker for their valuable comments on the various drafts of the manuscript. V. Sarich contributed to earlier drafts of this paper. I would also like to thank L. Brunker for her design of the phylogeny reproduced as Fig. 4, and C. Simmons for excellent drawings. Thanks also go to G. Simon for typing the manuscript.

Appendix

Retrospective on Hominoid Macromolecular Systematics

V. M. SARICH

Molecular Clocks and Hominoid Evolution—A Review and Evaluation of Objections

The molecular clock approach to problems in primate evolution had its beginnings in a seminar at the University of California, Berkeley conducted by Clark Howell and Sherwood Washburn in the fall of 1964. I was then a graduate student in the Anthropology Department and volunteered to do a presentation on what was then termed "systematic serology" as it might apply to primates. As it turned out, another member of the seminar was Andy Wilson, whose father had developed the trefoil Ouchterlony procedure and introduced Morris Goodman to it, and who himself had worked in Goodman's lab at Wayne State University. Andy and I then worked up a joint presentation. As one new to the field, I set myself the task of covering the existing literature (a rational undertaking in those distant days). It was from that survey that the idea that the hominid line originated only 4–5 m.y.a. ultimately derives. In a presentation to the Southwestern Anthropological Association in 1965 (the text of which was taken almost verbatim from the paper presented to the seminar), I stated:

> The development of the serological chronology, as is true for any chronology, must rely upon the discovery of some serological system whose variation or change is some definite function of time. The concept of such regularities in the evolutionary process must, at this late date, seem rather naive—yet a survey of the available biochemical evidence strongly indicates just such regularities.

V. M. SARICH • Departments of Anthropology and Biochemistry, University of California, Berkeley, California 94720.

. . . If then comparative serology is basically systematic in agreeing with current taxonomies, and those taxonomies are basically phylogenetic, then there is an almost inescapable conclusion to be drawn from such a situation: the essential variable governing the retention of antigenic similarity and appearance of antigenic diversity is time.

. . . If then the basic variable in structure change is time, we can proceed to the main question asked in this paper—what is the rate of change? Since in serological work we are limited to the end results of the evolutionary process; that is, living species, differential rates of evolution might give us a serological phylogeny which is distorted with respect to the actual one. For example, as illustrated in Fig. 5:

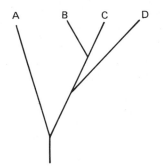

Fig. 5. A hypothetical cladogram or tree.

if species C has experienced a history of rapid change since its separation from the line leading to B, we might expect that from the perspective of species A or D, B and C would show different degrees of cross-reaction. In other words, different members of a monophyletic group would show different cross-reactions to a serum made against a species equally related to all of them phylogenetically.

Now this is a problem amenable to experimental testing—only unfortunately no one seems to have considered it and asked the relevant questions; therefore any tests I can make of the proposed hypothesis—that degree of change is some mathematical function of time—must use data which have not resulted from experiments designed to test the hypothesis. . . . All we can say at present is that there is no evidence to disprove the hypothesis that there is a regular rate of change in individual protein molecules through time.

. . . It would seem, then, that we may well be able to produce a truly phylogenetic taxonomy even in those groups where the fossil record is inadequate and morphological studies have not been able to resolve the issues. This is a situation particularly true of the primates, but hardly limited to them. . . . Comparative serology may also provide a check upon the interpretations given to fragmentary evidence from the fossil record. If, for example, the Hominidae became a distinct line only in the late Miocene, then rather obviously one cannot attribute an early Miocene fossil to the family.

All the basic ideas which have guided my work since (rate tests, a serological clock, testing fossil-based divergence times) were clearly and explicitly present in those two presentations. Indeed, while thinking about this section, it seemed to me that they were a little too fully formed to be seen as quite as original as I had come to think of them. In particular, there was the question of where the rate test concept so critical to making the clock a conclusion rather than an assumption had come from. If, for example, Goodman had seen its possibilities, there might have been little for Allan Wilson and me to do besides add a little quantitative gloss to the primate data set. So going back to the original manuscript and its bibliography, and discussing the point with Allan Wilson, I found what might have been the intellectual roots of the above in a paper by Margoliash (1963). To quote the most relevant passages from that paper:

> . . . it appears that the number of residue differences between the cytochromes C of any two species is mostly conditioned by the time elapsed since the lines of evolution leading to these two species originally diverged. If this is correct, the cytochromes C of all mammals should be roughly equally different from the cytochromes C of all birds. Since fish diverged from the main stem of vertebrates earlier than either birds or mammals, the cytochromes C of both mammals and birds should be equally different from the cytochromes C of fish. Similarly, all vertebrate cytochromes C should be equally different from the yeast protein. This is borne out by the comparison in Table 1.
>
> . . . If elapsed time is the main variable determining the number of accumulated substitutions, it should be possible to estimate roughly the period at which two lines of evolution leading to any two species diverged.

(Margoliash, 1963, p. 677)

So by 1965 I knew how to proceed; what remained was to get into the laboratory and do some work.

The laboratory became available through my meeting Allan Wilson, of the Biochemistry Department faculty, in the course of my seminar paper investigations, and Washburn recognized that this whole area might prove a fertile dissertation topic. I accepted Washburn's suggestion, Wilson accepted me as a member of his nascent laboratory, and a very profitable collaboration was under way. We began with the conviction that immunological comparisons of the products of a single locus would show sufficient internal consistency and resolving power to allow us to answer some of the questions raised in the seminar paper. The protein chosen was albumin and the comparative method microcomplement fixation (MC'F). As it still appears to many that our results are somehow albumin- and MC'F-specific, one should note here they have been amply corroborated by a good deal of later work, including the MC'F study of a second blood protein, transferrin, several DNA annealing studies, and, most recently, by restriction endonuclease and direct sequence comparisons of mitochondrial DNAs. In addition, there is nothing magical about the MC'F technique itself, as the original study could perfectly well have been carried out using classical quantitative precipitin analysis. It was simply that the MC'F approach offered somewhat greater resolving power, and, far more important, enormous savings in amounts of blood required.

The original goals were quite restricted and clear. We wanted to know whether the marked similarities among hominoid albumins were due to a slowdown in their rates of change, and if not, what sort of time scale for ape and human evolution could be extracted from the data. We already knew from the work of Boyden and his group at Rutgers and that of Goodman at

Wayne State that immunological comparisons of primate albumins did give similarities in line with taxonomic expectations. Relative to our albumin, those of the apes were most similar, then came those of the Old World monkeys, New World monkeys, "prosimians," and nonprimates, respectively. It should again be noted that "taxonomic expectations" imply some sort of time-dependent evolution. We then measured immunological distances using antisera made at Berkeley among a wide variety of primate albumins, applied the rate test concept to those data, and reported the results in a series of three papers published in 1966 and 1967 (Sarich and Wilson, 1966, 1967a,b). It is of some interest to note that one of that series which was logically required—because it presented the rate test concept and data—was rejected by *Science* as "not presenting anything new or unexpected." The other two, which *Science* did publish, were the first in the series, which was simply a technique paper, and the third, which applied the clock idea to the hominoids. The rejected second article, containing the only really original and critically necessary material, was finally published in the *Proceedings of the National Academy of Sciences*. One wonders if the continual accusation that we have somehow "assumed" the clock and imposed it upon the data stems in large part from this original misperception on the part of the reviewers *Science* used, as the article in the *Proceedings* has virtually never been referred to or critiqued. Ignoring that article does make it easier to assume that we assumed the clock—no doubt of that. In any case, we confirmed the taxonomic distance—immunological distance correlation, showed that various anthropoid albumins were similarly distant from those of two "prosimians," that those of the Old World monkeys and hominoids were about equally distant from those of several New World monkeys, and, finally, that all Old World monkey to hominoid albumin immunological distances were about the same. This told us that albumin evolution among the Anthropoidea did in fact provide a clock for dating within-anthropoid divergences. What remained was to calibrate the clock. In other words, what was the index of dissimilarity–time of divergence relationships? This, in 1966–67, was not at all obvious and the problem led to a good deal of soul-searching, as is evident in my dissertation. The problem was that the data base was so small. It was compounded by the fact that we had not yet developed sufficient ingenuity to circumvent that lack. Ultimately we came to rely on two observations made by Allan Wilson and others at Brandeis. First, MC′F indices of dissimilarity among a series of enzymes showed a log-linear relationship with the fossil-based divergence times for the species containing them; the same relationship held for indices of dissimilarity and number of amino acid sequence differences among a small series of human hemoglobins (Sarich, 1968). Thus we concluded that log(index of dissimilarity) ∝ number of amino acid differences ∝ time of separation, though we would not have been surprised to find some modification of this formula necessary in the future. It should be noted here that log(index of dissimilarity) × 100 equals immunological distance (Sarich, 1969). We finally concluded that a simple log-linear plot was the best bet, though we would not have been surprised to find it requiring modification as further testing was done.

   Four independent observations finally convinced me by about 1970 that the log-linear relationship we had teased out of a very meager data set in the latter part of 1967 was indeed correct. First, we plotted the *possible* divergence times (ranges based on the fossil record) against albumin immunological distances [log(index of dissimilarity)] for as many mammalian species pairs as we had data. The relationship was clearly linear. Second, we noted Goodman's demonstration that degree of spur formation in his double diffusion tests was linearly related to immunological distance. As spur formation is most reasonably interpreted as a linear function of surface sequence difference, so, too, must MC′F immunological distances. Third, immunological distances did apportion additively along derived cladograms. It was, and still is, difficult to see how this could be so unless sequence difference and immunological distance were linearly related. Finally and most importantly, the DNA annealing data of Kohne (1970) showed that ΔT_m was a linear function of albumin immunological distance for the same species pairs (it was known that ΔT_m was a linear function of the amount of sequence difference between the two DNA species). Thus, what was a tentative best guess in 1967 had developed into a virtual certainty in 1970—and remains so to this day. Since then we have added a very large body of albumin immunological comparisons, and an even larger number of transferrin immunological comparisons, and Benveniste and Todaro (1976), in particular, have markedly expanded the number of DNA annealing data points.

   All these new data have generally served to confirm the quantitative picture of anthropoid relationships presented in Sarich and Wilson (1967) and Sarich (1968). The only significant necessary modification involves the estimated time of separation between the gibbon line and that leading to the other hominoids. The albumins clearly underestimated this time, and a probable closer approximation to reality is made by adding the transferrin immunological and DNA annealing data (Sarich and Cronin, 1977). Otherwise the picture presented in Table 7 and Fig. 7 of Sarich (1968) does not look at all bad some 14 years later.

   All of this is appropriately emphasized in the molecular clock aspects of the early Berkeley work; it is with the clock and estimating primate times of divergence that we were mainly concerned. Given the demonstrated regularity in primate albumin evolution, albumin distances translated more or less directly into cladistic and temporal distances—and the details of cladistic analysis were simply deferred. It should also be noted that the immunological distances among the various hominoid albumins and those of the cercopithecines, the two groups of greatest interest to us at that time, were very small, and this fact should have precluded any detailed albumin comparisons done with the goal of obtaining high-resolution cladistic information. That is of course by way of hindsight; it took a good deal of wasted effort for us to develop a proper feel for the limitations of the immunological approach.

   Even so, the basic cladistic issues left unresolved by the early work are not all resolved even today, as discussed previously in this chapter. The basic human–chimpanzee–gorilla trichotomy is still that, or something very close to it, at the time of this writing. The position of the orangutan lineage was not cleanly resolved by the albumin work, and adding the transferrin data did not help significantly—we still obtained an essential three-way split among the gibbon, orangutan and human–chimpanzee–gorilla lines. The early DNA annealing studies of Hoyer *et al.* (1972) left the matter unresolved, and even now we have only a single one-way DNA annealing comparison between gorilla and orangutan which significantly associates the latter with the African apes and humans (Benveniste and Todaro, 1976). Finally, it might be noted that the pygmy chimpanzee was placed on the *Pan troglodytes* line as soon as the albumin immunological comparisons were done. *Pan paniscus* and *P. troglodytes* albumins are immunologically identical.

Molecular Clock: Evaluation of Objections

We have followed the attempts to question our molecular clock-based formulations of hominoid evolution with a certain degree of bemusement. These objections have clearly had their genesis in one or another of the following: (1) the molecular clock formulations do not accord with current interpretations of the primate fossil record—therefore they must be wrong; and (2) everyone knows that the evolutionary process has never shown temporal regularity—so why should it be any different at the level of the ultimate source of variation, the gene itself. This *prima facie* conviction that the molecular clock must be faulty in conception, execution, or both has tended to relieve the critics of having to see the implications of their specific criticisms; that is, as hypotheses to be tested in the world of real data. It has evidently been sufficient to think up a more or less plausible-sounding reason that the molecular clock approach was in some way faulty; those critics were then apparently relieved of any further responsibility—such as demonstrating that their objections had any viability on their own.

Indeed, the conviction that the macromolecular data had to be interpreted within cladistic and, particularly, temporal frameworks provided by contemporary paleontological opinion has, from the first, so channeled thinking into counterproductive areas that it has only been the existence of our small group at Berkeley that has kept the molecular approach alive at all. Most workers in the field, as is evident from the number of conferences and symposia on hominoid evolution in the last 15 years to which we have not been invited, would clearly have preferred for the whole idea to quietly go away. Otherwise it is evident that such long-overdue reappraisals of the meaning of the primate fossil record as have recently appeared would have been even longer in coming. The authors of most of these reappraisals seem to have found it impossible to admit, in print, that the molecular clock formulations just might have had something to do with changing their directions of thought—it was not, after all, the fossils that changed their morphology—but most of their readers will know better and make their assessments accordingly. Thus it is not so much the lack of attribution of ideas that bothers us (though we are not exactly ecstatic about it), but the clear and ever-present danger that the molecular clock formulations will see themselves accepted for precisely the same reasons that they were originally rejected. That is, they now agree with those interpretations of the fossil record changed to agree with them, where previously they were seen as in conflict with interpretations developed for the same fossil record prior to the time that there was any check possible on those interpretations. One is reminded of the continental drift controversy, which was more or less resolved in the 1960s—the same fossil record used to argue against moving continents in the 1950s is now put forward as support for the concept. This Janus-like aspect of much paleontological thinking is disturbing—and not just in relation to this minor controversy involving higher primate evolution and relationships. The main problem is that paleontologists have arrogated unto themselves the mantle of final arbiters of our understanding of the evolutionary process—in effect, if we cannot see it in the fossil record, it never really happened. One might hope that our experiences in the field of primate macromolecular studies will serve as an example as to how the very small portions of actual history preserved in the fossil record can be better worked into a better approximation of the actual history through a consideration of the far more complete bodies of data available from studies of extant organisms.

Morris Goodman laid the modern foundation for meaningful integration of macromolecular data into the development of evolutionary scenarios. At the same time, unfortunately, he also conceded to the paleontologists their coveted arbitrators mantle. This can be seen in his early interpretative attempts. Goodman found the immunological differences among hominoid serum proteins to be very small. This had to mean, then as now, one of two things: either the separation times among those hominoid lineages were very recent, or the rates of change along their protein lineages were retarded. Goodman accepted the 1960s paleontological opinion that the latter was the explanation of his observations. He wrote:

> The fact that the gibbons can be traced as a separate branch of the Hominoidea back into the Oligocene, 30–40 million years ago, highlights the significance of the finding of an extensive correspondence in antigenic correspondence among hominoid albumins . . . the fossil record suggests that the family Bovidae first emerged during the Miocene after extensive branching of the Hominoidea had already occured. Yet by the immunological plate technique chicken anti-beef albumin sera could detect divergences even with the subfamily Bovinae, kudu albumin diverging from beef albumin, whereas anti-human albumin sera failed to show any comparable divergences among hominoid albumins. Clearly, then, albumin evolution has been more rapid in the subfamily Bovinae than in the superfamily Hominoidea. This type of finding could be predicted by our theory, for the artiodactyls have an epitheliochorial placenta which, as previously noted, minimizes the possibility of transplacental immunizations.

(Goodman, 1963b, p. 226–227)

This sort of argument has set much of the subsequent pattern. First ask the paleontologists to put your data into context—in effect, to tell you what they mean—and then choose as ingeniously as possible among an almost unlimited set of potentially effective selective variables to "explain" your results. The fact that the alleged "slowdown" would produce an effect readily observable without reference to any external consideration was not seen by Goodman (which is hardly a criticism), but then has still not been accepted by him long after we pointed it out to him (which is a major criticism of his work).

In a similar, and even less excusable vein, the work of Kohne (1975) on annealing differences among higher primate DNAs is often referred to as showing this necessary-to-paleontologists slowdown in their macromolecular evolution. Yet all Kohne did was to accept as real some rather ancient divergence times among the hominoids (15–25 m.y.a. for human and chimpanzee) and some very short ones for some rodents (5–10 m.y.a. for rat and mouse) and then "measure" their respective "rates" of DNA evolution by dividing his actual DNA differences by his assumed divergence times. Not surprisingly, he found that human and chimpanzee DNAs had been changing much more slowly than those of rat and mouse (about 1/20 to 1/40th as fast). Given this "finding," the "explanation" was not hard to come by—Kohne opted for the shorter generation lengths among the rodents (conveniently close to 1/20 or 1/40th of those for human and chimpanzee. It seems to have escaped the attention of virtually everyone quoting Kohne as showing the slowdown of macromolecular evolution among the higher primates that the slowdown was in effect assumed when the above-cited divergence times were accepted. Though he could have tested his hypothesis without reference to the fossil record—and we (Allan Wilson and Sarich) pointed this out to him at excruciating length one very long morning in Berkeley in 1970—he chose not to, leading to a paper (Sarich 1972) in rebuttal.

In that same year, Lovejoy et al. (1972) published another approach which would accommodate the molecular data within the preferred paleontological framework with the help of the generation length idea. Here it was suggested that the longer generation lengths among early primates would so increase the number of amino acid and nucleotide substitutions among them as to render the rate tests we use insensitive among the higher primate lineages. Again, however, we see the usual pattern, as Lovejoy, et al. did not test the implications of their own hypothesis. The same assumptions leading to these very high rates of change early in primate history would necessitate that forms that continue to show very short generation lengths (here, tree shrews, or rats and mice) would then have to show markedly increased (several-fold) amounts of change using the rate tests supposedly insensitized by the generation length effect for higher primates. The data that would have allowed Lovejoy et al. (1972) to thus test their hypothesis directly were already in the literature, but they chose not to—it was enough to suggest another potential flaw in the molecular clock hypothesis giving these problematic divergence dates for the higher primates.

Another example in this genre of criticism, and the one which most strongly induced bemusement, was the effort of Benveniste and Todaro (1976). I should first note here that the criticism was indirect and the data presented are probably still the best quantitative assessment of genetic distances among higher primates. However, on p. 106 of that article, Benveniste and Todaro undertook to calculate divergence times among the Hominoidea using their DNA annealing differences, the molecular clock hypothesis, and, strangely enough, assumptions as to generation lengths at various times along the lineages involved. Not surprisingly, the generation lengths chosen, along with the assumption of an inverse proportionality between generation length and rate of DNA evolution, gave divergence times which would have made Elwyn Simons quite pleased. The real bemusement comes when one notes that on the lower left corner of the same page that presents the phylogenetic tree and divergence times resulting from the above exercise in the upper right corner is a table which shows that the various hominoid DNAs are equally distant from that of *Papio*. So Benveniste and Todaro actually and unconsciously tested the generation length hypothesis, found that it did not hold, and still used it to generate hominoid divergence times that fitted the paleontological party line then current. This was pointed out in an article published the following year (Sarich and Cronin, 1977).

Finally, Korey (1981) argues that, in effect, our divergence times among the Hominoidea might be biased low by a generation length effect which tends to hide actual substitution events. Now I pointed out to Korey at some length, both in person and by letter, that his proposed mechanism had at least two very testable consequences. First, rate tests should show less change along those lineages where its effect had been significant, and one ought to see more genetic variation (heterozygosity) along them as well. Again, such tests have been done, the data have long been in the literature, and the suggested effects have not been seen, yet it seems to be enough simply to go on producing "ingenious" explanations of the perceived conflict between the paleontological and molecular clock views of hominoid evolution. That one's "explanations" might require support from other than paleontological opinion seems not to be a serious consideration. That it might be the quality of fossil data and paleontological logic which are at fault seems to have escaped everyone's attention. Or at least it has not been seen as professionally healthy to turn one's attention to questions of that sort.

Another example of direct criticism was that of Uzzell and Pilbeam (1971). Here it was argued that the rate tests used to demonstrate the existence of molecular clocks were to a degree flawed by the fact that phenetic distances measured today as genetic differences between living species must be less than the sum of the number of actual nucleotide or amino acid substitutions along the two lineages involved since they last shared a common ancestor. Now this logic is impeccable, but it begs the questions of how large the effect might be, whether it be seen internally (that is, without reference to the fossil record), and, finally, what the implications for the molecular clock dates might be. Again, these matters were discussed by Sarich in a letter sent to Uzzell and Pilbeam in response to a prepublication copy of their manuscript, but they did not respond to them in the final article. The points have been discussed at some length (Sarich, 1973) and need not concern us unduly here. I simply note that a direct test of the magnitude of the proposed phenetic underestimation of phyletic distance (PUPD) effect had already been suggested in my doctoral dissertation (Sarich, 1967), carried out, shown to be insignificant at the immunological distances involved, and the results given to Uzzell and Pilbeam. These were ignored by them. The bemusing part comes when one considers the implications of the PUPD effect for our calculated times of divergence. As phyletic distance increases, phyletic underestimation will increase disproportionately (otherwise there is no effect); thus the ratio of actual phyletic distances for the, say, human–lemur and human–chimpanzee comparisons will be larger than measured. This would necessarily mean that the "actual" divergence times calculated for more recent separations (for example, human–chimpanzee) would be even more recent than those we obtained by ignoring, for good reasons, the PUPD effect. In other words, if Uzzell and Pilbeam had looked into the implications of their argument (or listened when I pointed them out), then they would have seen that it strengthened the case for a very recent separation between humans and the African apes. As it was, they played the game of misleadingly sowing doubt, knowing that the points made would, if carried to their logical conclusions, strengthen the opposing case.

Objections of a different sort were made by Read and Lestrel (1970) and Read (1975). Here it was argued that the scaling of immunological distance into phyletic distance used was not unequivocally defensible in terms of the known relationships between amino acid sequence difference and immunological distance. Though this argument could be dealt with in its own terms (Sarich, 1973), the simplest rebuttal is to point out that albumin immunological distances correlate very strongly with DNA annealing distances between the same species pairs (Sarich and Cronin, 1976), and it is known that the latter are linearly related to the actual proportion of sequence differences. Again, this was known as early as 1970 with Kohne's work [and pointed out in Sarich (1970)]; again Read ignored it in 1975, though, again, he was reminded of it by letter.

What, it might be asked here, has the modern pioneer in all this, Goodman, had to say about the molecular clock in relation to the immunological and DNA annealing data? The response has been peculiar. First, he has never discussed the DNA data in a clock context. Second, he has never discussed either his immunological data or ours in a clock context; that is, by doing rate tests on his own data or by assessing ours. What he has done is to continue to look for examples of nonclocklike accumulations of amino acid replacement, document those that do occur, and then show how silly the dates so obtained are. As Sarich and Cronin (1976) said:

> Inevitably some proteins will show accelerated change in some lineages and deceleration in others, and it may well be that hominoid globins are an example of the latter phenomenon. As we have already pointed out, if one looks at enough proteins in enough lineages one can readily accumulate a number of such cases. Each of these may be of appreciable intrinsic interest, but there can be no justification for calculating times of divergence in such cases without correcting the observed departures from regularity.

> (Sarich and Cronin, 1976, p. 156)

In other words, you have to show clocklike behavior before making use of the clock.

Finally, I note recent papers by Corruccini *et al.* (1979, 1980). Here it is argued that lineage by lineage comparisons of amounts of amino acid substitution along them for several different proteins show significant deviations from clock-like behavior. This I do not find surprising, especially when it is seen that the comparisons were done in a deterministic rather than stochastic manner. Even so, Corruccini *et al.* (1980) concede that ". . . it is noteworthy that the within-anthropoid analysis indicated linear evolution, and that all the macromolecules are very highly correlated with one another." They follow with the somewhat condescending final sentence: "Although it needs to be interpreted with very broad confidence intervals the molecular clock is of some value in calibrating the sequence of events in primate evolution" (p. 1218). Well, one might ask, "How much value?" "What are the confidence intervals?" For an allegedly quantitative statistical analysis, the conclusions are depressingly qualitative. This sort of thing contributes nothing but more confusion to the entire situation. Again there is the feeling engendered that because some uncertainty exists (it is, one needs to be reminded, a probabilistic, not metronomic, clock), we need not take the entire approach too seriously. The reader needs to be reminded that very few doubts concerning the branching order among the major primate lineages exist, that primate evolution is to be constrained to the Tertiary, and that the divergence times most at issue here are a long way and many separations down the line from the beginnings of the primate line. This tells us that it is not particularly productive to look at each separation between lineages in isolation—for the obvious reason that a phylogenetic framework as complex as that linking the living primates severely circumscribes the amount of give in the possible divergence times involved. When one adds the pattern of regularity observed for protein and nucleic acid evolution, the amount of give is reduced even more markedly. To stretch one lineage or set of lineages in time is to contract others, and one has to allow for this in considering the possible quantitative effects of minor departures from a linear relationship between genetic distance and time. Even more serious is the comment made that: "DNA evolves at a negative exponential (slowing) rate" (Corruccini *et al.*, 1980, p. 1218). This is pure invention. It has already been shown that no measurable difference in rates of change among higher primate DNA lineages can be seen using our standard rate tests. It was also shown some time ago that DNA annealing differences correlate very strongly ($r > 0.9$) with albumin immunological distances, as already mentioned. So what conceivable evidence could there be for this "slowdown"? As I must continue to note, it is difficult to avoid a strong degree of bemusement when dealing with this issue.

This section has clearly not been one of conciliation. It might be asked: "You would appear to have won most of your points, why not be magnanimous?" The question is, magnanimous to whom? It is not the personalities involved that are important. They can be left to the historians. The issue is not who is now seen as "right" and who as "wrong," it is why they were right or wrong. What do we learn from this whole affair? We will benefit only if we learn to discriminate productive from nonproductive or counterproductive ways of asking questions; learn how to judge the quality of an argument in evolutionary reconstruction; learn how not to be blinded by prevailing orthodoxy; and, above all, to recognize that the fossil record and its interpreters represent but one line of evidence contributing to our understanding of the evolutionary process in general and specific evolutionary events in particular. We want to learn, in Washburn's words, "to play the game of evolution" (Washburn, 1973).

References

Andrews, P. 1981. Species diversity and diet in monkeys and apes during the Miocene, in: *Aspects of Human Evolution* (C. B. Stringer, ed.), pp. 25–61, Taylor and Francis, London.

Andrews, P., and Cronin, J. E. 1982. The relationships of *Sivapithecus* and *Ramapithecus* and the evolution of the orang-utan. *Nature (Lond.)* **297:**541–546.

Avise, J. C. 1974. The systematic value of electrophoretic data. *Syst. Zool.* **23:**465–481.

Benveniste, R. E., and Todaro, G. J. 1976. Evolution of Type C viral genes: Evidence for an Asian origin of man. *Nature (Lond.)* **261:**101–108.

Bruce, E. J., and Ayala, F. J. 1979. Phylogenetic relationships between man and the apes: Electrophoretic evidence. *Evolution* **33:**1040–1056.

Brunker, L. 1980. A comparison of locomotor adaptations of the vervet and patas monkeys, and implications for early hominid evolution. *Am. J. Phys. Anthropol.* **52:**2–8.

Corruccini, R. S., Cronin, J. E., and Ciochon, R. L. 1979. Scaling analysis and congruence among anthropoid primate macromolecules. *Hum. Biol.* **51:**167–185.

Corruccini, R. S., Baba, M., Goodman, M., Ciochon, R. L., and Cronin, J. E. 1980. Non-linear macromolecular evolution and the molecular clock. *Evolution* **34:**1216–1219.

Cronin, J. E. 1975. *Molecular Systematics of the Order Primates*, Ph.D. Dissertation, University of California, Berkeley.

Cronin, J. E., and Meikle, W. E. 1979. The phyletic position of *Theropithecus:* Congruence among molecular, morphological, and paleontological evidence. *Syst. Zool.* **28:**259–269.

Cronin, J. E., and Meikle, W. E. 1982. Hominid and gelada baboon evolution: Agreement between molecular and fossil time scales. *Int. J. Primatol.* **3:**469–482.

Cronin, J. E., and Sarich, V. M. 1975. Molecular systematics of the New World monkeys. *J. Hum. Evol.* **4:**357–375.

Cronin, J. E., Cann, R., and Sarich, V. M. 1980. Molecular evolution and systematics of the genus *Macaca*, in: *The Macaques: Studies in Ecology, Behavior and Evolution* (D. Lindburg, ed.), pp. 31–51, Van Nostrand Rheinhold, New York.

Cronin, J. E., Boaz, N. T., Stringer, C. B., and Rak, Y. 1981. Tempo and mode in hominid evolution. *Nature (Lond.)* **292:**113–122.

Cronin, J. E., Sarich, V. M., and Ryder, O. 1983. Molecular evolution and speciation in the lesser apes in: *Biology of the Lesser Apes* (D. J. Chivers, H. Preuschoft, N. Creel, and W. Brockelman, eds.), Edinburgh University Press, Edinburgh (in press).

Delson, E. 1975. Evolutionary history of the Cercopithecidae. in: *Approaches to Primate Paleobiology* (F. S. Szalay, ed.). *Contrib. Primatol.* **5:**167–217.

Delson, E. 1977. Vertebrate paleontology, especially of non human primates, in China, in: *Paleoanthropology in the Peoples Republic of China*, pp. 40–65, National Academy of Sciences, Washington, D.C.

Ferris, S. D., Wilson, A. C., and Brown, W. M. 1981. Evolutionary tree of apes and humans based on cleavage maps of mitochondrial DNA. *Proc. Natl. Acad. Sci. USA* **78:**2432–2436.

Goodman, M. 1961. The role of immunochemical differences in the phyletic development of human behavior. *Hum. Biol.* **33:**131–162.

Goodman, M. 1963a. Serological analysis of the systematics of recent hominoids. *Hum. Biol.* **35:**377–436.

Goodman, M. 1963b. Man's place in the phylogeny of the primates as reflected in serum proteins, in: *Classification and Human Evolution* (S. L. Washburn, ed.), pp. 204–234, Aldine, Chicago.

Goodman, M. 1976. Towards a genealogical description of the Primates, in: *Molecular Anthropology* (M. Goodman and R. E. Tashian, eds.), pp. 321–353, Plenum, New York.

Goodman, M., and Moore, G. W. 1971. Immunodiffusion systematics of the primates. I. The Catarrhini. *Syst. Zool.* **20:**19–62.

Gould, S. J., and Eldredge, N. 1977. Punctuated equilibria: The tempo and mode of evolution reconsidered. *Paleobiology* **3:**115–151.

Greenfield, L. O. 1979. On the adaptive pattern of *"Ramapithecus"*. *Am. J. Phys. Anthropol.* **50:**527–548.

Howell, F. C. 1978. Hominidae, in: *Evolution of African Mammals* (V. J. Maglio and H. B. S. Cooke, eds.), pp. 154–248, Harvard University Press, Cambridge.

Hoyer, B. H., Van de Velde, N. W., Goodman, M., and Roberts, R. B. 1972. Examination of hominoid evolution by DNA sequence homology. *J. Hum. Evol.* **1:**645–649.

Huxley, T. H. 1863. *Evidence as to Man's Place in Nature,* Williams and Norgate, London. 159 pp.

Johanson, D. C., and White, T. D. 1979. A systematic assessment of early African hominids. *Science* **203:**321–330.

Johnson, S. C. 1981. Bonobos: Generalized hominid prototypes or specialized insular dwarfs? *Curr. Anthropol.* **22:**363–365.

Jungers, W., and Susman, R. 1981. A reply to S. C. Johnson. *Curr. Anthropol.* **22:**369–370.

Kay, R. F., Fleagle, J. G., and Simons, E. L. 1981. A revision of Oligocene apes of the Fayum Province, Egypt. *Am. J. Phys. Anthropol.* **55:**293–322.

King, M. C., and Wilson, A. C. 1975. Evolution at two levels in humans and chimpanzees. *Science* **188:**107–118.

Kohne, D. E. 1970. Evolution of higher organism DNA. *Quart. Rev. Biophys.* **3:**327–375.

Kohne, D. 1975. DNA evolution data and its relevance to mammalian phylogeny, in: *Phylogeny of the Primates* (W. P. Luckett and F. S. Szalay, eds.), pp. 249–264, Plenum, New York.

Korey, K. A. 1981. Species number, generation length, and the molecular clock. *Evolution* **35:**139–147.

Latimer, B. M., White, T. D., Kimbel, W. H., Johanson, D. C., and Lovejoy, C. O. 1981. The pygmy chimpanzee is not a living missing link in human evolution. *J. Hum. Evol.* **10:**475–488.

Liebhaber, S. A., Goossens, M., and Kan, Y. W. 1981. Homology and concerted evolution at the $\alpha 1$ and $\alpha 2$ loci of human α-globin. *Nature (Lond.)* **290:**26–29.

Lovejoy, C. O. 1981. The origin of man. *Science* **211:**341–350.

Lovejoy, C. O., Burstein, H., and Heiple, K. H. 1972. Primate phylogeny and immunological distance. *Science* **176:**803–805.

Margoliash, E. 1963. Primary structure and evolution of cytochrome C. *Proc. Natl. Acad. Sci. USA* **50:**672–679.

Martin, S. L., Zimmer, E. A., Kan, Y. W., and Wilson, A. C. 1980. Silent δ (delta) globin genes in Old World monkeys. *Proc. Natl. Acad. Sci. USA* **77:**3563–3566.

Miller, D. A. 1977. Evolution of primate chromosomes. *Science* **198:**1116–1124.

Nei, M. 1978. Estimation of average heterozygosity and genetic distance from a small number of individuals. *Genetics* **89:**583–590.

Nuttall, G. H. F. 1904. *Blood Immunity and Blood Relationship,* Cambridge University Press, Cambridge.

Patterson, C. 1981. Significance of fossils in determining evolutionary relationships. *Annu. Rev. Ecol. Syst.* **12:**195–223.

Pilbeam, D. 1979. Recent finds and interpretations of Miocene hominoids. *Annu. Rev. Anthropol.* **8:**333–352.

Read, D. W. 1975. Primate phylogeny, neutral mutations and molecular clocks. *Syst. Zool.* **24:**209–221.

Read, D. W., and Lestrel, P. E. 1970. Hominid phylogeny and immunology and rates of enzyme evolution in the Amphibia in relation to the origin of certain taxa. *Evolution* **20:**603–616.

Sarich, V. M. 1967. *A Quantitative Immunological Study of Evolution of Primate Albumins.* Ph.D. Dissertation, University of California, Berkeley.

Sarich, V. M. 1968. The origin of the hominids: An immunological approach, in: *Perspectives on Human Evolution,* Vol. 1 (S. L. Washburn and P. C. Jay, eds.), pp. 94–121, Holt, Rinehart and Winston, New York.

Sarich, V. M. 1969. Pinniped phylogeny. *Syst. Zool.* **18:**416–422.

Sarich, V. M. 1970. Primate systematics with special reference to Old World monkeys: A protein perspective, in: *Old World Monkeys* (J. R. Napier and P. H. Napier, eds.), pp. 175–226, Academic, New York.

Sarich, V. M. 1971. A molecular approach to the question of human origins, in: *Background for Man* (P. Dolhinow and V. M. Sarich, eds.), pp. 60–81, Little Brown, Boston.

Sarich, V. M. 1972. Generation time and albumin evolution. *Biochem. Genet.* **7:**205–212.

Sarich, V. M. 1973. Just how old is the hominid line? *Yearb. Phys. Anthropol.* **17:**98–112.

Sarich, V. M. 1977. Rates, sample sizes and the neutrality hypothesis for electrophoresis in evolutionary studies. *Nature (Lond.)* **265:**24–28.

Sarich, V. M., and Cronin, J. E. 1976. Molecular systematics of the primates, in: *Molecular Anthropology* (M. Goodman and R. E. Tashian, eds.), pp. 141–170, Plenum, New York.

Sarich, V. M., and Cronin, J. E. 1977. Generation lengths and rates of hominoid molecular evolution. *Nature (Lond.)* **269:**354–355.

Sarich, V. M., and Wilson, A. C. 1966. Quantitative immunochemistry and the evolution of primate albumins, Microcomplement fixation. *Science* **154:**1563–1566.

Sarich, V. M., and Wilson, A. C. 1967a. Rates of albumin evolution in primates. *Proc. Natl. Acad. Sci. USA* **58:**142–148.

Sarich, V. M., and Wilson, A. C. 1967b. Immunological time scale for hominoid evolution. *Science* **158:**1200–1203.

Sarich, V. M., and Wilson, A. C. 1973. Generation time and genomic evolution in primates. *Science* **179:**1144–1147.

Shea, B. 1981. A reply to S. C. Johnson. *Curr. Anthropol.* **22:**368–369.

Simons, E. L. 1970. The deployment and history of Old World monkeys (Cercopithecidae, Primates), in: *Old World Monkeys* (J. R. Napier and P. H. Napier, eds.), pp. 97–138, Academic, New York.

Simons, E. L. 1976. The fossil record of primate phylogeny, in: *Molecular Anthropology* (M. Goodman and R. E. Tashian, eds.), pp. 35–62, Plenum, New York.

Simons, E. L. 1981. Man's immediate forerunners. *Phil. Trans. R. Soc. Lond. B.* **292:**21–41.

Simons, E. L., and Fleagle, J. G. 1973. The history of extinct gibbon-like primates. *Gibbon and Siamang* **2:**121–148.

Slighton, J. L., Blechl, A. E., and Smithies, O. 1980. Human fetal Gγ and Aγ globin genes: Complete nucleotide sequences suggest that DNA can be exchanged between these duplicates genes. *Cell* **21:**627–638.

Szalay, F. S., and Delson, E. 1979. *Evolutionary History of the Primates*, Academic, New York. 580 pp.

Uzzell, T., and Pilbeam, D. 1971. Phyletic divergence dates of hominoid primates: A comparison of fossil and molecular data. *Evolution* **25:**615–635.

Walker, A. 1976. Splitting times among hominoids deduced from the fossil record, in: *Molecular Anthropology* (M. Goodman and R. E. Tashian, eds.), pp. 63–77, Plenum, New York.

Washburn, S. L. 1963. Behavior and human evolution, in: *Classification and Human Evolution* (S. L. Washburn, ed.), pp. 190–203, Aldine, Chicago.

Washburn, S. L. 1973. The evolution game. *J. Hum. Evol.* **2:**557–561.

White, T. D., Johanson, D. C., and Kimbel, W. H. 1981. *Australopithecus africanus:* Its phylogenetic position reconsidered. *S. Afr. J. Sci.* **77:**445–470.

Wilson, A. C., Carlson, S. S., and White, T. J. 1977. Biochemical evolution. *Annu. Rev. Biochem.* **46:**573–639.

Wooding, G. L., and Doolittle, R. F. 1972. Primate fibrinopeptide: Evolutionary significance. *J. Hum. Evol.* **1:**553–563.

Zihlman, A. L. 1981. A reply to S. C. Johnson. *Current Anthropol.* **22:**371–372.

Zihlman, A. L., Cronin, J. E., Cramer, D. L., and Sarich, V. M. 1978. Pygmy chimpanzee as a possible prototype for the common ancestor of humans, chimpanzees, and gorillas. *Nature (Lond.)* **275:**744–746.

Zimmer, E. A., Martin, S. L., Beverley, S. M., Kan, Y. W., and Wilson, A. C. 1980. Rapid duplication and loss of genes coding for the α chains of hemoglobin. *Proc. Natl. Acad. Sci. USA* **77:**2158–2162.

Cladistics and the Classification of the Great Apes

6

A. G. KLUGE

> In regard to classification and all the endless disputes about the "Natural System," which no two authors define in the same way, I believe it ought, in accordance to my heterodox notions, to be simply genealogical. But as we have no written pedigrees you will, perhaps, say this will not help much; but I think it ultimately will, whenever heterodoxy becomes orthodoxy, for it will clear away an immense amount of rubbish about the value of characters, and will make the difference between analogy and homology clear. The time will come, I believe, though I shall not live to see it, when we shall have very fairly true genealogical trees of each great kingdom of Nature.
>
> (Darwin, 1857, p. 104)

Introduction

Nineteenth century epistemologists identified the limits of exact knowledge and the detection of the laws of nature with the "natural system" of classification. In particular, William Whewell (1847), John Stuart Mill (1862), and W. Stanley Jevons (1874) stressed the importance of classification in the discovery of natural groups. Their contemporary, Charles Darwin, also argued that genealogy forms the basis for the natural system of classification of species, as evidenced by his letter (Darwin, 1857) to Thomas Huxley (see epigraph) and the repeated emphasis placed on genealogy in his widely read and influential

A. G. KLUGE • Museum of Zoology and Department of Ecology and Evolutionary Biology, The University of Michigan, Ann Arbor, Michigan 48109.

151

books *On The Origin of Species* (Darwin, 1859) and *The Descent Of Man* (Darwin, 1871).

The need for classification in biology has not diminished, and yet well-founded genealogical hypotheses are unavailable for the vast majority of plants and animals. Considering the preoccupation of human beings with their origins and destiny, it might seem reasonable to assume that a rigorously tested hominoid genealogical hypothesis exists. It does not! The absence is not due to lack of interest, nor to too few speculations, observations, or assertions. For example, Washburn (1973, p. 472) claimed "that man is most closely related to the African apes may now be considered a fact." *Homo*'s ancestry has been studied with living and extinct hominoids and with many data sources—anatomical, ontogenetic, behavioral, biomechanical, cytological, and molecular—and it is an understatement to say that the number of researchers exceeds the number of taxa involved (Pilbeam and Gould, 1974).

I believe our inability to discover the natural groups of hominoids, as for most other organisms, is the result of using overall similarity (phenetics), placing undue emphasis on fossils as common ancestors, and arbitrarily giving certain classes of observations more weight (Wiley, 1981). It is also apparent that the data analyses have rarely been conducted in a scientifically rigorous, testable, and repeatable manner (Popper, 1965, 1968). Tattersall and Eldredge (1977) and Eldredge and Tattersall (1975) also emphasized the lack of rationally determined genealogies, particularly *Homo*'s historical relationships to other hominids. Like these authors, I, too, will argue that discerning the genealogical classification for a difficult group of organisms such as hominoids requires a strictly cladistic methodology rather than phenetic (Sneath and Sokal, 1973) or syncretistic approaches (Mayr, 1969, 1974; Simpson, 1945, 1962, 1975).

The Cladistic Method

Any genealogical hypothesis involving terminal taxa (species *sensu* Simpson, 1962; Wiley, 1978) can be described as a nested set of monophyletic groups, where each set is characterized by all its members sharing at least one evolutionary novelty (Fig. 1). These shared derived features are the synapomorphies of Hennig (1966). Testing synapomorphies is straightforward. When the shared derived feature that defines a monophyletic group A + B is also found in the sister lineage to that group, C, then that novelty does not provide evidence for the original set hypothesis (Fig. 1). A condition shared by A + B is termed a symplesiomorphy when it is also exhibited by C. A shared primitive feature cannot provide evidence for the common ancestry of A + B; however, it may do so at a more inclusive level of monophyly (Fig. 1). Thus, symplesiomorphy and synapomorphy are relative, but only the latter can provide evidence of common ancestry.

The most commonly used method of estimating which of two or more similar attributes is derived (apomorphic) or primitive (plesiomorphic) is the

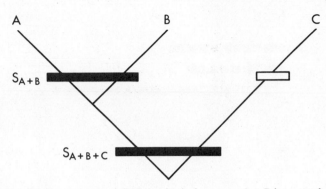

Fig. 1. Synapomorphy S_{A+B} supports the hypothesis that taxon A + B is monophyletic when the comparable condition in taxon C is plesiomorphic. Should the apomorphic condition be found in C, then the original synapomorphy no longer provides evidence for A + B, but it may do so at a more inclusive level, such as taxon A + B + C.

out-group comparison (Wiley, 1981). That condition occurring in the sister lineage to the group under study is considered primitive. The accuracy of the estimation is related to the reality of the monophyly hypothesized for the study group, as well as for that assemblage plus the out-group. Also, the transformation series observed during ontogeny can be employed in the estimation of relative primitiveness. Its use follows from the biogenetic law, as reformulated by Nelson (1978, p. 327): "given an ontogenetic character transformation, from a state observed to be more general to a state observed to be less general, the more general state is primitive, the less general derived."

It is not unusual to discover incongruent synapomorphies when studying most groups of species. Incongruent synapomorphies document alternate genealogical hypotheses (Fig. 2). Strictly following the methodological rule of parsimony, and choosing the simplest—most supported—proposition (Fig. 3), one can escape the impasse created by contradictory evidence. Some investi-

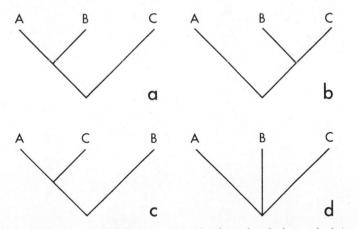

Fig. 2. Possible genealogical relationships are conveniently analyzed when only three taxa, A, B, and C, are considered because of the limited number of alternatives. There are as many as seven possible hypotheses when the taxa are species, because each can be ancestral to the others.

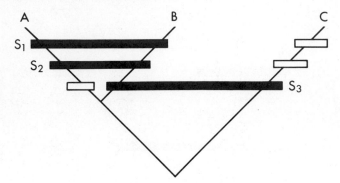

Fig. 3. Two synapomorphies S_1 and S_2 support the A + B monophyletic group, whereas synapomorphy S_3 supports another hypothesis of monophyly (B + C). Under these circumstances of conflicting evidence, S_3 is said to be incongruent with S_1 and S_2.

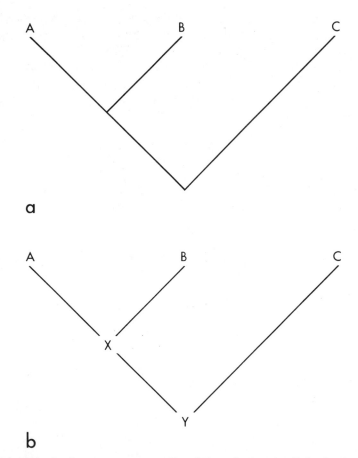

Fig. 4. The hypothesis of common ancestors X and Y emphasizes the distinction between cladograms (a) and phylogenetic trees (b). X and Y are not independent (e.g., when X = A, Y ≠ B). Y ≠ A, B, or X unless an evolutionary species concept is employed (Wiley, 1978).

gators differentially weight characters; however, differential weighting schemes usually involve unrealistic hidden assumptions. For example, most primatologists prefer hypotheses of hominoid relationships based on overall molecular similarity, rather than alternatives supported by morphological synapomorphies, presumably because "molecules are assumed to have a large amount of genetic information content directly correlated to taxonomic information content" (Frelin and Vuilleumier, 1979, p. 1). It is not clear what factual evidence the assumption is based on. The *ad hoc* hypothesis of homoplasy (parallelism and convergence) can always be used to give biological meaning to incongruent synapomorphies.

To delimit nested sets of monophyletic groups, termed cladograms when graphed, requires only two sets of assumptions (Wiley, 1981): (1) lineages split (speciation), and (2) new taxa are characterized by new features (apomorphies). Species of hybrid origin do not fulfill the first assumption (Hull, 1979; Wagner, 1980). Phylogenetic trees involve specific ancestor–descendant propositions and therefore make more assumptions than do cladograms (Fig. 4). The historical scenario painted by the syncretist–adaptationist is even more assumption-laden than the phylogenetic tree, and it is usually untestable (e.g., Lovejoy, 1981). Consistency with natural selection and overall plausibility appear to be important criteria guiding these authors in their choice among possible scenarios. Progress toward the identification of natural groups is likely to be most rapid when the fewest assumptions are made and the opportunity for testing always exists. My analysis of genealogy at the level of cladograms does not deny common ancestors, nor adaptation and associated micro- and macroevolutionary processes. It is pursued because it is repeatable and testable, and it leads efficiently to genealogical classification.

The Problem

The little progress made thus far toward the discovery of natural groups can be exemplified by living hominoids. The generally accepted hypothesis shown in Fig. 5 (Schwartz *et al.*, 1978) can be described as follows: *Gorilla* and *Pan* form a monophyletic group, and *Homo* is the sister lineage to *Gorilla* + *Pan*, and the more inclusive assemblage is (*Pongo* + *Gorilla* + *Pan* + *Homo*). *Hylobates* and *Symphalangus* are sister taxa, and that set, together with the Great Apes and *Homo*, form the monophyletic group usually designated Hominoidea. While most primatologists accept this hypothesis, they are inconsistent in recognizing (*Pongo* + *Gorilla* + *Pan*) as the Pongidae (or Ponginae), which is an unnatural, paraphyletic, group (Simpson, 1945, 1963). A particularly weak part of the generally accepted set of relationships, according to Delson *et al.* (1977), concerns the position of *Pongo*. These authors note that the hypothesis is based on biochemical and karyological evidence, morphological synapomorphies being "hard to define" (Delson *et al.*, 1977, p. 269; see also Bruce and Ayala, 1979). The fusion of the os centrale to the scaphoid

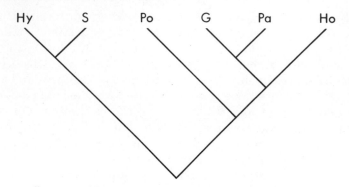

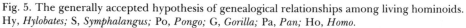

Fig. 5. The generally accepted hypothesis of genealogical relationships among living hominoids. Hy, *Hylobates;* S, *Symphalangus;* Po, *Pongo;* G, *Gorilla;* Pa, *Pan;* Ho, *Homo.*

supports the (*Gorilla* + *Pan* + *Homo*) grouping; however, a few *Pongo* exhibit that state as well (Corruccini, 1978). Other support may be found in the frontal and ethmoidal sinuses, but again there is a great deal of individual variation, and the exceptional development of the maxillary sinus, particularly in *Pongo,* tends to obscure the putative homologies (Cave, 1961; Cave and Haines, 1940).

　　If the relationship of *Pongo* is focused upon, and ones makes the reasonable assumptions that *Gorilla* + *Pan* is a natural group and that *Hylobates* + *Symphalangus* is the immediate out-group, then there are only four genealogical hypotheses possible (Fig. 6). These are: [*Pongo,(Gorilla,Pan)*], (*Pongo,Homo*), [*Homo, (Gorilla,Pan)*], and [*Pongo, Homo,(Gorilla,Pan)*]. The last possibility (Fig. 6d) is uninformative, in the sense that it does not go beyond the initial assumption of monophyly for the Great Apes + humans. The generally accepted molecular–karyological hypothesis is Fig. 6c. What synapomorphies do the biochemical and chromosome sources actually provide, can they be falsified, and can more parsimonious alternative relationships to Fig. 6c be suggested? The relationship of *Pongo* is a particularly interesting part of hominoid history, because of the emphasis placed on constancy of molecular evolution and the disparity between morphological and molecular data (King and Wilson, 1975; Cherry *et al.,* 1978). The generally accepted hypothesis (Fig. 6c) implies some rate heterogeneity and little congruence between molecules and morphology. Should the cladogram of either Fig. 6a or Fig. 6b be the natural classification, a more serious challenge to the molecular clock model would exist (Simons, 1976) and molecular–morphological incongruence would be exaggerated. Recently, the hypothesis of Fig. 6a was advocated by Tuttle (1975a), and it was explicitly put forward many years ago by Schultz (1930) and Keith (1931). I will examine only one aspect of the generally accepted hypothesis illustrated in Fig. 5, namely, the monophyletic group [*Homo+(Gorilla+Pan)*] (Fig. 6c). I will proceed by reviewing some of the biochemical studies, reanalyzing the major karyological evidence, and briefly setting forth other synapomorphies characterizing [*Pongo,+(Gorilla+Pan)*] as a highly corroborated monophyletic group. The (*Pongo+Homo*) hypothesis

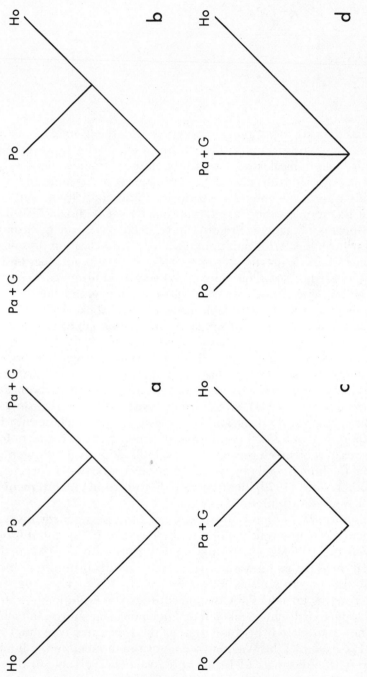

Fig. 6. Possible genealogical relationships among *Homo* and the Great Apes. Abbreviations are listed in the legend to Fig. 5. Pa + G is assumed to be monophyletic.

(Fig. 6b) is not considered further because it is supported by few synapomorphies.

Data Analysis

Molecular

Molecular data are thought to be consistent with the (*Gorilla*+*Pan*+*Homo*) monophyletic hypothesis (Fig. 6c; Corruccini *et al.*, 1979); however, several technical and philosophical issues cast doubt on the usefulness of most of this evidence in phylogenetic inference. The vast majority of the molecular studies of hominoids have used antigenic distances, immunodiffusion and micro-complement fixation, to assess taxonomic similarity (Goodman, 1975). More recently, estimates of genetic distance have been based on nuclear DNA hybridization (Kohne, 1975). Immunological and hybridization distances are inappropriate for phylogenetic inference because synapomorphies and symplesiomorphies are not separable. The use of such overall similarity measures would not be an issue if homogeneous rates of change and the absence of convergent evolution were reasonable assumptions (Mickevich, 1978). Further, genetic distance measures of similarity are questionable because they do not conform to the triangle inequality $[d(a,c) \leq d(a,b) + d(b,c)]$. Under such circumstances, negative branch lengths are produced, which is to say illogically that less than no evolution has occurred between two taxa. A negative branch length indicates nonmetric data and itself falsifies the hypothesis of a molecular clock (Farris, 1981). In addition, antisera do not give similar, let alone identical, reciprocal values, and Ferris *et al.* (1981) concluded for hominoids that there is about 25% error in measurement. The error term for the order Primates may be much greater. For example, Wilson and Prager (1974) reported that the human–baboon lysozyme distance is 66 in one direction and 127 in the other. Fitch (1976) identified a 15% reciprocal measurement error in primate DNA hybridization.

Nei's distance (*D*) has most often been used to assess genetic divergence with hominoid electrophoretic data (Bruce and Ayala, 1979), and it is usually analyzed with the UPGMA clustering algorithm (Sneath and Sokal, 1973). Nei's popular measure assumes a uniform rate of gene substitution, which has proven false for most organisms (Baverstock *et al.*, 1979). The Wagner tree method of grouping taxa does not assume constancy of evolutionary rate, and it might be applied to distances taken from any molecular source, but a best fit to data is not guaranteed, because that method assumes true metrics are employed. For example, the Wagner tree algorithm applied to the hominoid electrophoretic Nei distances of Bruce and Ayala (1979, Table 6) produced negative branch lengths [see Farris (1981) for a general critique]. Repeatability of Nei distances should also be of concern; considerably different values for *Homo–Pan* (0.39 and 0.62) have been reported (Wyles and Gorman, 1980, p. 67).

Goodman *et al.* (1979, p. 133) admitted that their gene lineage parsimony algorithm is not cladistic because "there can be no *a priori* specification of primitive and derived states; these states are assigned only after the nucleotide identities have been statistically sorted on the basis of how often they occur in the descendant sequences." Such sorting appears to lead to determining primitiveness on the basis of the commonality principle, which was dismissed as unreliable by Wiley (1981).

Amino acid and DNA–RNA base sequence data do not suffer from the problems of metricity and constancy of rate, reciprocity is not involved, and synapomorphies can be assessed, at least in the more thoroughly studied groups like hominoids (Farris, 1981). For example, the myoglobin molecule has been sequenced for *Homo, Pan, Gorilla, Pongo, Hylobates agilis,* and various lower primates (Romero-Herrera *et al.,* 1976, 1978). Three synapomorphies characterize the Hominoidea (sites 140-lysine, 144-serine, and 145-asparagine); one synapomorphy (23-glycine) supports a *(Gorilla+Pan+Homo)* hypothesis; and one (110-cysteine) a (Gibbon+*Gorilla+Pan+Homo*) group, excluding *Pongo.* By weight of all evidence (morphological and molecular), and even considering the myoglobin synapomorphies alone, the simplest interpretation of the site 110 substitution in *Pongo* is that it evolved independently (specifically, as an evolutionary reversal). The site 23 synapomorphy definitely corroborates the hypothesis of Fig. 6c; however, it is important to note that the evolution of glycine at that position is not a unique event. In fact it has evolved independently among birds and many other primate and nonprimate mammals.

Other molecular evidence for the *(Gorilla+Pan+Homo)* hypothesis of monophyly (Fig. 6c) comes from fibrinopeptide B amino acid sequences (Wooding and Doolittle, 1972). Serine at site 3 is characteristic of the African Great Apes and humans, while glycine is found in most other primates, including *Pongo.* Serine also occurs in the rhesus macaque, and there is an allelic serine in *Hylobates* (Mross *et al.,* 1970), reducing the significance of the putative site 3 synapomorphy. Only site 5 provides better support; phenylalanine appears in *Gorilla, Pan,* and *Homo,* and nowhere else among the primates.

The mitochondrial DNA (mtDNA) maps inferred from restriction endonuclease cleavage patterns are becoming increasingly popular sources of molecular data for phylogenetic studies. While such data can be expressed in point mutation units and analyzed cladistically, there appear to be some shortcomings to this approach. It seems the mtDNA evolves 5–10 times faster than does single-copy nuclear DNA, and while this difference in rate is seen as an advantage in discovering relationships among closely related species (Ferris *et al.,* 1981), users of mtDNA data must recognize that it is therefore not likely to be a conservative estimator [*viz.,* independently evolved mutations probably will be relatively common; see Fitch (1979)]. Also, the mtDNA of mammals, including hominoids, is about 16,500 base pairs long (Brown *et al.,* 1979), and only a tiny fraction of these are used in aligning the different species maps and in establishing monophyletic groups. In a hominoid study by Ferris *et al.* (1981), where all of the lineages considered in the present paper were mapped, 11 invariant sites were used for the mtDNA map alignments, and these involved 58 base pair sites (58/16,500 = 0.35%). Moreover, Ferris *et al.* recognized only 121 variable sites on the hominoid mtDNAs, and 79 of these were autapomorphic (variants found in one species). Of the remaining 42 cladistically useful, variable positions (~ 0.26% of the mtDNA base pairs), at least 24 must have exhibited one or more instances of independent evolution, assuming a minimum-length-tree hypothesis. The high degree of homoplasy, back mutations, and parallelisms was acknowledged by Ferris *et al.* (1981), and for the tree hypothesis that best fit the data, Fig. 6c here, there was a total of 25 such instances (67 − 42 = 25). The alternate tree hypothesis shown in Fig. 6a requires only four or five additional mutations (71 or 72; the exact number cannot be determined because of ambiguous coding in Ferris *et al.*'s data, their Table 6). I attach even less significance to such a small difference between alternative cladograms, because the mtDNA map of each species was based on one specimen, even though intraspecific variation was recognized by the authors. Subsequently, Ferris *et al.* (1982) documented considerable individual and geographic variation in the nucleotide sequence of mtDNA. Moreover, they found ape species to be 2–15 times more variable than *Homo*.

Recently, Brown *et al.* (1982) obtained a cleavage map of an 896-base-pair fragment of mtDNA from *Homo, Pan, Gorilla, Pongo*, and *Hylobates*. As before, it was shown that the maximum parsimony method produced nearly equally simple hypotheses; 37 evolutionary events were calculated for the equivalent of the hypothesis in Fig. 6a and 34 events for Fig. 6c. However, more compelling evidence for the [*Homo*+(*Gorilla*+*Pan*)] monophyletic group came from the analysis of the complete nucleotide sequence of the 896-base-pair fragment. Although the sequence is long, only three complete genes are included in the fragment, plus parts of two others (unidentified reading frame shifts). Moreover, only 81 of the 284 variable positions contained cladistic information pertaining to the relationship of *Pongo* (Figs. 6a and 6c). Autapomorphies accounted for much of the variation (*Hylobates,* 64; *Pongo,* 55; *Gorilla,* 22; *Pan,* 19; *Homo,* 15), 26 positions contained more than two nucleotide states and unique transformation series could not be deduced (W. Brown, personal communication), and two positions were intraspecifically variable (*Homo*). This left only 81 phylogenetically useful, unambiguous sites, which is less than 0.49% of the total mtDNA. This is not a rich sampling of the genotype–phenotype. Further, in the most parsimonious cladogram for the 81 characters (Fig. 6c), there were only 49 congruent synapomorphies, assuming *Hylobates* as the outgroup, whereas the remaining 32 supported various alternatives. Thus, as in the earlier paper by Ferris *et al.* (1981), considerable homoplasy was evident. Also relevant, in view of the extensive individual and geographic variation documented by Ferris *et al.* (1982), is the fact that each taxon, except *Homo,* was represented by a single clonal line, and the sequence for *Pan* was a composite "individual" (approximately two-thirds from a *Pan troglodytes* and one-third from a *Pan paniscus*). Finally, and perhaps most important, given the considerable variation, rapid rate of mtDNA evolution, and extensive homoplasy, only one member of the out-group was sequenced, and I have little confidence that many apomorphies were correctly deduced. Doubtless, primitive character state estimation would be improved by surveying several different hylobatids.

Karyotype

Chromosome number and morphology, including that based on various banding techniques, have been employed frequently in the study of hominoid relationships. According to Bruce and Ayala (1979), these data support the hypothesis of Fig. 6c. My review of the literature suggests that the observations are ambiguous, or perhaps better support the hypothesis that Great Apes form a monophyletic group (Fig. 6a). Consider the diploid chromosome number (Table 1). The human condition, $2n = 46$, is usually stated to have evolved from 48, like that found in Great Apes, by centric fusion of chromosome 2 (Dutrillaux *et al.*, 1975; Dutrillaux, 1979). To hypothesize $2n = 48$ as a synapomorphy for the Great Apes is also consistent with the distribution of diploid numbers. In this view, the 48 could have evolved from a karyotype with $2n = 46$, like that in *Homo,* by centric fission. To decide which state, 46 or 48, was the (*Pongo+Gorilla+Pan+Homo*) synapomorphy requires investigating the out-group. Obviously, hylobatids are the most relevant, but their diploid complement is variable, most with 44 (*Hylobates* [*Hylobates*]), others 50 or 52 (*Symphalangus syndactylus* and *Hylobates* [*Nomascus*] *concolor*). Confidently identifying the plesiomorphic state relative to the hylobatids requires that we go outside the Hominoidea. The sister lineage is not universally agreed upon, but most claim it to be among the Old World monkeys (Table 1). While the diploid number is variable in the Old World monkeys, 42–72, Chiarelli (1975, p. 123) concluded that 44, as in *Colobus* and *Hylobates,* "characterized the ancestor of Old World primates," and that the 50 in *Symphalangus* and 52 in *Hylobates* [*Nomascus*] *concolor* are derivatives by centric fission or polysomy. Eckhardt (1979, p. 44) also considered the lower diploid numbers to be primitive in the highly variable *Cercopithecus*. In addition to the colobine out-group comparison, Chiarelli's conclusion appears to be based on his observation (1972, pp. 94–95) that all *Hylobates* species with $2n = 44$ have only metacentric–submetacentric chromosomes and that one of these is marked by a large achromatic region, just as in the karyotype of all Old World monkeys. On the other hand, the karyotype of *Symphalangus syndactylus* includes a pair of acrocentrics that exhibit some achromatic filaments. If the achromatic conditions are homologous in the two hylobatid genera, then it must be concluded that at least one aspect of the *Symphalangus* karyotype is derived, perhaps by fission, although Chiarelli (1972, p. 97) speculated that a pericentric inversion was responsible.

The *Hylobates* [*Nomascus*] *concolor* karyotype has three small acrocentric

Table 1.   Somatic Chromosome Data on Old World Anthropoid Primates[a]

Genus (number of species)	2n	S–M	A
Macaca (13)	42	40	0
Papio (7)	42	40	0
Theropithecus (1)	42	40	0
Cercocebus (5)	42	40	0
Cercopithecus (18)	48–72	34–52	0–24
Erythrocebus (1)	54	36	16
Presbytis (4)	44	40	2
Pygathrix (1)	44	42	0
Rhinopithecus (1)	44	40	2
Nasalis (1)	48	46	0
Colobus (3)	44	42	0
Hylobates lar	44	42	0
Hylobates agilis	44	42	0
Hylobates moloch	44	42	0
Hylobates hoolock	44	42	0
Hylobates klossi	44	42	0
Hylobates [Nomascus] concolor	52	44	6
Symphalangus syndactylus	50	46	2
Pongo pygmaeus	48	26	20
Pan troglodytes	48	34	12
Pan paniscus	48	34	12
Gorilla gorilla	48	30	16
Homo sapiens	46	34	10

[a]Chiarelli (1975) and Chiarelli et al. (1979). The symbols stand for the following: 2n, diploid number, S–M, submetacentric–metacentric number; and A, acrocentric number.

chromosomes, two of which have an achromatic filament region, which also suggests a derived condition. The extreme similarity of some chromosome pairs in *Hylobates* [*Nomascus*] *concolor* implies duplication, another phenomenon suggesting that this species possesses a derived diploid number. Thus, it seems at least equally, if not more, consistent with the observed chromosome variation to treat $2n = 48$ as a synapomorphy of the Great Apes than it does to treat the human $2n = 46$ as a uniquely derived feature. The probable mechanism for the Great Ape synapomorphy is fission, the type of chromosome mutation that Imai and Crozier's (1980) recent theoretical arguments and empirical findings suggest has been the most common among all mammals [see also Seuanez (1979, p. 70)].

According to Seuanez (1979), the chromosomes of *Homo* and the African Great Apes fluoresce brilliantly when stained with quinacrine dihydrochloride (Q-banding), whereas that dye has no strong effect in the orangutan or gibbon. I believe there is good reason to question these observations as support for a (*Homo+Gorilla+Pan*) monophyletic assemblage. For example, homologous autosomes and sex chromosomes among the hominoids do not exhibit the same response to staining, and other types of banding techniques (e.g., C-banding) suggest incongruent synapomorphies (Seuanez, 1979, pp. 36–37, Chapter 6). Moreover, Seuanez (1979, pp. 37–38) attached little genealogical importance to the fluorescence, because those regions of the chromosomes so stained are considered genetically inert.

Several studies employing various banding techniques form the basis for identification of gene rearrangement and chromosome homologs among all Great Apes and *Homo* (Miller, 1977; Yunis *et al.*, 1980; Dutrillaux, 1975, 1979; Dutrillaux *et al.*, 1975; Seuanez, 1979). While there can be little doubt that human chromosome 2 was the only one involved in producing the diploid difference between these taxa ($2n = 46$ in *Homo;* $2n = 48$ in the Great Apes), the responsible mutation is unknown. Dutrillaux (1975, 1979) advocated a telomeric fusion of two acrocentric chromosomes ($2n = 48$ to $2n = 46$), although the opposite polarity for this transformation series, which would involve a fission, seems equally, if not more, plausible. Dutrillaux did not specify why his hypothesis was preferred, and either interpretation necessarily admits the existence of two narrowly separated centromeres in chromosome 2, with one being functionally supressed during cell division.

The same banding studies also have been used to identify comparable inversions among the Great Apes and human, although the conclusions arc ambiguous. Contributing to the uncertainity is the absence of similarly detailed maps of hylobatid chromosomes (Dutrillaux, 1979; Seuanez, 1979, pp. 65–69; Turleau and de Grouchy, 1979). Such characterizations are essential if apomorphic and plesiomorphic macromutation states are to be distinguished among the Great Apes and human. Also contributing to the ambiguity is the fact that not all inversion synapomorphies are congruent with each other, and there exists the prospect of discovering more such mutations with improved technology. For example, Dutrillaux (1975; see also Dutrillaux, 1979; Seuanez, 1979, pp. 73–77) identified "all" pericentric and paracentric inversions, telomeric fusions, translocations and translations among the Great Apes and human, and from his observations, Vogel *et al.* (1976) prepared a complete karyotype distance matrix. These raw distances lead to the conclusion that *Homo* and *Pan* are sister lineages; however, Yunis *et al.* (1980) recently discovered three new pericentric differences between *Homo* and *Pan troglodytes* which were overlooked by Dutrillaux, and they alter his opinion. Moreover, the taxonomic sequence preferred by Dutrillaux, *Gorilla* to *Pan* to *Homo,* based on inversion similarities of only certain chromosomes (2, 4, 5, 7, 12, 17) led him to an "unorthodox" hybridization hypothesis between *Gorilla* and *Pan.* An alternative sequence of *Homo* to *Gorilla* to *Pan* requires fewer mutations and no hybridization. Thus, I conclude that the diploid number supports placing the Great Apes in a monophyletic group, while only questionable information is available from gene rearrangements. The chromosome information is particularly disappointing considering the number of cytotaxonomic studies of hominoids. Perhaps no other primate research better documents the need for a strictly cladistic methodology (Seuanez, 1979, pp. 71–77).

Urethrovaginal Septum

According to Bolk (1907), Eckstein (1958), and Hill (1951), the urethrovaginal area of *Pongo, Gorilla,* and *Pan* differs markedly from the more primitive condition found in *Homo.* In the Great Apes, the vaginal vestibule is deep and possesses irregular margins formed by the labia minora. The vestibule narrows abruptly to a short anteroposterior cleft, into which the urethra and vagina open (Fig. 7b). A urethral papilla (prominence) is absent. In *Homo, Hylobates* (and presumably *Symphalangus;* Matthews, 1946), and Old World monkeys (Eckstein, 1958), the vestibule is much shallower, its margins are regular, the urethra and vaginal openings are separated by a distinct septum, both empty directly into the vestibule, and the urethral meatus is located within a raised area suggesting a papilla or prominence (Fig. 7a). Thus, urethrovaginal morphology suggests that the Great Apes form a monophyletic group.

Labia Majora

Many contradictory observations exist in the literature on the degree of development of the labia majora in hominoids (Hill, 1958). It seems that most of the reported variation is due to individual differences in age, stage in the

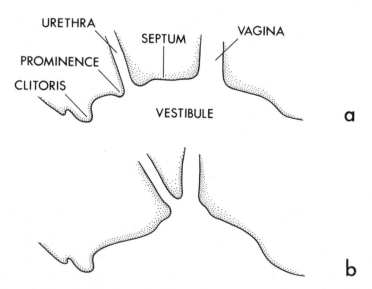

Fig. 7. Diagrammatic midsagittal sections illustrating two types of urethrovaginal anatomy in hominoids. (a) *Homo* and *Hylobates;* (b) *Pan, Gorilla,* and *Pongo.*

reproductive cycle, and nutritional state. Nevertheless, the conditions are remarkably similar in sexually mature hylobatids and *Homo* (Schultz, 1927). The labia majora form relatively distinct, narrow, cutaneous folds peripheral to the labia minora. The two sets of folds are easily distinguished because the latter form the preputial (anterior) covering of the clitoris. On the other hand, in sexually mature Great Apes, swollen, well-defined labia majora are absent. Such cutaneous folds are present in these taxa only in foetal life, subadults, and exceptionally obese caged individuals. According to Wislocki (1932), possession of large labia majora is the primitive character state. This is supported by the ontogenetic sequence in the Great Apes, and the fact that such folds are present in adult platyrrhines and many catarrhines.

Pelvic Inlet and Newborn Sizes

It has long been recognized that birth in the Great Apes is relatively easy compared to other primates (Table 2; Schultz, 1949; Harms, 1956; Leutenegger, 1972, 1973). Some of the infant size and adult female pelvic diameter variation may be independent(Leutenegger, 1970, 1974) and interpreted as separate synapomorphies corroborating the (*Pongo+Gorilla+Pan*) group. Table 3 indicates that Great Apes have relatively smaller infants, according to newborn weight, than do *Homo, Hylobates,* and other higher primates. Surprisingly, the relative size of the pelvic ring is also smaller in the Great Apes than in other hominoids and monkeys.

Another variable related to pelvic inlet and newborn sizes is the degree of sexual differentiation in the ischiopubic distance. With few exceptions, among all primates, the Great Apes have the least developed sexual dimorphism in the pelvis (Leutenegger, 1974), whereas it is strongly differentiated in *Homo*. While the index may be independent of much of the birth canal variation, the

Table 2. **Average Pelvis Inlet Diameter of Adult Females and Size of Newborns in Representative Primates**[a]

		Adult females			Newborns		
	N	Pelvic inlet breadth	Symphysion– promontorium	N	Head length	Head breadth	Shoulder breadth
Ateles geoffroyi	7	54.4	90.3	8	66.9	52.2	50.4
Macaca mulatta	41	50.9	67.7	28	66.3	50.7	49.2
Nasalis larvatus	15	51.8	71.5	1	64.0	49.0	59.0
Hylobates lar	87	55.9	78.7	6	64.2	52.7	51.2
Pongo pygmaeus	26	102.5	149.6	4	84.1	74.9	81.8
Pan troglodytes	29	98.0	149.5	9	83.0	71.0	84.9
Gorilla gorilla	10	122.6	175.7	4	97.0	79.0	92.0
Homo sapiens	10	121.6	112.9	10	123.8	98.5	118.3

[a]Schultz (1949).

Table 3. Relative Adult Pelvic Inlet Breadth and Relative Newborn Weight in Representative Primates[a]

| | N | Pelvic breadth[b] | | Newborn weight[c] |
		$\bar{X}$	Orv	$\bar{X}$
Ateles geoffroyi	7	51.8	50.5–53.3	7.0
Macaca mulatta	41	59.3	53.4–67.5	6.7
Nasalis larvatus	15	49.3	46.5–56.4	4.6
Hylobates lar	87	57.5	52.1–68.7	7.5
Hylobates moloch	7	52.0	48.1–58.1	—
Pongo pygmaeus	26	42.9	39.2–46.1	4.1
Pan troglodytes	30	41.0	36.8–47.0	4.0
Gorilla gorilla	15	37.5	34.8–41.6	2.4
Homo sapiens	50	49.5	47.2–52.2	5.5

[a]Schultz (1949).
[b]The percentage relation between the greatest breadth of the pelvic inlet and the greatest pelvic breadth.
[c]The average weight of newborns as a percentage of average weight of adult females.

lack of correlation is not well documented, and I have not recognized it as an independent synapomorphy.

Glans Penis

Adult hylobatids and *Homo*, like Old World monkeys, have a well-developed glans penis, set off dorsally from the corpora cavernosa by a prominent corona glandis. The glans is scarcely differentiated in *Pongo*. It is not demarcated dorsally by a corona glandis. The glans is not differentiated at all in *Pan* (Hill, 1958). While the glans is moderately well developed in adult *Gorilla*, the corona glandis is absent (Raven, 1950) or defined dorsally by only a "relatively feeble" shallow border (Hill and Matthews, 1949, p. 373). Thus, I consider the reduced nature of the corona glandis in adult Great Apes to be a synapomorphy. I hypothesize this state to be derived on the basis of its well-developed nature in hylobatids and Old World monkeys and relatively prominent appearance in some juvenile Great Apes. The relative reduction during ontogeny is most conspicuous in *Gorilla* (Hill and Matthews, 1950).

Form of Hand-Walking

The African Great Apes have long been recognized as knuckle-walkers, a unique form of quadrupedal locomotion (Tuttle, 1975b). Several recent studies have correlated various shared anatomical and biomechanical novelties with that form of locomotion [Tuttle and Basmajian, 1974; Jenkins and Fleagle, 1975; see review in Tuttle (1975a)]. Many *Pongo* also employ closed hands during locomotion, either on the ground or on large branches well above ground, although versatile arboreal climbing is their most common mode of travel (Wallace, 1902, p. 45; Tuttle, 1977). Unlike the African Great Apes, orangutans do not always walk with flexed fingers [palmigrade postures of one or both hands have been observed (Tuttle, 1967)]. *Pongo*'s distinctly different biomechanical features have caused Tuttle and others to refer to its closed-hand quadrupedal locomotion as fist-walking. While fist-walking has not yet been diagnosed anatomically, most authorities interpret fist- and knuckle-walking to be part of the same flexed-finger transformation series. Also, according to Tuttle (1969*a*, p. 960), "there are no features in the bones, ligaments, or muscles of [*Homo*] that give evidence for a history of knuckle-walking"; however, Corruccini (1978) countered that claim by citing the fusion of the os centrale and scaphoid. Unlike young Great Apes, humans show no sign of pongid fist- or knuckle-walking during infantile quadrupedalism (Schultz, 1936). Some infants clench their fists, but they do not place them on the backs of the proximal phalanges or radial sides down as do orangutans (R. Tuttle, personal communication). I interpret the human infant's purely palmigrade form of locomotion, like that of certain catarrhines, as further evidence that the closed-hand form of walking is a synapomorphy of the Great Apes alone.

Flexor Pollicus Longus Tendon

According to Straus (1942*a,b,* 1949) and Tuttle (1970, 1975*a*), the deep, long flexor of the thumb is vestigial or absent in the Great Apes, whereas in humans, hylobatids, and many cercopithecine monkeys it is well developed. Also, the relative mass of the intrinsic thumb muscles is greatly reduced in the apes; in humans, hylobatids, and many cercopithecines it is relatively large (Straus, 1949; Tuttle, 1969*b*, 1972, pp. 279–282). Both the small size of the muscles of the thenar eminence and the reduction of the long pollical flexor tendon may document the same synapomorphy, namely the reduced nature of the thumb in the Great Apes.

Tendons of Toe Flexor Muscles

The tendons of the flexor hallucis longus and flexor digitorum longus are fused in prosimians and evenly distributed to the five digits. Fusion and an even distribution to some toes characterizes platyrrhines, catarrhines, hylobatids, and humans. Only in the Great Apes are the two sets of tendons markedly discrete and totally unevenly distributed, the hallucis serving toes I, III, and IV and the digitorum serving II and V; the tendon for the great toe is nearly always absent in *Pongo* (Straus, 1949; Tuttle, 1970). The unique separation in the Great Apes is almost certainly derived.

Foramen Intercondyloideum

Schultz (1930) summarized the variability of the foramen intercondyloideum, a perforation in the olecranon fossa of the humerus, in the Hominoidea. The foramen is usually absent in juveniles, and more adult females than males possess the opening. The foramen is usually present in adult Great Apes (males 37–76%; females 43–84%), but it only occurs in 10% or less of the hylobatids. The frequencies in humans are more like the presumed primitive condition in hylobatids (males 5–7%; females 15–28%), except for two samples which overlap the Great Ape range of variation. Further documentation of the extent of between-population variation in *Homo* is required before the presence of the foramen intercondyloideum can be fully endorsed as a Great Ape synapomorphy.

Epiphysial Union

Schultz (1944) stated that all Great Apes show a tendency toward relatively early closure of the proximal epiphysis of the humerus, a synapomorphy, whereas it is late appearing in Old World monkeys, gibbons, and *Homo* [the *Symphalangus* condition is unknown (Schultz, 1973)]. The out-group criterion suggests that *Homo*'s late closure is plesiomorphic.

Cervical Vertebral Spines

The relative lengths of the dorsal spines of the cervical vertebrae vary considerably among hominoids (Schultz, 1961). In Old World monkeys, hylobatids, and humans, the spines are short, whereas in the Great Apes, particularly among males, they are much longer. According to Slijper (1946, p. 109), the increased demands made on the cervical muscles and ligaments are positively reflected in the length of the cervical spines. In turn, the length of the postcondylar part of the occiput is inversely correlated with the development of the muscles and ligaments. The derived nature of the cervical spinal trait is inferred from both out-group (hylobatids and Old World monkeys) and ontogenetic criteria.

Lumbar Vertebral Number

According to Schultz (1961), the number of lumbar vertebrae in Old World monkeys varies from six to nine. The intrageneric mode is usually seven. He also noted that in hylobatids and humans the mode of five is relatively primitive compared to the mode of four that characterizes the Great Apes (Table 4). As expected, this Great Ape synapomorphy is also reflected in

Table 4.   Number of Vertebrae in Hominoids[a]

	Sample size	Number of lumbar vertebrae						
		3		4		5		6
Hylobates	319			16	3	242	10	48
Symphalangus	29	1	1	12	2	13		
Pongo	127	8	6	99	3	11		
Pan	162	65	11	86				
Gorilla	81	28	2	50	1			
Homo	125			6	3	112	4	

[a]Schultz (1961). Partially fused vertebrae are shown between columns.

the relative weight of the lumbar portion of the vertebral column (Schultz, 1961, p. 30) and the modest mobility of the back when climbing, walking, or sitting (Slijper, 1946).

Thoracic Shape

Adult Great Apes have a distinctly funnel-shaped thorax, whereas it is more barrel-like in all other hominoids and Old World monkeys. Both the out-group comparison and the ontogenetic sequence in the Great Apes (Schultz, 1968) suggest that the Great Ape condition is derived.

Distance between Iliac Crest and Last Rib

The anthropoid apes exhibit little space between the pelvis and the last rib, in contrast the large distance in all other hominoids and Old World monkeys. This apparent (*Pongo+Gorilla+Pan*) synapomorphy is not entirely explained by their shortened lumbar region, because the difference is not totally offset in individuals with exceptional numbers of trunk (thoracolumbar) vertebrae (Schultz, 1936).

Prognathism

Schultz (1960, 1962, 1973), among many primatologists, has drawn attention to the relatively much larger face of the Great Apes, which can only be due in part to a size allometry. The prosthion–basion and nasion–prosthion distances relative to the basal length of the skull (nasion–basion) clearly expresses this pongid synapomorphy (Table 5). Hylobatids and humans, like most Old and New World monkeys, do not have significantly protruding chins as adults. The large face in a few monkeys (e.g., *Alouatta*) is more in the

Table 5.   Degree of Prognathism in Adult Hominoids[a]

	Sex	N	P–B/N–B	N–P/N–B
Hylobates moloch	m	10	1.23	0.55
	f	7	1.21	0.52
Hylobates lar	m	25	1.20	0.56
	f	30	1.18	0.53
Symphalangus	m	10	1.29	0.60
	f	8	1.27	0.58
Pongo	m	10	1.64	1.04
	f	10	1.56	0.97
Pan	m	25	1.39	0.89
	f	25	1.36	0.86
Gorilla	m	50	1.46	0.95
	f	30	1.37	0.91
Homo	m	21	0.97	0.69
	f	21	0.99	0.69

[a]Schultz (1962). The symbols stand for the following: P–B, prosthion–basion; N–B, nasion–basion; N–P, nasion–prosthion.

snout. The prognathism seen in *Macaca* and *Cercopithecus* is best interpreted as convergent to the Great Apes, because few if any other characters corroborate that assemblage as monophyletic.

Simian Shelf

A backward-projecting plate of bone, the simian shelf, unites the two halves of the mandible at the base of the symphysial region in some primates. It is almost always well-developed in adult Great Apes and is also present in some adult Old World monkeys. It is absent in platyrrhines, many catarrhines, hylobatids, and humans (Straus, 1949). Schultz (1969) illustrated an adult male chimpanzee with only a weakly developed shelf, but that general region of its mandible appears to be unusual and may represent a developmental abnormality that affected the entire symphysial region. The condition of *Australopithecus afarensis* (particularly specimen A.L. 400-1a) is questionably a simian shelf. While "the inferior transverse torus is low and rounded rather than shelf-like," there exists "a moderate superior or transverse torus" (Johanson and White, 1979). I interpret the presence of a large simian shelf in the Great Apes as a synapomorphy, because it is absent in early developmental stages of all primates, adult hylobatids, and many adult catarrhines.

Deciduous Dentition

In Old World monkeys, hylobatids, and humans, the sequence of eruption of the deciduous teeth is middle incisors, lateral incisors, first molars, canines, and second molars. The canines nearly always appear last in the Great Apes (Schultz, 1956), which I interpret as a shared apomorphic condition. There is little information on variation in the sequence of eruption of deciduous teeth (S. M. Garn, personal communication), and future research could easily change my treatment of this character.

Discussion

How accurate is the phenotypic evidence supporting the Great Ape monophyletic group (Fig. 6a), and does each synapomorphy represent an independent assessment? The first question can be answered in two ways, both of which involve further tests. First, the observations that led me to each hypothesis of synapomorphy require additional study. Anatomists and cytologists must check the accuracy of the many authors' observations that served as my source of specimen information. While I consciously avoided highly variable traits, and those about which there was unresolvable disagreement in the literature, I have no doubt that some of the proposed synapomorphies will require redefinition. Perhaps some will have to be eliminated altogether because I misinterpreted the primary literature. Second, further observations on the distribution of the traits in out-groups and their time of appearance in ontogeny are necessary to test my conclusion as to which condition is apomorphic. Some of the hypothesized synapomorphies I recognize may be judged symplesiomorphies.

The second question posed above is not so easily answered. When dealing with only one or a few grossly characterized steps in a transformation series, and when all of that variation is focused on testing a single monophyletic group (Fig. 6), the lack of independence may appear to be greater than it really is. The existence of correlation can be due to the reality of the monophyletic group that the synapomorphies corroborate, or it might represent the expected outcome of genetic linkage, pleiotropy, or biomechanical and developmental constraints having little to do with common ancestry. In the present study, a factor that might be responsible for the lack of independence among the synapomorphies is the presumed common form of locomotion and posture that the Great Apes share. If, for example, all Great Apes tended to locomote in a similar manner, and many of the synapomorphies were necessary products of that common condition, then I would have to doubt the extent of the evidence corroborating the hypothesis of Fig. 6a, perhaps even to the degree that those few remaining incongruences would be hypothesized as independent evolutionary events. At least some of the synapomorphies that I have recognized seem unlikely necessary correlates of a common pattern of locomotion and posture (e.g., karyotype, glans penis, simian shelf, deciduous dentition). The remaining synapomorphies might be functionally related, but is there really a common pattern of locomotion and posture among the Great Apes with which a necessary correlation might exist? On this subject there seems to be no general agreement among the functional morphologists [in particular, see the review by Tuttle (1977)]. In fact, according to Slijper (1946, pp. 63–64), the locomotor and postural patterns of hominoids are incongruent with the cladogram I claim to be strongly corroborated (Fig. 6a).

Another factor that might affect the lack of independence among the synapomorphies is the process of paedomorphosis. If *Homo* were generally paedomorphic (neotenic), as claimed by Gould (1977), then its many traits interpreted as symplesiomorphies might be, in reality, autapomorphies. However, the more general problem is that the hypothesis that a single heterochronic change has simultaneously affected many traits (an epiphenomenon), or even that a single trait is paedomorphic, requires some prior conclusion of ancestor–descendent relationships, *viz.*, a genealogical hypothesis (Wiley, 1980). A conclusion of heterochrony, like a hypothesis of convergence, is an *a posteriori* judgement of process used to explain one or more incongruent synapomorphies. Thus, in the present study, many synapomorphies support the hypothesis that the Great Apes form a monophyletic group (Fig. 6a), and to invoke paedomorphosis as an explanation for the conditions exhibited by *Homo* is to make *ad hoc* explanations where none are called for; to do so is to decrease the simplicity of the genealogical hypothesis (Fink, 1982; Kluge, 1982).

Which hypothesis of hominoid relationships should be adopted (Fig. 6a or Fig. 6c)? In view of the contradictory nature of the available evidence, I suggest that neither possibility be ignored. Certainly, it would be unscientific to accept only certain biochemical and chromosome data and to continue to "prove" *Homo*'s close relationship to the African Great Apes with that evidence. No doubt, science will be best served by directing future research toward falsification of all hypotheses. Also, considering the amount and variety of data examined to date, it seems unlikely that either competing cladogram (Fig. 6a or Fig. 6c) will ever receive overwhelming corroboration. I believe there are two issues here. First, how does one judge the best fit to all data, when the evidence comes in such disparate forms as nucleotides, karyotypes, number of vertebrae, and pattern of locomotion? The qualitative differences in the units of observation seem so great as to defy a common synthesis. Second, the sheer number of possible characters at the DNA basepair level of study necessarily favors the molecular approach. At least, I find it difficult to imagine equivalent numbers of putative synapomorphies being characterized from morphology or behavior. Thus, I suggest it may be more fruitful in the future to judge the truth of the alternative hypotheses in terms of consilience (Kluge, 1982). As Whewell wrote many years ago (1847, Volume 2, p. 469): "the consilience of inductions takes place when an induction, obtained from one class of facts, coincides with an induction, obtained from another *different class*. This consilience is a test of the truth of the theory in which it occurs" (my italics).

A truly "different class" of evidence corroborating the Great Ape monophyletic hypothesis (Fig. 6a) comes from the study of parasites. A cladistic analysis of pinworms, genus *Enterobius* (13 species and 31 binary characters), indicated that *E. buckleyi* (host *Pongo*), *E. lerouxi* (host *Gorilla*), and *E. an-thropopitheci* (host *Pan*) form a monophyletic assemblage (Brooks and Glen, 1982). Moreover, *E. vermicularis* is the sister lineage to that monophyletic group of pinworms, and it inhabits *Hylobates* and *Homo* (D. Brooks, personal communication). Only six of the 31 pinworm traits studied were incongruent with the best fitting cladogram. This is a high consistency (Kluge and Farris, 1969), and all of the predicted homoplasy is found in the parasites infecting colobines. Figure 6a provides the simplest explanation for these parasite relationships—namely host–parasite coevolution—whereas Fig. 6c forces us to adopt various *ad hoc* explanations for the co-occurences.

Should the Great Ape assemblage be accepted as a monophyletic group, with *Homo* being its sister lineage, two major considerations must be confronted. Such a phylogenetic history denotes a clear lack of congruence between molecular and other, more traditional, kinds of data. The molecular information may not provide good evidence of common ancestry, because of associated technical difficulties and/or the predisposition toward independent evolution (convergence, parallelism, and evolutionary reversal). As discussed earlier, there is evidence for both. The other major consideration involves the age of the lineage leading to *Homo*. A time of 15–20 m.y.a. for such a divergence is suggested (Read, 1975; Uzzell and Pilbeam, 1971; Szalay and Delson, 1979). Perhaps this does not require an unrealistic revision in our thinking, because even *Homo*'s special form of bipedality seems to have existed much earlier than previously suspected, as inferred from Pliocene hominid footprints [3.6–3.8 m.y.a. (White, 1980)].

My view of hominoid relationships, contrary to Washburn's (1973, p. 472), is that much research remains to be done, and that we are little closer to the natural system of classification of man and his ancestors than when Charles Darwin wrote to T. H. Huxley in 1857.

ACKNOWLEDGMENTS

Many persons offered unpublished data, literature sources, and valuable criticisms, although not all endorsed my phylogenetic methods or my conclusions of primate relationships. I especially wish to thank Dan Brooks, Wes Brown, Ken Creighton, Steve Dobson, J. S. Farris, Dan Fisher, Stanley M. Garn, Phil Gingerich, David Hull, Mary Mickevich, Phil Myers, Marsha Robertson, Russ Tuttle, and Dave Wake.

References

Baverstock, P. R., Cole, S. R., Richardson, B. J., and Watts, C. H. S. 1979. Electrophoresis and cladistics. *Syst. Zool.* **28**:214–219.

Bolk, L. 1907. Beitrage zur Affen-Anatomie, IV, Zur Entwicklung und vergleichende Anatomie des Tractus urethrovaginalis der Primaten. *Z. Morphol. Anthropol.* **10**:250–316.

Brooks, D. R., and Glen, D. 1982. Pinworms and primates: A case study of coevolution. *Proc. Helminthol. Soc. Wash.* **49**:76–85.

Brown, W. M., George, M., Jr., and Wilson, A. C. 1979. Rapid evolution of animal mitochondrial DNA. *Proc. Natl. Acad. Sci. USA* **76**:1967–1971.

Brown, W. M., Prager, E. M., Wang, A., and Wilson, A. C. 1982. Mitochondrial DNA sequences of primates: Tempo and mode of evolution. *J. Molec. Evol.* **18**:225–239.

Bruce, E. J., and Ayala, F. J. 1979. Phylogenetic relationships between man and the apes: Electrophoretic evidence. *Evolution* **33**:1040–1056.

Cave, A. J. E. 1961. The frontal sinus of the *Gorilla. Proc. Zool. Soc. Lond.* **136**:359–374.

Cave, A. J. E., and Haines, R. W. 1940. The paranasal sinuses of the anthropoid apes. *J. Anat.* **74**:493–523.

Cherry, L. M., Case, S. M., and Wilson, A. C. 1978. Frog perspective on the morphological difference between humans and chimpanzees. *Science* **200**:209–211.

Chiarelli, B. 1972. The karyotypes of the gibbons. *Gibbon and Siamang* **1**:90–102.

Chiarelli, B. 1975. The study of primate chromosomes, in: *Primate Functional Morphology and Evolution* (R. Tuttle, ed.), pp. 103–127, Mouton, Paris.

Chiarelli, B., Koen, A. L., and Ardito, G. 1979. Primate chromosome atlas, in: *Comparative Karyology of Primates* (B. Chiarelli, A. L. Koen, and G. Ardito, eds.), pp. 211–281, Mouton, New York.

Corruccini, R. S. 1978. Comparative osteometrics of the hominoid wrist joint, with special reference to knuckle-walking. *J. Hum. Evol.* **7**:307–321.

Corruccini, R. S., Cronin, J. E., and Ciochon, R. L. 1979. Scaling analysis and congruence among anthropoid primate macromolecules. *Hum. Biol.* **51**:167–185.

Darwin, C. R. 1857. Letter to T. H. Huxley—Sept. 26, 1857, *More Letters of Charles Darwin* (F. Darwin, ed.), Volume 1, p. 104, Appleton, New York.

Darwin, C. R. 1859. *On the Origin of Species by Means of Natural Selection, or the Preservation of Favoured Races in the Struggle for Life.* John Murray, London.

Darwin, C. R. 1871. *The Descent of Man, and Selection in Relation to Sex.* John Murray, London.

Delson, E., Eldredge, N., and Tattersall, I. 1977. Reconstruction of hominid phylogeny: A testable framework based on cladistic analysis. *J. Hum. Evol.* **6**:263–278.

Dutrillaux, B. 1975. Sur la nature et l'origine des chromosomes humains. *Monogr. Ann. Genet.* **1975**:41–71.

Dutrillaux, B. 1979. Chromosomal evolution in primates: Tentative phylogeny from *Microcebus murinus* (Prosimian) to man. *Hum. Genet.* **48**:251–314.

Dutrillaux, B., Rethore, M.-O., and Lejeune, J. 1975. Comparison du caryotype de l'orangoutang (*Pongo pygmaeus*) a celui de l'homme, du chimpanze et du gorille. *Ann. Genet.* **18**:153–161.

Eckhardt, R. B. 1979. Chromosome evolution in the genus *Cercopithecus*, in: *Comparative Karyology of Primates* (B. Chiarelli, A. L. Koen, and G. Ardito, eds.), pp. 39–46, Mouton, New York.

Eckstein, P. 1958. Internal reproductive organs. *Primatologia* **3**:542–629.

Eldredge, N., and Tattersall, I. 1975. Evolutionary models, phylogenetic reconstruction, and another look at hominid phylogeny, in: *Approaches to Primate Paleobiology* (F. Szalay, ed.), pp. 218–242, Karger, Basel.

Farris, J. S. 1981. Distance data in phylogenetic analysis, in: *Advances in Cladistics* (V. A. Funk and D. R. Brooks, eds.), pp. 3–23, New York Botanical Garden, New York.

Ferris, S. D., Wilson, A. C., and Brown, W. M. 1981. Evolutionary tree for apes and humans based on cleavage maps of mitochondrial DNA. *Proc. Natl. Acad. Sci. USA* **78**:2432–2436.

Ferris, S. D., Brown, W. M., Davidson, W. S., and Wilson, A. C., 1982. Extensive polymorphism in the mitochrondrial DNA of apes. *Proc. Natl. Acad. Sci. USA* **78**:6319–6323.

Fink, W. L. 1982. The conceptual relationship between ontogeny and phylogeny. *Paleobiology* **8**:253–263.

Fitch, W. M. 1976. Molecular evolutionary clocks, in: *Molecular Evolution* (F. J. Ayala, ed.), pp. 160–178, Sinauer, Sunderland, Massachusetts.

Fitch, W. M. 1979. Cautionary remarks on using gene expression events in parsimony procedures. *Syst. Zool.* **28:**375–379.

Frelin, C., and Vuilleumier, F. 1979. Biochemical methods and reasoning in systematics. *Z. Zool. Syst. Evol.* **17:**1–10.

Goodman, M. 1975. Protein sequence and immunological specificity; Their role in phylogenetic studies of primates, in: *Phylogeny of the Primates* (W. P. Luckett and F. S. Szalay, eds.), pp. 219–248, Plenum, New York.

Goodman, M., Czelusniak, J., Moore, G. W., Romero-Herrera, A. E., and Matsuda, G. 1979. Fitting the gene lineage into its species lineage: A parsimony strategy illustrated by cladograms constructed from globin sequences. *Syst. Zool.* **28:**132–163.

Gould, S. J. 1977. *Ontogeny and Phylogeny,* Harvard University Press, Cambridge.

Harms, J. W. 1956. Schwangerschaft und geburt. *Primatologia* **1:**661–722.

Hennig, W. 1966. *Phylogenetic Systematics,* University of Illinois Press, Chicago.

Hill, W. C. O. 1951. The external genitalia of the female chimpanzee, with observations on the mammary apparatus. *Proc. Zool. Soc. Lond.* **121:**133–145.

Hill, W. C. O. 1958. External genitalia. *Primatologia* **3:**630–704.

Hill, W. C. O., and Matthews, L. H. 1949. The male external genitalia of the *Gorilla,* with remarks on the os penis of other Hominoidea. *Proc. Zool. Soc. Lond.* **119:**363–378.

Hill, W. C. O., and Matthews, L. H. 1950. Supplementary note on the male external genitalia of *Gorilla. Proc. Zool. Soc. Lond.* **120:**311–316.

Hull, D. L. 1979. The limits of cladism. *Syst. Zool.* **28:**416–440.

Imai, H. T., and Crozier, R. H. 1980. Quantitative analysis of directionality in mammalian karyotype evolution. *Am. Nat.* **116:**537–569.

Jenkins, F. A., Jr., and Fleagle, J. G. 1975. Knuckle-walking and the functional anatomy of the wrists in living apes, in: *Primate Functional Morphology and Evolution* (R. H. Tuttle, ed.), pp. 213–227, Mouton, Paris.

Jevons, W. S. 1874. *The Principles of Science: A Treatise on Logic and Scientific Method,* Macmillan, London.

Johanson, D. C., and White, T. D. 1979. A systematic assessment of early African hominids. *Science* **203:**321–330.

Keith, A. 1931. *New Discoveries Relating to the Antiquity of Man,* Norton, New York.

King, M. C., and Wilson, A. C. 1975. Evolution at two levels in humans and chimpanzees. *Science* **188:**107–116.

Kluge, A. G. 1982. The relevance of parsimony to phylogenetic inference, in: *The Estimation of Evolutionary History: Proceedings of a Workshop on the Theory and Application of Cladistic Methodology* (T. Duncan and T. F. Stuessy, eds.), pp. 1–23, Columbia University Press, New York.

Kluge, A. G., and Farris, J. S. 1969. Quantatative phyletics and the evolution of anurans. *Syst. Zool.* **18:**1–32.

Kohne, D. E. 1975. DNA evolution data and its relevance to mammalian phylogeny, in: *Phylogeny of the Primates* (W. P. Luckett and F. S. Szalay, eds.), pp. 249–261, Plenum, New York.

Leutenegger, W. 1970. Beziehungen zwischen der Neugeboreneugrosse und dem Sexualdimorphismus am Becken bei simischen Primaten. *Folia Primatol.* **12:**224–235.

Leutenegger, W. 1972. Newborn size and pelvic dimensions in *Australopithecus. Nature (Lond.)* **240:**568–569.

Leutenegger, W. 1973. Maternal–fetal weight relationships in primates. *Folia Primatol.* **20:**280–293.

Luetenegger, W. 1974. Functional aspects of pelvic morphology in simian primates. *J. Hum. Evol.* **3:**207–222.

Lovejoy, C. O. 1981. The origin of man. *Science* **211:**341–350.

Matthews, L. H. 1946. Notes on the genital anatomy and physiology of the gibbon (*Hylobates*). *Proc. Zool. Soc. Lond.* **116:**339–364.

Mayr, E. 1969. *Principles of Systematic Zoology,* McGraw-Hill, New York.

Mayr, E. 1974. Cladistic analysis or cladistic classification. *Z. Zool. Syst. Evol.* **12:**94–128.

Mickevich, M. F. 1978. Taxonomic congruence. *Syst. Zool.* **27:**143–158.

Mill, J. S. 1862. *A System of Logic, Ratiocinactive and Inductive, Being a Connected View of the Principles of Evidence, and the Methods of Scientific Investigation*, Parker, Son, and Bourn, London.

Miller, D. A. 1977. Evolution of primate chromosomes. *Science* **198**:1116–1124.

Mross, G. A., Doolittle, R. F., and Roberts, B. F. 1970. Gibbon fibrinopeptides: Identification of a glycine–serine allelism at position B-3. *Science* **170**:468–470.

Nelson, G. 1978. Ontogeny, phylogeny, paleontology, and the biogenetic law. *Syst. Zool.* **27**:324–345.

Pilbeam, D., and Gould, S. J. 1974. Size and scaling in human evolution. *Science* **186**:892–901.

Popper, K. R. 1965. *Conjectures and Refutations: The Growth of Scientific Knowledge*, Harper and Row, New York.

Popper, K. R. 1968. *The Logic of Scientific Discovery*, Harper and Row, New York.

Raven, H. C. 1950. Regional anatomy of the *Gorilla*, in: *The Anatomy of the Gorilla* (W. K. Gregory, ed.), pp. 15–195, Columbia University Press, New York.

Read, D. W. 1975. Primate phylogeny, neutral mutations, and "molecular clocks." *Syst. Zool.* **24**:209–221.

Romero-Herrera, A. E., Lehmann, H., Castillo, O., Joysey, K. A., and Friday, A. E. 1976. Myoglobin of the orangutan as a phylogenetic enigma. *Nature (Lond.)* **261**:162–164.

Romero-Herrera, A. E., Lehmann, H., Joysey, K. A., and Friday, A. E. 1978. On the evolution of myoglobin. *Phil. Trans. R. Soc. Lond.* **283**:61–163.

Schultz, A. H. 1927. Studies on the growth of *Gorilla* and of other higher Primates with special reference to a foetus of *Gorilla* preserved in the Carnegie Museum. *Mem. Carnegie Mus.* **11**:1–86.

Schultz, A. H. 1930. The skeleton of the trunk and limbs of higher primates. *Hum. Biol.* **2**:303–438.

Schultz, A. H. 1936. Characters common to higher primates and characters specific for man. *Q. Rev. Biol.* **11**:259–283, 425–455.

Schultz, A. H. 1944. Age changes and variability in gibbons; A morphological study of a population sample of a man-like ape. *Am. J. Phys. Anthropol.* **2**:1–129.

Schultz, A. H. 1949. Sex differences in the pelves of primates. *Am. J. Phys. Anthropol.* **7**:401–423.

Schultz, A. H. 1956. Postembryonic age changes. *Primatologia* **1**:887–964.

Schultz, A. H. 1960. Age changes and variability in the skulls and teeth of the Central America monkeys *Alouatta*, *Cebus* and *Ateles*. *Proc. Zool. Soc. Lond.* **133**:337–390.

Schultz, A. H. 1961. Vertebral column and thorax. *Primatologia* **4**:1–66.

Schultz, A. H. 1962. Metric age changes and sex differences in primate skulls. *Z. Morphol. Anthropol.* **52**:239–255.

Schultz, A. H. 1968. The recent primates, in: *Perspectives on Human Evolution* (S. L. Washburn and P. C. Jay, eds.), pp. 122–195, Holt, Rinehart and Winston, New York.

Schultz, A. H. 1969. The skeleton of the chimpanzee. *Chimpanzee* **1**:50–130.

Schultz, A. H. 1973. The skeleton of the Hylobatidae and other observations on their morphology. *Gibbon and Siamang* **2**:1–54.

Schwartz, J. H., Tattersall, I., and Eldredge, N. 1978. Phylogeny and classification of the primates revisited. *Yearb. Phys. Anthropol.* **21**:95–133.

Seuanez, H. N. 1979. *The Phylogeny of Human Chromosomes*, Springer-Verlag, New York.

Simons, E. L. 1976. The nature of the transition in the dental mechanism from pongids to hominids. *J. Hum. Evol.* **5**:511–528.

Simpson, G. G. 1945. The principles of classification and a classification of mammals. *Bull. Am. Mus. Nat. Hist.* **85**:1–350.

Simpson, G. G. 1962. *Principles of Animal Taxonomy*, Columbia University Press, New York.

Simpson, G. G. 1963. The meaning of taxonomic statements, in: *Classification and Human Evolution* (S. L. Washburn, ed.), pp. 1–31, Aldine, Chicago.

Simpson, G. G. 1975. Recent advances in methods of phylogenetic inference, in: *Phylogeny of the Primates* (W. P. Luckett and F. S. Szalay, eds.), pp. 3–19, Plenum, New York.

Slijper, E. J. 1946. Comparative biologic–anatomical investigations on the vertebral column and spinal musculature of mammals. *Verh. K. Ned. Akad. Wet. Natuurk.* **42**:1–128.

Sneath, P. H. A., and Sokal, R. R. 1973. *Numerical Taxonomy*, Freeman, San Francisco.

Straus, W. L., Jr. 1942a. The homologies of forearm flexors: Urodeles, lizards, mammals. *Am. J. Anat.* **70:**281–316.

Straus, W. L., Jr. 1942b. Rudimentary digits in primates. *Q. Rev. Biol.* **17:**228–243.

Straus, W. L., Jr. 1949. The riddle of man's ancestry. *Q. Rev. Biol.* **24:**200–223.

Szalay, F. S., and Delson, E. 1979. *Evolutionary History of the Primates,* Academic, New York.

Tattersall, I., and Eldredge, N. 1977. Fact, theory, and fantasy in human paleontology. *Am. Sci.* **65:**204–211.

Turleau, C., and de Grouchy, J. 1979. Comparison of banding patterns in man and primates and their evolutionary signficance, in: *Comparative Karyology of Primates* (B. Chiarelli, A. L. Koen, and G. Ardito, eds.), pp. 47–57, Mouton, New York.

Tuttle, R. H. 1967. Knuckle-walking and the evolution of hominoid hands. *Am. J. Phys. Anthropol.* **26:**171–206.

Tuttle, R. H. 1969a. Knuckle-walking and the problems of human origins. *Science* **166:**953–961.

Tuttle, R. 1969b. Quantitative and functional studies on the hands of the Anthropoidea. I. The Hominoidea. *J. Morphol.* **128:**309–364.

Tuttle, R. H. 1970. Postural, propulsive, and prehensile capabilities in the cheiridia of chimpanzees and other great apes. *Chimpanzee* **2:**167–253.

Tuttle, R. 1972. Relative mass of cheiridial muscles in catarrhine primates, in: *The Functional and Evolutionary Biology of Primates* (R. Tuttle, ed.), pp. 262–291, Aldine-Atherton, Chicago.

Tuttle, R. 1975a. Parallelism, brachiation, and hominoid phylogeny, in: *Phylogeny of the Primates* (W. P. Luckett and F. S. Szalay), pp. 447–480, Plenum, New York.

Tuttle, R. 1975b. *Socioecology and Psychology of Primates,* Mouton, Paris.

Tuttle, R. H. 1977. Naturalistic positional behavior of apes and models of hominid evolution, 1929–1976, in: *Progress in Ape Research* (G. H. Bourne, ed.), pp. 277–296, Academic, New York.

Tuttle, R. H., and Basmajian, J. V. 1974. Electromyography of forearm musculature in *Gorilla* and problems related to knuckle-walking, in: *Primate Locomotion* (F. A. Jenkins Jr., ed.), pp. 293–347, Academic, New York.

Uzzell, T., and Pilbeam, D. 1971. Phyletic divergence dates of hominoid primates: A comparison of fossil and molecular data. *Evolution* **25:**615–635.

Vogel, F., Kopun, M., and Rathenberg, R. 1976. Mutation and molecular evolution, in: *Molecular Anthropology* (M. Goodman and R. E. Tashian, eds.), pp. 13–33, Plenum, New York.

Wagner, W. H., Jr. 1980. Origin and philosophy of the groundplan-divergence method of cladistics. *Syst. Bot.* **5:**173–193.

Wallace, A. R. 1902. *The Malay Archipelago: The Land of the Orang-utan, and the bird of paradise. A narrative of Travel, with Studies of Man and Nature,* Macmillan, New York.

Washburn, S. L. 1973. Primate studies and human evolution, in: *Nonhuman Primates and Medical Research* (G. H. Bourne, ed.), pp. 467–485, Academic, New York.

Whewell, W. 1847. *The Philosophy of the Inductive Sciences, Founded upon Their History,* Parker, London.

White, T. D. 1980. Evolutionary implications of Pliocene hominid footprints. *Science* **208:**175–176.

Wiley, E. O. 1978. The evolutionary species concept reconsidered. *Syst. Zool.* **27:**17–26.

Wiley, E. O. 1980. Phylogenetic systematics and vicariance biogeography. *Syst. Bot.* **5:**194–220.

Wiley, E. O. 1981. *Phylogenetics. The Theory and Practice of Phylogenetic Systematics,* Wiley, New York.

Wilson, A. C., and Prager, E. M. 1974. Antigenic comparison of lysozymes, in: *Lysozyme* (E. F. Osserman, R. E. Canfield, and S. Beychock, eds.), pp. 127–141, Academic, New York.

Wislocki, G. B. 1932. On the female reproductive tract of the gorilla, with a comparison of that of other primates. *Contrib. Embryol. Carnegie Inst. Wash.* **23:**165–204.

Wooding, G. L., and Doolittle, R. F. 1972. Primate fibrinopeptides: Evolutionary significance. *J. Hum. Evol.* **1:**553–563.

Wyles, J. S., and Gorman, G. C. 1980. The albumin immunological and Nei distance correlation: A calibration for the saurian genus *Anolis* (Iguanidae). *Copeia* **1:**66–71.

Yunis, J. J., Sawyer, J. R., and Dunham, K. 1980. The striking resemblance of high resolution G-banded chromosomes of man and chimpanzee. *Science* **208:**1145–1148.

Evidence from Craniodental Morphology

III

New Interpretations of the Phyletic Position of Oligocene Hominoids

7

J. G. FLEAGLE AND R. F. KAY

Introduction

In attempts to trace the ancestry of living hominoids, the Miocene apes from Europe, Asia, and sub-Saharan Africa have long played a central role. First described over 125 years ago, the genera *Dryopithecus* and *Pliopithecus* were recognized from their initial descriptions as an ancestral pongid and hylobatid, respectively, a situation which Darwin (1871) acknowledged and which, until most recently, was universally accepted.

Since the first decade of this century, fossil evidence for the hominoid antecedents of these taxonomically diverse and geographically widespread Miocene apes has come almost exclusively from the (?Eocene–) Oligocene sediments of the Fayum province of Egypt. When Schlosser (1910) described the first fossil anthropoids from Egypt, he recognized this relationship by naming one genus *Propliopithecus*. Since Schlosser's original description, numerous additional species have been described and dozens of new fossil specimens have been recovered from these same deposits (e.g., Simons, 1967, 1972; Kay *et al.*, 1981). Over the past 70 years, many divergent hypotheses concerning relationships with later apes and humans have been proposed for these early anthropoids. Various Fayum species have been identified at one time or another as the earliest hominid, pongid, hylobatid, Old World

J. G. FLEAGLE • Department of Anatomical Sciences, Health Sciences Center, State University of New York, Stony Brook, New York 11794. R. F. KAY • Department of Anatomy, Duke University Medical Center, Durham, North Carolina 27710.

monkey, the common ancestor of all hominoids, or the common ancestor of all catarrhines including Old World monkeys, apes, and humans. Many of the relationships were advanced when the Oligocene species were known only from a single type specimen. Now most are known from dozens of jaws, as well as cranial and skeletal material. With the greatly increased fossil record of these early anthropoids, paleoprimatologists can better evaluate the phyletic relationship of the Fayum apes with respect to later ape and human evolution.

Systematics and Interrelationships of the Fayum Apes

The first recognized ape jaws from the Fayum Province were collected by Richard Markgraf, apparently in the spring of 1907 (Gingerich, 1978). Schlosser (1910) named two taxa, *Propliopithecus haeckeli* and *Moeripithecus markgrafi*, each on the basis of a single specimen (see Figs. 1, 2). Simons (1967) suggested that the two are congeneric and dropped *Moeripithecus*, a move which has been followed by all later workers. The distinctiveness of the two species is generally accepted. Unfortunately, no information survives as to the stratigraphic position of either type specimen.

The type specimen of *Oligopithecus savagei* (Simons, 1962) was recovered from the Lower Fossil Wood Zone in 1961. This animal is still known only from the type specimen preserving $\bar{C}$–M_2 and a recently discovered unworn lower molar. It resembles extant catarrhines in having two rather than three

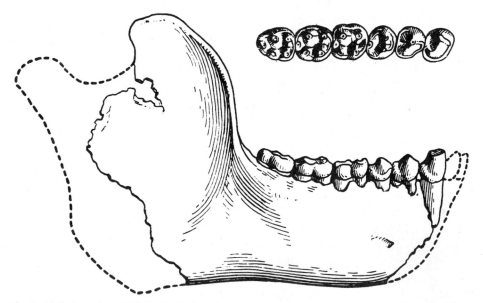

Fig. 1. Right lateral and left occlusal view of the type (SNM 12638) jaw and teeth of *Propliopithecus haeckeli* Schlosser (X c.2.0). [From Gregory (1916).]

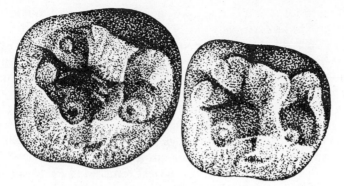

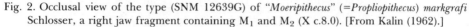

Fig. 2. Occlusal view of the type (SNM 12639G) of *"Moeripithecus"* (=*Propliopithecus*) *markgrafi* Schlosser, a right jaw fragment containing M_1 and M_2 (X c.8.0). [From Kalin (1962).]

premolars, the anterior of which bears a well-developed wear facet for contact with an interlocking upper canine. On this basis, and because it has a relatively deep jaw, Simons (1972) has taken the view the *Oligopithecus* should be tentatively linked to the Fayum apes. He notes that the Upper Fossil Wood Zone parapithecid *Parapithecus grangeri,* which he feels stands closer to Old World monkeys, still retained three premolars, implying that the two-premolared condition of the Cercopithecidae was achieved separately at a later time. However, *Oligopithecus* lacks the peculiar derived molar occlusal pattern found in other Fayum apes, parapithecids, and all later catarrhines (Kay, 1977a). Thus, it is equally plausible that *Oligopithecus* is much more primitive and is an offshoot of the ancestral Old World anthropoid group that gave rise to parapithecids and apes, and that reduction from three to two premolars by loss of P_2^2 occurred in parallel in this genus (Szalay and Delson, 1979). An alternative view is that *Oligopithecus* is an adapid (Gingerich, 1977b).

Simons (1965) described two species of ape from the Upper Fossil Wood Zone, *Aegyptopithecus zeuxis* and *Aeolopithecus chirobates.* The former, based on four specimens, including an edentulus jaw, which had been collected for the American Museum of Natural History in 1906, is morphologically distinct from *Propliopithecus* in several taxonomically important ways (Kay *et al.*, 1981). In all specimens of *Aegyptopithecus zeuxis* M_2 is much larger than M_1, whereas in all *Propliopithecus* species M_1 is about the same size as M_2. The lower molar occlusal surfaces of *Aegyptopithecus* are compressed buccal-lingually, and the crown margins flare outward, giving the molars a bulbous appearance. In *Propliopithecus* species typically the lower molar crowns are not compressed and the crowns are more steep-sided. *Aegyptopithecus* also differs from *Propliopithecus* species in that the premolar cingulum tends to be less developed. *Aegyptopithecus* lacks a lingual cingulum on P_4, whereas a P_4 lingual cingulum is invariably preserved in *Propliopithecus*. The first upper molar of *Propliopithecus* is consistently broader than that of *Aegyptopithecus*. Finally, lower incisors in *Propliopithecus* are shorter and of smaller caliber in relation to M_1 size than are those of *Aegyptopithecus*.

On the basis of a single lower jaw, a second species of ape was described by Simons (1965) as *"Aeolopithecus" chirobates* (Fig. 3). Because of chemical erosion [probably due to digestive enzymes in the gut of a crocodile (Fisher, 1981)], few features are preserved on the teeth of the type, but these seemed at the time to indicate an ape quite distinct from other known species. The type of *"Aeolopithecus"* is much smaller than contemporaneous *Aegyptopithecus zeuxis*. The large canines and premolar heteromorphy of the type initially set it apart from *Propliopithecus haeckeli*. Recent collections of material from Quarries I and M show, however, that the type of *"Aeolopithecus" chirobates* is a male; the females of this species have small canines (Fleagle *et al.*, 1980; Kay *et al.*, 1981; see Fig. 4). This species has been best referred to *Propliopithecus* (Szalay and Delson, 1979; Kay *et al.*, 1981). It resembles *P. haeckeli* and is distinctively different from *Aegyptopithecus* in having similar sized first and second molars with marginally placed lower molar cusps. In addition, lower molar crowns are steep-sided, and premolar cingula are relatively well developed.

Gingerich (1978) tentatively suggested that, should more material of *P. chirobates* be found, it might prove to be conspecific with *P. haeckeli*. Currently available samples of the two taxa do not show sufficient overlap in morphology to confirm this. In premolar morphology, *P. chirobates* is more derived than *P. haeckeli*, which it closely resembles in overall size. The shape of the P_3 is oval and compressed in both presumed males and presumed females of *P. chirobates* (Fig. 5), leading to premolar heteromorphy similar to that of *Aegyptopithecus;* in *P. haeckeli*, the sole known specimen of which is probably a female, P_3 is practically round in occlusal outline. Our sample of *P. chirobates*

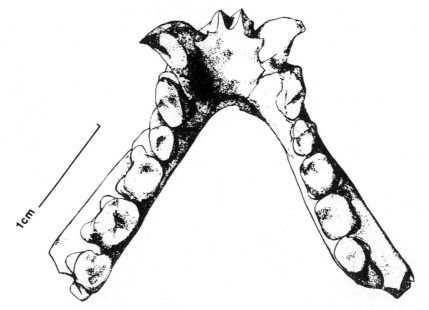

Fig. 3. Type mandible of *Aeolopithecus* (= *Propliopithecus*) *chirobates*, CGM 26923 (X c.2.0). [From Simons (1965).]

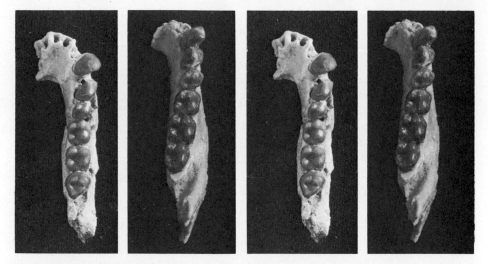

Fig. 4. Stereophotographs of mandibles of *Propliopithecus chirobates* from the Upper Fossil Wood Zone, Jebel el Qatrani Formation. Left, DPC 1029, a male. Right, DPC 1103, a female. (X c.1.5). [From Kay *et al.* (1981).]

mandibles shows that their mandibular corpora average considerably shallower than in *P. haeckeli*. Even so, jaw depth varies considerably in sexually dimorphic species; in body-size dimorphic species, jaw depth under M_{1-2} is usually greater in males than in females (Fleagle *et al.*, 1980). While the molars in the type specimen of *P. haeckeli* (a female as judged from canine size) are smaller than any in the sample of *P. chirobates*, its mandible is 30% deeper under M_{1-2} than are females of *P. chirobates*. Thus, in various slight differences, *P. haeckeli* is distinct from all known *P. chirobates* in having a less sectorial P_3 and a very deep mandible. We prefer to maintain two species at present.

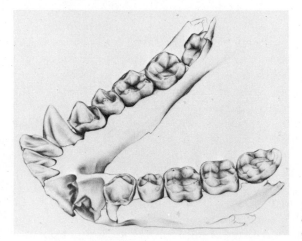

Fig. 5. *Propliopithecus chirobates*, oblique lateral view of DPC 1069 (X c.l.8).

We also cannot demonstrate that "*Moeripithecus*" *markgrafi* and *Propliopithecus haeckeli* are conspecific. In molar size and M_1:M_2 proportions, "*M.*" *markgrafi* resembles *P. haeckeli*, but the molar crowns of the former are very compressed (see Fig. 2), a feature of similarity shared with *Aegyptopithecus*. The larger samples of *P. chirobates* show that this degree of morphological distinctiveness would not be anticipated in a single species and we conclude that "*M.*" *markgrafi* is a valid species. The question of whether "*Moeripithecus*" is a valid genus is more of a problem, given the lack of information about most of its anatomy. In a few known features, this species bridges the gap between *Aegyptopithecus* and *Propliopithecus*, with a few morphological similarities to the former and the size proportions of the latter. If more material of "*Moeripithecus*" were to become available, it might be necessary to conclude that it was congeneric with *Aegyptopithecus* on grounds of molar morphology. Such a step is not warranted at present, given the very incomplete nature of the type, and we take the more conservative course of including "*M.*" *markgrafi* within *Propliopithecus*.

The presently available ape material from the Fayum is thus divisible into four species, *Propliopithecus haeckeli*, *P. chirobates*, *P. markgrafi*, and *Aegyptopithicus zeuxis*. *Aegyptopithicus zeuxis* and *P. chirobates* are known to be contemporaries of one another; the geological provenence of the other taxa is unknown. How are the four interrelated? Table 1 summarizes the distribution of 13 character states discussed more extensively elsewhere (Kay *et al.*, 1981). In terms of overall similarity, *Aegyptopithecus zeuxis* and *Propliopithecus haeckeli* are at opposite poles: the two differ in every characteristic of the dentition that shows variation among the four species. *Propliopithecus chirobates* falls in an intermediate position, but shares more features with *Propliopithecus haeckeli* than with *Aegyptopithecus zeuxis*. The dentition of *P. markgrafi* is so poorly known that it can be compared with the other three taxa in three characters only.

To assess correctly the phylogenetic relationships among these forms, it would be necessary to know the morphology of their last common ancestor. This would permit determination of which features of similarity result from the shared retention of primitive characteristics from the common ancestor and which result from the acquisition of new characters at various stages subsequent to the initial speciation event. Only the latter similarities might indicate relative relatedness within the group. If one assumes that other Oligocene anthropoids (parapithecids and *Oligopithecus*) demonstrate the primitive condition for the apes, all of the character states shown by *P. haeckeli* are probably primitive for the four species, with the possible exception of the degree of cingulum development (characters 6 and 10 in Table 1). If this interpretation is correct, by far the most parsimonious cladogram of the three best known forms shows an initial split of *P. haeckeli* from the common ancestor of *P. chirobates* and *A. zeuxis*, followed by the separation of the latter two. If *P. haeckeli* proves to be primitive in all respects, and if it turns out to be geologically older than the other species, it would be an ideal ancestor for the others. Given the poor state of knowledge about *P. markgrafi*, it is not possible

Table 1. Morphological Comparisons of Fayum Apes

	Propliopithecus haeckeli	Propliopithecus chirobates	Propliopithecus markgrafi	Aegyptopithecus zeuxis
1.	—	M^1 smaller than M^2	—	M^1 much smaller than M
2.	—	M^x conules absent	—	M^2 conules variably present
3.	P_3 rounded: no mesial–buccal flare	P_3 oval with well-developed mesial–buccal flare	—	P_3 oval with well-developed mesial–buccal flare
4.	—	C–P_m sexual size differences	—	C–P_m sexual size differences
5.	P_4 trigonid broad buccal–lingually	P_4 trigonid broad buccal–lingually	—	P_4 trigonid compressed buccal–lingually
6.	P_4 lingual cingulum present	P_4 lingual cingulum variable	—	P_4 lingual cingulum is absent
7.	$M_1 = M_2$ in mesial–distal length	M_1 averages 5% shorter (M–D) than M_2	M_1 5% shorter (M–D) than M_2	M_1 averages 17% shorter (M–D) than M_2
8.	Mandibular corpus very deep for female	Mandibular corpora broad, shallow under M_{1-2} when sexual dimorphism is controlled	—	Mandibular corpora broad, shallow under M_{1-2} when sexual dimorphism is controlled
9.	Molar crowns broad, steep-sided	Molar crown broad, steep-sided	Molar crown compressed, sides flare	Molar crown compressed, sides flare
10.	Molar buccal cingulum well developed	Molar cingulum moderate	Lower molar cingulum weak	Lower molar cingulum weak
11.	M_3 distal fovea absent	M_3 distal fovea small to lacking	—	M_3 with large distal fovea
12.	—	I_1 broad, low-crowned	—	I_1 narrow, high-crowned

to decide which are its closest relations. If a well-developed molar cingulum were primitive for the last common ancestor of Fayum apes, a possible link with *A. zeuxis* is implied; otherwise, *P. markgrafi* is almost the same as *P. chirobates.*

Evidence for Relationships with Later Anthropoids

In his initial description of *Propliopithecus haeckeli*, Schlosser (1910, 1911) suggested that the species could represent an ancestral gibbon through the lineage *Propliopithecus → Pliopithecus → Hylobates*. He argued that *P. haeckeli* was gibbon-like largely through its similarities to *Pliopithecus*, especially in having molars with five rounded cusps located around the periphery of the teeth. On the other hand, he suggested that the small canine, deep jaw, and relatively broad anterior premolar were hominid-like features. At the same time, Schlosser was very impressed with many "primitive" similarities which he found between *Propliopithecus* and cebids—especially in size, in the shape of the jaw, and the morphology of the premolars. However, because of the catarrhine dental formula, he never made much of these resemblances.

Largely because of a lack of any new fossil material attributed to this genus for over 50 years, Schlosser's two main interpretations have persisted down to the present and are even found in some recent textbooks. Over the years, various authorities have adopted one, the other, or both of these positions. For example, Keith (1923) argued that since *Propliopithecus* was gibbon-like, this was possible corroborating evidence that the earliest apes were brachiators. In contrast, some authors offered the human-like nature of the canines and premolars in *Propliopithecus* as well as its generally primitive molars and mandible as evidence for a very early separation of humans from the ape lineage (Osborn, 1927; Simons, 1965; Pilbeam, 1967; Kinzey, 1971; Kurten, 1972; Wood-Jones, 1929). This phylogenetic scheme had the decided advantage of not requiring a reduction of canine size in hominid evolution. Others, such as Gregory (1922), Kalin(1961), Simons (1967, 1972), Delson and Andrews (1975), Andrews (1978), and Szalay and Delson (1979) argued that this taxon was probably best interpreted as a generalized ancestral hominoid which was broadly ancestral to most later apes, including the dryopithecines.

Schlosser's other ape species, *"Moeripithecus" markgrafi,* has had an equally diverse history of opinions as to its affinities. Schlosser originally suggested that his species might be an early ancestor of Old World monkeys. Abel (1931) felt that *"Moeripithecus" markgrafi* was related to *Apidium* and probably also to later cercopithecoids. Kalin (1961, 1962) considered the possibility of cercopithecoid affinities for this species, but decided otherwise. Most recent workers (Simons, 1967, 1972; Delson, 1975; Szalay and Delson, 1979) have tended to regard this taxon as just another species of *Propliopithecus* with no specific affinities to particular later taxa.

In 1965, Simons described *Aeolopithecus* as a possible early hylobatid because of the large canines, reduced third molar, supposed posterior shallowing of the mandibular corpus, and a high, deep genial fossa. He tentatively supported this relationship in a series of later publications (Simons, 1967, 1972; Simons *et al.*, 1978), as did other workers (e.g., Howell, 1967), with the proviso that additional material of this species might necessitate a reevaluation of this relationship. As described above, additional specimens have indeed increased the knowledge of this species considerably and now cast considerable doubt as to its hylobatid affinities. Such hylobatid features as the posteriorly shallowing mandible and the marked reduction of M_3 in the type specimen are artifacts of the poorly preserved type specimen, and the large canines are found only in the males (Fleagle *et al.*, 1980; Kay *et al.*, 1981). Thus, unlike extant gibbons, *P. chirobates* had sexually dimorphic canines. Now that the molar morphology is known, this species is clearly very close to and possibly even conspecific with *Propliopithecus haeckeli*, and its phyletic affinities are best considered in conjunction with that taxon.

When Simons (1965) first described *Aegyptopithecus zeuxis*, he argued that it was more closely allied with *Proconsul* species from the Miocene of East Africa and with the later pongids and hominids than with either the small apes from Africa, *Pliopithecus*, or hylobatids, primarily on the basis of its relatively large M_3. He later expanded the argument that *Aegyptopithecus zeuxis* was clearly more closely related to *Dryopithecus* sp. from the Miocene and to the Great Apes than to any other group of anthropoids (Simons, 1974a). Howell (1967) carried this possibility to the extreme by suggesting synonymy between *Aegyptopithecus* and *Dryopithecus*. In direct contrast, Andrews (1970) argued that *Aegyptopithecus zeuxis* was very similar to *Limnopithecus legetet* and *Dendropithecus macinnesi*. Indeed he even provisionally assigned specimens from Rusinga Island (early Miocene, Kenya) to a new species of *Aegyptopithecus* before reassigning them to *Dendropithecus macinnesi* (Andrews, 1978).

More recently Delson and Andrews (1975), Delson (1977), and Szalay and Delson (1979) have argued that, on the basis of shared primitive features, all of the Fayum hominoids are most closely allied to *Pliopithecus* from Europe and *Dendropithecus macinnesi* from East Africa and that they should all be placed in a primitive family of hominoids, the Pliopithecidae, which is the sister taxa of all other hominoids. Likewise, Simons and Fleagle (1973) and Fleagle and Simons (1978a) noted the numerous similarities (in primitive features) between *Aegyptopithecus* and *Pliopithecus* (but not with *Dendropithecus*).

Although many authors have noted the many ways in which *Aegyptopithecus* (and *Propliopithecus*) are far more primitive than any Miocene to Recent catarrhines (Gregory, 1922; Simons and Fleagle, 1973; Fleagle and Simons, 1978a, 1979; Fleagle, 1980; Delson and Andrews, 1975; Delson, 1975, 1977; Szalay and Delson, 1979), only a few (Remane, 1965; Groves, 1972; Fleagle and Simons, 1978b; Fleagle, 1980; Kay *et al.*, 1981; Andrews, 1981; Fleagle and Rosenberger, 1983) have actually argued that the Oligocene taxa are probably ancestral to both cercopithecoid monkeys and later hominoids.

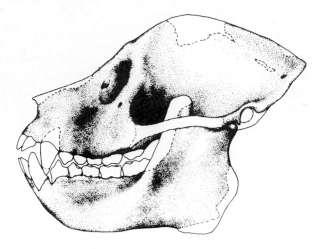

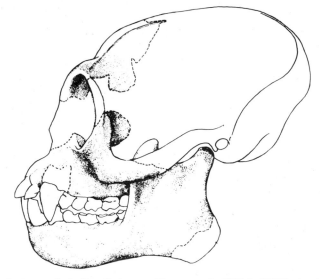

Fig. 6. Lateral views of skulls of *Aegyptopithecus zeuxis* (CGM 40237) (top) and *Pliopithecus vindobonensis* (bottom), based on material described by Zapfe (1960). (X c.0.75).

In light of current knowledge about the very primitive dental, cranial, and skeletal morphology of the Oligocene hominoids *Propliopithecus* and *Aegyptopithecus*, there is no reason to believe that any single group of extant hominoids (either hylobatids, hominids, or pongids) can be traced back to an Oligocene divergence (Fleagle *et al.*, 1980). However, the issue of the relationship between these Oligocene hominoids and the numerous Miocene apes from Africa and Europe is very complex and controversial and can only be evaluated by a careful review of the available cranial, dental, and skeletal material of Oligocene and early Miocene species.

Cranial Morphology

The best material available for cranial comparisons with *Aegyptopithecus zeuxis* are the partial skulls (Figs. 6–8) of *Proconsul africanus* (East African early Miocene) (Le Gros Clark and Leakey, 1951; Napier and Davis, 1959; Davis and Napier, 1963) and *Pliopithecus vindobonensis* (Czechoslovakian middle Miocene) (Zapfe, 1960). The skull of *Aegyptopithecus* (CGM 40237; Fig. 6, 8) stands out for its primitiveness with respect to all later Anthropoidea. The large facial component of the skull, combined with the small brain case, the somewhat dorsally facing orbits, the sharp angulation of the nuchal region with the cranial vault, and the extreme postorbital constriction are reminiscent of the skulls of Eocene adapids, such as *Leptadapis*, and of some of omomyids, such as *Rooneyia*. On the other hand, *Aegyptopithecus* shows an anthropoid grade of organization in having a fused mandibular symphysis with both superior and inferior transverse tori (unlike Eocene–Oligocene adapids, which lack the superior transverse torus, even when the symphysis is fused) (see Figs. 9, 10). *Aegyptopithecus* also exhibits an advanced stage in postorbital closure, very close to the condition seen in extant gibbons, whereas

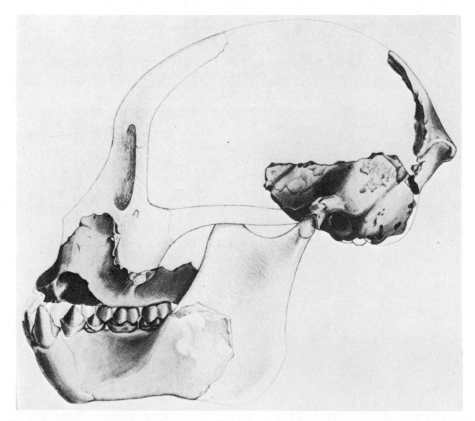

Fig. 7. Lateral view of reconstructed skull of *Proconsul africanus*. [From Davis and Napier (1963).] Note that illustration has been reversed to facilitate comparison with Fig. 6 (X c.0.75).

no known Eocene primate has advanced beyond the "prosimian" grade of postorbital configuration.

A detailed comparison of the cranial morphology of *Aegyptopithecus zeuxis*, *Pliopithecus vindobonensis*, and *Proconsul africanus* (Figures 6–8, Table 2) shows that most of the similarities between *Aegyptopithecus* and *Pliopithecus* are primitive for the apes in general, whereas *Pliopithecus* shows a number of derived features with *Proconsul*. This suggests that the Miocene taxa may have shared

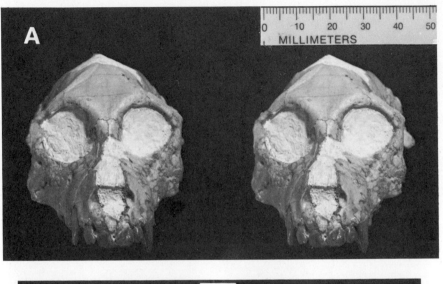

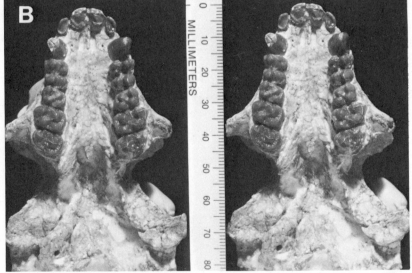

Fig. 8. Stereophotographs of *Aegyptopithecus zeuxis* skull (CGM 40237) (various magnifications, scales provided). A, frontal view; B, palatal view; C, posterior view; D, top view; E, right lateral view of ear region.

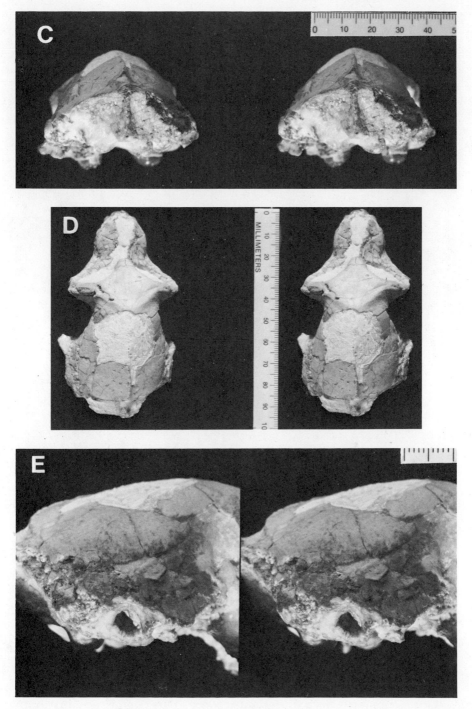

Fig. 8. (*continued*)

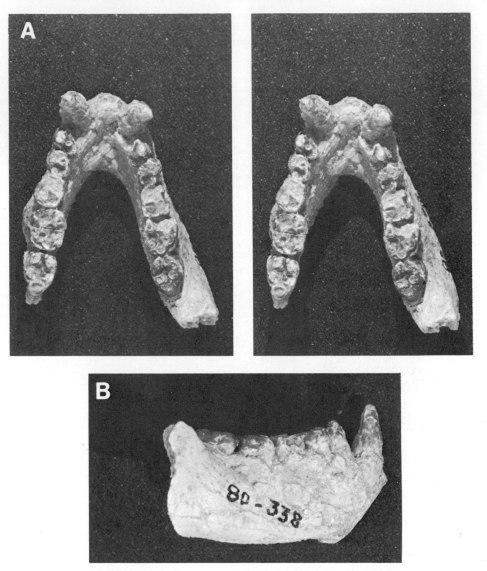

Fig. 9. A, stereophotograph of occlusal view of mandible of *Aegyptopithecus zeuxis*, female, newly recovered (1980) from Quarry M, Upper Fossil Wood Zone. B, Lateral view of the same (X c.1.5).

a more recent common ancestor with one another in their derivation from a form like *Aegyptopithecus*.

Details of the facial skeleton of *Aegyptopithecus* (Fig 8A) are considerably at variance with those of *Pliopithecus* and *Proconsul*. The premaxilla of *Aegyptopithecus* is uniquely large for an anthropoid. This bone is very broad anteroposteriorly, particularly at its dorsal end, much in the same way as is seen among Eocene–Oligocene adapids (Simons, 1972; Gingerich, 1973) and

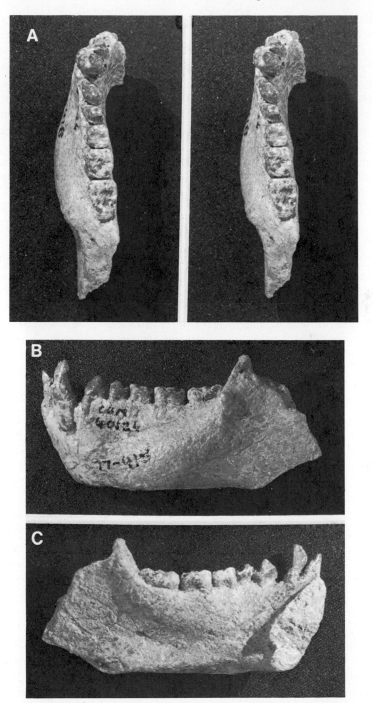

Fig. 10. A, stereophotograph of occlusal view of mandible of *Aegyptopithecus zeuxis*, male, (CGM 40137) from Quarry M, Upper Fossil Wood Zone. B, Lateral view, and C, medial view of the same (X c.1.2).

omomyids, such as *Rooneyia* (Table 2). (Here and throughout, the terms dorsal, ventral, anterior, and posterior are used in relation to the Frankfurt Horizontal.) Anterodorsally it forms a sharp-edged margin for the nasal aperture. The infraorbital surface of the maxilla is smooth, lacking a pronounced canine fossa or a lip-like projection of the inferior orbital margin. The distance between the inferior orbital margin and the ventral root of the zygomatic arch increases slightly proceeding laterally. The inferior orbital margin is slightly posterior to the root of the zygomatic arch. The root of the zygomatic arch is above M^2. The nasal bones are broken away, leaving only the interorbital portion. However, it is clear from the configuration of the maxillae and the nasal aperture that the nasal bones were long and formed a very acute angle with the Frankfurt Horizontal. The breadth between the orbits, taken as a proportion of the P^3-M^3 length, is 0.30, indicating a relatively large interorbital breadth, comparable to that of gibbons, most extant colobines, and Eocene primates as well.

By comparison with *Aegyptopithecus zeuxis*, the premaxillae of both *Pliopithecus* and *Proconsul* are very narrow anteroposteriorly at their dorsal ends. The overall configuration of the premaxilla in the latter genera is similar to that found among extant catarrhines. The infraorbital surface of the maxilla of *Proconsul* resembles that of *Aegyptopithecus* in lacking a pronounced canine fossa or a well developed lip-like forward projection of the inferior orbital margin. Additionally, the inferior orbital margin is posterior to the plane of the root of the zygomatic arch. It differs from *Aegyptopithecus* in that the distance between the inferior orbital margin and the root of the zygomatic arch shallows laterally. In *Proconsul* the root of the zygomatic arch is above M^{1-2}.

The conformation of the maxilla in the infraorbital region in *Pliopithecus* is distinct from both *Aegyptopithecus* and *Proconsul* in that the inferior orbital margin forms a lip-like projection anteriorly. As a result, the inferior orbital margin of *Pliopithecus* is situated well anterior to the root of the zygomatic arch. *Pliopithecus* is also different from either of the other forms in that the vertical distance between the inferior orbital margin and the root of the zygomatic arch becomes much greater laterally. These features are not seen in *Proconsul* but occur in recent hylobatids, colobines, and some cercopithecines.

In *Pliopithecus*, the root of the zygomatic arch is above M^{1-2}. The confiruration and size of the surrounding bones give some indication of the size and orientation of the nasal bones of *Pliopithecus* even though they are missing. The greater part of the nasals are preserved in *Proconsul*. In both of the Miocene forms, the nasals must have been considerably smaller than in *Aegyptopithecus* and well within the range of modern hominoids. The long axis of the nasals of *Pliopithecus* must have been oriented quite obliquely with respect to the Frankfurt Horizontal. Those of *Proconsul* were apparently oriented somewhat more acutely than in *Pliopithecus* but were not as horizontal as in *Aegyptopithecus*. In the latter respect, the reconstructions of Le Gros Clark and Leakey (1951) and Davis and Napier (1963) are quite different.

Table 2. Comparison of the Cranial Anatomy of *Aegyptopithecus zeuxis*, *Pliopithecus vindobonensis*, *Proconsul africanus*, and a Sample of Eocene Omomyids and Adapids, Including *Notharctus*, *Rooneyia*, *Adapis*, and *Tetonius*

Eocene/Oligocene "prosimians"	Aegyptopithecus	Pliopithecus	Proconsul
Premaxilla large	Premaxilla large	Premaxilla small	Premaxilla small
Infraorbital surface of maxilla smooth	Infraorbital surface of maxilla smooth	Infraorbital surface projecting, lip-like	Infraorbital region smooth
Variable	Infraorbital maxilla depth increases slightly laterally	Infraorbital maxilla depth increases greatly laterally	Infraorbital maxilla depth decreases laterally
Above either M_1 or M_2	Root of zygomatic arch above M_2	Root of zygomatic arch above M_2	Root of zygomatic arch above M_2
Nasals long, projecting	Nasals probably long, projecting	Nasals probably short, projecting	Nasals short, not projecting
Extremely large interorbital breadth	Large interorbital breadth	Extremely large interorbital breadth	Large interorbital breadth
Palate straight-sided or widens posteriorly	Palate widens posteriorly	—	Palate widens posteriorly
Orbit size variable	Small orbits	Orbits large	Orbits intermediate in size between *Pliopithecus* and *Aegyptopithecus*
Orbits face slightly laterally and dorsally	Orbits face slightly laterally and dorsally	Orbits face forward	—
Small brain compared to modern Anthropoidea	Small brain compared to modern Anthropoidea	Possibly modern sized brain	Brain in modern anthropoid size range
?Large visual cortex	Large visual cortex	—	Large visual cortex
Large olfactory bulbs	Small olfactory bulbs	—	Small olfactory bulbs
Marked postorbital constriction	Marked postorbital constriction	Little postorbital constriction	Little postorbital constriction
Variable	No external auditory tube	External auditory tube incomplete ventrally	Well-developed tubular ectotympanic

The interorbital breadth, taken as a ratio of P^3–M^3 length, is 0.38 in *Proconsul*, approximately the same as in *Aegyptopithecus*. The same index is 0.60 in *Pliopithecus*, illustrating the extremely great distance between the orbits in that taxon.

The palate of *Aegyptopithecus* (Fig. 8B) widens posteriorly such that the distance between the lingual margins of the M^3's is 1.33 times greater than the distance between the lingual margins of the P^3's. Data presented by Andrews (1978) show that the palates of Miocene African dryopithecines except for *Proconsul major* widen posteriorly but not to the degree seen in *Aegyptopithecus*. The palate of *Pliopithecus* is not well enough preserved for one to be certain of its shape. [Andrews (1978) provided measurements of the palate of *Pliopithecus*, apparently based on the reconstruction of Zapfe (1960). Unfortunately, the reconstructed palate in question is mostly made of plaster.]

The orbits of *Aegyptopithecus* (Figs. 8A, 8D) are quite small by comparison with extant cercopithecoids, using skull length or P^3–M^3 length as a standard for comparison (Kay and Simons, 1980). The ratio of orbit diameter to P^3–M^3 length is 0.63. The orbits of *Pliopithecus* are relatively much larger, comparable to those of many extant catarrhines. The ratio of orbit size to P^3–M^3 length is 0.89 for *Pliopithecus*. The orbits of *Proconsul* are proportionately smaller than those of *Pliopithecus* but larger than those of *Aegyptopithecus*. The orbit to P^3–M^3 ratio is 0.74. The relatively small orbit in *Proconsul africanus* compared with *Pliopithecus* is not out of line for the larger size of the former and the negatively allometric trend in orbit size among Old World anthropoids (Kay and Cartmill, 1977). However, the orbits of *Aegyptopithecus* are surprisingly small for an Old World anthropoid.

In *Aegyptopithecus*, the lateral orbital margin is more posterior than the medial orbital margin, and the superior orbital margin is more posterior than the inferior orbital margin. Thus, the orbits are directed somewhat laterally and dorsally. In *Proconsul*, the orbits are badly distorted so that their orientation is indeterminant. The orbits of *Pliopithecus* face forward in a manner similar to modern catarrhines.

The brain size of *Aegyptopithecus*, as estimated by Radinsky (1977) and allometrically corrected for body size, shows that this species had a smaller brain than extant anthropoids and within the range of extant strepsirhines (Kay and Simons, 1980). *Aegyptopithecus* was also more primitive than extant anthropoids in having comparatively smaller frontal lobes (Radinsky, 1973). On the other hand, the brain of *Aegyptopithecus* had a larger visual cortex and smaller olfactory lobes than is the case for most "prosimians," and within the modern anthropoid range.

The brain of *Proconsul africanus* (see Falk, this volume, Chapter 9) was apparently within the modern anthropoid range (Gingerich, 1977a). All aspects of the morphology of the endocast resemble extant Old World anthropoids (Radinsky, 1977). No information is available about the configuration of the available parts of the inside of the brain case of *Pliopithecus*. A hint that *Pliopithecus* may have an essentially modern-sized Old World anthropoid brain is indicated by the lack of any great postorbital constriction. The small-brained *Aegyptopithecus* skull has a marked postorbital constriction and relatively flat-lying frontal bone. The large-brained *Proconsul* more closely resembles *Pliopithecus* in this feature.

The skull of *Aegyptopithecus* has well-developed ridges marking the origin of the temporalis muscles (Figs. 8C–8E). These converge posteriorly and medially to form a pronounced sagittal crest. The anteriormost extent of the sagittal crest is found in the same coronal plane as the maximum postorbital constriction. The crest continues posteriorly to inion, where it is continuous with flange-like nuchal crests. The pronounced development of this cresting is unique for an anthropoid of comparable body size, especially considering that this individual was probably a young adult (the canines are not quite fully erupted). This development is more likely to reflect the small size of the brain case than to be an indication of comparatively better developed jaw or neck musculature than seen in extant catarrhines. This conclusion is drawn because the overall size of the nuchal plate, the temporalis fossa, and the surface area of the origin of the temporalis muscles are similar to those of comparably sized Old World monkeys [5000–5500 g (Gingerich, 1977a; Kay and Simons, 1980)]. The foramen magnum of *Aegyptopithecus* is located well under the cranial vault in a fashion similar to the position in extant cercopithecines.

The temporal lines of *Pliopithecus* converge to form a very low midsagittal ridge well back on the skull. These crests again diverge before reaching the back of the skull. Laterally they are confluent with the nuchal crest. The only specimen of *P. africanus* with the relevant areas preserved is a juvenile female, so the condition of temporal lines on the adult cranium is indeterminate.

The ectotympanic bone of *Aegyptopithecus* is fused solidly to the petrosal bone, but is unique among adult Old World Anthropoidea in not being prolonged outward as a tube (Simons, 1972; see Fig. 8E). By contrast, *Proconsul* had a well-developed tubular ectotympanic (Davis and Napier, 1963). *Pliopithecus* is intermediate in this respect (Szalay and Delson, 1979); short ectotympanic flanges are prolonged laterally from anterior and posterior parts of the petrosal bulla, but the ventral portion is incompletely ossified (Zapfe, 1960, Figs. 28–30).

The mandibular tooth rows of *Aegyptopithecus* (Fig. 9A) diverge posteriorly, giving the jaws a V-shaped appearance in occlusal view. The ratio of breadth between P_4's to breadth between M_2's is 0.74 in one female *Aegyptopithecus*. A similar ratio is present in the type of *Propliopithecus chirobates*. The tooth rows of East African Miocene apes diverge posteriorly also, as do those of *Pliopithecus* (Andrews, 1978; Zapfe, 1960). In the Fayum apes, the ratio of mandibular depth under M_{1-2} to M_2 length is extremely high for males by comparison with males of Miocene ape species [comparative data from Andrews (1978) and Zapfe (1960)]. Two male *Aegyptopithecus zeuxis* have a ratio of depth to M_2 length of 3.03; two male *Propliopithecus chirobates* have a ratio of 3.57. Three male *Pliopithecus vindobonensis* average 2.17 and a *P. antiquus* male has 2.38 (Zapfe, 1960). Some male *Proconsul nyanzae* and *P. major* approach the relative jaw depth of male Fayum apes, but the males of some other early Miocene species have shallower jaws.

The mandibular symphysis of both *Aegyptopithecus zeuxis* (Fig. 10C) and *Propliopithecus chirobates* is buttressed by well-developed superior and inferior transverse tori. This condition is found also in *Dendropithecus macinnesi* (Miocene), but East African *Proconsul* and *Limnopithecus* usually lack an inferior transverse torus (Andrews, 1978). *Pliopithecus* tends to have a well-developed inferior transverse torus; in *Pliopithecus vindobonensis* the superior torus is quite variable (Zapfe, 1960).

Male *Aegyptopithecus zeuxis* mandibles show an extremely heavy scar for the origin of the anterior belly of the digastric muscle. This scar extends on the ventral surface of the mandibular corpora from near the midline of the symphysis back to the middle of M_2. Such a broad origin for digastric muscle is typical of extant catarrhines (Hill, 1966; Raven, 1950).

Dental Morphology

All of the Fayum and Miocene ape taxa are known from dental remains, allowing a broader range of comparison than was possible with cranial material. The most complete reviews of the dental anatomy of Miocene apes on which the present comparison is based are the studies of Andrews (1978), Fleagle and Simons (1978a), and Zapfe (1960).

Andrews places considerable emphasis on central incisor shape and height as diagnostic criteria for recognition of East African Miocene taxa. In particular, *Dendropithecus macinnesi* is said to have strongly mesial-distally compressed, high-crowned incisors by comparison with the contemporaneous taxa. Using Andrews' (1978) criteria, the lower central incisor of *Propliopithecus chirobates* (Fig. 11) is comparatively broad and low-crowned, resembling most East African Miocene taxa; those of *Aegyptopithecus zeuxis* are very narrow and high-crowned, resembling *Dendropithecus*. However, the sample sizes on which Andrews' conclusions are based are very small and the variability in the measurements is very high. The largest sample of lower central incisors for a Miocene ape consist of six specimens of *Proconsul* (*Rangwapithecus*) *gordoni; Dendropithecus macinnesi* is represented by two specimens only, and unaccountably one specimen is eliminated from all of Andrews' sample statistics. The shape indices within each species vary widely. One sample of three I_1's of *Limnopithecus legetet* has buccal-lingual/mesial-distal ratios ranging from 104 to 148. This exceeds the range of species means for all six Miocene taxa analyzed by Andrews. In fact, the only statistically significant difference in lower incisor shape which emerges from Andrews' study is that *Proconsul* (*Rangwapithecus*) apparently had comparatively narrow incisors. [This would be consistent with a possibly more folivorous diet for this species (Kay, 1977b).] There are no associated upper central incisors for any of the Fayum apes.

The upper canines of the Fayum species (one each of *Aegyptopithecus zeuxis* and *Propliopithecus chirobates*) are nearly round in cross section at the base. The lower canines of these species are more bilaterally compressed. In

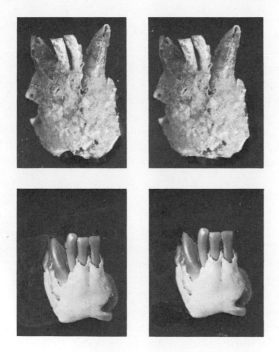

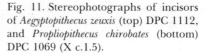

Fig. 11. Stereophotographs of incisors of *Aegyptopithecus zeuxis* (top) DPC 1112, and *Propliopithecus chirobates* (bottom) DPC 1069 (X c.1.5).

Pliopithecus and East African Miocene hominoids, upper canine breadth is 70–80% of canine length (in the occlusal plane) and the lower canines are also bilaterally compressed. Delson and Andrews (1975) argued that "bilaterally compressed" upper and lower canines were primitive for Old World Anthropoidea, and that dryopithecines are characterized by rounded upper canines, but the evidence of the Fayum apes suggests that the upper canines of ancestral apes may have been more circular in cross section.

The upper premolars and M^{1-2} of *Propliopithecus* and *Aegyptopithecus* are quite broad buccal-lingually, M^1 of the former more so than that of the latter. The ratio of P^4 breadth to length exceeds 1.50 for these species; the ratio of M^2 breadth to length is greater than 1.20. In this respect, the Fayum species resemble species of *Micropithecus, Pliopithecus, Limnopithecus, Dendropithecus*, and *Proconsul. Proconsul (Rangwapithecus) gordoni* and *P. (R.) vancouveringi* differ from all contemporary and earlier apes in having much narrower upper premolars and nearly square molars [data from Andrews (1978), Fleagle and Simons (1978a), and Zapfe (1960)]. In both of these respects, *P. (Rangwapithecus)* very closely resembles later *Sivapithecus* and extant Great Apes.

The lower fourth premolar of the Fayum apes tends to be broad: the ratio of buccal-lingual breadth to mesial-distal length is greater than 1.00. In this respect there is a close similarity to *Proconsul*. On the other hand, P_4 tends to be longer (breadth/length < 100) in *Pliopithecus, Limnopithecus, Dendropithecus*, and *Proconsul (Rangwapithecus)*.

As noted above, the M_2 is about 5% longer (mesial-distally) than M_1 in *Propliopithecus*. In *Aegyptopithecus*, M_2 is much longer than M_1. *Micropithecus,*

Limnopithecus, Dendropithecus, Pliopithecus, Proconsul, and *P.* (*Rangwapithecus*) all have much longer M_2's than M_1's, resembling *Aegyptopithecus.*

The upper premolar and molar lingual cingula of *Aegyptopithecus zeuxis* and *Propliopithecus chirobates* are very strongly developed. In African Miocene apes, the upper lingual cingulum is best developed on the upper molars and is progressively weaker toward the anterior upper premolar. The cingulum development reaches forward to include P^3 in *Proconsul* (*Rangwapithecus*), which resembles the Fayum apes in this respect. It is absent from P^3 but strongly developed on P^4–M^2 in *Dendropithecus, Pliopithecus* and *Limnopithecus.* It is poorly developed on P^4 and well developed on M^{1-2} in *Proconsul.*

Postcranial Morphology

In the past 5 years, increasing numbers of skeletal elements attributable to the Oligocene hominoid have been recovered and described [Fleagle *et al.* (1975); Conroy (1976*a,b*); Fleagle and Simons (1978*b*, 1982); reviewed in Fleagle (1980)], including most of the humerus and ulna, a hallucial metatarsal, and a phalanx of *Aegyptopithecus zeuxis.* In addition to their value for reconstructing the locomotor habits of these early anthropoids, these skeletal elements provide important phyletic information of the relationship between Oligocene and later hominoids.

The humeri attributed to *A. zeuxis* (Fig. 12) are more primitive than the humeri of any extant higher primate, either platyrrhine or catarrhine (Fleagle and Simons, 1978*b*, 1982). The morphology of tuberosities, the bicipital groove, the deltoid plane, the brachialis flange, and the shaft as a whole are more "prosimian"-like than anthropoid-like. The distal aspect of the humerus in both *A. zeuxis* and *P. chirobates,* however, is more comparable to that of extant platyrrhines in the shape of articular surfaces and the possession of an entepicondylar foramen. This latter feature, primitive for mammals as a group, is found in *Pliopithecus* and about half the genera of extinct platyrrhines (often variably), but is absent in all Miocene to Recent catarrhines, including all extant cercopithecines, and hominoids except as a rare variation in *Homo.*

The relatively wide trochlea, large medial epicondyle, and thin supracapitular region in the humeri of Fayum species are apparently primitive characteristics which link them with platyrrhines and later hominoids and contrast with the derived conditions seen in cercopithecoids. The Fayum taxa and *Pliopithecus* lack the spool-shaped trochlea seen in extant hominoids, and to a lesser degree in *Proconsul africanus.*

The ulna attributed to *A. zeuxis* (Fleagle *et al.,* 1975; Preuschoft, 1975; Conroy, 1976*a;* Schön-Ybarra and Conroy, 1978) is similar to the same bone in the living platyrrhines such as *Alouatta* and unlike the ulnae of either extant cercopithecines or hominoids. Morphologically it is similar to (but more robust than) the ulna of *Pliopithecus* and neither shares any clearly derived features with any group of later catarrhines.

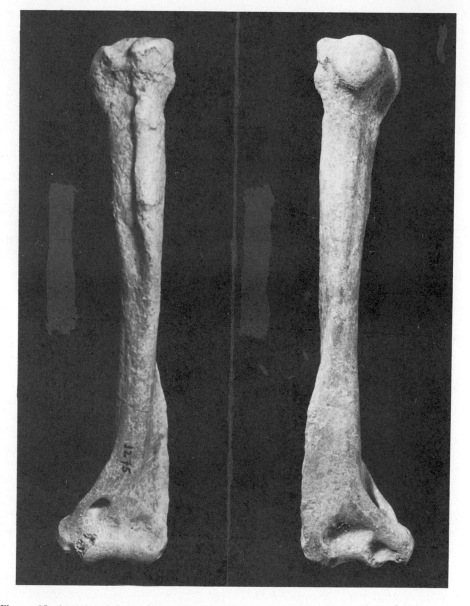

Figure 12. Anterior (left) and posterior (right) views of the humerus of *Aegyptopithecus zeuxis* (DPC 1275). Approximate size.

The first hallucial metatarsal attributed to *A. zeuxis* clearly demonstrates the presence of an opposable grasping hallux in this species (Preuschoft, 1975) and also shows a facet for a prehallux bone. A prehallux and the associated facet is not present in hominids, living pongids, or cercopithecoids, but is found in many platyrrhines, in *Hylobates*, in *Pliopithecus*, and in *P. africanus* (Conroy, 1976*b*). Its presence in *A. zeuxis* is certainly a primitive anthropoid feature. The phalanges are similar to those of other arboreal anthropoids with no remarkable features.

Summary of Morphological Features

In all known characteristics, with the possible exception of very small orbit size, the cranial anatomy of *Aegyptopithecus zeuxis* appears to present what might be expected of a primitive anthropoid. In fact, other than the diagnostic anthropoid features of postorbital closure and a fused mandibular symphysis, the cranial anatomy is very similar to that of adapids and advanced omomyids like *Rooneyia*. *Aegyptopithecus zeuxis* does not share any derived features of the cranium, suggesting a closer relationship to either *Pliopithecus*, *Proconsul africanus*, or any other single Miocene taxon. On the other hand, *Pliopithecus* and *Proconsul* share such advanced anthropoid cranial characteristics as dorsally reduced premaxillae, short, nonprojecting nasals, reduced postorbital constriction, relatively shallow mandibular rami, and at least partial development of a tubular ectotympanic bone. In this last feature, *Proconsul* is apparently more advanced than is *Pliopithecus*.

Dentally, both *Aegyptopithecus* and *Propliopithecus* are very similar to African Miocene hominoids and *Pliopithecus*. Lower incisor morphology is comparable. The lower canines of the Fayum species are very similar to those of Miocene apes, but the upper canines are more rounded (less bilaterally compressed) than in any of the Miocene species considered here. The upper premolars and molars of the Fayum apes are very broad buccal-lingually, which is true for the Miocene apes except *Proconsul* (*Rangwapithecus*). M_2 is longer than M_1 in *Aegyptopithecus* and the Miocene apes. *Propliopithecus* in this regard differs strikingly in having M_2 about the same length as M_1. All taxa have a well-developed lingual cingulum on the upper molars, and in many species there is a comparable development of the cingulum on the upper premolar also.

In postcranial anatomy (see Fleagle, this volume, Chapter 11) the Fayum hominoids are completely primitive with respect to both living hominoids and cercopithecoids (or even platyrrhines in many ways). There are no derived features linking the Egyptian genera with any one Miocene taxon or with either extant catarrhine groups. Indeed, the Fayum genera show a primitive condition that could be ancestral to both groups. *Pliopithecus* is more "advanced" than *Aegyptopithecus* in the more gracile nature of the humerus and ulna, but retains the primitive entepicondylar foramen and generally primitive limb anatomy. All of the dryopithecines from the early Miocene of East Africa, including *Dendropithecus*, lack this foramen, as do all extant catarrhines. *Proconsul africanus* appears to be even more like a modern hominoid in that it has a distal humeral articulation like a living ape and also apparently a very chimpanzee-like foot and shoulder (Walker and Pickford, this volume, Chapter 12). However, both *Pliopithecus* and *Proconsul africanus* lack the characteristic hominoid wrist morphology with a very reduced ulnar styloid.

In summary, the preponderance of anatomical evidence indicates that

Pliopithecus and the early Miocene hominoids from Kenya and Uganda, were more like modern hominoids and more derived than *Aegyptopithecus*. The Fayum apes are both sufficiently primitive and geologically old enough to have been the ancestors of all later hominoids.

So primitive is the anatomy of *Aegyptopithecus* and *Propliopithecus* that these Oligocene anthropoids might easily be ancestral to all later catarrhines (Fleagle and Simons, 1978*b;* Fleagle, 1980; Kay *et al.*, 1981). Miocene to Recent catarrhines (both hominoids and cercopithecoids) share a number of probably derived features not present in *Aegyptopithecus*, including a tubular ectotympanic, shortened nasals, reduced premaxillae, and enlarged brains, as well as loss of the entepicondylar foramen and numerous modifications in the shape of the humerus. Having ape-like molars would not necessarily rule out the Fayum "apes" from Old World monkey ancestry. Certainly hominoid molars more closely resemble the primitive primate molar pattern than do those of Old World monkeys.

Other considerations have nevertheless led many authors to exclude the Fayum apes from the ancestry of Old World monkeys and to regard them as specially related to Miocene–Recent hominoids alone. There are possibly derived features of the molar dentitions which link *Aegyptopithecus* and *Propliopithecus* with Miocene hominoids and would tend to rule them out of the ancestry of Old World monkeys. The upper molar lingual cingulae are particularly well developed in Fayum apes and the M_{1-2} hypoconulids are very large and centrally or buccally placed. These features are the reverse of what would be expected in a cercopithecoid ancestor. A second, but independent, argument against deriving Old World monkeys from the Fayum "apes" is that the Parapithecidae, particularly *Parapithecus grangeri* from the Upper Fossil Wood Zone, show a considerable degree of derived dental similarity to the Cercopithecidae. Recently recovered dental specimens show that, although *Parapithecus grangeri* still had three premolars, the first of these was considerably reduced, and the lower molars are extremely high-crowned and incipiently bilophodont (Simons, 1970, 1972, 1974*b*), Unlike *Apidium*, upper molars of *Parapithecus* are quadrate, with four main cusps. This, and a combination of other derived dental features reviewed by Kay (1977*a*), support Simons' (1970) case that the parapithecids are ancestral to Old World monkeys. However, like the Fayum apes, parapithecids retain a number of very primitive cranial and postcranial features which suggest that they antecede the divergence of hominoids and cercopithecoids (Cartmill *et al.*, 1981; Delson and Andrews, 1975; Szalay and Delson, 1979; Fleagle, 1980; Fleagle and Rosenberger, 1983).

Phylogeny and Systematics of Fayum Anthropoids

Any phylogeny linking Fayum primates with later catarrhines must involve considerable parallelism. The "prosimian"-like features in the humerus and skull of *Aegyptopithecus* must have been lost independently in platyrrhines and catarrhines. Furthermore, if *Aegyptopithecus* is a phyletic ape [that is, specially related to apes alone; as advocated by Simons (1967, 1972), Simons and Pilbeam (1972), and Szalay and Delson (1979)], then the few osteological features proposed which unite extant catarrhines (e.g., tubular ectotympanic; loss of entepicondular foramen) are necessarily also parallelisms. Alternatively, if *Aegyptopithecus* and the other Fayum anthropoids phyletically precede the hominoid–cercopithecoid divergence, then the dental similarities between *Parapithecus* and cercopithecoids (Simons, 1974*a;* Kay, 1977*a*) are parallelisms. On the present evidence, we support the latter phylogeny because of the lack of convincing derived features linking *Aegyptopithecus* and *Propliopithecus* uniquely with later hominoids and their lack of many apparently derived features seen in all modern catarrhines.

If *Aegyptopithecus* and *Propliopithecus* are really primitive catarrhines belonging to a group ancestral to both cercopithecoids and living hominoids, rather than uniquely related to later apes, to what higher taxon of higher primates should they be assigned? They clearly should not be regarded as cercopithecoids; all fossil and living Old World monkeys are a very coherent group showing numerous uniquely derived features in their dentition and skeleton (Szalay and Delson, 1979). *Aegyptopithecus* and *Propliopithecus* show none of these derived features and do not even show any inkling of special cercopithecoid affinities.

There are two remaining taxonomic alternatives. They can be left in the Hominoidea with the clear implication that Hominoidea is a wastebasket taxon, or they can be placed in a new taxon for noncercopithecoid, nonhominoid catarrhines. [Szalay and Delson (1979) have used a comparable taxon for the parapithecids.] We prefer to leave the Oligocene apes in the Hominoidea for several practical reasons.

First, and most important, is the very obvious fact that *Aegyptopithecus* and *Propliopithecus* are merely the oldest of numerous "dental apes" from the Oligocene and Miocene of Africa and Eurasia whose phyletic position *vis à vis* later apes is debatable. *Pliopithecus* shares only a few more derived features with living catarrhines than do the Oligocene taxa, and even the dryopithecines share very few clearly derived dental, cranial, or skeletal features with living pongids, hylobatids, or hominids. Expanding the Hominoidea to include *Aegyptopithecus* and *Propliopithecus* eliminates the additional difficulties of deciding what to do with the other dental apes who grade ever so tenuously into a more modern-looking hominoid appearance.

Second, a broad definition of the Hominoidea recognizes what anatomists have long realized (e.g., Wood-Jones, 1929; Le Gros Clark, 1934), that in many aspects of dental and skeletal anatomy, cercopithecoids are equally or more derived from the ancestral anthropoid condition than are apes. Furthermore, Old World monkeys appear in the fossil record "full-blown" in recognizably modern form, but it is impossible to identify a likely ancestor for cercopithecoids among the "dental apes" of either the Oligocene or Miocene (Delson, 1975).

Within the Hominoidea, the most primitive taxa can reasonably be grouped in a single family, the Pliopithecidae (Remane, 1965; Szalay and Delson, 1979). In contrast with the latter authors, we would recognize two subfamilies: Propliopithecinae for the more primitive Fayum genera such as *Aegyptopithecus* and *Propliopithecus*, and Pliopithecinae for the somewhat more advanced *Pliopithecus*. *Dendropithecus* is more closely allied with the other dryopithecines of East Africa, and should be placed with them (Fleagle and Simons, 1978a).

Summary

Four species of "dental apes" currently are recognized from the Fayum (Oligocene, Egypt). *Aegyptopithecus zeuxis* (Simons, 1965) and *Propliopithecus chirobates* (Simons, 1965) from the Upper Fossil Wood Zone are both known from numerous jaws and various skeletal elements. *Propliopithecus haeckeli* (Schlosser, 1910) and *Propliopithecus markgrafi* (Schlosser, 1910) are known only from their type specimens collected early this century from an unknown stratigraphic level and possible isolated teeth from Quarry G (Simons, 1972). Dentally, these taxa are similar to later hominoids, particularly *Pliopithecus* from the Miocene of Europe and *Proconsul, Limnopithecus,* and *Dendropithecus* from East Africa. In skeletal anatomy they share many primitive features with *Pliopithecus.* In contrast, in the cranial and skeletal anatomy they are more primitive than any later catarrhine and share no obviously derived features with any group of living catarrhines, cercopithecoid monkeys, hylobatids, pongids, or hominids. They are thus suitable phyletic ancestors for all later catarrhines. They should be retained within the superfamily Hominoidea as a separate subfamily Propliopithecinae of the primitive hominoid family Pliopithecidae. This arrangement implies that Hominoidea are the primitive catarrhines and that cercopithecoids are derived from a hominoid ancestor.

ACKNOWLEDGMENTS

The research reported here was supported by NSF grant BNS77-25921 and BNS79-24149 to John G. Fleagle, BNS77-08939 to Richard F. Kay, and BNS77-20104, BNS80-16206, and Smithsonian Foreign Currency Grant FC0869600 and FC80974 to Elwyn L. Simons. The field work was carried out with the cooperation of the Egyptian Geological Survey and Mining Authority, and especially Ragi Eissa, Darwish el Far, Abed El Ghani Ibrahim, Bahay Issawi, Baher el Khashab, Galal Ali Moustafa, M. F. el Ramly, and Rushdi Said. We also thank W. L. Jungers for comments on the manuscript. We thank the curators of Mammals at the American Museum of Natural History, the Museum of Comparative Zoology, and the Smithsonian Institution for allowing us to study the primate specimens in their care. We are especially grateful to Prof. E. L. Simons for allowing us to participate in the Fayum expeditions under his direction.

References

Abel, O. 1931. *Die Stellung des Menschen im Rahmen der Wirbeltiere*, G. Fisher, Jena.

Andrews, P. 1970. Two new fossil primates from the Lower Miocene of Kenya. *Nature (Lond.)* **288:**537–540.

Andrews, P. 1978. A revision of the Miocene Hominoidea of East Africa. *Bull. Br. Mus. (Nat. Hist.) Geol.* **30**(2):85–244.

Andrews, P. 1981. Species diversity and diet in monkeys and apes during the Miocene, in: *Aspects of Human Evolution* (C. Stringer, ed.), pp. 25–62, Taylor and Francis, London.

Cartmill, M., MacPhee, R. D. E., and Simons, E. L. 1981. Anatomy of the temporal bone in early anthropoids, with remarks on the problem of anthropoid affinities. *Am. J. Phys. Anthropol.* **56:**3–21.

Conroy, G. C. 1976*a*. Primate postcranial remains from the Oligocene of Egypt. *Contrib. Primatol.* **8:**1–134.

Conroy, G. C. 1976*b*. Hallucial tarsometatarsal joint in an Oligocene anthropoid, *Aegyptopithecus zeuxis. Nature (Lond.)* **263:**684–686.

Darwin, C. 1871. *The Descent of Man, and Selection in Relation to Sex*, John Murray, London.

Davis, P. R., and Napier, J. (1963). A reconstruction of the skull of *Proconsul africanus* (R.S. 51). *Folia Primatol.* **1:**20–28.

Delson, E. 1975. Toward the origin of the Old World monkeys, in: Evolution des Vertébrés— Problèmes Actuels de Paléontologie. *Colloq. Int. Cent. Nat. Rech. Sci.* **218:**839–850.

Delson, E. 1977. Catarrhine phylogeny and classification: Principles, methods and comments. *J. Hum. Evol.* **6:**433–459.

Delson, E., and Andrews, P. 1975. Evolution and interrelationships of the catarrhine primates, in: *Phylogeny of the Primates* (W. P. Luckett and F. S. Szalay, eds.), pp. 405–446, Plenum, New York.

Fisher, D. C. 1981. Crocodilian scatology, microvertebrate concentrations and enamel-less teeth. *Paleobiology* **7**(2):262–275.

Fleagle, J. G. 1980. Locomotor behavior of the earliest anthropoids: A review of the current evidence. *Z. Morphol. Anthropol.* **71:**149–156.

Fleagle, J. G., and Rosenberger, A. L. 1983. Cranial morphology of earliest anthropoids, in: *Morphologie Evolutive, Morphogenese du Crane et Anthropogenese* (M. Sakka, ed.) CNRS, Paris.

Fleagle, J. G., and Simons, E. L. 1978*a*. *Micropithecus clarki*, a small ape from the Miocene of Uganda. *Am. J. Phys. Anthropol.* **49:**427–440.

Fleagle, J. G., and Simons, E. L. 1978*b*. Humeral morphology of the·earliest apes. *Nature (Lond.)* **273:**705–707.

Fleagle, J. G., and Simons, E. L. 1979. Anatomy of the bony pelvis of parapithecid primates. *Folia Primatol.* **31:**176–186.

Fleagle, J. G., and Simons, E. L. 1982. The humerus of *Aegyptopithecus zeuxis*, a primitive anthropoid. *Am. J. Phys. Anthropol.* **59**(2):175–194.

Fleagle, J. G., Simons, E. L., and Conroy, G. C. 1975. Ape limb bone from Oligocene of Egypt. *Science* **189:**135–137.

Fleagle, J. G., Kay, R. F., and Simons, E. L. 1980. Sexual dimorphism in early anthropoids. *Nature (Lond.)* **287:**328–330.

Gingerich, P. D. 1973. Anatomy of the temporal bone in the Oligocene anthropoid *Apidium* and the origin of the Anthropoidea. *Folia Primatol.* **19:**329–337.

Gingerich, P. D. 1977*a*. Correlation of tooth size and body size in living hominoid primates, with a note on relative brain size in *Aegyptopithecus* and *Proconsul. Am. J. Phys. Anthropol.* **47:**395–398.

Gingerich, P. D. 1977*b*. Radiation of Eocene Adapidae in Europe. *Géobios, Mém. Spéc.* **1**:165–182.

Gingerich, P. D. 1978. The Stuttgart collection of Oligocene primates from the Fayum Province of Egypt. *Paleontol. Z.* **52**:82–92.

Gregory, W. K., 1916. Studies on the evolution of the primates. *Bull. Amer. Mus. Nat. Hist.* **35**:239–355.

Gregory, W. K. 1922. *The Origin and Evolution of the Human Dentition*, Williams and Wilkins, Baltimore. 548 pp.

Groves, P. 1972. Systematics and phylogeny of gibbons. *Gibbon and Siamang* **1**:1–80.

Hill, W. C. O. 1966. *Primates. Comparative Anatomy and Taxonomy.* Volume VI, *Catarrhini, Cercopithecoidea*, Edinburgh University Press, Edinburgh. 757 pp.

Howell, F. C. 1967. Recent advances in human evolutionary studies. *Q. Rev. Biol.* **42**:471–513.

Kalin, J. 1961. Sur les Primates de l'Oligocène inférieur d'Egypte. *Ann. Paleontol.* **47**:1–48.

Kalin, J. 1962. Über *Moeripithecus markgrafi* Schlosser und die phyletischen Vorstufen der Bilophodontie der Cercopithecoidea. *Bibl. Primatol.* **1**:32–42.

Kay, R. F. 1977*a*. The evolution of molar occlusion in the Cercopithecidae and early catarrhines. *Am. J. Phys. Anthropol.* **46**:327–352.

Kay, R. F. 1977*b*. Diets of early Miocene African hominoids. *Nature (Lond.)* **268**:628–630.

Kay, R. F., and Cartmill, M. 1977. Cranial morphology and adaptations of *Palaechthon nacimienti* and other Paromomyidae (Plesiadapoidea, ?Primates), with a description of a new genus and species. *J. Hum. Evol.* **6**:19–53.

Kay, R. F., and Simons, E. L. 1980. The ecology of Oligocene African Anthropoidea. *Int. J. Primatol.* **1**:22–37.

Kay, R. F., Fleagle, J. G., and Simons, E. L. 1981. A revision of the Oligocene apes from the Fayum Province, Egypt. *Am. J. Phys. Anthropol.* **55**:293–322.

Keith, A. 1923. Man's posture: Its evolution and disorders. *Br. Med. J.* **1**:451–454.

Kinzey, W. 1971. Evolution of the human canine tooth. *Am. Anthropol.* **73**:680–694.

Kurten, B. 1972. *Not from the Apes*, Vantage Books, New York.

Le Gros Clark, W. E. 1934. *Early Forerunners of Man*, Balliere, London. 296 pp.

Le Gros Clark, W. E., and Leakey, L. S. B. 1951. The Miocene Hominoidea of East Africa. *Fossil Mammals of Africa* (Br. Mus. Nat. Hist.) **1**:1–117.

Napier, J., and Davis, P. R. 1959. The fore-limb skeleton and associated remains of *Proconsul africanus. Fossil Mammals of Africa* (Br. Mus. Nat. Hist.) **16**:1–69.

Osborn, H. F. 1927. Recent discoveries relating to the origin and antiquity of man. *Palaeobiologica* **1**:189–202.

Pilbeam, D. R. 1967. Man's earliest ancestors. *Sci. J.* **3**(2):47–53.

Preuschoft, H. 1975. Body posture and mode of locomotion in fossil primates: Method and example—*Aegyptopithecus zeuxis*, in: *Proceedings from the Symposia of the Fifth Congress of the International Primatological Society 1974*, pp. 345–359, Japan Science Press, Tokyo.

Radinsky, L. 1973. *Aegyptopithecus* endocasts: Oldest record of a pongid brain. *Am. J. Phys. Anthropol.* **39**:239–248.

Radinsky, L. 1977. Early primate brains: Facts and fiction. *J. Hum. Evol.* **6**:79–86.

Raven, H. C. 1950. *The Anatomy of the Gorilla*, Columbia University Press, New York. 259 pp.

Remane, A. 1965. Die Geschichte der Menschenaffen, in: *Menschliche Abstammungslehre* (G. Heberer, ed.), pp. 249–309, Fischer, Stuttgart.

Schlosser, M. 1910. Über einige fossile Saugetiere aus dem Oligocan von Ägypten. *Zool. Anz.* **34**:500–508.

Schlosser, M. 1911. Beiträge zur kenntnis der Oligozänen Landsäugetiere aus dem Fayum, Ägypten. *Beitr. Paläontol. Oesterreich-Ungarns Orients* **24**:51–67.

Schön-Ybarra, M., and Conroy, G. C. 1978. Nonmetric features in the ulna of *Aegyptopithecus, Alouatta, Ateles, Lagothrix. Folia Primatol.* **29**:178–195.

Simons, E. L. 1962. Two new primate species from the African Oligocene. *Postilla* (Peabody Mus. Nat. Hist., Yale Univ.) **64**:1–12.

Simons, E. L. 1965. New fossil apes from Egypt and the initial differentiation of the Hominoidea. *Nature (Lond.)* **205**:135–139.

Simons, E. L. 1967. The earliest apes. *Sci. Am.* **217**(6):28–35.

Simons, E. L. 1970. The deployment and history of Old World monkeys (Cercopithecidae, Primates), in: *Old World Monkeys* (J. R. Napier and P. H. Napier, eds.), pp. 97–137, Academic, New York.

Simons, E. L. 1972. *Primate Evolution,* Macmillan, New York. 322 pp.

Simons, E. L. 1974a. The relationships of *Aegyptopithecus* to other primates. *Ann. Geol. Surv. Egypt* **4:**149–156.

Simons, E. L. 1974b. *Parapithecus grangeri* (Parapithecidae, Old World Higher Primates): New species from the Oligocene of Egypt and the initial differentiation of the Cercopithecoidea. *Postilla* (Peabody Mus. Nat. Hist., Yale Univ.) **166:**1–12.

Simons, E. L., and Fleagle, J. G. 1973. The history of extinct gibbon-like primates. *Gibbon and Siamang* **2:**121–148.

Simons, E. L., and Pilbeam, D. 1972. Hominoid paleo-primatology, in: *The Functional and Evolutionary Biology of Primates* (R. Tuttle, ed.), pp. 36–62, Aldine-Atherton, Chicago.

Simons, E. L., Andrews, P., and Pilbeam, D. R. 1978. Cenozoic apes, in: *Evolution of African Mammals* (V. J. Maglio and H. B. S. Cooke, eds.), pp. 120–146, Harvard University Press, Cambridge.

Szalay, F. S., and Delson, E. 1979. *Evolutionary History of the Primates,* Academic, New York. 580 pp.

Wood-Jones, F. 1929. *Man's Place among the Mammals,* Arnold, London. 371 pp.

Zapfe, H. 1960. Die primatenfunde aus der Miozänen spaltenfüllung von Neudorf an der March (Děvinská Nová Ves), Tschechoslowakei. Mit Anhang: Der Primatenfund aus dem Miozän von Klein Hadersdorf in Niederosterreich. *Schweiz. Palaeontol. Abh.* **78:**1–293.

Maxillofacial Morphology of Miocene Hominoids from Africa and Indo-Pakistan

8

S. C. WARD AND D. R. PILBEAM

Introduction

The contact of the Afro-Arabian Plate with Eurasia around 17 million years ago (m.y.a.) is associated with profound changes in Miocene faunal communities. At or about this time, a hominoid appeared in East Africa that was anatomically quite different from the *Proconsul* species complex that had been endemic there for over 6 million years. These differences involved elements of occlusal design, thickness of molar enamel caps, and gnathic buttressing, which showed a general increase in robusticity. By the mid 1970s, this combination of features had come to be regarded as exclusively characteristic of australopithecines or their immediate ancestors. Thus the presence of a hominoid with thick enamel and robust maxillae at Fort Ternan seemed to provide the oldest evidence of hominids (Simons, 1968; Andrews and Tekkaya, 1976; Simons and Pilbeam, 1978). The later discovery and diagnosis of *Australopithecus afarensis* from Pliocene deposits in Ethiopia and Tanzania (Johanson *et al.*, 1978; Johanson and White, 1979) tended to further strengthen the argu-

S. C. WARD • Department of Anthropology, Kent State University, Kent, Ohio 44242 and Department of Anatomy, Northeast Ohio Universities College of Medicine, Rootstown, Ohio 44272.   D. R. PILBEAM • Department of Anthropology, Harvard University, Cambridge, Massachusetts 02138.

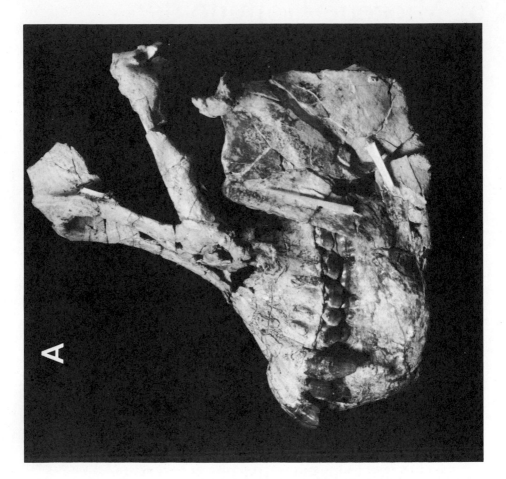

Fig. 1. (A) GSP 15000, a partial *Sivapithecus indicus* face from locality 410 near Kaulial Village, Pakistan.   (B) Three-quarter anterior view of GSP 15000 (center), *Pongo* (right), and *Pan* (left). Note the similarity in the form of premaxilla and supraorbital region of *Sivapithecus indicus* and *Pongo*.

ment that an animal like *Ramapithecus* from East Africa and Indo-Pakistan was ancestral to the earliest undoubted hominids.

By the end of the 1970s, new Miocene hominoid fossils had been recovered from localities in Pakistan (Pilbeam *et al.*, 1977, 1980), Turkey (Andrews and Tobien, 1977; Andrews and Tekkaya, 1980), Greece (de Bonis *et al.*, 1974), Arabia (Andrews, Hamilton and Whybrow, 1978), and China (Xu and Lu, 1979, 1980). Along with these specimens, new data concerning Miocene climatic conditions, composition of floral and faunal communities, and widespread use of various dating techniques had markedly expanded the scope of hominoid paleontology. As perhaps should have been expected, an augmented fossil record and improved calibration of biostratigraphic successions has made it necessary to revise earlier views of hominoid evolution. It appears to us that the taxonomic position of *Ramapithecus* is not as certain as was previously thought. *Ramapithecus* and *Sivapithecus* are similar to each other in a number of ways. Both have thick molar enamel caps, and gnathic buttressing systems are similar. They differ in canine dimensions, maxillary pneumatiziation, and molar size. While *Ramapithecus* conforms to most expectations of what a hominid ancestor should look like, *Sivapithecus* does not. Recently, Andrews and Tekkaya (1980) noted that the maxilla of *S. meteai* (MTA 2125) is similar in a number of respects to that of living *Pongo*. This observation is supported by preliminary observations on a new partial face of *S. indicus* recovered in the Potwar region of Pakistan (Pilbeam, 1981, 1982) (Fig. 1A). This specimen (GSP 15000) is similar in many respects to the Turkish maxilla and confirms that the maxilla of *Sivapithecus* is in fact quite similar to that of living orangutans, especially in the premaxillary and supraorbital regions (Fig. 1B). On the other hand, the mandibles of *Sivapithecus* do not show any detailed affinities with either *Pongo* or *Australopithecus*. They are however, very similar to the mandibles of *Ramapithecus*.

It is clear, then, that the recently expanded sample of *Sivapithecus* maxillae has complicated earlier reconstructions of hominid phylogeny in which thick molar enamel was assumed to be a derived hominid trait. Since the maxillae and mandibles of *Sivapithecus* and *Ramapithecus* appear to be more similar to each other than either is to *A. afarensis*, it becomes necessary to invoke parallelism in order to account for the distribution of enamel thickness in Miocene hominoids and Pliocene hominids. Parallel evolution in African and Asian hominoid lineages would also be demonstrated if *Sivapithecus* and *Ramapithecus* could be shown to postdate the divergence of the African and Asian hominoid radiations. This would be documented if the later Miocene hominoids of Asia were shown to possess a uniquely shared and derived trait or traits with living orangutans. In this chapter, we report the results of recent observations on the maxillary alveolar process of early and later Miocene hominoids, focusing specifically on the Asian radiation. Following the suggestion of our colleague Dr. Andrew Hill, we shall refer to *Ramapithecus*, *Sivapithecus*, *Ouranopithecus*, and *Gigantopithecus* collectively as the "ramamorphs." We recognize in these specimens a complex of features that distinguishes them from the earlier and more primitive hominoids from East

Africa, which we will designate the dryomorphs. These terms are convenient linguistic devices that summarize the geographic, temporal, and morphological limits of the two groups, while avoiding uncritical and informal use of subfamily (ramapithecines) and family (ramapithecids) designations which connote biological relationships that are at the moment unclear.

The Subnasal Alveolar Process

African Patterns

Ramamorph and dryomorph hominoids differ considerably in the topographic relationships of their premaxillary regions, as do living African apes and orangutans. These differences are based primarily on the disposition of the incisive canal and fossa, and their relationship to the hard palate. In all *Proconsul* species from the early Miocene of East Africa, the incisive fossa is a transversely broad basin that opens directly into the oral cavity (KNM-SO 700, KNM-ME 1, KNM-RU 1803, UMP 62-11) (Fig. 2). This configuration also characterizes the small early Miocene dryomorph, *Dendropithecus* (KNM-SO 417). The subnasal alveolar process appears as a flattened oval in sagittal section, and the distance from alveolare to nasaspinale is relatively short. Because the anterior edge of the hard palate is retracted distally from nasaspinale, no true incisive canal is present. This morphological pattern is rather similar to the subnasal morphology of living cercopithecines.

Unfortunately, the subnasal region is not preserved in any African middle Miocene specimens, although the Tayassuidae and elephants from Moroto could suggest a capping age for that site of 14 million years. As we have just noted, the Moroto palate (UMP 62-11) presents a subnasal pattern that is similar to the well-dated early Miocene hominoids recovered in Kenya. However, the premaxilla is well represented from the Hadar Formation. The Sidi Hakoma Member of this Formation overlies a basalt which approaches 4 m.y.a. (J. L. Aronson, personal communication). Hominids attributed to *Australopithecus afarensis* (Johanson et al., 1978; Johanson and White, 1979) have a subnasal configuration that is similar to living chimpanzees (Fig. 2). In sagittal section, the premaxillary region forms an average angle of 34° with the alveolar plane (Ward et al., 1982). The nasoalveolar clivus as measured from alveolare to nasaspinale is long, and projects a considerable distance into the nasal cavity. One consequence of this posterior position of nasaspinale is the formation of an incisive canal. The subnasal alveolar process overrides the anterior edge of the hard palate, producing an overlapping relationship between these two elements. As is the case in the African dryomorphs, the incisive fossa is transversely broad. However, the deflection of the palate beneath nasaspinale results in a sharp drop from the posterior pole of the clivus into the incisive fossa.

It appears, then, that two African subnasal "patterns" can be identified in

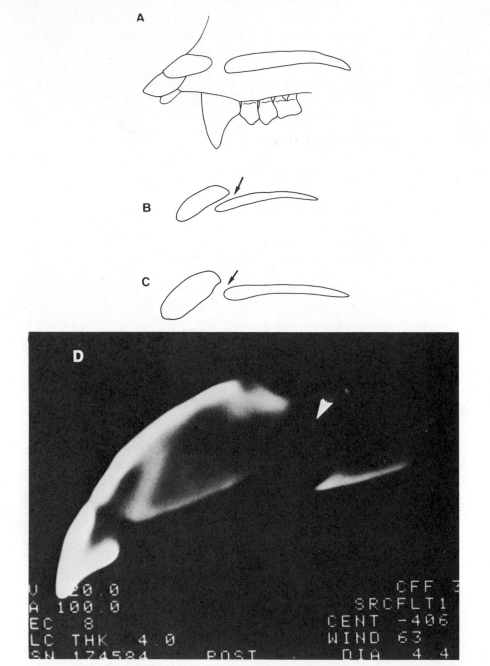

Fig. 2. Midsagittal sections through the premaxillae and palates of (A) *Proconsul major* (UMP 62-11); (B) *Pan troglodytes;* (C) *Gorilla;* (D) *Australopithecus afarensis* (computerized tomogram). Arrows indicate the incisive fossa.

the early Miocene hominoids and the Pliocene hominids. The primary differences in the dryomorph and hominid patterns involve the length of the nasoalveolar clivus and the topographic relationship of this element to the hard palate. The hominids have a well-defined incisive canal and the dryomorphs do not. Both groups share a broad incisive fossa, but this feature appears to be relatively wider in the dryomorphs. Given the present evidence, we think it is likely that the subnasal pattern presented by *A. afarensis* was derived from something like the dryomorph condition. Later clade differentiation in the australopithecines involved some modifications in the subnasal region (Ward *et al.*, 1983). *Australopithecus africanus* from the South African cave localities does not show as great a posterior projection of the nasoalveolar clivus into the nasal cavity as do the Hadar australopithecines, and their vomeronasal contact is not as extensive. The robust clade as represented by *A. robustus* and *A. boisei* shows an increase of the subnasal angle of the clivus to the alveolar plane and a marked amount of vertical thickening of both the subnasal alveolar process and the anterior one-third of the hard palate. All of these variations can reasonably be derived from *A. afarensis*.

The suite of features that constitute australopithecine subnasal morphology is also clearly recognizable in the living African apes. The nasoalveolar clivus of both chimpanzees and gorillas projects well back into the nasal cavity and drops sharply into the incisive fossa, which is transversely broad. The fossa is divided into two chambers by the vomeronasal contact (Fig. 3). The hard palate is deflected beneath nasospinale, and as a consequence, an incisive canal is formed. *Pan* and *Gorilla* differ in the angle of the clivus with respect to the alveolar plane, and in the geometry of the subnasal alveolar process in midsagittal section. The clivus angle of gorillas is more vertical than is the case for chimpanzees. The subnasal section in *Pan* is an elongated oval, while in *Gorilla* the section is somewhat more rectangular. In a few cases, especially in males, the gorilla palate is not deflected beneath nasospinale. This occurs when the nasoalveolar clivus is acutely angled, and the resulting configuration is reminiscent of the early Miocene dryomorphs. In general, the subnasal pattern characteristic of chimpanzees bears the closest resemblance to *A. afarensis* of any known hominoid.

The Asian Pattern

Premaxillary morphology of the ramamorph hominoids recovered from "Chinji"- and "Nagri"-aged sediments in the Siwaliks is quite distinct from the early Miocene dryomorph condition. In both *Sivapithecus* sp. and *Ramapithecus punjabicus* the nasoalveolar clivus intersects the alveolar plane at a shallow angle. Unlike the situation for *Proconsul* sp. and the australopithecines, the clivus arcs posteriorly into the nasal cavity without terminating at a broad and deep incisive fossa. The fossa is a narrow depression just behind nasospinale. The incisive canal opens into the floor of the fossa as a narrow cleft (Fig. 4). The very narrow canal continues anteriorly beneath the subnasal alveolar

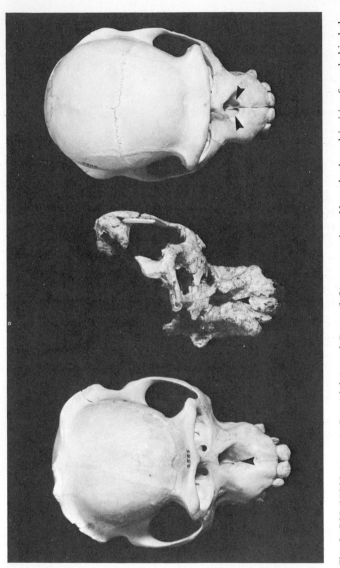

Fig. 3. GSP 15000 (center), *Pan* (right), and *Pongo* (left); vertex view. Note the broad incisive fossa behind the premaxilla in the chimpanzee skull. The narrow incisive fossa of *Pongo* and *Sivapithecus* is situated somewhat more posteriorly in the nasal cavity (arrows).

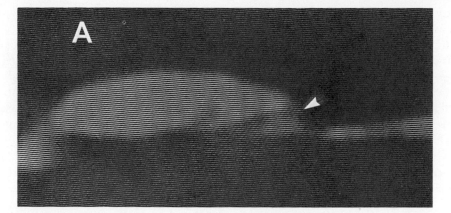

Fig. 4. (A) Sagittal computerized tomogram through the subnasal alveolar process of GSP 15000. The opening of the narrow incisive canal just below the posterior terminus of the nasoalveolar clivus is indicated by the arrow. (B) Sagittal section through the subnasal alveolar process of *Pongo*.

process, terminating at a small incisive foramen on the palate. As a consequence of these relationships the palate and the nasoalveolar clivus appear to contact each other, rather than being offset as is typical of the Pliocene hominids and living African pongids. Another correlate of the Asian subnasal pattern is a differential distribution of palatal thickness. The ramamorph hard palate is usually thin anteriorly, and becomes thicker posteriorly (GSP 15000, GSP 11704, GSP 9977). The reverse is true of the Pliocene hominids, in which the palate tends to be thickest just posterior to the incisive fossa (Fig. 5).

The only reasonably complete *Ramapithecus* premaxillary region is preserved in the Haritalyangar palate (YPM 13799). As is well known, this specimen is broken lateral to the midline. Although the incisive canal and fossa are not present, the subnasal region is sufficiently preserved to show that it is quite similar to *Sivapithecus* in its overall configuration. It is certainly unlike

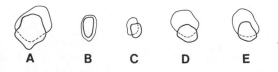

Fig. 5. Transverse sections through the maxillar canine roots of (A) *Proconsul major* (KNM-SO 585); (B) *P. nyanzae* (KNM-RU 1677E); (C) *P. africanus* (KNM-RU 2049); (D) *Sivapithecus indicus* (GSP 8925); (E) the large hominoid canine from Fort Ternan (KNM-FT 39). All sections are oriented as the roots would appear in their sockets. The larger section is through the first and middle thirds of the root, and the smaller section is through the root at approximately two-thirds of the distance from cervix to apex.

any of the known dryomorphs, Pliocene australopithecines, or African pongids.

Premaxillary morphology of living orangutans is very similar to that of *Sivapithecus*. The labial surface of the nasoalveolar clivus passes posteriorly into the nasal cavity without a sharp drop into the incisive fossa. Also similar to *Sivapithecus*, the incisive fossa of *Pongo* is a narrow slit posterior to and slightly below nasospinale. The palate is not deflected beneath the subnasal alveolar process, and is usually thin anteriorly. Unlike *Sivapithecus*, there may be more variation in the projection angle of the clivus. Older orangutans often have a more vertically oriented clivus than do younger individuals (Greenfield, 1979), but this metric variability does not affect the qualitative relationships of the premaxilla and hard palate.

The occurrence of two premaxillary morphs in Miocene hominoids is useful in assessing the affinities of other specimens, such as the Rudabánya palates (Kretzoi, 1975) and the recently described *S. meteai* palate (MTA 2125) from the Mount Sinap series in Turkey (Andrews and Tekkaya, 1980). While Simons (1976) initially attributed *Rudapithecus* (Rud, 12) to *Ramapithecus* sp., the premaxillary region of this specimen is clearly dryomorph in character. The clivus is abbreviated, and the incisive fossa is a relatively broad basin that opens into the oral cavity. There is a slight suggestion of an offset relationship between the premaxilla and the anterior edge of the hard palate. From these observations we conclude that Rud 12 is more closely related to the African dryomorphs than it is to the later Miocene hominoids of Asia. The Turkish maxilla (MTA 2125) has the typical Asian subnasal pattern, and conforms in all particulars to the ramamorphs recovered in Indo-Pakistan. We have also recently learned that the relatively complete, but badly crushed *Sivapithecus* cranium from Lufeng also has an Asian subnasal pattern (Bai Cheuh Lei, personal communication).

The distribution of subnasal patterns* among living and Miocene hominoids suggests that *Sivapithecus* and *Ramapithecus* are more similar to *Pongo* than they are to *A. afarensis* or the living African apes. If we are correct in arguing that the subnasal pattern that characterizes *Sivapithecus*, *Ramapithecus*, and *Pongo* indicates a close phylogenetic relationship, then the last common ancestor of the Asian and African hominoids must predate the oldest ramamorph hominoid.

Canine Implantation

The hominoid canine root is in topographic association with the anterior premolar, the nasal cavity, and the maxillary sinus. The proximity of the canine alveolus to each of these neighboring structures and spaces is a func-

---

*It is interesting to note that the Lesser Apes, the Hylobatidae, are characterized by the dryomorph subnasal pattern. This would indicate that the dryomorph condition is probably the primitive ancestral morph for the Hominoidea.

tion of its length, volume, and axial alignment. These factors, as well as the degree of external rotation of the root, can influence the appearance of the canine pillar and margins of the nasal aperture. There are several variations with respect to canine implantation in the dryomorph hominoids from East Africa, apparently due to differences in root morphology and size. The Chinji–Nagri ramamorphs, on the other hand, manifest only one canine implantation pattern despite a range of probable body sizes spanning *P. major* through *P. africanus*.

The maxillary canine root of the commonly recognized *Proconsul* species is implanted in a pyramidal mass of bone, bounded by the premaxilla, nasal cavity, maxillary sinus, and oral vestibule. This region is relatively massive in the Moroto palate (UMP 62-11) assigned to *P. major* (Simons *et al.*, 1978; Andrews, 1978) and also in the holotype of *P. nyanzae* (BMNH M 16647). On the basis of canine crown dimensions, both specimens are probably male. Several maxillary specimens attributed to *P. africanus* by Andrews (1978) (BNMH M 32363; KNM-RU 1705) may include both males and females. In this species the bone housing the canine root is less massive, and the septum separating the alveolus from the nasal cavity is considerably thinner than in *P. major* and *P. nyanzae*. The latter species are similar in the expression of a moderately bulging vestibular cortex draped over the large canine roots. Greater surface relief is evident in *P. nyanzae*, since its bladelike root causes the overlying cortical plate to drop away over the root system of P3. The maxillary canine roots of the (presumably) male *P. major* from Moroto, Napak, and Songhor are, in contrast, plump and tusklike. They show no evidence of the labiolingual compression which characterizes the upper canines of *P. nyanzae* and *P. africanus*. As a consequence the canine region of the maxilla is quite inflated. One effect of this configuration is a distal prolongation of the canine pillar over P^3. The canine fossa is thereby much diminished, producing overall surface relief on the facial surface of the maxilla.

In addition to influencing the expression of surface topography, canine root morphology is a useful sorting criterion in identifying taxonomic groupings. Male *P. major* canines are quadrangular in section and are deeply excavated by mesial and distal longitudinal grooves, while the maxillary canines of *P. nyanzae* and *P. africanus* are mediolaterally compressed. The root surfaces of *P. nyanzae* are usually devoid of longitudinal grooves, with only the distal groove being incipiently expressed. Although the canines of *P. africanus* are smaller, they tend to exhibit more surface relief, with a well-developed mesial groove incising the root from the cervix to within a short distance of the apex (Fig. 5).

Some problems in clarifying the affinities of the large East African dryomorph sample of isolated canines have been recently discussed by Bosler (1981). Identifying female *P. major* canines is especially difficult, since they are probably not unlike those of *P. nyanzae*, both metrically and anatomically. In addition, we believe that at least one maxillary canine tooth attributed to *P. nyanzae* (KNM-FT 39) (Andrews, 1978) is more similar to *Sivapithecus* than it is to any of the African dryomorphs (see p. 223).

The canine pillar of all known *Sivapithecus* maxillae is quite unlike that of

Proconsul. It is delimited anteriorly by a concavity overlying the alveolus of the lateral incisor, and posteriorly by a well-developed canine fossa (GSP 11704, GSP 9977, GSP 11708, GSP 15000). In frontal view, the canine juga converge toward the midline. These relationships are the product of four factors: root volume, axial disposition, root rotation, and topographic relations of the maxillary sinus.

Maxillary canine root volume of the largest *Sivapithecus* individuals is roughly similar to that of *P. nyanzae.* Differences in root morphology between the two genera are limited to torsion of the primary root axis with respect to the crown, and longitudinal curvature from cervix to apex. *Sivapithecus* canine roots are roughly quadrangular in transverse section, the labial surface being broader than the lingual. The mesial surface is flat, while the distal surface may be indented by a shallow longitudinal depression. As noted above, this pattern is also represented in the large isolated canine from Fort Ternan (Fig. 5).

The maxillary canine roots of *Sivapithecus* are also curved longitudinally. The proximal two-thirds of the root establish its primary or dominant orientation. Above this level, the root reverses this axis and twists bucally. The apical third of the root also curves distally, bringing its apex into approximation with the maxillary sinus.

From the topographic relationships of the canine socket of YPM 13799, in addition to casts made of its empty alveolus, one sees that the canine root morphologies of *Ramapithecus* and *Sivapithecus* are quite similar. The root of YPM 13799 is quite unlike that of any of the known dryomorphs in its reduced mesial grooving and the presence of torsion and curvature patterns similar to those described for *Sivapithecus.* It is also of interest that the isolated canine usually associated with the Fort Ternan maxilla (KNM-FT 46, 47) is unlike the canine that was housed in the empty socket of YPM 13799, since the former tooth has well-developed longitudinal grooves.

Another component of maxillary canine implantation that readily distinguishes ramamorphs from the *Proconsul* species is increased medial angulation. In frontal projection, the canine alveoli tilt strongly toward the midline. The degree of medial inclination is greater in *Sivapithecus* than in *Ramapithecus.* *Proconsul* canines also converge toward the midline—but at a very shallow angle. This contrast is also evident in living pongids. *Pongo* shows pronounced medial angulation of its canine roots in both males and females. This relationship is the primary determinant of the pyramidal circumnasal region so characteristic of orangutans and ramamorphs.

Also contributing to strongly developed canine juga in *Pongo, Sivapithecus,* and *Ramapithecus* is the pronounced external rotation of the canine root. All ramamorphs are characterized by outwardly rotated roots, and the degree of rotation is apparently independent of body size. Transverse computed tomography through the canine socket of *Sivapithecus* and YPM 13799 reveal the same alignment of the canine root. Lateral root rotation influences canine jugum morphology by causing the vestibular cortex of the alveolar process to rise over the convex union of the mesial and buccal root surfaces (Fig. 6).

In orangutans, the maxillary canines are also externally rotated. However, the degree of rotation is always greater in males than in females. In addition, neither male or female orangutan canines are rotated to the degree observed in ramamorphs. However, the canine jugum of *Pongo* is produced in the same fashion as that of *Sivapithecus*. The thin alveolar plate is draped over the rotated root mass, producing a bulge on the facial surface of the maxilla.

As we have previously noted, the large maxillary canine from Fort Ternan (KNM-FT 39) does not conform to the morphological pattern typical of *P. nyanzae* canines. As Simons (1981) has recently noted, this specimen is more similar to the canines of *Sivapithecus* than to any of the East African dryomorphs. We agree with Simons (1981) that FT 39 is likely to be a large, probably male *Sivapithecus* tooth, and this position is supported by the collection of several new isolated ramamorph canines from the Potwar Plateau in Pakistan (GSP 11003, GSP 8925, GSP 13167). Crown, root, and sectional morphology of these specimens are similar to KNM-FT 39. Using the mesial developmental groove to orient the Asian and Fort Ternan canines, it is also clear that the Fort Ternan specimen was inclined medially when viewed in frontal projection, much like the canines of the ramamorphs and *Pongo*. It is also apparent from this orientation that KNM-FT 39 was externally rotated within its alveolar process. It thus appears possible that a large ramamorph was present at Fort Ternan contemporaneously with *R. wickeri*.

Dryomorph hominoids differ considerably from their Asian counterparts in their complete lack of canine rotation. Their roots, which tend to be elliptical in section, are aligned more or less along the axis of the alveolar process, or are in fact rotated slightly internally. While the biomechanical implications of this configuration are unclear, its role in canine jugum formation is straightforward. The broad, flat buccal root surface provides an unbroken surface upon which lies the buccal cortical plate of the maxilla. Only a modest bulge in the cortex indicates the presence of the canine alveolus. A similar pattern characterizes chimpanzees and gorillas. Both animals are sexually dimorphic in canine dimensions, yet males and females are similar in terms of canine implantation. Female *Pan* and *Gorilla* canines present three surfaces in transverse section. A flattened mesial root surface gives way to a sharply convex distal-lingual surface. The buccal surface is the largest of three, and terminates at an abrupt flexure which marks the union of the buccal and mesial surfaces. The major axis of any transverse section through the root is anteroposteriorly aligned.

The topographic relations and sectional geometry of the male chimpanzee canine is similar to that of the female. All sections are mesiodistally aligned. The canine root of male gorillas is more massive than that of the female, and its major axis either is mesiodistal or, in many specimens, is internally rotated. These relationships produce canine jugum topography that is similar to that of the *Proconsul* species: unrotated canines fail to raise a substantial bulge in the buccal plate of the alveolar process.

At the moment, both the ontogenetic and adaptive bases of canine rotation are unclear. Wallace (1978) has suggested that variations in fetal premaxillary fusion times might account for differences in the extent of external

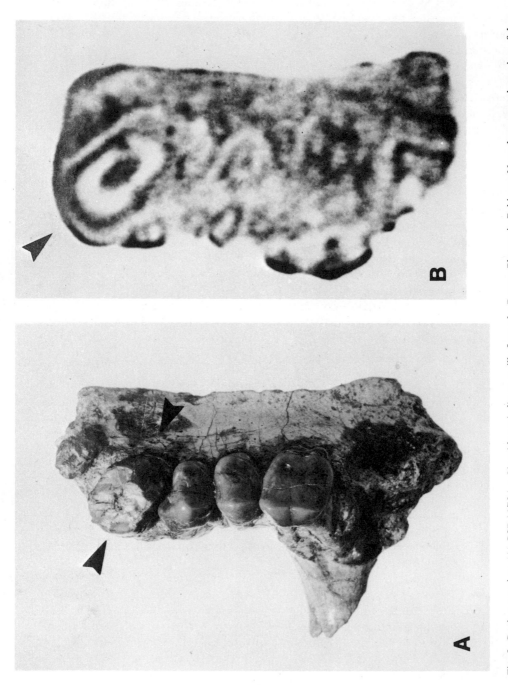

Fig. 6. Canine rotation. (A) GSP 11704, a *Sivapithecus indicus* maxilla from the Potwar Plateau in Pakistan. Note the external rotation of the canines. (B) A transverse computerized tomogram at mid-canine root level of GSP 11704. (C) Transverse computerized tomogram of YPM 13799. The empty canine socket is indicated (arrows). (D) Transverse CT image of an adult male orangutan.

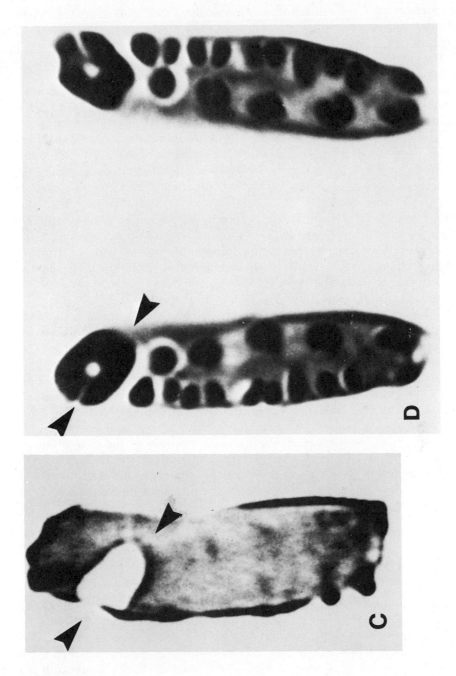

Fig. 6. (continued)

rotation that characterize the canines of *A. africanus* and *A. robustus*. While early fusion of the premaxillary suture may in fact promote canine rotation in australopithecines, it is not clear how premaxillary ontogeny would effect the rotational configuration of Miocene hominoid canines. However, the adaptive significance of canine rotation may be more amenable to empirical observation. We have observed in a large series of transverse computed tomograms of living and fossil hominoid maxillae that a rotated canine root, especially in its cervical one-third, effectively extends the length of the maxillary alveolar process by providing more room for the premolars and molars (see Fig. 6). Thus it might be expected that thickly enameled and/or megadont hominoids would have outwardly rotated canines in order to accommodate their more voluminous teeth. This mechanism would also mitigate the need to elongate the maxilla in a lineage undergoing selection for thick enamel and dental enlargement. Potentially complicated rearrangements of buttressing systems would thereby be avoided.

Our hypothesis seems to be supported by the presence of canine rotation in Miocene ramamorphs and Pliocene hominids. The thickly enameled and possibly megadont ramamorphs all have markedly rotated canines. In addition, their canine roots tend to be quadrangular, a geometry that probably also conserves space within the alveolar process. *Australopithecus afarensis* also has rotated canines, megadont molars, and thick postcanine enamel caps. It seems likely that by rotating the canine, australopithecines and ramamorphs were able to add more enamel to their premolars and molars without any other major modifications in facial design.

In contrast, the early Miocene dryomorphs had thin, or moderately thick, enamel and relatively uncrowded molars. As we have shown, their canines are aligned along the axis of the tooth rows. An interesting intermediate condition appears to characterize living orangutans. The molar enamel of *Pongo* seems to be thicker than that of *Pan* and *Gorilla*, but it does not appear to be large with respect to body size. The maxillary canines of the orangutan are externally rotated, but to a lesser extent than those of the ramamorphs or Pliocene hominids. There also appears to be some sexual dimorphism with respect to canine rotation in orangutans, since males always have more highly transversely positioned canines than do females.

The Postcanine Alveolar Process

An attribute of palatofacial architecture that Simons (1976) interpreted as a shared derived character linking *Ramapithecus* and *Australopithecus* is the relative robusticity of the maxillary alveolar process. Simons correctly noted that the alveolar recess of the maxillary sinus invades the alveolar process in all adult hominoids (Cave and Haines, 1940; Wegner, 1936; Sicher and Du-Brul, 1980). The extent of pneumatization may be so great that in many older individuals the sinus drops into the furcations of the premolar and molar root systems. The exposed alveoli are visible in the sinus floor as bulges. There appear to be no differences between males and females with respect to the topographic relations of the maxillary sinus in living apes and humans. However, this may not be true of Miocene hominoids and Pliocene hominids. *Ramapithecus* maxillae (YPM 13799; GSI D-185) appear to be quite robust in coronal sections taken through the postcanine alveolar process. This "robusticity" is a function of the level of the sinus floor. In most *Sivapithecus* specimens (GSP 9977, GSP 15000, GSP 11786) the alveolar recess of the sinus drops from a high position over the distal premolar directly into the molar root system. In *Ramapithecus*, on the other hand, the alveolar recess does not invade the alveolar process to the same extent. Its floor is confined to the level of the molar root apices. Thus when *Sivapithecus* and *Ramapithecus* are compared, the former presents what appears to be a more "lightly" constructed alveolar process (Figs. 7 and 8).

A similar pattern of diversity characterizes *A. afarensis*. A small maxilla (A.L. 199) appears quite robust in coronal sections distal to the canine, for the same reason *Ramapithecus* does. The alveolar recess of the sinus does not encroach upon the molar root systems. In contrast, the sinus of larger specimens (A.L. 200-1a, A.L. 333w-1) deeply excavates the alveolar process, and the apices of the molar and premolar roots are visible as bulges in the sinus floor (Ward *et al.*, 1982). In section, these specimens appear more gracile than A.L. 199.

The contours of the sinus floor itself are also rather variable in ramamorphs and hominids. These variations are expressed primarily in the complexity of septum development in the alveolar recess. Some *Sivapithecus* maxillae have strongly developed loculi that divide the sinus floor into two or three chambers. The most anterior chamber tends to be the largest and deepest. It becomes increasingly narrow toward the canine alveolus, and terminates over the root system of P^4. If transverse septa are present, the depth of the sinus floor changes abruptly from front to back, but in most cases is deepest over the first molar and shallowest over the third molar (GSP 15000, GSP 9977). Several specimens show very little evidence of septum formation in the maxillary sinus (GSP 11704, GSP 11786), but are similar to those that do have septa in presenting a deep alveolar recess anteriorly that becomes shallower posteriorly.

Ramapithecus maxillae (YPM 13799; GSI D-185) show a pattern of anterior narrowing similar to that of *Sivapithecus*, and have no septa. However, there are not sufficient specimens to determine if all *Ramapithecus* maxillae lack transverse septa. As Andrews and Walker (1976) have noted, the maxil-

A   **B**

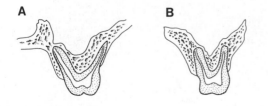

Fig. 7. Coronal sections through the postcanine alveolar process of *Sivapithecus indicus*. (A) GSP 11786. (B) GSP 9977. The sections are taken through the first molars.

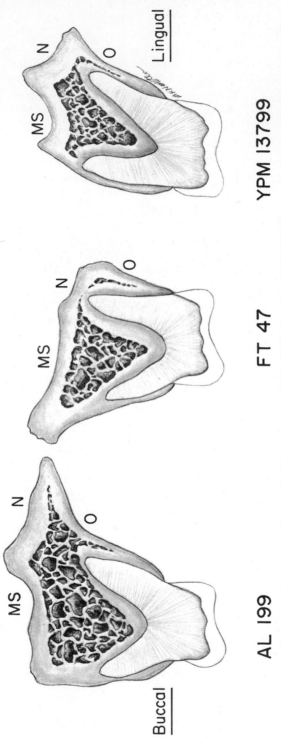

AL 199   FT 47   YPM 13799

Fig. 8. Reconstructions of maxillary sinus relationships in *R. punjabicus* (YPM 13799), *R. wickeri* (KNM-FT 47), and *A. afarensis* (AL-199). These smaller specimens lack extensive invasion of the maxillary sinus into the alveolar process, and therefore appear relatively "robust." Abbreviations: O, oral cavity; N, floor of the nasal cavity; MS, floor of the maxillary sinus.

lary alveolar process of KNM-FT 46, 47 does not extensively invade the alveolar process, nor does it have any septa. In this respect the Fort Ternan maxilla is very similar to that of *R. punjabicus*.

Maxillary sinus morphology of the dryomorph hominoids has been discussed by Andrews (1978). The most common pattern occurring in *Proconsul* species is a narrow, troughlike sinus that is in direct proximity to the dental alveoli. Its floor is not complicated by partitions and chambers. Some of the smaller maxillae (*P. Rangwapithecus*) demonstrate a more aggressive pattern of pneumatization, and transversely wide sinuses (KNM-SO 7000). These specimens are more similar to chimpanzees and gorillas than are the maxillae of the larger *Proconsul* species.

In addition to influencing the robusticity of the maxilla, the maxillary sinus also contributes to the formation of the canine fossa. In the African apes, the sinus extends anteriorly to approximate, and in some cases partially surround, the canine alveolus. This condition is most advanced in chimpanzees, as evidenced by extensions of the maxillary sinus into the hard palate and premaxilla (Cave and Haines, 1940). The sinus does not develop as far anteriorly in orangutans, and seldom encroaches upon the canine root. In all the hominoids, including humans, the maxillary sinus also grows into the root of the zygomatic arch. With increasing age, this zygomatic recess may extend into the zygomatic bone itself. Orangutans and the African apes differ slightly in the extent to which the zygomatic root is inflated by the sinus. The zygomaticoalveolar crest of chimpanzees and gorillas is generally broader than it is in orangutans. As a result, the surface of the maxilla distal to the canine fossa is only slightly indented in the African apes, and therefore the canine prominence is only moderately expressed. In contrast, the maxillary sinus of *Pongo* does not inflate the zygomatic root. Coupled with the fact that the sinus does not greatly inflate the maxilla anterior to the zygomatic root either, the canine fossa of orangutans is more highly developed.

The relationship of both the canine alveolus and zygomatic root in Miocene ramamorphs are more similar to *Pongo* than they are to any other hominoid. The maxillary sinus of *Sivapithecus* does not extend beyond the anterior premolar, and in most cases terminates over P^4. The midface is consequently fairly robust anterior to the canine. There is also a tendency for the facial surface of the maxilla to "collapse" inward over the premolar roots in the absence of an underlying sinus. This condition is most pronounced in GSP 15000, which also is one of the dentally oldest specimens from the Potwar Plateau. Recession of the maxillary cortex is so great that the buccal roots of the premolars are implanted in a shelflike extension of the alveolar process (Fig. 9). This configuration is very similar to that occurring in Pleistocene and Recent baboon maxillae, which lack a true maxillary sinus.

Sivapithecus is also similar to *Pongo* in the morphology of its zygomatic root. The zygomaticoalveolar crest is relatively thin when compared to the African apes, and its anterior flare above the alveolar crest is also similar to the condition in living orangutans. The combination of an anteriorly diminishing maxillary sinus, zygomatic flare, and rotated canine roots produces a

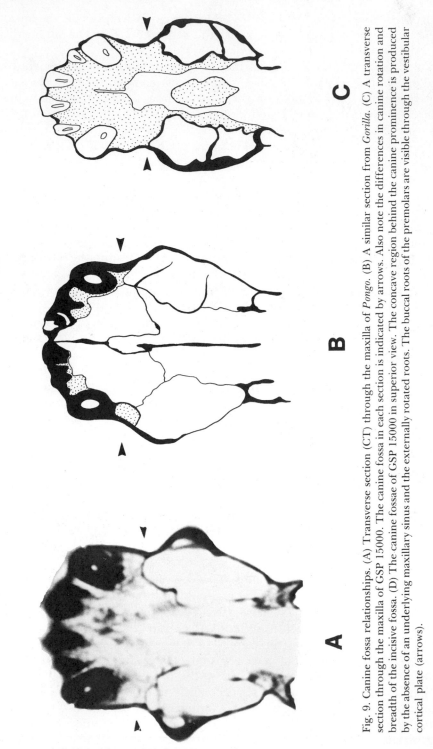

Fig. 9. Canine fossa relationships. (A) Transverse section (CT) through the maxilla of *Pongo*. (B) A similar section from *Gorilla*. (C) A transverse section through the maxilla of GSP 15000. The canine fossa in each section is indicated by arrows. Also note the differences in canine rotation and breadth of the incisive fossa. (D) The canine fossae of GSP 15000 in superior view. The concave region behind the canine prominence is produced by the absence of an underlying maxillary sinus and the externally rotated roots. The buccal roots of the premolars are visible through the vestibular cortical plate (arrows).

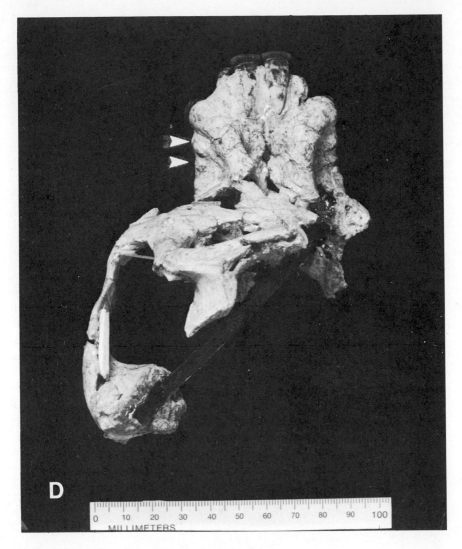

Fig. 9. (*continued*)

well-developed canine fossa. Smaller ramamorphs (*R. punjabicus* and *R. wickeri*) also have canine fossae, probably for the same reasons.

Miocene Hominoids and Human Origins

Asian and African Miocene hominoids are clearly divergent in the organization of their maxillary alveolar processes. However, specimens from both groups do show similarities to the earliest australopithecines. The subnasal region of *A. afarensis* is most similar to living chimpanzees, and we think it probable that it is derived from something like the *Proconsul* pattern. On the other hand, the subnasal region of *Ramapithecus* and *Sivapithecus* is very similar to that of living orangutans. In contrast, certain features related to canine implantation are similar in the Hadar australopithecines, *Ramapithecus,* and *Sivapithecus,* while the *Proconsul* species exhibit patterns of canine alignment and angulation similar to those of the African apes. Maxillary pneumatization characteristics are also shared by the later Miocene Asian hominoids and *A. afarensis,* in that large individuals tend to have alveolar processes that are deeply excavated by the maxillary sinus, while smaller individuals show little encroachment by the sinus onto the molar root systems. Interestingly, the Fort Ternan maxilla (KNM-FT 46, 47) is similar to both *Ramapithecus* (YPM 13799; GSI D-185) and small *A. afarensis* (A.L. 199) in the construction of its postcanine alveolar process (Fig. 8).

The fact that the front of the maxilla clearly segregates Asian and African hominoids, while the posterior components do not, offers some interesting possibilities in reconstructing ape and human ancestry. An important issue that the fossil record may be complete enough to resolve is the taxonomic position of *Ramapithecus* and its role in human origins. If, as Lewis, Simons, and others have maintained since 1937, *R. punjabicus* is a hominid, in the strictest sense, it must then be ancestral to the earliest australopithecines and not to any extant apes. Alternatively, and in a less restrictive sense, *Ramapithecus* might be considered a hominid if it is ancestral to all African hominoids, but postdates the last common ancestor of African and Asian hominoids. Finally, if *R. punjabicus* can be shown to be part of the radiation from which *Pongo* evolved, then it cannot in either sense be considered a hominid. In order to resolve this issue, the following questions must be answered: first, what is the taxonomic relationship between *Ramapithecus* and *Sivapithecus?*; and second, how can the dental and facial features of Miocene and Recent hominoids best be used in phylogeny reconstruction?

Ramapithecus and Sivapithecus

In 1961, Simons revived a long-dormant dispute concerning the affinities of *Ramapithecus*. In that and subsequent papers (Simons 1961, 1964, 1968, 1976, 1978, 1979) Simons supported Lewis' (1934, 1937) view that the dental and gnathic anatomy of *Ramapithecus, Bramapithecus,* and *Sugrivapithecus* were more "humanlike" than any early hominid then known. This position was criticized by Hrdlička (1935) when he noted that dental anatomy and details of mandibular structure of the known Asian hominoids could easily be interpreted as being more ape-like than human. This position was in turn forcefully rejected by Simons (1968). An important point that was lost in this extended debate was the nature of the relationship obtaining between *Ramapithecus* and *Sivapithecus.*

Beginning early in this century, Pilgrim (1910, 1915, 1927) diagnosed a series of large hominoid teeth as belonging to at least two species of *Siva-pithecus*. For some reason, *Sivapithecus* was never a popular taxon among American paleontologists, who preferred to allocate larger Siwalik hominoids to *Dryopithecus* (Brown *et al.*, 1924). This view was sustained by Simons and Pilbeam (1965, 1971), who found no basis for segregating most European, African, and Asian hominoids at the genus level. This taxonomic scheme reflected the differences that Simons and Pilbeam felt at that time distinguished *Ramapithecus* and *Dryopithecus* (*Sivapithecus*). However, by the end of the 1970s, accumulating specimens from localities ranging from Greece to China generated some revisions in Miocene hominoid taxonomy. Simons (1976) and Pilbeam (Pilbeam *et al.*, 1977; Pilbeam, 1979) resurrected *Siva-pithecus*, recognizing that enamel thickness and palatofacial buttressing distinguished these forms from the early Miocene hominoids from East Africa. Recent reviews of the later Miocene fossil hominoids by Greenfield (1979), Kay (1982), and Kay and Simons (this volume, Chapter 23) have not only sustained the validity of *Sivapithecus*, but have also proposed that *Sivapithecus* and *Ramapithecus* are closely related either at the genus or species levels, and have included all thickly enameled Miocene hominoids in *Sivapithecus*.

We have shown that the structure and topographic relations of their alveolar processes support the hypothesis that *Ramapithecus* and *Sivapithecus* are closely related. The only significant difference between them other than size involves the extent of maxillary sinus invasion into the alveolar process. A rather similar pattern of midfacial pneumatization also characterizes *A. afarensis*. Taken together, however, the basic organization of the ramamorph maxillary alveolar process is sufficiently distinctive to show that *Ramapithecus* and *Sivapithecus* are more similar to each other than either is to the Pliocene hominids. We think it less likely, though, that they are conspecific, but they may be congeneric. Recent collecting in the Potwar Plateau has produced specimens with size differences that could exceed what may reasonably be expected to be incorporated by one species. Until associated postcranial and gnathic elements are recovered, we prefer to avoid synonmizing *Ramapithecus* and *Sivapithecus* at the species level, though future collecting should soon fully justify their synonymy at the genus level.

Enamel Thickness

The adaptive importance in Miocene hominoids and early hominids of cheek teeth with thick enamel was first clearly articulated by Simons and Pilbeam (1972). Recent observations on primate molar enamel thickness (Gantt, 1977, 1981; Molnar and Gantt, 1977; Kay, 1981) have confirmed the basic assumption made by Simons and Pilbeam that some primates have thicker enamel caps on their molars than others, that this variation in thickness can influence how tooth wear accumulates, and that there is probably some relationship between enamel thickness and the composition of primate diets. The presence of very thick enamel on the molars of *Ramapithecus* therefore became the most compelling evidence linking this later Miocene hominoid with thickly enameled *Australopithecus*. A correlate of this hypothesized ancestor–descendent relationship was the assumption that *Ramapithecus* must have been a terrestrial forager that consumed hard, resistant food items found on the ground, such a niche being that inferred for *Australopithecus*. Kay (1981) has noted that there are in fact severe difficulties in breaking open certain arboreal food resources in order to take advantage of the high-energy carbohydrates within. Kay (1981) has also shown that these large seeds and nuts are important dietary constituents of arboreal monkeys that possess relatively thick molar enamel. It thus appears that enamel thickness is a poor predictor of substrate preference.

Another misleading assumption concerning enamel thickness is that it is either thick or thin. In fact, some Miocene and extant hominoids have enamel of intermediate thickness. Gorillas and chimpanzees have thin molar enamel, but orangutans and humans possess enamel caps that are intermediate in thickness between the ramamorphs and robust australopithecines on the one hand, and African apes on the other. Thus, for comparative purposes, we suggest that hominoid molar enamel be characterized in at least three categories, thin, intermediate, or thick. It was initially assumed that the primitive condition for living hominoids was thin enamel (Simons and Pilbeam, 1972). This meant that thick enamel was a derived trait shared by the ramamorphs and australopithecines. However, it now seems possible that the last common ancestor of Asian and African hominoids possessed either thick or moderately thick molar enamel caps. This would mean that thin enamel in *Pan* and *Gorilla* would be a derived feature. Several possible cladistic reconstructions based on changes in enamel thickness are presented in Fig. 10. Assuming that relative enamel thicknesses are correctly interpreted, arbitrary weights can be assigned to each change in thickness. In this simple exercise, we assign a value of 0.5 for each qualitative change from one thickness category to another. If we assume that the last common ancestor of all post-middle Miocene hominoids had thin enamel, then a minimum of 3.0 changes would be necessary to account for the present distribution of enamel thickness in apes and humans. If this last common ancestor had thick enamel, contemporary hominoid enamel thickness patterns could be explained by invoking a minimum of 2.0 changes. Other possibilities present themselves. The most interesting segment of the cladogram is the branch between the last common ancestor of Asian and African hominoids (LCA_1) and the last common ancestor of the African Great Apes and hominids (LCA_2). On the basis of the known evidence, this form could have had thick, intermediate, or thin enamel. The absence of a fossil record for late Miocene, Pliocene, or Pleistocene chimpanzees and gorillas presents problems. We think it possible that LCA_2 had thick or intermediate enamel, since the only late Miocene African hominoid dental remains (Ngorora, Lukeino, Lothagam) have thick enamel caps.

Despite the multiple alternatives that can be derived from cladistic reconstructions based on enamel thickness, some simplification of the data can be achieved concerning the ramamorphs and australopithecines. If enamel

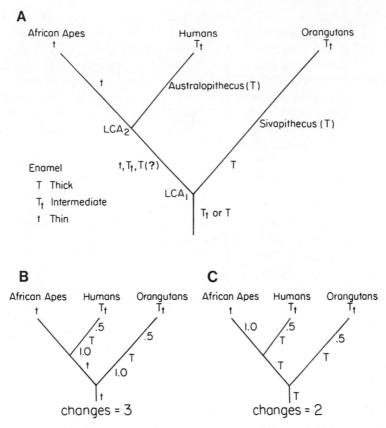

Fig. 10. Alternative phyletic pathways with regard to hominoid enamel thickness. T designates thick enamel, T_t moderately thick enamel, and t thin enamel. Enamel thicknesses of humans and the Great Apes are shown in A. This figure also indicates that the last common ancestor of Asian and African hominoids (LCA₁) may have had thick or moderately thick enamel. In addition, the last common ancestor of African apes and hominids (LCA₂) could have had thick, thin, or moderately thick enamel. By assigning arbitrary weights to each presumed shift in enamel thickness, it appears that a phylogeny based on thick enamel for both last common ancestors would be the most parsimonious reconstruction. However, other alternatives are possible (B and C). While thick enamel could be a primitive retention in early hominids and ramamorphs, it could also have evolved in parallel between the two lineages.

thickness and megadonty in *Ramapithecus*, *Sivapithecus*, and *A. afarensis* are *either* retained primitive traits or have evolved in parallel, then it can be argued that *Ramapithecus* and *Sivapithecus* are not necessarily hominids. If our interpretation of the value of premaxillary morphology in assessing phylogenetic relationships is sustained, then *neither Ramapithecus nor Sivapithecus* can be hominids. It would also then be clear that enamel thickness and presumably associated gnathic buttressing systems are the result of parallel evolution. Both the fossil record and molecular biology support the possibility of fairly extensive parallel evolution in hominoid phylogeny. Recently reported results of DNA restriction mapping (Ferris *et al.*, 1981) are consistent with earlier observations based on DNA hybridization (Beveniste and Todaro, 1976) showing that humans and African apes are more closely related to each other than either is to *Pongo*. These procedures yield branching sequences that are consistent with fossil-based phylogenies positing a split of the lineage ancestral to orangutans prior to the last common ancestor of the living African apes and hominids. If this is true, then the dental mechanism of the ramamorph hominoids and early hominids was subject to parallel trends. This position is supported by recent collecting in the Chinji Formation in Pakistan (Raza, personal communication). We are presently preparing these new specimens for description; one of them (GSP 16075) is a partial maxilla in which the premaxillary region is fairly complete. It clearly has an Asian subnasal pattern, as well as canine root morphology and implantation features that are all characteristic of *Sivapithecus*. It is probable that the rocks from which this specimen was recovered are between 12 and 13 million years old. We have preliminary evidence that the earliest sediments of the Chinji sequence in the Potwar Plateau are about 14 million years old. It is evident, then, that the ramamorph subnasal and canine segments of the alveolar process have been established since the late–middle Miocene, and may prove to be older. We conclude that thick enamel, rotated canines, robust maxillae, and perhaps megadont molars are not necessarily shared derived features linking the ramamorphs and early hominids, but can just as easily be shown to be the result of parallel evolution.

References

Andrews, P. J. 1978. A revision of the Miocene Hominoidea of East Africa. *Bull. Br. Mus. (Nat. Hist.) Geol.* **30:**85–224.

Andrews, P. J., and Tekkaya, I. 1976. *Ramapithecus* in Kenya and Turkey, in: *Le Plus Anciens Hominides*, (P. V. Tobias and Y. Coppens, eds.), pp. 7–21, Colloque VI, IX Union Internationale des Sciences, Prehistoriques et Protohistoriques, Nice. CNRS, Paris.

Andrews, P. J., and Tekkaya, I. 1980. A revision of the Turkish Miocene hominoid *Sivapithecus meteai*. *Paleontology* **23:**85–95.

Andrews, P. J., and Tobien, H. 1977. A new Miocene locality in Turkey with evidence on the origin of *Ramapithecus* and *Sivapithecus*. *Nature (Lond.)* **268:**699–701.

Andrews, P. J., and Walker, A. C. 1976. The primate and other faunas from Fort Ternan, Kenya, in: *Human Origins: Louis Leakey and the East African Evidence* (G. L. Isaac and E. R. McCown, eds.), pp. 279–304, Benjamin, Menlo Park, California.

Andrews, P., Hamilton, W. R., and Whybrow, P. J. 1978. Dryopithecines from the Miocene of Saudi Arabia. *Nature (Lond.)* **274:**249–250.

Benveniste, R., and Todaro, G. J. 1976. Evolution of type C viral genes: Evidence for an Asian origin of man. *Nature (Lond.)* **261:**101–108.

Bosler, W. 1981. Species groupings of early Miocene dryopithecine teeth from East Africa. *J. Hum. Evol.* **10:**151–158.

Brown, B., Gregory, W. K., and Hellman, M. 1924. On three incomplete anthropoid jaws from the Siwaliks, India. *Am. Mus. Novit.* **130:**1–9.

Cave, A. J. E., and Haines, R. W. 1940. The paranasal sinuses of the anthropoid apes. *J. Anat.* **72:**493–523.

De Bonis, L., Bouvrain, G., Geraads, D., and Melentis, J. 1974. Première decouverete d'un Primate hominoide dan le Miocene Superieur de Macedoine. *C. R. Acad. Sci. Paris D* **278:**3063–3066.

Ferris, S. D., Wilson, A. C., and Brown, W. M. 1981. Evolutionary tree for apes and humans based on cleavage maps of mitochondrial DNA. *Proc. Natl. Acad. Sci. USA* **78:**2432–2437.

Gantt, D. G. 1977. *Enamel of Primate Teeth,* Ph.D Dissertation, Washington University, St. Louis. 403 pp.

Gantt, D. G. 1981. Enamel thickness and Neogene hominid evaluation. *Am. J. Phys. Anthropol.* **54:**222 (Abstract).

Greenfield, L. 1979. On the adaptive pattern of *"Ramapithecus". Am. J. Phys. Anthropol.* **50:**527–548.

Hrdlička, A. 1935. The Yale fossils of anthropoid apes. *Am. J. Sci.* **29:**39–40.

Johanson, D. C., and White, T. D. 1979. A systematic reassessment of early African hominids. *Science* **203:**321–330.

Johanson, D. C., White, T. D., and Coppens, Y. 1978. A new species of the genus *Australopithecus* from the Pliocene of Eastern Africa. *Kirtlandia* **28:**1–14.

Kay, R. F. 1981. The nut-crackers—A new theory of adaptations of the Ramapithecinae. *Am. J. Phys. Anthropol.* **56:**141–152.

Kay, R. F. 1982. *Sivapithecus simonsi:* A new species of Miocene hominoid, with comments on the phylogenetic status of the Ramapithecinae. *Int. J. Primatol.* **3:**113–174.

Kretzoi, M. 1975. New ramapithecines and *Pliopithecus* from the lower Pliocene of Rudabánya in Northeastern Hungary. *Nature (Lond.)* **257:**578–581.

Lewis, G. E. 1934. Preliminary notice of new man-like apes from India. *Am. J. Sci.* **27:**161–179.

Lewis, G. E. 1937. Taxonomic syllabus of Siwalik fossil hominoids. *Am. J. Sci.* **34:**139–147.

Molnar, S., and Gantt, D. G. 1977. Functional implications of primate enamel thickness. *Am. J. Phys. Anthropol.* **46:**447–454.

Pilbeam, D. 1979. Recent finds and interpretations of Miocene hominoids. *Annu. Rev. Anthropol.* **8:**333–352.

Pilbeam, D. 1981. New fossil hominoid from Pakistan. *Am. J. Phys. Anthropol.* **54:**263 (Abstract).

Pilbeam, D. 1982. New hominoid skull material from the Miocene of Pakistan. *Nature (Lond.)* **295:**232–234.

Pilbeam, D., Meyer, G. E., Badgley, C., Rose, M. D., Pickford, M. H. L., Behrensmeyer, A. K., and Shah, S. M. I. 1977. New hominoid primates from the Siwaliks of Pakistan and their bearing on hominoid evolution. *Nature (Lond)* **270:**689–695.

Pilbeam, D., Rose, M. D., Badgley, C., and Lipschutz, B. 1980. Miocene hominoids from Pakistan. *Postilla* (Peabody Mus. Nat. Hist., Yale Univ.) **181:**1–94.

Pilgrim, G. E. 1910. Notices of new mammalian genera and species from the tertiaries of India. *Rec. Geol. Surv. India.* **40:**63–71.

Pilgrim, G. E. 1915. New Siwalik primates and their bearing on the question of the evolution of man and the Anthropoidea. *Rec. Geol. Surv. India* **45:**1–74.

Pilgrim, G. E. 1927. A *Sivapithecus* palate and other primate fossils from India. *Mem. Geol. Surv. India (Palaeontol. Ind.)* **14:**1–26.

Sicher, H., and DuBrul, E. L. 1980. *Oral Anatomy,* Mosby, St. Louis.

Simons, E. L. 1961. The phyletic position of *Ramapithecus. Postilla* (Peabody Mus. Nat. Hist., Yale Univ.) **57:**1–9.

Simons, E. L. 1964. On the mandible of *Ramapithecus. Proc. Natl. Acad. Sci. USA* **51:**528–535.

Simons, E. L. 1968. A source for dental comparison of *Ramapithecus* with *Australopithecus* and *Homo. S. Afr. J. Sci.* **64:**92–112.

Simons, E. L., 1976. The nature of the transition in the dental mechanism from pongids to hominids. *J. Hum. Evol.* **5:**500–528.

Simons, E. L. 1978. Diversity among the early hominids: A vertebrate paleontologist's viewpoint, in: *Early Hominids of Africa* (C. Jolly, ed.), pp. 543–566, Duckworth, London.

Simons, E. L. 1979. L' origine des hominides. *Recherche (Paris)* **10:**260–267.

Simons, E. L. 1981. Man's immediate forerunners. *Phil. Trans. R. Soc. Lond. B* **292:**21–41.

Simons, E. L., and Pilbeam, D. 1965. Preliminary revision of the Dryopithecinae (Pongidae, Anthropoidea). *Folia Primatol.* **3**:81–152.

Simons, E. L., and Pilbeam, D. 1971. A gorilla sized ape from the Miocene of India. *Science* **173**:23–27.

Simons, E. L., and Pilbeam, D. 1972. Hominoid paleoprimatology, in: *The Functional and Evolutionary Biology of Primates* (R. Tuttle, ed.), pp. 36–62, Aldine, Chicago.

Simons, E. L., Andrews, P., and Pilbeam, D. 1978. Cenozoic apes, in: *Evolution of African Mammals* (V. J. Maglio and H. B. S. Cooke, eds.), pp. 120–146, Harvard University Press, Cambridge.

Wallace, J. 1978. Evolutionary trends in the early hominid dentition, in: *Early Hominids of Africa* (C. Jolly, ed.), pp. 285–310, Duckworth, London.

Ward, S. C., Johanson, D. C., and Coppens, Y. 1982. Subocclusal morphology and alveolar process relationships of hominid gnathic elements from the Hadar Formation: 1973–1977 collections. *Am. J. Phys. Anthropol.* **57**:605–630.

Ward, S. C., Kimbel, W. H., and Pilbeam, D. 1983. Subnasal alveolar morphology and the systematic position of *Sivapithecus. Am. J. Phys. Anthropol.* **61**(1).

Wegner, R. N. 1936. Sonderbildungen der Kieferhole bei Anthropoiden. *Anat. Anz.* **83**:161–193.

Xu, Q., and Lu, Q. 1979. The mandibles of *Ramapithecus* and *Sivapithecus* from Lufeng, Yunnan. *Vertebr. Palasiat.* **17**:1–13.

Xu, Q., and Lu, Q. 1980. The Lufeng ape skull and its significance. *China Reconstructs* **29**:50–57.

A Reconsideration of the Endocast of *Proconsul africanus*
Implications for Primate Brain Evolution

9

D. FALK

Introduction

Description of the cortical sulcal pattern of the early Miocene *Proconsul africanus* skull (BMNH M 32363) from Rusinga Island, Kenya has had a checkered history. Le Gros Clark and Leakey (1951) described the sulcal markings on the endocranial aspect of the frontal and parietal regions of the right side of the *Proconsul* skull. They suggested that *Proconsul* had a relatively small frontal lobe and a sulcal pattern that appeared to be cercopithecoid-like, rather than hominoid-like. Radinsky (1974) redescribed the sulcal pattern of the same specimen from an endocast that extended slightly further caudally than the area described by Le Gros Clark and Leakey (1951). Radinsky noted that the endocast resembled the brains of modern gibbons more than those of cercopithecoids and concluded that the *Proconsul* endocast therefore represented a hominoid rather than a cercopithecoid sulcal pattern. This latter assessment was based on comparison of the *Proconsul* sulcal pattern with sulcal patterns from a small number of cercopithecoid and gibbon specimens. Since then, large numbers of cercopithecoid (Falk, 1978*a,b;* Radinsky, 1979), ceboid

D. FALK • Department of Anatomy and Caribbean Primate Research Center, University of Puerto Rico, Medical Science Campus, San Juan, Puerto Rico 00936.

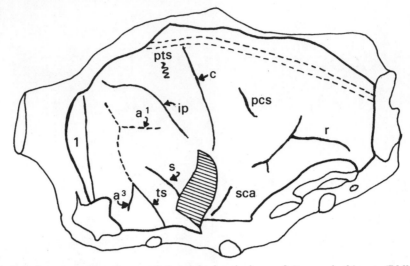

Fig. 1. Sulcal pattern of endocast of the right hemisphere of *Proconsul africanus* (BMNH M 32363), who lived approximately 18 m.y.a. See text for discussion. Abbreviations of sulci after Connolly (1950): a^1, superior parallel; a^3, anterior occipital; *c*, central; *ip*, intraparietal; *l*, lunate; *pcs*, precentral superior; *pts*, postcentral superior; *r*, rectus; *s*, Sylvian fissure; *sca*, subcentral anterior; *ts*, superior temporal. Shading indicates damaged area (Le Gros Clark & Leakey, 1951). Not to scale.

(Falk, 1980*b*, 1981), pongid (Radinsky, 1975, 1977, 1979; Falk, 1980*a*), and hominid (Falk, 1979*b*, 1980*a*, 1980*c;* Radinsky, 1979) endocasts and/or brains have been studied and cladistic analysis has been applied to the study of sulcal patterns (Falk, 1979*a*). The purpose of this chapter is to redescribe the partial endocast of *Proconsul* (BMNH M 32363) a third time, in light of recent findings, and to discuss the implications of its sulcal pattern for primate brain evolution.

Figure 1 illustrates the right lateral view of the somewhat crushed endocast from the *Proconsul africanus* specimen, which was studied at the British Museum of Natural History in London. This figure was prepared from a plaster copy of the endocast. The lunate and postcentral superior sulci seen in Fig. 1 were not figured in either Le Gros Clark and Leakey's (1951) or Radinsky's (1974) illustrations. However, except for these and other minor details, such as the degree to which certain sulci are shaded, Fig. 1 is in general agreement with the sulcal pattern published by Radinsky. Thus it is my interpretation of the sulcal pattern, rather than the basic sulcal pattern itself, that differs from previous descriptions of the *Proconsul* specimen.

The Proconsul Sulcal Pattern

Frontal Lobe

Three sulci appear in the frontal lobe of *Proconsul* (i.e., rostral to the level of the central sulcus; see Fig. 1): the precentral superior (*pcs*), subcentral anterior (*sca*), and rectus (*r*) sulci (Radinsky, 1974). Radinsky suggested that the sulcus identified as the subcentral anterior in his illustration of the *Proconsul* endocast might possibly be the opercular sulcus (*io*). However, *io* forms part of the superior limiting sulcus of the insula (Connolly, 1950, p. 54) and I think the sulcus in question in the *Proconsul* endocast is located too far caudally for this identification. I therefore agree with Radinsky's decision to label the sulcus as the subcentral. The *sca* and *pcs* sulci are present frequently in gibbons, the Great Apes, both subfamilies of Old World monkeys, and the larger brained New World monkeys (Table 1).

The frontal lobe of *Proconsul* reproduces a laterally curved rectus sulcus (*r*) that gives off a rostromedially directed spur from its middle portion and another smaller, laterally directed spur in its caudal extremity (Fig. 1). *Proconsul's* frontal lobe is noteworthy for its lack of an arcuate sulcus (*arc*) that is separate from and surrounds the caudal end of *r* (Radinsky, 1974). Presence of *arc* typifies all genera of extant Old World monkeys and *Cebus* in the New World (Falk, 1979a). During primate evolution, there appears to have been a shift from a longitudinally oriented sulcul pattern in frontal/parietal regions (i.e., similar to the coronolateral sulcus of "prosimians") to a transversely oriented central sulcus as seen in anthropoids (Radinsky, 1975). Thus a simple rectus sulcus (*r*) which most closely resembles the simple, longitudinal orientation of the rostral portion of the "prosimian" coronolateral sulcus appears to be relatively primitive, whereas addition of a transversely oriented

Table 1. Sulcal Pattern of *Proconsul* Compared to Sulcal Patterns of Extant Anthropoids[a]

Sulci	Proconsul	Ceboids Small-brained	Ceboids Large-brained	Cercopithecoids Colobines	Cercopithecoids Cercopithecines	Hominoids Gibbons	Hominoids Great Apes
ip, s, ts	x	x	x	x	x	x	x
c, l, r	x	d	x	x	x	x	x
pcs, sca	x	o	x	x	x	x	x
pts	d	o	x	x	x	x	x
a^1	x	o	o	o	x	o	x
a^3	x	o	o	o	o	x	x
arc[c]	o	o	o[b]	x	x	o	x
fs	o	o	x	o	x	x	x
fo	o	o	x	x	x	x	x
io	o	o	o	o	o	x	x

[a]Sulci listed above line are present on the *Proconsul* endocast; those listed below the line are absent from the frontal lobe of *Proconsul*. Abbreviations of sulci: *arc*, arcuate; *fs*, superior frontal; *fo*, orbitofrontal; *io*, opercular that are present on lateral surface of brain. Other sulci as in Fig. 1. An x means that sulcus is present frequently if not always, o means the sulcus is rarely if ever present, and d means that the sulcus is represented by a dimple.
[b]An exception to this entry is *Cebus*, in whom *arc* is typically present. Presence and absence of sulci determined from the following sources: *Proconsul* (this report); ceboids (Falk, 1980b); cercopithecoids (Falk, 1978b); gibbons and the Great Apes (Connolly, 1950).
[c]In Connolly's (1950) abbreviations, *arc* = *pci* + *h*.

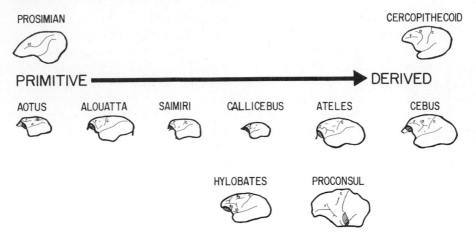

Fig. 2. Frontal lobe of *Proconsul* (reversed from right to left view for comparative purposes) compared to frontal lobes of New World monkeys. Top: typical "prosimian" and cercopithecoid frontal lobe sulcal patterns [after Radinsky (1975)]. Note shift from primitive, longitudinally oriented coronolateral sulcus (*cl*) in lower primates to rectus sulcus (*r*) surrounded by a transversely oriented arcuate sulcus (*a*) and a transversely oriented central sulcus (*c*) in higher primates such as cercopithecoids. Frontal lobe patterns of individual extant New World monkey specimens [after Falk (1980*b*)] are arranged from left to right from the most primitive to the most derived. The frontal lobe sulcal pattern of *Proconsul* resembles that of the *Ateles* specimen, i.e., it has not achieved the fully derived state found in all genera of Old World monkeys and present only in *Cebus* in the New World. Note also that the typical gibbon frontal lobe [after Connolly (1950)] retains the primitive condition. See text for discussion. Not to scale.

arcuate sulcus represents the derived condition (Fig. 2). Sulcal patterns of New World monkeys are much more variable than sulcal patterns of Old World monkeys. Character states can be found in ceboids that span from the very primitive rectus seen in some *Aotus* specimens to the fully derived rectus/arcuate complex that is typical of *Cebus*.

Figure 2 shows that the pattern of *r* and lack of a separate *arc* of *Proconsul* resembles the sulcal pattern of New World monkeys such as *Ateles*, whose lateral portion of frontal lobe is intermediate between the primitive condition seen in the *Aotus* specimen and the derived condition which is typical for only *Cebus* in the New World. Thus the frontal lobe of *Proconsul* is primitive compared to those of all Old World monkeys and *Cebus* and derived compared to "prosimians" and many New World monkeys (Fig. 2). The rectus sulcus of *Proconsul* resembles *r* of gibbons (Radinsky, 1974), which have retained the primitive condition (i.e., lack of *arc*).

Other Sulci

The postcentral superior (*pts*) appears as more of a dimple than a discrete sulcus. It was detected visually and by palpation. The lunate (*l*), intraparietal (*ip*), caudal portion of Sylvian (*s*), and lateral portion of superior temporal (*ts*) sulci are all clearly present on the *Proconsul* endocast (Fig. 1). These sulci appear frequently in ceboids, cercopithecoids, gibbons, and the Great Apes (Table 1).

The caudal end of *ts* appears to be curved forward (Radinsky, 1974) although this sulcus (a^1 = superior parallel sulcus) is not as distinct as most of the others on the endocast. A^1 is typically present in the Great Apes, but not in gibbons or ceboid monkeys (Connolly, 1950; Falk, 1979*a*, 1980*b*). Connolly (1950) shows an a^1 for various cercopithecines, but not for colobine Old World monkeys. This fact is verified by my endocast collection. For example, a^1 is present in five of the ten *Papio* hemispheres which reproduce the caudal end of *ts*. This sulcus is rarely present in colobine endocasts.

A caudolaterally directed spur from *ts* (a^3 = anterior occipital sulcus) is clearly present on the *Proconsul* endocast (Radinsky, 1974). It is also frequently present on hominoid brains, including those of gibbons. As noted earlier (Radinsky, 1974), this feature is rarely if ever seen in cercopithecoid endocasts/brains, nor is it present in ceboid endocasts/brains (Table 1).

Sulci Missing from Frontal Lobe

As noted above, *arc* is a derived feature present in the Great Apes and cercopithecoids but lacking in gibbons, ceboid monkeys (except *Cebus*), and *Proconsul* (Table 1). A frontalis superior sulcus (*fs*) is also derived for primates (Radinsky, 1974), present in all apes and many monkeys, and lacking in *Proconsul* (Radinsky, 1974). Two additional sulci are present in all apes but lacking in the frontal lobe of *Proconsul:* the fronto-orbital sulcus (*fo*) (also present in many monkeys) and the opercular sulcus (*io*), which forms part of the superior limiting sulcus of the insula and is not developed on lateral hemispheres of monkeys (Connolly, 1950, p. 54). Since these sulci are present in gibbon brains that are smaller than the *Proconsul* endocast, allometry does not account for their presence (see below). Therefore *fo* and *io*, like *fs*, seem to be associated with differential expansion of cortex, i.e., these sulci appear to be derived in extant anthropoids.

Brain Size

Radinsky (1974) estimated the endocranial volume of *Proconsul* to be 150 cm³, but later he reconsidered and decided that no reasonably accurate estimate of volume could be made, because of the "crushed and incomplete nature of the endocast" (Radinsky, 1979). I agree that it would be foolish to attempt an exact estimate of endocast volume on a partial, crushed specimen. Nevertheless, I have compared *Proconsul* to endocasts of adult primates in my collection (representing all genera of extant monkeys) in an effort to determine general size and shape. The length of the frontal lobe as well as the length of its orbital edge suggest that the *Proconsul* endocast is large. In these respects, *Proconsul* seems to resemble *Papio* more than any other genus of monkey (in his 1974 paper, Radinsky suggested that the *Proconsul* endocast was about the size of a *Papio* endocast). However, the frontal lobe of *Proconsul* seems very flattened compared to frontal lobes of *Papio* and other extant monkeys (due either to artificial crushing or to real shape differences). Nevertheless, the *Proconsul* endocast is clearly larger than the two *Hylobates* and three *Ateles* endocasts in my collection. Therefore, even by today's standards, the *Proconsul* endocast appears to be large—i.e., it is definitely larger than endocasts of the biggest brained ceboids (the atelines) and the smallest brained hominoids (gibbons). The size of the *Proconsul* endocast also compares favorably with endocast size of the larger brained cercopithecoids; it definitely appears larger than endocasts of adult macaques and (distortion and incompleteness aside) *may* even approximate the still larger size of baboon endocasts. However, as stated above, an exact estimate of the *Proconsul* endocast volume is precluded by the condition of the specimen (Radinsky, 1979). Unfortunately, without a numerical estimate of brain size, one cannot reliably estimate the important statistic of relative brain size.

Discussion and Conclusions

Three features of sulcal pattern were earlier (Radinsky, 1974) believed to align *Proconsul* with hominoids (particularly gibbons) rather than cercopithecoids: (1) a laterally curved sulcus rectus and concomitant lack of a separate arcuate sulcus, (2) a forward curved caudal end of the superior temporal sulcus (a^1 of Fig. 1), and (3) a caudolaterally directed spur from the superior temporal sulcus (a^3 of Fig. 1). As discussed above, lack of *arc* in *Proconsul* and gibbons is a shared primitive feature (see Table 1). According to the tenets of cladistic analysis (Delson, 1977), only shared derived traits can be used to align two taxa into sister groups; thus, the shared primitive lack of *arc* in *Proconsul* and gibbons does *not* indicate that *Proconsul* was a hominoid. It merely indicates that gibbons are primitive in this feature.

Presence of a^1 on the *Proconsul* endocast is a shared derived characteristic with the cercopithecine subfamily of Old World monkeys and the Great Apes, but not gibbons (Table 1). Table 1 shows that the only derived sulcus shared exclusively by *Proconsul* and all hominoids is a^3. However, since there is only one such feature, parallel evolution cannot be ruled out as an explanation for its presence in both groups.

Table 1 also reveals that, of the 11 sulci present on the *Proconsul* endocast, nine are common to anthropoids: *c, ip, l, pcs, pts, r, s, sca,* and *ts.* [Elsewhere, I have attributed lack or dimpling of some of these sulci in the smallest brained ceboids to allometry. See Falk (1979a, 1980b) for details.] *Proconsul* shares with anthropoids certain similarities of sulcal pattern which differ from the typical "prosimian" pattern [the reader is referred to Radinsky (1968, 1970) for thorough discussion of lower primate sulcal patterns]: *Proconsul* and anthropoids are characterized by a well-marked transverse *c*. Instead of exhibiting a "prosimian"-like coronolateral sulcus, *Proconsul* and extant anthropoids are characterized by discrete *ip* and *r*. The primary visual cortex of *Proconsul* and anthropoids, but not "prosimians," is relatively expanded and consequently delineated by *l*. Smaller sulci, such as *sca* and *pcs*, are present in *Proconsul* and anthropoids but not in lower primates. Thus, as confirmed by Table 1, the sulcal pattern of *Proconsul* is basically anthropoid-like. It is also important to establish what sulci are *not* present on the *Proconsul* endocast. According to Radinsky (1974), lack of a frontal sulcus (*fs*) is the only primitive feature evident on the endocast. However, in addition to lack of *fs*, lack of *arc* is a second primitive feature of the endocast (see above). Table 1 also lists two additional derived sulci, *fo* and *io*, that are present in hominoids, including gibbons, but lacking in *Proconsul*. Thus, *Proconsul* lacks a total of four derived features in its frontal lobe alone that appear in the frontal lobes of the Great Apes (and three of which are also shared by gibbons; see Fig. 2). Since the *Proconsul* endocast clearly appears to be larger than endocasts of many cercopithecoids and gibbons, *allometric considerations cannot account for lack of these sulci in Proconsul.*

According to Radinsky (1974, p. 26), "except for the lack of a frontal sulcus, I see no reason to consider the *Dryopithecus* [=*Proconsul*] endocast primitive compared to modern hominoids." However, to me the endocast of *Proconsul* appears to reproduce basic anthropoid features (first nine sulci in Table 1) in its sulcal pattern. *Proconsul* is more primitive than extant anthropoids because its endocast lacks four derived sulci that characterize the latter's frontal lobes in varying degrees (Table 1), and its endocast is not as convoluted as one would expect for a modern anthropoid endocast of roughly equivalent size (see above). The *Proconsul* endocast shares only one derived sulcus (a^1) with some cercopithecines and all the Great Apes and shares only one derived sulcus (a^3) exclusively with gibbons and the Great Apes. Presence of two derived features in *Proconsul* that are distributed differently across subsets of apes (and monkeys) is not strong enough evidence to offset the remainder of Table 1, i.e., evidence that strongly suggests that *Proconsul* is a primitive anthropoid (i.e., derived compared to lower primates but lacking certain derived features of extant anthropoids) in its sulcal pattern.

There are no derived features on the *Proconsul* endocast that preclude it from being ancestral to the brains of the Great Apes. However, the situation with gibbons is more problematic. For *Proconsul* to be ancestral to gibbons, there would have to have been a reduction in brain size during gibbon evolution, loss of one derived sulcus (a^1), and acquisition of three other derived sulci: *fs*, *fo*, and *io*.

Similarly, for *Proconsul* to be ancestral to ceboids there would have to have been a reduction in brain size, loss of two derived sulci (a^1, a^3), and acquisition in larger ceboids of *fs* and *fo*. To be ancestral to cercopithecoids, descendants of *Proconsul* would have to have exhibited some reduction in brain size, loss of a^1 and a^3, acquisition of *arc* and *fo*, and later acquisition of a^1 and *fs* in the cercopithecine subfamily. (See Table 1.)

Tabulating the above information, the simplest phylogenetic explanation to account for *all* sulcal patterns listed in Table 1 is to hypothesize a recent common ancestor of *Proconsul* on the one hand and the common ancestor of extant anthropoids on the other hand. The brain of this ancestor would resemble the *Proconsul* endocast except that (1) it would lack a^1 and a^3 and (2) its volume would be smaller. In other words, *the endocast of the common ancestor of Proconsul and all extant anthropoids would appear like the endocast in Fig. 1 except that it would be smaller and a^1 and a^3 would be missing.* If one accepts *Proconsul* as directly ancestral to the Great Apes, on the other hand, then one must deal with the above-mentioned problems associated with the ancestry of gibbons plus potential dilemmas (depending on how the other groups are treated) to do with the necessity of postulating repeated secondary size reduction as well as repeated parallel loss of the same derived sulci. However, no matter how one interprets the phylogenetic history of anthropoid brain evolution, one fact remains clear: the endocast of *Proconsul* appears to be basically primitive for anthropoids in its succal pattern.

Implications for Primate Brain Evolution

The oldest record of a primitive anthropoid sulcal pattern is provided by three incomplete endocasts of the Fayum Oligocene primate *Aegyptopithecus zeuxis* (Radinsky, 1973, 1974), which lived approximately 27 m.y.a. Although the complete sulcal pattern remains unknown, Radinsky has shown that *Aegyptopithecus* was characterized by central, sylvian, superior temporal and lunate sulci (Table 1) as well as an apparent lack of sulci on preserved portions of the frontal lobes. Since the *Aegyptopithecus* endocast(s) is much smaller than that of *Proconsul* (Radinsky, 1973), allometry could account for the fact that *Aegyptopithecus* manifests fewer frontal lobe sulci than *Proconsul*. Thus, one cannot make a definitive statement about the relationship between *Aegyptopithecus* and *Proconsul* based on sulcal patterns except to say that there is nothing to rule out the possibility of *Aegyptopithecus* as an ancestor of *Proconsul*.

Walker (Pilbeam and Walker, 1968) described an endocast of the Napak frontal bone from early Miocene deposits of Uganda. He considered the Napak frontal to be cercopithecoid-like rather than hominoid-like. Radinsky (1974) notes that the fragmentary Napak endocast reproduces only *pcs* and *r* and that it apparently lacked *arc*. Radinsky therefore suggested that the Napak frontal, like the endocast of *Proconsul*, represented a hominoid on the basis of affinities with gibbons. Based on the foregoing analysis, I would reclassify the sulcal pattern of the Napak frontal as primitive anthropoid.

Prior to this investigation, it was believed that the oldest, most complete record of anthropoid brains of *modern* appearance was provided by the 18-million-year-old *Proconsul* endocast (Radinsky, 1974). Although *Proconsul* may have been a "dental ape," this analysis suggests that neurologically, *Proconsul* was a primitive anthropoid. Interestingly, the oldest known record of endocasts that reproduce sulcal patterns similar to those of modern Great Apes were produced in South African australopithecine skulls (Falk, 1980*a*)! It appears, then, that during pongid as well as hominid evolution, evolution of the external morphology of the cerebral cortex lagged strikingly behind dental/skeletal evolution (see also Radinsky, 1973, 1974, 1975, 1979). Since this is the case for two separate groups of fossil primates (hominids and pongids), "neurological lag" may represent a general trend in primate, or even mammalian, brain evolution.

ACKNOWLEDGMENTS

I thank authorities at the British Museum (Natural History), London for access to the *Proconsul* specimen (BMNH M 32363) and Leonard Radinsky for loaning me a copy of its endocast. I am grateful to Lydia Warlop for typing the manuscript, Juan Mussenden for preparing the illustrations, and Robert Eaglen and Leonard Radinsky for criticism of the manuscript. The views expressed in this chapter are solely my own.

References

Connolly, C. J. 1950. *External Morphology of the Primate Brain,* Thomas, Springfield, Illinois.

Delson, E. 1977. Catarrhine phylogeny and classification: Principles, methods and comments. *J. Hum. Evol.* **6:**433–459.

Falk, D. 1978a. Brain evolution in Old World monkeys. *Am. J. Phys. Anthropol.* **48:**315–320.

Falk, D. 1978b. External Neuroanatomy of Old World monkeys (Cercopithecoidea). *Contrib. Primatol.* **15:**1–95.

Falk, D. 1979a. Cladistic analysis of New World monkey sulcal patterns: Implications for primate brain evolution. *J. Hum. Evol.* **8:**637–645.

Falk, D. 1979b. On a new australopithecine partial endocast. *Am. J. Phys. Anthropol.* **50:**611–614.

Falk, D. 1980a. A reanalysis of the South African australopithecine natural endocasts. *Am. J. Phys. Anthropol.* **53:**525–539.

Falk, D. 1980b. Comparative study of the endocranial casts of New and Old World Monkeys, in: *Evolutionary Biology of the New World Monkeys and Continental Drift* (R. L. Ciochon and A. B. Chiarelli, eds.), pp. 275–292, Plenum, New York.

Falk, D. 1980c. Hominid brain evolution: The approach from paleoneurology. *Yearb. Phys. Anthropol.* **23:**93–107.

Falk, D. 1980d. Language, handedness and primate brains: Did the australopithecines sign? *Am. Anthropol.* **82:**72–78.

Falk, D. 1981. Sulcal patterns of fossil *Theropithecus* baboons: Phylogenetic and functional implications. *Int. J. Primatol.* **2:**57–69.

Le Gros Clark, W. E. and Leakey, L. S. B. 1951. The Miocene Hominoidea of East Africa. *Fossil Mammals of Africa* (Br. Mus. Nat. Hist.) **1:**1–117.

Pilbeam, D., and Walker, A. 1968. Fossil monkeys from the Miocene of Napak, north-east Uganda. *Nature (Lond.)* **220:**657–660.

Radinsky, L. B. 1968. A new approach to mammalian cranial analysis, illustrated by examples of prosimian primates. *J. Morphol.* **124:**167–180.

Radinsky, L. B. 1970. The fossil evidence of prosimian brain evolution, in: *The Primate Brain* (C. Noback and W. Montagna, eds.), pp. 209–224, Appleton-Century-Crofts, New York.

Radinsky, L. B. 1973. *Aegyptopithecus* endocasts: Oldest record of a pongid brain. *Am. J. Phys. Anthropol.* **39:**239–248.

Radinsky, L. B. 1974. The fossil evidence of anthropoid brain evolution. *Am. J. Phys. Anthropol.* **41:**15–28.

Radinsky, L. B. 1975. Primate brain evolution. *Am. Sci.* **63:**656–663.

Radinsky, L. B. 1977. Early primate brains: Fact and fiction. *J. Hum. Evol.* **6:**79–86.

Radinsky, L. B. 1978. Evolution of brain size in carnivores and ungulates. *Am. Nat.* **112:**815–831.

Radinsky, L. B. 1979. *The Fossil Record of Primate Brain Evolution. 49th James Arthur Lecture, 1979.* American Museum of Natural History, New York.

The Enamel of Neogene Hominoids 10

Structural and Phyletic Implications

D. G. GANTT

Introduction

The present view of primate evolution is based almost completely on dental remains. This nearly total reliance on teeth by paleoanthropologists in their attempts to reconstruct phylogenies and to interpret functional parameters makes it imperative that every piece of information be evaluated. Teeth are the most abundant fossil remains because they are the hardest and most highly calcified elements of the mammalian skeleton. Thus, they are best suited to survive the processes of death, scavenging, and burial. They are also least susceptible to the processes of fossilization. The teeth are an unique organ, constructed of three calcified tissues: enamel, dentin, and cementum.

The tooth crown of primates, as in most mammals, is completely covered by enamel. This tissue is the hardest and most highly calcified tissue in the body and provides protection for the underlying softer layers of dentin and cementum. The major function of enamel is to serve as a masticatory surface. Changes in tooth morphology result from either minor changes due to localized thickening of the enamel or major alterations in cusp and groove

D. G. GANTT • Institute of Dental Research, University of Alabama School of Dentistry, University Station, Birmingham, Alabama 35294.

patterns. The configuration of the enamel crown provides the most definitive conclusions regarding the zoological position and the affinities of a given species. However, morphological and odontometric analyses of the enamel crown have neither resolved problems nor provided definitive answers regarding phylogenies and adaptation, especially when isolated teeth are evaluated. The analyses of these remains have often led to conflicting theories and interpretations. To provide for a more complete understanding of adaptation and phylogeny, new parameters and methods of analysis must be applied.

Dental tissues, in this chapter, are analyzed at three levels: (1) gross analysis—at which morphological and odontometric studies are conducted (magnifications of I× to 500×); (2) architectural analysis—at this level, finer features of the dental tissues, such as enamel prism orientation and direction, are studied in thin sections and fractured and ground sections (magnifications of 500× to 2000×; Figs. 1, 2, 14, and 19; and (3) ultrastructural analysis—this level of analysis is beyond the limits of the light microscope and requires the use of the scanning electron (SEM) and transmission electron (TEM) microscopes (magnifications of 2000× to 100,000× or greater; Figs. 3, 6, 13, 15–18, 20–22, and 24).

The study of dental structures must involve analysis on all three levels before evolutionary changes in tooth form and function can be fully understood. At present all of the knowledge and data are derived from level 1 analyses, although during the past decade a few studies have been conducted at level 2. Molnar and Ward (1975) and Molnar (1976) conducted studies on the microstructural quality of nonhuman primate teeth showing that a gradation existed. A few microstructural defects occur in Old and New World monkeys, which increase in the pongids and are nearly universal in humans. More recently, Poole and Shellis (see Lavelle *et al.,* 1977) attempted to describe the calcified dental tissues of primates. In reviewing the literature, they found that no comparative study had been conducted; therefore, it became necessary to conduct a survey of a number of primate species to provide the basic data. Although their sample size was limited, they concluded: "With regard to taxonomy, our conclusion is that tooth structure is of value. The dental tissues of certain species are sufficiently characteristic to be readily recognized, and a particular combination of characters may well allow an unknown specimen to be assigned. . . . Establishment of the full taxonomic potential of tooth structure must, however, await a truly comprehensive survey" (Lavelle *et al.,* 1977, p. 266).

Several studies by the author (Gantt, 1977, 1979*a,b,c,* 1980, 1981*a,b*) have revealed a number of significant differences within the order Primates in the thickness of enamel, enamel prism patterns, and microstructure. It becomes clear that, apart from taxonomic implications, scrutiny of dental tissues of primates highlights a whole series of unsolved problems. Of special interest and importance are those concerning the relationship between function and structure of a tissue and between structure and its development.

Furthermore, studies on other mammalian orders have also verified the importance of architectural analysis. Von Koenigswald (1977, 1980) has

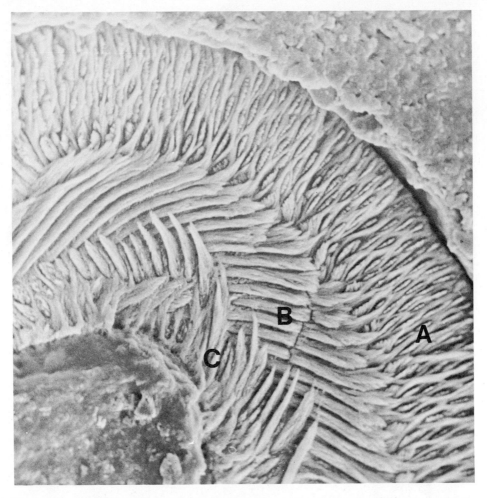

Fig. 1. The enamel structure of the arvicolid rodent molar correlated with the stress pattern during mastication on its leading and trailing edges. The molars of the Arvicolidae are differentiated from those of other rodents by the occurrence of the lamellar enamel within the chewing surface, along with radial and tangential enamels. (A) The radial enamel area, (B) tangential enamel, (C) lamellar enamel. [Courtesy of W. von Koenigswald.]

shown that three enamel types were present in the Arvicolidae, Villafranchian rodents. He demonstrated that the changes in enamel architecture corresponded to the forces of mastication (Fig. 1).

   Also, Remy (1976), in a study of tooth histology and the Equoidea, was able to supply new arguments in favor of the Palaeotheriidae being considered as a single family opposite the Equidae. Remy concluded, based on the data obtained from tooth histology (architectural analysis) and morphological analysis, that it appeared suitable to place all European Paleogene Equoidea into the family of Palaeotheriidae except *Hyracotherium*. *Hyracotherium* may be remote because it stands at the bottom and approximates the ancestral stock of both equoid phyla in the primeval family, Hyracotheriidae.

In addition, Shellis and Poole (1979) have documented that the incisor enamel of the aye-aye, *Daubentonia madagascariensis*, is unique among mammals. The enamel prism decussation (crossing over) is in the form of a spiral (Fig. 2); therefore, a fragment of enamel from the aye-aye can easily be distinguished from rodent and other primate incisors.

The third level, ultrastructural analysis, was attempted as early as 1849 by Tomes, although he was unable to correctly describe the enamel prism, due to the limitations of the light microscope. It was not until the advent of the

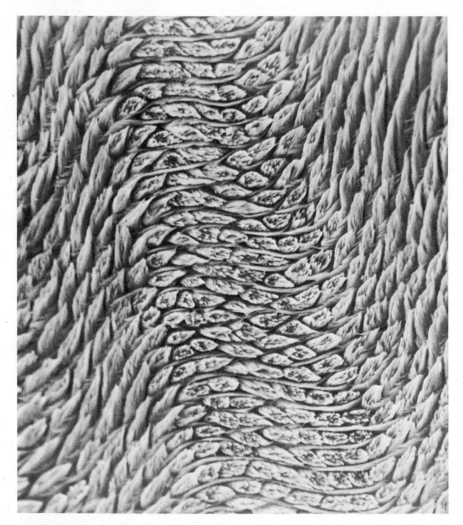

Fig. 2. The enamel architecture in the continuously growing incisors of *Daubentonia madagascariensis*, the aye-aye, revealing that the arrangement of prisms is totally different from that seen in the incisors of rodents. [Courtesy of Peter Shellis and David Poole.]

scanning electron microscope (SEM) and transmission electron microscope (TEM) that researchers could begin to understand the development and ultrastructural features of enamel, especially the enamel prism. Poole and Brooks (1961) have identified the arrangement of crystallites within human enamel prisms, and Meckel (1971) has defined the "keyhole," a human enamel microstructure (see the section on enamel microanatomy and refer to Fig. 3). However, the work of Boyde has provided a more complete understanding of enamel structures in mammals (Boyde, 1964, 1965a,b, 1967, 1969a,b, 1970, 1971, 1975, 1976a,b, 1978; Boyde and Jones, 1972).

Recent studies have documented that specific enamel prism patterns do exist and can be used as a taxonomic indicator (Boyde, 1978; Gantt, 1977, 1979a,b,c; Gantt et al., 1977; Lavelle et al., 1977; Shellis and Poole, 1979; Remy, 1976). However, while a few researchers have reported different prism patterns in a particular species [Shellis and Poole, in Lavelle et al., (1977)]; others have even suggested that prism morphology contains no information on phylogenetic relationships (Vrba and Grine, 1978a,b). Some species-specific differences obtained by Shellis and Poole were found to be acid etching artifacts or due to the method of analysis (Shellis and Poole, personal communication). On the other hand, the differences obtained by Vrba and Grine (1978a,b) were the result of a number of variables: (1) lack of stereoanalysis; (2) differential effects of acid etchants; (3) type of fossilization; and (4) prism direction. To be more specific, these four variables were not adequately considered (Gantt, 1979b,c), but must be evaluated before correct interpretation can be made. These variables and associated problems of interpretation are as follows:

1. The first variable involves methods of analysis. The main problem in studying enamel is that its structures must be seen in three dimensions. The importance of stereoanalysis of enamel structures cannot be overemphasized. Boyde (1964, 1975) has published extensively on this subject and on the methods and techniques of applying stereoanalysis to the study of enamel prism patterns.

2. The differential effects of acid etchants is the second variable. Vrba and Grine (1978a,b) used a dilute solution of hydrochloric acid (10% HCl) to prepare their specimens for analysis. Acid treatment is necessary to etch the specimen in order to remove the prismless layer (approximately 10 μm) and to more fully reveal the organization of the prisms and crystallite arrangement. Also, etching is of potential interest in the field of mammalian taxonomy, because enamel is the only tissue which is virtually fossilized before death.

An intensive study of the qualitative and quantitative aspects of enamel etching with acid and EDTA on human teeth has brought to light certain inadequacies and the presence of artifacts as the result of specific etchants and/or lengths of etching time (Boyde et al., 1978). The data obtained revealed that the etchants, including the 10% hydrochloric acid solution used by Vrba and Grine (1978a,b), will give rise to artifacts of one sort or another. Among the important artifacts of etching are the generation of either the open honeycomb pattern or its converse in which the prism boundary crack is widened. In addition, problems of the growth of the new phase (reprecipitation of crystallites) due to acid etchants must be considered. The results of this study (Boyde *et al.*, 1978) led the authors to conclude that a dilute solution of phosphoric acid (0.74 M = 0.5% vol/vol of concentrated, 85% acid) was most suitable for etching polished enamel surfaces to study crystallite orientation and prism organization. This method has been successfully used with etch times of 30–60 s (Boyde, 1978; Gantt, 1979*b,c*).

3. The product of fossilization is the third variable. Fossilization subjects the enamel to a variety of chemical and physical agents. These agents change the surface chemical and physical composition. Therefore, treatment of the tooth surface with acid etchants will produce greatly differing prism patterns due to the remineralization products, that is, the differences in mineral content of the crystals and effects of differential etching. The fossil enamel may already be highly etched or its acid resistance altered. Acid resistance may be higher, related to a high fluoride content, or lower, due to a high carbonate content, than in the natural state. The specimens examined by Vrba and Grine (1978*a,b*) were recovered from a limestone fissure and had a high carbonate fossilization matrix; therefore, they were highly susceptible to etching (see Fig. 19B for example of the fossilization zone).

4. A fourth variable which must be considered is the position of the prisms in the plane of view and prism decussation (crossing over). Studies of the nature and development of prism decussation in mammalian enamel document that the greatest extent of decussation is in the middle portions of the tooth (Boyde, 1969*a*). It has also been shown that the position of the prism bundles varies from the cemento–enamel junction to the cusp apex (Boyde, 1964, 1969*b*). The results of these studies document the importance of prism direction and position for the correct interpretation and description of prism organization and cross-sectional area (Gantt, 1979*c*).

These four variables—stereoanalysis, differential effects of acid etchants, type of fossilization, and prism direction—must be considered together before correct interpretations can be made of any species' enamel prism pattern.

The Microanatomy of Enamel

Enamel is a unique tissue of ectodermal origin. Its structural composition is that of hydroxyapatite crystals. The hydroxyapatite of enamel is similar to the structure of synthetic and geological apatites. It consists of calcium, phosphate, and hydroxyl ions in a stoichiometric ratio: 10 Ca:6 (PO_4):2 OH. Each crystallographic unit cell consists of 18 ions, arranged as hydroxyapatite (Weatherell, 1975). These apatite crystals vary in thickness (10–40 nm), in width (30–60 nm), and in length (100–1000 nm). They are arranged into elongated prisms or rods, comparable to the atoms in a crystal lattice. The prisms and the crystals are regularly arranged and, until the advent of the

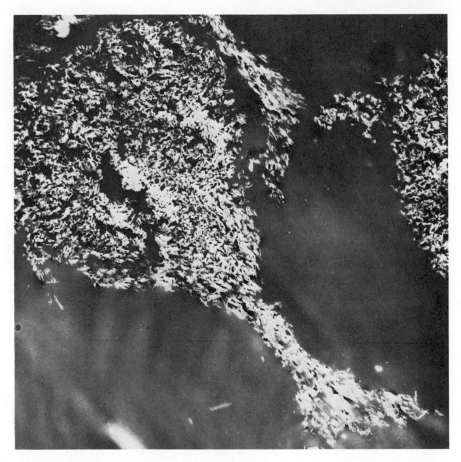

Fig. 3. A transmission electron microphotograph of a developing enamel prism ("keyhole") from *Homo sapiens* taken at 35,000×. Note the large "winged" or "fishtail" process of the tail portion of this prism. [Courtesy of D. Poole.]

scanning electron microscope, they were beyond the resolution capabilities of the optical microscope (2000×; see Figs. 3–5). The maturation of these crystallites is still questionable, but their orientation has been defined by Poole and Brooks (1961) in human enamel. The crystallites are arranged parallel to the prism axis, with a gradual transition to an oblique angle of 20–45° in specific areas [see Figs. 4 and 5, and Meckel (1971), Osborn (1974), and Poole and Brooks (1961)].

The prisms are oriented at right angles to the dentin surface and follow an undulating course throughout the thickness of the enamel. Prism decussation does vary from species to species, as has been shown by Boyde (1969*a*). However, within a given species, prism decussation remains constant, although each prism has its own shape and direction, and this direction may change abruptly. In general, the prisms are approximately horizontal in the cervical and central parts of the crown. Near the incisal edge or tip of the

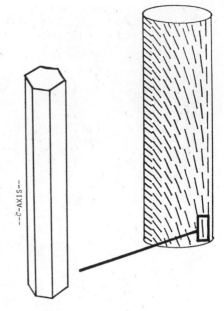

Fig. 4. Crystallite orientation within the enamel prism. In the human "keyhole" enamel prism, hydroxyapatite crystals are arranged parallel to the prism axis with a gradual transition to an oblique angle of 20–45°.

cusps they change gradually to an increasingly oblique direction, obtaining a vertical position at the ends or tips of the cusps in enamel crowns. These changes account for the different patterns and also account for the variations in prism patterns over the tooth surface.

Observations of the enamel prisms through the light microscope reveal the presence of an interprismatic substance which separates the prisms. Recent investigations utilizing the transmission and scanning electron microscopes have shown what had previously been considered as interprismatic material to be actually the *tail* of each prism (Helmcke, 1967; Meckel, 1971; Osborn, 1974). Therefore, it was suggested that the term *interprismatic substance* be dropped in favor of *interprismatic region* (Boyde, 1965b; Meckel, 1971; Osborn, 1968, 1974).

Another feature of the prism observed through the light microscope is the prism sheath which surrounds the prism and is usually incomplete on its cervical side. Within each prism, the ordering of the crystals is distinct from the adjacent prism or prisms (see Figs. 4 and 5); thus, there is a boundary between contiguous enamel prisms. At these boundaries there appear to be distinct changes in crystal orientation which are related to the position of the prism sheath (Helmcke, 1967). Frank and Nalbandian (1967) suggest that the prism sheath is a portion of the organic matrix about 100 nm wide which fails to calcify and is continuous with the prism body as well as the interprismatic region. Osborn (1974) disagrees and suggests that the prism sheath is an optical artifact due to crystal orientation.

The analysis of the shape of the prism has provided for a better understanding of the inner structures of the prisms. The "keyhole"-shaped prism pattern in human enamel was proposed by Meckel *et al.* (1965a,b) based on the analysis of prism crystallites by Poole and Brooks (1961). Subsequent studies

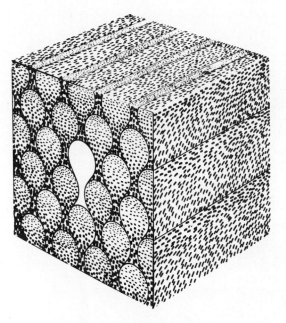

Fig. 5. A model of the human "keyhole" enamel prism pattern after Poole and Brooks (1961). Note the wide tail process described as "winged" or "fishtail." Also note the hydroxyapatite crystal arrangement.

by Boyde (1964, 1969a,b, 1971), Meckel (1971), and Osborn (1967, 1968, 1970, 1974) have documented the crystallite arrangement within each keyhole-shaped prism (see Figs. 3–6). These investigators have also demonstrated that previously described patterns observed in human enamel were the result of sections cut at different planes. The *head* of the enamel prism is near the occlusal and/or incisal surface, while the *tails* point cervically. The prisms are approximately 5 μm across the head and 9μm in length. The crystallite orientation within any single prism is coaxial with the prism in the extreme portion, but tipped more and more away from the prism axis in moving toward the tail of each prism (Meckel *et al.*, 1965a,b). Although there are other patterns which appear in human enamel, the "keyhole" pattern appears most often and is considered as the "norm" [see Figs. 3, 5, and 6, and Boyde (1964, 1971) and Meckel (1971)].

Enamel Prism Patterns

The structural unit of enamel is the prism, which has remained basically the same since the evolution of mammals from the reptiles approximately 150 million years ago (m.y.a.) The development of mammalian tooth is seen as a marked increase in the thickness of enamel in the form of *prismatic* enamel (Poole, 1973). This structural unit is common to all extant mammals, although, just as enamel crown patterns vary, so do the enamel prism patterns.

Tomes (1849) was the first to attempt an investigation of the taxonomic possibilities of enamel prism patterns in marsupials and rodents. Later, Carter (1922) studied the prism patterns in a series of nonhuman primates, both extant and extinct forms. The study revealed different patterns, which led

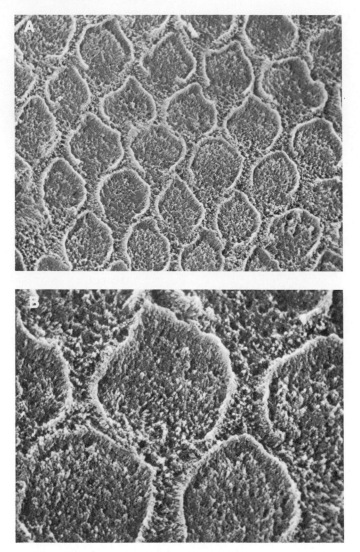

Fig. 6. Enamel prism pattern of *Homo*. The SEM microphotographs of the enamel prism pattern seen in both *Homo sapiens* and *Homo erectus* were taken at 2000× (A) and 5000× (B). The enlargement (B) clearly shows the wide tail processes described as the "winged processes" or "fishtails."

Carter to propose a separation of the Lemuroidea from the Lorisoidea and Tarsioidea. These studies, as well as others, had little if any affect upon theoretical schemes of primate taxonomy. The reason for this, perhaps, is the lack of understanding concerning the ultrastructure of enamel in humans and other mammals. It was not until the advent of the scanning electron microscope that researchers could begin to understand the development and ultrastructural features of enamel.

The work of Boyde has done much to provide a better and more complete understanding of enamel structures in mammals. Analyses with the SEM of developing and mature enamel have documented that specific enamel prism patterns do exist and can be used as a taxonomic indicator (Boyde, 1964, 1965a,b, 1971, 1978). These basic patterns describe the developmental surface of the enamel and not the mature outer surface. Figure 7 illustrates that each pattern has a variety of subpatterns or variants (Boyde, 1964, 1969a).

Pattern 1 prisms are most common in the Insectivora, Sirenia, and some of the Chiroptera. This pattern is also found in some Lemuriformes and in human early cuspal enamel. Boyde (1964, 1969a) suggests that in the developing enamel, the Tomes process (the secretory part of the ameloblast) lies in a pit in the floor of the developing enamel and forms the central portion of the prism pattern. Thus, in Pattern 1 prisms the pits are nearly straight cylinders with flat floors, being separated from one another by a continuous interprismatic region which formed the mutual walls of the pit at the formative surface and by the sites of adjacent Tomes processes [Boyde, 1976b (p. 341, Figs. 7 and 10A)]. Pattern 1 prisms are the result of the ameloblast movement, which is nearly perpendicular to the developing surface, which causes the crystal orientation to be the same in the centers of the interprismatic and prismatic regions (see Figs. 12 and 13).

Pattern 2 prisms are found in most ungulates, marsupials, and rodents. In Pattern 2 prisms the ameloblasts are not perpendicular to the surface as they are in the Pattern 1 prism. In the developing enamel, the Tomes process lies in a pit which enter the surface obliquely so that the floor of a pit becomes tilted and forms the missing part of the wall [Boyde, 1976b (p. 341, Figs. 9 and 10B)]. In Pattern 2 prisms, the floor is in the cervical side of the pit, with the ameloblast aligned in rows parallel with the longitudinal axis of the tooth. The

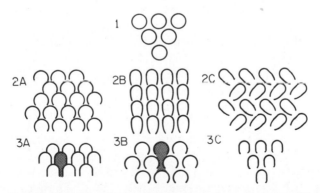

Fig. 7. The developmental patterns in enamel that have been documented by Boyde (1964, 1965b, 1976b). Pattern 1 is most commonly found in the Insectivora, Sirenia, and some of the Chiroptera, as well as in some Lemuriformes and in human early cuspal enamel. Pattern 2 is found in most ungulates, marsupials, and rodents. Pattern 3 prism patterns are predominately found in ape and human enamel, and in the latter group a variant, the "keyhole" enamel prism, is found (see text for further explanation).

crystallites in the floor of the pits grow parallel with the direction of movement of the ameloblasts, while the majority of the crystals in the interrow sheets grow perpendicularly to the tops of the interpit walls, which are in the general plane of the developing surface. Two ameloblasts are considered to make each prism, with two predominant crystal orientations—those in the prisms and those in the interprismatic sheets.

Pattern 3 prisms are most commonly found in human enamel, being described as the "keyhole" enamel prism pattern. The Pattern 3 prisms pits are inclined to the general plane of the ameloblast matrix. The pits are arranged in transverse rows, though the row axis is rotated in areas where the net cell movement has a lateral as well as a cuspal component. The developmental enamel floor is the cervical side of the pit, which merges with the interpit wall next to it instead of joining the next cervical wall as in Pattern 2 prisms [Boyde, 1976*b* (p. 341, Figs. 9 and 10C)]. There exists an extreme range of crystal orientations resulting from the cell movements, but the orientation in the interprismatic (winged process or tail) regions of the prisms merges gradually into that in the prism body proper. In the mature prisms, the tail regions are what were the interpit walls and contain the crystals that grew mostly in relation to the tops of the interpit walls and thus normal to the general plane of the developing surface [Boyde, 1976*b* (p. 342, Fig. 15)]. The crystals in the prism body (head region) centers are parallel with the prism direction, so that there must be a span of orientations between the body and the tail at least as great as the angle of the prisms with respect to the developing surface. However, the crystals in the most cuspal side of the prism may diverge cuspally toward the next prism by some 5 or 10°; and those few that grow in the limited amount of new matrix added along the long cuspal wall of the pits (i.e., in the far cervical, fan-shaped part of the tails of the prisms) diverge cervically toward that surface, so that the total fan angle of orientations from top of head to bottom of tail of Pattern 3 prisms may exceed 60° [Boyde, 1976*b* (p. 343, Figs. 3, 5, and 6)]. Each prism is considered to be formed by the contributions of four ameloblasts.

Although three basic patterns have been documented by Boyde, several variations have also been identified [see Fig. 1 of Boyde (1964)]. The range and extent of variation have not been fully documented. In addition, there have been few attempts to correlate the prism patterns of the fully mineralized enamel surface to the developmental prism patterns documented by Boyde (1964, 1965*a,b*, 1976*b*).

Of importance to our consideration of Neogene hominoid evolution is the analysis of the variants of Pattern 3 prisms. Figures 5, 7, and 15 illustrate the variants of the type 3 prisms. Pattern 3A prisms have no regions which can properly be defined as interprismatic (Boyde, 1964). These prisms have been referred to as "tadpole-shaped" by Shellis and Poole (1979) because of their slender "tails" [Boyde (1964) refers to this portion of the prism as the "winged process"]. The "tail" or "winged process" region contains the crystallites with the maximum deviation from the prism axis. The Pattern 3B prisms differ from those of Pattern 3A in that the crystallite orientation boundary planes often extend through more than half a circle as seen in transverse section [i.e., a transition toward Pattern 1, although "prism" and "interprismatic" regions are still continuous with each other (Boyde, 1964)].

The prism outlines range from a half circle in Pattern 3A to an almost closed circle in Pattern 3C. In addition to the extent of the prism outlines, a very important feature is the extent of the interprismatic regions. Interprismatic enamel is clearly recognizable in 3C, much less so in 3B, and in 3A it is difficult to allocate any part of the structure to the interprismatic region. The 3C prisms are smaller than 3B or 3A prisms; in fact they approach the size of the circular prisms of Pattern 1.

Boyde (1964) considers the differences between the variants of Pattern 3 to lie in the extent of the planes of abrupt change in crystallite orientation and the separation between them [refer to Figs. 3, 5–7, 15–17, and 20–22 of this chapter and Fig. 2.25 of Boyde (1964), p. 89)]: "In Pattern 3A it seems logical to conceive of the regions between adjacent prism boundaries in a horizontal direction as 'winged processes' belonging to the region circumscribed by the neighbouring (occlusally) prism boundary. However, in Pattern 3C one more naturally imagines the prism boundaries to be complete, (cylinders, i.e. circular in cross-section) when the remaining regions outside 'the prisms' may be described as 'interprismatic'" (Boyde, 1964, p. 89).

Hypotheses to Be Tested

Evolutionary Changes in Tooth Form and Function

The bulk of the primate fossil record consists of teeth, especially the molars. This fact dictates that most attempts to explain or to reconstruct the primate fossil record must resort to concepts developed within the field of functional dental morphology.

Recently, a number of concepts and theories have been proposed which seek to functionally explain the relationship between the molar form, occlusion, and dietary preference. The earlier studies (and some recent studies) in functional dental morphology attempted to explain molar form and occlusal relationships from an analysis of molar wear features (Butler, 1972, 1973; Gingerich, 1974; Mills, 1955, 1963, 1973). However, major technological innovations have recently permitted *in vivo* cineradiographic, cinefluorographic, and stress analyses of mandibular movements during mastication (Hiiemae and Kay, 1972, 1973; Hylander, 1977, 1979; Kay, 1973; Kay and Hiiemae, 1974). These and similar studies have provided new data which have permitted a much clearer understanding of jaw movement, molar wear patterns, and molar occlusion.

Based on these new data, an effort has been made to relate dietary specializations to specific molar designs (Kay, 1973, 1974, 1975, 1977, 1978; Kay and Cartmill, 1977; Kay and Hiiemae, 1974; Kay and Hylander, 1979; Seligsohn, 1977). Studies of the masticatory behavior in a variety of mammals reveal that the tooth design and resultant wear patterns are related to the patterns of jaw movements and types of food eaten (Kay, 1973). Therefore, mastication is considered to comprise two successive phases: "puncture-crushing," in which food is pulped and the teeth do not come into contact; and "chewing," in which food is reduced by a variety of methods (Hiiemae and Kay, 1973). Through the investigation of both feeding behaviors and the interaction of the food bolus with the dentition, it has become apparent that the physical properties of food have been the primary factors in the evolutionary selection of dental features (Hiiemae and Kay, 1973; Jolly, 1970; Kay, 1973, 1975, 1978; Rensberger, 1973, 1975; Rosenberger and Kinzey, 1976).

Perhaps the most significant changes in tooth form and function are evident in the hominoids. The initial split and subsequent radiation of the Neogene hominoids are considered to be the result of an adaptation to new diets that required increased crushing and grinding. This adaptation led to changes in crown morphology and an increase in the thickness of enamel, as well as changes in enamel ultrastructure (Gantt, 1977, 1979a,b; Gantt et al., 1977; Jolly, 1970; Molnar and Ward, 1977; Pilbeam, 1978; Simons, 1976; Simons and Pilbeam, 1972).

Studies of the dental complex of the robust australopithecines have led investigators to suggest that the robust australopithecines are unique in that they have relatively small incisors and canines and large molars and premolars with both thick enamel and flattened occlusal surfaces. This morphology is suggestive of a chewing apparatus especially designed to generate and dissipate large forces during powerful postcanine biting or chewing (Crompton and Hiiemae, 1969; Gantt, 1977; Hylander, 1979; Jolly, 1970; Kay, 1973, 1977; Kay and Hiiemae, 1974; Pilbeam, 1972; Robinson, 1956; Simons, 1976, 1977; Wolpoff, 1973). However, there presently exist no quantitative data to support either the suggestion that hominids have thicker enamel than other hominoids or that the robust autralopithecines have thicker enamel than other African hominoids. Kay (1981) has recently published on the relative thickness of enamel in a number of hominoid species. This study was based on a comparison of wear facets and dentin exposure to my published data (Gantt, 1977) on enamel thickness in primates. I am not in agreement with Kay's conclusions and am presently preparing a comprehensive evaluation of the thickness of enamel in the Neogene hominoids.

The differences in enamel prism patterns are considered to be the result of functional alterations of prism packing and crystallite arrangement. For example, Boyde (1964) has documented several variants for each developmental prism pattern type (see Fig. 7). In Pattern 3, there are at least three variants (3A, 3B, and 3C). These variants could be explained by and would have to be accompanied by slight modifications of the shape of the developing enamel surface; they should show a different type of transverse fracture surface. The degree of dove-tailing of the common interpit wall-derived component (the prism tail) between the prism heads would mean that the type 3B prisms (like that found in *Homo*) would part in a transverse fracture plane by rupture of the tail crystals under tension, in addition to separation along the prism boundary planes (Boyde, 1976a). The tails of the type 3A (like that found in *Pan, Gorilla,* and *Pongo*) prisms might tear out in a more intact condition requiring less energy to exploit the separation of the prism boundary discontinuities (Boyde, personal communication; see Fig. 8). If this speculation is correct, the related differences in structure of the fractured surface should be best and most easily seen by study in the SEM.

Taxonomy and Relationship of Extinct Primates

Phylogeny is an attempt to reconstruct the evolutionary history of a group of organisms and to include both cladistic and anagenetic information. Classification can be considered as essentially a means of conceptualization, communication, and storage of information about the group of organisms (Luckett and Szalay, 1978). As such, a phylogeny can only be expected to reflect history to the extent that actual historical records, i.e., fossils, are available to document successive stages of evolutionary change (Gingerich and Schoeninger, 1977). Because they are relatively rare and are often only represented by dental remains, fossils may not provide sufficient information to permit phyletic reconstruction. The only alternative is to make a detailed comparative study of the living representatives of the group of interest and then to arrange them in the most parsimonious arrangement. These data, together with those obtained from the analyses of the fossil specimens, will allow maximum parsimony to be achieved.

In this chapter, the "cladistic" (Hennig, 1966) approach will be used. I consider the methods of phylogenetic inference, which include multiple-character analyses of both fossil and extant taxa, coupled with a consideration of the temporal and paleogeographic distribution of fossils, to offer the best approach for interpreting phylogeny. This "cladistic" method attempts to group taxa on the basis of the number of shared and derived characters. The branching sequences of a phylogeny are determined by phylogenetic analysis of as many characters as feasible in order to distinguish among similarities due to shared ancestral features, shared derived features, convergences, or parallelisms. That is, the relative states of primitive vs. derived characters are determined by the distribution of character states in higher categories, while relatively rare and uniquely acquired character states are generally considered to be derived (Luckett and Szalay, 1978; Schwartz *et al.,* 1978).

The data obtained in this study, together with the existing data concerning primate evolution, will provide for a much more complete understanding of dental evolution. To accomplish this, a number of existing theories and hypotheses concerning hominoid evolution were tested. However, the significance of the data obtained from the study of fossil dentitions may not be fully evaluated or understood until the extant taxa are studied to determine the range of variation and distribution of these characters.

The Hominid/Hominoid Differentiation

Within the past 5 years, major new discoveries of fossil hominoids have significantly reshaped conceptualizations of hominoid evolution (Andrews, 1978; de Bonis and Melentis, 1976; Pilbeam, 1978; Pilbeam *et al.*, 1977). Not only have these new discoveries revealed a much larger variety of species and a wider adaptive radiation, they have also led scientists to question the previous established relationships among these taxa and the modern species of Hominoidea. The Neogene hominoids presently represent at least three groups: (1) the Hylobatidae, (2) the Dryopithecidae, and (3) the Ramapithecidae (Pilbeam, 1978; Pilbeam *et al.*, 1977). However, Simons is not in agreement with this taxonomic arrangement (Simons, 1976, 1977; Simons *et al.*, 1978).

Analysis of the enamel structure and the thickness of enamel in a number of these Neogene hominoids documents that distinct differences do exist. Preliminary analysis of the dryopithecids documents that *Limnopithecus* is distinct from *Proconsul*. In the ramapithecids, distinct enamel prisms were obtained in *Sivapithecus indicus*, *S. sivalensis*, *Gigantopithecus*, and *Ramapithecus*. Also, all of these forms have thick enamel.

The Hominid Status of Ramapithecus

In 1961, Simons described a maxilla which he considered to be the earliest hominid remain. This specimen was assigned to genus *Ramapithecus* and Simons has continued to add new specimens to this taxon to provide support for his claim (Simons, 1961, 1976, 1977). Recently, a number of researchers have questioned the hominid status of *Ramapithecus,* including Pilbeam (1978). The most outspoken criticism of the hominid status of *Ramapithecus* is that of Greenfield (1979). Greenfield states that "the morphological and adaptive patterns suggested by the known parts of '*Ramapithecus*' and *Sivapithecus* are virtually identical. Both are reconstructed as dryopithecines. . . . Consequently, the taxon '*Ramapithecus*' was formally synonymized with *Sivapithecus*. . . ." (Greenfield, 1979, p. 544).

The Status of the Pongidae

Based on molecular data, a number of researchers have questioned the relationship of the extant pongids, especially *Pan* and *Gorilla*, to the Neogene hominoids. These molecular data suggest a recent split of *Pan* and *Gorilla* from the hominids with an estimated divergence date of 4–5 m.y.a. (Goodman, 1974; Sarich and Cronin, 1976). However, the molecular data document that *Pongo* has a much earlier time of divergence. New discoveries of the face of *Sivapithecus meteai* document a number of derived characters which are shared only with *Pongo* among the pongids; therefore, *Pongo*'s ancestry may extend back into the late Miocene (Andrews and Tekkaya, 1980).

Methods and Procedures

Scanning Electron Microscopy

The methods and procedures used in the analysis of enamel ultrastructure follow those described by Gantt (1979c), except that the specimens are embedded in a clear, removable plastic, which allows for a more precise control of preparation of a facet of specific size and depth. The facet is then polished to 0.25 μm finish and etched in a 0.074 M solution of phosphoric acid (H_3PO_4) for 60 s. This procedure has been used at the British Museum in London and at the MRC Dental Unit in Bristol with great success and allows the removal of only a few micrometers of enamel.

To correctly understand the differences in prism shape and direction, a number of specimens were serially sectioned or polished in a stepwise fashion to remove a few micrometers at a time. SEM analysis was conducted on these specimens to document the changes in enamel prism patterns and the size and shape of prisms. HCl and H_3PO_4 solutions were used to delineate specific features resulting from the differential effects of these acids on calcified tissues [see Figs. 1, 2, and 14, and Boyde *et al.* (1978)].

Transverse fracture analysis was conducted to determine prism pattern resistances to stress. Boyde (1976a) has demonstrated that in the "keyhole" enamel prism pattern (developmental type 3B) the crystallites of the tail region fracture under tension. My preliminary studies of the enamel pattern in the pongids (developmental type 3A) indicate that the crystallites of the tails tear out in a more intact condition, requiring less energy to exploit the separation of the prism boundary discontinuities (see Fig. 8).

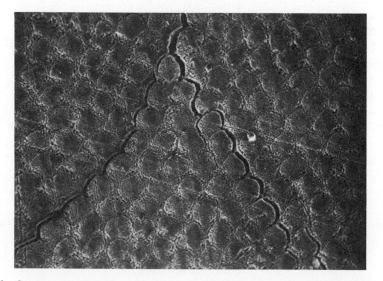

Fig. 8. The fracture pattern of type 3A prisms in a microphotograph of *Gorilla* taken at 1000×.

Thin Section Analysis

The methods and procedures involved in the analysis of thin sections and the assessment of the thickness of enamel are described in more detail by Gantt (1977) and Molnar and Gantt (1977). In the fossil samples two alternative approaches are used:

1. If the specimens are fractured in a plane that allows analysis of the thickness of enamel, they are measured in a light microscope and/or photographed. This allows a greater amount of material to be analyzed.

2. A 50-μm thick diamond wire is used to cut through the center of the cusps, which removes no more than 75 μm of enamel (far less than the 0.1-mm standard error in measuring the teeth), thereby allowing the SEM analysis of two halves of the specimen for thickness of enamel and architectural analysis. This method is currently being used by Andrews (personal communication) in the study of the Paşalar Neogene hominoids. The specimens are then glued back with minimal damage.

Architectural Analysis

Transverse and longitudinal sections of enamel were made in order to evaluate prism orientation, prism decussation, and crystallite arrangement in the enamel. The sections were then etched with either 10% HCl for 2–3 s or 0.074 M H_3PO_4 for 60 s. The use of different acid etchants produces different results due to the differential effects of etching, thereby highlighting various structures [see Figs. 1, 2, and 14, and Boyde *et al.* (1978)].

Procedures for Enamel Thickness Analysis

Once the morphological analysis is completed, the teeth are individually prepared for sectioning. The tooth is placed in absolute alcohol for 1 h, then in acetone for 1 h, and then in a peelaway plastic embedding mold (see Fig. 9). The orientation of the specimen is critical to ensure that a buccal–lingual section be cut through the centers of the cusps, parallel to the long axis of the tooth (see Figs. 9 and 10). An expoxy resin (Buehler, Ltd. #20-8130-032) is poured in the mold, placed in vacuum for 10 min, and then allowed to cure for 8 h at room temperature. This material has proven a ready medium for sectioning, with the Isomet thin sectioning saw. The technique used is the same as described by Gillings and Buonocore (1959, 1961a,b) and by Gantt (1977).

To estimate the thickness of the section desired, knowledge of the thickness of the blade is necessary (this thickness plus the thickness of the desired thin section). The teeth are sectioned with a diamond saw blade, approximately 350 μm thick. Each tooth is positioned so the cut can be made through

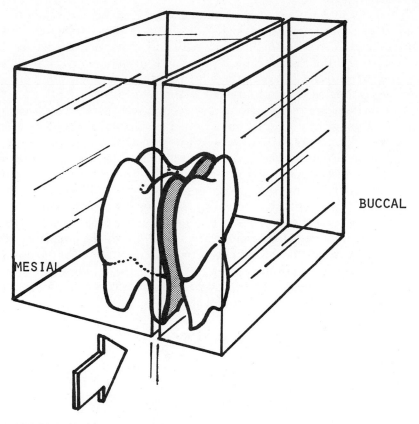

BUCCAL

MESIAL

DIRECTION OF CUT

Fig. 9. Model of an embedded molar and plane of section. This diagram indicates the direction of the cut so as to ensure that the centers of the cusps are cut in a buccal–lingual plane.

the centers of the cusps, in order to outline the contours of the dentinal horn and because enamel in this area has been shown to be the first to wear away. Kay (1973) showed that a large proportion of the force of mastication is directed perpendicular to the tips of the cusps, which results in the enamel in this area being worn away first. A subsequent study of the developmental aspects of dental wear in cercopithecoids has verified these observations (Gantt, 1976, 1977, 1979*d;* Kay, 1977, 1978; Molnar and Gantt, 1977).

The embedded tooth is then sectioned with a diamond saw through the centers of the cusps in the coronal plane (buccal to lingual direction; Fig. 9). The thin sections are approximately 100 μm in thickness. Several sections are cut from each tooth to ensure that the plane of section is, as nearly as possible, parallel to the long axis of the tooth, through the center of the cusps (Fig. 9). Using this technique, thin sections (90–110 μm) of good quality are produced, useful at low or high magnifications without grinding or polishing.

Fossil specimens that are too valuable to be sectioned are cut in the same plane as previously described, but with a 50-μm diamond wire. This procedure allows analyses of the two halves of the tooth with the reflected light microscope and the SEM. The loss of tissue is in the area of 50 μm, far less than the 0.1 mm measurement error. In addition, broken fragmentary remains are evaluated to determine the enamel thickness.

Cut sections are mounted on 75 × 25 mm glass slides with Kleermount (Carolina Biological) and enlarged 25 times by means of a projecting microscope (Bausch and Lomb). Tracings of the projected images are made from the upper surface of the sections, thus eliminating tracing errors due to surface slope in each section. As the measurements of the tracings are made to the nearest 0.025 mm, they have an accuracy of ±25μm (Gillings and Buonocore, 1961a). On the average, two sections from each tooth are traced.

To ensure that the same relative orientation for each tracing of the thin sections is maintained, the sections are aligned with the phantom occlusal plane perpendicular to the long axis of the tooth and the odontoblast process in the cusp tips parallel with the long axis (Fig. 10). Analysis is made of 16 linear and two area measurements taken from tracings of thin sections of the posterior dentition (see Fig. 11 and Table 1).

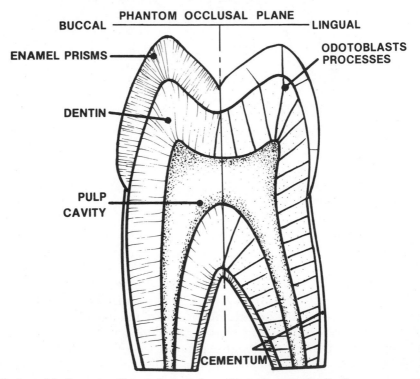

Fig. 10. A model of a section (buccal–lingual) through a molar revealing the fine structures of the dental tissues.

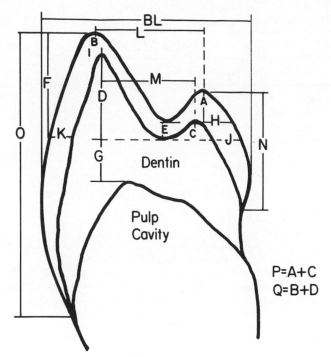

Fig. 11. A model of the thickness of enamel illustrating the positions of each measurement (see Table 1 for a description of each measurement).

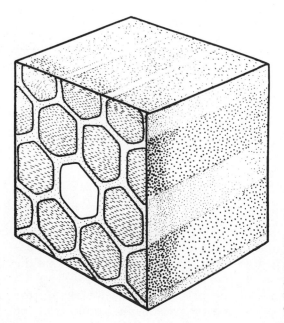

Fig. 12. Model of the Pattern 1 prisms most commonly found in the Insectivora, Sirenia, and some Chiroptera and Lemuriformes as well as occurring in the developmentally early human enamel and around the cusp tips.

Table 1. Posterior Dentition Variable Identification of Terms Utilized in Fig. 11

Variable	Variable name
MD	Total length of tooth, mesial to distal direction
BL	Total width of tooth, buccal to lingual direction
A	Enamel thickness at the buccal cusp
B	Enamel thickness at the lingual cusp
C	Dentinal horn height at the buccal cusp
D	Dentinal horn height at the lingual cusp
E	Enamel thickness in the groove between the cusps (not in the fissure)
F	Occlusal plane height (taken at the DEJ[a] of the enamel groove to the highest cusp)
G	Distance from DEJ of the enamel groove to the pulp cavity (this measurement proved to be too unpredictable, and thus was removed from subsequent analyses)
H	Buccal enamel width (taken at the dentinal horn)
I	Lingual enamel width (taken at the dentinal horn)
J	Buccal enamel width (taken at the DEJ of the enamel groove)
K	Lingual enamel width (taken at the DEJ of the enamel groove)
L	Enamel cusp width (the distance from the centers of the cusp tips)
M	Dentinal horn width (the distance from the centers of the horn tips)
N	Buccal cusp height (the distance from the CEJ[b] to the cusp tip)
O	Lingual cusp height (the distance from the CEJ to the cusp tip)
P	Buccal functional enamel thickness (A + C)
Q	Lingual functional enamel thickness (B + D)
R	Area of enamel
S	Area of dentin
HH	Radial enamel thickness at point H
II	Radial enamel thickness at point I
JJ	Radial enamel thickness at point J
KK	Radial enamel thickness at point K
EA	Radial enamel thickness in the groove buccal side
EB	Radial enamel thickness in the groove lingual side

[a]DEJ, Dentino-enamel junction.
[b]CEJ, Cemento-enamel junction.

Results

Scanning Electron Microscopy of Extant Taxa

The analysis of the extant taxa through the use of the SEM revealed a number of important facts which are necessary for the correct evaluation of prism patterns. First, it is necessary to limit the area of analysis to the mid-lateral portions of the crown. It is in this area of the enamel crown that the prisms display a consistent pattern. The prism pattern is the result of the angle made between the long axes of the prisms and ameloblasts during development. Boyde (1964) has shown that the angle is the resultant of the lateral slope of the prism in its zone and the vertical inward and cervical slope

Fig. 13. Microphotograph of the type 1 enamel prism pattern seen in *Ursus deningeri* (the Pleistocene cave bear) from Petralona, Greece.

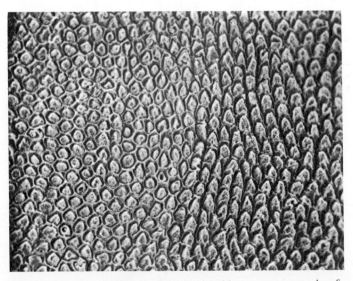

Fig. 14. Microphotograph of *Pongo* illustrating the transition zone progressing from "hexagonal" or "circular" (Pattern 1) prisms to the "tadpole"-shaped (Pattern 3A) prisms. The area of analysis is along the cusp apex. Note that 10% HCl was used to highlight this transition. The left side of this figure represents the occusal surface direction, whereas the right represents the cervical (or root) direction.

against the normal plane of the surface of the developing enamel. In humans as well as in apes, round prisms are found over the cusp tips (this pattern, type 1, is seen in all the enamel of those mammals in which the prisms do not slope very much against the enamel–dentine junction, nor form into decussating zones such as found in Chiroptera, Insectivora, Cetacea, and Sirenia) [see Figs. 7, 12, and 13, and Boyde (1964)]. Analysis of the extant hominoids shows a marked transition from the round prisms, similar to type 1, to the type 3 prism patterns (see Fig. 14; type 1, as seen in carnivores, is, however, quite different—see Figs. 12 and 13).

The second fact resulting from this study is the consistency of pattern

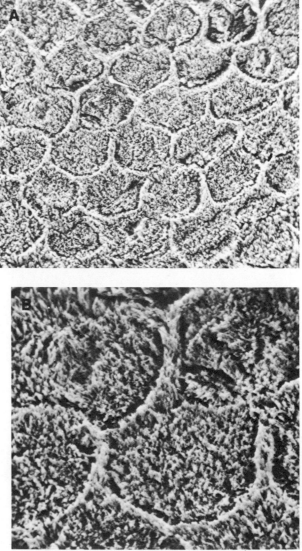

Fig. 15. Enamel prism pattern in *Pongo*. The SEM microphotographs of the enamel prism patterns seen here were taken at 2000× (A) and 5000× (B). Note in A that the "tadpole"-like tails are pointing upward instead of downward, due to the SEM chamber-size limitations, which prevented orienting the specimen in its correct position, tails down.

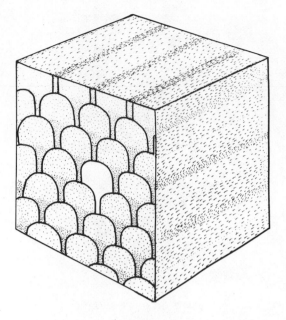

Fig. 16. Model of the Pattern 3A prisms seen in *Gorilla, Pan, Pongo,* and *Hylobates.*

types. Human enamel, which has been extensively studied, has a type 3B or "keyhole" enamel prism pattern (Figs. 3 and 5–7), while the extant apes display a type 3A (Figs. 7, 15, and 16).

The third fact relates to the effects of acid etching. In Fig. 17 the prisms are highly etched with crystallite relationships not definable; however, in Figs. 18–22, the prism patterns are distinct, with definable crystallite relationships, as the result of using the acid etch procedures described by Boyde *et al.* (1978) and Gantt (1979*c*).

Scanning Electron Microscopy of Extinct Taxa

The fossil taxa evaluated are listed in Table 2 (Hylobatidae were excluded from this study due to the lack of fossil forms). They are generally more difficult to study and to obtain distinct prism patterns for, due to the effects of fossilization. In Fig. 19B the effects of fossilization can be clearly seen; this is a longitudinal section through a specimen of *Ramapithecus* from the Siwaliks of India (Fig. 19A). The outer surface of the enamel reveals a layer of varying width and degree of change due to fossilization. This area must be removed before analysis of the prism patterns can be determined.

Three patterns were obtained from the set of Neogene hominoids. The classical "keyhole," or type 3B, prism pattern was obtained in the hominids: *Australopithecus robustus, A. africanus, A. afarensis, Homo erectus,* and *H. sapiens* (see Figs. 3, 5–7, and 20).

The second pattern, type 3A, was obtained in all of the Miocene hominoids (*Ramapithecus, Sivapithecus, Ouranopithecus, Proconsul,* and *Gigantopithecus*

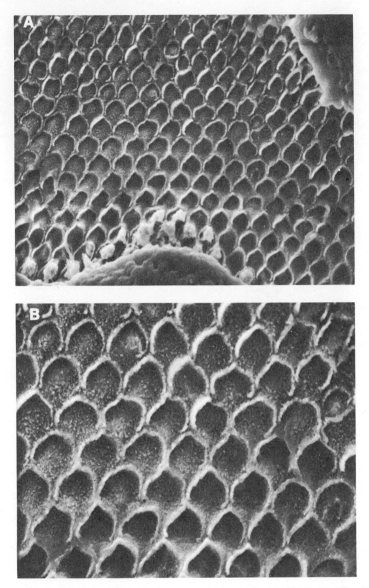

Fig. 17. Enamel prism pattern of *Ramapithecus punjabicus* (AMNH 19565-B). The SEM micro-photographs of the prism pattern of *R. punjabicus* were taken at 1000× (A) and 2000 (B). The specimen was etched with 10% HCl for 2.5 min and was illustrated in Gantt *et al.* (1977).

bilaspurensis) as well as in the Plio-Pleistocene hominoids, *Gigantopithecus blacki*, and *Pongo* (see Figs. 7, 15–18, 21, and 22).

The third pattern, which has been reported from *Sivapithecus sivalensis*, is that of a "hexagonal" prism shape. This pattern was obtained on one specimen only. Reanalysis of this specimen (YPM 8926) reveals that the facet had inadvertently been placed on the cusp, and, as expected, a round or hexagonal-shaped prism was obtained.

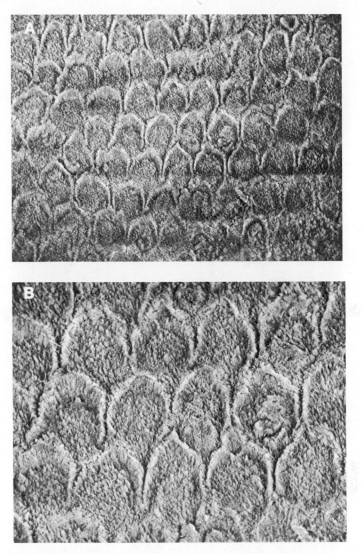

Fig. 18. Enamel prism pattern of *Ramapithecus punjabicus* (AMNH 19565-B). This SEM micro-photograph is of the same *R. punjabicus* specimen as presented in Fig. 17, here also taken at 1000× (A) and 2000× (B), but this time the specimen was polished and etched with 0.074 M solution of H_3PO_4 for 60 s. Note the crystallite pattern in both microphotographs (A and B), as well as the narrow tails of each prism as documented in B, which represents the type 3A prism pattern.

Enamel Thickness

Data obtained from the analyses of thin sections, longitudinally cut speci-mens, and broken or fractured specimens provided perhaps the most surpris-ing results. Enamel thickness in extant apes followed patterns previously described by Gantt (1977, 1979*a,b*) and Molnar and Gantt (1977). These data

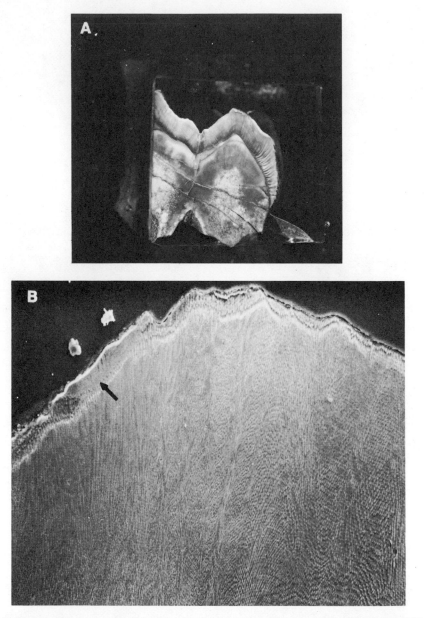

Fig. 19. The profile, cut buccal-lingually, of enamel in *Ramapithecus punjabicus* (AMNH 19565-B). Note the thick enamel covering seen in A (~4×). In B, the microphotograph (250×) is of the face of the enamel taken from the sectioned specimen (A) to illustrate the fossilization zone, as indicated by the arrow.

document that among the living apes, *Pongo* has the thickest enamel, while *Gorilla* has the thinnest relative to body size (see Fig. 23).

Thin enamel appears to be related to dietary specializations of the taxa. The folivorous taxa all fall below the regression line, having thin enamel, while the frugivorous to graminivorous forms have thick enamel (Gantt, 1977). Certainly significant is the position of *Homo* relative to the other hominoids. Humans have thicker enamel than do any other living primate (Gantt,

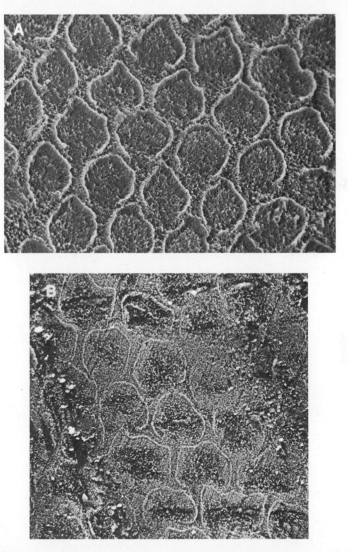

Fig. 20. The "keyhole" or type 3B enamel prism pattern. (A) Microphotograph of the type 3B, "keyhole" pattern seen in both *Homo sapiens* and *Homo erectus* taken at 2000×. Note the wide "winged" or "fishtail" processes. (B) Microphotograph (2000×) from a specimen of "*Pithecanthropus erectus*" (*Homo erectus*) from Java. Note as in A the long and wide "winged" or "fishtail" processes of the type 3B prism pattern.

Table 2.   Neogene Hominoid Teeth Analyzed in This Study

Africa
Proconsul africdnus
   KNM-SO 901 M_1
   KNM-SO 1078 M^2
Proconsul major
   KNM-SO 933 M^1
   Napak I 21834   Lower M
   KNM-SO 953 M_1
Proconsul nyanzae
   KNM-SO 962 M_2
   KNM-RU 1839 M_2
   KNM-RU 1720 M^1
   KNM-KA 6 M^2
Australopithecus sp.   Lower M, Swartkrans
Australopithecus afarensis M^3, Garusi
Australopithecus africanus?   Lower M, Koobi Fora

Asia
Sivapithecus indicus
   GSP 11536/7   Left infant mandible, dm_1, dm_2, M_2
   GSP 10500   Left M^2
   GSP 6999   Right I^1
   BMNH M 13364   Right M_3
Ramapithecus punjabicus
   AMNH 19565-B   Right M_1
Sivapithecus/Ramapithecus
   GSP 12000   Right M^1
   GSP 5018   Left M^2
   GSP 7618   Left immature mandible P_3, P_4, with an un-
                                         erupted M_2
   GSP 7308   Right M^1
   BP 1310   Upper M
   YPM 12865   Lower M
   YPM 8926 M_3
Ramapithecus brevirostris
   YPM 13799   Holotype, maxillary fragment with P^{3-4},
                                         M^{1-2}

Gigantopithecus blacki
   Specimen 2   Right M_3
   (von Koenigswald, 1952)
Gigantopithecus bilaspurensis
   GSP 7144   Left M_2 or M_1
   GSP 13163   Lower M
Ouranopithecus macedoniensis
   RPl-?   Two lower M
Homo erectus erectus   Lower M, Java
Homo erectus pekinensis   Three lower M, China
Pongo pygmaeus   Five lower M, China, Java, Sumatra

Europe
Homo erectus?   Right M^2, Petralona

1977, 1979*a,b;* Molnar and Gantt, 1977). Among the extant hominoids, *Pongo* most closely approaches the thickness of enamel seen in humans. Studies of the Pleistocene orangutan from China and Java reveal that they also had thick enamel, comparable to that seen in living humans (a detailed study of the thickness of enamel in the Neogene hominoids, especially *Pongo,* is in progress; also see Gantt, 1981*a*). Analysis of a few hominoid specimens (*Ramapithecus, Sivapithecus* and *Proconsul*) also documents that they, too, had thick enamel [see Fig. 23, and Gantt (1981*c*)].

Comparison of the thickness of enamel in similar sized hominoids (e.g., *Ramapithecus* and *Proconsul africanus*) reveals that the thickness of enamel is identical, contrary to Kay (1981). Evidence does not support the division of Miocene hominoids into "thin" (ancestral, early Miocene forms) and "thick" (descendant, middle to late Miocene forms).

Discussion

Enamel Prism Patterns

The data presented here concerning the prism patterns of the Neogene hominoids are the result of 5 years of study. A variety of scanning scopes have been used, which makes comparisons of the results difficult. Negatives of the microphotographs of some of the specimens were not available for reproduction because they are the property of the British Museum of Natural History and MRC Dental Unit, University of Bristol.

The first description of hominoid prism patterns was presented in 1977 (Gantt, 1977; Gantt *et al.,* 1977). In our discussion we pointed out that "interpretation of these structural differences is at present conjectural" and that "prism patterns appear to be potentially very interesting for functional analysis and, perhaps, eventually for phylogenetic and taxonomic purposes" (Gantt *et al.,* 1977, p. 1157). Our purpose in writing that article was to stimulate interest in this approach and to assist us in obtaining needed extant and extinct dental specimens.

Our preliminary data indicated that *Ramapithecus* had a prism pattern which looked remarkably similar to that seen in humans, the "keyhole" prism pattern (see Figs. 5 and 17). That same year, Shellis and Poole (see Lavelle *et al.,* 1977, Chapter 5, pp. 197–279) completed the first comparative study of the calcified dental tissues of living primates. Their analyses of the prism patterns of hominoids were limited to a single specimen each of *Gorilla gorilla* (M) and *Pan troglodytes* (M1). They compared their results to the developmental patterns of Boyde (1971), but did not compare their findings to any of the observed variants which were described by Boyde (1964) (see also Fig. 7). For humans they obtained the Pattern 3 or "keyhole" prism pattern (Lavelle *et al.,* 1977, p. 206, Fig. 29), which shows a striking resemblance to the prism pattern illustrated in our article on *Ramapithecus* (Gantt *et al.*, 1977, p. 1156, Fig. 3; and this chapter, Fig. 17). For *Gorilla,* Shellis and Poole obtained a Pattern 3 also, but noted that the prisms differed in that they were "closely packed tadpole-shaped prisms" (Lavelle *et al.*, 1977, p. 221, Fig. 38). However, a different pattern was obtained in *Pan,* which they reported as a "circular or polygonal [Pattern 1]" prism pattern (Lavelle *et al.*, 1977, p. 222, Fig. 39). This pattern is also like the pattern observed for the extant apes, especially for *Pongo,* as was illustrated in Gantt *et al.* (1977, p. 1156, Fig. 2).

In 1978 I attended the Third International Symposium on Tooth Enamel sponsored by the National Institute of Dental Research. At this symposium, I had the opportunity to discuss my research, and the problems I was encountering with Dr. Alan Boyde. Dr. Boyde and his colleagues had just completed a comprehensive study of the quantitative and qualitative aspects of etching enamel with acid and EDTA (Boyde *et al.*, 1978). The results of that study indicated that hydrochloric acid (HCl) was not the etchant of choice to study prism patterns, but rather a dilute solution of phosphoric acid (H_3PO_4) should be used. Based on this I subsequently stated that "due to the recent study by Boyde *et al.* (1978) of acid etchants and the subsequent development of a new technique to analyze and to describe enamel patterns (see Gantt, 1979*c*), previous patterns described by myself and others are questionable" (Gantt, 1979*b*).

During the spring and summer of 1979, I was able to work in England with Dr. Boyde in London and with Drs. Shellis and Poole in Bristol, during which time we were able to compare data and to evaluate the results of our study as well as to discuss the differences obtained. During this period we were able to determine that the "hexagonal" or "circular" shape obtained in a number of the extant apes by both myself and Shellis and Poole was the result of patterns obtained from the occlusal surface and the areas along the cusp tips. In Fig. 14 one can clearly see the transition from a "hexagonal" or "circular" shaped prism (Pattern 1) to the "tadpole-shaped" prism described by Shellis and Poole (Lavelle *et al.*, 1977). The problem in using these descriptive terms is that they represent two different views: (1) one which is concerned with the general shape and (2) one which counts the sides of each prism to determine its description. Therefore, to eliminate confusion, all descriptions of prism patterns will be related to the developmental patterns described by Boyde (1964) and illustrated in Fig. 7.

As a result of this collaboration, we were able to determine that the pattern which best describes the extant apes is Pattern 3A (see Figs. 15 and 16). This pattern was obtained in *Gorilla* by Shellis and Poole and by myself in *Gorilla, Pan,* and *Pongo,* and by Boyde (personal communication).

Therefore, based on the analyses of extant apes by myself, by Shellis and Poole, and by Boyde, it is now possible to correctly describe the prism pattern of extant apes, that is, Pattern 3A. This pattern is consistently obtained if the following procedures are employed:

1. A highly polished facet is placed on the *midlateral* crown (incorrectly referred to as the midcervical crown).

2. The facet is etched with a 0.074 M = 0.5% vol/vol of concentrated, 85% phosphoric acid (H_3PO_4) (Boyde *et al.*, 1978; Gantt, 1979*c*).

3. Stereoanalysis is used to evaluate prism patterns.

4. The prism patterns are related to the developmental patterns described by Boyde (1964) (see the present chapter, Fig. 7).

Vrba and Grine's Results

Although there is a large volume of literature concerning the significance of enamel prism studies to paleontology and mammalogy [see Boyde (1964) as well as the references in the present chapter], Vrba and Grine (1978*a,b*) have questioned the importance of this approach. In a study of the enamel prism patterns of South African australopithecines and extant hominoids, they reported that in all hominid and pongid specimens a Pattern 3 was found to predominate on all surfaces examined and that on the occlusal surfaces Pattern 1 was found (Vrba and Grine, 1978*a*, p. 890). In their Fig. 2b (Vrba and Grine, 1978*a*, p. 891) and Fig. 4 (Vrba and Grine, 1979*b*, p. 126) one can clearly see the transition zone from the round cuspal enamel prism (Pattern 1) to the more open Pattern 3 prisms, as is illustrated in Fig. 14 of this chapter. Immediately clear from an analysis of Vrba and Grine's (1978*a,b*) illustrations is that the specimens show a great deal of differential etching. None of the microphotographs clearly reveals the crystallite relationships as is documented in this chapter or in my past publications (Gantt, 1979*b,c*, 1980). One figure (Vrba and Grine, 1978*a*, p. 891, Fig. 3b) is even out of focus. The results led these authors to "suggest that gross prism morphology contains no information on phylogenetic relationships of hominoid species" (Vrba and Grine, 1978*a*, p. 891). Furthermore, they concluded, "we believe that enamel prism structure cannot be utilized in either phylogenetic or taxonomic deductions within the Hominoidea" (Vrba and Grine, 1978*b*, p. 126). These statements are incorrect and premature.

In 1978, I informed Dr. Vrba that I no longer was using the acid etching techniques as described in my 1977 publication and suggested that a dilute H_3PO_4 solution should be used as described by Boyde *et al.* (1978). In addition, I informed her that, based on the analysis of a fragment of enamel from *Australopithecus robustus* from Swartkrans, the South African australopithecine material may be difficult to analyze due to its high carbonate component. The South African australopithecine material was recovered from travertine deposits in which the fossilized specimens were subjected to a great deal of calcium carbonate replacement. Therefore, etching has a greater differential effect. Furthermore, the fossilized specimens are removed from their matrix by etching away the travertine in nitric acid. Thus the specimens are already highly etched!

I applaud their attempt to study the enamel prism patterns of the australopithecines, but deplore their lack of scientific judgement by abandoning this approach before the data could fully be evaluated. Neither Vrba nor

Grine is a specialist in comparative enamel histology or with the SEM. The voluminous literature on this subject [e.g., Boyde (1964, 1965a,b, 1967, 1969a,b, 1970, 1971, 1975, 1976a,b, 1978); Osborn (1970, 1974); Poole (1973); Remy (1976); Shellis and Poole (1979); International Symposiums on Tooth Enamel I (1964), II (1969), III (1978)] clearly indicates both the difficulties in studying enamel and the significance of enamel prism studies. Therefore, I suggest that their conclusions should be abandoned.

Enamel Prism Patterns of Neogene Hominoids

The prism patterns obtained on the sample of Neogene hominoids (Table 2) are the results of collaboration with a great number of colleagues from around the world. SEM facilities have been used for this study in Germany; London and Bristol, England; and Harvard, Tallahassee, and Birmingham, in the U.S. The problems with this approach stem from the differences in the SEM used, and comparison of these microphotographs is often difficult, especially in the application of Fourier analysis in order to determine size and shape differences.

Using the techniques and procedures outlined in this chapter (see also Gantt, 1979c; Boyde *et al.*, 1978) I have obtained consistent and reproducible results. A reanalysis of the specimen of *Ramapithecus punjabicus* [AMNH 19565-B (Gantt *et al.*, 1977)] clearly reveals the crystallite relationship and documents a type 3A pattern (see Figs. 7 and 16–18). This pattern was also obtained on the type specimen, *Ramapithecus brevirostris* (YPM 13799).

Analysis of *Sivapithecus indicus, Ouranopithecus macedoniensis, Gigantopithecus bilaspurensis,* and *G. blacki* documents a type 3A prism pattern in all these taxa (see Figs. 7, 16, and 21). A number of other specimens were considered to either represent *Ramapithecus punjabicus* or *Sivapithecus sivalensis.* These specimens were lumped together (see Table 2). Analysis of these specimens revealed a type 3A prism pattern like that obtained for *Sivapithecus indicus.* On one specimen (YPM 8926), a different pattern was obtained, but upon reanalysis, it was found that the polished facet had inadvertently been placed on the cusp. The pattern obtained is a "hexagonal" prism described by the author for *Sivapithecus* (Gantt, 1979c). This pattern results from the bending and folding of the ameloblast sheet, and as predicted by Boyde (1964), the prisms in this area take on a circular to round appearance.

Prism patterns obtained on the African hominoids *Proconsul africanus, P. major,* and *P. nyanzae* reveal a type 3A prism pattern like that seen in the Asian hominoids of the Miocene (see Figs. 7, 16, and 22). In Fig. 22B one can clearly distinguish the crystallites. The long, slender, and narrow *tail* of each prism may be easily seen and compared to the wide, "wing-like" or "fishtail" process of the *tails* in the type 3B prisms (see Figs. 21, 22A, and 22B and compare these to Figs. 6 and 20).

Several subfossil and fossil orangutans were studied, and they provided a

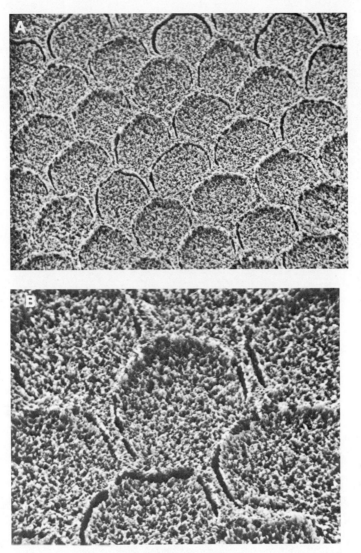

Fig. 21. The enamel prism pattern typical of the Eurasian Miocene hominoids. The microphotographs of the 3A prism pattern seen in *Sivapithecus* were taken at 2000× (A) and 5000× (B). The enlargement B reveals the crystallite arrangement and the narrow "tadpole"-like processes of each type 3A prism.

clue to a possible phyletic link with the Asian Miocene hominoids, especially *Sivapithecus indicus* and *Ouranopithecus*. The prism pattern obtained for these specimens is type 3A (see Figs. 15 and 16) and appears to be remarkably similar in size and shape to that of *Sivapithecus indicus* (see Fig. 21; note that the tails of *Pongo* in Figs. 15, 16 are pointed up instead of down due to the limitations of the chamber size of the S4-10 Cambridge Stereoscan preventing the correct positioning of the specimen). This is of special interest, since *Pongo*

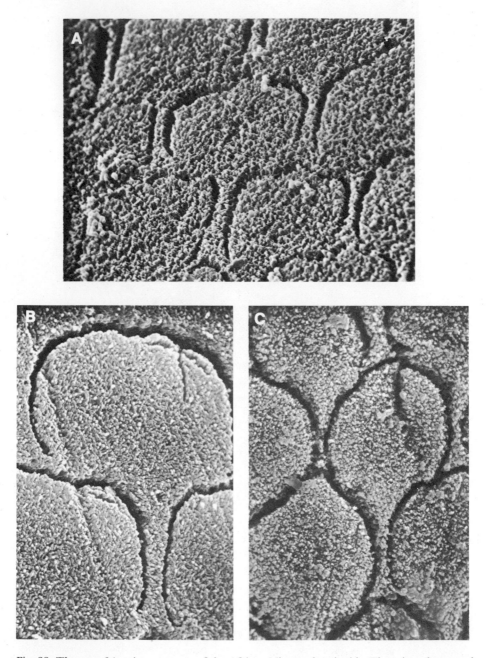

Fig. 22. The type 3A prism patterns of the African Miocene hominoids. The microphotographs of a type 3A prism from *Proconsul major* were taken at 5000× (A) and 7000× (B) and illustrate the crystallite arrangement as well as the "tadpole"-like tails. (C) Microphotograph (5000×) of a Miocene hominoid from Saudia Arabia which represents an as yet undetermined lineage. [Specimen courtesy of Peter Andrews.]

is believed to have diverged at an earlier date than the African apes (Goodman, 1974; Sarich and Cronin, 1976).

Enamel Prism Patterns of Hominids

The material which is considered to represent the hominids consists of two samples, *Australopithecus* and *Homo erectus*. In *Australopithecus,* the material was limited to a few fragments, although it appears that a type 3B prism pattern or the "keyhole" prism pattern is the characteristic pattern for this taxon (Gantt, 1981*b*). However, it would be necessary to study a more complete sample to clearly document their prism pattern. The sample of middle Pleistocene (*Homo erectus*) hominids was much more complete and included specimens from Java, China, and Greece. Analysis of these specimens clearly documented the typical human "keyhole" enamel prism pattern [see Figs. 3, 5–7, and 20; the type 3B prism pattern (Gantt *et al.,* 1980)].

Enamel Thickness in Hominoids

The thickness of enamel has recently become an important factor, especially as it relates to the nature of the transition in the dental mechanism from pongids to hominids (Simons and Pilbeam, 1972; Simons, 1976; Kay, 1981). Analysis of a number of primate species as well as a few fossil hominoids reveals a surprising result. There is a close correlation between the thickness of enamel and body size among the extant primates, the exception being *Homo sapiens* (see Fig. 23). Humans have much thicker enamel than any

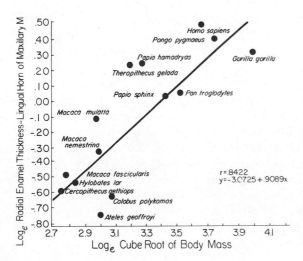

Fig. 23. A regression analysis of the relationship between body size and enamel thickness in selected primate species. In particular note the position of *Gorilla gorilla,* with its relatively very thin layer of enamel. See Gantt (1981*c*) for further discussion and information regarding the data analyzed in this figure.

extant primate (Gantt, 1976, 1977, 1979a,b; Molnar and Gantt, 1977). Among the extant apes, there is another interesting result; *Gorilla* has "thin" enamel when compared to its body size [see Fig. 23 and Gantt (1976, 1977, 1981a) and Molnar and Gantt (1977)]. Based on the results of this and other studies (Gantt, 1979a,b, 1981a,c), it appears that highly folivorous forms have "thin" molar enamel, while the more frugivorous to graminivorous forms have "thick" enamel (see Fig. 23). Analysis of the thickness of enamel in *Ramapithecus punjabicus* (AMNH 19565-B), *Sivapithecus indicus,* and several fossil orangs reveals that they all have thick enamel compared to either *Pan* or *Gorilla* (a detailed study of the thickness of enamel in the Neogene hominoids is currently underway).

This preliminary analysis suggests that at least by the middle Miocene, hominoids had evolved "thick" enamel and that *Pongo* is probably derived from that radiation. The evidence does not indicate that *Ramapithecus* has thicker enamel than the other Miocene hominoids. In Fig. 19A one can clearly see the thick layer of enamel which covers the crowns of these Miocene hominoids.

Conclusions

The data obtained in this study of enamel prism patterns and enamel thickness clearly indicate that distinct differences do exist among extant and extinct hominoids. Enamel prism patterns in the extant hominoids reveal two patterns, a type 3A (see Figs. 7, 14–17, and 24) in the apes (*Pan, Gorilla, Pongo,* and *Hylobates*) and a type 3B (see Figs. 3, 5–7, and 20) in *Homo sapiens.* Analysis of fracture patterns reveals that, in the extant apes, prisms are pulled apart intact (Fig. 8), while in human enamel the "winged" process or "fishtail" is ruptured (Boyde, 1976a, and personal communication). These data indicate that a functional difference exists between the two variants of Pattern 3 (3A and 3B). Boyde (personal communication) suggests that the "winged" processes of type 3B allow the enamel to be subjected to greater stress before cracking. This, together with the evidence of an increase in the thickness of enamel in humans as compared to extant apes, supports this hypothesis.

Concerning the question of ape and human ancestry, the data presented here suggest that, of the extant apes, *Pongo* has an ancestry extending back to the middle Miocene hominoids of Eurasia. The prism pattern and the thick enamel seen in both extant and extinct orangutans most closely approximate what is seen in *Sivapithecus indicus* and *Ouranopithecus.* Furthermore, *Ourano-pithecus* is characterized by enamel wrinkling like that seen in the orangutans. This, together with the data from molecular studies (Goodman *et al.,* 1976), new fossil finds [a skull of *Sivapithecus* recently recovered by Pilbeam which most closely resembles *Pongo* (Pilbeam, personal communication)], and the recent analysis of the palate and lower face of *Sivapithecus meteai* (Andrews and Tekkaya, 1980), supports this proposal.

The status of *Ramapithecus* is still uncertain. Based on the analysis of prism patterns and the thickness of enamel, *Ramapithecus* cannot be distinguished from *Sivapithecus sivalensis* as Simons (1976) and others have earlier proposed. Rather, *Ramapithecus* may best be considered to represent a "small"-canine radiation, while *Sivapithecus indicus* and *Ouranopithecus* represent a "large"-canine radiation (the new finds from China also appear to have large, conical canines). There is no reason to suggest that *Ramapithecus* is not a possible hominid ancestor based on present data. What is in question is its generic taxonomic status and its distinctiveness from that of *Sivapithecus sivalensis*, a point most recently raised by Greenfield (1979). Morphologically they are exceedingly similar forms and could very well represent sexual dimorphs of the same species, or alternatively, two closely related species.

The African apes differ from both humans and the orangutans. Both *Pan* and *Gorilla* have a type 3A prism, unlike humans, and both have a reduced thickness of enamel, unlike humans and orangutans. This reduction in the thickness of enamel is carried to an extreme in *Gorilla* (see Fig. 23) and is considered to be a derived feature for the African apes. The type 3A prism pattern is considered to be primitive for the Neogene hominoids, whereas the marked increase in thickness of enamel and the "keyhole" enamel prism (type 3B) pattern seen in *Australopithecus*, *Homo erectus*, and *Homo sapiens* are considered derived. Therefore, the "keyhole" enamel prism and the marked increase in enamel thickness are unique features in human evolution, just as is bipedality.

A final question concerns the position of the hylobatids, *Hylobates* and *Symphalangus*, within the Hominoidea. Though they are known to share a number of unique postcranial specializations with the Great Apes, in many ways their cranial anatomy resembles the primitive catarrhine morphotype. An analysis of the enamel thickness in *Hylobates lar* shows that it is characterized by relatively thin enamel (Fig. 23). Since the gibbon diet is primarily frugivorous in nature, the apparent correlation between frugivory and "thick" enamel in the higher primates is not supported by this example. A recently completed analysis of the enamel prism pattern in *Hylobates lar* shows that it exhibits a type 3A prism pattern (Fig. 24). This firmly establishes *H. lar* as a member of the Hominoidea and excludes it from any possible relationship with the Cercopithecoidea. It also indicates that the origin of the Hylobatidae was probably closely connected with the cladogenesis of the Hominoidea as a whole and with the evolution of the type 3A enamel prism pattern in particular.

A cladogram is presented in Fig. 25 to illustrate the pattern of distribution within the Neogene to Recent Hominoidea of the derived characters discussed in this chapter and elsewhere. The hominoids are grouped together into nested sets on the basis of common possession of derived character states. The sharing of derived character states indicates an evolutionary relationship. The nature of the evolutionary relationships is postulated in the classification presented in Table 3. This classification is based on the consideration of anatomical and molecular data which emphasize the striking similarities be-

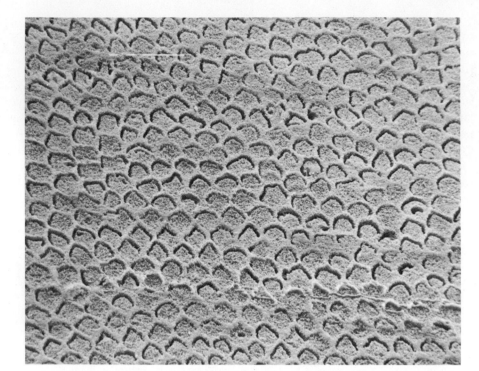

Fig. 24. Enamel prism pattern of *Hylobates lar*. The SEM microphotograph of this enamel prism pattern was taken at 1000× and documents a type 3A prism pattern as seen in the other living apes.

tween the African apes and humans and the striking differences between these forms and orangutans (Gantt, 1981*c;* Goodman, 1974).

A New Interpretation of Ape and Human Ancestry

Ramapithecus and *Sivapithecus sivalensis* are considered to represent the same taxon or two closely related species of a "small" canine radiation to which *Gigantopithecus* is related. The "large"-canine radiation is represented by *Sivapithecus indicus, Ouranopithecus,* and *Pongo* (the taxonomic status of such taxa as *Ankarapithecus, Bodvapithecus,* and the Chinese material is still questionable, although they all appear to be "large"-canine forms). These taxa are assigned to the family Pongidae based on molecular evidence, facial morphology, thickness of enamel, enamel wrinkling (found in *Ouranopithecus, Pongo,* and the recently discovered Miocene hominoids from China), and paleoecological data (Andrews and Tekkaya, 1980; de Bonis, this volume, Chapter 24; Goodman, 1974; Gantt, 1981*c;* Pilbeam, 1978; Pilbeam *et al.,* 1977).

The African Neogene hominoids have recently undergone much revision (Andrews, 1978; Simons *et al.,* 1978; Szalay and Delson, 1979; this volume,

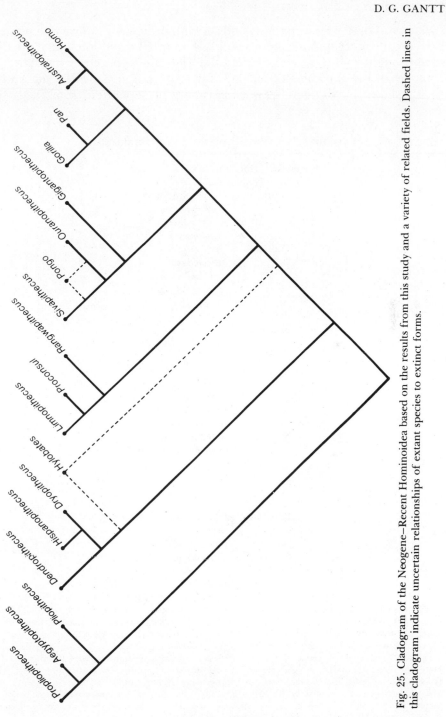

Fig. 25. Cladogram of the Neogene–Recent Hominoidea based on the results from this study and a variety of related fields. Dashed lines in this cladogram indicate uncertain relationships of extant species to extinct forms.

various authors). The numerous finds of both previously known and new taxa have greatly increased our knowledge of hominoid evolution and at the same time brought many previously held theories and hypotheses into question. Analyses of the thickness of enamel and prism patterns provide data which clearly show that the genus *Proconsul* cannot be separated from *Sivapithecus*

Table 3.   A Classification of the Hominoidea[a]

Superfamily: Hominoidea
   Family: Pliopithecidae
      Subfamily: Propliopithecinae
         Propliopithecus
         Aegyptopithecus
         Pliopithecus
      Subfamily: Dryopithecinae
         Micropithecus
         Dendropithecus
         Hispanopithecus
         Dryopithecus
   Family: Hylobatidae
      Subfamily: Hylobatinae
         Hylobates
         Symphalangus
   Family: Pongidae
      Subfamily: Proconsulinae
         Limnopithecus
         Proconsul
         Rangwapithecus
      Subfamily: Ramapithecinae
         Sivapithecus (= *Ramapithecus*)
         Pongo
         Ouranopithecus
         Gigantopithecus
   Family: Hominidae
      Subfamily: Paninae
         Gorilla
         Pan
      Subfamily: Homininae
         Australopithecus
         Homo

[a]A diagnosis of the family Pliopithecidae follows that of Szalay and Delson (1979, pp. 435–437), who view the taxa of this family as being grouped together based on their numerous primitive retained characters. The subfamily Dryopithecinae is constructed to contain the "thin"-enameled hominoids of Europe. The small early Miocene hominoids of East Africa, *Dendropithecus* and *Micropithecus*, are tentatively placed in this subfamily. The "small" hominoids of East Africa are currently undergoing a revision by Peter Andrews and Terry Harrison; therefore, the correct placement of these forms remains uncertain. Furthermore, Simons *et al.* (1978) have suggested that *Dendropithecus* and *Micropithecus* might lie within or near the ancestry of the Hylobatidae. Pongidae has priority [see Article 23 (d)(i) of the Rules of Nomenclature]; therefore two subfamilies are constructed to contain the East African early to middle Miocene hominoids (Proconsulinae) and the middle to late Miocene hominoids of East Africa and Eurasia together with the Plio-Pleistocene hominoids of Asia (Ramapithecinae). The family Hominidae is divided into two subfamilies: (1) the Paninae, which contains the African Great Apes (*Pan* and *Gorilla*), and (2) the Homininae (see the text for a discussion of these two subfamilies).

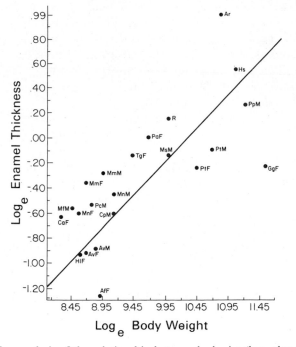

Fig. 26. A regression analysis of the relationship between body size (log$_e$ plotted on the X axis) and enamel thickness (log$_e$ plotted on the Y axis) which includes several fossil hominoids. The first letter identifies the genus, second letter the species, and the third letter the sex (M, male; F, female). Af, *Ateles fusciceps;* Ar, *Australopithecus robustus;* Av, *Alouatta villosa;* Ca, *Cercopithecus aethiops;* Cp, *Colobus polykomos;* Gg, *Gorilla gorilla;* Hs, *Homo sapiens;* Hl, *Hylobates lar;* Mf, *Macaca fascicularis;* Mm, *Macaca mulatta;* Mn, *Macaca nemestrina;* Ms, *Mandrillus sphinx;* Pa, *Papio anubis;* Pc, *Presbytis cristatus;* Pp, *Pongo pygmaeus;* Pt, *Pan troglodytes;* R, *Ramapithecus* sp. and *Proconsul africanus;* and TG, *Theropithecus gelada.* Note: Weights taken from Kay (1973) for all primates except the fossils. The estimated weights of the fossils were based on ranges of 40–80 lb for *Ramapithecus* and *Proconsul africanus,* and 80–140 lb for *Australopithecus.* Although these fossil species were not used in the statistical analysis, they are placed on the regression line plot based upon a body weight of 50 lb for *Ramapithecus* and *Proconsul,* and 110 lb for *Australopithecus.* To correctly evaluate fossil forms, enamel thickness should be plotted against tooth size (Gantt, in preparation). Fortunately, in this comparison of *Ramapithecus* and *Proconsul* identical sized teeth were selected which revealed an enamel thickness of identical dimensions. These two fossil species are plotted as R.

(=*Ramapithecus*) (see Fig. 26) as had previously been suggested (de Bonis, this volume, Chapter 24; Kay, 1981; Kay and Simons, this volume, Chapter 23; Pilbeam, 1976; Pilbeam *et al.*, 1977; Simons, 1976; Szalay and Delson, 1979). Therefore, it follows that the "thick" enamel characteristic of many Miocene to Recent large hominoids (excluding *Pan* and *Gorilla*) might be regarded as a shared primitive character complex. This provides support for the hypothesis that the last common ancestors of the middle to late Miocene hominoids of Asia and Africa, the Ramapithecinae, were the Proconsulinae (see Fig. 2). The subfamily Proconsulinae contains the early to middle Miocene (~22–15 m.y.a.) hominoids of East Africa (*Proconsul, Rangwapithecus,* and *Limnopithecus*). The other small East African hominoids of this period are still poorly known (e.g., *Micropithecus*), except for *Dendropithecus.* The affinities of the small hominoids of East Africa are uncertain although some suggest that they might lie in or near the ancestry of the Hylobatidae (Simons *et al.,* 1978). In Fig. 25 I have attempted to indicate this proposed link between *Dendropithecus* and *Hylobates* (see Table 3).

The European hominoids of the middle to late Miocene (~12–8 m.y.a.) *Dryopithecus* and *Hispanopithecus* are poorly represented and not described in detail (e.g., *Hispanopithecus*); however, from an analysis of casts of a number of isolated upper and lower teeth, it appears that *Hispanopithecus* is a small and distinct genus. Both taxa appear to have thin enamel, based on an evaluation of wear facets and chipped and/or broken enamel specimens. These taxa are therefore grouped with small hominoids of East Africa (*Dendropithecus* and *Micropithecus*) in the subfamily Dryopithecinae. They can be described as having thin enamel; the dental sizes in these species range from smaller than that of a gibbon to that of a siamang (enamel prism analysis is pending). Because of the retention of primitive traits, the Dryopithecinae are grouped with the Propliopithecinae in the family Pliopithecidae, as defined by Szalay and Delson (1979, pp. 435–437) (see Fig. 25 and Table 3).

Therefore, I feel that the classical classification of:

> Superfamily Hominoidea
> Family Hominidae
> *Homo*
> Family Pongidae
> *Pan*
> *Gorilla*
> *Pongo*
> Family Hylobatidae
> *Hylobates*
> *Symphalangus*

is no longer feasible due to the overwhelming evidence obtained from morphological, molecular and biochemical, behavioral, paleontological, and paleoecological studies. Based on an analysis of these data, together with the data presented here, I feel that the following classification is the most parsimonious. A phylogenetic classification of the extant Hominoidea which accommodates the present evidence on hominoid phylogeny and conforms to the principles of Hennig (1966) is presented in Table 3 (also see Fig. 25). In this classification I have placed the "thick"-enamel hominoids of East Africa and Eurasia into the family Pongidae (see Fig. 26). I feel that the Ramapithecinae, especially the "large"-canine forms (*Ouranopithecus* and/or *Sivapithecus indicus*), are specifically related to *Pongo* for the following reasons:

1. Large, broad canines.
2. Thick molar enamel. The fossil orangutans of the Pleistocene of China have thicker enamel than the modern form. They are in the same range as the Ramapithecinae—contrary to Kay and Simons (this volume, Chapter 23).

3. Enamel prism morphology identical to *Sivapithecus* (see Figs. 14–17 and 21).

4. Facial morphology. The evidence presented by Andrews and Tekkaya (1980) and Andrews (this volume, Chapter 17) clearly documents a *Sivapithecus–Pongo* relationship. [I do not accept Kay and Simons' interpretation of this relationship. If one considers the morphological pattern seen in *Pongo* as primitive, then the features it shares with *Pan* and *Gorilla* as described by Kay and Simons (this volume, Chapter 23) are primitively retained features.]

5. Molecular and biochemical data clearly establish that *Pongo* diverged prior to *Pan* and *Gorilla*. *Pan* and *Gorilla* are more closely related to *Homo* than either is to *Pongo*. Following Goodman (1974), I accept his proposal regarding the placement of *Pan, Gorilla,* and *Homo* in the family Hominidae.

6. Fossil evidence documents that *Pongo* extends back into the Pleistocene of China. This evidence consist of isolated teeth of large forms with thick enamel. (There are no fossil remains of *Pan* or *Gorilla!*)

7. Enamel wrinkling is an inherited characteristic which clearly distinguishes the postcanine teeth of *Pongo* from those of *Pan, Gorilla,* and *Homo*. This unique characteristic is found among the Ramapithecinae, in *Ouranopithecus,* and in the newly discovered Chinese specimens.

8. Behavioral data also indicate that *Pongo* may have been more territorial in the past and that its present arboreal specializations are secondarily derived (Andrews, this volume, Chapter 17; Smith and Pilbeam, 1980).

The African apes, *Pan* and *Gorilla,* are placed within the family Hominidae as proposed by Goodman (1974) and based on the new finds of *A. afarensis,* which show a surprising resemblance to *Pan* (White *et al.,* this volume, Chapter 29). This remarkable cranial resemblance, together with the data from molecular biology, has led some to suggest that *Pan* (i.e., *Pan paniscus*) is an archetype for the ancestor of living hominids (Goodman, 1974; Zihlman *et al.,* 1978). However, *Pan* and *Gorilla* are here separated from the other known hominids (*Australopithecus* and *Homo*) into the subfamily Paninae, for the following reasons:

1. *Pan* and *Gorilla* have a uniquely derived locomotor system—knuckle-walking, as compared to the bipedality of *Australopithecus* and *Homo*.

2. Both *Pan* and *Gorilla* have reduced enamel thickness as compared to *Pongo* or other pongids; whereas *Australopithecus* has markedly increased thickness of enamel, especially among the robust form. Furthermore, *Gorilla* has significantly reduced its thickness of enamel if allometric effects are considered. This reduction is viewed as a secondary specialization to folivory (see Figs. 23 and 26).

This study is by no means a final conclusion. Many questions remain unanswered; of special interest and importance are those concerning the relationships between function and structure, and between structure and its development. Data on prism patterns and enamel thickness are limited to a very few primate species and are almost nonexistent in other mammalian orders. Studies must be conducted to determine the range of variation of prism patterns, prism size, and density. Moreover, architectural analysis, developmental analysis, the study of enamel thickness, hardness testing, chemical analysis, and further methods and procedures need to be used to provide additional data. This information will be of assistance in the attempt to reconstruct phylogenies.

It is the author's hope that this study will stimulate interest in structural and ultrastructural analyses of dental tissues, especially enamel. But a word of caution—this is a new field of research, with many questions to be answered and new techniques to be tried. Those who wish to undertake this type of analysis must be prepared to provide themselves with a thorough understanding of enamel histology and SEM techniques as well as a willingness to keep an open mind.

Acknowledgments

The success of this project is due to the support of my colleagues, especially Prof. G. H. R. von Koenigswald, Prof. David Pilbeam, and Dr. Peter Andrews, who provided me not only with extant and extinct hominoid dental materials and financial support, but also the encouragement to continue with this research effort. I would also like to thank Dr. Alan Boyde of University College, London, and Drs. David Poole and Peter Shellis of the MRC Dental Unit, University of Bristol, for the opportunity to work with them and for sharing their knowledge of enamel histology and ultrastructure.

Much of this research has required technical assistance, for which I am grateful: the SEM staff of the Senkenberg Museum, Frankfurt/Main; Donald Claugher, Director of the SEM facilities, British Museum of Natural History, London; and Dr. William Miller, Director of the SEM facilities, Florida State University.

I wish to acknowledge financial support from the L. S. B. Leakey Foundation, the Foundation for Research into the Origins of Man (FROM), and the Alexander von Humboldt-Stiftung Foundation.

References

Andrews, P. 1978. Taxonomy and relationships of fossil apes, in: *Recent Advances in Primatology*, Volume 3, *Evolution* (D. J. Chivers and K. A. Joysey, eds.), pp. 43–56, Academic, London.

Andrews, P., and Tekkaya, I. 1980. A revision of the Turkish Miocene hominoid *Sivapithecus meteai. Palaeontology* **23**:85–95.

Boyde, A. 1964. *The Structure and Development of Mammalian Enamel*, Ph.D. Dissertation, University of London.

Boyde, A. 1965a. The structure of developing mammalian dental enamel, in: *Tooth Enamel* (R. W. Fearnhead and M. V. Stack, eds.), pp. 163–167, Wright, Bristol.

Boyde, A. 1965b. The development of enamel in mammals, in: *Calcified Tissues* (J. J. Backwood and M. Owen, eds.), pp. 276–280, Springer-Verlag, New York.

Boyde, A. 1967. The development of enamel structure. *Proc. R. Soc. Med.* **60**:13–18.

Boyde, A. 1969a. Electron microscopic observation relating to the nature and development of prism decussation in mammalian dental enamel. *Bull. Group Int. Rech. Sc. Stomat.* **12**:151–207.

Boyde, A. 1969*b*. Correlation of ameloblast size with enamel prism pattern: Use of scanning electron microscope to make surface area measurements. *Z. Zellforsch.* **93:**583–593.

Boyde, A. 1970. The contribution of the scanning electron microscope to dental histology. *Apex* **4:**15–21.

Boyde, A. 1971. Comparative histology of mammalian teeth, in: *Dental Morphology and Evolution* (A. A. Dalberg, ed.), pp. 81–93, University of Chicago Press, Chicago.

Boyde, A. 1975. Some aspects of the photogrammetry of SEM images. *Photogrammetric Record* **8:**408–445.

Boyde, A. 1976*a*. Enamel structure and cavity margins. *Operative Dentistry* **1:**13–28.

Boyde, A. 1976*b*. Amelogenesis and the structure of enamel, in: *Scientific Foundations of Dentistry* (B. Cohen and I. R. H. Kramer, eds.), pp. 335–352, Heineman Medical Books, London.

Boyde, A. 1978. Development of the structure of the enamel of the incisor teeth in the three classical subordinal groups of the Rodentia, in: *Development Function and Evolution of Teeth* (P. M. Butler and K. A. Joysey, eds.), pp. 43–58, Academic, London.

Boyde, A., and Jones, S. J. 1972. Scanning electron microscopic studies of the formation of mineralized tissue, in: *Developmental Aspects of Oral Biology* (S. Slavkin and R. Bavetta, eds.), pp. 243–274, Academic, New York.

Boyde, A., Jones, S. J., and Reynolds, P. S. 1978. Quantitative and qualitative studies of enamel etching with acid and EDTA. *Scanning Electron Microscop.* **1978**(II):991–1002.

Butler, P. 1972. Some functional aspects of molar evolution. *Evolution* **26:**474–483.

Butler, P. 1973. Molar wear facets of early tertiary North American primates, in: *Craniofacial Biology of Primates* (M. R. Zingeser, ed.), pp. 474–483, Karger, Basel.

Carter, J. T. 1922. The structure of the enamel in the primates and some other mammals. *Proc. R. Zool. Soc. Lond.* **1922**:599–608.

Crompton, A. Q. and Hiiemae, K. 1969. Functional occlusion in tribosphenic molars, *Nature* (*Lond.*) **222:**678–679.

De Bonis, L., and Melentis, M. 1976. Les Dryopithécinés de Macédoine (Grèce): Leur place dans l'évolution des Primates hominoides du Miocène, in: *Les Plus Anciens Hominidés* (P. V. Tobias and Y. Coppens, eds.), pp. 28–38, Colloque VI, IX Union Internationale des Sciences Prehistoriques, Nice. CNRS, Paris.

Frank, R. M., and Nalbandian, J. 1967. Ultrastructure of amelogenesis, in: *Structure and Chemical Organization of Teeth* (A. E. Miles, ed.), Volume 1, pp. 399–466, Academic, London.

Gantt, D. G. 1976. Enamel thickness: Its functional and possible phyletic implication. *Am. J. Phys. Anthropol.* **44:**179–180.

Gantt, D. G. 1977. *Enamel of Primate Teeth: Its Thickness and Structure with Reference to Functional and Phyletic Implications.* Ph.D. Dissertation, Washington University, St. Louis.

Gantt, D. G. 1979*a*. Comparative enamel histology of primate teeth, in: Proceedings of the Third International Symposium on Tooth Enamel, *J. Dent. Res.* **58**(Special Issue B):1002–1003.

Gantt, D. G. 1979*b*. Taxonomic implications of primate dental tissues. *J. Biol. Buccale* **7:**149–156.

Gantt, D. G. 1979*c*. A method of interpreting enamel prism patterns. *Scanning Electron Microscop.* **1979**(II):491–496.

Gantt, D. G. 1979*d*. Patterns of dental wear and the role of the canine in Cercopithecinae. *Am. J. Phys. Anthropol.* **51:**353–360.

Gantt, D. G. 1980. Implications of enamel prism patterns for the origin of New World Monkeys, in: *Evolutionary Biology of the New World Monkeys and Continental Drift* (R. L. Ciochon and A. B. Chiarelli, eds.), pp. 201–217, Plenum, New York.

Gantt, D. G. 1981*a*. Enamel thickness and Neogene hominoid evolution. *Am. J. Phys. Anthropol.* **54:**222.

Gantt, D. G. 1981*b*. Notes on the enamel prism pattern of the Garusi and Eyasi hominids, in: *Die Archaeologischen und Anthropologischen Ergegnisse der Kohl-Larsen Expeditionen in Nordtanzania, 1933–1939* (R. R. R. Protsch, ed.), Volume V, *The Palaeoanthropological Finds of the Pliocene and Pleistocene:* Part I, *Monograph Garusi;* Part II, *Monograph Eyasi*, pp. 150–153, Archaeological Venatoria, W. Kohlhammer, Stuttgart.

Gantt, D. G. 1981*c*. Neogene hominid evolution: A tooth's inside view, in: *Teeth-Form, Function and Evolution* (B. Kurtén, ed.), pp. 107–120, Columbia University Press, New York.

Gantt, D. G., Pilbeam, D., and Steward, G. 1977. Hominoid enamel prism patterns. *Science* **198:**1155–1157.

Gantt, D. G., Xirotiris, N., Kurten, B., and Melentis, J. K. 1980. The Petralonia dentition— Hominid or cave bear? *J. Hum. Evol.* **9:**483–487.

Gillings, B., and Buonocore, M. 1959. An apparatus for the preparation of thin serial sections of undecalcified tissues. *J. Dent. Res.* **38:**1156–1165.

Gillings, B., and Buonocore, M. 1961*a*. An investigation of enamel thickness in human lower incisor teeth, *J. Dent. Res.* **49:**105–118.

Gillings, B., and Buonocore, M. 1961*b*. Thickness of enamel at the base of pits and fissures in human molars and bicuspids. *J. Dent. Res.* **40:**119–133.

Gingerich, P. 1974. Dental function in the Paleocene primate *Plesiadapis,* in: *Prosimian Biology* (R. D. Martin, G. A. Doyle, and A. C. Walker, eds.), pp. 533–541, Duckworth, London.

Gingerich, P., and Schoeninger, M. 1977. The fossil record and primate phylogeny. *J. Hum. Evol.* **6:**483–505.

Goodman, M. 1974. Biochemical evidence on hominid phylogeny, *Annu. Rev. Anthropol.* **3:**203–228.

Goodman, M., Tashian, R. E., and Tashian, J. H. 1976. *Molecular Anthropology,* Plenum, New York. 550 pp.

Greenfield, L: O. 1979. On the adaptive pattern of "*Ramapithecus*". *Am. J. Phys. Anthropol.* **50:**527–531.

Helmcke, J. G. 1967. Ultrastructure of enamel, in: *Structural and Chemical Organization of Teeth* (A. E. W. Miles, ed.), Volume II, pp. 135–163, Academic, London.

Hennig, W. 1966. *Phylogenetic Systematics,* University of Illinois Press, Urbana. 263 pp.

Hiiemae, K. M., and Kay, R. F. 1972. Trends in the evolution of primate mastication. *Nature (Lond.)* **240:**486–487.

Hiiemae, K., and Kay, R. F. 1973. Evolutionary trends in the dynamics of Primate mastication, in: *Craniofacial Biology of Primates* (M. R. Zingeser, ed.), Volume 3, pp. 28–64, Karger, Basel.

Hylander, W. L. 1977. *In vivo* bone strain in the mandible of *Galago crassicaudatus, Am. J. Phys. Anthropol.* **46:**309–326.

Hylander, W. L. 1979. The functional significance of Primate mandibular form. *J. Morphol.* **160:**223–240.

Jolly, C. J. 1970. The seed-eaters: A new model of hominid differentiation based on a baboon analogy. *Man* **5:**5–26.

Kay, R. F. 1973. *Mastication, Molar Tooth Structure, and Diet in Primates,* Ph.D. Dissertation, Yale University, New Haven.

Kay, R. F. 1974. Jaw movement and tooth use in recent fossil primates. *Am. J. Phys. Anthropol.* **40:**227–256.

Kay, R. F. 1975. The functional adaptations of primate molar teeth. *Am. J. Phys. Anthropol.* **43:**195–215.

Kay, R. F. 1977. The evolution of molar occlusion in Cercopithecidae and early catarrhines. *Am. J. Phys. Anthropol.* **45:**227–256.

Kay, R. F. 1978. Molar structure and diet in extant Cercopithecidae, in: *Development Function and Evolution of Teeth* (P. Butler and K. Joysey, eds.), pp. 309–339, Academic, London.

Kay, R. F. 1981. The nut-crackers—A new theory of the adaptations of the Ramapithecinae. *Am. J. Phys. Anthropol.* **55:**141–151.

Kay, R. F., and Cartmill, M. 1977. Cranial morphology and adaptation of *Palaechthon nacimienti* and other Paromomyidae (Plesiadapoidea, ?Primates), with a description of a new genus and species. *J. Hum. Evol.* **6:**19–53.

Kay, R. F., and Hiiemae, K. M. 1974. Mastication in *Galogo crassicaudatus,* A cinefluorographic and occlusal study, in: *Prosimian Biology* (R. D. Martin, G. A. Doyle, and A. C. Walker, eds.), pp. 501–530, Duckworth, London.

Kay, R. F., and Hylander, W. L. 1979. The dental structure of mammalian folivores with special reference to Primates and Phalageroidea (Marsupialia), in: *The Biology of Arboreal Folivores* (G. G. Montgomery, ed.), pp. 173–191, Smithsonian Institute Press, Washington, D.C.

Kay, R. F., Fleagle, J. G., and Simons, E. L. 1981. A revision of the Oligocene apes of the Fayum Province, Egypt. *Am. J. Phys. Anthropol.* **55:**293–321.

Lavelle, C. L. B., Shellis, R. P., and Poole, D. F. G. 1977. *Evolutionary Changes to the Primate Skull and Dentition,* Chapter 5: The calcified dental tissues of primates, pp. 197–279, Thomas Springfield, Illinois.

Luckett, W. P., and Szalay, F. S. 1978. Clades versus grades in primate phylogeny, in: *Recent Advances in Primatology,* Volume 3, *Evolution* (D. J. Chivers and K. A. Joysey, eds.), pp. 227–237, Academic, London.

Meckel, A. H. 1971. The keyhole concept of enamel microstructure, in: *Chemical and Physiology of Enamel,* pp. 25–42, University of Michigan Press, Ann Arbor, Michigan.

Meckel, A. H., Griebstein, W. J., and Neal, R. J. 1965a. Ultrastructure of fully calcified human dental enamel, in: *Tooth Enamel* (M. V. Stack and R. W. D. Fearnhead, eds.), pp. 160–162, John Wright, Bristol.

Meckel, A. H., Griebstein, W. J., and Neal, R. J. 1965b. Structure of mature human dental enamel as observed by electron microscopy. *Arch. Oral Biol.* **10:**775–783.

Mills, J. R. E. 1955. Ideal dental occlusion in the Primates. *Dent. Pract.* **6:**47–61.

Mills, J. R. E. 1963. Occlusion and malocclusion of the teeth of Primates, in: *Dental Anthropology* (D. Brothwell, ed.), pp. 29–51, Pergamon, London.

Mills, J. R. E. 1973. Evolution of mastication in primates, in: *Craniofacial Biology of Primates* (M. R. Zingeser, ed.), pp. 23–36, Karger, Basel.

Molnar, S. 1976. The microstructure of Primate teeth, in: *Orofacial Growth and Development* (A. A. Dahlberg and T. M. Graber, eds.), pp. 57–61, Mouton, The Hague.

Molnar, S. 1977. On the hominid masticatory complex: Biomechanical and evolutionary perspectives. *J. Hum. Evol.* **6:**551–568.

Molnar, S., and Gantt, D. G. 1977. Functional implications of primate enamel thickness. *Am. J. Phys. Anthropol.* **56:**447–454.

Molnar, S., and Ward, S. C., 1975. Mineral metabolism and micro-structural defects in primate teeth. *Am. J. Phys. Anthropol.* **43:**3–18.

Molnar, S., and Ward, S. C. 1977. On the hominid masticatory complex: Biomechanical and evolutionary perspectives. *J. Hum. Evol.* **6:**551–568.

Osborn, J. W. 1967. Three-dimensional reconstruction of enamel prisms. *J. Dent. Res.* **46:**1412–1419.

Osborn, J. W. 1968. An evaluation of previous assessments of prism directions in human enamel. *J. Dent. Res.* **47:**217–222.

Osborn, J. W. 1970. The mechanism of prism formation in teeth: A hypothesis. *Calc. Tiss. Res.* **5:**115–132.

Osborn, J. W. 1974. Variation in structure and development of enamel, in: Dental Enamel: Development, Structure and Caries. *Oral Sci. Rev.* **3:**1–84.

Pilbeam, D. R. 1972. *The Ascent of Man,* pp. 49–61, Macmillan, New York.

Pilbeam, D. R. 1978. Rearranging our family tree. *Hum. Nat.* **1**(6):38–45.

Pilbeam, D. R., Meyer, G. E., Badgley, C., Rose, M. D., Pickford, M. H. L., Behrensmeyer, A. K., and Shah, S. M. I. 1977. New hominoid primates from the Siwaliks of Pakistan and their bearing on hominid evolution. *Nature (Lond.)* **270:**689–695.

Poole, D. F. G. 1973. Phylogeny of tooth tissues: Enameloid and enamel in recent vertebrates, with a note on the history of cementum, in: *Structural and Chemical Organization of Teeth* (A. E. W. Miles, ed.), pp. 111–149, Academic, London.

Poole, D. F. G., and Brooks, A. W. 1961. The arrangement of crystallites in enamel prisms. *Arch. Oral Biol.* **5:**14–26.

Remy, J. A. 1976. *Étude Comparative des Structures Dentaires Chez les Palaeotheriidae et Divers Autres Périssodactyles Fossiles,* Thesis, Université Louis Pasteur, Strasbourg.

Rensberger, J. M. 1973. An occlusal model for mastication and dental wear in herbivorous mammals. *J. Paleontol.* **47:**515–528.

Rensberger, J. M. 1975. Function in the cheek tooth evolution of some hypsodont geomyoid rodents. *J. Paleontol.* **49:**10–22.

Robinson, J. T. 1956. The dentition of the Australopithecinae. *Transvaal Mus. Mem.* **9**:1–79.

Rosenberger, A. L., and Kinzey, W. G. 1976. Functional patterns of molar occlusion in platyr-rhine primates. *Am. J. Phys. Anthropol.* **45**:281–298.

Sarich, V. M., and Cronin, J. E. 1976. Molecular systematics of the Primates, in: *Molecular Anthropology* (M. Goodman and R. E. Tashian, eds.), pp. 141–170, Plenum, New York.

Schwartz, J. H., Tattersall, I., and Eldredge, N. 1978. Phylogeny and classification of primates revisited. *Yearb. Phys. Anthropol.* **21**:95–133.

Seligsohn, D. 1977. Analysis of species-specific molar adaptations in Strepsirhine Primates. *Contrib. Primatol.* **11**:1–116.

Shellis, R. P., and Poole, D. F. G. 1979. The arrangement of prisms in the enamel of the anterior teeth of the aye-aye. *Scanning Electron Microscop.* **1979**(II):481–490.

Simons, E. L. 1961. The phyletic position of *Ramapithecus*. *Postilla* (Peabody Mus. Nat. Hist., Yale Univ.) **57**:1–9.

Simons, E. L. 1976. The nature of the transition in the dental mechanism from pongids to hominids. *J. Hum. Evol.* **5**:511–528.

Simons, E. L. 1977. *Ramapithecus*. *Sci. Am.* **236**(5):28–35.

Simons, E. L., and Pilbeam, D. R. 1972. Hominoid paleoprimatology, in: *The Functional and Evolutionary Biology of Primates* (R. Tuttle, ed.), pp. 36–62, Aldine-Atherton, Chicago.

Simons, E. L., Andrews, P., and Pilbeam, D. R. 1978. Cenozoic apes, in: *Evolution of African Mammals* (V. J. Maglio and H. B. S. Cooke, eds.), pp. 120–146, Harvard University Press, Cambridge.

Smith, R. J., and Pilbeam, D. R. 1980. Evolution of the orang-utan. *Nature (Lond.)* **284**:447–448.

Szalay, F. S., and Delson, E. 1979. *Evolutionary History of the Primates*, pp. 434–502, Academic, New York.

Tomes, J. 1849. On the structure of the dental tissues of marsupial animals, and more especially of the enamel. *Phil. Trans. R. Soc. Lond.* **150**:743–745.

Von Koenigswald, G. H. R. 1952. *Gigantopithecus blacki* von Koenigswald, A giant fossil hominoid from the Pleistocene of southern China. *Anthropol. Pap. Am. Mus. Nat. Hist.*, **43**:291–326.

Von Koenigswald, W. 1977. *Micmonys* cf. *reidi* aus der villafranchischen Spaltenfullung Schambach bei Treuchtlingen. *Mitt. Bayer. Staatssamml. Paleaontol. Hist. Geol.* **17**:197–212.

Von Koenigswald, W. 1980. Schmelzstruktur und Morphologie in den Molaren der Arvicolidae (Rodentia). *Abh. Senckenb. Naturforsch. Ges.* **539**:1–129.

Vrba, E. S., and Grine, F. E. 1978a. Australopithecine enamel prism patterns. *Science* **202**:890–892.

Vrba, E. S., and Grine, F. E. 1978b. Analysis of South African Australopithecine enamel prism patterns. *Proc. Electron Microscop. Soc. S. Afr.* **8**:125–126.

Weatherell, J. A. 1975. Composition of dental enamel. *Brit. Med. Bull.* **31**:115–119.

Wolpoff, M. H. 1973. Posterior tooth size, body size, and diet in South African gracile australopithecines. *Am. J. Phys. Anthropol.* **39**:375–394.

Zihlman, A. L., Cronin, J. E., Cramer, D. L., and Sarich, V. M. 1978. Pygmy chimpanzee as a possible prototype for the common ancestor of humans, chimpanzees, and gorillas. *Nature (Lond.)* **275**:744–746.

Evidence from Postcranial Morphology IV

Locomotor Adaptations of Oligocene and Miocene Hominoids and Their Phyletic Implications

11

J. G. FLEAGLE

Introduction

For most of this century, discussions of ape and human ancestry have frequently centered on questions of skeletal morphology and locomotor adaptations. The proponents of a close evolutionary relationship between humans and living apes have always based much of their argument on the many "brachiating" adaptations found in the musculoskeletal systems of all living hominoids (e.g., Keith, 1923; Gregory, 1934; Washburn, 1968). At the same time, their opponents emphasized the many ways in which apes showed specialized adaptations for brachiation which were not found in humans (e.g., Wood-Jones, 1929, 1948; Straus, 1949). The discovery of skeletal remains for many fossil apes from the Miocene provided what Le Gros Clark (1950) and many others initially perceived as a broadly synthetic solution to the issue of ape–human relationships, but in the long run only complicated the arguments and recalibrated the debate (Fleagle and Jungers, 1982). Thus, for the

J. G. FLEAGLE • Department of Anatomical Sciences, Health Sciences Center, State University of New York, Stony Brook, New York 11794.

past three decades, the debate has shifted from mainly neontological comparisons of living apes and humans to true evolutionary investigation of the morphological and temporal relationships among the fossil apes, living apes, and humans.

These considerations have profited greatly from new knowledge about the earliest hominoids from the Oligocene of Egypt and by new studies of the Miocene species. At the same time they have long been complicated and frustrated not only by the still incomplete nature of the fossils themselves, but by a diversity of initial phylogenetic assumptions. In this chapter, I will summarize current knowledge about the skeletal anatomy of Oligocene and Miocene hominoids, providing what I consider a representative summary of the locomotor adaptations of the best known taxa, and discuss the implications of this material for our understanding of ape and human evolution.

Oligocene Hominoids

Although there is increasing evidence for the existence of fossil anthropoids from the Eocene Pondaung fauna of Burma, the earliest higher primates for which we have any information about the postcranial skeleton are those from the Oligocene deposits in the Fayum Province of Egypt. Many of the taxa from these deposits were known from dental remains (of unknown stratigraphic level) in the first decade of this century (Osborn, 1908; Schlosser, 1910, 1911), but skeletal elements attributed to these early anthropoids have only been described in the last decade [e.g., Fleagle *et al.* (1975), Conroy (1976a), and Preuschoft (1975); reviewed in Fleagle (1980)]. At present we have skeletal material attributable to four species—two parapithecids (*Apidium phiomense* and *Parapithecus grangeri*) and two hominoids (*Aegyptopithecus zeuxis* and *Propliopithecus chirobates*). Although the parapithecids are certainly critical for understanding the origin of all anthropoids and the initial appearance of catarrhines (and possibly Old World monkeys), they are by general consensus collateral to the phyletic mainstream of ape and human evolution and will not be discussed here. At the same time, while there is certainly debate about the exact phyletic position of *Aegyptopithecus* and *Propliopithecus* relative to later groups of catarrhines (e.g., Simons, 1972; Szalay and Delson, 1979; Fleagle and Kay, this volume, Chapter 7), there is a similar consensus that these taxa are suitable and likely ancestors for the later dryopithecines of East African early Miocene (Simons and Pilbeam, 1972). As such they provide our only truly historic evidence of the primitive morphology that subsequently gave rise to later apes and humans.

Aegyptopithecus zeuxis

Aegyptopithecus zeuxis, from the Upper Fossil Wood Zone of the Jebel Qatrani Formation, is the early hominoid species for which we have the greatest information about skeletal anatomy. At present, several humeri, an ulna, a

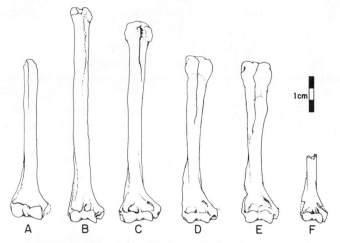

Fig. 1. Anterior view of the right humerus of (A) *Proconsul africanus;* (B) *Dendropithecus macinnesi;* (C) *Pliopithecus vindobonensis;* (D, E) *Aegyptopithecus zeuxis;* (F) *Propliopithecus chirobates.* (A and D are reversed drawings of the left humerus.)

metatarsal, several phalanges, and a talus are known for this species, which was probably the size of a large howling monkey (6–8 kg.). Most of this material has been recovered in the last 5 years, by ongoing expeditions from Duke University and the Egyptian Geological Survey.

The humerus (Figs. 1D, 1E, 2D, 2E) is the most complete (and most recently recovered) limb element known from this species and is particularly useful for phyletic studies because of the numerous humeri known for other fossil anthropoids. Two virtually complete humeri were recovered in 1980. The maximum length of the humerus in this species is about 140 mm, the size of the humerus in the neotropical howling monkey *Alouatta* sp., and consider-

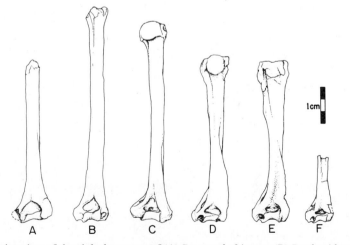

Fig. 2. Posterior view of the right humerus of (A) *Proconsul africanus;* (B) *Dendropithecus macinnesi;* (C) *Pliopithecus vindobonensis;* (D, E) *Aegyptopithecus zeuxis;* (F) *Propliopithecus chirobates.* (A and D are reversed drawings of the left humerus.)

ably shorter than the same bone in the Miocene hominoids *Pliopithecus, Den-dropithecus,* or *Proconsul africanus.* The humerus of *A. zeuxis* is quite robust, and similar to that for many arboreal cercopithecoids of the same size.

Proximally, the greater and lesser tuberosities are both quite large and laterally placed with respect to the articular surface of the head, similar to the situation seen in most arboreal quadrupeds. In this regard, the humerus differs from that of either the more cursorial pronograde primates (baboons, patas monkeys, vervets), in which the greater tuberosity is much larger and anteriorly placed in relation to the head (Jolly, 1967), or the more suspensory apes and atelines, in which the lesser tuberosity is rotated anteriorly and the articular surface enlarged medially. Concomitantly, the bicipital groove is relatively wide and shallow, and the relatively small oval-shaped head faces posteriorly. The deltopectoral crest is quite pronounced and extends nearly 40% down the shaft; laterally, however, there is neither a flat deltoid plane nor a deltotriceps crest as seen in most later cercopithecoids and many ceboids. Rather, the lateral and posterior surfaces of the upper portion of the shaft look very much like that region in extant and fossil "prosimians." On the medial side, below the lesser tuberosity there is a pronounced crest for the insertion of Teres major, a feature commonly found in arboreal quadrupeds including many "prosimians" and ceboids.

On the lower half of the shaft there is a broad, pronounced brachialis flange extending well up the posterolateral surface, a feature found in many arboreal "prosimians" and to a lesser extent in some of the more arboreally quadrupedal cebids, such as *Cebus apella.* Medially, there is an entepicondylar foramen, above the large medial epicondyle. The medial epicondyle projects medioposteriorly at an average angle of 23° as in many arboreal, quadrupedal cebids. The posterior surface of this process has a deep pit (Conroy, 1976a) for the attachment of the ulnar collateral ligament. The distal articular region is distinctly ceboid-like. There is a relatively large, rounded capitulum separated by a low ridge from the wide trochlea. The trochlea has distinct but not flaring distal-medial and posterolateral lips. The olecranon fossa is wide and shallow. The distal articular region lacks the derived features of either extant cercopithecoids or extant hominoids (Fleagle and Simons, 1978).

The ulna of *Aegyptopithecus zeuxis* (Figs. 3C, 4C) has been described, analyzed, and discussed many times during the past decade (Fleagle *et al.,* 1975; Conroy, 1976a; Preuschoft, 1975; Schön-Ybarra and Conroy, 1977). The single ulnar specimen is complete except for the posterior surface of the olecranon process and distal-most part of the shaft. The olecranon process is

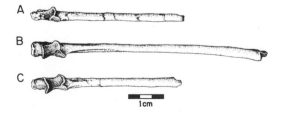

Fig. 3. Anterior view of the right ulna of (A) *Dendropithecus macinnesi;* (B) *Pliopithecus vindobonensis;* (C) *Aegyptopithecus zeuxis.*

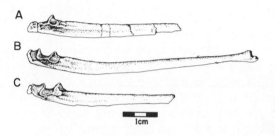

Fig. 4. Lateral view of the right ulna of (A) *Dendropithecus macinnesi;* (B) *Pliopithecus vindobonensis;* (C) *Aegyptopithecus zeuxis.*

quite long relative to the length of the shaft, comparable to the condition seen in *Alouatta* or *Varecia.* All evidence suggests that it is projected proximally as in arboreal quadrupeds rather than posteriorly as in more terrestrial species. The sigmoid cavity is relatively broad and shallow. The radial notch is small and not excavated as in cercopithecoids and terrestrial quadrupeds. The shaft is relatively robust, with a slight anterior concavity and a more pronounced posterior convexity.

Like the humerus, the ulna of *A. zeuxis* is clearly that of a slowly moving, relatively short-limbed arboreal quadruped (Bown *et al.,* 1982). The long olecranon and low, wide sigmoid notch are almost certainly primitive features for anthropoids. Like the humerus, the ulna shows no derived features that would link it uniquely with either later cercopithecoids or hominoids.

The hindlimb of *Aegyptopithecus* is known thus far from only a hallucial metatarsal and, most recently, a slightly damaged left talus. The talus is slightly smaller than expected for a primate the size of *A. zeuxis,* so it probably belonged to a female of this very highly dimorphic species (Fleagle *et al.,* 1980). The bone is strikingly similar to that bone described for the Miocene apes of East Africa—*Dendropithecus, Limnopithecus,* and *Proconsul.* The superior articular surface is relatively high and curved. The lateral side is higher than the medial, a characteristic of arboreal quadrupeds (Fleagle, 1977). The neck is relatively long. Medially, at the base of the neck is a very distinct, cup-shaped facet for the medial malleolus. The head is badly eroded, but appears to have been much broader medially than laterally.

The hallucial metatarsal is relatively robust, suggesting a powerful grasping hallux (Preuschoft, 1973). It also shows an articular surface for a pre-hallux bone, a primitive feature found in most platyrrhines but lost in Old World monkeys, pongids, and hominids (Conroy, 1976b). Several phalanges have also been attributed to *Aegyptopithecus.* They are curved, indicating grasping digits, but lack either the elongation of living apes or the robustness seen in terrestrial monkeys. They suggest arboreal habits (Preuschoft, 1975).

Propliopithecus chirobates

Propliopithecus (=*Aeolopithecus*) *chirobates* is the only other Oligocene hominoid for which any limb bones are known. Several partial humeri (Figs. 1F, 2F), a fragmentary ulna, a relatively complete tibia, and several partial calcanei have been attributed to this taxon. These bones are all indicative of arboreal quadrupedal habits. The forelimb remains are similar to those of *A. zeuxis* in the general features preserved. The tibia is relatively short for the size of the articular surfaces and shows evidence of very distal insertions for many hip muscles crossing the knee—a characteristic of arboreal quadrupeds. Both the distal articulation of the tibia and the calcaneus show arboreal quadrupedal, rather than leaping adaptations. The ankle of *Propliopithecus* is very similar to that of *Pliopithecus* from the Miocene of Europe and suggests possibilities of hindlimb suspension (Fleagle and Simons, 1982a). From these few remains there is little in the postcranial skeleton that distinguishes this taxon from *Aegyptopithecus zeuxis* either adaptively or phylogenetically except for size.

The most striking aspect of the skeletal anatomy of the Oligocene anthropoids is the remarkably primitive nature of the forelimb bones. In a more detailed analysis of the humerus, Fleagle and Simons (1982a) compare *A. zeuxis* with *Alouatta* and the Eocene "prosimian," *Notharctus,* following Gregory (1920). In nine of 12 features, the Oligocene ape is more primitive than *Alouatta* and usually intermediate in form between the two taxa. Only in aspects of the distal articular surface is *Aegyptopithecus* similar to modern anthropoids. The ulna is likewise most similar to large platyrrhines and lemurs but offers few distinctive characters suitable for phyletic "grading". The similarities which *A. zeuxis* and *P. chirobates* show to later catarrhines are usually with colobines or hominoids rather than cercopithecines. In all cases, these similarities are almost certainly primitive features, usually related to an arboreal quadrupedal mode of locomotion. Unique derived features related to either rapid pronograde quadrupedalism as seen in many cercopithecoids or advanced suspensory behavior as seen in extant hominoids are totally absent.

Miocene Hominoids

Pliopithecus vindobonensis

Partial skeletons from three individuals of this species were recovered in the late 1940s from a limestone fissure in Czechoslovakia, approximately 17 million years old. Since the first detailed descriptions by Zapfe (1958, 1960), the skeletal anatomy of this species and its phyletic implications have been the subject of numerous, often apparently conflicting analyses (e.g., Simons and Fleagle, 1973; Corruccini *et al.,* 1975, 1976; Ciochon and Corruccini, 1977; Groves, 1972; Andrews and Groves, 1975; Morbeck, 1976). In his initial description, Zapfe emphasized the many ways in which *Pliopithecus* resembled "prosimians" and platyrrhines, with occasional resemblances to hominoids. The main similarities to cercopithecids were in overall proportions. The following summary of the skeletal morphology largely follows Zapfe's original (1958, 1960) description.

The shoulder girdle is known from two complete clavicles, and a partial scapula preserving the glenoid fossa, and part of the acromion. The clavicle is quite straight and robust, more closely resembling hylobatids than either cercopithecoids, large ceboids, or "prosimians." On the scapula, the glenoid fossa is oval, as in many colobines or large platyrrhines, rather than round, as in either *Ateles* or living hominoids—certainly that of a quadruped animal. The acromion, however, is quite robust, is broad, and overhangs the shoulder joint, suggesting a large distal attachment origin for both trapezius and deltoid muscles, and some suspensory abilities for *Pliopithecus*.

Zapfe (1958, 1960) noted that, overall, the humerus (Figs. 1C, 2C) is primitive, with some noteworthy similarities to "prosimians." The head of the humerus is much less globular than in extant hominoids. Like *Aegyptopithecus*, *Pliopithecus* shows neither the enlarged anteriorly placed greater tuberosity seen in the more cursorial quadrupedal primates nor the anteriorly rotated lesser tuberosity of extant hominoids. Rather, it shows the more generally primitive condition of large medially and laterally placed tuberosities separated by a wide, shallow intertubercular sulcus as seen in colobines, platyrrhines, many "prosimians," and *Aegyptopithecus*. Ciochon and Corruccini (1977) performed a multivariate phenetic analysis of the shoulder joint of *Pliopithecus* and found the greatest similarity with arboreal quadrupeds such as *Presbytis* or *Cebus* rather than with hominoids or cercopithecines. The shaft is relatively straight, rounded in cross section, and slender, with a robustness less than that of most cercopithecines but comparable to *Ateles* or *Alouatta*. There is a distinct deltopectoral crest extending about one-third of the distance down the anterior border of the shaft (*contra* Zapfe), but laterally there is no pronounced deltotriceps crest and hence no flat deltoid plane as in most cercopithecoids. In this primitive morphology *Pliopithecus* resembles *Aegyptopithecus*, some platyrrhines, and most "prosimians."

Distally, there is a moderately developed brachialis flange (not so extensive as in *Aegyptopithecus*) and a pronounced supinator crest on the lateral side. On the medial side is a large entepicondylar foramen—another primitive feature commonly found in "prosimians," some platyrrhines, in *Aegyptopithecus*, and only rarely, as an anomaly, among extant catarrhines. The distal articular surface and the shallow olecranon fossa are virtually identical to the same region in *Aegyptopithecus zeuxis* and larger platyrrhines such as *Alouatta* or *Lagothrix*. There is a medially directed medial epicondyle, a characteristic of arboreal climbing forms; a rounded capitulum; and a broad trochlea with a very low anteromedial and posterolateral margins. The olecranon fossa is wide and shallow.

The humerus of *Pliopithecus*, and of *A. zeuxis*, looks like that of a very primitive arboreal quadruped with many "prosimian" and platyrrhine features and no derived features characteristic of either later hominoids or cercopithecoids. It differs from the Oligocene species in showing adaptive differences suggesting more suspensory (as opposed to quadrupedal) abilities (e.g., less robust shaft, smaller brachialis flange, lower margins on trochlea, more medially directed medial epicondyle).

The ulna of *Pliopithecus vindobonensis* (Figs. 3B, 4B) is completely preserved in two individuals. It is strikingly similar to the same bone in large platyrrhines, some "prosimians," and *Aegyptopithecus zeuxis* (Zapfe, 1958, 1960; Fleagle *et al.*, 1975). The olecranon process is relatively short, comparable to that in *Ateles,* and is directed proximally as in arboreal quadrupeds rather than posteriorly as in more terrestrial monkeys. The sigmoid cavity is broad and shallow, with a low coronoid process and a small, not excavated, radial notch, much as in *Aegyptopithecus zeuxis* and the larger platyrrhines. The shaft is relatively slender—again, much like *Ateles,* with a distinct anterior concavity and no pronounced interosseus crest. Distally there is a large prominent styloid process as in most nonhominoid primates. The ulna of *Pliopithecus* is remarkably like the same bone in *Ateles* and shows adaptations for arboreal quadrupedalism and some suspensory abilities. In all respects it looks like a longer, more slender, more suspensory version of the same bone in *Aegyptopithecus zeuxis* (Fleagle *et al.*, 1975). Two complete radii are known for *P. vindobonensis.* Like the ulna, this bone is relatively long and slender, with marked similarity to large platyrrhines (Zapfe, 1958, 1960).

Much of the hand is known for *Pliopithecus.* The metacarpals and phalanges are relatively short and "monkey-like" compared with those of extant hominoids. The capitate is relatively broad, with a very distinct "spiral facet" as in arboreal quadrupedal species (Jenkins and Fleagle, 1975; Fleagle, 1977; Jenkins, 1981). Unlike extant hominoids, the wrist of *Pliopithecus* has an ulnar-triquetral articulation.

Various parts of the axial skeleton are known from *Pliopithecus.* The sternum is relatively broad in shape, less than the condition seen in living hominoids, but more so than in many extant platyrrhines, "prosimians," or cercopithecoids; it has five distinct segments as in cercopithecoids and ceboids. The lumbar region is fairly long, with six or seven vertebrae. The lumbar vertebrae were relatively broad as in extant hominoids and large cebids, and possess the accessory articular processes found in all primates except extant hominoids. *Pliopithecus* had three sacral vertebrae. On the basis of the shape of the vertebral centra and the overall morphology of the sacrum, Zapfe (1958, 1960) argued that *Pliopithecus* was tail-less. Ankel (1965, 1972), however, suggested from the relative dimensions of the sacral canal at the beginning and end of the bone that this taxon had a tail of moderate length. In any case, it is clear that *Pliopithecus* did not have a long sacrum like extant hominoids.

From the pelvis, only a small part of the right ilium of a single individual is known for *Pliopithecus.* The bone appears to be relatively narrow, with none of the iliac expansion characteristic of extant hominoids.

The several complete femora show numerous similarities to hylobatids and large platyrrhines. The rounded femoral head is set on a long neck extending well above the head of the greater trochanter—a characteristic feature of suspensory forms and arboreal quadrupeds which allows considerable abduction and hip mobility. In contrast, among leaping or cursorial, quadrupedal species the less spherical femoral head is set on a shorter, stouter neck which lies below the greater trochanter. McHenry and Corruccini (1976) also found morphological similarities to *Hylobates* in the femora of *Pliopithecus* and other Miocene hominoids, but felt that it was not possible to make functional inferences from this region.

The femoral shaft is relatively long and slender, with a robustness comparable to that of gibbons but much less than that in most cercopithecoids. The patellar groove is very broad and shallow—a characteristic of most platyrrhines and hominoids (except *Homo*). This condition is characteristic of suspensory species and contrasts with the condition seen in leapers or terrestrial quadrupeds (Fleagle, 1977). The thin, circular patella is very gibbon-like. Likewise, the femoral condyles are relatively broad, shallow anteroposteriorly, and subequal in size, as in *Hylobates* or Ateles. The femur of *Pliopithecus* is clearly that of a primate using its hindlimbs in suspension rather than leaping or fast running.

The tibia of *Pliopithecus* is more gibbon-like than like that of a terrestrial quadruped and also lacks pronounced muscle markings. The shaft is very similar to the same bone in *Ateles*. Distally, there is a relatively large medial malleolus.

The talus is very similar to the same bone in *Hylobates*, with a long neck and a relatively high, narrow trochlea, with low subsequent medial and lateral edges—a morphological configuration allowing a wide range of flexion and extension but minimal lateral stability at the ankle. It shows neither the pronounced lateral trochlear ridge seen in the trochlea surface of most arboreal quadrupeds nor the high trochlea margins seen in leapers (Fleagle, 1977). At the same time the trochlear surface is not as flat nor the neck as short as in extant pongids. Like *Aegyptopithecus*, *Hylobates*, and cercopithecoids, there is a distinct cup-shaped depression on the medial side of the base of the neck where the medial malleolus articulates in full flexion of the ankle (Le Gros Clark and Leakey, 1951).

The calcaneus of *Pliopithecus* is extremely similar to the same bone in *Hylobates*, *Ateles*, *Pongo*, and other suspensory species. The calcaneal tuberosity is very short and very deep—a morphology that provides virtually no leverage for either leaping or quadrupedal running, but allows control of the ankle in extreme positions of flexion during suspension. The bone is also extremely broad, with a pronounced peroneal tubercle laterally and a wide sustentaculum medially. The inferior surface of the sustentaculum has a very large groove for the flexor hallucis longus, indicating very strong grasping abilities for the foot.

Zapfe (1958, 1960) noted that the metatarsals and phalanges were very similar to those of hylobatids. The phalanges are long relative to the short but slender metatarsals, indicating arboreal habits. Schultz (1963) noted that the proportions of the foot in *Pliopithecus* were intermediate between those of "a macaque" and *Hylobates*. The hallucial metatarsal shows evidence of a prehallux bone (*contra* Lewis, 1972a), a primitive sesamoid in the tendon of the Peroneus longus found in many platyrrhines, gibbons, and *Aegyptopithecus* (Conroy, 1976b).

In virtually all features, the hindlimb of *Pliopithecus* is that of an arboreal, suspensory-quadrupedal primate much like *Ateles* or *Brachyteles*. There are no indications of functional adaptations to more quadrupedal or leaping behavior and no suggestion of terrestrial quadrupedalism. The hindlimb elements show numerous similarities to living gibbons not seen in the forelimb. Functionally, they are almost certainly related to hindlimb suspensory behavior; phyletically, they are likely to be primitive features.

The intermembral index of *Pliopithecus* is 94, similar to that for some macaques, baboons, and large cebids such as *Alouatta* or *Lagothrix*. Zapfe (1958, 1960) argued from this and from the fact that *Pliopithecus* was found in a fissure fill that it was macaque-, baboon-, or cercopithecoid-like in locomotor habits and frequently came to the ground. Like most such indices, however, the intermembral index of *Pliopithecus* is very misleading when taken out of context of the size of the animal. Compared with living catarrhines of the same size, *Pliopithecus* has both long forelimbs and long hindlimbs (Jungers *et al.*, 1982). Functionally, all of the evidence from the skeletal anatomy of *Pliopithecus vindobonensis* [including the multivariate analysis of Ciochon and Corruccini (1977) and those of McHenry and Corruccini (1976)] suggests an arboreal, suspensory, monkey-like animal similar to *Ateles* without a prehensile tail.

Phyletically, the postcranial anatomy of *Pliopithecus vindobonensis* is very primitive compared with living catarrhines, either hominoids or cercopithecoids. In most features of the forelimb, *Pliopithecus* is platyrrhine-like or even "prosimian"-like. There are no derived features of the forelimb which clearly link the Miocene species with either group of living catarrhines—both of which show distinct, presumably derived, characters in forelimb morphology. The primitive nature of the *Pliopithecus* forelimb is emphasized by the many similar primitive features it shares with *Aegyptopithecus*.

The hindlimb of *Pliopithecus* is more intriguing and difficult to interpret phyletically. In many aspects of the hindlimb, especially the femur, talus, and calcaneus, *Pliopithecus* is very similar to living gibbons. Are these features primitive catarrhine features, parallelisms, or primitive hominoid synapomorphies? There is very little evidence on the question and I will defer it until after a discussion of the Miocene apes from Kenya.

At least eight species of dryopithecine hominoids are known from the early Miocene of Kenya and Uganda (e.g., Andrews, 1981), providing evidence of an extraordinary adaptive radiation. While many are known mainly from teeth and from only isolated, tentatively referred limb elements, two species—*Dendropithecus* (formerly *Limnopithecus*) *macinnesi* and *Proconsul africanus*—are known from relatively complete skeletons that allow direct comparisons with other fossil apes.

Dendropithecus macinnesi

Originally described as *Limnopithecus macinnesi,* this species has long been recognized as the most gibbon-like of the early Miocene apes from Kenya and has usually been placed in the Hylobatidae (Le Gros Clark and Leakey, 1951;

Simons, 1972; Andrews, 1978; Andrews and Simons, 1977). The limb bones are from several associated individuals found on Rusinga Island in 1948.

The humerus (Figs. 1B, 2B) is the most complete bone. It lacks only the proximal articular epiphysis, which presumably was not yet fused. The estimated length of the bone is 193 mm, approximately the size of the humerus in several species of *Ateles*. Proximally, very little remains of the greater or lesser tuberosities. From the shape of the proximal end of the shaft, Le Gros Clark and Thomas (1951) have estimated the torsion of the humeral head at 108°— considerably less than that of either *Ateles* or any extant hominoid, but similar to what is found in many extant cercopithecoids and platyrrhines. On the medial side, the shaft shows a marked crest extending inferiorly from the lesser tuberosity approximately one-quarter of the way down the shaft to the insertion of Teres major. Pronounced development of this crest is certainly a primitive hominoid feature, as it is found in most "prosimians," platyrrhines, *Aegyptopithecus, Pliopithecus,* and to a lesser extent in other hominoids. It is rare in cercopithecoids. A relatively narrow bicipital groove (*contra* Le Gros Clark and Thomas, 1951) separates the medial crest from the distinct deltopectoral crest. The shape and position of the bicipital groove in *Dendropithecus* are similar to those seen in *Ateles,* approximating those of living hominoids and contrasting with the condition seen in *Aegyptopithecus, Pliopithecus,* most platyr- rhines, and cercopithecoids. In contrast with both *Aegyptopithecus* and *Pliopithecus,* the humerus of *Dendropithecus* also shows a flattened deltoid plane delimited medially by the deltopectoral crest and laterally by a distinct de- ltotriceps crest. The crest is, however, much less developed than that found in extant cercopithecoids. In robustness, the shaft is much thicker than in extant gibbons, more gracile than in most extant cercopithecoids, and very similar to that of *Ateles.* Distally, there is a marked supinator crest, but not broad brachialis flange as seen in either *Aegyptopithecus* or *Pliopithecus.* Like all extant catarrhines, *Dendropithecus* lacks an entepicondylar foramen. The distal artic- ulation is very similar to that in *Ateles* and *Pliopithecus,* with a rounded capitu- lum and a broad trochlea with a low anteromedial and posterolateral lips. There is a very low ridge separating the capitulum from the trochlea. The medial epicondyle is large and projects medially as in both *Ateles* and *Pliopithecus.* As Le Gros Clark and Thomas (1951) noted, in all aspects, the humerus of *Dendropithecus* is virtually identical with that of *Ateles,* indicating that this species was probably a very suspensory arboreal quadruped.

Only the proximal half of the ulnar shaft is known for *Dendropithecus* (Figs. 3A, 4A). On the basis of comparison with other species, Le Gros Clark and Thomas (1951) estimated the total length of the bone at about 215 mm. The proximal end of the sigmoid cavity and the olecranon process are both badly eroded, but it seems likely that the humerus had an olecranon process similar in length to that of *Ateles.* The sigmoid cavity is relatively narrow, with a small, laterally facing radial facet much as in *Pliopithecus* or *Ateles* and smaller than that in any cercopithecoid. The ulnar shaft is very narrow, without pronounced muscle attachments or a strong interosseus line as seen in most cercopithecoids.

The radius, like the ulna, is missing its distal one-third. The head is oval in shape and the radial neck is very long as in hominoids and *Ateles* but not in cercopithecoids. The shaft of the radius is very slender and bowed.

The forelimb of *Dendropithecus macinnesi* is clearly much more primitive and much less elongated than that of extant hylobatids. It is also lacking any specializations which would link it specifically with living cercopithecoids. At the same time it is in some features more advanced than the comparable elements in either *Aegyptopithecus* or *Pliopithecus*. The loss of the entepicondy-lar foramen, reduction of the brachialis flange, and development of a flat deltoid plane with a distinct deltotriceps crest are all features linking *Dendro-pithecus* with living anthropoids.

The hindlimb of *Dendropithecus** is known only from several fragmentary long bones (all lacking articular surfaces), a partial calcaneus, and a broken talus. Le Gros Clark and Thomas (1951) noted that in the shape and projec-tion of the lesser trochanter as well as the straightness and cylindrical shape of the shaft, the femur resembles that of hylobatids more than that of cer-copithecoids. The robustness of the femur appears to have been greater than in gibbons and less than in most cercopithecines or *Ateles*.

Only the shaft is known of the tibia. It is very compressed laterally. Le Gros Clark and Thomas (1951) felt it was cercopithecoid-like, rather than gibbon-like. It is very long, but not a very diagnostic bone. The fibula is also not very helpful.

Le Gros Clark and Thomas (1951) argued that the cuboid facet in the partial calcaneus of *D. macinnesi* was cercopithecoid-like, while the relatively short lever-arm/load-arm ratio was gibbon-like. The broken talus of *D. macin-nesi* shows gibbon-like features in the shape of the trochlea surface. They argued incorrectly that the cup-like depressions on the anterior surface of the medial malleolus facet was a cercopithecoid feature not found in hylobatids.

The hindlimb of *Dendropithecus* offers much less information than the forelimb because of the lack of any relatively complete elements. However, the hindlimb was clearly very long for the size of the animal (Jungers *et al.*, 1982), and shows no obviously cercopithecoid-like features.

In their initial description and analysis of the limbs of *Dendropithecus macinnesi*, Le Gros Clark and Thomas (1951) noted that in overall proportion and in many detailed features, the Miocene ape was certainly less advanced than modern gibbons. Indeed, the main feature which *Dendropithecus macin-nesi* shares with gibbons is length and slenderness of the limbs. In this respect, *Dendropithecus* lies intermediate between extant cercopithecoids and extant hylobatids and is very similar to *Ateles* (Le Gros Clark and Thomas, 1951). However, in morphological details, the Miocene Ape shows few features that could be considered characteristic of either cercopithecoids or hylobatids.

---

*McHenry and Corruccini (1976) analyzed a small hominoid femur from Songhor (KNM-SO 1011). As their analysis revealed, the fossil is very gibbon-like in the high neck angle, rounded head, and small intertrochlea fossa. Because of its gibbon-like appearance, they argued it was probably from *Dendropithecus macinnesi*. The fossil is very much smaller than *D. macinnesi* from Rusinga and almost certainly does not belong to that taxon.

Proconsul sp.

Postcranial elements are known for three species of the larger ape from the East African early Miocene ape, *Proconsul*. The largest species, *Proconsul major*, is known from a talus, a calcaneus, the distal part of the tibia, and a lumbar vertebra (Le Gros Clark and Leakey, 1951; Walker and Rose, 1968; Rose, this volume, Chapter 15). More postcranial elements are known for *Proconsul nyanzae*, but many of these have not yet been completely described (Le Gros Clark and Leakey, 1951; Szalay and Delson, 1979; Rose, this volume, Chapter 15). As a result of recent work by Walker and Pickford, *Proconsul africanus* is now known from a nearly complete juvenile skeleton (McKean, 1981; Walker and Pickford, this volume, Chapter 12). Since other authors in this volume discuss the limb skeleton of *Proconsul* sp. in considerable detail, I shall confine my treatment to a broader comparison of this genus, and especially *P. africanus*, with the Oligocene and smaller early Miocene species described above.

As most workers have properly emphasized (e.g., Napier and Davis, 1959; Morbeck, 1975; Feldesman, 1982; Rose, this volume, Chapter 15; Walker and Pickford, this volume, Chapter 12), the forelimb of *Proconsul africanus*, including the humerus (Figs. 1A, 2A), cannot be easily characterized by comparison with extant catarrhines. It is clearly unlike that of either living apes or living monkeys and shows a suite of probably primitive features that demand a more careful, broadly comparative approach to understanding its locomotor habits and its proper phyletic position. Comparison with Oligocene and other early Miocene apes is probably more straightforward. Like *Pliopithecus* and *Dendropithecus*, *P. africanus* was basically an arboreal quadruped. Particularly diagnostic anatomical regions such as the medial epicondyle and the olecranon are unfortunately not known for adults of this species; however, several aspects of its anatomy suggest that species of *Proconsul* were probably less suspensory than were either *Pliopithecus* or *Dendropithecus* (but see Aiello, 1981). *Proconsul africanus* apparently had both absolutely and relatively shorter forelimbs than either *Pliopithecus* or *Dendropithecus* even though the latter were absolutely much smaller animals (Figs. 1, 2). Also, the ulna shows possible indications of dorsal rather than ventral concavity, suggesting more quadrupedal habits. For the larger *P. nyanzae*, the relatively narrow trochlea and posteriorly directed olecranon process (Szalay and Delson, 1979, p. 481) suggest even more quadrupedal, possibly even terrestrial habits.

The wrist and hand of *P. africanus* certainly show greater similarity to those of arboreal quadrupedal monkeys (including cebids) than to those of extant apes (Jenkins and Fleagle, 1975; Napier and Davis, 1959; Schön and Ziemer, 1973; McHenry and Corruccini, this volume, Chapter 13; Rose, this volume, Chapter 15; *contra* Lewis, 1972*b*).

The hindlimb of *Proconsul* remains more poorly known and less extensively studied than the forelimb; however, the features indicative of extreme hip mobility (rounded head on a high neck, low greater trochanter) found in many of the smaller apes are also found in *Proconsul*. The knee is poorly documented, but the ankle and foot of *Proconsul* are now very well known.

The fibular side of the leg was apparently well developed as in extant apes and the foot showed a large grasping hallux and relatively long metatarsals and phalanges indicative of arboreal grasping habits.

Overall, the larger *Proconsul* was, I suspect, a more quadrupedal animal than was either *Pliopithecus* or *Dendropithecus*. There was probably also a considerable range of locomotor adaptation among the different-sized members of the genus, the larger species being more terrestrially inclined than the smaller. Like living apes, however, all apparently retained the primitive arboreal grasping abilities of the hindlimb, and as a group lacked the rapid running or leaping abilities of living cercopithecoids.

Discussion

The available skeletal material of Oligocene and early Miocene hominoids discussed above provides a necesarily incomplete, but consistent picture of early catarrhine locomotor evolution during this period. All of the forelimb elements for *Aegyptopithecus zeuxis* indicate that this species was an arboreal quadruped with locomotor affinities most comparable to the neotropical monkey *Alouatta* sp. or the large lemur *Varecia variagatus*. It was neither a terrestrial nor a rapidly running arboreal, quadruped in the manner of many extant cercopithecoid monkeys. Rather, it was a heavily muscled, slow moving, quadrupedal, climbing species. The only known elements of the hindlimb—a talus, a metatarsal, and a phalanx—are consistent with this interpretation. Likewise, the few skeletal remains of *Propliopithecus chirobates* indicate that this small Oligocene species was an arboreal quadruped.

Pliopithecus vindobonensis, from the Miocene of Europe is, in its forelimb skeleton, very much a gracile version of *Aegyptopithecus*. Although probably basically an arboreal quadruped, it certainly engaged in more suspensory behavior than did *Aegyptopithecus*. The suspensory adaptations of *Pliopithecus* are especially obvious in the hindlimb, which is very similar to that of living hylobatids. The best locomotor analogs to *Pliopithecus* are large cebids like *Lagothrix* and *Ateles* rather than extant cercopithecoids. There is no anatomical evidence for either rapid running or terrestrial adaptations in the limb skeleton of *Pliopithecus*. All evidence points toward a more suspensory, climbing adaptation.

The skeletal remains of *Dendropithecus macinnesi* from the early Miocene of Kenya also suggest a quadrupedal, long-limbed, suspensory climber much like the neotropical *Ateles*. The limb bones and few articular surfaces all suggest gracile, slender bones with flexible joints adapted for mobility. Again, there are no indications of either terrestrial adaptations or rapid arboreal running or leaping.

There has certainly been more debate over the locomotor adaptations indicated by the partial juvenile forelimb of *Proconsul africanus* than about the locomotion of any other fossil ape. Description and analysis of the remainder of the skeleton (Walker and Pickford, this volume, Chapter 12) both amplify and complicate our understanding of this species. Nevertheless, the presently available analyses permit a relatively coherent picture of the animal's likely behavior. There is no reliable evidence that *Proconsul africanus* was "a brachiator" or any more suspensory than any of the smaller fossil apes from that time span, such as *Pliopithecus* or *Dendropithecus*. Although *Proconsul africanus* was certainly much larger than *Dendropithecus* and the articular surfaces of the forelimb are much larger, the absolute lengths of the limbs are shorter, and although immature, are most unlikely to have grown much more. Thus, it was a much shorter-limbed species, an adaptation one expects in an arboreal quadruped but not in either a suspensory or a terrestrially quadrupedal species. There is also little evidence that *P. africanus* was particularly inclined toward terrestrial quadrupedalism. The impression one gets is of an arboreal quadruped with some suspensory and climbing abilities, as indicated by the chimpanzee-like nature of the elbow, leg, and foot, but no obvious adaptation for either brachiation (as in the Lesser Apes or orangutans), knuckle-walking, or fast terrestrial quadrupedalism as in baboons or patas monkeys. The larger *Proconsul* species (e.g., *P. nyanzae*) were probably more terrestrial in their habits, but very little skeletal material attributed to the larger taxa has been described or analyzed.

In comparing the locomotor adaptations of the Oligocene and Miocene hominoids with those of living monkeys and apes, it is clear that none of the fossils show either the extreme suspensory adaptations of living hominoids or extreme adaptation for either arboreal or terrestrial quadrupedalism or leaping seen among extant cercopithecoids. In that sense, they were all more generalized than the extant forms. At the same time, they were probably more effective arboreal quadrupeds than are living apes and more suspensory than are living cercopithecoids (see Rose, this volume, Chapter 15). As has been frequently emphasized, the best adaptive models are the larger New World monkeys. Within this adaptive zone, there were certainly trends and differences. *Aegyptopithecus* was almost certainly slower, more limited to branch quadrupedalism, and much less suspensory than any of the Miocene genera. Within the Miocene taxa, I would suggest that *Pliopithecus* and *Dendropithecus* were more suspensory than *Proconsul* sp. based on limb proportions and available joint morphology (but see Aiello, 1981).

Hominoid Phylogeny

The description of the skeletal anatomy and the interpretations of locomotor adaptations of fossil apes from the Oligocene and Miocene summarized above are not substantially different from those offered by either the original describers or the vast majority of scientists analyzing these remains over the past 20 years. Yet, based on the same fossils some workers have argued that these animals were "incipient" brachiators or knuckle-walkers phyletically related to extant ape genera, while others have argued that they were quadrupedal monkeys just like extant cercopithecoids with no particular phyletic or behavioral relationships to living apes. Why have the skeletal anatomy and locomotor adaptations of these animals encouraged such apparent disagreement and confusion? The contradictions and the disagreement have come largely, I believe, from the very different phylogenetic hypotheses and evolutionary assumptions that have underlain the analysis and interpretation of fossil apes during the past three decades.

For most of the scientists who originally described the limb skeleton of fossil apes (Le Gros Clark and Thomas, 1951; Le Gros Clark, 1950; Napier and Davis, 1959; Zapfe, 1958, 1960) and for many others, the phylogenetic positions of the fossil taxa were largely predetermined by analysis of the dentition. For these authors, analysis of the skeletal anatomy provided very little information about the phyletic position of the respective taxa in hominoid evolution; rather, it provided evidence about the locomotor adaptations of a particular lineage at a point in time. Thus, for Le Gros Clark the limbs of *Limnopithecus* showed that Miocene gibbons had not yet attained the specializations of the recent member of the family. Zapfe (1958, p. 454) was even more explicit in summarizing, "Prior to this discovery, it was highly unlikely, to say the least, that this humerus or auditory region would have been recognized and properly identified as anthropoid remains."

More subtle, but no less pervasive has been the way in which views of the phyletic relationships of extant taxa influenced conclusions about the phyletic position and the habits of the fossils. Both Le Gros Clark and Thomas (1951) and Zapfe (1958, 1960) compared their fossils with a wide range of extant primates. The former authors found *Dendropithecus* to be virtually identical with *Ateles,* while Zapfe (1958, 1960) noted numerous platyrrhine and "prosimian" features in the skeleton of *Pliopithecus*. Despite finding all of these primitive features, their phyletic discussions were still largely limited to deciding whether these Miocene "apes" were more like gibbons or like cercopithecoid monkeys—the presumed ancestral catarrhine condition. Very few authors have seriously considered the phyletic likelihood that both extant hominoids and extant cercopithecoids show derived morphologies relative to the ancestral catarrhine (i.e., two premolared) anthropoids (e.g., Fleagle and Simons, 1978a). Rather, most people have tended to place a very literal phylogenetic interpretation on the "*scala natura*" of primates which Huxley (1863) and Le Gros Clark (1934, 1950) emphasized. Thus, most studies of supposed fossil apes have almost invariably tried to interpolate them somewhere *between* presumably ancestral living cercopithecoid monkeys and derived hominoid apes.

This is most obvious in Le Gros Clark and Thomas (1951), who, after comparing *Dendropithecus* with *Ateles* throughout their monograph, conclude that this fossil was like a cercopithecoid monkey! Corruccini *et al.* (1976) generally used similar reasoning. Although they briefly discuss the possibility that cercopithecoids may be derived from a hominoid ancestor (also see Delson, 1975; Szalay and Delson, 1979), they nevertheless continue, "Presumably a hominoid ancestor of cercopithecoids would still retain the primitive generalized mammalian postcranial pattern, just as extant cercopithecoids do to-day." They further argue that, "Although Pilbeam (1969) and Simons (1972) infer that dryopithecines, if quadrupedal, were cebid-like quadrupeds as opposed to cercopithecoid-like, this makes no difference *vis-à-vis* their differences from hominoids. Thus, consideration of whether dryopithecines were more similar to New World or Old World monkeys is immaterial" Corruccini *et al.*, 1976, p. 218).

It is certainly true that if one is only posing a simple dichotomous question—Are these fossils like modern apes or different?—the issue of the primates with which they are compared becomes less critical. However, if we are interested in ascertaining the correct phylogenetic position of the fossil "apes," it is critical that we make appropriate comparisons.

Theoretically, the most appropriate way of investigating the phyletic affinities of the Oligocene and early Miocene hominoids would be to identify the unique derived features of catarrhines, cercopithecoid monkeys, and extant hominoids and then examine the fossils *vis à vis* these features. As students of primate evolution have long realized (e.g., Gregory, 1910; Le Gros Clark, 1939; Szalay, 1977; Luckett, 1980), distinguishing between primitive and derived features is both the most important aspect of phylogeny reconstruction and the most difficult. When we add to this the incompleteness of the fossils themselves, we are left with a very difficult task indeed. Fortunately, there are a few aspects of skeletal anatomy where we can make relatively good assessments among numerous taxa. The skeletal region that has been analyzed most exhaustedly for the fossil apes is the elbow—a region in which most workers feel it is possible to identify distinctive, derived morphologies in cercopithecoid monkeys and extant hominoids, and an aspect of the skeleton that is well represented in the fossil record (e.g., Washburn, 1968; McHenry and Corruccini, 1975; Feldesman, 1982). Figure 5 is a phylogeny of humeral morphology. Despite considerable diversity, the humerus and ulna in extant platyrrhines are relatively generalized by either mammalian or primate standards, and almost certainly retain the primitive anthropoid or higher primate morphology (e.g., Szalay and Dagosto, 1980). The trochlear surface usually has only a slight medial lip anteriorly, inferiorly, and posteriorly. There is also a relatively low lip posterolaterally. There is usually a shallow sulcus separating the trochlea from the very rounded capitulum. Above the medial epicondyle there is often an entepicondylar foramen. Laterally there is often a moderately developed brachialis flange with a distinct supinator crest extending well up the shaft. Posteriorly, the olecranon fossa is shallow and wide. The ulna in ceboids generally has a shallow, relatively broad sigmoid notch with a gently sloping surface. The radial notch is a relatively small laterally facing facet in most species. The olecranon is moderately long. The shaft is usually deeper than wide, and is gently formed, and the styloid process is long. The primitive platyrrhine morphology could easily be transformed into either that found among extant hominoids or that found among extant cercopithecoids.

The humero-ulnar joint in living pongids and humans is quite distinctive

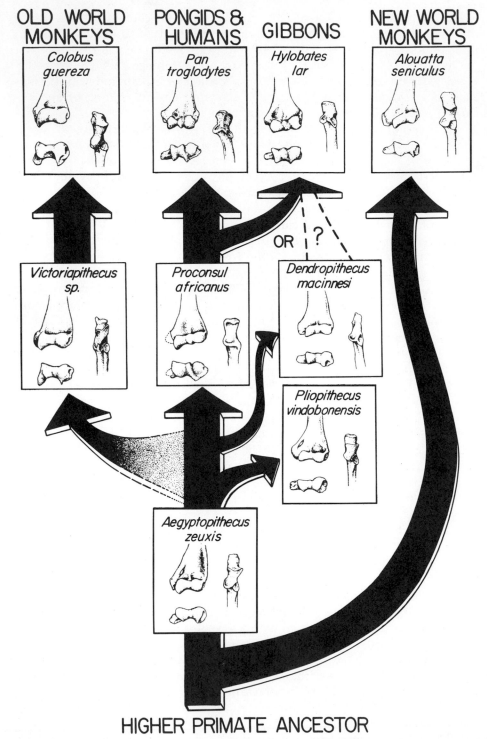

Fig. 5. A phylogeny of Oligocene and early Miocene catarrhines based on the morphology of the elbow region.

and clearly derived with respect to all other mammals (Gregory, 1943; Washburn, 1968; Jenkins, 1973). The most distinctive area is the deep trochlea, with distinct margins both medially and laterally extending almost 270° around the articular surface and divided by a deep gutter. This trochlear morphology is reflected in the ulna by a pronounced median keel. The capitulum in hominoids is spherical and separated from the lateral lip of the trochlea by another gutter. The medial epicondyle of all extant hominoids is large and projects medially. The lateral epicondyle is well formed. In addition to the trochlear keeling, the ulna shows a small lateral facet for the articulation of the radius and a very short olecranon process. The shaft of the ulna is bowed but lacks pronounced interosseus markings. Distally, there is a reduced styloid process.

The elbow joint of gibbons shows the same basic design seen in pongids and humans. However, the trochlear "guttering" and associated "keel" on the trochlear notch of the ulna are less marked than in pongids. Also, the capitulum is less spherical and the brachialis flange is more reduced. It is difficult to determine the extent to which the differences gibbons show from pongids represent primitive conditions indicative of a earlier split from the ancestral ape stock or specialized features associated with their unique locomotor adaptations (Feldesman, 1982; Fleagle, 1983).

The humero-ulnar joint of cercopithecoids is equally distinctive, but in very different ways. In the distal articular surface of the humerus in Old World monkeys, the medial lip of the trochlea is pronounced anteriorly and inferiorly, but not posteriorly. Posteriorly, there is a pronounced lateral lip, which forms the lateral edge of the deep olecranon fossa. On the anterior and inferior surfaces, the trochlea grades smoothly into the capitulum, which lacks either the spherical shape or the separation from the capitulum as seen in apes. The medial epicondyle is always relatively short and posteriorly directed, often very much so. The ulna in cercopithecoids has a narrow sigmoid cavity with the anterior and posterior parts barely connected. The coronoid process is relatively high and the radial notch is large and often deeply excavated. The olecranon process is longer than that in extant apes. Like extant hominoids, all cercopithecoids lack an entepicondylar foramen. The distinctive elbow features of Old World monkeys are present in the earliest known fossil monkeys from Africa, *Victoriapithecus* (Szalay and Delson, 1979).

Although clearly catarrhine in their dental formula, the fossil hominoids *Aegyptopithecus, Propliopithecus, Pliopithecus,* and *Dendropithecus* all look much more like platyrrhines in their skeletal anatomy than like either group of extant catarrhines (see also Simons and Fleagle, 1973; Fleagle and Simons, 1978a; Szalay and Delson, 1979; Feldesman, 1982). They all lack any unique derived features linking them with either group of extant catarrhines and the first three still retain an entepicondylar foramen, a feature lost in both later groups. The relatively uniform platyrrhine-like morphology of these early fossil catarrhines from three different areas and a time span of approximately 10 million years is very strong evidence that this is the primitive elbow morphology for Old World higher primates, which are nevertheless catarrhine in their dental formula. A phyletic position ancestral to groups of modern catarrhines suggested by the forelimb morphology of *Aegyptopithecus* and *Plio-pithecus* is also supported by the morphology of the ear region. *Aegyptopithecus* clearly lacks the tubular external auditory meatus found in both groups of extant catarrhines, and in *Pliopithecus,* this structure is only partially developed (Simons, 1967; Zapfe, 1960; Szalay and Delson, 1979; Cartmill *et al.,* 1981; Fleagle and Kay, this volume, Chapter 7).

Compared with *Aegyptopithecus* and *Pliopithecus,* both *Dendropithecus* and *Proconsul* are more modern-looking in the morphology of their humerus and ulna. Both lack an entepicondylar foramen and both show indications of a lateral trochlear ridge. This development of a lateral trochlear margin and the associated "gutter" between that and the medial lip is particularly pronounced in *P. africanus.* The same articular phenomenon is also evident in the low median keel in the trochlear notch of the ulna in that species. In this regard, *Proconsul africanus* looks more like a living ape than does any other early Miocene or Oligocene hominoid (see also Feldesman, 1982). It shows, incipiently, the same elbow features which first appear fully developed in the late Miocene apes of Europe and Asia (see Rose, this volume, Chapter 15; Morbeck, this volume, Chapter 14).

The more advanced appearance of the elbow region in *Proconsul* generally accords with the evidence from the remainder of the skull and skeleton. For *Dendropithecus* the evidence is admittedly thin and based largely on the fact that this genus appears in its dentition to be part of a single radiation of primitive hominoids in the early Miocene of East Africa (Simons and Fleagle, 1973; Fleagle and Simons, 1978*b*). There is, however, good evidence that *Proconsul africanus* is more advanced and modern-looking than the earlier *Aegyptopithecus* or its European contemporary *Pliopithecus.* The tubular external auditory meatus is fully formed as in modern catarrhines and the brain apparently shows ape-like conditions (Radinsky, 1974). Most of the "prosimian"-like features which were so apparent in the shaft and head of the humerus of *Aegyptopithecus* and *Pliopithecus* are not obvious in *Dendropithecus* and *Proconsul.* At the same time it is also clear that *P. africanus* (and *Dendropithecus*) lacks many uniquely derived features of the skull, dentition, wrist, and hand of extant apes. *Proconsul africanus* was probably more than just a "dental ape," but was still only a "formative ape" (Feldesman, 1982).

Recognition that a basically platyrrhine-like, rather than a cercopithecoid-like, forelimb morphology is primitive for catarrhines means that many of the characteristic forelimb features of extant hominoids are probably primitive catarrhine features. These would include the development of a moderate brachialis flange, a large medially projecting medial epicondyle, and a spherical capitulum separated from the trochlea by a shallow groove. These are features which earlier workers used to link the fossils with extant apes.

If we try to reconstruct an early catarrhine phylogeny for other parts of the skeleton, the fossil evidence becomes much more limited. Few hindlimb elements are known for the Oligocene taxa. However, all suggest a divergent grasping hallux and arboreal quadrupedal habits. The various Miocene taxa are known from numerous, albeit usually unassociated hindlimb elements.

The most striking aspect of all of the Miocene material from medium-sized and small "apes" is the recurrence of climbing and suspensory adaptations. In the femur, all tend to have a rounded femoral head set on a high neck, which extends well above the greater trochanter. Where the distal end of the femur is known, there is a broad, shallow patellar groove and shallow, symmetric condyles—all features found in more suspensory platyrrhines and hominoids. The cercopithecoid femur with a high greater trochanter, relatively restricted femoral head, and relatively deep patellar groove is certainly a derived condition.

The talus and calcaneus are other areas that have been discussed extensively with very conflicting functional and phyletic conclusions. Wood (1973), like Le Gros Clark and Leakey (1951) and Preuschoft (1973), argued that in ankle morphology the African dryopithecines were cercopithecoid-monkey-like, while Lisowski *et al.* (1976) argued that they were probably more suspensory. Preuschoft's (1973) argument that the pronounced basal calcaneal tubercle seen in most of the smaller dryopithecine fossils from East Africa (and *Pliopithecus*) is an adaptation for terrestrial quadrupedal running with the heel off the ground is contradicted by the fact that such a prominent basal tubercle is found not in cercopithecoid monkeys and baboons, but in more suspensory anthropoids such as *Alouatta, Ateles, Hylobates,* and *Pongo.* The same is true of the very broad sustentaculum and short lever arm (e.g., Zapfe, 1960).

The extent to which somewhat suspensory hindlimbs (Stern, 1971) represent the ancestral condition for both modern catarrhine groups—hominoids and cercopithecoids—cannot be confirmed at present. We need more hindlimb material for the Oligocene hominoids (see Fleagle and Simons, 1982a). However, the consistent findings of numerous suspensory adaptations in the skeleton of *Pliopithecus,* which in many cranial and skeletal features seems to precede the cercopithecoid–hominoid divergence, and in many of dryopithecines from East Africa is certainly suggestive. Such a hindlimb structure would certainly agree functionally with available evidence for the forelimb of these early catarrhines. It would also suggest that many of the suspensory, climbing features of extant hominoids are just further developments of the primitive early catarrhine adaptation.

ACKNOWLEDGMENTS

I am grateful to the curators of the American Museum of Natural History (New York), The Museum of Comparative Zoology (Cambridge), The United States National Museum (Washington, D.C.), and the Naturhistorisches Museum (Vienna) for permission to study the fossils and extant primates in their collections. The Oligocene fossils discussed in this paper were collected through research grants BNS 77-20104, BNS 80-16206, FC 0869600, and FC 80974 to Elwyn L. Simons and the cooperation of the Egyptian Geological Survey. I thank Dr. Simons and Dr. Baher el Khashab of the Cairo Geological

Museum for access to fossils under their care. Drs. M. D. Rose, A. C. Walker, W. L. Jungers, J. T. Stern, and R. L. Ciochon offered many helpful discussions regarding the contents of the manuscript. Lucille Betti drafted the figures and Judy Nimmo typed many versions of the manuscript. This research was supported by NSF research grants BNS 77-25921 and BNS 79-24149.

References

Aiello, L. C. 1981. Locomotion in the Miocene Hominoidea, in: *Aspects of Human Evolution* (C. B. Stringer, ed.), pp. 63–98, Taylor and Francis, London.

Andrews, P. 1978. A revision of the Miocene Hominoidea of East Africa. *Bull. Br. Mus. (Nat. Hist.) Geol.* **30:**85–244.

Andrews, P. 1981. Species diversity and diet in monkeys and apes during the Miocene, in: *Aspects of Human Evolution* (C. B. Stringer, ed.), pp. 25–62, Taylor and Francis, London.

Andrews, P., and Groves, C. 1975. Gibbons and brachiation. *Gibbon and Siamang* **4:**167–218.

Andrews, P., and Simons, E. L. 1977. A new African Miocene gibbon-like genus, *Dendropithecus* (Hominoidea, Primates) with distinctive postcranial adaptations: Its significance to origin of Hylobatidae. *Folia Primatol.* **28:**161–170.

Ankel, F. 1965. Der canalis sacralis als indikator für die lange der primaten. *Folia Primatol.* **3:**263–276.

Ankel, F. 1972. Vertebral morphology of fossil and extant primates, in: *The Functional and Evolutionary Biology of Primates* (R. H. Tuttle, ed.), pp. 223–240, Aldine, Chicago.

Bown, T. M., Kraus, M. J., Wing, S. L., Fleagle, J. G., Tiffany, B., and Simons, E. L. 1982. The Fayum forest revisited. *J. Hum. Evol.* **11(7):**603–632.

Cartmill, M., McPhee, R. D., and Simons, E. L. 1981. Anatomy of the temporal bone in early anthropoids, with remarks on the problem of anthropoid origins. *Am. J. Phys. Anthropol.* **56:**3–22.

Ciochon, R. L., and Corruccini, R. S. 1977. The phenetic position of *Pliopithecus* and its phylogenetic relationship to the Hominoidea. *Syst. Zool.* **26:**290–299.

Conroy, G. C. 1976a. Primate postcranial remains from the Oligocene of Egypt. *Contrib. Primatol.* **8:**1–134.

Conroy, G. C. 1976b. Hallucial tarsometatarsal joint in an Oligocene anthropoid, *Aegyptopithecus zeuxis*. *Nature (Lond.)* **262:**684–686.

Corruccini, R. S., Ciochon, R. L., and McHenry, H. M. 1975. Osteometric shape relationships in the wrist joint of some anthropoids. *Folia Primatol.* **22:**209–226.

Corruccini, R. S., Ciochon, R. L., and McHenry, H. M. 1976. The postcranium of Miocene hominoids: Were dryopithecines merely "dental apes"? *Primates* **17:**205–223.

Delson, E. 1975. Toward the origin of the Old World monkeys, in: Evolution des Vertébrés—Problèmes Actuels de Paléontologie. *Colloq. Int. Cent. Nat. Rech. Sci.* **218:**839–850.

Feldesman, M. R. 1982. Morphometric analysis of the distal humerus of some Cenozoic catarrhines: The late divergence hypothesis revisited. *Am. J. Phys. Anthropol.* **59:**173–95.

Fleagle, J. G. 1977. Locomotor behavior and skeletal anatomy of sympatric Malaysian leaf-monkeys (*Presbytis obscura* and *Presbytis melalophos*). *Am. J. Phys. Anthropol.* **46:**297–308.

Fleagle, J. G. 1980. Locomotor behavior of the earliest anthropoids. A review of the current evidence. *Z. Morphol. Anthropol.* **71:**149–156.

Fleagle, J. G. 1983. Are there any fossil gibbons?, in: *Biology of the Lesser Apes* (D. J. Chivers, H. Preuschoft, N. Creel, and W. Brockelman, eds.), Edinburgh University Press, Edinburgh. (in press).

Fleagle, J. G., and Jungers, W. L. 1982. Fifty years of higher primate phylogeny, in: *Fifty Years of Physical Anthropology in North America* (F. Spencer, ed.), pp. 187–230, Academic, New York.

Fleagle, J. G., and Simons, E. L. 1978a. Humeral morphology of the earliest apes. *Nature (Lond.)* **276:**705–707.

Fleagle, J. G., and Simons, E. L. 1978b. *Micropithecus clarki,* a small ape from the Miocene of Uganda. *Am. J. Phys. Anthropol.* **49:**427–440.

Fleagle, J. G., and Simons, E. L. 1982a. Skeletal remains of *Propliopithecus chirobates* from the Egyptian Oligocene. *Folia Primatol.* **39:**161–177.

Fleagle, J. G., and Simons, E. L. 1982b. The humerus of *Aegyptopithecus zeuxis,* a primitive anthropoid. *Am. J. Phys. Anthropol.* **59**(2):175–194.

Fleagle, J. G., Simons, E. L., and Conroy, G. C. 1975. Ape limb bone from the Oligocene of Egypt. *Science* **189:**135–137.

Fleagle, J. G., Kay, R. F., and Simons, E. L. 1980. Sexual dimorphism in early anthropoids. *Nature (Lond.)* **257:**328–330.

Gregory, W. K. 1910. The orders of mammals. *Bull. Am. Mus. Nat. Hist.* **27:**3–524.

Gregory, W. K. 1920. On the structure and relations of *Notharctus,* an American Eocene primate. *Mem. Am. Mus. Nat. Hist. N.S.* **3:**51–243.

Gregory, W.K. 1934. *Man's Place Among the Anthropoids,* Oxford Press, London. 119 pp.

Gregory, W. K. 1943. The humerus from fish to man. *Am. Mus. Novit.* **1400:**1–54.

Groves, C. P. 1972. Systematics and phylogeny of gibbons. *Gibbon and Siamang* **1:**1–80.

Huxley, T. H. 1863. *Evidence as to Man's Place in Nature,* Williams and Norgate, London. pp. 159.

Jenkins, F. A., Jr. 1973. The functional anatomy and evolution of the mammalian humero-ulnar articulation. *Am. J. Anat.* **137:**281–298.

Jenkins, F. A., Jr. 1981. Wrist rotation in primates: A critical adaptation for brachiators. *Symp. Zool. Soc. Lond.* **48:**429–451.

Jenkins, F. A., Jr., and Fleagle, J. G. 1975. Knuckle-walking and the functional anatomy of the wrist in living apes, in: *Primate Functional Morphology and Evolution* (R. H. Tuttle, ed.), pp. 213–227, Mouton, The Hague.

Jolly, C. J. 1967. The evolution of baboons, in: *The Baboon in Medical Research* (H. Vagtborg, ed.), Volume 2, pp. 427–457, University of Texas Press, Austin.

Jungers, W. L., Fleagle, J. G., and Simons, E. L. 1982. Limb proportions and skeletal allometry in fossil catarrhine primates. *Am. J. Phys. Anthropol.* **57:**1200 (Abstract).

Keith, A. 1923. Man's posture: Its evolution and disorders. *Br. Med. J.* **1923**(1):451–454, 499–502, 545, 548, 587–590, 624–626, 669–672.

Le Gros Clark, W. E. 1934. *Early Forerunners of Man,* Balliere, London.

Le Gros Clark, W. E. 1939. *The Scope and Limitation of Physical Anthropology,* British Association for the Advancement of Science/Ballantine Press, London, pp. 1–24.

Le Gros Clark, W. E. 1950. Fossil apes and men. *Spectator* **1950**(Jan. 13):38–39.

Le Gros Clark, W. E., and Leakey, L. S. B. 1951. The Miocene Hominoidea of East Africa. *Fossil Mammals of Africa* (Br. Mus. Nat. Hist.) **1:**1–117.

Le Gros Clark, W. E., and Thomas, D. P. 1951. Associated jaws and limb bones of *Limnopithecus macinnesi. Fossil Mammals of Africa* (Br. Mus. Nat. Hist.) **3:**1–27.

Lewis, O. J. 1972a. The evolution of the hallucial tarsometatarsal joint in the Anthropoidea. *Am. J. Phys. Anthropol.* **37:**13–34.

Lewis, O. J. 1972b. Osteological features characterizing the wrists of monkeys and apes, with a reconsideration of this region in *Dryopithecus (Proconsul) africanus. Am. J. Phys. Anthropol.* **36:**45–58.

Lisowski, F. P., Albrecht, G. H., and Oxnard, C. E. 1976. African fossil tali: Further multivariate morphometric studies. *Am. J. Phys. Anthropol.* **45:**5–18.

Luckett, W. P. 1980. The suggested evolutionary relationships and classification of tree shrews, in: *Comparative Biology and Evolutionary Relationships of Tree Shrews* (W.P. Luckett, ed.), pp. 3–34, Plenum, New York.

McHenry, H. M., and Corruccini, R. S. 1975. Distal humerus in hominoid evolution. *Folia Primatol.* **23:**227–244.

McHenry, H. M., and Corruccini, R. S. 1976. Affinity of Tertiary hominid femora. *Folia Primatol.* **26:**139–150.

McKean, K. 1981. A new ancient ape. *Discover* **2**:52–55.

Morbeck, M. E. 1975. *Dryopithecus africanus* forelimb. *J. Hum. Evol.* **4**:39–46.

Napier, J. R., and Davis, P. R. 1959. The forelimb skeleton and associated remains of *Proconsul africanus*. *Fossil Mammals of Africa* (Br. Mus. Nat. Hist.) **16**:1–69.

Osborn, H. F. 1908. New fossil mammals from the Fayum Oligocene, Egypt. *Bull. Am. Mus. Nat. Hist.* **24**:265–272.

Pilbeam, D., 1969. Tertiary Pongidae of East Africa: Evolutionary relationships and taxonomy. *Bull. Peabody Mus. Nat. Hist.* (Yale Univ.) **31**:1–185.

Preuschoft, H. 1973. Body posture and locomotion in some East African Miocene dryopithecinae, in: *Human Evolution* (M. Day, ed.), pp. 13–46, Symposium of the Society for the Study of Human Biology, Volume 11, Barnes and Noble, New York.

Preuschoft, H. 1975. Body posture and mode of locomotion in fossil primates—Method and example: *Aegyptopithecus zeuxis*, in: *Proceedings from the Symposia of the Fifth Congress of the International Primatological Society 1974*, pp. 346–359, Japan Science Press, Tokyo.

Radinsky, L. C.1974. The fossil evidence of anthropoid brain evolution. *Am. J. Phys. Anthropol.* **41**:15–27.

Schlosser, M. 1910. Über einige fossile Säugertiere aus dem Oligocän von Ägypten. *Zool. Anz.* **34**:500–508.

Schlosser, M. 1911. Beiträge zur kenntnis dër Oligozänen Landsäugetiere aus dem Fayum: Ägypten. *Beitr. Paläontol. Oesterreich-Ungarns Orients* **6**:1–227.

Schön, M. A., and Ziemer, L. K. 1973. Wrist mechanism and locomotor behavior of *Dryopithecus (Proconsul) africanus*. *Folia Primatol.* **20**:1–11.

Schön-Ybarra, M. A., and Conroy, G. C. 1977. Non-metric features in the ulna of *Aegyptopithecus, Alouatta, Ateles* and *Lagothrix. Folia Primatol.* **29**:178–195.

Schultz, A. H. 1963. Relations between the lengths of the main parts of the foot skeleton in primates. *Folia Primatol.* **1**:150–171.

Simons, E. L. 1967. The earliest apes. *Sci. Am.* **217**(6):28–35.

Simons, E. L. 1972. *Primate Evolution: An Introduction to Man's Place in Nature*, Macmillan, New York. 322 pp.

Simons, E. L., and Fleagle, J. G. 1973. The history of extinct gibbon-like primates. *Gibbon and Siamang* **2**:121–148.

Simons, E. L., and Pilbeam, D. R. 1972. Hominoid paleo-primatology, in: *The Functional and Evolutionary Biology of Primates* (R. Tuttle, ed.), pp. 36–62, Aldine-Atherton, Chicago.

Stern, J. T. 1971. Functional myology of the hip and thigh of cebid monkeys and its implications for the evolution of erect posture. *Bibl. Primatol.* **14**:1–318.

Straus, W. L. 1949. The riddle of man's ancestry. *Q. Rev. Biol.* **24**:200–223.

Szalay, F. S. 1977. Ancestors, descendants, sister-groups, and testing of phylogenetic hypotheses. *Syst. Zool.* **26**:12–18.

Szalay, F. S., and Dagosto, M. 1980. Locomotor adaptations as reflected in the humerus of Paleogene primates. *Folia Primatol.* **34**:1–45.

Szalay, F. S., and Delson, E. 1979. *Evolutionary History of the Primates*, Academic, New York. 580 pp.

Walker, A., and Rose, M. D. 1968. Fossil hominoid vertebra from the Miocene of Uganda. *Nature (Lond.)* **217**:980–981.

Washburn, S. L. 1968. *The Study of Human Evolution*, Condon, Oregon System of Higher Education, Eugene, Oregon. 48 pp.

Wood, B. 1973. Locomotor affinities of hominoid tali from Kenya. *Nature (Lond.)* **246**:45–46.

Wood-Jones, F. 1929. *Man's Place Among the Mammals*, Longmans, Green & Co., New York.

Wood-Jones, F. 1948. *Hallmarks of Mankind*, Williams and Wilkins, Baltimore.

Zapfe, H. 1958. The skeleton of *Pliopithecus (Epipliopithecus) vindobonensis* Zapfe and Hürzeler. *Am. J. Phys. Anthropol.* **16**:441–455.

Zapfe, H. 1960. Die primatenfunde aus der Miozänen spaltenfüllung von Neudorf an der March (Děvinská Nová Ves), Tschechoslowakei. Mit Anhang: Der Primatenfund aus dem Miozän von Klein Hadersdorf in Niederosterreich. *Schweiz. Palaeontol. Abh.* **78**:1–293.

New Postcranial Fossils of *Proconsul africanus* and *Proconsul nyanzae*

12

A. C. WALKER AND M. PICKFORD

Introduction

Our understanding of early Miocene hominoid postcranial material has been built upon a fossil record that consists, for the most part, of unassociated fragments of different individuals from different sites. Although many skeletal parts have been known from small to large *Proconsul* species, major problems have arisen due to lack of knowledge of limb proportions and to difficulties in assigning postcrania to species based on teeth and jaws. Because of this our viewpoint has been heavily biased by interpretations based on the associated partial skeleton of *P. africanus* (Napier and Davis, 1959). Recently two things happened to change this situation. The first was the recognition that more parts of the associated skeleton, KNM-RU 2036, were still in blocks of matrix in the National Museums of Kenya and that more are likely still to be on the site, locality R114, Rusinga Island. This has led to the recognition of many more parts of this individual and their preparation from their matrix. Importantly, much of the hindlimb skeleton is now known. The second was the finding of the major part of an associated foot and leg skeleton that can be reasonably attributed to *P. nyanzae* at site R1-3, Rusinga Island. These new finds allow us to do several things:

    1. The limb proportions of *P. africanus* can be almost perfectly calculated for one individual.

A. C. WALKER • Department of Cell Biology and Anatomy, The Johns Hopkins University, School of Medicine, Baltimore, Maryland 21205. M. PICKFORD • National Museum of Kenya, P.O. Box 40658, Nairobi, Kenya.

2. The estimate of body weight for this individual is now on a much sounder foundation than before.

3. The hypotheses concerning locomotor adaptations that were based on the forelimb can now be checked against the hindlimb.

4. The body weight of *P. nyanzae* can be estimated, based on the size of the leg and foot skeleton and other reasonably attributed parts.

5. Comparisons can be made between the two species.

Regional Stratigraphy and Geochronology

Both localities R114 and R1-3 (see Fig. 1) are in the Hiwegi Formation (Van Couvering, 1972). This unit unconformably overlies the Rusinga Agglomerate dated at 19.6 million years (Van Couvering and Miller, 1969) and is unconformably overlain by the Kulu Formation, the Kiangata Agglomerate, and the Lunene Lava. In the vicinity of R1-3, the Hiwegi Formation is underlain by a localized unit, the Ombonya Formation. Van Couvering and Miller (1969) suggest an age of about 17.5–18.0 million years for the Hiwegi Formation on the basis of its position between the dated Rusinga Agglomerates (19.6 million years) and Lunene Lava (16.5 million years), taking into account various unconformable surfaces in the sequence. Pickford (1981) supported this

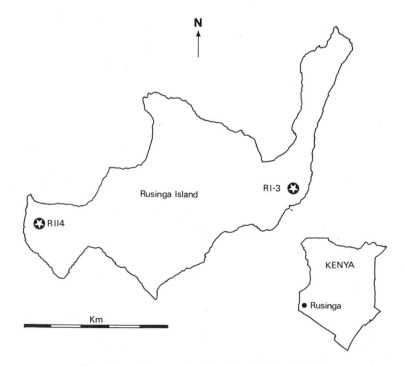

Fig. 1. Locality map showing sites R114 and R1-3.

inference on the basis of the more advanced aspect of the fauna compared with the assemblages at Songhor, Koru, and Napak, which are dated at about 19.0–19.5 million years. Nevertheless, it is desirable to refine this chronology, since the Rusinga fauna is the standard upon which so many comparisons are based.

Depositional Environments

Locality R114.   The sequences both above and below the fossiliferous calcereous grit at R114 are fluviatile. Sand-grade clasts predominate, but beds with cobbles up to 10 cm in diameter occur below the site, as do several clay intercalations and well-laminated silts. The strata are cross-bedded throughout, are bedded on a 1 mm to 1 cm scale, and show numerous cut-and-fill episodes, clay drapes, and swale and swell structures. Lateral continuity of strata indicates a low-lying flattish depositional environment such as occurs in the lower reaches of braided stream systems or on floodplains.

The clasts, which comprise both volcanic debris and fenitised basement fragments, are subrounded and suggest a modicum of fluviatile transportation. Abundant fossil wood characterizes the strata underlying the fossil level.

The fossil bed itself is not now well exposed, and may have been dug away in 1951. Abundant blocks are scattered about the site. They are comprised of calcareous "grit" of coarse sand to fine conglomerate grade. From what little that can be inferred from the site without further excavation, it seems that the grit overlays a clay drape which thickens laterally where it still contains a tragulid skeleton. In the immediate vicinity of the site, the clay drape is thin and green. Laterally it becomes thicker and brownish, and bones from it are coated in a red-brown layer. Bones from the grit are whitish.

According to Whitworth's field notes, he thought the deposit was "an agglomerate pipe," while Leakey (fieldnotes, December 16, 1951) "Found it to be a filling of agglomerate in a pothole in the rock." Both these statements suggest that the deposit was more or less circular in plan and had steep sides, yet a photograph in the Kenya National Museum Archives suggests a deposit only about 0.3 m thick with a gently sloping lower contact. Until the outcrop is cleaned up (it is now full of silt and blocks of grit) it will not be possible to confirm or disprove the "pothole" hypothesis. Recent impressions gained from examining the surface features of the area suggest that the deposit was about 10 × 5 m in extent and about 0.3–0.5 m thick, with a gently sloping basal profile.

Nevertheless, whether the fossiliferous grit accumulated in a pothole or in a shallow depression, there is little doubt that the site formed in a fluviatile setting of floodplain character. Subsequent to its deposition it was calcified and vegetation became established, as shown by abundant rootlet casts on naturally weathered surfaces of the grit.

Locality R1-3.   The stratum at R1-3 from which the *Proconsul nyanzae* leg and foot skeleton came was likewise deposited on a floodplain. The bed which yielded the specimen is a red, silt-grade, limey stratum a few centimeters thick, but of relatively widespread occurrence. It appears to have been an overbank accumulation in a thick sequence of floodplain strata comprising sheet conglomerates up to 1 m thick, intercalated with silts and sands. The whole sequence is characterized by cross-bedding and cut-and-fill features. The lateral continuity of many of the beds indicates deposition in an area of low relief.

Taphonomy

Locality R114.   Many of the fossils at R114 occurred as articulated or semiarticulated skeletons, each quite separate from its neighbors in the deposit. This suggests that the skeletons of mammals and reptiles alike were still enclosed in their skins or were held together by ligaments when buried and that subsequent mixing or interpenetration of skeletal elements did not occur, such as took place in the Meswa Bridge locality (Pickford and Andrews, 1981), where an immense tangle of bones resulted.

The *Proconsul* skeleton suffered some warping, especially to the anterior mandibular dentition and the long bones of the forelimb. Alignment of warped areas in the forelimbs suggests that the forearm was fully flexed and lying parallel to the humerus so that both were warped by the same mechanism. This in turn suggests that the forearm skeletons had dried out and that shrinkage of the muscles and tendons had resulted in complete flexing of the forelimb prior to burial.

Another interesting aspect of the R114 assemblage is that most of the larger mammals were juveniles. Three pigs, one *Proconsul*, two carnivores, and two ruminants were definitely subadult, while the remainder are too incomplete for an age to be assigned. The micromammals all seem to have been adults. There is carnivore damage on the proximal femur of the *Proconsul* specimen. Therefore, carnivore accumulation cannot as yet be ruled out.

Locality R1-3.   The *Proconsul nyanzae* foot is missing large parts of the calcaneum, talus, and distal tibia as well as some phalanges. The talus also has a distinct puncture on the tibial facets. The damage was evidently sustained prior to burial and seems to be consistent with carnivore activity. *Postmortem* damage is unusual for Rusinga fossils, in that the bone (especially hollow bones, such as the tibia) had broken into a number of fragments each of which was unwarped but displaced from its neighbors. The lack of warping has allowed excellent reconstruction to be made by extraction of each fragment from its matrix followed by its proper realignment against its neighbors. The constituent parts of the foot were not in articulation when found, but seem to have drifted apart during burial in the sediment. The drifting was localized, so that all the elements were within a few centimeters of each other.

The holotype of *Proconsul nyanzae* is the only other specimen from Rusinga with comparable postburial damage, comprising clean breaks and displacement without warping (Whybrow and Andrews, 1978). It is just possible that this facial skeleton and the foot came from the same layer and that they belong to the same individual.

Paleoenvironments

The gastropod assemblage at R114 is rather poor, but the presence of *Maizania* suggests that the immediate area was unlikely to be forested. *Maizania* has never been found at the Koru sites, which contain abundant snails of rain-forest aspect. It is, however, common at most sites on Rusinga in association with other gastropods that in modern assemblages occur in dry-forest to open-woodland habitats (e.g., Kibwezi Forest and Tsavo in Kenya).

The flamingo and *Brachyodus* indicate that surface water was likely to be found nearby, a suggestion supported by the fluviatile depositional environment.

At R1-3, a much more comprehensive gastropod assemblage was obtained, but it was derived from a number of beds spanning an unknown time period. Nevertheless, the assemblage has the appearance of being a natural assemblage of a limited number of environmental types and could be duplicated in woodland areas in lowland Kenya today, such as Kibwezi, through which flowed rivers flanked by denser forest. Rainfall data from Verdcourt (1963) suggests a mean annual rainfall of between 660 mm (26 in.) and 1140 mm (45 in.), while altitudinal limits of between 900 m (3000 ft) and 1210 m (4000 ft) are suggested by the cooccurrence of *Ligatella, Maizania,* and *Burtoa.* At present *Burtoa* and *Ligatella* only occur together above 3000 ft and below 4000 ft, these being the lower and upper altitudinal limits, respectively.

The common occurrence of Cryptodesmid millipedes is suggestive of dry woodland areas, the group being common today in *Commiphora* woodland areas such as Tsavo and near Isiolo. The lizard *Gerrhosaurus* is evidently an arid to semiarid country taxon. The relatively abundant hypsodont taxa *Myohyrax* and *Kenyalagomys* may indicate that grass was an important local food resource. Additionally, the relatively great proportion of large mammals would be unusual in a thick forest (Bourliere, 1963). The picture that emerges is that the R1-3 area sampled a fauna which lived in or near a floodplain and that the local paleoenvironment was moderately dry and most likely contained open woodland to perhaps dry "Kibwezi-like" forest (actually a continuous single-canopy woodland). It is unlikely that thick rain forest occurred nearby. In addition, the marked rarity of aquatic fossils at R1-3 suggests that the area lay in the dry floodplain facies of the depositional basin.

The R1-3 assemblage differs markedly from the Koru rain-forest assemblages on the one hand and the wet-floodplain/swamp assemblages of Karungu and Chianda on the other.

Table 1. Faunal List from Site R114, Rusinga Island

Mollusca	*Maizania*	4
	Limicolaria	1
	Small species	1
Chordata		
Reptilia	Varanidae	1–2 Skeletons
	Ophidia	1 Large skeleton
		1 Tiny skeleton
Aves	Phoenicopteridae	1 Individual
Mammalia	*Proconsul africanus*	1 Skeleton
	Paraphiomys pigotti	4 Skeletons
	Paraphiomys stromeri	11 Skeletons
	Proheliophobius leakeyi	1 Skeleton
	Kenyalagomys rusingae	5 Skeletons
	Hyaenodon pilgrimi	1 Skeleton
	?Teratodon sp.	1 Skeleton
	Kichechia sp.	1 Skeleton
	Rhinocerotidae	1 Bone
	Suid *nov. gen. et sp.*	3 Skeletons
	Libycochoerus jeanelli	1 Specimen
	Brachyodus aequatorialis	1 Specimen
	Walangania africanus	1 Skeleton
	Dorcatherium pigotti	2 Skeletons

Associated Faunas

The fauna from R114 is from a single lithologic unit, the so-called "pothole" fill or "agglomerate pipe" of Whitworth's field notes (Whitworth, 1953). Many of the individuals were articulated or nearly so and most of the mammals, except the microfauna, were juveniles with mixed dentition or incompletely erupted permanent dentitions. Table 1 is a list of materal known to have come from the site, but is incomplete, since abundant fossil material remains in the field and in blocks in various museums.

The fauna from R1-3 proved impossible to sort out on a bed by bed basis, so the following compilation necessarily yields a composite faunal list from many areas of exposures containing numerous beds within the same general area as the stratum from which the *P. nyanzae* leg skeleton was derived. The latter bed is poorly fossiliferous. The entire area of R1-3 has yielded in excess of 3000 fossils. The faunal list is given in Table 2.

Descriptions of the New Specimens

KNM-RU 2036

Complete descriptions of the previously known parts are to be found in Napier and Davis' classic monograph (1959). These are not repeated here, except when there is new information on broken or distorted parts (see Fig. 2).

Table 2. Faunal List from Site R1–3, Rusinga Island

Taxa		Number of Specimens
Arthropoda		
Insecta	Woodlice	
	Millipedes	
	Centipedes	
	Cocoons and brood cells of various sorts	
Mollusca	*Ligatella*	152
	Trochonanina	80
	Cerastus	57
	Maizania	266
	Limicolaria	282
	Burtoa	59
	Bloyetia	10
	Homorus	228
	Thapsia	13
	Urocyclidae	28
	Edentulina	4
	Gonaxis	8
	Gulella leakeyi	3
	Tayloria	13
	Other helicoid snails	173
Chordata		
Reptilia	Amphisbaenidae	⎫
	Lacertidae (including Gerrhosaurs)	⎬ 15
	Ophidia	⎪
	Varanidae	⎭
	Chelonia	3
	Crocodilia	2
Aves	Phoenicopteridae	1
Mammalia	*Propotto leakeyi*	1 possibly from R1–3
	Rhynchocyon rusingae	⎫
	Rynchocyon clarki	⎪
	Gymnurechinus leakeyi	⎪
	Gymnurechinus camptolophus	⎪
	Gymnurechinus songhorensis	⎪
	Galerix africanus	⎬ Common
	Amphechinus rusingensis	⎪
	Myohyrax oswaldi	⎪
	Kenyalagomys rusingae	⎪
	Kenyalagomys minor	⎭
	Progalago	Rare
	Dendropithecus macinnesi	⎫
	Proconsul africanus	⎬ 61
	Proconsul nyanzae	⎪
	Limnopithecus legetet	⎭
	Trilophodont	11
	Prodeinotherium hobleyi	8
	Kichechia zamanae	⎫
	Anasinopa leakeyi	⎪
	Hyaenodon pilgrimi	⎬ Moderately common
	Hyaenodon andrewsi	⎪ for carnivores
	Metapterodon kaiseri	⎭

(continued)

Table 2—*(Continued)*

Taxa	Number of Specimens
Myorycteropus africanus	8
Megalohyrax championi	101
Meroehyrax bateae	Rare
Diamantomys leuderitzi	
Paraphiomys pigotti	Several hundred
Paraphiomys stromeri	
Anomaluridae	2
Vulcanisciurus	Rare
Megapedetes pentadactylus	32
Chalicotherium rusingense	Rare
Aceratherium acutirostratum	
Dicerorhinus leakeyi	38
Brachypotherium henzelini	
Brachyodus aequatorialis	10
Hyoboops africanus	3
Suid *nov. gen. et sp.*	
Suid *nov. gen. kijivium*	68
Libyeochoerus jeanelli	
Dorcatherium parvum	
Dorcatherium pigotti	
Dorcatherium chappuisi	Several hundred
Propalaeoryx nyanzae	
Canthumeryx sirtensis	2
Walangania africanus	Rare

Right Humerus and Scapula (Ex KNM-RU 3630A) (Fig. 3).   Part of the right scapula has the spine, acromion, and most of the supraspinous and some of the infraspinous regions preserved. The medial part is distorted, but the lateral part is seemingly undistorted. The spine is well developed, reflected cranially, giving a deep supraspinous fossa. It is strongly buttressed inferiorly, so that its root merges into the infraspinous fossa gradually. Medially there is a "trapezius tubercle" that is always found close to the vertebral border. Thus nearly all the spine is preserved. The deltoid area of the acromion is small and triangular, being restricted to the lateral part of the spine. The notch between the glenoid and the base of the spine is deep. The triceps insertion is close to the glenoid, but the teres area is marked by a strong double keel. From glenoid to vertebral border the estimated length is 57 mm.

This fragmentary specimen is quite remarkable in that it looks, in the parts preserved, like a diminutive version of a chimpanzee scapula.

Associated with the scapula is about 90 mm of the proximal right humerus from the epiphyseal plate distally. The two bones are in such tight contact (see Fig. 3) that they cannot be separated without damage. The piece of the humerus is badly distorted distally, but not proximally. It is bent posteromedially around the scapula in its distal part. The epiphyseal plate has a strong posterolateral–anteromedial crest (the anterior part is abraded). The question of whether the humerus of *P. africanus* was retroflexed or not, an issue of some importance in deciding its locomotor role, cannot ever be settled with the two humeri from this individual, since both are distorted in the region that would show it.

Napier and Davis tried to estimate the degree of torsion of the head of the left humerus fragment, but they did not have much bone to base this on. It is now clear that there was considerable medial torsion of the head on the shaft, so that the articular surface faced posteromedially.

The impression for the insertion of the latissimus dorsi tendon is very clear and so an estimate can be made for the distance between the proximal

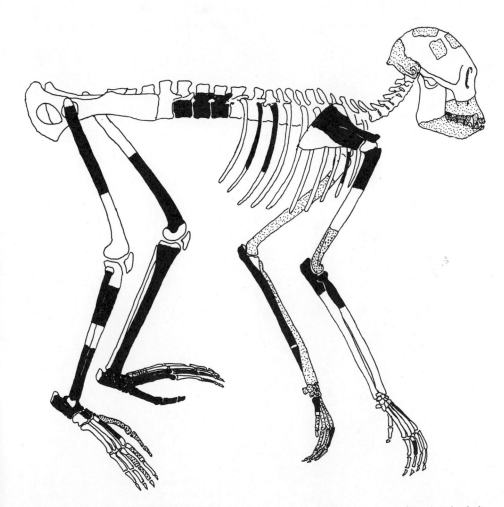

Fig. 2. Reconstruction of KNM-RU 2036, showing parts described by Napier and Davis (stippled) and parts described here (solid).

epiphysis and the impression. Reconstructing the proximal epiphysis using several different primate models gives a mean estimate of 35 mm. Using the left humerus length from the latissimus dorsi impression to the distal epiphysis maximum distal extent, an estimate of 167 mm is obtained for the length of the humerus. Napier and Davis had to obtain an estimate for the missing proximal portion of the left humerus and it is not surprising that their estimate of 182 mm is rather large.

*Right Radius (*Ex KNM-RU 3630C). About 50 mm of the proximal right radius, including the head, is preserved. Napier and Davis, whose left radius was missing the head, allowed 2 mm for the epiphyseal thickness. A value of 4 mm can now be seen to be correct. The total length of the radius can now be reliably given as 158 mm. Napier and Davis commented upon the disposition of the broken left bicipital tuberosity, estimating it to conform more closely to the cercopithecine condition. This region is complete on the newly discovered

Fig. 3. Right partial scapula and right proximal humerus of KNM-RU 2036, showing warping of the humerus. Dimensions given in the text.

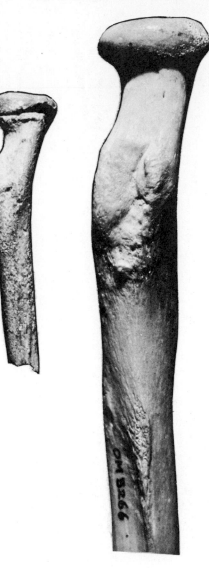

Fig. 4. Right proximal radius of KNM-
RU 2036 (left) compared with the same
part of an adult *Pan troglodytes* (right).
Dimensions given in the text.

right radius and the relationship of the tuberosity to the interosseous crest
and the nutrient foramen cannot be distinguished, except for the size, from
that seen in modern *Pan* radii (see Fig. 4).

   *Proximal Left Ulna (Ex KNM-RU 3629A) and Proximal Right Ulna (Ex
KNM-X 38).* The new piece of left ulna is an undistorted 120-mm-long
portion from the olecranon epiphyseal surface distally. The coronoid process
and the anterior olecranon are sheared off together with a portion of the

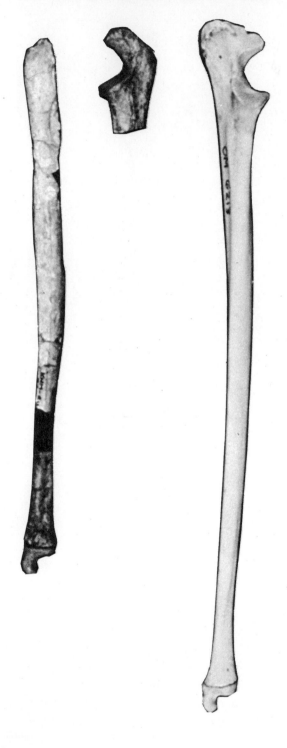

Fig. 5. Left ulna shaft and proximal right ulna of KNM-RU 2036 compared with (right) ulna of adult *Papio cynocephalus*. Dimensions given in the text.

anterior surface of the shaft. Close inspection of the bone in the region of the sigmoid notch shows outer cortical bone, and the distance from the deepest part of the notch to the posterior surface of the shaft can thus be estimated. The distal 40 mm of the shaft is intact. Napier and Davis described the distal part of this same bone. The two pieces do not quite fit, as there is a gap of about 13 mm as judged from the radius length. The total length of the ulna (minus the proximal epiphysis) is estimated at 176 mm. The shaft is gently bowed, convex laterally in basal (posterior) view. The posterior profile is almost straight for the upper two-thirds, but very gently concave posteriorly in the lower third. In midshaft the bone is oval in outline, 10 mm anteroposteriorly and 7 mm mediolaterally. The portion missing from the proximal end was split through the depth of the notch, but the proximal end of the right ulna was found among specimens with no locality data. This had been suspected of belonging to this individual, but without the left proximal ulna, or even a radial head, this was impossible to verify. Thus, nearly the whole of the ulnar morphology is known from one side or the other. The proximal epiphysis is not known. The olecranon process was relatively longer than in the living Great Apes, as judged from the 4-mm proximal projection without the epiphysis. The sigmoid notch is saddle-shaped and almost symmetric about both longitudinal and mediolateral axes. The anterior olecranon projection is 7 mm and the coronoid projection about 8 mm anterior to the deepest part of the notch. From the deepest part of the notch to the posterior border of the shaft is 11 mm. The radial facet is not deeply excavated into the coronoid as in Old World monkeys and apes. There is a strong cresentic ridge running distally from the medial side of the coronoid that is presumably the medial limit of attachment of the brachialis tendon. Immediately distal to the coronoid there is a strong keel that becomes continuous with the anterior border of the shaft. This is 13 mm deep to the posterior border at 10 mm distal to the coronoid process. The keel is set more laterally than sagittally. The ulnar distal epiphysis is known for the right side, but adds nothing to what is already known. The parts are shown in Fig. 5.

Hand Bones.   The left hand skeleton has had new elements added: the first metacarpal shaft and the proximal phalanx of the third digit, both known previously from their proximal epiphyses. The proximal phalanx of the third digit is nearly identical to that described by Napier and Davis for the fourth. It is only fractionally larger than their estimated length in their reconstruction. It is 31 mm long. The head of the first metacarpal is missing, although old glue on the distal break suggests that it was once present. What remains makes the bone complete from the proximal epiphysis for a length of 17.5 mm. It is quite stout, oval in section, is waisted about 10 mm along the shaft, and is seen expanding in width toward the head. Most surprisingly, the opponens flange is easily discernible, a very unusual feature in such a small primate. Napier and Davis gave a very tentative length of 25 mm for the first metacarpal. This is almost exactly the value that we determined, based on extrapolation of the distal expansion, and can now be taken as a more certain estimate.

A reconstruction of the hand skeleton and forearm of *P. africanus* appears in Fig. 6.

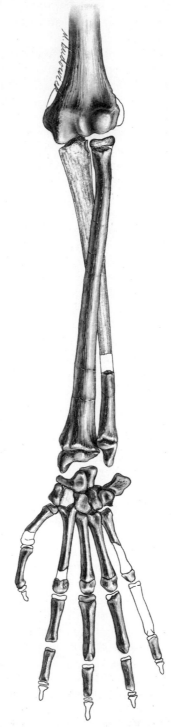

Fig. 6. Reconstruction of left forearm and hand skeleton of KNM-RU 2036.

*Right (*Ex KNM-RU 3611B & 3629B) *and Left (*Ex KNM-RU 3628C) *Femora.* The proximal half of the right femur is missing the epiphyses of the head and greater trochanter. The neck shows tooth marks made by carnivores. The anterior surface of the neck is flattened. The neck is short and not set at the high angle that is more typical of living Great Apes. The shaft is circular in cross section throughout most of its length. The midshaft diameters are 12 mm mediolaterally by 11 mm anteroposteriorly. The linea aspera is wide, with one superior limb running from anterior to the lesser trochanter and the other from the gluteal tuberosity. This latter swings diagonally across the midline to make a relatively deep groove on the medial side of the linea. The intertrochanteric crest is not sharply defined, and runs at an angle of about 45° from lesser to greater trochanters. The distal half of the left femur is missing its distal epiphysis. The oblique part with its deep groove is seen proximally, so an estimate of total femoral length can be made by using both femoral parts. It is estimated at 195 mm. The distal shaft is slightly bowed, convex anteriorly, whereas the proximal shaft is almost straight. The distal metaphysis is widely flaring medially and laterally with a shallow suprapatellar notch and slightly more buildup of bone lateral to the notch (see Fig. 7).

*Left Tibia (*Ex KNM-RU 2036BA & 3687 & part in block with foot) *and Right Tibia (*Ex KNM-RU 3611D, 3611E, 3611S and 3630V). Most of the left tibia is known. It is missing its proximal epiphysis and a small part of the distal shaft. The shaft is very straight and compressed mesiolaterally in the upper half, becoming more rounded in section distally. Midshaft diameters are 11 mm anteroposteriorly and 9 mm mediolaterally. The upper part has a slight posterior retroflexion. There is a well-marked pit on the anteromedial surface of the shaft that marks the insertion of sartorius, gracilis, and semitendinosus muscles. This lies about 35 mm distal to the proximal epiphyseal line. A similar elongated pit is found in *Pan.* Distally the tibia is quite robust, with a large malleolus and strong tibialis posterior groove. Laterally the distal shaft has two strong bony ridges, between which is the area of attachment of the interosseous ligament. The articular surface is 15 mm broad anteriorly and 10 mm posteriorly, with a marked division between the larger medial part and the lateral part. The malleolar part is set vertically against the medial part. An estimated length is 179 mm.

*Left Fibular Midshaft Fragment and Left Distal Epiphysis (*from block with foot)*, Right Proximal Shaft of Fibula (Minus Epiphysis) (*Ex KNM-RU 3630L)*, and Right Distal Shaft.* A surprisingly stout bone with big epiphyses, sharply edged posteriorly and rounded anteriorly, and compressed mesiodistally. Midshaft diameters are 4.5 mm mediolaterally and 7.0 mm anteroposteriorly. Distally it is very flaring to accommodate the large epiphysis. This has a very marked pit for the posterior tibiofibular ligament, a strong peroneal groove, and a wide, flat articular surface for the talus. The estimated length is 170 mm.

Foot Skeleton. When Napier and Davis described the forelimb bones, they also described several bones of the foot. These included the medial cuneiform and the metatarsal and phalanges of the right foot and a few other fragments. The foot skeleton of the right foot has been improved by the

Fig. 7. Proximal right and distal left parts of femora of KNM-RU 2036. Dimensions given in the text.

addition of a nearly complete calcaneum, a complete talus, and most of the navicular. Much of the skeleton of the left foot, together with the tibial and fibular epiphyses, was contained in a block of matrix (Field No. 1507 1951). It includes most of the calcaneum, the talus, the navicular, and all three cuneiforms. The metatarsals include virtually complete ones for the first, second, and third and parts of the fourth and fifth. Several parts of phalanges are known, including the proximal phalanx of the hallux (see Fig. 8).

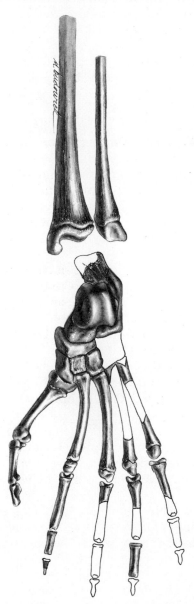

Fig. 8. Reconstruction of left lower leg and foot skeleton of KNM-RU 2036.

*Talus (*right Ex KNM-RU 4347*).* Bones similar to these have been known for many years from the East African Miocene sites. The trochlear surface is asymmetric, with a larger and higher lateral margin than the medial. It is 12 mm wide anteriorly and 7 mm posteriorly. The intervening sulcus is relatively deep. The lateral facet is steep, with only a little flare in its deepest part, and a strongly formed ligament pit. The medial facet is developed only anteriorly, is tear-shaped in outline, and has a deep pit for the tibial malleolus excavated into it anteriorly. This excavation extends onto the neck. There is a deep ligament pit immediately posterior to this excavation. The posterior calcaneal facet is a symmetric cylindrical one, 14 mm long by 9 mm wide, separated by a wide sinus from the anterior facet which is L-shaped. There is no strong change in profile between the navicular facet of the head and the anterior facet. The neck is moderately long and the head almost hemispherical.

*Calcaneum (*right Ex KNM-RU 3636*).* The tuber is damaged on the most complete specimen, but was placed obliquely, its long axis superolateral to inferomedial. It has a salient crest running from it to the posterior margin of the posterior talar facet. The base is only moderately convex transversely and is built out below the sustentaculum medially and as a strong peroneal tubercle laterally. The posterior facet is a simple cylindrical one about 12 mm by 9 mm. The anterior facets are joined, the sustentacular part being oval, with its long axis anteroposterior and the anterior part more nearly circular. The cuboid facet has a very deep excavation medially and a surrounding flatter area, especially laterally and inferiorly.

*Navicular (*both sides now known*).* A stout bone with well-defined facets for the cuneiforms. They are set in an arc, with that for the lateral cuneiform meeting the cuboid facet at 90°. The middle anteroposterior length is 8 mm.

*Medial Cuneiform (*both sides now known*).* This was described by Napier and Davis for the right side. It is a short, deep bone with an anteroposterior length of 13.0 mm and a height of 15.0 mm.

*Middle Cuneiform (*left only known*).* This bone is practically square in superior outline, with an anteroposterior length of 9 mm and a breadth of 7.5 mm. It is twisted along its length so that the proximal facet is twisted clockwise to the distal one in proximal view.

*Lateral Cuneiform (*left only known*).* Complete except for the tubercle. The dorsal surface is rectangular in outline (11.5 anteroposteriorly by 7.0 mm mediolaterally). The configuration of the three cuneiforms means that the second metatarsal base was set in by a depth of 4.0 mm between the two outer bones.

*First Metatarsal (*both sides known*).* This element was described by Napier and Davis. A single, elongated, large sesamoid was found floating in the matrix of the foot block. It is likely that this is associated with the head of the left first metatarsal, but the possibility that it is a peroneal sesamoid cannot be ruled out.

*Hallucial Phalanges (*both sides known*).* These were described by Napier and Davis. The proximal left is distorted.

*Second Metatarsal (*left complete and shaft only of right*).* A very stout bone with a rugged base and a single articulation for the third metatarsal. The shaft is circular in section and noticeably curved, concave medially. The head is compressed mediolaterally and twisted relative to the base by an angle of about 45°. The length is 49.0 mm.

*Third Metatarsal (*left only*).* Complete but damaged at the base. The base was clearly smaller than that of the second metatarsal. Its shaft, however, is thicker and the head less compressed. Shaft torsion is also less than seen in the second, but the head is still clearly twisted to face the hallux.

*Fourth Metatarsal (*left only*).* Proximal and distal fragments are known. Only half of the head is preserved.

*Fifth Metatarsal (*left only*).* Proximal and distal fragments are known. The base is damaged, but shows signs of having had a moderately developed styloid process.

Pedal Phalanges. Described by Napier and Davis. They are more slender and more tapered distally than those of the hand. Assignment to digit is not possible except in the obvious case of the first.

Three Lumbar Vertebrae. These are quite similar in general configuration to those reported by Walker and Rose (1968) for *Proconsul major* from Moroto, Uganda. One is represented by the body and the pedicle bases only. The other two are more complete, each having the laminae and each part of a transverse process. The best preserved has both posterior articular facets and the base of the spine. All are missing their epiphyses, but one has a fragment still attached by matrix. The bodies are 17–19 mm long, with a small ventral keel and gentle ventrolateral hollowings. The bodies are kidney-shaped in anterior and posterior views. Strong pits on each side of the anterior edge of the spine are for the attachment of the ligamenta flava. The small, oval posterior articular facets are angulated at about 45° to the sagittal plane. The styloid apophyses are represented by only the slightest traces.

KNM-RU 5872. Left Distal Tibia, Fibula, and Partial Foot Skeleton Attributed to Proconsul nyanzae (Figs. 9 and 10)

Tibia. This bone was cracked into many pieces, which were pressed against each other and recemented by matrix. The pieces were taken apart and realigned. There was no plastic deformation and the bone has now been given its original shape. Only the proximal portion, weathered and squashed, has been left in the original field state. The shaft is only moderately compressed mediolaterally (estimated midshaft diameters, 18 mm anteroposteriorly, 14.0 mm mediolaterally). There is a patch of roughened bone on the medial side of the shaft at midlength that most probably represents a healed wound. The distal part of the soleal line is seen at the proximal part of the bone. There is a single nutrient foramen about 138 mm above the talar surface. The anterolateral muscular (interosseous) line runs from the anterior part of the distal fibular surface and keeps well anterior for the distal 70 mm

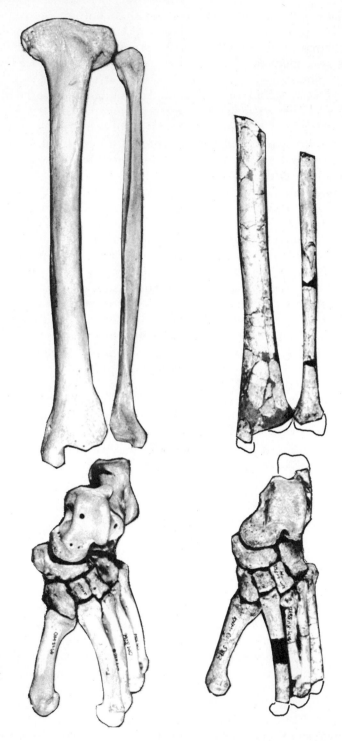

Fig. 9. Tibia, fibula, tarsal, and metatarsal bones of *Pan troglodytes* (left) and KNM-RU 5872 (right).

before swinging posteriorly, then anteriorly at the proximal break. The distal surface and malleolus are badly eroded and only a small part of the articular surface remains. There is a tibialis posterior groove about 5 mm wide. The distal part of the shaft is triangular in cross section, being flat anteriorly and medially and more curved, convexly, posteromedially.

Fibula. Part of the fibula was recovered next to the tibia *in situ.* About 140 mm of its length remains, but the pieces are not in complete alignment because some small chips of bone are missing. The distal part has had the epiphysis chewed away and the posterior cortical bone splintered off. The shaft is very stout. At the midshaft it is 12.5 mm anteroposteriorly and 7.5 mm mediolaterally. In section it has a teardrop outline, with a sharp anterior crest and a rounded posterior margin and the medial surface flatter than the lateral. The distal metaphysis is very expanded and triangular in section.

Calcaneum. The tubercle was destroyed by a carnivore before death. The posterior talar facet is mostly present. It is a simple cylindrical surface about 11 mm wide. The anterior facet is double, with a small isthmus joining the two parts. The posterior part is teardrop-shaped in outline, its apex anterior. It is 10.0 mm wide by 14.0 mm long. The outline of the anterior facet is subrectangular and is 10.0 mm wide by 8.0 mm long. The sustentaculum is large and some 7.0 mm thick, with a flexor hallucis longus groove about 8 mm wide. The cuboid facet is deeply indented for the cuboid process inferomedially. The indentation has a hemispherical surface and is surrounded by flatter regions which are even convex laterally. The total facet is 20.0 mm wide by 18 mm high.

Talus. Only the lateral part of the trochlea and part of the head remain. A 10.0 mm by 6.0 mm oval, deep ligament pit lies posterior to the largely vertical fibular facet. There is a clear carnivore tooth mark that breaks the surface of the facet. Another talus (KNM-RU 1743) is known from Rusinga that is identical in preserved parts and which fits the talar facet of the calcaneum and navicular reasonably well. It is slightly larger than the broken associated bone. Both medial and lateral ligament pits are especially large and pronounced. This talus was described by Le Gros Clark and Leakey (1951).

Navicular. This is a stout bone with well-defined facets for the cuneiforms, that for the lateral being joined to that for the cuboid at a little less than 90°. The thickness at the middle part is 11.0 mm.

Medial Cuneiform. This is a tall bone with a large hallucial facet. It is 20.0 mm long and 24.5 mm high. The facet for the first metatarsal is in a spiral form, with the greatest lateral extension inferiorly and the greatest medial extension superiorly.

Middle Cuneiform. The proximal articular facet is badly damaged, but if the bone is articulated with its neighbors it can be seen that little bone has been lost. The dorsal surface has a length of 13.0 mm and a width of 10.0 mm. The facet for metatarsal II is concave, with a diagonal long axis running superomedially to inferolaterally.

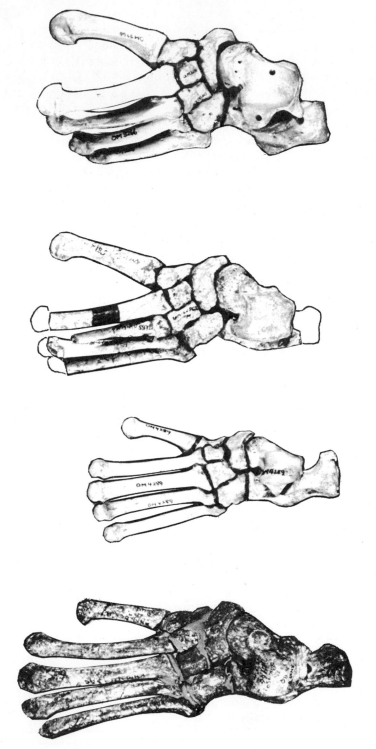

Fig. 10. Tarsal and metatarsals of, from left to right, *Paracolobus chemeroni*, *Papio cynocephalus*, KNM-RU 5872, and *Pan troglodytes*. This comparison was made to avoid problems of comparison between species of greatly different size. *Paracolobus chemeroni*, although extinct, has postcranial adaptations that show that it was as arboreal as most living colobines (M. G. Birchette, Jr., personal communication).

Lateral Cuneiform.   This rectanguloid bone has a dorsal surface 17.0 mm long by 13.0 mm wide and is 16.5 mm deep. There is only a single superior facet for the middle cuneiform. There is a small distal and a larger proximal cuboid facet. The tubercle is very stout.

Cuboid.   This bone, the first known cuboid for an East African Miocene hominoid, is missing the part that would articulate with the distolateral part of the calcaneum and the inferior parts of the facets for metatarsals IV and V. The dorsal surface is 22.0 long and 17.0 mm wide. There is a very strong calcaneal process that projects proximally from the surrounding flatter articular surface. The process arises quite far laterally. The proximal and distal margins converge laterally, giving the bone a wedge-shaped outline in dorsal view. The peroneus longus groove runs parallel to the plane of the metatarsal facets. The facet for the fifth metatarsal does not extend much onto the lateral border.

First Metatarsal.   This element is complete and is 57.5 mm long. It is a large, stout bone with a cylindrical proximal facet, 11.0 wide superiorly and 17.0 mm deep. The distal articulation shows two strong sesamoid grooves and the keel between them faces laterally and inferiorly when the bone is articulated with the medial cuneiform.

Second Metatarsal.   The head and a 10-mm piece of shaft are missing. This bone has a very stout base (dorsal surface 11.0 mm wide) and has an estimated length of 70 mm. The shaft is rounded dorsally, but keeled on the plantar surface. It is curved along its length so as to be concave medially.

Third Metatarsal.   This bone is somewhat crushed and its facets have to be taken from those on the neighboring elements.

Fourth Metatarsal.   This is missing only the head. There is a large superior facet and a tiny inferior facet on a small pedestal of bone for articulation with metatarsal III. There is a single, large concave facet for metatarsal V. The width at the base is 11.0 mm. Near the head the shaft is becoming markedly mediolaterally compressed. Its estimated length is 73 mm.

Fifth Metatarsal.   This, too, is missing only the head. It bears a small, convex facet for the cuboid and a stout styloid process. The shaft becomes compressed mediolaterally near the head. The estimated length is 73.0 mm. When these last two metatarsals are articulated with the cuboid, the dorsal surface of the cuboid overhangs the concave metatarsal facets such that the cuboid/metatarsal articulation is dorsiflexed.

Proximal Phalanx.   This is the distal part of a phalanx from near the base. It is a long, gently curved bone with rather parallel sides and a flattened plantar surface with strong flexor sheath attachments. The head is symmetrically condyloid. The estimated length is 45 mm. The width is 8.4 mm and the thickness 5.5 mm at midshaft.

Middle Phalanx.   This bone is nearly complete and is of the right size to articulate with the proximal phalanx. It is only gently curved and has its tendon sheath markings placed proximally. It is 29.0 mm long and is 7.3 mm wide and 4.5 mm thick at the midshaft.

Discussion

Although it will take some time to complete an analysis of all the new skeletal elements, some preliminary observations can be made. The limb proportions of the single individual KNM-RU 2036 can be given almost with certainty. In Napier and Davis' discussion of the forelimb proportions, they had to rely on an estimate of the length of the humerus. We now have a much better estimate of this length and it is perhaps not surprising that unlike the first estimate, where the brachial index was less than that for all primates except *Gorilla* and *Homo,* the new estimate of 96.4 lies closer to many other extant primate brachial indices. It is particularly close to the values recorded by Napier and Napier (1967) for *Pan,* and had this new material been available to the original describers, it might have been used to demonstrate a greater resemblance to chimpanzees. However, although the brachial index is well within the range recorded for *Pan,* the crural and intermembral indices are outside those recorded for *Pan.* The crural index of 91.8 lies within the ranges for quadrupedal primates of diverse locomotor types, as does the intermembral index of 86.9. In fact the overall proportions are not like those of any extant primate, resembling most, perhaps, those of certain macaques. Since the interrelationships between the indices are not given in most standard compilations of limb proportions, but only in the means and ranges for each index, and since the species of certain genera (e.g., *Macaca*) are dealt with as one, it is not possible to say more at present than that, with the exception of the forelimb proportions, the limb proportions of this specimen were not like those of living hominoids.

Most estimates of body weight for fossil specimens are made by calculating a relationship between the size of some anatomical part and body weight in living primates and then using that relationship to estimate the weight of the extinct species. In this case the overall size and robustness of the fossil skeleton was compared with that of the skeletons of extant species until the best match could be made. Among the skeletons available, male *Colobus guereza* skeletons gave a close match. These have a body weight of about 11 kg (M. G. Birchette, personal communication). Using regression coefficients generated from analyses of lower second molar length and body weights of extant primates, Gingerich (1977) estimated the body weight of *Proconsul africanus* as an average of 27.4 kg, with the 95% confidence interval ranging from 16.1 to 34.0 kg. We would regard this as a very high value, given what we now know of the postcranial skeleton of KNM-RU 2036, an individual whose third molars were erupting and several epiphyses of which were fused or nearly fused. Although Gingerich (personal communication) has revised his body weight estimate for *Proconsul africanus* downward, the reader is referred to Smith (1981) for a discussion of some problems inherent in the study of allometry. The body weight of the individual represented by the lower leg and foot (KNM-RU 5872) cannot be estimated with such certainty. In size it conforms well with that of several female *Pan troglodytes* specimens that we examined and was much larger than the largest male *Papio cynocephalus* we could find. A body weight of as much as 40 kg is not, then, out of the question for this individual. Certainly the weight of this individual was much greater than three times that of the *P. africanus* individual and it may safely be assumed that they represent different species.

Many previously unattributed postcranial fragments of *Proconsul* can now be placed in their respective species based on the proportions of the KNM-RU 2036 parts. This will make attempts at reconstructing postcranial adaptations in the Miocene forms much easier, since many of these bones can now be used in a composite way to assess the size and shape variability that seemed bewildering before. Of particular interest is the preliminary observation that *P. nyanzae* appears to be, in the parts known, a larger *isometrically* scaled version of *P. africanus*. Despite the apparent large size difference, no major shape or proportional differences have yet been found between parts that can reasonably be attributed to these two species.

The forelimb skeleton of *Proconsul africanus* described by Napier and Davis has been at the center of debate over the origins of hominoid locomotor adaptations. To a large extent, the questions being asked and the methods devised to answer them have resulted in the creation of a morphological (and sometimes behavioral) dichotomy—monkey-like or ape-like? This issue is dealt with by Rose (this volume, Chapter 15). In the absence of a living primate whose postcranial skeleton is identical to that of *Proconsul*—in which case we would have no alternative, following the assumption of matching form with function, but to attribute the same behavior seen in the living species to that of the extinct species—we can still make some general statements about the locomotor capabilities of species of *Proconsul*. In general the robustness of the skeleton is something of a surprise. For such a small primate, the limb bones of the *P. africanus* individual seem quite strongly built, with the articular surfaces relatively large. The scapula and proximal humerus suggest that ranges of motion at the shoulder joint were large. Similarly, ranges of elbow joint motion seem to have been large and the range of supination and pronation of the forearm fairly large in all elbow joint positions. The radius is a stout, strut-like bone, while the ulna is relatively straight, leaving little space between them for a wide interosseous membrane. There are no overt signs in the hand of any special postures, such as knuckle-walking or digitigrade ones, and the safest assumption is that the hand was used mostly in a palmigrade way. In the hindlimb, in the absence of information about the hip and knee joints, the striking feature is the strength of the fibula and the hallux—both indicating strong grasping and climbing capabilities, since flexor hallucis longus arises from the fibula. The overall impression is one of a powerful, relatively slow-moving arboreal quadruped that was not much constrained by joint and ligament morphology to a limited range of postures.

It is instructive to note that had the foot and lower leg skeletons of these two species of *Proconsul* been available to us before the forelimb bones, our perceptions of both the postcranial adaptations and the phylogenetic implications would have been very different. It has been the consensus that the postcranial skeleton of *Proconsul* was, in many features, reminiscent of those of certain New and Old World monkeys. It now seems that in other respects there are many features that are also found in living apes, particularly *Pan*. Without dealing with the complex issue of whether these features are primitive or derived, it appears that in the forelimb there is a gradient of features along the limb such that features that are more hominoid-like are found more proximally. In the hindlimb the gradient is reversed, with more hominoid-like features more distally. It is quite likely that much of the remainder of the KNM-RU 2036 skeleton will be recovered from the site. Of great interest will be the pelvis, unknown for the early Miocene hominoids. The number of vertebrae in the sacrum will be a clear indication of whether major hominoid characteristics had developed in the pelvic skeleton of *Proconsul*. We look forward to revising this chapter when new parts are found.

ACKNOWLEDGMENTS

We thank Richard Leakey and the Trustees of the National Museums of Kenya for permission to study the new material. We thank the following individuals for help and discussion: P. Andrews, M. G. Birchette, J. G. Fleagle, L. Jacobs, M. G. Leakey, R. E. F. Leakey, D. R. Pilbeam, and M. D. Rose.

References

Bourliere, F. 1963. Observations on the ecology of some large African mammals, in: *African Ecology and Human Evolution* (F. C. Howell and F. Bourliere, eds.), pp. 43–64, Wenner-Gren Foundation, New York.

Gingerich, P. D. 1977. Correlation of tooth size and body size in living hominoid primates, with a note on relative brain size in *Aegyptopithecus* and *Proconsul. Am. J. Phys. Anthropol.* **47**:395–398.

Le Gros Clark, W. E., and Leakey, L. S. B. 1951. The Miocene Hominoidea of East Africa. *Fossil Mammals of Africa* (Br. Mus. Nat. Hist.) **1**:1–117.

Napier, J. R., and Davis, P. R. 1959. The forelimb skeleton and associated remains of *Proconsul africanus. Fossil Mammals of Africa* (Br. Mus. Nat. Hist.) **16**:1–70.

Napier, J. R., and Napier, P. H. 1967. *A Handbook of Living Primates*, Academic, London. 456 pp.

Pickford, M. 1981. Preliminary Miocene mammalian biostratigraphy for Western Kenya. *J. Hum. Evol.* **10**:73–98.

Pickford, M., and Andrews, P. 1981. The Tinderet Miocene sequence in Kenya. *J. Hum. Evol.* **10**:11–34.

Smith, R. J. 1981. On the definition of variables in studies of primate dental allometry. *Am. J. Phys. Anthropol.* **55**:323–329.

Van Couvering, J. A. 1972. *Geology of Rusinga Island and Correlation of the Kenya Mid-Tertiary fauna*, Ph.D. Dissertation, University of Cambridge. 208 pp.

Van Couvering, J. A., and Miller, J. A. 1969. Miocene stratigraphy and age determinations, Rusinga Island, Kenya. *Nature (Lond.)* **221**:628–632.

Verdcourt, B. 1963. The Miocene non-marine Mollusca of Rusinga Island, Lake Victoria and other localities in Kenya. *Palaeontographica A* **121:**1–37.

Walker, A., and Rose, M. D. 1968. Fossil hominoid vertebrae from the Miocene of Uganda. *Nature (Lond.)* **217:**980.

Whitworth, T. 1953. A contribution to the geology of Rusinga Island. *Q. J. Geol. Soc. Lond.* **109:**75–96.

Whybrow, P. J., and Andrews, P. 1978. Restoration of the holotype of *Proconsul nyanzae. Folia Primatol.* **30:**115–125.

The Wrist of *Proconsul africanus* and the Origin of Hominoid Postcranial Adaptations

13

H. M. McHENRY AND R. S. CORRUCCINI

Introduction

One of the most important discoveries in hominoid evolutionary studies is the fossil forelimb of *Proconsul africanus* (KNM-RU 2036), discovered in 1951 from the early Miocene deposits of Rusinga Island, Kenya. Napier and Davis (1959, p. 1) describe the importance of the fossil:

> The discovery of the fore-limb bones of *P. africanus* is therefore an event of considerable moment, for they constitute the oldest and most complete skeleton of the hominoid fore-limb so far known. It is clear that these bones belong to one of the most significant periods of primate evolution; a period when the generalized catarrhine stock was emerging from a prolonged phase of arboreal quadrupedalism with its limited opportunities for adaptive radiation and was entering upon a phase that would provide a diversity of environmental opportunity leading ultimately to the emergence of four distinct patterns of locomotion among the Anthropoidea: (1) arboreal quadrupedalism, (2) terrestrial quadrupedalism, (3) brachiation, (4) terrestrial bipedalism.

H. M. McHENRY • Department of Anthropology, University of California, Davis, California 95616. R. S. CORRUCCINI • Department of Anthropology, Southern Illinois University, Carbondale, Illinois 62901.

Although they conclude that "there is no denying the quadrupedal habit of *P. africanus*" (p. 61), they stress the generalized pattern of the limb with many examples of preadaptations for brachiation (e.g., elongation of deltoid and absence of a strong deltotriceps crest).

Since this conclusion there have been numerous reinterpretations of *Proconsul*'s locomotor mode. While most have reaffirmed the original describers' conclusion that the fossil was a monkey-like quadruped (Corruccini *et al.*, 1975, 1976; Morbeck 1972, 1975, 1977; O'Connor, 1975, 1976; Preuschoft, 1973; Schön and Ziemer, 1973), some have proposed knuckle-walking (Conroy and Fleagle, 1972; Zwell and Conroy, 1973) or brachiation (Lewis, 1971, 1972).

Recent stationary radiographic and cineradiographic studies of the hominoid wrist reveal a great deal of new information about structural adaptations for knuckle-walking and forelimb suspension. Jenkins and Fleagle (1975) find numerous features in the carpal architecture of chimpanzee and gorilla which appear to be adaptations to knuckle-walking, in contrast to the suspensory adapted wrist of the Asian apes. They note particularly the extension limiting and stabilizing features of the knuckle-walkers (e.g., concavoconvex proximal articular surface of the scaphoid, volar inclination of distal articular surface of the radius, concavoconvex facet for the os centrale on the capitate), in contrast to the adaptive shift toward a ball-and-socket proximal carpal joint and configurations in the midcarpal joint for a wider range of movements in the Asian apes. They also note that *Macaca* has many of the extension limiting and stabilizing features characteristic of knuckle-walkers. Examination of casts of *Proconsul* reveals for them a pattern more like their quadrupedal group (chimpanzee, gorilla, and macaque) than their forelimb suspensory group (orangutan and gibbon). Corruccini (1978*a*) shows that the anatomical features discussed by Jenkins and Fleagle (1975) and others can be quantified and used to discriminate the wrist adaptations of knuckle-walkers and Asian apes. He also shows that even with traits designed only to distinguish among hominoids, the wrist morphology of *Macaca* is quite distinct from the quadrupedal apes (chimpanzee and gorilla). The purpose of our chapter is to determine if the wrist morphology of *Proconsul* has any special functional similarity to the modern knuckle-walking apes [as suggested by Conroy and Fleagle (1972) and Zwell and Conroy (1973)] or is closer to the pattern seen in *Macaca*.

The issue of *Proconsul*'s adaptations has wide implications. If the fossil forelimb has many of the unique hominoid specializations, then an early Miocene divergence of the evolutionary lineages leading to extant species does not have to imply extensive parallel evolution of the hominoid forelimb. If the fossil does not possess these specializations, then either there occurred extensive parallel evolution after an early Miocene divergence or hominoids shared a later common ancestor which possessed the unique hominoid forelimb adaptations. The latter alternative is certainly more likely if one is sympathetically inclined toward the molecular clock (Sarich and Cronin, 1977).

Materials and Methods

A full description of the 30 measurements used in this study including discussion of their functional significance is given in Corruccini (1978a). They include four of the lunate [(1) height of triquetral facet, (2) mediolateral width, (3) medial and distal facet angulation, (4) mediolateral diameter of the radial articular surface], five of the scaphoid [(5) anteroposterior diameter of capitate facet, (6) length of tubercle, (7) angulation of tubercle, (8) mediolateral diameter of radial facet, (9) diameter of concave portion of proximal facet], five of the capitate [(10) relative length, (11) mediolateral diameter of head, (12) neck constriction, (13) length of ridge between lunate and scaphoid facets, (14) length of concave portion of os centrale facet], three of the hamate [(15) length of triquetral facet, (16) proximal anteroposterior diameter, (17) triquetral facet constriction], eight of the radius [(18) volar slant of distal articular surface, (19) depth of scaphoid facet, (20) anteroposterior diameter of scaphoid facet, (21) depth of mediolateral curvature of the distal articular surface, (22) mediolateral breadth of distal articular surface, (23) proximodistal height of ulnar facet, (24) angulation between ulnar and carpal articular planes, (25) difference between the mediolateral diameter of the scaphoid and lunate facets], and five of the ulna [(26) cartilage depth, (27) styloid process projection, (28) semilunar facet mediolateral width, (29) minimum anteroposterior width of the distal ulnar notch, and (3) relative distal radial facet curvature]. All of the measurements described above are illustrated in Fig. 1.

One-hundred and six extant catarrhine wrist joints are measured, including 16–20 each for *Pan, Gorilla, Pongo, Homo, Hylobates,* and *Macaca,* in addition to the fossil *Proconsul africanus* (KNM-RU 2036). The original specimen of this fossil is examined with the help of a binocular dissecting scope. The problem of relying on casts instead of examining original specimens is acute, especially when the investigator is examining the detailed morphology of joint surfaces to reconstruct movement potentials. Morbeck (1977) explains how some very serious mistakes in interpretations have been made about the *Proconsul* wrist because of reliance on casts. In our own study, we found it nearly impossible to discern, for example, the boundaries of the triquetral facet on the lunate or the extent of the ridge between the scaphoid and lunate facets on the head of the capitate even on the relatively accurate casts now produced in Kenya. A binocular scope with oblique light on the original specimens was imperative for accurately taking most of our measurements and observations.

The measurements are converted to dimensionless shape variables by subtracting the standard size (average normalized) variable of each individual from each normalized measurement and adjusting the resulting variables to their residual from the regression line relating size to shape, using slopes calculated within the African apes as described in Corruccini (1978b). We consider *a priori* that within the broader definition of the genus *Pan* (i.e., *P.*

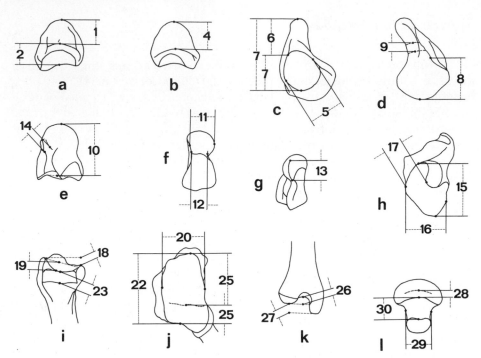

Fig. 1. Measurements as described in the text. a: Lunate, medial side. b: Lunate, lateral side. c: Scaphoid, distal side. d: Scaphoid, proximal side. e: Capitate, lateral side. f: Capitate, dorsal side. g. Capitate, palmar side. h: Hamate, proximal side. i: Radius, medial side. j: Radius, distal side. k: Ulna, medial side. l: Ulna, distal side.

gorilla and *P. troglodytes*) there are no important functional differences (Tuttle, 1967) but only superficial variations due to size. Therefore in this analysis we are testing specific hypotheses of shape similarity between groups, with size-related effects such as those connecting chimpanzees and gorillas removed. Standard multivariable procedures are followed, including the calculation of Sokal's *d*, canonical variates, and Mahalanobis *D* (Blackith and Reyment, 1971; Sneath and Sokal, 1973).

Results

Lunate. The original describers of *Proconsul africanus* state that the lunate ". . . cannot be distinguished from that of *Cercopithecus* or *Papio*," noting that the angle between the scaphoid and radial facets is much wider than it is in apes and more like monkeys, and the overall shape is flat like monkeys, not round like apes (Napier and Davis, 1959, p. 45). This assessment is confirmed by our results: Figure 2 shows the z-scores for the four lunate measurements of the African apes (shaded), *Macaca* and *Proconsul* (also shaded to emphasize

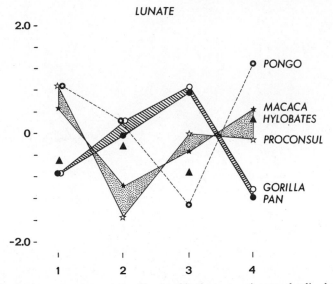

Fig. 2. Means for the lunate measurements discussed in the text as size-standardized z-scores. The ordinate is calibrated in standard deviation units. *Proconsul africanus* and *Macaca* are shaded to emphasize how they are similar.

their similarity), *Pongo*, and *Hylobates*. In each of the four traits *Proconsul* is much closer to *Macaca* than to the African or Asian apes (as well as *H. sapiens*, which is not shown in Fig. 2). The triquetral facet in the fossil is relatively large (measurement 1), which contrasts sharply with its relatively small size in the knuckle-walking apes. The capitate and hamate facets (measurement 2) are relatively small in *Macaca* and *Proconsul*, contrasting with all of the apes. The bending between the medial and distal lunate facets (measurement 3) is greatest in African apes (probably reflecting dorsal compressive forces), intermediate in *Macaca* and *Proconsul*, and least in the Asian apes. The radial articular surface (measurement 4) is greatest in *Pongo*, intermediate in *Macaca* and *Proconsul*, and is smallest in *Gorilla* and *Pan*. Jenkins and Fleagle (1975, p. 223) point out that "in the two African apes, as in the macaque, the proximal articular surface of the scaphoid and lunate are roughly similar in size," whereas "in *Pongo*, the articular surface of the lunate makes a much greater contribution to the proximal carpal joint than does the scaphoid." This, they feel, may be part of "... an adaptive shift toward a ball-and-socket mechanism, specifically for providing additional rotary capability," whereas "the African apes, like the cercopithecoids, retain a less specialized biaxial joint" (Jenkins and Fleagle, 1975, p. 223).

   Scaphoid. Napier and Davis (1959, p. 44) find that the *Proconsul africanus* scaphoid "... differs profoundly from that of *Pan*," with an intermediate tubercular length and an oblong capitate facet extending laterally as it does in monkeys and not in apes. The os centrale is not fused as it is in African apes and in humans. Again our measurements confirm the original describers'

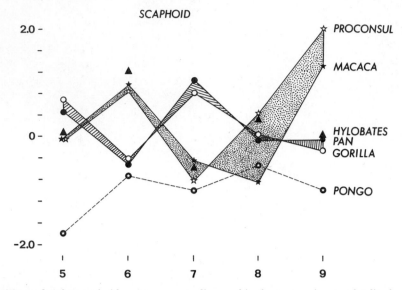

Fig. 3. Means for the scaphoid measurements discussed in the text as size-standardized z-scores. The ordinate is calibrated in standard deviation units. *Proconsul africanus* and *Macaca* are shaded to emphasize how they are similar.

assessment (Fig. 3). The size of the capitate facet (measurement 5) is relatively large in *Pan* and *Gorilla*, intermediate in *Macaca, Proconsul,* and *Hylobates,* and very small in *Pongo,* reflecting the absence of os centrale fusion. The relative length of the scaphoid tubercle (measurement 6) is great in *Hylobates, Macaca,* and *Proconsul* and relatively smaller in the Great Apes. The angulation of the scaphoid tubercle (measurement 7) is greatest in the knuckle-walking apes and smaller in the rest. The mediolateral diameter of the proximal (radial) articular surface of the scaphoid (measurement 8) is relatively large, as it is in the African Great Apes and *Hylobates,* and not reduced as in *Pongo* and *Macaca.* The extent of concave articular surface on the proximal articular facet (measurement 9) is relatively large in *Proconsul* and *Macaca,* which contrasts sharply with its very limited extent in *Pongo.* Jenkins and Fleagle (1975, pp. 216–217) note that "mechanically, the concave part of the proximal scaphoid surface, in contact with the distal radius, appears to limit extension at the proximal carpal joint. Hinge movement ceases as the radius contacts the concave part of the scaphoid."

 Capitate. Our five measurements of the *Proconsul* capitate reveal a pattern more like *Macaca* than any hominoid (Fig. 4). This is especially true of capitate neck construction, which is less pronounced in *Proconsul* and *Macaca* than it is in the Great Apes. Lewis (1969, 1972, 1973) postulates that the capitular neck configuration is related to a locking of the scaphoid around the capitular neck in hominoids. This view has been refuted by Corruccini *et al.* (1975), Jenkins and Fleagle (1975), O'Connor (1975), and Schön and Ziemer (1973). Tuttle (1967) proposes that the restricted capitular neck is related to the limiting of dorsal and lateral intercarpal movement in knuckle-walking.

The ridge on the capitular head separating the lunate and scaphoid facets (measurement 13) is less well developed in *Proconsul* than in the African knuckle-walking apes although there is some damage in the fossil along the crest. Certainly the two articular facets are not smoothly confluent as they are in the Asian apes, where they approach what Jenkins and Fleagle (1975) describe as a ball-and-socket joint in the midcarpal area. The size of the concave articular surface on the posterolateral part of the scaphoid facet of the capitate (measurement 4) is also most like *Macaca* in *Proconsul* and least like the Asian apes, in which the os centrale facet is completely convex. The stationary radiographic and cineradiographic studies of Jenkins and Fleagle (1975) show that the convex portion of the capitular os centrale facet limits extension in the midcarpal region. The orientation of the scaphoid (os centrale) facet (not measured in our analysis) is dorsal in *Proconsul* and *Macaca* and not oblique, as it is in African apes, nor radial, as it is in the Asian apes.

In sum, the capitular morphology of *Proconsul* is very similar to *Macaca* and differs sharply from all hominoids, especially the Asian apes.

Hamate. The choice of measurements in this study was guided by consideration of contrasts between knuckle-walking and non-knuckle-walking apes. In general shape the *Proconsul* hamate is more like that of a monkey, as Napier and Davis (1959) point out, particularly in its small "hook." Corruccini *et al.* (1975) show that the relative height of the hamate "hook" in *Proconsul* is about two standard deviations less it is in hominoids but in the range of variation of most monkeys. Napier and Davis (1959) also note the slight medial orientation of the metacarpal V facet, which they point out is more like

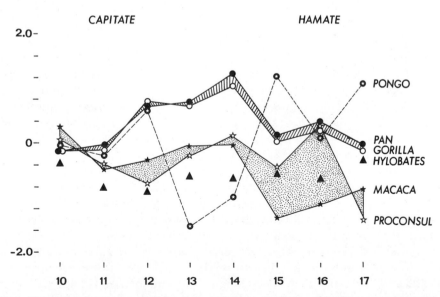

Fig. 4. Means for the capitate and hamate measurements discussed in the text as size-standardized z-scores. The ordinate is calibrated in standard deviation units. *Proconsul africanus* and *Macaca* are shaded to emphasize how similar they are.

Papio and *Cercopithecus* than *Pan.* O'Connor (1976) shows that this facet for metacarpal V also extends more dorsally than is true of hominoids, allowing greater dorsiflexion.

The relative length of the triquetral facet (measurement 15, Fig. 4) is greatest in *Pongo,* less in the African apes, and smallest in *Macaca. Proconsul* is intermediate between *Macaca* and the African apes. The relative proximal breadth (measurement 16) of *Proconsul* is similar to the Great Apes and unlike *Macaca,* in which the hamate is relatively narrower. The constriction of the triquetral facet (measurement 17) is greater in *Macaca* and *Proconsul* than in the apes.

Lewis (1972) notes what appears to be a hominoid characteristic of *Proconsul* in its concavoconvex triquetral facet, although O'Connor (1975, 1976) shows that this is a highly variable trait in cercopithecoids and therefore not diagnostic. Jenkins and Fleagle (1975) point out that *Macaca, Gorilla,* and *Proconsul* have a triquetral facet which is oriented obliquely, unlike the Asian apes. They also note the extensive lunate articulation in the Asian apes, which is only slight in *Proconsul, Macaca,* and the African apes.

Radius.   Napier and Davis (1959, p. 34) find "the distal articular surface of the fossil radius is morphologically indistinguishable from that of all cercopithecoids except for the styloid process, which is relatively bulkier as a result of the greater expansion of the distal extremity in the fossil form." They note particularly that the distal articular surface is flat, which ". . . differs profoundly from the deeply concave surface in *Pan.*"

Our analysis generally confirms the original describers' assessment of the distal radius. The volar slant (measurement 18, Fig. 5), which is so conspicuous in the apes, is much less so in *Proconsul* and *Macaca.* This is related to the absence of a prominent dorsal margin which acts to resist shearing stress dorsally (Jenkins and Fleagle, 1975). The depth of the scaphoid facet as measured from the plane formed by the volar and dorsal margins (measurement 19) is relatively small in *Pongo,* greater in African apes, and greatest in *Proconsul, Macaca,* and *Hylobates,* contrary to Napier and Davis' interpretation. *Proconsul* is uniquely broad in the volar–dorsal diameter of the scaphoid facet (measurement 20). The degree of curvature from radial to ulnar apsects of the distal articular surface (measurement 21) is greater in the Asian apes than the African apes. *Proconsul* and *Macaca* are more like *Pongo* in this respect. Jenkins and Fleagle (1975, p. 223) suggest that the greater side-to-side curvature in the Asian apes ". . . may represent an adaptive shift toward a ball-and-socket mechanism, specifically for providing additional rotary capability," but this interpretation is seemingly negated by the appearance of *Macaca* and *Proconsul.* The mediolateral breadth of the distal articular surface (measurement 22) is greatest in *Pongo* and *Proconsul,* least in *Hylobates.* In this respect *Proconsul* and *Macaca* are rather different. *Macaca* and *Proconsul* are quite similar in the height and angulation of the ulnar facet, however (measurements 23 and 24, respectively). The height of the ulnar facet is relatively small in knuckle-walking apes. The angulation between this facet and the lunate facet is least in *Pongo* and greatest in *Macaca* and *Proconsul.* As noted by

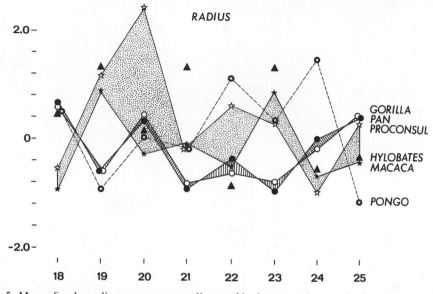

RADIUS

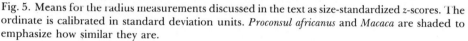

Fig. 5. Means for the radius measurements discussed in the text as size-standardized z-scores. The ordinate is calibrated in standard deviation units. *Proconsul africanus* and *Macaca* are shaded to emphasize how similar they are.

Jenkins and Fleagle (1975), the relative size of the lunate and scaphoid facets differs among the apes, with the two facets being about equal in knuckle-walkers and *Macaca,* and the scaphoid is smaller in *Pongo* (measurement 25). Although the line between the two facets is difficult to see in *Proconsul,* examination under a binocular dissecting microscope reveals that the two facets are about equal in size, thus constituting one trait that the fossil has in common with *Pan.*

   Ulna.   The *Proconsul* distal ulna is clearly unlike that of any modern hominoid. As Napier and Davis (1959, p. 35) point out, "the articular portion of the lower end of the ulna in *P. africanus* corresponds in extent and definition to that in catarrhine monkeys." They note especially the absence of the fossa for attachment of the intra-articular disc. The relative size of the styloid process is over two standard deviations larger in *Proconsul* than in the Great Apes (Corruccini *et al.,* 1975). Morbeck (1975) postulates that the prominent styloid possibly limits radial deviation of the antebrachial complex. The styloid bears a distinct facet for the triquetrum on its dorsolateral surface and another (fainter) facet for the pisiform (Morbeck, 1975, 1977). The head is narrow and transversely elongated, like that of monkeys and not like that of hominoids (Morbeck, 1975, 1977; O'Connor, 1975). This crescent shape of the hominoid ulnar head allows greater pronation-supination (O'Connor, 1975). Our results generally confirm these findings (Fig. 6).

   There are several other important traits of the *Proconsul* distal ulna that were not measured here. Lewis (1972) argues that the presence of a bifasicular palmar radiocarpal ligament in *Proconsul* is a hominoid trait, but Morbeck

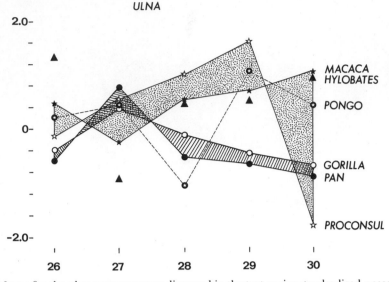

Fig. 6. Means for the ulna measurements discussed in the text as size-standardized z-scores. The ordinate is calibrated in standard deviation units. *Proconsul africanus* and *Macaca* are shaded to emphasize how similar they are.

(1975) points out that this is not diagnostically hominoid since it often occurs in cercopithecoid monkeys. The dorsal tubercle of the radius lies more on the ulnar side in *Proconsul* and monkeys than is true of hominoids (Corruccini *et al.*, 1975; Morbeck, 1975; O'Connor, 1975). This is related to the position of the extensor pollicis longus tendon, which is characteristically above the lunate facet in palmigrade monkeys and *Proconsul* and above the scaphoid facet in hominoids.

Multivariate Analysis. Since the wrist is a complex and integrated structure, it is somewhat artificial to dissect it into individual measurements with-

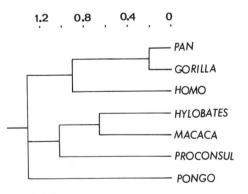

Fig. 7. Dendrogram of unweighted variable-group clustering of Sokal's *d*, with 0.05 S.D. tolerance. The co-phenetic correlation coefficient is +0.95.

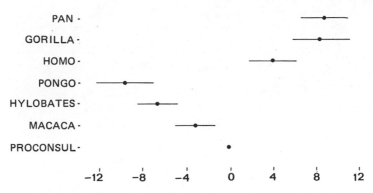

Fig. 8. The 95% confidence limits along the first canonical variate.

out studying them as a complete, intercorrelated system. We therefore subject the 30 traits to multivariate analyses.

Figure 7 gives the results of an unweighted variable-group clustering of Sokal's d with 0.05 S.D. tolerance for cluster admittance. The co-phenetic correlation coefficient is +0.95. The tightest cluster is that of *Pan* and *Gorilla*, which is no surprise, given the special knuckle-walking adaptations of their wrists. Next to join the African ape cluster is *Homo sapiens*, an indirect demonstration of our close genetic (and perhaps functional) affinities. The wrist of *Proconsul* is most similar to *Macaca* and *Hylobates*. *Pongo* is unique and equidistant from these two clusters.

The canonical variate analysis produces four variates with eigenvalues greater than one, with the first accounting for over 64% of the total discrimination. This variate maximally separates the African and Asian apes, with *Proconsul* and *Macaca* falling between. Figure 8 gives the 95% confidence limits along this axis. Jenkins and Fleagle (1975) suggest that generalized terrestriality is the ancestral condition, and that knuckle-walking is very little different from that condition. However, the extreme positions of knuckle-walking and suspensory apes on canonical variate one indicate that none of the extant hominoid taxa well respresents the ancestral morphotype. The intermediate position of *Proconsul* and *Macaca* and functional interpretation indicate that their wrists are more generalized, likely indicating a morphocline whose *middle* represents the pleisiomorph pole, rather than one or the other end.

Canonical variate two (accounting for 16.6% of total discrimination) acts to separate maximally *Pongo* at one extreme and *Proconsul* and *Macaca* at the other (Table 1). The third variate (11.3%) isolates the uniqueness of the human wrist. Again *Macaca* and *Proconsul* are closely approximated and do not approach the *Homo* projection at all. The fourth variate (6.3%) separates the gibbon wrist from all others, especially from *Proconsul* and *Macaca*, which occupy the opposite extreme. The fifth variate has an eigenvalue below one and accounts for only 0.9% of the total discrimination. It is of interest because it is on this variate that the uniqueness of the *Proconsul* wrist is isolated.

Table 1.   Mean Projections on Canonical Variates

	I	II	III	IV	V
Percent of total discrimination	64.7%	16.6%	11.3%	6.3%	0.9%
Pan	8.81	0.05	2.62	0.19	0.07
Gorilla	8.35	0.05	2.60	−0.05	−0.23
Homo	3.93	−2.88	−5.55	−0.45	−0.02
Pongo	−9.49	−4.48	1.87	2.22	−0.06
Hylobates	−6.61	1.87	0.74	−4.09	0.19
Macaca	−3.18	6.83	−1.64	2.54	−0.64
Proconsul	−0.17	7.04	−1.76	5.48	8.00

The Mahalanobis D values in general follow the same pattern as those for Sokal's d. The human wrist is closer to that of a chimpanzee (9.99) than to that of an orangutan (15.64). *Proconsul* is closer to *Macaca* (9.62) than to *Pan* (15.50). The African ape wrists are separated by only 1.40.

Discussion

These results do not support the contention that the *P. africanus* forelimb was adapted for knuckle-walking as suggested by several authors (e.g., Conroy and Fleagle, 1972; Zwell and Conroy, 1973; Jenkins and Fleagle, 1975). Our traits were selected specifically to isolate the peculiarities of the knuckle-walking wrist, particularly from the wrists of the Asian apes. Although the fossil approaches the knuckle-walking configuration in a few traits, in every case the macaque wrist is closer still to the *Proconsul* pattern. The similarities between the wrist of the African apes and *Proconsul* are merely retentions of generalized traits of a common ancestor (plesiomorphic) and not shared derived features (apomorphic). The same is true for other traits as well, such as the prominent deltopectoral crest, the posteriorly directed medial epicondyle, the anterior convexity of the humeral shaft, and the medial slope of the inferior articular surface of the humerus (cited by Conroy and Fleagle, 1972). Multivariate analyses of 16 traits describing the articular surface and general shape of the distal humerus show no special affinity between *Proconsul* and the Great Apes. One of the most diagnostic osteological features of the knuckle-walking forelimbs is the presence of dorsal ridges on the distal metacarpals (Tuttle, 1967), a trait conspicuously absent in the well-preserved *P. africanus* specimen. Preuschoft (1973) shows that the joint surfaces and shaft cross sections of the *Proconsul* metacarpals are adapted to palmigrade quadrupedalism. He notes especially the fact that the distal epiphyses are broad volarly and narrow distally, which is just the opposite of the condition found in modern knuckle-walkers. Zwell and Conroy (1973) find multivariate affinity between *Proconsul* and *Pan;* however, reanalysis of the same data by more suitable statistical procedures shows that the fossil's closest affinities are with *Cercopithecus* and *Presbytis* (Corruccini, 1975). Although Jenkins and Fleagle (1975) note four key features in which the *Proconsul* wrist resembles modern knuckle-walkers and *Macaca*, we show in this study that in these and other features the macaque is easily more similar to the fossil, while there are important differences between *Macaca* and knuckle-walkers.

Napier and Davis (1959) place great emphasis on the "brachiating" traits of *Proconsul*, although they extend the term "brachiation" to include monkeys such as *Presbytis*. In fact almost all of the features that they attribute to brachiation can be matched with equivalent traits in monkeys. The postcranium of *Proconsul africanus* is not in any way intermediate between modern apes and monkeys despite the fact that its dentition is so clearly hominoid (with the possible exception of the distal humerus).

The old error is assuming that hominoids are derived or "advanced" in all traits, dentition, crania, and postcrania. It now appears that the hominoid dental and cranial complexes as seen in the early Miocene are similar to the ancestral morphotype of all catarrhines and that cercopithecoids are derived (not primitive) in these patterns. The opposite appears to be true of the postcrania: the generalized quadrupedalism typical of monkeys is plesiomorphic and the hominoid pattern is apomorphic. This mosaic evolution fits the model proposed by Temerin and Cant (1983) for the cercopithecoid–hominoid divergence: early Miocene catarrhines were frugivores with hominoid-like dentitions and quadrupedal monkey-like postcrania. Later in the Miocene a divergence took place, with monkeys becoming more folivorous in dental adaptation but remaining quadrupedal and the hominoids specializing in frugivory by adapting to movement within and between terminal branch locations using forelimb suspension.

Conclusion

Our results add to the accumulating documentation of the lack of hominoid postcranial specializations in the apes of the early part of the Miocene, a fact not predicted by many before the East African discoveries described by Le Gros Clark and Leakey (1951), Le Gros Clark and Thomas (1951), Napier and Davis (1959), and others. All early Miocene hominoid postcranial fragments so far described in the literature plus several dozen undescribed fossils that have been examined by one of us (H.M.M.) show no uniquely hominoid adaptation, but are precisely what one might expect as the primitive catarrhine morphotype without brachiating, knuckle-walking, or digitigrade quadrupedal specializations. When did the hominoid postcranial adaptations arise? The fossil evidence is still sparse, but new material from Rudabánya (Kretzoi, 1975, 1976) and the Siwaliks (Pilbeam *et al.*, 1977, 1980). shows that by middle Miocene times some aspects of the elbow at least were beginning to show the pattern characteristic of modern Great Apes. The Siwalik specimen GSP

12271, for example, is indistinguishable from *P. gorilla* in an eight-variable discriminant analysis. The distinctively hominoid trait of short ulnar olecranon process is present in the old specimen of "*Austraicopithecus*," which dates from the middle Miocene. However, the middle Miocene *Pliopithecus* postcrania, known from two well-preserved although incomplete specimens, are decidedly un-hominoid (Zapfe, 1958, 1960). *Oreopithecus* stands as one of the few whose elbow complex is clearly hominoid, but whose teeth are so unlike hominoid dentitions that most would exclude the fossil from this superfamily. Perhaps the ancestors of modern hominoids evolved their distinctive postcranial adaptations in reaction to the differentiation of the Cercopithecoidea, an event relating to the exploitation of a new adaptive niche (Temerin and Cant, 1983). If this were so, we would predict that the first appearance of the distinctive cercopithecoid dentition would also be when the first hominoid postcranial adaptations occurred.

References

Blackith, R. E., and Reyment, R. A. 1971. *Multivariate Morphometrics*, Academic, London. 412 pp.

Conroy, G. C., and Fleagle, J. G. 1972. Locomotor behavior in living fossil pongids. *Nature (Lond.)* **237:**103–104.

Corruccini, R. S. 1975. Multivariate analysis in biological anthropology: Some considerations. *J. Hum. Evol.* **4:**1–19.

Corruccini, R. S. 1978*a*. Comparative osteometrics of the hominoid wrist joint, with special reference to knuckle-walking. *J. Hum. Evol.* **7:**307–321.

Corruccini, R. S. 1978*b*. Relative growth and shape analysis. *Homo* **28:**222–226.

Corruccini, R. S., Ciochon, R. L., and McHenry, H. M. 1975. Osteometric shape relationships in the wrist of some anthropoids. *Folia Primatol.* **24:**250–274.

Corruccini, R. S., Ciochon, R. L., and McHenry, H. M. 1976. The postcranium of Miocene hominoids: Were dryopithecines merely "dental apes"? *Primates* **17:**205–223.

Jenkins, F. A., Jr., and Fleagle, J. G. 1975. Knuckle-walking and the functional anatomy of the wrists in living apes, in: *Primate Functional Morphology and Evolution* (R. H. Tuttle, ed.), pp. 213–227, Mouton, The Hague.

Kretzoi, M. 1975. New ramapithecines and *Pliopithecus* from the Lower Pliocene of Rudabánya in north-eastern Hungary. *Nature (Lond.)* **257:**578–581.

Kretzoi, M. 1976. Emberré Válás Es az Australopithecinak. *Anthropol. Kozl.* **20:**3–11.

Le Gros Clark, W. E., and Leakey, L. S. B. The Miocene Hominoidea of East Africa. *Fossil Mammals of Africa* (Br. Mus. Nat. Hist.) **1:**1–117.

Le Gros Clark, W. E., and Thomas, D. P. 1951. Associated jaws and limb bones of *Limnopithecus macinnesi*. *Fossil Mammals of Africa* (Br. Mus. Nat. Hist.) **3:**1–27.

Lewis, O. J. 1969. The hominoid wrist joint. *Am. J. Phys. Anthropol.* **30:**251–68.

Lewis, O. J. 1971. Brachiation and the early evolution of the Hominoidea. *Nature (Lond.)* **230:**577–578.

Lewis, O. J. 1972. Osteological features characterizing the wrists of monkeys and apes, with a reconsideration of this region in *Dryopithecus (Proconsul) africanus*. *Am. J. Phys. Anthropol.* **36:**45–58.

Lewis, O. J. 1973. The hominoid os capitatum with special reference to the bones from Sterkfontein and Olduvai Gorge. *J. Hum. Evol.* **2:**1–12.

Morbeck, M. E. 1972. *A Re-examination of the Forelimb of the Miocene Hominoidea*, Ph.D. Dissertation, University of California, Berkeley.

Morbeck, M. E.1975. *Dryopithecus africanus* forelimb. *J. Hum. Evol.* **4**:39–46.

Morbeck, M. E. 1977. The use of casts and other problems in reconstructing the *Dryopithecus* (*Proconsul*) *africanus* wrist complex. *J. Hum. Evol.* **6**:65–78.

Napier, J. R., and Davis, P. R. 1959. The forelimb skeleton and associated remains of *Proconsul africanus*. *Fossil Mammals of Africa* (Br. Mus. Nat. Hist.) **16**:1–69.

O'Connor, B. L. 1975. The functional morphology of the cercopithecoid wrist and inferior radioulnar joints, and their bearing on some problems in the evolution of the Hominoidea. *Am. J. Phys. Anthropol.* **43**:113–122.

O'Connor, B. L. 1976. *Dryopithecus* (*Proconsul*) *africanus:* Quadruped or non-quadruped? *J. Hum. Evol.* **5**:279–283.

Pilbeam, D. R., Meyer, G. E., Badgley, C., Rose, M. D., Pickford, M. H. L., Behrensmeyer, A. K., and Shah, S. M. I. 1977. New hominoid primates from the Siwaliks of Pakistan and their bearing on hominoid evolution. *Nature* (*Lond.*) **270**:689–695.

Pilbeam, D. R., Rose, M. D., Badgley, C., and Lipschutz, B. 1980. Miocene hominoids from Pakistan. *Postilla* (Peabody Mus. Nat. Hist., Yale Univ.) **181**:1–94.

Preuschoft, H. 1973. Body posture and locomotion in some East African Dryopithecinae, in: *Human Evolution* (M. Day, ed.), pp. 13–46, Symposium of the Society for the Study of Human Biology, Volume 11, Barnes and Noble, New York.

Sarich, V. M., and Cronin, J. E. 1977. Molecular systematics of the primates, in: *Molecular Anthropology* (M. Goodman and R. E. Tashian, eds.), pp. 141–170, Plenum, New York.

Schön, M. A., and Ziemer, L. K. 1973. Wrist mechanism and locomotor behavior of *Dryopithecus* (*Proconsul*) *africanus*. *Folia Primatol.* **20**:1–11.

Sneath, P. H. A., and Sokal, R. 1973. *Numerical Taxonomy. The Principles and Practice of Numerical Classification,* Freeman, San Francisco.

Temerin, L. A., and Cant, J. G. H. 1983. The evolutionary divergence of Old World monkcys and apes. *Am. Nat.* (in press).

Tuttle, R. H. 1967. Knuckle-walking and the evolution of hominoid hands. *Am. J. Phys. Anthropol.* **26**:171–206.

Zapfe, H. 1958. The skeleton of *Pliopithecus* (*Epipliopithecus*) *vindobonensis*, Zapfe and Hürzeler. *Am. J. Phys. Anthropol.* **47**:5–66.

Zapfe, H. 1960. Die Primatenfunde aus der miozänen spaltenfüllung van Neudorf an der March. *Schweiz. Palaeontol. Abh.* **78**:1–293.

Zapfe, H. 1960. Die primatenfunde aus der Miozänen spaltenfüllung von Neudorf an der March (Děvínská Nová Ves) Tschechoslowakei. *Schweiz. Palaeontol. Abh.* **78**:1–293.

Zwell, M., and Conroy, G. C. 1973. Multivariate analysis of the *Dryopithecus africanus* forelimb. *Nature* (*Lond.*) **244**:373–375.

Miocene Hominoid Discoveries from Rudabánya

14

Implications from the Postcranial Skeleton

M. E. MORBECK

Introduction

Recent discoveries and new interpretations of hominoids living 8–14 million years ago (m.y.a.), including species from several geographical regions, have generated interest in reevaluating the nature of the middle to late Miocene hominoid radiation. Current evidence from both the fossil record and biochemical studies of modern taxa suggest that this time period was crucial for hominoid and, perhaps, hominid differentiation (Zihlman *et al.*, 1978; Sarich and Cronin, 1976, 1977; McHenry and Corruccini, 1980). Functional and phylogenetic affinities of primarily dental and some craniofacial specimens from Hungary, Turkey, Greece, China, and Pakistan, in addition to earlier discoveries from India and Kenya, indicate a far more complex pattern of hominoid/hominid evolution than once assumed (Kretzoi, 1975; Andrews and Tobien, 1977; Pilbeam *et al.*, 1977, 1980; Pilbeam, 1978; Vasishat *et al.*, 1978; Simons, 1976, 1977, 1978; Simons and Pilbeam, 1978; Qi, 1979; Xu and Lu, 1979; Jia, 1980; Kay, 1981; Andrews and Cronin, 1982; see also Lewis, 1934, 1937, Simons, 1961, 1964, 1968, 1972; Simons and Pilbeam,

M. E. MORBECK • Department of Anthropology, University of Arizona, Tucson, Arizona 85721.

1965; Prasad, 1964, 1969; Conroy and Pilbeam, 1975; Khatri, 1975; Leakey, 1962, 1967, 1968; Le Gros Clark and Leakey, 1951; Andrews, 1971; Walker and Andrews, 1973; Andrews and Walker, 1976). Pilbeam (1978), for example, formerly a staunch supporter of the special role of the middle Miocene *Ramapithecus* as the first hominid (Pilbeam, 1966, 1968, 1972; Simons and Pilbeam, 1978), no longer promotes this genus, "ramapithecines," or "ramapithecids," as *the* definitive candidates for the earliest hominids (but see Simons, 1977). He now envisions several related groups of middle–late Miocene hominoids shifting into new habitats and exhibiting dietary changes and does not link them phylogenetically to extant taxa. Historically, not all researchers have seen middle–late Miocene hominoids like *Ramapithecus* as the key to human evolution (Hrdlička, 1935; Frisch, 1967; Genet-Varcin, 1969; Eckhardt, 1972; Vogel, 1975; Frayer, 1976; Sarich, 1971). Recent discoveries have provided fuel for these old and some new controversies. Greenfield (1979), for instance, revitalizes the case for dryopithecine affinities of these fossil populations by focusing on the new data as well as on specimens in established collections. Controversies over taxonomic status of fossil groups, no doubt, will continue.

Paleoanthropologists recognize the importance of understanding biological, inferred behavioral, and environmental interrelationships in reconstructing adaptations of past populations. Interpretation of Miocene hominoids now emphasizes functional implications of feeding mechanisms, inferred dietary patterns, and possible ecological niches. Here, postcranial elements can be important in understanding positional and manipulative behaviors associated with critical maintenance activities such as feeding.

Skeletal remains from the Hadar Formation, Ethiopia show that by at least 3½ million years ago, early hominids had morphological capabilities for bipedalism (Johanson *et al.*, 1976; Lovejoy, 1979, 1981; Johanson and White, 1979; McHenry and Temerin, 1979). The few associated postcranial remains from earlier Miocene hominoid taxa, especially *Dryopithecus* or *Proconsul* from East Africa, although not necessarily related to hominids, suggest quadrupedalism with a weight-bearing wrist complex (Morbeck, 1975, 1977; O'Connor, 1976; Schön and Ziemer, 1973; Corruccini *et al.*, 1975; but see Lewis, 1972; see also Napier and Davis, 1959; Walker and Pickford, this volume, Chapter 12; Rose, this volume, Chapter 15), and an angulated humeral shaft probably related to sustaining compressive forces (Preuschoft, 1973). Middle to late Miocene data, when well-dated associated specimens are discovered, may provide the clues to understanding the evolutionary history of the Hominoidea.

Interpretation of biobehavioral capabilities and, perhaps eventually, taxonomic classification must include postcranial data. Unassociated postcranial remains representing middle to late Miocene hominoids have been recovered recently from Rudabánya, Hungary (Kretzoi, 1975, and in preparation) and the Potwar Plateau, Pakistan (Pilbeam *et al.*, 1977, 1980; Rose, personal communication). The isolated humerus from Fort Ternan (cf. *Dryopithecus;* Andrews and Walker, 1976) and other isolated limb bones, primarily from Europe, also may represent the complexity of hominoid evolutionary history within this time period [see Morbeck (1972) for review of hominoid (assumed dryopithecine) limb remains]. This chapter focuses on some of the postcranial remains from Rudabánya, Hungary and comparative contemporary and fossil morphological data from the middle–late Miocene of Africa, Europe, and Asia.

Rudabánya, Hungary

The Rudabánya lignite and clay deposits in northeastern Hungary have yielded an extensive flora and fauna, including many primate fossils. Three hominoid species have been described briefly in English by Kretzoi (1975) on the basis of over three dozen craniofacial and dental specimens. In addition, 34 primate postcranial elements have been recovered from the deposits. Hominoidea identified and described by Kretzoi include: (1) *Pliopithecus hernyáki* Kretzoi 1974; (2) *Bodvapithecus altipalatus* Kretzoi 1974; and (3) *Rudapithecus hungaricus* Kretzoi 1967.

The various localities at Rudabánya have produced different fossil assemblages. For instance, over 90 species of plants and a useful mollusc assemblage have been recovered from Locality 3. Preliminary analysis of botanical data suggests a Mediterreanean-subtropical climate with elements of different local zones represented in the deposits. However, there is little evidence for primates at Locality 3 and most other species recovered here are not found at other localities, in particular, Locality 2. Of the several areas sampled, most of the primate fossils and all of the important postcranial specimens were recovered from Locality 2. This locality now is preserved as a Hungarian Nature Protection Site.

Chronometric dates are not available for these Rudabánya deposits and the localities can only be described in terms of faunal/floral relationships. The "*Hipparion* fauna of archaic character" (Kretzoi, 1975) indicates a maximum age of 12.5 million years or perhaps 11.5 million years (Berggren and Van Couvering, 1974; Szalay and Delson, 1979). Pilbeam *et al.* (1979), however, suggest that the first *Hipparion* reached Europe less than 11 m.y.a. The present faunal list and references to Pannonian strata suggest that these localities may be several million years younger than original estimates [E. Lindsay, personal communication (1980)]. Rudabánya hominoids apparently are several million years younger than the Fort Ternan *Ramapithecus wickeri* and older than *Ramapithecus, Sivapithecus,* and *Gigantopithecus* from the Potwar Plateau, Pakistan.

Phylogenetic relationships among these European hominoids and those from India, Pakistan, Kenya, and other regions have yet to be determined. The *Rudapithecus* and *Bodvapithecus* specimens have been interpreted by Kretzoi (1975, and in preparation) as "ramapithecines" and "pongo-hominids." *Rudapithecus* is seen as a short-faced hominoid illustrating a trend toward hominization with a subparabolic dental arcade, relatively small incisors, canines that to not project far below the tooth row, and brachydont cheek teeth with no cingula (e.g., Rud 1, Rud 12, Rud 15, Rud 17). *Bodvapithecus* is larger than *Rudapithecus* and characterized by larger canines and relatively hypsodont, thick-enameled cheek teeth with marked cingula (e.g., Rud 7, Rud 14, Rud 44a,b). Although other researchers (Pilbeam *et al.*, 1977; Szalay and Delson, 1979) link "*Rudapithecus*" with *Ramapithecus* and "*Bodvapithecus*" with *Sivapithecus* on the basis of the morphology of the jaws and teeth [see also Simons (1976) and, for a different view, Greenfield (1979)], Kretzoi continues to distinguish the Hungarian taxa generically from other Eurasian and African forms [Kretzoi, personal communication (1980)].

Table 1. Primate Postcranial Remains from Rudabánya, Hungary
(*ca.* 11.5 m.y.a.)[a]

Shoulder	Elbow	Wrist	Hand
	Rud 53 Unassociated distal humerus with articular surfaces		Rud 28, 29, 30, 31, 32, 33, 34, 35, 36, 37, 38, 39, 40, 41, 42, 43, 54, 56, 59, 60, 63, 74 V11983 (1977)[b];
	Rud 22 Unassociated proximal ulna with partial articular surfaces		Unassociated metacarpals and phalanges of either hand or foot
	Rud 66 Unassociated proximal radial head		

Hip	Knee[c]	Ankle	Foot
Rud, 23, 55 Unassociated proximal femora with articular surfaces	Rud 24 Unassociated distal femur with articular surfaces but lacking part of the lateral condyle Rud 25 Unassociated proximal tibia with articular surfaces Rud 26 Unassociated patella (?)	Rud 27, 72[d] Unassociated tali with trochlear articular surface; Rud 27 lacks head	Rud 57 Metatarsal I fragment Rud 28, 29, 30, 31, 32, 33, 34, 35, 36, 37, 38, 39, 40, 41, 42, 43, 54, 56, 59, 60, 63, 74 V11983 (1977)[b]; Unassociated metatarsals and phalanges of either hand or foot

[a]Taxa: *Pliopithecus hernyáki:* Lower jaw, fragments and teeth. *Rudapithecus hungaricus:* Upper and lower jaw, fragments and teeth. *Bodvapithecus altipalatus:* Upper and lower jaw, fragments and teeth.
[b]This specimen has a Rudabánya field number only; no Rud number has been assigned as yet.
[c]It is likely that the knee elements are not primate. They may be Carnivora. See text.
[d]It is possible that Rud 72 is not a primate.

Postcranial remains from Rudabánya include elements from the elbow, hand, hip, knee, ankle, and foot complexes (Table 1). As with postcranial material from other middle–late Miocene sites, including the Pakistan localities, none of these specimens is associated directly with dental remains. Joint surfaces and, in one case, a link or shaft specimen provide data on structure and movement capabilities of the larger hominoids, that is, *Rudapithecus* or *Bodvapithecus*. Most postcranial specimens are hand and foot bones not discussed here. Many are fragmentary, but a wide variation in size suggests sampling of different hominoid taxa at Rudabánya.

Upper Limb—Elbow

The bones of the elbow complex recovered from Locality 2 probably represent the large primates, *Rudapithecus* or *Bodvapithecus*, both defined as hominoid on the basis of dental characteristics. Specimens include: (1) Rud 53, a distal left humerus from the lignite; (2) Rud 22, a proximal right ulna from the gray clays that is broken midway through the trochlear notch and distal to the radial notch; and (3) Rud 66, a proximal radial epiphysis also from the clay. The oval shape of the Rud 66 radius presents a problem in assessing structural affinities, since many primates and some nonprimates are similar in both size and shape. The evidence from the humerus and ulna, however, suggests that at least one, and perhaps both, of the large Rudabánya primates is similar in overall structural pattern to contemporary pongids and, to a lesser extent, humans.

Rud 53/Humerus.   Rud 53 is a complete distal humerus, including the medial and lateral epicondyles as well as the trochlea, olecranon fossa, and capitulum. Its bi-epicondylar breadth is 56.0 mm. Rud 53 shows slight anterior–posterior distortion and some abrading of the capitulum.

The trochlea is broad relative to the capitulum. The anterior trochlear breadth is 20.9 mm, while the capitular breadth is narrower, at 19.9 mm. Its trochlear index (ratio of trochlear breadth to capitular breadth × 100) is 105.

The midportion of the trochlea is narrow in both proximodistal and anteroposterior directions. Both its medial and lateral edges are prominent. The lateral edge of the trochlea separates the trochlear articular surface from that of the capitulum and wraps distally to form the lateral edge of the posterior trochlea and olecranon fossa (Figs. 1A, 1B).

The olecranon fossa is triangular. Its lateral wall is steep-sided (Fig. 1C). On the anterior aspect, the Rud 53 coronoid fossa is deep and well defined (Fig. 1A).

Within the humeroradial joint, the capitulum is bulbous and, as discussed above, separated from the trochlea. The radial fossa is shallow. In posterior view (Fig. 1C), the area lateral to the olecranon fossa is broad, accounting for about 41% of the bi-epicondylar breadth. As I have shown previously (Morbeck, 1976), this breadth does not necessarily reflect the role of the capitulum

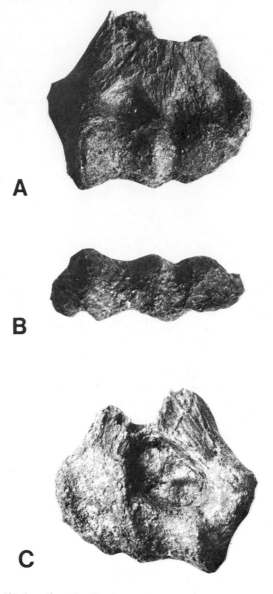

Fig. 1. Rud 53: (A) Anterior, (B) distal, (C) posterior view. Approximately actual size.

in the elbow complex. For example, here the capitulum represents only approximately 35% of the Rud 53 bi-epicondylar breadth.

The medial epicondyle is small and oriented posteromedially. Its size relative to the bi-epicondylar breadth and its proximal position suggest a possible break at the epiphysis. The lateral epicondyle, on the other hand, is within the size range of modern pongids and *Homo*.

Rud 22/Ulna.   Rud 22 is a proximal right ulna with the distal half of its trochlear notch, an almost complete radial notch, coronoid process, and part of the ulnar shaft. The ulna is broken at the approximate midpoint of the humero-ulnar joint surface and lacks the proximal joint surface and ole-cranon process (Figs. 2A, 2B, 2C).

The trochlear notch is segmented, with a proximodistal ridge separating its medial and lateral aspects. Its linear breadth at the proximal break is 12.8 mm. Unfortunately, there is no information concerning the proximal aspect of the trochlear notch or the size or structure of the olecranon process (and the functionally related role of the triceps and related muscles). Nor can the angle of the orientation of the proximal ulna relative to its shaft be recon-structed with any certainty.

The radial notch, functioning within the proximal radio-ulnar joint, is U-shaped, deep proximodistally, and slightly concave but not deeply incised. Radial notch breadth is 10.0 mm, while its height is 9.5 mm (Fig. 2C).

Rud 66/Radius.   Only maximum and minimum breadths can be taken for Rud 66. Maximum diameter is 16.5 mm, while its minimum breadth is 15.0 mm, yielding a radial head index of 110. Morphologically, the radius appears to be a primate (Fig. 2D).

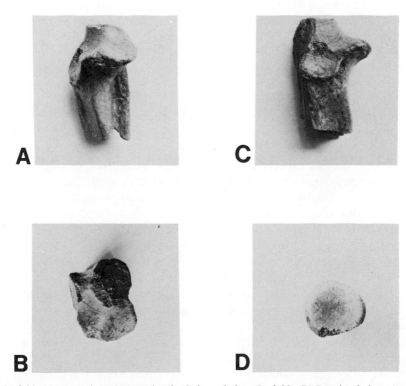

Fig. 2. Rud 22: (A) Anterior, (B) proximal, (C) lateral view. Rud 66: (D) Proximal view. Approx-imately actual size.

Lower Limb—Hip

Rud 23, Rud 55/Femoral Heads.   Rud 23 is a well-preserved head and part of the neck of a left femur. Rud 55 is a partial, poorly preserved femoral head that approximates Rud 23 in size. Head diameters of Rud 23 are superior–inferior 22.2 mm and anterior–posterior 22.4 mm, and, for the more fragmentary Rud 55, superior–inferior 20.8 mm and anterior–posterior 20.5 mm. In both specimens, the articular surface curves to form a sphere. Only in Rud 23 is the extent of the articular surface and position of the fovea for the ligamentum teres evident. Here, the articular surface extends slightly onto the posterior aspect of the femoral neck. On the femoral head, the fovea for ligamentum teres is visible in posterior view.

Lower Limb—Knee

The knee is represented at Rudabánya by three specimens. These include: (1) Rud 24 a distal femur; (2) Rud 25, a proximal tibia and part of the tibial shaft; and (3) Rud 26, a possible patella. Similar size, shape of articular facets, and provenience suggest that Rud 24 and 25 represent one individual. Although Kretzoi (1975) refers these fossils to ?*Rudapithecus*, their status as a primate is questioned. The suite of morphological features, particularly those related to the patellar facet, extent of surface area of the femoral condyles, and proximal tibia, are unlike extant primates.

Rud 24/Femur.   Rud 24 is a distal femur preserving an almost complete patellar facet, complete medial condyle, intercondylar region, and broken lateral condyle. The femur is separated from its shaft close to the proximal extent of the patellar groove.

The most striking feature, at least if compared to primates, is the long, narrow patellar groove. It is narrow, deep, with a U-shaped profile, and elongated proximodistally. The medial edge is apparently more prominent than its lateral counterpart. The lateral border, however, is weathered and/or may have been damaged by chewing. Preliminary metrical analysis of Rud 24 compared to a broad range of primate and nonprimate genera confirms its unusual morphology. The so-called patella (Rud 26) is short and thick anteroposteriorly. Compared to the Rud 24 patellar groove, it is broad. Little else can be determined from this questionable element.

The femoral condyles are asymmetric. Bicondylar breadth is 46.4 mm. The intercondylar area is broad. The posterior aspect of the lateral condyle is broken. Both the medial condylar articular surface and that inferred from the lateral condyle exhibit a short surface area associated with knee flexion or extension.

Rud 25/Tibia.   The facets on the proximal tibia, Rud 25, are asymmetric, with its lateral facet larger and convex and the medial facet narrower mediolaterally and slightly concave. Articular surfaces extend posteriorly on both facets. The tibial tuberosity is broken and the shaft is broken just distal to the tuberosity. A mediolaterally compressed tibial shaft (83 mm), lacking its distal end, may be associated with Rud 25. It has few distinguishable features.

Lower Limb—Ankle

Rud 27/Talus. Rud 27 is the body of a left talus with articular surfaces for the tibia and fibula within the talocrural joint. Its trochlear facet, lateral and medial aspects, and, inferiorly, the posterior facet of the subtalar joint are complete. There is some evidence of weathering or chewing activity on the trochlear facet. The talar head and most of the neck are missing. However, the head appears to have been directed anteromedially.

A second talus, Rud 72, has been recovered recently and assigned tentatively to primates (cercopithecoid?) or to carnivores (viverrid?) by Kretzoi [personal communication (1980)]. This talus is pieced together from fragments and includes a talar head. However, the fragmentary nature of the body precludes detailed comparison with Rud 27.

The trochlear facet on Rud 27 is complete but chipped posteriorly (length, 21 mm). The anterior breadth is wider than its posterior breadth (anterior 14.2 mm, posterior 8.3 mm). The ratio of trochlear facet breadth to its length indicates an extremely narrow facet, at least compared to those of extant primates. The facet is concave throughout its length, although the narrow posterior surface is flatter. There is no defined demarcation of the articular surface edge. The oblique groove for the flexor hallucis longus is shallow.

The lateral border of the trochlear facet is pronounced. It contributes laterally to the steep-sided lateral aspect of the body associated with the distal fibula and medially to an inclined trochlear facet. The height of the lateral border, taken from a plane defined by the most inferior points of the posterior subtalar facet and the small area of articular surface of the anterior facet, is greater than that of the medial edge (maximum height of lateral facet, 14.1 mm; medial facet, 10 mm). The lateral aspect of the body is larger than the medial surface. Ligament attachment areas, for the most part, are well defined. Without the talar head and other ankle and foot bones, especially the tibia and calcaneus, precise orientation of talocrural articular surfaces is impossible. However, the talocrural plane is "tilted" or sloped from a high lateral side to a lower medial side.

The inferior aspect of the body includes the posterior part of the subtalar joint. The facet for the calcaneus is concave, with an even curvature over its length. Its long axis diverges from that of the trochlear facet. Only a small portion of the anterior facet for the calcaneus is present and little else can be determined from the inferior surface.

Structure

The Rudabánya specimens show interesting morphological affinities with extant taxa. The humerus and ulna, although representing different individuals of different sizes, are similar in structure to extant Great Apes and humans. Rud 53, also, is similar to *Pan* and *Homo* in overall size. Rud 66, conversely, differs markedly in its radial head shape from most modern large hominoids. Little morphological detail can be ascertained from the isolated femoral heads. The knee is equivocal. The morphology of the talus, however, indicates a habitually inverted foot.

The overall structural pattern seen in the Rud 53 distal humerus is similar to that of a spectrum of living large hominoids. These features include: (1) a broad trochlea relative to the capitulum, with its midportion constricted and edges prominent, (2) a trochlear lateral edge that separates the trochlea and capitulum anteriorly and inferiorly and wraps distally to form the lateral edge of the posterior trochlea and meet the olecranon fossa, (3) a deep coronoid fossa and shallow radial fossa, (4) bulbous capitulum, and (5) on its posterior aspect, a broad area lateral to the olecranon fossa. The only major difference is in the small size of the medial epicondyle.

The segmented trochlear notch of Rud 22 and deep radial notch also are structurally similar to the large hominoids. The concavoconvex humero-ulnar articulation provides a "good fit," thus providing transverse stability of the elbow. The narrow, deep hominoid radial notch, bulbous capitulum, and pronounced lateral trochlear edge, in combination with an almost circular radial head in living taxa, allows considerable forearm rotation in any degree of flexion or extension.

The radius, on the other hand, is more typical of a structural pattern seen in quadrupedal monkeys, especially cercopithecoids. The oval shape of the radial head of these monkeys usually is tilted with respect to the radial shaft. It is combined with a concave and anteriorly extended radial notch of the ulna which is incised into the proximal shaft, thus providing joint stability in a habitually flexed and pronated forearm position.

Morphological affinities of the femoral heads and the knee are difficult to assess. The proximal femora are similar to many primates, with few distinguishing features. The knee, in contrast, is different from most primates. The long, narrow patellar facet and limited surface area condyles suggest a habitually flexed knee with limited movement capabilities. This morphology cannot be matched in living primates and is more similar to some carnivores.

The steep, slanted, narrow trochlear facet of the talus with its high lateral edge indicates a habitually inverted ankle joint. This morphological feature separates it from bipedal hominids that show a flatter, broader articular surface that is perpendicular to parallel tibial shafts. Based on modern forms, it is suggestive of an arboreal lifestyle.

Size and Proportions

The Rudabánya postcranial material, in general, represents a broad range of sizes, similar to extant large cercopithecoids, pongids, and, for the distal humerus, humans. Here, I focus on the elbow joint of the upper limb, that is, the distal humerus and proximal ulna, Rud 53 and Rud 22, respectively. Middle–late Miocene comparative data for these articular elements are known from Africa (Fort Ternan humerus), Europe (Klein Hadersdorf ulna), and Asia (Pakistan distal humeri). In addition, Oligocene and Miocene hominoid forelimb data are available for comparison, as are data from Pliocene and Pleistocene Hominidae. The other joint complex, for which there is more than one bone, is the Rudabánya knee. Metrical analysis is not yet completed. In addition, given its uncertain taxomonic status as a primate, it is best to postpone detailed discussion of these fossils. Hip and ankle data are fragmentary and are considered here only in a general way.

Linear and joint surface area measurements have been collected for all Rudabánya specimens discussed here. Surface area data have not been analyzed fully with a broad comparative sample and are not considered in this chapter. Raw linear measurements indicate size; indices and scaled data describe proportions for extant primates (Tables 2 and 3). Multivariate analyses using a variety of data transformations (work in progress) show both size and shape affinities with contemporary and fossil hominoids. Most work on size and proportions has focused on Rud 53 and, secondarily, on Rud 22.

Rudabánya 53 is reminiscent of contemporary pongids and *Homo* in its general size and most proportions (Table 4). It falls within the range of *Pan paniscus* for 15 of the 16 linear measurements taken on distal humeri, for which comparative data on 11 anthropoid genera are available (Tables 2 and 3). Over one-half (9/16) of Rud 53 measurements fall within respective 95% confidence intervals for the genus *Pan*, with seven in the 95% confidence intervals for the pygmy chimp, *Pan paniscus*, and for *Homo sapiens*. Thirteen of the Rud 53 direct measurements are included in the *Pongo* range, but only three within respective 95% confidence intervals. Interestingly, the medial epicondyle breadth, a derived value calculated by subtracting the trochlea–lateral epicondyle from the bi-epicondylar breadth, lies outside the ranges for all pongids and *Homo*.

The humerus falls outside of the size ranges computed for all but two values in *Gorilla* and all but one in *Hylobates*. Among the nonhominoids compared in this investigation, Rud 53 is in a size range for six *Papio* values

Table 2. Primate Genera Used in Comparative Samples for Metric Analyses[a]

Hominoidea	Cercopithecoidea
Homo	*Presbytis*
Gorilla	*Nasalis*
Pan	*Cercopithecus*
P. troglodytes	*Papio*
P. paniscus	Ceboidea
Pongo	*Saimiri*
Hylobates	*Ateles*

[a]Metric data were collected from osteological specimens housed in the following institutions: American Museum of Natural History; Arizona State Museum; British Museum (Natural History); Musée Royal de l'Afrique Centrale; Museum of Comparative Zoology (Harvard); Powell-Cotton Museum; Smithsonian Institution.

Table 3.   Linear Measurements Used in Metric Analyses[a]

Humerus	Humerus (Cont.)
Bi-epicondylar breadth	Olecranon fossa breadth
Articular breadth	Olecranon fossa height
Distal articular breadth	Olecranon fossa depth
Capitulum breadth	Lateral posterior breadth
Trochlea breadth	Ulna
Capitulum–medial epicondyle	Trochlear notch breadth
Trochlea–lateral epicondyle	Trochlear notch A–P thickness
Medial trochlea thickness	Radial notch breadth
Capitulum thickness	Radial notch height
Medial trochlea height	Radius
Capitulum height	Maximum diameter head
Posterior trochlea breadth	Minimum diameter head

[a]Measurements [defined in Morbeck (1972)] used in these analyses are those that are available for the Rudabánya elbow specimens.

(related primarily to the size of the medial aspect of the trochlea and the capitulum) and in the range for only one other monkey, *Nasalis* (olecranon depth).

The trochlear index and the lateral posterior breadth index indicate size relationships or proportions within the distal humerus. Exhibiting a trochlear index of 105, that is, the trochlea is broader than the capitulum, Rud 53 lies

Table 4.   Humeral Bi-epicondyle, Trochlea, and Capitulum Breadths

Genus	N	Bi-epicondylar breadth mean, mm	Trochlea breadth mean, mm	Capitulum breadth mean, mm
Homo	20	55	19	20
Gorilla	20	90	33	28
Pan				
P. troglodytes	25	60	22	23
P. paniscus	16	55	19	20
Pongo	28	65	24	22
Hylobates	20	27	9	11
Presbytis	21	23	7(19)[a]	10(20)[a]
Nasalis	18	31	8	16
Cercopithecus	31	21	6	8
Papio	18	38	11	16
Saimiri	15	13	3	5
Ateles	16	29	7(13)[a]	13(13)[a]
Rud 53	1	56.0	20.9	19.9
KNM-FT 2751	1	44.6	17.2	16.0
GSP 12271	1	—	—	27.4[b]
GSP 6663	1	—	—	18.3[b]

[a]Adjusted sample size for which data are available.
[b]Rose's "transverse diameter capitulum" (in Pilbeam *et al.*, 1980) is not equivalent to "capitulum breadth" as used here. His figures for GSP 12271 and GSP 6663 are higher than those reported here.

below the average for *Gorilla* (117) and *Pongo* (109) and above that of *Pan* (96) and *Homo* (97). Trochlear indices of contemporary pongids and *Homo* are greater than those of hylobatids and monkeys studied (Morbeck, 1976).

In posterior view, the surface lateral to the olecranon fossa is as wide as in most hominoids. The value of 41% of bi-epicondylar breadth is greater than the means, but falls within the ranges for extant hominoids, including *Hylobates*. It is wider relative to bi-epicondylar breadth than in any monkey sampled, a proportion that also reflects the structures of the components of the bi-epicondylar breadth, including the lateral posterior surface and the medial epicondyle (Morbeck, 1976).

Scaled measurements, that is, measurements expressed as a ratio of an average measurement for an animal, yield more information on proportions and shape. Average measurements are calculated for Rud 53 and a comparative pongid/*Homo* sample, both using and eliminating epicondylar dimensions. The small size of the Rudabánya medial epicondyle is suspect (i.e., it could be broken at the epiphysis) and data were run both ways. Twelve non-overlapping measurements were used, with ten for the second comparison.

In many cases results were similar to those observed with raw measurements. For example, Rud 53 capitular breadth falls within the ranges for *Homo* and *Pan* whether raw measurements or either set of scaled data are used. On the other hand, it lies within the range of *Pongo* using raw data and data scaled with epicondyles, but not in the range of the results of scaled data with epicondyles. In general, however, Rud 53 falls within fewer ranges using scaled data of either category. Raw measurements can be misleading with respect to understanding proportions or shape.

It is also possible that no overlap occurs in results of the different analyses. Using raw measurements, for example, capitular height and olecranon height each lie within the range of *Gorilla,* but neither falls with *Gorilla* using scaled data. Rud 53 trochlear breadth and height, posterior trochlear breadth, olecranon breadth, and lateral epicondyle breadth are each included within the *Gorilla* ranges for scaled data, but not on the basis of raw measurements.

Discriminant function analyses of logged raw measurements and scaled values of 15 measurements of the distal humerus defined here (olecranon depth is not used in these analyses due to sampling problems), derived values, and, in some cases, indices calculated from these data separate contemporary Hominoidea genera from other groups. The results of discriminant function analyses using different data transformation techniques show interesting patterns of similarities and differences (Hagaman and Morbeck, 1982). Rud 53 falls within a group that includes humans, chimpanzees, and orangutans when interpolated into a discriminant functional analysis using both logged raw measurements and scaled data.

Rud 22, with its segmented trochlear notch and characteristic U-shaped deep radial notch, is smaller than living pongids. Of the four measurements for which comparative data are available (i.e., anteroposterior depth trochlear notch, mediolateral breadth trochlear notch, radial notch breadth, and radial notch height), the Rudabánya ulna is below the range for all Great Apes. The trochlear notch measurements also suggest generally smaller size than in humans since the articular breadth is well below the *Homo* range and the proximal anteroposterior breadth falls at the low end of the human size range. Both the radial notch breadth and height fall within a 95% confidence interval for *Homo*, yet the fossil radial notch index (breadth/height × 100) is at the low end of the human range and outside of its 95% confidence interval (Table 5).

Size relationships between Rud 53 and Rud 22 do not suggest that these specimens belong to a single individual. Multiple regression analysis shows that given all 16 measurements and all comparative taxa or only the Pongidae, the Rud 53 humerus data predict an ulna that is larger than Rud 22. Using all humeral measurements, a predicted ulnar trochlear notch is 19.0 mm wide, while the observed breadth in Rud 22 is 12.8 mm. An even larger ulnar trochlear notch is predicted by using only the humeral trochlear breadth in pongids, 20.9 mm. The Rudabánya ulna from the clays clearly is too small for the humerus from the lignites.

On the radial side, the predicted value for the maximum radial head breadth using the capitulum as a basis for calculations is 21.3 mm, yet the observed value for Rud 66 is only 16.5 mm. The same relationship holds true for predicting minimum radial head breadth, where the expected value derived from all humeral measurements in the pongid subsample is 4.9 mm greater than the observed value in Rud 66. The Rudabánya radial epiphysis is too small for the humerus.

When the ulna is used to predict the size of the radial head, Rud 66 again is smaller than expected. However, the radius and ulna are closer to the expected sizes than are the humerus and either the radius or ulna. These size

Table 5. Ratio of Radial Notch Breadth to Height × 100

Genus	N	Mean	Range	95 Percent confidence interval
Homo	20	141	102–187	129–153
Gorilla	20	126	85–178	115–138
Pan				
P. troglodytes	25	137	89–176	128–145
P. paniscus	16	153	121–199	141–166
Pongo	28	147	83–179	139–154
Hylobates	20	168	129–250	156–180
Presbytis	20	156	128–180	149–163
Nasalis	18	137	111–181	126–147
Cercopithecus	28	180	137–232	172–189
Papio	19	196	143–244	185–208
Saimiri	15	101	72–128	92–111
Ateles	14	130	108–156	123–138
Rud 22	1	105		
Klein Hadersdorf	1	85.5		

Table 6. Femoral Head Diameter (Anterior–Posterior) in mm

Genus	N	Mean	Range	95 Percent confidence interval
Gorilla	20	46	34–54	43–49
Pan				
P. troglodytes	25	33	28–38	33–34
P. paniscus	16	30	23–34	29–32
Pongo	28	34	28–41	33–36
Hylobates	20	15	12–17	15–16
Presbytis	21	14	13–16	14–15
Nasalis	18	20	17–24	19–21
Cercopithecus	29	12	10–15	12–13
Papio	17	23	18–26	22–24
Saimiri	15	6	6–7	6–7
Ateles	16	16	14–19	16–17
Rud 23	1	22.4		
Rud 55	1	20.5		
BMNH M 16331 (Maboko)	1	23.1		
Eppelsheim	1	22.5[a]		
GSP 11867	1	26.2[b]		
GSP 13929	1	24.5[b]		
GSP 15782	1	26.8[b]		
GSP 9894	1	20.0[b]		

[a]Measurement from cast.
[b]From M. D. Rose.

relationships suggest that Rud 53, Rud 22, and Rud 66 represent different individuals. The morphology of Rud 22 and Rud 66 suggests the same conclusion.

Lower limb data, although fragmentary, indicate that in terms of size the femoral heads, Rud 23 and Rud 55, fall below the means for pongids sampled here (Table 6). Rud 23 lies within the size range for *Pan paniscus*, and the large cercopithecoids, *Papio* and *Nasalis*, but within a 95% confidence interval for only *Papio*. Rud 55 is closest in size to *Nasalis* within this sample.

The knee data are difficult to interpret structurally. The bicondylar breadth of Rud 24, the distal femur, is 46.4 mm; thus is below the means for all pongids, but within the ranges of *Pan paniscus*, *Pongo*, and *Papio*, respectively (Table 7). Comparisons of patellar groove shape show that the patellar surface on the femur is longer and narrower than for all anthropoids. Other data, including those on femora and tibiae of many nonprimates, have not yet been analyzed fully.

Rud 27 is smaller than extant pongids, although its trochlear facet length (21 mm) is within range of both *Pan* and *Pongo*. Its midpoint breadth (10.4 mm), however, falls below the means for apes and within the range, but not within 95% confidence intervals, of monkeys sampled here. The ratio of its breadth to length (×100) lies below all taxa investigated here (Table 8).

Table 7.   Distal Femur Breadth (mm)

Genus	N	Mean	Range	95 Percent confidence interval
Gorilla	20	85	65–103	79–91
Pan				
P. troglodytes	25	61	53–69	60–63
P. paniscus	16	57	44–67	54–60
Pongo	28	57.5	46–69	55–60
Hylobates	20	25	21–29	24–26
Presbytis	21	25	22–28	24–26
Nasalis	18	33	29–40	31–35
Cercopithecus	29	22	7–27	20–23
Papio	17	38	30–46	36–41
Saimiri	15	12	10–13	11–12
Ateles	16	30	26–34	28–31
Rud 24	1	46.4		
Eppelsheim	1	38.5[a]		

[a]Measurement from cast.

In summary, the Rudabánya postcranial bones indicate a variety of size and structural affinities when compared to modern primate taxa. The elbow components show similarities in structure and, especially with the distal humerus, size to extant large Hominoidea. The hip elements (the femoral heads) can only be discussed in terms of size, that is, approximately as large as large monkeys among living primates. The knee is problematic. The small talus with its tilted proximal surface indicates habitual ankle and foot positions similar to those of arboreal taxa.

Table 8.   Talar Index (Breadth/Length × 100)

Genus	N	Mean	Range	95 Percent confidence interval
Gorilla	18	69	55–86	64–73
Pan				
P. troglodytes	22	72	59–123	66–78
P. paniscus	16	82	66–151	70–94
Pongo	27	86	60–109	81–91
Hylobates	19	70	60–80	68–73
Presbytis	19	83	65–96	79–87
Nasalis	18	80	72–96	77–83
Cercopithecus	24	79	68–94	76–81
Papio	16	82	69–100	78–86
Saimiri	14	71	60–82	67–74
Ateles	13	76	66–84	72–80
Rud 27	1	49.5		

Middle–Late Miocene Comparative Data

Postcranial data are available for middle–late Miocene probable hominoids from Africa, Europe, and Asia (Tables 9–12). Unfortunately, these specimens suffer from the same fate as those from Rudabánya. None are associated directly with cranial and/or dental remains that would provide a more conclusive statement of taxonomic affinities. The postcranial data considered here include: (1) limb bones from Maboko Island and Fort Ternan in Kenya; (2) limb bones from Austria, France, and Germany in western Europe; and (3) limb bones from the Potwar Plateau in Pakistan. The associated skeletal material from *Oreopithecus,* which is close in age to some of the Pakistan specimens [at least less than 10 million years old (Pilbeam *et al.,* 1977; Delson, 1979], is not considered here. Delson (1979) and Szalay and Delson (1979) consider this European taxon to be a cercopithecoid. *Pliopithecus,* another European form (Zapfe, 1960), also is not discussed in detail here. A complete study, which is not possible here, should survey the past 30 million years, including Oligocene and early Miocene data, these European taxa, and Pliocene and Pleistocene Hominidae postcranial data.

African Middle Miocene Postcranial Evidence

Maboko Island, Kenya

The Maboko Island fauna includes small samples of several hominoids, including isolated teeth of *Ramapithecus wickeri* and jaws and teeth of several dryopithecines as well as *Limnopithecus legetet* (Andrews, 1978; Simons and Pilbeam, 1978). Several individuals representing *Dryopithecus* (or *Proconsul*) *nyanzae* and *D. vancouveringi* have been identified, while *D. africanus* is questionable (Andrews, 1978). Cercopithecoids also are reported from this site (Simons and Delson, 1978; Simons *et al.,* 1978). Paleoecological analysis suggests woodland and floodplain habitats (Evans *et al.,* 1981). The Maboko faunal sample includes some of the same species recovered from Fort Ternan (Van Couvering and Van Couvering, 1976; Van Couvering, 1980; Andrews *et al.,* 1981).

The Maboko deposits may be as old as 14–15 million years, a tentative date based on faunal similarities to Fort Ternan and low faunal similarity with earlier sites such as Rusinga (Van Couvering and Van Couvering, 1976; Andrews *et al.,* 1981; Pickford, 1981). Van Couvering and Van Couvering (1976) also suggest a date earlier than Fort Ternan based on geological evidence, perhaps 15+ million years.

Primate postcranial remains recovered in 1933 by Archdeacon Owen are both unassociated and fragmentary, in most cases lacking joint surfaces (Table 9). Upper limb specimens include only a fragmentary clavicle and a left

Table 9.   Primate Postcranial Remains from Maboko Island, Kenya (*ca.* 14.0–15.0 m.y.a.)[a]

Shoulder	Elbow	Wrist	Hand
BMNH M 16335 Unassociated clavicle fragment	—	—	—
	BMNH M 16334 Unassociated shaft humerus lacking articular surfaces	—	—

Hip	Knee	Ankle	Foot
BMNH M 16331 Unassociated proximal femur with articular surfaces	—	—	—
	BMNH M 16330 BMNH M 16332-3 Unassociated shafts, femora lacking articular surfaces		

[a]Taxa: identified using dental remains: *Ramapithecus wickeri:* Isolated teeth. *Dryopithecus/Proconsul nyanzae:* Three individuals. *Dryopithecus/Proconsul vancouvering:* ?*Dryopithecus/Proconsul africanus. Limnopithecus legetet:* Teeth.

humeral shaft that is missing both shoulder and elbow joint surfaces. Several femoral fragments, including a proximal femur with its articular head, are known for the lower limb (Le Gros Clark and Leakey, 1951). The limb bones probably represent a large primate, the most likely candidates being one of the dryopithecines—for example, according to Le Gros Clark and Leakey (1951), *Proconsul* or *Sivapithecus,* and probably the former taxon. Maboko postcranial specimens usually are assigned to *Dryopithecus* (or *Proconsul*). Even with additional primate dental remains now known from the site, these specimens still are problematic.

BNMH M 16335/Clavicle and M 16334/Humerus.   Shoulder and elbow morphology cannot be determined from these fragmentary remains. Le Gros Clark and Leakey (1951) describe BNMH M 16335, "associated" with the humeral and femoral remains, as the middle portion of a clavicle lacking articular surfaces (fragment length 50.7 mm). Ten years ago Alan Walker suggested that the so-called "large hominoid clavicles" from the early Miocene site of Rusinga Island, also described by Le Gros Clark and Leakey (1951; Le Gros Clark, 1952), probably are crocodile femora (Walker, cited in Morbeck, 1972; Walker, 1980). The Maboko clavicle fragment contributes little to middle Miocene hominoid morphology.

The humeral shaft has been pieced together from several fragments and is laterally displaced at its distal end. Its overall length is 210 mm. BNMH M 16334 is broken just below the humeral head and above the upper edge of the olecranon fossa at its distal end. The upper shaft is triangular in cross section, exhibiting a prominent deltopectoral crest with forward bending of the shaft. There is a suggestion, from the shaft morphology based on the posterior position of the buttress of the head, that the humeral head was directed posteriorly. However, neither shoulder nor elbow morphology can be reconstructed with certainty and thus cannot be compared directly with the Rudabánya forelimb specimens.

BNMH M 16331, BNMH M 16330-3/Femora. The Maboko sample includes a left proximal femur (BNMH M 16331) with an almost complete femoral head, part of the greater trochanter, the lesser trochanter, and part of its proximal shaft. In addition, the shafts of both a right and left femur (BNMH M 16330-3) have been recovered.

Le Gros Clark and Leakey (1951) describe M 16331 as possessing an almost spherical head with the fovea for the ligamentum teres below its center region. In addition, the articular surface extends distally on its posterior surface, slightly more than the extent of posterior extension observed in Rud 23. The Maboko femoral head approximates Rud 23 in size: anterior–posterior breadth 23.1 mm and superior–inferior breadth 22.8 mm (my measurements) which are similar to Le Gros Clark and Leakey's (1951) measurements of A–P 23 mm and S–I 22.6 mm. The greater trochanter in BNMH M 16331 lies distal to the femoral head. A major portion is missing. The lesser trochanter projects medially as a flattened extension of the shaft, with an oval facet facing proximally.

The right shaft, BNMH M 16332-3 (length fragment 266 mm), and left broken shaft (BNMH M 16330) (166.3 mm) provide information on the link between the hip and the knee joints. The right femoral shaft corresponds with BNMH M 16331 where similar surfaces are present. The posterior surface is smooth, with some evidence for muscle attachment, but no linea aspera. The distal portion is distorted and broken proximal to the articular surfaces within the knee.

BNMH M 16330 shows a relatively straight shaft with anterior–posterior distortion and some missing surfaces on its posterior aspect. It is broken below the lesser trochanter and above the condyles and patellar surface. The remaining portion is similar in structure to BNMH M 16332-3. Accurate distal reconstruction is impossible for both diaphyses.

Fort Ternan, Kenya

Several hominoids have been identified at Fort Ternan, a middle Miocene site in Kenya dated radiometrically to 12.5–14.0 m.y.a. Only one probable primate postcranial specimen, an unassociated, incomplete right humerus with distal articular surfaces, has been published from the extensive collections (Table 10). The Fort Ternan hominoids, described from remains of jaws

**Table 10.   Primate Postcranial Remains
from Fort Ternan, Kenya (14.0 m.y.a.)**[a]

Shoulder	Elbow	Wrist	Hand
	KNM-FT 2751 Unassociated distal humeral shaft and articular surfaces		

[a]Taxa: *Ramapithecus wickeri:* Seven facial/dental specimens with upper and lower jaw fragments and teeth. *Dryopithecus* cf. *nyanzae:* six isolated teeth. *Dryopithecus* cf. *africanus:* three isolated teeth. *Limnopithecus legetet:* Fourteen teeth plus jaw fragments with teeth. Humerus may be "*D.* cf. *nyanzae*" but possibly *Ramapithecus* or nonhominoid.

and teeth, include: (1) *Ramapithecus wickeri;* (2) possibly two species of *Dryopithecus, D.* cf. *africanus* and *D.* cf. *nyanzae;* and (3) *Limnopithecus legetet* (Andrews and Walker, 1976; Leakey, 1962; Andrews, 1971, 1978; Walker and Andrews, 1973).

Fort Ternan is significant among the East African Miocene sites since it represents the first fauna in this area to be dominated by artiodactyls (Gentry, 1970; Churcher, 1970). Geological and faunal evidence suggest aquatic, forest, woodland, as well as floodplain and even grassland habitats (Andrews *et al.,* 1979; Walker, 1974; Shipman *et al.,* 1981; Evans *et al.,* 1981). The sediments with fauna have been dated by chronometric techniques to approximately 14.0 m.y.a. (Bishop *et al.,* 1969; Bishop, 1971; Andrews and Walker, 1976; Shipman *et al.,* 1981). New localities with fossils have been discovered recently by Pickford (1980).

KNM-FT 2751/Humerus.   The Fort Ternan primates are reviewed and described in detail by Andrews and Walker (1976). The isolated humerus, KNM-FT 2751, has been grouped tentatively with the dryopithecines, *Dryopithecus* cf. *nyanzae,* by these researchers; however, since it is not associated with teeth, it may be a ramapithecid or may well warrant another taxon. The shaft of KNM-FT 2751, which is broken well below the humeral head, is relatively straight, oval in cross section, and narrowed anteroposteriorly throughout its length. KNM-FT 2751 is smaller than Rud 53 and shows a different overall structural pattern from that of the Hungarian specimen. Its bi-epicondylar breadth is 44.6 mm, while the trochlea is 17.2 mm and the capitulum is 16.0 mm wide.

Like the Rudabánya humerus, its trochlea is broader than its capitulum, with a trochlear index of 107.5, near both the Rud 53 value and the *Pongo* mean and well within the range for all living hominoids and some colobines (Morbeck, 1976). The trochlea is separated from the capitulum, but unlike that of Rud 53, its lateral edge is bulbous and not defined sharply. In addition, its medial portion is rounded.

The most striking differences relate to the posterior aspect of the humerus and the epicondyles. The olecranon fossa is deep and its lateral edge is formed by a pronounced flange of bone creating an extremely deep lateral wall. The area lateral to the fossa is slightly concave and very broad, accounting for almost one-half of the bi-epicondylar breadth.

The medial epicondyle is small relative to its bi-epicondylar breadth and, unlike Rud 53, is markedly bent posteriorly. The lateral epicondyle is large and pronounced anteriorly. The supracondylar crest associated with brachioradialis and extensor carpi radialis longus and brevis is prominent. The shaft itself is flattened anteroposteriorly, especially on its distal posterior aspect.

European Middle Miocene Postcranial Evidence

Klein Hadersdorf, Austria

Unassociated limb bones from Klein Hadersdorf, Austria include a right humeral shaft and almost complete ulna (Table 11). These specimens, originally called *Austriacopithecus weinfurteri* by Ehrenberg over 40 years ago, have been described in detail by Zapfe (1960). Fossils from Klein Hadersdorf are perhaps 14 million years old (Szalay and Delson, 1979; Zapfe, 1960, "Miocene"). Although not associated with dentition, the isolated upper limb bones often are assumed to be *Dryopithecus fontani* (Simons and Pilbeam, 1965; Pilbeam and Simons, 1971). Szalay and Delson (1979), however, recently have suggested that on both morphological and geological/geographical evidence, these postcranial specimens may be *Sivapithecus darwini*. Part of their argument centers on the morphology of the proximal ulna, in particular, a "clear reduction of the olecranon process as would be *expected* [my emphasis] for a *Sivapithecus* elbow" (Szalay and Delson, 1979, p. 490). Assigning these upper limb remains to a particular taxon, however, is premature. It certainly cannot be accomplished on the basis of what one expects to find in the fossil record.

Humerus. This fragmentary right humerus includes the main part of the shaft (length 188.7 mm). It is missing both its proximal and distal extremities and thus, like the Maboko and Saint Gaudens humeri, provides no conclusive evidence for shoulder and elbow morphology. The diaphysis is broken just proximal to the deltopectoral crest region and above the distal articular surfaces, but preserving part of the olecranon fossa.

The proximal shaft exhibits a prominent deltopectoral crest and a corresponding triangular cross section. This accounts for much of the forward convexity of the shaft. The distal shaft flattens anteroposteriorly and broadens mediolaterally. The supracondylar crest on its lateral aspect is well defined. The posterior surface exhibits the proximal portion of olecranon fossa. It appears to have been "deep" and a foramen may have been present.

Ulna. The almost complete Klein Hadersdorf ulna (length 209 mm) includes the olecranon process and the trochlear notch within the humero-

Table 11. Primate Postcranial Remains from Western Europe

	Shoulder	Elbow	Wrist	Hand
Klein Hadersdorf, Austria (ca. 14.0 m.y.a.) *"Austriacopithecus weinfurteri"*: No cranial, facial, dental remains.	Unassociated shaft humerus lacking articular surfaces	Unassociated ulna with proximal articular surfaces		

	Shoulder	Elbow	Wrist	Hand
St. Gaudens, France (ca. 12.5 m.y.a.) *Dryopithecus fontani:* Jaws and teeth.	Unassociated shaft humerus lacking articular surfaces			

	Hip	Knee	Ankle	Foot
Eppelsheim, Germany (ca. 11 m.y.a.) *"Eppelsheim femur"*: No cranial, facial, dental remains.	Unassociated, reconstructed femur with articular surfaces			

ulnar joint and the radial notch within the proximal radio-ulnar joint. Its distal radio-ulnar surface, the ulnar head, and distal shaft are missing.

Proximal ulna size approximates that of Rud 22. However, these ulnae differ in structure. The Klein Hadersdorf trochlear notch is flattened medio-laterally. It does not have the distinctive proximodistal ridge that separates medial and lateral facets and is characteristic of the Rud 22 specimen and of extant large hominoids. The radial notch, although not deeply incised, shows widening at its proximal edge and a shallow concavity on its joint surface.

Unlike Rud 22, the olecranon process is present. Its posterior surface is complete, while the anterior surface and the most proximal portion of the trochlear notch are damaged. The existing olecranon process, at least in its posterior aspect, is short and apparently bent medially.

Saint Gaudens, France

Dryopithecus fontani from Saint Gaudens, France was the first *Dryopithecus* to be described (Lartet, 1856). A partial humerus, probably representing a young individual, was recovered from the same site as a fragmentary lower jaw with teeth (Table 11). Apparently, *D. fontani* is the only hominoid known from this site (Pilbeam and Simons, 1971). Faunal correlation of European sites suggests an age of 12.5 million years (Szalay and Delson, 1979) to 14 million years (Pilbeam and Simons, 1971).

The fragmentary humerus includes most of the diaphysis. The humeral head and distal articular surfaces are missing and the shaft appears to be compressed anteroposteriorly. The distal surface includes part of the olecranon fossa.

In addition to Lartet's brief discussion (1856), Le Gros Clark and Leakey (1951) compare the Saint Gaudens humerus to that from Maboko Island, suggesting similarities especially in the weak development of the supracondylar crest. Pilbeam and Simons (1971) speculate on medial humeral head orientation, capitular development, and trochlear features, and compare the shaft favorably with a *Pan paniscus* individual. Their assessment must be extremely tentative since, not only is the fossil incomplete and their comparative sample inadequate, but form–function relationships of shaft morphology and articular surface size and shape are not defined clearly (Morbeck, 1976).

The overall shaft is slender. The proximal portion, unlike those from Maboko and Klein Hadersdorf, shows little development of the deltopectoral crest and is oval in cross section. The distal shaft exhibits mediolateral broadening and a weakly developed supracondylar crest, but with the distal extremity, it tells little about elbow size and structure.

Eppelsheim, Germany

The affinities of the relatively complete, but isolated, right femur from Eppelsheim, purported to be approximately 11 million years old (Szalay and Delson, 1979), have been controversial for more than a century. Since its discovery in 1820, it has been described and/or renamed several times, linking it with the gibbons, the dryopithecines, or separating it as a distinct taxon (McHenry and Corruccini, 1976). In a recent review of fossil primate taxonomy, Szalay and Delson (1979) tentatively refer it to *Dryopithecus (Dryopithecus)*, as did Simons and Pilbeam (1965). McHenry and Corruccini (1976), based on their morphometric analysis of primate proximal femora, suggest phenetic affinities with *Pliopithecus vindobonensis* and *Hylobates*. As with other Miocene postcranial remains that are not associated with cranial and dental elements, it is best not to assign this specimen to a particular taxon.

The Eppelsheim femur includes most of its femoral head, neck, greater and lesser trochanters, shaft, and distal articular surfaces (Table 11). Kretzoi (1975; and personal communication 1980) suggests that reconstruction of its shaft has added to its overall length. I have not had the opportunity to study the original specimen. Observations discussed here are based on a cast and reports in the literature, and therefore must be considered to be provisional in nature.

Within the hip complex, the femoral head is not quite spherical and has a deep fovea for the ligamentum teres. The articular surface on its posterior surface apparently extends on the femoral neck, a feature evident on the cast and noted by McHenry and Corruccini (1976). As described by Le Gros Clark and Leakey (1951) and earlier workers, the neck exhibits a posterior tubercle. The greater trochanter appears to lie distal to the proximal aspect of the femoral head. The shaft is straight and lacks strong muscular markings. It is only moderately widened at its distal extremity.

The distal articular surfaces articulating within the knee complex show some damage. The patellar surface appears to be broad; its medial and lateral edges are subequal in height. The condyles are close in size, asymmetric, and only slightly divergent.

Asian Middle–Late Miocene Postcranial Evidence

Potwar Plateau, Pakistan

The rich fauna from the middle–late Miocene deposits in Pakistan includes at least three hominoid genera as well as other primates (Pilbeam *et al.,* 1977, 1979, 1980). The evidence from jaws, teeth, and postcranial remains at several localities, especially Locality 311 in the Sethi Nagri area and Localities 182, 260, and 317 in the Khaur area, is crucial for understanding hominoid differentiation during this time period. *Ramapithecus punjabicus, Sivapithecus indicus,* and *Gigantopithecus bilaspurensis* are identified from these remains. *Gigantopithecus,* however, is known from only a single molar fragment (GSP 7144, left M_1) and possible, but unassociated, postcranial remains. Another species of *Sivapithecus, S. sivalensis,* also may be present (Pilbeam *et al.,* 1977). However, Pilbeam *et al.* (1979) do not list *S. sivalensis* among the hominoid species found during 1974–1978 seasons. In a more recent summary, Pilbeam *et al.* (1980) describe specimens without reference to specific taxa so as to lessen controversy about hominoid classification and phylogeny. Cercopithecoids and lower primates also have been identified from several localities (Pilbeam *et al.,* 1979).

The Pakistan localities in biostratigraphic Zones 5 and 6 (Barry and Lindsay, in Pilbeam *et al.,* 1979), where the majority of important hominoids have been recovered and where taphonomic and paleoecological studies have been undertaken, apparently indicate a mosaic of habitats, including woodland, bush, and grassland, associated with past river systems (Badgley and Behrensmeyer, in Pilbeam *et al.,* 1979). Badgley and Behrensmeyer describe these localities as dominated by "medium-sized" mammals with evidence for both open country and woodland adaptations. In addition, they suggest that bone concentrations in Locality 260, a site that has yielded *Ramapithecus* and *Sivapithecus* fossils as well as several primate postcranial specimens, may be the result of predator activities. The mixed nature of the paleoecological data suggesting a broad environmental grain may be broadly similar to patterns of community diversity at other middle–late Miocene sites.

Important hominoid sites include: (1) Locality 311, in the Sethi Nagri area, Zone 5 primarily with isolated teeth, at least one of which has been identified tentatively as *Gigantopithecus,* and several postcranial elements; and (2) in the Khaur area, Zone 6 localities 182, 260, 317, and others with *Ramapithecus* and *Sivapithecus* remains. Zones 5 and 6 are estimated to be 9 million years old on the basis of lithostratigraphic, paleomagnetic, and faunal sequence data by Barry and Lindsay (in Pilbeam *et al.,* 1979). However, more detailed analysis and correlation of these data with a calibrated magnetic polarity time scale completed by Tauxe (1979) suggests a younger estimate of 8 million years for the Khaur area hominoids. The middle–late Miocene hominoid sites with primate postcranial fossils therefore span at least 6 million years.

Postcranial fossils representing three size classes have been recovered from several localities in Pakistan, including those with *Ramapithecus, Siva-pithecus,* and *Gigantopithecus* specimens (Table 12). As pointed out by Pilbeam *et al.* (1977) and Rose (Rose, personal communication; Pilbeam *et al.,* 1980), none of the postcranial specimens is associated directly with cranial or dental remains. However, the three size classes may correspond with taxa, for example, the larger representing *Gigantopithecus,* the middle-size class from *Sivapithecus,* and the smaller fossils representing *Ramapithecus* (Pilbeam *et al.,* 1977, 1979). Postcranial elements for the upper limb include a fragmentary humeral and radial shaft and elbow components within different size classes.

Table 12. Primate Postcranial Remains from Potwar Plateau, Pakistan (*ca.* 8.0–9.0 m.y.a.)[a]

Shoulder	Elbow	Wrist	Hand
	GSP 12271 Unassociated distal humerus with partial articular surfaces	GSP 17119 Capitate	GSP 6664 Unassociated proximal phalanx of thumb
	GSP 6663 Unassociated distal humerus with partial articular surfaces		GSP 13168, 15783, 15784, 6666, 17154 Unassociated phalanges either of hand or foot
GSP 13606 Unassociated shaft humerus lacking articular surfaces			
	GSP 7611 Unassociated proximal radius lacking epiphysis and radial shaft lacking distal articular surface		

(*continued*)

Table 12 (*Continued*)

Hip	Knee	Ankle	Foot
GSP 11867, 13929, 6178, 15782, 9894; Unassociated proximal femora with articular surfaces		GSP 10785b Unassociated talus fragment	GSP 4664, 17606, 17152 Unassociated calcaneus fragments
			GSP 6454, 17118 Unassociated cuneiforms
GSP 12654 Unassociated proximal shaft femur with partial neck and greater trochanter			GSP 14046, 14045 Unassociated distal metatarsal I and II phalanges
	GSP 13420 Unassociated shaft femur lacking articular surfaces		GSP 13168, 15783, 15784, 6666; Unassociated phalanges of either hand or foot

[a]Taxa: *Ramapithecus punjabicus:* Jaws and teeth. *Sivapithecus indicus:* Jaws and teeth. *Sivapithecus sivalensis?:* Questionable taxon. *Gigantopithecus bilaspurensis:* One tooth. *Presbytis sivalensis? Macaca paleindicus.*

Recently discovered material includes an as yet unpublished capitate (Rose, personal communication). The lower limb is represented by femoral heads, partial femoral shafts, isolated tarsal bones, and part of the hallux.* In addition, several unassociated phalanges have not been sorted as to hand or foot body segments. Primate postcranial fossils from the Potwar Plateau are described formally by Rose (in Pilbeam *et al.,* 1980).

GSP 12271/Humerus.   GSP 12271 is a distal fragment of a right humerus from Locality 311. Classed as a large-sized specimen (it is considerably larger than Rud 53) and provisionally as *Gigantopithecus,* the lateral side of this humerus is almost complete, including the distal anterior and posterior shaft and lateral epicondyle, capitulum, and lateral border of the trochlea, and part of the olecranon fossa.

The capitulum, which is missing its proximolateral surface anteriorly, is bulbous and narrows distally. The articular surface extends to the posterior aspect. It is clearly separated from the trochlea as indicated by the prominent lateral border of the trochlea. Only a portion of the posterior trochlea is evident, and its lateral edge merges with the steep lateral surface of the olecranon fossa. The distal humeral shaft is deep anteroposteriorly, as is the olecranon fossa. The posterior shaft surface area lateral to the olecranon fossa

*In addition to the capitate (GSP 17119) now described by Rose, two incomplete calcanei (GSP 17606 and GSP 17152) have been recovered.

is broad and oriented horizontally. The supracondylar crest is well defined and the lateral epicondyle is pronounced. Muscle and ligament attachment areas are apparent, especially an elongated facet on the epicondyle for the forearm extensor muscles.

GSP 6663/Humerus. A second distal right humerus, also from Locality 311, classed by Rose with medium-size specimens, includes most of the capitular surface, part of the lateral border of the trochlea, and much of the posterior trochlea. The capitulum is damaged proximally, but is well rounded and clearly separated from the trochlea. The trochlear lateral edge is well defined where present on the distal and posterior aspects. It apparently wraps distally toward the lateral border of the olecranon fossa. Little else can be determined from these surfaces, and the remainder of the humerus is missing.

GSP 13506/Humerus. GSP 13506, a humeral shaft lacking both proximal and distal ends, from Locality 260, may be a primate. Rose (in Pilbeam *et al.*, 1980) includes it within his medium-sized category and suggest that it is within a size range predicted for GSP 6663. This locality has yielded both *Ramapithecus* and *Sivapithecus* as well as other mammals. The shaft, a right humerus, is broken below the head proximally but includes much of the olecranon fossa and associated shaft.

Overall shape is reminiscent of that seen in the Fort Ternan humerus, although GSP 13506 is larger. The shaft broadens distally and is narrower anteroposteriorly. Its posterior surface is relatively flat throughout its length. The supracondylar crest is well defined and extensive. The lateral posterior aspect is damaged, but the olecranon fossa is triangular and appears to have a steep-sided lateral wall. In inferior view, the broken medial epicondyle region appears to be bent posteriorly.

GSP 7611/Radius. The small (according to Rose's size categories) juvenile left radius from Locality 317 is pieced together from fragments of the shaft. Hominoids identified from this locality include both *Ramapithecus* and *Sivapithecus* (Pilbeam *et al.*, 1977). The radial head and distal articular surfaces are missing. The proximal epiphyseal surface is oval and apparently tilted slightly. The bicipital tuberosity is not pronounced but is well defined. A defined interosseous border runs distally from the level of the bicipital tuberosity. Although the shaft widens distally, there is no evidence for details of the distal joint complex.

GSP 11867, 13929, 6178, 15782, 9894/Femora. These unassociated femoral heads from Localities 260 (GSP 15782, 9894), 317 (11867), 221 (13929), and 224 (6178) provide some evidence of hip morphology. Recognized hominoids from the localities in which these lower limb specimens were recovered include *Ramapithecus* and *Sivapithecus* at Localities 260 and 317 and, according to Pilbeam *et al.* (1977), only several specimens of possible *Ramapithecus* at Localities 221 and 224.

These specimens are roughly the same size but in varying states of preservation. The joint surface is spherical, with articular surface extending on the neck anteriorly and posteriorly in two specimens (Rose, in Pilbeam *et al.*, 1980). The fovea for the ligamentum teres is apparent in all specimens.

GSP 12654, 13420/Femora.   GSP 12654 is a proximal left femoral shaft from Locality 260 with part of the neck and greater trochanter preserved. The fragmentary long bone shaft, GSP 13420 from the same locality, also lacks articular surfaces. Rose (in Pilbeam *et al.*, 1980) suggests that neck measurements on GSP 12654 match those of GSP 13929 and GSP 11867 from other localities. Two shaft fragments make up GSP 13420, also from Locality 260, each approximately 60 mm in length. The proximal fragment, which is broken at the level of the lesser trochanter, is smooth and oval in cross section. The second fragment presents problems in identification and interpretation. Rose, however, describes this fragment as a distal shaft broken just proximal to the epicondylar areas. Together, these fragments provide some information about distance between the lip and knee, but little else.

GSP 10785b/Talus.   This small body of a right talus from Locality 310 includes the trochlear facet, the lateral surface, and posterior calcaneal facet. Pilbeam *et al.* (1977) tentatively classify this specimen as a possible *Ramapithecus*, but list no cranial or dental *Ramapithecus* remains from Locality 310 in the Khaur area.

The concave, trochlear facet is wedge shaped, with its anterior portion wider than its posterior aspect. In posterior view, part of the flexor hallucis longus groove is present. The medial and lateral borders of the trochlear facet apparently are subequal in height. Due to the fragmentary nature of the fossil, this is difficult to determine accurately. The lateral surface of the body is smooth except for ligament attachment areas. Its counterpart on the medial side is missing, as is the talar head. Inferiorly, the posterior talocalcaneal facet is almost complete. This concave, elongate facet diverges from the main orientation of the trochlear facet.

GSP 4664/Calcaneus.   GSP 4664, from Locality 182, which has yielded both *Ramapithecus* and *Sivapithecus* fossils, is small and within range for the talus from Locality 310. This right calcaneus is extremely eroded, exhibiting only part of the talar articulation, part of the posterior extension, and the sustentaculum tali. GSP 4664 is too eroded to describe with accuracy. The posterior facet for the talus is convex and evenly curved. The sustentaculum tali is weathered on its edges. Details of the extent of articular surface are difficult to assess. The anterior, medial, and lateral aspects of GSP 4664 are eroded but show some evidence for muscle and ligament attachments.

Discussion

Middle-late Miocene upper limb and, in particular, elbow data for probable hominoids fall into at least two structural groups. The Rudabánya humerus and ulna and the less complete GSP 12271 and GSP 6663 distal humeri from the Potwar Plateau show structural similarities with the extant large hominoids. The Fort Ternan humerus and the Klein Hadersdorf ulna differ significantly. In terms of size, GSP 12271 is the largest, GSP 6663 and Rud 53 are close in size, and KNM-FT 2751 is considerably smaller (but still larger than earlier Miocene hominoids). Table 4 indicates comparative size relationships.

The middle–late Miocene humeral shafts also show at least two distinctive structural patterns. The immature Saint Gaudens specimen and the Potwar Plateau shaft both have relatively straight shafts, the latter very similar to that from Fort Ternan. The Maboko Island and Klein Hadersdorf humeri, which are close in overall length, both exhibit a forward angulated shaft and prominent deltopectoral crest. Although the shapes are different, behavioral interpretation is confused by recognition of potential bone–muscle interactions and skeletal plasticity, perhaps greater than in joints, and the unknown influences of expressed positional behavior in the life history of these individuals on the proximal shaft shape.

The known femoral heads are apparently similar in structure. Each fossil, including Rud 23 and Rud 55, is smaller in anteroposterior head diameter than the mean values for extant pongids. The Maboko femur and three of the Pakistan specimens are large enough to fall within the range of *P. paniscus*, but not within its 95% confidence interval. The larger monkeys sampled here, *Nasalis* and *Papio* (means equal 20 and 23 mm, respectively) have ranges that include the Rudabánya, Maboko, Eppelsheim, and the smaller Pakistan femoral heads (Table 6). Although *Pliopithecus vindobonensis* is not discussed here in detail, it is interesting that the anteroposterior head diameters for three femoral specimens, one right and two left sides, are 17.5 mm, that is, smaller than the other fossils investigated here but slightly larger than *Hylobates* and most monkeys in this sample.

The Maboko femoral shafts, Eppelsheim reconstructed diaphysis, and Rudabánya tibial shaft, if primate, provide limited data on lower limb morphology. The Eppelsheim distal femur, unlike that from Rudabánya, is reminiscent of the East African early Miocene unassociated distal femora. Although the Rudabánya femur distal breadth is within the range for *Pan paniscus, Pongo,* and *Papio* (but below pongid means) sampled here (Table 7), the long, narrow patellar surface alone of Rud 24 distinguishes it from known large fossil and extant anthropoids.

The talar data are also limited, with only the Rudabánya and Pakistan specimens known for the middle–late Miocene time period, except, of course, that of the associated *Pliopithecus* skeletons. Both Rud 27 and GSP 10785b are fragmentary, exhibiting primarily the trochlear facet and associated structures. Based on trochlear length, the Pakistan talus is about three-quarters the size of Rud 27. The *Pliopithecus* individual is even smaller. Rud 27 facet length (21 mm) falls within the ranges of *Pan* and *Pongo* (and is close to *Nasalis* and *Papio*), but its narrow breadth produces a talar index lower than all taxa sampled (Table 8). The high lateral border clearly separates it from bipedal hominids, but taxonomic affinities for either specimen remain unknown.

Explaining the morphological variation observed in the fossil record is crucial for interpreting past lifestyles and for determining taxonomic status. Although the data are incomplete, middle–late Miocene probable hominoid postcranial remains indicate that more than one morphological pattern exists, suggesting different movement capabilities and/or expressed behaviors. For example, the elbow components from Rudabánya and the distal humeri from the Potwar Plateau are similar to those of extant large hominoids. Movement capabilities emphasizing a stable humero-ulnar joint in flexion and extension and considerable rotation within the proximal radio-ulnar and humeroradial joints are most likely also are similar. Positional capabilities can only be confirmed with associated data from other joints and body segments. Expressed locomotor and postural behavior, however, is impossible to reconstruct with fossil data.

Sorting morphological variation between and within taxa is difficult. Between-group variation reflecting species' positional adaptation, perhaps, is easier. However, it is complicated by unknown ranges of variation due to age, sex, and individual life histories as well as to species-defined differences. The Saint Gaudens humerus, for example, may be an immature individual. At the present time, there is no way of ascertaining if an adult of this taxon would exhibit a similar humeral shaft configuration or exhibit a greater development of the deltopectoral crest possibly related to longer term bone–muscle interactions during its life history.

Sexual dimorphism and the inability to identify size variation due to sex in these postcranial data pose additional problems in sorting morphological variation. For example, the recent recognition of sexually dimorphic species based on increased sample sizes in Oligocene anthropoids from the Fayum deposits in Egypt and the shift in taxonomic classification for some earlier finds (e.g., *Propliopithecus haeckeli*) alter many of the previously held views of Oligocene hominoid evolution (Fleagle *et al.*, 1980). New fossils also may have solved some arguments (or created new ones) about variation due to sexual dimorphism in Plio-Pleistocene hominids [e.g., KNM-ER 406 and KNM-ER 732 (R. E. F. Leakey, 1973; Walker and Leakey, 1978)]. Unfortunately, even in dental remains for the middle–late Miocene hominoid sample, the possibility of sexual dimorphism is not appreciated fully (e.g., variation observed in the Rudabánya specimens).

Studies of modern primate positional behavior and postcranial anatomy show that the life history of individuals within species and between closely related species influence skeletal structure (Fleagle, 1977; Ward and Sussman, 1979). Skeletal variation produced by variation in habitual positional behavior and substrate preference reflecting the mechanical demands placed on the functioning organism has only recently been appreciated by anatomists and paleoanthropologists. Attacking the morphology– behavior– environment interface in living primates from many directions (Gomberg *et al.*, 1979; Fleagle, 1979) and understanding interactions in living taxa will provide a more comprehensive framework for interpreting fossil taxa.

The major problem in trying to sort morphological variation in the middle–late Miocene hominoid postcranial sample is that the data are incomplete. There are no associated skeletons and few complete bones. The sample is small and geographically widespread.

In the past, incomplete fossils have led to problems in reconstruction, in interpretation of structure and function, and in taxonomic classification. Reconstruction of shaft morphology from joints—for example, estimating distances between joints or muscle lever actions relative to joint centers and interpreting joint morphology from shafts—must be approached with extreme caution. Pilbeam and Simons (1971) reconstructed the orientation of the humeral head from the position of the bicipital groove and inferred a hominoid-like distal humerus from the distal posterior shaft, neither of which should be attempted given the status of one immature diaphysis. Reconstructions of jaws and teeth have also been the results of grand leaps from fossil morphology to potential evolutionary scenarios (cf. the history of *Ramapithecus* discussions).

It is even more dangerous to assume that because one region is similar, then overall morphological and behavioral patterns are equivalent to those observed in modern groups and/or to expect particular morphologies for given taxa. Speculations concerning bipedalism or tool use based on dentition, fortunately, are no longer prevalent in the literature. However, as mentioned above, Szalay and Delson (1979) use a morphological feature, a small olecranon process, to justify placement of the Klein Hadersdorf humerus and ulna in a taxonomic classification, *Sivapithecus darwini,* for which no associated postcranial data are recognized but for which they expect to find certain features. Ideas and models derived from personal scenarios cannot be confused with the process of taxonomic classification.

The middle–late Miocene fossil data, in general, and the postcranial evidence, in particular, do not yet allow the luxury of secure classification and detailed interpretations of hominoid evolutionary history. In describing how an anatomist should proceed when studying fossil limb bones, Day (1979) suggests that one should start with the original specimens, initially identify elements, and study the anatomy in a comparative context, including, if possible, measurement and biomechanical analysis. Detailed analysis of postcranial remains, especially if fossil bones can be placed in a functional complex, a regional complex, or whole skeleton context, as shown in Day's model, can lead to both a taxonomic judgement and a locomotor or positional adaptation judgment. The discussion of postcranial fossils from Rudabánya and the middle–late Miocene of Africa, Europe, and Asia fits within the early stages of analysis, primarily as a result of the incomplete nature of the fossil record. Original material has been identified and the anatomy studied using a comparative method. As Day (1979, p. 252) points out, the comparative method works well when "(1) there is sufficient comparative material of good quality, and (2) the recognized features or combinations of features are securely linked with only one observed type of locomotion." Among primates, anatomists recognize morphological patterns in joints and associated body segment as linked with particular movement capabilities (e.g., range of flexion or rotation) and, often, locomotor/postural capabilities (e.g., hominid bipedalism, pongid knuckle-walking). The Rudabánya humerus and ulna are recognized as similar to those of large extant hominoids and assumed to have had similar movement capabilities, but no conclusions are possible concerning the assessment of positional capabilities or taxonomic classification. The comparative fossil data are not of sufficient quality to allow a complete analysis. The patterns of variation that do appear (e.g., Rudabánya and Potwar Plateau distal humeri versus Fort Ternan humerus; Saint Gaudens humerus versus Maboko Island and Klein Hadersdorf humeri) may well prove to be functionally and taxonomically important. Eventually, with the discovery of associated, well-dated postcranial data, anatomists and paleoanthropologists will be able to test the tentative ideas about this sorting of variation and come closer to a taxonomic and phylogenetic judgment for the middle–late Miocene postcranial data.

ACKNOWLEDGMENTS

I thank Miklós Kretzoi for permission to study the Rudabánya fossils and for his good company during my visits to Hungary. I am grateful to the late L. S. B. Leakey and R. E. F. Leakey (National Museums of Kenya), D. Brothwell and T. Molleson [British Museum (Natural History)], D. Pilbeam (Harvard University) and M. Rose (College of Medicine and Dentistry of New Jersey), and H. Zapfe (Paläontologisches Institut der Universität, Vienna) for making fossils in their care available to me for study. The directors and staffs of many institutions, including the American Museum of Natural History, Arizona State Museum, British Museum (Natural History), Musée Royal de l'Afrique Centrale, Museum of Comparative Zoology (Harvard), National Museums of Kenya, Powell-Cotton Museum, and Smithsonian Institution, have been helpful in collecting comparative data. I am grateful for their generosity. I thank Roberta Hagaman for her statistical analyses of both fossil and comparative metric data. Mike Rose has been extremely helpful in discussing the postcranial fossils from the Potwar Plateau, Pakistan. This research was supported in part by the Wenner-Gren Foundation for Anthropological Research, the University of Arizona, the Research Council of the University of Massachusetts/Amherst, and a Special Career and NDEA Title IV Fellowships at the University of California, Berkeley. This support is gratefully acknowledged.

References

Andrews, P. 1971. *Ramapithecus wickeri* mandible from Fort Ternan, Kenya. *Nature (Lond.)* **231:**192–194.

Andrews, P. 1978. A revision of the Miocene Hominoidea of East Africa. *Bull. Br. Mus. (Nat. Hist.) Geol.* **30**(2):85–224.

Andrews, P., and Cronin, J. E. 1982. The relationships of *Sivapithecus* and *Ramapithecus* and the evolution of the orang-utan. *Nature (Lond.)* **297:**541–546.

Andrews, P., and Tobien, H. 1977. New Miocene locality in Turkey with evidence on the origin of *Ramapithecus* and *Sivapithecus*. *Nature (Lond.)* **268:**699–701.

Andrews, P., and Walker, A. C. 1976. The primate and other fauna from Fort Ternan, Kenya, in *Human Origins: Louis Leakey and the East African Evidence*, (G. L. Isaac and E. R. McCown, eds.), pp. 279–304, Benjamin, Menlo Park, California.

Andrews, P., Lord, J. M., and Nesbit-Evans, E. M. 1979. Patterns of ecological diversity in fossil and modern mammalian faunas. *Biol. J. Linn. Soc.* **11**:177–205.

Andrews, P., Meyer, G., Pilbeam, D., Van Couvering, J. A., and Van Couvering, J. A. H. 1981. The Miocene of Maboko Island, Kenya: Geology, age, taphonomy and palaeontology. *J. Hum. Evol.* **10**:35–48.

Berggren, W. A., and Van Couvering, J. A. 1974. The late Neogene: Biostratigraphy, geochronology and paleoclimatology of the last 15 million years in marine and continental sequences. *Palaeogeogr. Paleoclimatol. Palaeoecol.* **16**:1–215.

Bishop, W. W. 1971. The late Cenozoic history of East Africa in relation to hominoid evolution, in: *The Late Cenozoic Glacial Ages* (K. K. Turekian, ed.), pp. 493–537, Yale University Press, New Haven, Connecticut.

Bishop, W. W., Miller, J. A., and Fitch, F. J. 1969. New potassium–argon age determinations relevant to the Miocene fossil mammal sequence in East Africa. *Am. J. Sci.* **267**:669–699.

Churcher, C. S. 1970. Two new upper Miocene giraffids from Fort Ternan, Kenya, East Africa: *Palaeotragus primaevus* n. sp. and *Samotherium africanum* n. sp., in *Fossil Vertebrates of Africa*, Volume 2 (L. S. B. Leakey and R. J. G. Savage, eds.), pp. 1–105, Academic, London.

Conroy, G., and Pilbeam, D. R. 1975. *Ramapithecus:* Review of its hominid status, in: *Paleoanthropology, Morphology and Paleoecology* (R. Tuttle, ed.), pp. 59–86, Mouton, The Hague.

Corruccini, R., Ciochon, R., and McHenry, H. 1975. Osteometric shape relationships in the wrist joint of some anthropoids. *Folia Primatol.* **24**:250–274.

Day, M. 1979. The locomotor interpretation of fossil primate postcranial bones, in: *Environment, Behavior, and Morphology: Dynamic Interactions in Primates* (M. E. Morbeck, H. Preuschoft, and N. Gomberg, eds.), pp. 245–258, Gustav Fischer, New York.

Delson, E. 1979. *Oreopithecus* is a cercopithecoid after all. *Am. J. Phys. Anthropol.* **50**:431–432.

Eckhardt, R. 1972. Population genetics and human evolution. *Sci. Am.* **226**(1):94–103.

Evans, E. M., Van Couvering, J. A. H., and Andrews, P. 1981. Palaeoecology of Miocene sites in western Kenya. *J. Hum. Evol.* **10**:99–116.

Fleagle, J. 1977. Locomotor behavior and skeletal anatomy of sympatric Malaysian leaf-monkeys (*Presbytis obscura* and *Presbytis melalophos*). *Yearb. Phys. Anthropol.* **20**:440–453.

Fleagle, J. 1979. Primate positional behavior and anatomy: Naturalistic and experimental approaches, in: *Environment, Behavior, and Morphology: Dynamic Interactions in Primates* (M. E. Morbeck, H. Preuschoft, and N. Gomberg, eds.), pp. 313–325, Gustav Fischer, New York.

Fleagle, J., Kay, R., and Simons, E. L. 1980. Sexual dimorphism in early anthropoids. *Nature (Lond.)* **287**:328–330.

Frayer, D. 1976. A reappraisal of *Ramapithecus. Yearb. Phys. Anthropol.* **18**:19–30.

Frisch, J. 1967. Remarks on the phyletic position of *Kenyapithecus. Primates* **8**:121–126.

Genet-Varcin, E. 1969. *À la Recherche du Primate Ancête de l'Homme*, Boubée, Paris. 336 pp.

Gentry, A., 1970. The Bovidae (Mammalia) of the Fort Ternan fossil fauna, in: *Fossil Vertebrates of Africa*, Volume 2 (L. S. B. Leakey and R. J. G. Savage, eds.), pp. 243–323, Academic, London.

Gomberg, N., Morbeck, M. E., and Preuschoft, H. 1979. Multidisciplinary research in the analysis of primate morphology and behavior, in: *Environment, Behavior, and Morphology: Dynamic Interactions in Primates* (M. E. Morbeck, H. Preuschoft, and N. Gomberg, eds.), pp. 5–21, Gustav Fischer, New York.

Greenfield, L. O. 1979. On the adaptive pattern of "*Ramapithecus.*" *Am. J. Phys. Anthropol.* **50**:527–548.

Hagaman, R., and Morbeck, M. E. 1982. Data transformations in multivariate morphometric analyses. *J. Hum. Evol.* (in press).

Hrdlička, A. 1935. The Yale fossils of anthropoid apes. *Am. J. Sci.* **29**:34–40.

Jia, L.-P. 1980. *Early Man in China*, Language Press, Beijing.

Johanson, D. C., and White, T. 1979. A systematic assessment of early African hominids. *Science* **203**:321–330.

Johanson, D. C., Lovejoy, C. O., Burstein, A. H., and Heiple, K. G. 1976. Functional implications of the Afar knee joint. *Am. J. Phys. Anthropol.* **44**:188 (Abstract).

Kay, R. F. 1981. The nut-crackers—A new theory of the adaptations of the Ramapithecinae. *Am. J. Phys. Anthropol.* **55**:141–151.

Khatri, A. P. 1975. The early fossil hominids and related apes of the Siwalik Foothills of the Himalayas: Recent discoveries and new interpretations, in *Paleoanthropology, Morphology and Paleoecology* (R. Tuttle, ed.), pp. 31–58, Mouton, The Hague.

Kretzoi, M. 1975. New ramapithecines and *Pliopithecus* from the Lower Pliocene of Rudabánya in north-eastern Hungary. *Nature (Lond.)* **257:**578–581.

Lartet, E., 1856. Note sur un grand singe fossile qui se rattache au groupe des singes supérieurs. *C. R. Acad. Sci. Paris* **43:**219–228.

Leakey, L. S. B. 1962. A new lower Pliocene fossil primate from Kenya. *Ann. Mag. Nat. Hist.* **4:**689–696.

Leakey, L. S. B. 1967. An early Miocene member of Hominidae. *Nature (Lond.)* **213:**155–163.

Leakey, L. S. B. 1968. Lower dentition of *Kenyapithecus africanus*. *Nature (Lond.)* **217:**827–830.

Leakey, R. E. F. 1973. Australopithecines and hominines: A summary of the evidence from the early Pleistocene of eastern Africa. *Symp. Zool. Soc. Lond.* **33:**53–69.

Lewis, G. E. 1934. Preliminary notice of new man-like apes from India. *Am. J. Sci.* **27:**161–179.

Lewis, G. E. 1937. Taxonomic syllabus of Siwalik fossil anthropoids. *Am. J. Sci.* **34:** 139–147.

Le Gros Clark, W. E. 1952. Report on fossil hominoid material collected by the British–Kenya Miocene Expedition, 1949–1951. *Proc. Zool. Soc. Lond.* **122:**273–286.

Le Gros Clark, W. E., and Leakey, L. S. B. 1951. The Miocene Hominoidea of East Africa. *Fossil Mammals of Africa* (Br. Mus. Nat. Hist.) **1:**1–117.

Lewis, O. J. 1972. Osteological features characterizing the wrists of monkeys and apes, with a reconsideration of this region in *Dryopithecus (Proconsul) africanus*. *Am. J Phys. Anthropol.* **36:**45–58.

Lovejoy, C. O. 1979. A reconstruction of the pelvis of AL-288 (Hadar Formation, Ethiopia). *Am. J. Phys. Anthropol.* **50:**460 (Abstract).

Lovejoy, C. O. 1981. The origin of man. *Science* **211:**341–350.

McHenry, H., and Corruccini, R. 1976. Affinities of Tertiary hominoid femora. *Folia Primatol.* **26:**139–150.

McHenry, H., and Corruccini, R., 1980. Late Tertiary hominoids and human origins. *Nature (Lond.)* **285:**397–398.

McHenry, H., and Temerin, A. 1979. The evolution of hominid bipedalism: Evidence from the fossil record. *Yearb. Phys. Anthropol.* **22:**105–131.

Morbeck, M. E. 1972. *A Re-Examination of the Forelimb of the Miocene Hominoidea*, Ph.D. Dissertation, University of California, Berkeley. 205 pp.

Morbeck, M. E. 1975. *Dryopithecus africanus* forelimb. *J. Hum. Evol.* **4:**39–46.

Morbeck, M. E. 1976. Problems in reconstruction of fossil anatomy and locomotor behaviour: The *Dryopithecus* elbow complex. *J. Hum. Evol.* **5:**223–233.

Morbeck, M. E. 1977. The use of casts and other problems in reconstructing the *Dryopithecus (Proconsul) africanus* wrist complex. *J. Hum. Evol.* **6:**65–78.

Napier, J., and Davis, P. 1959. The forelimb skeleton and associated remains of *Proconsul africanus*. *Fossil Mammals of Africa* (Br. Mus. Nat. Hist.) **16:**1–69.

O'Connor, B. L. 1976. *Dryopithecus (Proconsul) africanus:* Quadruped or nonquadruped? *J. Hum. Evol.* **5:**279–283.

Pickford, M. 1980. Recent discoveries in Middle Miocene rocks of western Kenya, *L. S. B. Leakey Foundation News* **17:**7–8.

Pickford, M. 1981. Preliminary Miocene mammalian biostratigraphy for western Kenya. *J. Hum. Evol.* **10:**73–97.

Pilbeam, D. R. 1966. Notes on *Ramapithecus*, the earliest known hominid, and *Dryopithecus*. *Am. J. Phys. Anthropol.* **25:**1–5.

Pilbeam, D. R. 1968. The earliest hominids. *Nature (Lond.)* **219:**1335–1338.

Pilbeam, D. R. 1972. *The Ascent of Man: An Introduction to Human Evolution*, Macmillan, New York.

Pilbeam, D. R. 1978. Rearranging our family tree. *Hum. Nat.* **1**(6):38–45.

Pilbeam, D. R., and Simons, E. L., 1971. Humerus of *Dryopithecus* from Saint Gaudens, France. *Nature (Lond.)* **229:**406–407.

Pilbeam, D. R., Meyer, G. E., Badgley, C., Rose, M. D., Pickford, M., Behrensmeyer, A. K., and Shah, S. M. I. 1977. New hominoid primates from the Siwaliks of Pakistan and their bearing on hominoid evolution. *Nature (Lond.)* **270:**689–695.

Pilbeam, D. R., Behrensmeyer, A. K., Barry, J. C., and Shah, S. M. I. 1979. Miocene sediments and faunas of Pakistan. *Postilla* (Peabody Mus. Nat. Hist., Yale Univ.) **179:**1–45.

Pilbeam, D. R., Rose, M. D., Badgley, C., and Lipschutz, B. 1980. Miocene hominoids from Pakistan. *Postilla* (Peabody Mus. Nat. Hist., Yale Univ.) **181:**1–94.

Prasad, K. N. 1964. Upper Miocene anthropoids from the Siwalik Beds of Haritalyangar, Himachal Pradesh, India. *Palaeontology* **7:**124–134.

Prasad, K. N. 1969. Critical observations on the fossil anthropoids from the Siwalik system of India. *Folia Primatol.* **10:**288–317.

Preuschoft, H. 1973. Body posture and locomotion in some East African Miocene Dryopithecinae, in: *Human Evolution* (M. Day, ed.), pp. 13–46, Symposium of the Society for the Study of Human Biology, Volume 11, Barnes and Noble, New York.

Qi, G. 1979. Pliocene mammalian fauna of Lufeng, Yunnan. *Vertebr. Palasiat.* **17:**14–22.

Sarich, V. 1971. A molecular approach to the question of human origins, in: *Background for Man* (P. Dolhinow and V. Sarich, eds.), pp. 60–81, Little, Brown and Company, Boston.

Sarich, V. M., and Cronin, J. E. 1976. Molecular systematics of the primates, in: *Molecular Anthropology* (M. Goodman and R. E. Tashian, eds.), pp. 141–170, Plenum, New York.

Sarich, V. M., and Cronin, J. E. 1977. Generation length and rates of hominoid molecular evolution. *Nature (Lond.)* **269:**354.

Schön, M., and Ziemer, L. 1973. Wrist mechanism and locomotor behaviour of *Dryopithecus* (*Proconsul*) africanus. *Folia Primatol.* **20:**1–11.

Shipman, P., Walker, A., Van Couvering, J. A., Hooker, P., and Miller, J. 1981. The Fort Ternan hominoid site, Kenya: Geology, age, taphonomy and paleoecology. *J. Hum. Evol.* **10:**49–72.

Simons, E. L. 1961. The phyletic position of *Ramapithecus*. *Postilla* (Peabody Mus. Nat. Hist., Yale Univ.) **57:**1–9.

Simons, E. L. 1964. On the mandible of *Ramapithecus*. *Proc. Natl. Acad. Sci. USA* **51:**528–535.

Simons, E. L. 1968. A source for dental comparisons of *Ramapithecus* with *Australopithecus* and *Homo*. *S. Afr. J. Sci.* **64:**92–112.

Simons, E. L. 1972. *Primate Evolution: An Introduction to Man's Place in Nature*, Macmillan, New York.

Simons, E. L. 1976. The nature of the transition in the dental mechanism from pongids to hominids. *J. Hum. Evol.* **5:**511–528.

Simons, E. L. 1977. *Ramapithecus*. *Sci. Am.* **236**(5):28–35.

Simons, E. L. 1978. Diversity among the early hominids: A vertebrate paleontologist's viewpoint, in: *Early Hominids of Africa* (C. J. Jolly, ed.), pp. 543–566, St. Martin's Press, New York.

Simons, E. L., and Delson, E. 1978. Cercopithecidae and Parapithecidae, in: *Evolution of African Mammals* (V. J. Maglio and H. B. S. Cooke, eds.), pp. 100–119, Harvard University Press, Cambridge, Massachusetts.

Simons, E. L., and Pilbeam, D. R. 1965. Preliminary revision of the Dryopithecinae (Pongidae, Anthropoidea). *Folia Primatol.* **3:**81–152.

Simons, E. L., and Pilbeam, D. R. 1978. *Ramapithecus* (Hominidae, Hominoidea), in: *Evolution of African Mammals* (V. J. Maglio and H. B. S. Cooke, eds.), pp. 147–153, Harvard University Press, Cambridge, Massachusetts.

Simons, E. L., Andrews, P., and Pilbeam, D. R. 1978. Cenozoic apes, in: *Evolution of African Mammals* (V. J. Maglio and H. B. S. Cooke, eds.), pp. 120–146, Harvard University Press, Cambridge, Massachusetts.

Szalay, F., and Delson, E. 1979. *Evolutionary History of the Primates*, Academic, New York.

Tauxe, L. 1979. A new date for *Ramapithecus*. *Nature (Lond.)* **282:**399–401.

Van Couvering, J. A. 1980. Community evolution in East Africa during the Late Cenozoic, in: *Fossils in the Making: Vertebrate Taphonomy and Paleoecology* (A. K. Behrensmeyer and A. Hill, eds.), pp. 272–298, University of Chicago Press, Chicago.

Van Couvering, J. A. H., and Van Couvering, J. A. 1976. Early Miocene mammal fossils from East Africa: Aspects of geology, faunistics and paleoecology, in: *Human Origins: Louis Leakey and the East African Evidence* (G. L. Isaac and E. R. McCown, eds.), pp. 155–196, Benjamin, Menlo Park, California.

Vasishat, R. N., Gaur, R., and Chopra, S. R. K. 1978. Geology, fauna and palaeoenvironments of Lower Sivalik deposits around Ramnagar, India. *Nature (Lond.)* **275:**736–737.

Vogel, C. 1975. Remarks on the reconstruction of the dental arcade of *Ramapithecus*, in: *Paleoanthropology, Morphology and Paleoecology* (R. H. Tuttle, ed.), pp. 87–98, Mouton, The Hague.

Walker, A. C. 1974. Excavation of the Miocene fossil site of Fort Ternan, Kenya 1974. Manuscript.

Walker, A. C. 1980. Functional anatomy and taphonomy, in: *Fossils in the Making: Vertebrate Taphonomy and Paleoecology* (A. K. Behrensmeyer and A. Hill, eds.), pp. 182–196, University of Chicago Press, Chicago.

Walker, A. C., and Andrews, P. 1973. Reconstruction of the dental arcades of *Ramapithecus wickeri*. *Nature (Lond.)* **244:**313–314.

Walker, A. C., and Leakey, R. E. F. 1978. The hominids of East Turkana. *Sci. Am.* **239**(2):54–66.

Ward, S., and Sussman, R. 1979. Correlates between locomotor anatomy and behavior in two sympatric species of *Lemur*. *Am. J. Phys. Anthropol.* **50:**575–590.

Xu, Q., and Lu, Q., 1979. The mandibles of *Ramapithecus* and *Sivapithecus* from Lufeng, Yunnan. *Vertebr. Palasiat.* **17:**1–13.

Zapfe, H. 1960. Die primatenfunde aus der Miozänen spaltenfüllung von Neudorf an der March (Děvínská Nová Ves), Tschechoslowakei. Mit anhang: Der primatenfund aus dem Miozän von Klein Hadersdorf in Niederösterreich. *Schweiz. Palaeontol. Abh.* **78:**1–293.

Zihlman, A. L., Cronin, J. E., Cramer, D. L., and Sarich, V. M. 1978. Pygmy chimpanzees as a possible prototype for the common ancestor of humans, chimpanzees and gorillas. *Nature (Lond.)* **275:**744–746.

Miocene Hominoid Postcranial Morphology

15

Monkey-like, Ape-like, Neither, or Both?

M. D. ROSE

Introduction

It is not the purpose of this chapter to provide definitive answers to any of the questions asked in its title, even though various aspects of these questions have formed a large part of the lively debate that has been conducted in recent years concerning Miocene hominoid postcrania. The material available for investigation has been increased significantly recently by new specimens from the early Miocene of East Africa [KNM-RU 2036C and KNM-RU 5872 specimens (Walker and Pickford, this volume, Chapter 12)], the later Miocene of Rudabánya, Hungary [Rud specimens (Morbeck, this volume, Chapter 14)], and the Potwar Plateau of Pakistan [most GSP specimens (Pilbeam *et al.*, 1980)]. The main purpose of this chapter is to make some general comments on functional features of the morphology of some Miocene hominoid postcrania and on possible positional capabilities consistent with those features. Similarities to and differences from features of Miocene species evident in the postcrania of groups of living higher primates will be made purely in terms of function. Attention will be directed toward the larger bodied Miocene hominoids. Original specimens of all the East African and Asian material have been examined. The European material has been examined in cast form.

M. D. ROSE • Department of Anatomy, New Jersey Medical School, University of Medicine and Dentistry of New Jersey, 100 Bergen St., Newark, New Jersey 07103.

The Forelimb

Shoulder. The partial scapula, and proximal humeral shaft and metaphysis, KNM-RU 2036CH, of *Proconsul africanus* from Rusinga Island show a number of distinctive features. In posterior view the scapular spine is set fairly obliquely and the lateral half of the spine inclines somewhat cranially from its root. The acromion extends somewhat lateral to the cranially inclined glenoid fossa. As evident from the metaphysial surface, there was a greater amount of medial torsion of the humeral head than estimated by Napier and Davis (1959) for the KNM-RU 2036AH specimen. Details of the proximal shaft and the region of deltoid insertion indicate that there was some anterior inclination of the shaft, with relatively well developed deltotriceps and deltopectoral crests. The deltoid insertion may not have extended as far distally on the shaft as estimated by Napier and Davis. The proximal shaft features are present in the dryopithecine humeri from Maboko, BMNH M 16334 (Le Gros Clark and Leakey, 1951), and Klein Hadersdorf (Zapfe, 1960). There was evidently considerable mobility within the shoulder region of *P. africanus*. The scapular features imply that overhead positions of the forelimb could be achieved fairly easily. The relative importance of deltoid as an abductor at the glenohumeral joint is equivocal, as is the positioning of the scapula on the thorax. This leaves the functional significance of the humeral head torsion explicable in more than one way. The torsion would be an obvious consequence if the scapula was relatively dorsally placed (Le Gros Clark, 1959). However, if the scapula was more laterally placed, the torsion might imply that some medial rotation of the forelimb at the glenohumeral joint was associated with commonly used forelimb positions.

No living primates show a pattern of features of the shoulder region that closely corresponds to that described above. The cercopithecoid, and especially the cercopithecine, shoulder region is specialized mainly in terms of features relating to parasagittal movements of the limb as a whole. Similarities with the Miocene hominoid pattern are only evident in the proximal humeral region, and not in features related to general mobility of the shoulder girdle. Some larger cebids, and living hominoids in which shoulder mobility is high, show similarities in scapular features and in the torsion of the humeral head, but the proximal humerus of living hominoids shows a different pattern of features.

Elbow and Forearm. The elbow region of *P. africanus* has received considerable attention (e.g., Napier and Davis, 1959; McHenry and Corruccini, 1975; Corruccini *et al.*, 1976; Morbeck, 1975, 1976). The humero-ulnar joint of *P. africanus* can be reconstructed from the specimens KNM-RU 2036AH and AK (distal humeri) and CF (proximal ulna). The spool-shaped humeral trochlea is relatively broad compared with the capitular region and has a fairly even transverse curvature throughout its extent. A moderately developed medial trochlear lip extends posteriorly and a sharply defined lateral lip continues as the lateral margin of a quite deep olecranon fossa. As seen in the proportions of the trochlear surface of the proximal ulna, the articular surface is relatively long compared to its width. The expanded olecranon beak of the ulna extends somewhat anteriorly and articulates proximolaterally with the lateral lip of the humeral trochlea and lateral wall of the olecranon fossa. The coronoid beak extends slightly further anteriorly than the olecranon beak and is moderately expanded where it engages medially with the medial lip of the humeral trochlea. In side view the articular surface of the coranoid beak slopes anterodistally.

Flexion–extension is obviously the predominant movement allowed at the joint. The combination of a moderately protuberant olecranon beak and fairly deep olecranon fossa indicates that considerable extension was allowed. Napier and Davis (1959) suggest a range of up to 180° on the basis of humeral morphology. The slope of the coronoid beak suggests that a common working position of the joint may have been one of semiflexion, when this surface would be aligned horizontally. Preuschoft (1973) reaches similar conclusions on the basis of other morphological features of the KNM-RU 1786 *Proconsul nyanzae* ulna. The morphology of the medial and lateral sides of the joint provide stabilization against medially or laterally directed forces throughout the flexion–extension range (Napier and Davis, 1959; Jenkins, 1973). Conjunct movements are minimal in this type of joint. A broad area for the origin of brachialis indicates that powerful flexion at the humero-ulnar joint may have been possible. A complete ulnar olecranon process is present on the KNM-RU 1786 *P. nyanzae* ulna. The process is mediolaterally broad, quite short proximodistally, and inclined somewhat posteriorly. Triceps insertion was thus quite extensive, indicating that it may have been a quite powerful muscle. Posterior inclination of the olecranon process has been discussed by Jolly (1967), who relates it to efficient triceps action in elbow positions of nearly full extension. Harrison (1982) suggests that this and other features of the elbow region in *Proconsul* species indicate some terrestrial activity. As far as the olecranon process is concerned, this morphology is not incompatible with efficient triceps action at more flexed elbow positions, or as a synergist or fixator, for example, during activities in which the limb is held in an overhead position.

The KNM-RU 1786 *P. nyanzae* ulna is similar to that of *P. africanus* in most features of the trochlear surface. Although it is eroded, it is possible that the coronoid beak extended further anteriorly than in *P. africanus,* indicating that more extended elbow positions may have been common. However, the inclination of the surface is similar to that of *P. africanus,* indicating that more flexed elbow positions were also used. The Klein Hadersdorf ulna shows similar features. The possibly hominoid ulna KMN-FT 3381 from Fort Ternan has a coronoid beak that extends even further anteriorly and its articular surface slopes more anteriorly than anterodistally. These features are shared by the Rud 22 ulnar fragment. The emphasis in these specimens is therefore more on extension at the humero-ulnar joint. The later Miocene humeral specimens Rud 53, the medium-sized GSP 6663 and 13606, and large-sized GSP 12271 have an even broader and more spool-shaped trochlea, and better defined medial and lateral trochlear lips than the earlier Miocene specimens.

The olecranon fossa is relatively deep and its lateral wall is aligned so as to make extensive contact with the lateral part of the olecranon beak. These features, which are most marked in GSP 12271, indicate an emphasis on extension, and the stability of the humero-ulnar joint in this position.

The humeral capitulum of *P. africanus* is directed as much distally as anteriorly, has a strong mediolateral curvature that is more pronounced laterally, a fairly even anteroposterior curvature, and is separated from the lateral lip of the trochlea by a pronounced groove. The proximal radial head of KNM-RU 2036CE is complete and the dished area for articulation with the central part of the capitulum has a circular margin, surrounded by a flatter articular rim that is more extensive anterolaterally, giving the head an oval outline in proximal view. The head is slightly tilted proximomedially to distolaterally. There was thus good contact and free movement between the joint surfaces during forearm pronation–supination at any position of elbow flexion–extension. Contact would be maximal in full pronation, when the anterolateral area of the radial head would engage the more medial part of the capitulum, the tilt of the head having been effectively eliminated by medial movement of the distal radius. The articular area on the radial head for the proximal radio-ulnar joint extends completely round the head and is proximodistally narrowest dorsolaterally. The corresponding radial notch on the ulna faces anterolaterally. As with the humeroradial joint, the proximal radio-ulnar joint would be most stable in a position of full pronation. This position was thus probably a common working position of the forearm.

The form of the capitulum in other large Miocene hominoids largely conforms to the pattern described above. While the lateral trochlear region shows some peculiarities in the KNM-FT 2751 humerus (Morbeck, this volume, Chapter 14), it is clearly separated from the globular capitulum. These features are even more pronounced in the later Miocene specimens GSP 6663 and 12271, and Rud 53. This may indicate that the radial head was more circular in outline and less tilted than in *P. africanus*. The shape of the radial head of Rud 66, certainly approaches circularity (see Fig. 2D in Morbeck, this volume, Chapter 14). The *P. nyanzae* ulna KNM-RU 1786, the Fort Ternan ulna KNM-FT 3381, and the Klein Hadersdorf ulna all have radial notches similar in form to that of *P. africanus*. The radial notch of Rud 22 faces completely laterally.

As with the shoulder region, morphological patterns of the elbow and forearm regions are not closely matched by that of any group of living higher primates. The humero-ulnar joint of cercopithecoids, especially cercopithecines, show a number of specializations that indicate functional differences. The humeral articular surface is mediolaterally narrow and is buttressed mostly to withstand medially directed forces tending to adduct the ulna. Laterally directed forces are ultimately withstood by the shape of the humeroradial joint and the annular ligament of the proximal radio-ulnar joint (Washburn, 1951). Due to spiraling of the articular surface, flexion is accompanied by a conjunct medial translation of the ulna and attached proximal radius. Extension is limited mostly by the form of the ulnar olecranon beak. Larger cebids show some similarities in this region, especially to the morphology of the ulnar trochlea of earlier Miocene hominoids. It is the living hominoids that share most resemblances with the Miocene hominoids (Morbeck, 1975, 1976, and this volume, Chapter 14). Similarities are least to the earlier Miocene hominoids, where differences in the proportions of the articular surfaces, the degree of mediolateral curvature of the trochlea, and the degree of development of the lateral trochlear lip are evident (McHenry and Corruccini, 1975). Similarities are greatest to the partial humerus, GSP 12271, for features that can be compared (Pilbeam *et al.*, 1980; McHenry and Corruccini, this volume, Chapter 13). These similarities all relate to an extensive flexion–extension range that includes marked extension, with mediolateral stability throughout the range, but especially in extension. The situation with the humeroradial and radio-ulnar joints is more complex. The ability of the radius to spin on the capitulum at different positions of the flexion–extension range of the humero-ulnar joint, the range of pronation–supination possible, and habitual working positions of the elbow and forearm are all involved. In cercopithecoids the morphology indicates that a habitual working position is one of semiflexion at the elbow and full pronation of the forearm. In this position the oval tilted head of the radius is in a close-packed position at both the humeroradial and the proximal radio-ulnar joints. An anteriorly facing radial notch, bringing the radius into a position almost completely anterior to the ulna in pronation, is an additional feature in this complex.

Cercopithecoids differ from Miocene hominoids in the shape of the capitulum, which is flatter distally and medially in cercopithecoids, and in the positioning of the radial notch of the ulna. Larger cebids generally show more similarities than cercopithecoids to the Miocene hominoids. Similarities include a more globular capitulum and a more laterally facing radial notch, as well as details of radial head shape. Activities involving a fully pronated forearm are implied in the morphology of all three groups. Cercopithecoids are additionally specialized for stability at the humeroradial and proximal radioulnar joints in positions of semiflexion of the elbow. The relationships of the orientation of the radial notch of the ulna to pronation–supination range is not clear, but it is certain that the range is relatively great in cebids, where the notch faces anterolaterally, and extensive in hominoids, where it faces laterally (O'Connor and Rarey, 1979). In living hominoids, of course, the circular radial head, completely surrounded by articular cartilage, and the extensive distal radio-ulnar articular surfaces are the main factors relating to this range. The extent of the articular surface for the distal radio-ulnar joint is slightly greater in *P. africanus* than in most cercopithecoids and cebids (Corruccini *et al.*, 1975; Robertson, 1979). The indications are therefore that the pronation–supination range in *P. africanus* was at least as great as in some cebids. The orientation of the radial notch of Rud 22 and the form of the radial head in Rud 66 suggest that the range may have been even greater in those individuals.

One interpretation of some of the features of the shoulder, elbow, and forearm mentioned above is that during the later stages of the stance phase of quadrupedal locomotion in *P. africanus* the forelimb was placed so that the forearm was in full pronation and the semiflexed elbow pointed posterio-laterally, due to medial rotation at the glenohumeral joint. Grand (1968*a*) has investigated this type of forelimb use in *Alouatta caraya*. In later Miocene hominoids the elbow may have been held in a more extended position during the stance phase. All Miocene hominoids could probably place the forearm over a reasonable pronation–supination range during a wide range of elbow positions. There are numerous activities other than quadrupedal progression in which these capabilities might be employed.

Wrist and Hand.   The morphology of the wrist and to a lesser extent the hand of *P. africanus* has received considerable attention. The general consensus is that to the extent that quadrupedal progression was used, that quadrupedalism was palmigrade (e.g., McHenry and Corruccini, this volume, Chapter 13, and references therein). Two points concerning palmigrady are worth discussing. The first relates to the fact that the prime requirement in palmigrady is for adequate contact to be made between the palm and the substrate, and the second relates to the fact that in all types of quadrupedalism the joints of the wrist and hand take up their close-packed positions toward the end of the stance phase, as the propulsive thrust is made against the substrate. In cercopithecoids the palm of the hand does not reach a fully horizontal position when the forearm reaches the end of its limited pronation range. Wrist rotation is necessary to bring about a more horizontal palm position. This rotation mainly involves the trapezoid and capitate in the distal row, and the scaphoid and centrale in the proximal row. Jenkins (1981) has demonstrated that in *Macaca mulatta* at least, the embrasure formed by the trapezoid and capitate is wide dorsally and narrower ventrally. Midcarpal pronation, during which the scaphoid and centrale rotate dorsally, is relatively free. Midcarpal supination is limited by the centrale engaging in the capitate–trapezoid embrasure. During quadrupedal progression passive midcarpal pronation produced by body weight acting during hand placement would effectively complete the rotation started by forearm pronation. Midcarpal supination occurring at the pushoff stage would lock the centrale and bring the "beak-like" process of the scaphoid into contact with the trapezium (Napier and Davis, 1959). This rotation lock would combine with extension locks within the carpus and between the carpus and metacarpus (O'Connor, 1975, 1976). In *Ateles* the midcarpal supination locking mechanism is lacking due to the fact that supination is extensive and is used during suspensory activities (Jenkins, 1981). During quadrupedal activities carpal locking is presumably achieved mainly by close packing in extension.

Schön and Ziemer (1973) have documented such a complex in *Ateles* and *Alouatta*. They point out similarities between it and a locking complex that can be reconstructed for the incomplete KNM-RU 2036 *P. africanus* carpus. Thus, in *P. africanus* at least, rotation required to bring the palm into contact with the substrate could probably have been generated within the forearm, and locking of the carpus could have been achieved mostly by close packing in extension. The presence or absence of an additional supination lock in *P.*

africanus is problematic, as there are no trapezoid or centrale specimens. Evidence from the form of the KNM-RU 2036M *P. africanus* capitate is indirect and equivocal. The GSP 17119 capitate (Rose, 1982) shares some features of head shape with *Ateles* and so provides minimal evidence that a midcarpal supination lock may not have been present. The absence of the "beak-like" process on the KNM-RU 2036Q scaphoid also suggests the absence of a rotatory lock. If relatively free midcarpal rotation was present in *P. africanus,* it would be consistent with the rotatory capabilities suggested above for the shoulder and forearm. It should be emphasized that most groups of living hominoids, suspensory, knuckle-walking, and quadrumanous, possess rotatory capabilities within the carpus, in at least some positions of flexion–extension of the carpus, that are considerably greater than those being discussed here, and that in these groups, as well as in cercopithecoids and various cebids, different patterns of rotatory and extension locking take place (Jenkins and Fleagle, 1975; Lewis, 1977; Jenkins, 1981).

Abduction–adduction within the wrist of *P. africanus* probably took place mostly at the midcarpal joint (Morbeck, 1975, 1977), although Robertson (1979) presents evidence from distal radius and lunate features indicating that there was an adduction (ulnar deviation) set to the hand of *P. africanus.* This is consistent with the possible positioning of the forelimb during quadrupedal activities that has been described above. Napier and Davis (1959) found a mixture of features within the metacarpus and phalanges of *P. africanus,* none of which are inconsistent with grasping in association with arboreal activities, a conclusion also reached by Preuschoft (1973). The New World monkeys are the only group of living higher primates showing similarities to *P. africanus* in the form of the trapezium–first metacarpal joint. The large proximal pollicial phalanx GSP 6664 shows features indicative of a powerful grasping capability. As with the large humeral fragment GSP 12271, it is the large living hominoids that share most features with this specimen.

In summary, the forelimb of the earlier Miocene hominoids is characterized by having moderately well-developed rotatory powers in all segments, a relatively mobile shoulder girdle, an extensive range of movement at the elbow, and a freely mobile hand. The forelimb is well designed for palmigrade quadrupedal activities performed in a fairly flexed-limbed way, and the general mobility suggests that a wide range of other general activities could have been used. There is no evidence of functional features related to highly specialized quadrupedal, suspensory, saltatorial, or other specialized positional activities.

The Hindlimb

Hip and Knee.   The proximal femur shows a number of distinctive features that are common to most of the larger Miocene hominoids. These include a generally quite spherical femoral head with an articular surface that extends onto the neck anteriorly, and especially posteriorly, where there is a smooth transition between head and neck. These features are found in KNM-SO 399 from Songhor, probably from *P. africanus*, BMNH M 16331 (Le Gros Clark and Leakey, 1951), probably a *Proconsul* species, KNM-RU 1753 (Le Gros Clark and Leakey, 1951) of *P. nyanzae*, and GSP 11867 and 13929 from the medium-sized Potwar group. A relatively highly inclined neck is present in the KNM-SO 399, BNMH M 16331, and KNM-RU 1753 femora, and can be inferred for the medium-sized GSP 12654 specimen. The functional implications are that movement was relatively free about all axes at the hip joint, and that abduction and lateral rotation were favored. A tubercle on the posterior aspect of the femoral neck, present on KNM-SO 399, BNMH M 16331, KNM-RU 1753, and the Eppelsheim femur, has also been interpreted, by Walker in Napier (1964), as being related to a lateral rotatory capability.

Information on distal femoral morphology is sparse for Miocene hominoids. The KNM-RU 3709 and KNM-RU 5527 *P. nyanzae* distal femora are both incomplete and distorted. Some indication as to the proportions of the distal femur can be gained from the KNM-RU 2036 CQ *P. africanus* specimen, which lacks the epiphysis, and from the medium-sized GSP 13420 specimen, where some of the supracondylar region is present. The distal end of the Eppelsheim femur shows similarities with KNM-RU 5527. The articular region is fairly narrow mediolaterally, the condyles are subequal in size, and extend distally to the same extent. The patellar groove is relatively wide. Harrison (1982) suggests that for the KNM-RU 5527 specimen this combination of features is at least in part related to a placement of the knees directly beneath the hip joints in the quadrupedal posture and with the hip joints in the neutral position. The incomplete proximal tibial metaphysis of *P. africanus*, KNM-RU 2036CR, generally confirms the proportional features of the *P. nyanzae* distal femur. The angulation of the metaphysial surface in side view indicates that the surface would have been horizontal in a position of semiflexion. These features of hip and knee are consistent with a hindlimb position in which the hip is flexed, abducted, and laterally rotated, and the semiflexed knee points anterolaterally. This position is illustrated by Grand (1968b) for *Alouatta caraya* and is the hindlimb equivalent of the forelimb position during quadrupedal activities described above. A number of studies have concluded that the femur of Miocene hominoids is not clearly matched by that of any living primate (Le Gros Clark and Leakey, 1951; Napier, 1964; McHenry and Corruccini, 1976; Pilbeam *et al.*, 1980; Harrison, 1982). The pattern in cercopithecoids, which, as with the forelimb, relates to parasagittal movement, is similar for the distal femur, while cebids and large living hominoids share features relating to general hip mobility.

Ankle and Foot. The ankle region is relatively well represented for *Proconsul* species. *Proconsul africanus* specimens include a distal fibula, KNM-CA 1834 from Koru; distal tibiae KNM-RU 2036BA and CN and complete tali KNM-RU 2036BF and CO, KNM-RU 1744, and KNM-RU 1745. *Proconsul nyanzae* specimens include distal tibiae KNM-RU 1939 and KNM-RU 5872 and complete tali KNM-RU 1743 (Le Gros Clark and Leakey, 1951) and KNM-RU 1896 (Preuschoft, 1973). *Proconsul major* specimens include a distal tibia from Napak, Napak I 58, currently in the care of the British Museum (Natural History), and a complete talus KNM-SO 389 (Le Gros Clark and Leakey, 1951). Partial tali from later Miocene sites are Rud 27 and GSP 10785b. Features relevant to the talocrural joint include a relatively deep talar trochlea with more or less equally developed medial and lateral lips, with moderate wedging of the articular surface in dorsal view. The lateral trochlear lip is emphasized in Rud 27 (Morbeck, this volume, Chapter 14). The medial and lateral surfaces of the talar body, articulating with the fibular and tibial maleoli, are relatively steep-sided. A distinctly developed, almost globular articular surface on the medio-anterior part of the tibial maleolus is matched by an area on the junction of the medial side of the talar body with the neck.

Details of the action of the talocrural joint in higher primates are complex (Barnett and Napier, 1952; Conroy, 1976; Lewis, 1980a). They center around the degree to which conjunct abduction and inversion of the foot as a whole accompany dorsiflexion and mechanisms by which a stable position in dorsiflexion are achieved. These movements and positions are all involved during the later stages of the stance phase of quadrupedal climbing and walking. The features mentioned above are consistent with a reasonable degree of abduction and not a great deal of inversion (except in Rud 27) accompanying dorsiflexion, and good stability in dorsiflexion, when the trochlea and the maleolar-talar "subjoint" are both fully engaged.

The proportions within the tarsus in different groups of living primates vary considerably. This is reflected in the proportions of individual bones and in the amount of articular contact between neighboring bones. However, in all groups in which arboreal activities are at all important, the mechanisms for making effective contact with the substrate and for stabilization of the tarsus tend to be similar (Lewis, 1980b,c). The virtually complete tarsus of *P. nyanzae*, KNM-RU 5872, and the almost complete tarsus of *P. africanus*, KNM-RU 2036, are used as the basis for the following discussion. Additional specimens include the *P. major* calcanei KNM-SO 390 (Le Gros Clark and Leakey, 1951) and KNM-SO 969 (Preuschoft, 1973), the small GSP 4664 calcaneus, and the medium-sized intermediate cuneiform GSP 6454 and lateral cuneiform GSP 17118 (Rose, 1982).

The calcaneus is relatively elongated, especially in its more distal part. There are confluent facets for the anterior talocalcaneal joint on the sustentaculum, and the axis of the subtalar joint is directed distomedially. The orientation of the facets of the posterior talocalcaneal joint are such that the calcaneus is advanced to make more complete contact with the cuboid during the considerable inversion movement taking place at the subtalar joint. There is extensive contact between the head of the talus and the navicular, especially laterally. The calcaneocuboid joint is characterized by a well-developed ventral beak on the cuboid, which articulates with a correspondingly deep surface on the distal calcaneus and close packs when the forefoot supinates about the hindfoot to complete inversion taking place at the subtalar joint. There are relatively extensive areas of contact between the cuboid and the navicular, and between the calcaneus and the navicular. These contacts allow the navicular to move on the talus during inversion–eversion, and with the cuboid during supination–pronation, and also allow the calcaneus to fully contact the navicular when supination and inversion are combined. This pattern of articulation seems to be a consequence of the relative proximodistal proportions of the tarsus as much as being functionally required.

The three cuneiforms are all moderately elongated proximo-distally. There is extensive contact distally between the medial and intermediate cuneiforms, minimal distal contact between the intermediate and lateral cuniform, and only dorsal contact between the lateral cuneiform and the cuboid. GSP 6454 and 17118 show similar features, except that in each case they are relatively shorter proximodistally than the earlier Miocene specimens. *Proconsul* metatarsals and such phalanges as are known are relatively gracile. The articulation between the metatarsus and the tarsus is relatively mediolaterally narrow. The first metatarsal–medial cuneiform joint indicates that hallux faced as much plantarward as laterally in its neutral position. The medium-sized partial hallux GSP 14046 is more robust than those of *Proconsul* species, and was capable of grasping powerfully.

Lisowski *et al.* (1976), Lewis (1980*a*), and Harrison (1982) have all noted that there are no consistent patterns of similarities between the morphology of the talus in living primates and the tali of Miocene hominoids, although the various authors have emphasized different partial sets of similarities in each case. This inconsistency is evident when comparisons are made to features of the tarsus and metatarsus of *Proconsul* species. The tarsus is relatively proximodistally longer in cercopithecoids and shorter in living hominoids. Cercopithecoids share features of the talocrural joint relating to dorsiflexion stability, but not features relating to the inversion set of the joint. Hominoids are the only group of living primates with a similar calcaneocuboid joint, while both cercopithecoids and hominoids differ from *Proconsul* species in details of the cubonavicular and calcaneonavicular joints, and in the pattern of articulations in the distal tarsus. More similarities are evident between cercopithecoids and *Proconsul* species in the metatarsus and in the digits, although the hallux faces more laterally than plantarward in cercopithecoids, and more plantarward than laterally in living hominoids. There are more resemblances between the hallux and cuneiforms of large living hominoids and the late Miocene Potwar specimens than with the *Proconsul* specimens. However, the more "monkey-like" Rud 27 talus maintains the overall inconsistency of the comparative picture.

Conclusions as to overall hindlimb function are similar to those made for the forelimb of larger Miocene hominoids. General mobility within the limb is high and features relating to very specialized hindlimb function are lacking. The limb is well adapted for arboreal quadrupedal activities during which, as with the forelimb, a fairly flexed limb position may have been used. The morphology is also compatible with the utilization of a range of other general positional activities.

Conclusions and Discussion

The general conclusion of this study, that the functional morphology of Miocene hominoids relates to generalized arboreal quadrupedalism, does not differ from the assessment made by Napier and Davis (1959) for *P. africanus*, or many of the studies referred to above. Rather more material has been considered, and aspects of the function of some regions has been considered in a somewhat different way than in some previous studies. When considered *in toto* the Miocene material shows some interesting features. Size variation between different species at any particular time does not involve major differences in functional morphology. Thus the comparable features of the *Proconsul* species are remarkably consistent. Similarly, although the sample sizes are extremely small and only limited comparisons can be made, the medium- and large-sized Potwar material and the Rudabánya specimens do not show fundamental differences. Similarly, differences between earlier and later Miocene hominoids are mostly of degree rather than of kind. The basic plan upon which these variations were made was evidently a very successful one. This success was based on a generalized morphology underlying generalized capabilities.

The differences in degree between earlier and later Miocene species as shown in the elbow region relate to increased movement in the elbow and forearm and a greater independence of elbow and forearm function in the later Miocene forms. These features are shared with some living apes. Were later Miocene hominoids therefore more "ape-like" than earlier species, and were the earlier species more "monkey-like"? These labels have purposely been avoided in this chapter, except in its title, for a number of reasons. While cercopithecoids are similar to Miocene hominoids in showing a number of variations on a basic morphological plan, and while this plan relates to generalized capabilities, the morphology is nevertheless a relatively specialized one. The morphology is a conservative one, and as it relates to quadrupedal progression, it is expressed in predominantly parasagittal limb movement. Cercopithecoids are thus in a sense "specialized generalists" rather than "generalized generalists." Similarities with Miocene hominoids are greatest in those parts of the limb skeleton that are most generalized, for example, the wrist, and least in specialized regions, such as the humero-ulnar joint. Similarly, living hominoids share features, for example, in the shoulder, elbow, and hip, that relate to general abilities. Thus, in terms of special features Miocene hominoids are neither (cercopithecoid) "monkey-like" nor "ape-like." To the extent that some cebids are morphologically and behaviorally quite generalized, the early Miocene hominoids are more "New World monkey-like" than "Old World monkey-like." While information on the morphology and behavior of living primates is of great importance in understanding functional features of any fossil species, the importance of these features may be lost if labels such as "monkey-like" and "ape-like" are applied to them in any but very clearly defined and limited ways. In some ways it is better to ask how particular living primates resemble Miocene hominoids, as has been done above, rather than *vice versa*.

Many of the studies of Miocene hominoid postcrania have been concerned as much with phylogeny as with function, and in this context the demonstration of features that are, for example, more "monkey-like" than "ape-like" may be important. However, relative similarity does not necessarily correspond to absolute similarity. There are of course very important questions concerning features that are primitive or derived for various groups of catarrhine primates involved with these points, but they are not the main concern of this chapter. Miocene hominoids were a highly successful group, both in time and space, and were primarily "Miocene hominoid-like" rather than like any contemporary group.

ACKNOWLEDGMENTS

I would like to thank D. Pilbeam for making the Potwar postcranial material available to me for investigation, R. E. F. Leakey for permission to study the Miocene hominoid material in the collections of the Nairobi National Museum, P. Andrews for permission to study the Miocene hominoid material in the collections of the British Museum (Natural History), and M. E. Morbeck for providing me with casts of the Rudabánya postcranials.

I have learned much from discussions relevant to this chapter with P. Andrews, J. G. Fleagle, T. Harrison, F. A. Jenkins, Jr., M. Marzke, M. E. Morbeck, D. Pilbeam, and A. C. Walker.

I would also like to thank R. Ciochon, A. Hill, B. Jacobs, L. Jacobs, M. Marzke, M. E. Morbeck, and C. Rose, who have at various times provided encouragement, assistance, and hospitality, or have shown great patience during the course of this study.

The research reported here was supported by a grant from the L. S. B. Leakey Foundation.

References

Barnett, C. H., and Napier, J. R. 1952. The axis of rotation at the ankle joint. *J. Anat.* **86:**1–9.

Conroy, G. C. 1976. Primate postcranial remains from the Oligocene of Egypt. *Contrib. Primatol.* **8:**1–134.

Corruccini, R. S., Ciochon, R. L., and McHenry, H. M. 1975. Osteometric shape relationships in the wrist joint of some anthropoids. *Folia Primatol.* **24:**250–274.

Corruccini, R. S., Ciochon, R. L., and McHenry, H. M. 1976. The postcranium of Miocene hominoids: Were dryopithecines merely "dental apes"? *Primates* **17:**205–223.

Grand, T. 1968a. Functional anatomy of the upper limb. *Bibl. Primatol.* **7:**104–125.

Grand, T. 1968b. The functional anatomy of the lower limb of the howler monkey (*Alouatta caraya*). *Am. J. Phys. Anthropol.* **28:**163–182.

Harrison, T. 1982. *Small-Bodied Apes from the Miocene of East Africa*, Ph.D. Dissertation, University of London.

Jenkins, F. A., Jr. 1973. The functional anatomy and evolution of the mammalian humero-ulnar articulation. *Am. J. Anat.* **137:**281–298.

Jenkins, F. A., Jr. 1981. Wrist rotation in primates: A critical adaptation for brachiators. *Symp. Zool. Soc. Lond.* **48:**429–451.

Jenkins, F. A., Jr., and Fleagle, J. G. 1975. Knuckle walking and the functional anatomy of the wrist in living apes, in: *Primate Functional Morphology and Evolution* (R. H. Tuttle, ed.), pp. 213–227, Mouton, The Hague.

Jolly, C. J. 1967. The evolution of baboons, in: *The Baboon in Medical Research* (H. Vagtborg, ed.), Volume II, pp. 23–50, University of Texas Press, Austin.

Le Gros Clark, W. E. 1959. *The Antecedents of Man,* Edinburgh University Press, Edinburgh.

Le Gros Clark, W. E., and Leakey, L. S. B. 1951. The Miocene Hominoidea of East Africa. *Fossil Mammals of Africa* (Br. Mus. Nat. Hist.) **1:**1–117.

Lewis, O. J. 1977. Joint remodelling and the evolution of the human hand. *J. Anat.* **123:**157–201.

Lewis, O. J. 1980a. The joints of the evolving foot. Part I. The ankle joint. *J. Anat.* **130:**527–543.

Lewis, O. J. 1980b. The joints of the evolving foot. Part II. The intrinsic joints. *J. Anat.* **130:**833–857.

Lewis, O. J. 1980c. The joints of the evolving foot. Part III. The fossil evidence. *J. Anat.* **131:**275–298.

Lisowski, F. P., Albrecht, G. H., and Oxnard, C. E. 1976. African fossil tali: Further multivariate morphometric studies. *Am. J. Phys. Anthropol.* **45:**5–18.

McHenry, H. M., and Corruccini, R. S. 1975. Distal humerus in hominoid evolution. *Folia Primatol.* **23:**227–244.

McHenry, H. M., and Corruccini, R. S. 1976. Affinities of Tertiary hominoid femora. *Folia Primatol.* **26:**139–150.

Morbeck, M. E. 1975. *Dryopithecus africanus* forelimb. *J. Hum. Evol.* **4:**39–46.

Morbeck, M. E. 1976. Problems in reconstruction of fossil anatomy and locomotor behaviour: The *Dryopithecus* elbow complex. *J. Hum. Evol.* **5:**223–233.

Morbeck, M. E. 1977. The use of casts and other problems in reconstructing the *Dryopithecus* (*Proconsul*) *africanus* wrist complex. *J. Hum. Evol.* **6:**65–78.

Napier, J. R. 1964. The evolution of bipedal walking in the hominids. *Arch. Biol.* (*Liege*) **75:**673–708.

Napier, J. R., and Davis, P. R. 1959. The forelimb skeleton and associated remains of *Proconsul africanus*. *Fossil Mammals of Africa* (Br. Mus. Nat. Hist.) **16:**1–69.

O'Connor, B. L. 1975. The functional morphology of the cercopithecoid wrist and inferior radioulnar joints, and their bearing on some problems in the evolution of the Hominoidea. *Amer. J. Phys. Anthropol.* **43:**113–122.

O'Connor, B. L. 1976. *Dryopithecus* (*Proconsul*) *africanus:* Quadruped or non-quadruped? *J. Hum. Evol.* **5:**279–283.

O'Connor, B. L., and Rarey, K. E. 1979. Normal amplitudes of radioulnar pronation and supination in several genera of anthropoid primates. *Am. J. Phys. Anthropol.* **51:**39–44.

Pilbeam, D. R., Rose, M. D., Badgley, C., and Lipschutz, B. 1980. Miocene hominoids from Pakistan. *Postilla* (Peabody Mus. Nat. Hist., Yale Univ.) **181:**1–94.

Preuschoft, H. 1973. Body posture and locomotion in some East African Miocene Dryopithecinae, in: *Human Evolution* (M. Day, ed.), pp. 13–46, Symposium of the Society for the Study of Human Biology, Volume 11, Barnes and Noble, New York.

Robertson, M. 1979. Positional behavior in *Dryopithecus* (*Proconsul*) *africanus*. Paper presented at the 78th Annual Meeting of the American Anthropological Association.

Rose, M. D. 1982. Hominoid postcranial specimens from the middle Miocene Chinji Formation, Pakistan. *Nature* (*Lond.*) (in press).

Schön, M. A., and Ziemer, L. K. 1973. Wrist mechanisms and locomotor behavior of *Dryopithecus* (*Proconsul*) *africanus*. *Folia Primatol.* **20:**1–11.

Washburn, S. L. 1951. The analysis of primate evolution with particular reference to the origin of man. *Cold Spring Harbor Symp. Quant. Biol.* **15:**67–77.

Zapfe, H. 1960. Die primatenfunde aus der Miozänen spaltenfüllung von Neudorf an der March (Děvínská Nová Ves), Tschechoslowakei. *Schweiz. Palaeontol. Abh.* **78:**1–293.

Evidence from Paleoenvironmental Studies

V

Sequence and Environments of the Lower and Middle Miocene Hominoids of Western Kenya

16

M. PICKFORD

Introduction

A widespread hypothesis in anthropological science is that at some stage in the evolution of humans there must have been an adaptive shift from an arboreal forest existence to a terrestrial savannah one. The reasoning behind this idea seems to be that most extant apes are forest-dwelling, leading to the conclusion that this is probably the primitive (plesiomorphic) hominoid adaptation. Humans, in contrast, are adapted to savannah settings, and this represents a highly derived (apomorphic) condition among hominoids.

Because the hypothesis was erected using extant situations, paleoanthropologists are unable to provide adequate tests using data relating to modern conditions. Therefore, in order to test the hypothesis, the fossil record must be examined. Such a testing program requires several basic types of information: (1) the sequence in which hominoids lived, (2) their environment, and (3) the pathways by which they evolved. Even so, a number of assumptions must be made, the most important of which is that the fossil record is repre-

M. PICKFORD • National Museum of Kenya, P. O. Box 40658, Nairobi, Kenya.

sentative of what actually happened in the past. In a continent such as Africa, the fossil record of large areas is too incomplete or nonexistent for the paleoenvironment to be both adequately and uniformly reconstructed.

The available data largely relate to the fossil record of East Africa, since this is the only part of the continent where a long and continuous record of Neogene strata (containing hominoids) exists. West Kenya, including Baringo, has the most complete record of strata from late Oligocene (23 m.y.a.) to Recent times and is the only one in which such a long sequence of hominoid species can be demonstrated on superpositional grounds.

Lithostratigraphy and Biostratigraphy

Until recently only two chronological faunal groupings were recognized in the Miocene of West Kenya. The most persistent grouping was the "lower Miocene" one, comprising a supposed unit fauna labeled "Rusingan" by Van Couvering and Van Couvering (1976), but having its conceptual roots in Kent's (1944) Miocene Lake. In this grouping were placed Rusinga (as a kind of typical fauna), Songhor, Koru, Karungu, Chianda, Mfwangano, Ombo, and Maboko.

The second and younger grouping was recognized in the early 1960s with recovery of a varied fauna from Fort Ternan. In the reassessment of Miocene sites triggered by the discovery of Fort Ternan, Maboko was pulled out of the "Rusinga-like" assemblage and moved upward toward Fort Ternan. It is interesting to note, however, that as early as 1948, Hopwood was claiming a status of "Helvetian" (middle Miocene) for some of the Maboko proboscideans. But so pervasive was the "unit fauna" hypothesis that Hopwood's and other data were either ignored or put down to haphazard collecting or to the presence of an undetected unconformity in the Maboko succession. This kind of juggling was in many ways a consequence of the sketchy regional mapping techniques employed before 1967. The real basis for superposition began with Van Couvering's (1972) work on Rusinga and the regional mapping in Baringo District by the East African Geological Research Unit (Bishop *et al.*, 1971). This work has continued with the recent mapping in Tinderet (Pickford and Andrews, 1981). Details concerning the mapping of the major sequences are available elsewhere (Pickford, 1981) and a summary of the main points of concern to paleoanthropologists is given here.

Results presented here draw upon collections made as long ago as 70 years and as recently as this year. Some 30,000 fossil mammals from over 100 discrete fossil localities have been examined (see Fig. 1). They derive from three geologic provinces, the oldest in the Tinderet region (Pickford and Andrews, 1981), the second in the Rangwa area (Van Couvering, 1972), and the third associated with the Plateau Phonolite Group (King and Chapman, 1972). Six distinct faunal assemblages are consistently recognizable in the three provinces, and these have been termed "faunal sets," numbered from I

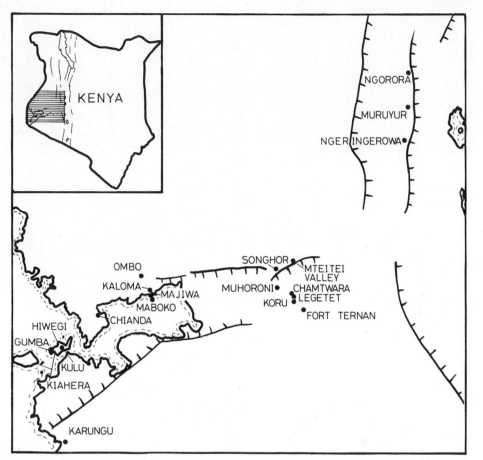

Fig. 1. Index map of lower and middle Miocene collecting localities in West Kenya. Geological boundaries of the Kenya Rift Valley System indicated by hachured (fault) lines.

(oldest) to VI (youngest) (Pickford, 1981). Together, the six sets span much of the lower and middle Miocene and all have yielded hominoids and/or monkeys.

The bare framework of the biostratigraphy of West Kenya is presented in Table 1; the details are available in a separate publication (Pickford, 1981; see also Fig. 2, this Chapter). Representation of hominoids and other primates within the West Kenya sequence is sketchy. Hiwegi, Songhor, and Chamtwara contain large samples of hominoids, while most other units have yielded less adequate samples. It is perhaps significant that most units yielded new hominoid taxa when first collected. It seems that the chances are high that a new locality in a new age bracket will yield a new taxon. If this is in fact the case, then the total diversity of hominoids in the East African Miocene record is a long way from being fully appreciated.

An important feature is the existence of a calibrated sequence spanning significant portions of the Miocene. The most advanced hominoids in the

Table 1. Temporal Distribution of Miocene Primates in West Kenya

Faunal set	Age, m.y.a.	Stratigraphic units	Hominoidea		Dominant spp.	Common spp.	Rare spp.
			Specimens	Taxa			
VI	10.5–8	Ngeringerowa	0	0	—	—	Monkeys
V	12–10.5	Ngorora	4	2	—	—	Ramapithecine, small-bodied apes, monkeys
IV	14.5–12	Fort Ternan[a]	31	4 or 5	*Pliopithecus*, *Ramapithecus wickeri*[a]	—	*?Proconsul* sp.
		Muruyur	1	—	—	—	*?Proconsul* sp.
III	16–14.5	Moboko Formation[a]	33	4 or 5	Monkeys	Ramapithecine two spp. (*Sivapithecus africanus*[a], *Ramapithecus* sp.)	*?R. vancouveringi*, *?L. legetet*, *?Proconsul nyanzae*

II	18.5–16	Kulu Formation	4	2	—	—	D. macinnesi, P. nyanzae
		Hiwegi Formation[a]	319	5 or 6	D. macinnesi, P. africanus, P. nyanzae[a]	—	L. legetet, ?R. gordoni
		Kiahera Formation	17	5	D. macinnesi, P. africanus, P. nyanzae	—	?R. gordoni, R. vancouveringi
I	20–18.5	Songhor[a]	252	7	Songhor-type assemblages, Taxon A, R. gordoni[a]	P. major[a], D. songhorensis[a]	R. vancouveringi, L. legetet, D. macinnesi
		Chamtwara[a]	280	6	Chamtwara-type assemblages	—	—
		Legetet Formation	92	5	L. legetet,[a] M. clarki	D. macinnesi, D. songhorensis	P. africanus[a] (Xenopithecus)[a]
		Koru Formation	33	6	—	P. major	—
Pre-I?	?23	Muhoroni agglomerates	11	1	—	?P. major or ?P. nyanzae	—

[a]Type locality for taxon.

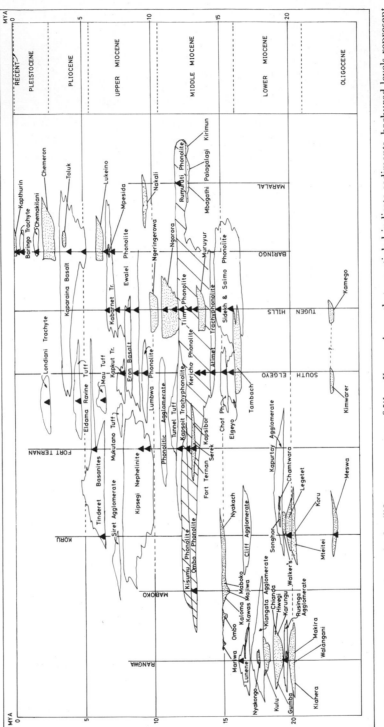

Fig. 2. Correlations between the main fossiliferous sequences of West Kenya. Areas that are stipled indicate sediments, hachured levels represent plateau phonolites, and black triangles show position of radiometric dates.

lower part of the succession consist of various *Proconsul* species, whereas in the middle portions they are ramapithecines. In the Plio-Pleistocene come australopithecines and *Homo*. Although the details of the transitions require considerably more attention, the sequential basis exists upon which these studies can proceed.

Temporal Relationships of Ramapithecines

There is virtually no chronological overlap between the East African ramapithecine sample and that from the Indian subcontinent [Ngorora (one specimen) and Chinji (about ten specimens) are the closest in time]. The Majiwa/Fort Ternan ramapithecine samples are separated by some 5–6 million years from the main Potwar sample (Pilbeam *et al.*, 1977).

It is also apparent that there is not much sample overlap between proconsulines and ramapithecines as now known. There is an appreciable decrease in the number of proconsuline specimens after Rusinga (Hiwegi). Although they may persist until Fort Ternan, in all collections from Faunal Sets III, IV, V, and VI, there are fewer proconsulines than ramapithecines. Prior to set III, no ramapithecines have yet been recognized in West Kenya.

Depositional Environments

In West Kenya, the study of relationships between faunal assemblages and the depositional environments in which they are found reveals consistent well-developed patterns. In general terms, "watery" depositional environments (lake, lake margin, swamp, wet and dry floodplain) yield a large proportion of forms such as *Brachyodus, Hyoboops, Diamantohyus, Prodeinotherium,* and Rhinocerotidae, while "dry" ones (subaerial) yield most or all specimens of taxa such as *Bathyergoides, Afrocricetodon, Prochrysochloris, Rangwapithecus,* and *Dorcatherium songhorensis*. There are, of course, some wide-ranging species (*Diamantomys, Paraphiomys, Miohyrax, Walangania*) which occur at most localities, but even these seem to be more common in selected localities (*Paraphiomys* and *Miohyrax* in wet floodplain; *Walangania* and *Diamantomys* in drier subaerial situations). Similarly, primate species occur preferentially with particular depositional environments (Table 2).

The determination of depositional type is strictly a sedimentological exercise. Some revision of published ideas concerning sedimentary facies has been necessary, since almost all previous concepts assumed deposition in or near lakes (Kent, 1944; Shackleton 1951; Whitworth, 1953; Oswald, 1914). It is now evident that in proximity to positive relief settings, such as occurred at Tinderet and Rangwa, most of the eruptives and sediments accumulated above local base level. Local tectonic influences (doming, paravolcanic basins, fault-bounded basins, and volcanic dams) may have given rise to basins in which sediments accumulated, but these were often short-lived and rapidly filled.

At Tinderet, most fossils occur in residual soils developed on subaerial tuffs, agglomerates, and carbonatite extrusives. The Rangwa fossil localities are mostly in reworked volcanogenic material predominantly in floodplain settings (e.g., R1–3, R5). Minor lacustrine facies are present in the sequence [Gumba, Kulu (partim)], but yielded few of the mammals. In addition, a lake margin facies is recognized on Rusinga, where paleotopographic considerations suggest that "finger lakes" (drowned river valleys) with steep walls were receiving coarse to fine debris around their edges (in the Kulu Formation). The bulk of the Rangwa succession is pyroclastic, however. At Karungu, the sediments are of fluviatile origin with significant paludal and overbank elements. There is little sedimentary evidence to suggest lacustrine conditions, although algal coatings on pebbles indicate standing water at intervals. Soil formation occurred at Karungu, and several "handpans" with rootcasts are preserved. The main fossiliferous sediments at Karungu probably constitute a wet floodplain facies with localized channel, overbank, and swamp environments.

The Maboko Formation of middle Miocene age also represents floodplain deposition. The three outcrops, at Maboko, Majiwa, and Kaloma, record interesting facies variations. Part of the Maboko sequence accumulated close to surface water since it yields crocodile and abundant chelonian remains (*Trionyx*, Pelomedusidae) in silts and clays. At Majiwa there are few chelonians. In contrast to Maboko it yields abundant terrestrial gastropods. At Kaloma, further to the north, the sequence contains four to six well-developed 0.5-m-thick calcretes containing abundant terrestrial gastropods and few aquatic specimens. The preliminary analysis is that the three areas preserve different surface facies in a single floodplain sequence. The main fossil beds at Maboko were probably swamp to riverside; Majiwa started out as lateral to the channel at the edge of the calcrete field, but became close to the meander axis in the upper half of the sequence. Kaloma seems always to have been lateral to the main channel system and developed a fine sequence of superposed stage 4 calcretes (Leeder, 1975). The possible role of carbonatite activity at Homa and Ruri in the development of calcretes at Majiwa has yet to be ascertained, but if Leeder's data are applicable in the Maboko Formation, the Kaloma area underwent infrequent sediment accretion (minimum 10,000 years) compared with areas closer to the meander belt (i.e., Maboko and Majiwa). However, the Kaloma stage 4 calcretes are considerably thinner than any mentioned by Leeder and are closer to those in the Serengeti area described by Hay and Reeder (1978), where carbonatites provide a very high ambient concentration of carbonate.

At Fort Ternan deposition of the fossil beds seems to have been in fluviatile to subaerial settings, and there is a strong paleosol overprint.

Table 2 summarizes the depositional environments and their primate assemblages. The rather prominent pattern between the species and depositional categories suggests that fossils found in each locality were derived from

Table 2. Depositional Environments of Hominoid Material: Western Kenya

			Specimens				
			Flood plain				
	Lake	Lake margin	Wet	Dry	Channel	Subaerial	Total
P. africanus	2	5	37	42	1	18	104
P. nyanzae	—	5	24	44	—	—	73
P. major	—	—	—	—	2	83	85
R. gordoni	—	—	—	3	—	71	74
R. vancouveringi	—	—	1	5	—	3	9
D. macinnesi	1	5	29	89	—	23	124
D. songhorensis	—	—	—	—	—	46	46
L. legetet	1	1	1	8	—	121	132
M. clarki	—	—	—	—	—	79	79
Taxon A	—	—	—	—	—	103	103
Pliopithecus sp.	—	—	—	—	—	20+	20+
Small ramapithecine	—	—	—	5	—	—	5
Large ramapithecine	—	—	7	3	—	5	15
Monkeys	—	—	100	2	—	—	102

areas within the bounds of that facies (i.e., local) and that mixing of fossils from different areas was not a significant factor. This, in turn, suggests that most individuals died and presumably lived within the limits of that facies.

The distribution of primates *vis-à-vis* the depositional environments is of much interest (Table 2), and as with other mammals, some seem to have been restricted in their depositional types, while others were apparently more widespread.

There is a prominent difference between floodplain and subaerial assemblages. In general, larger bodied species are dominant in the floodplain deposits, while smaller bodied species predominate in the subaerial localities. *Proconsul major*, however, the largest of the hominoids, occurs only in the subaerial category. Ramapithecines are most common in floodplain deposits, but are known also in subaerial deposits at Fort Ternan. Monkeys make their earliest appearance in the local record in floodplain deposits at Maboko.

Paleoenvironments

The sequence of fossil deposits allows time-related studies to be made on faunal groups associated with the hominoids. Examination of gastropod assemblages is of particular relevance to those interested in documenting time-related environmental changes, since the Miocene species are close to, if not actually similar to, living species (Verdcourt, 1963). Detailed studies of the molluscan assemblages have begun, but were initiated so recently that it would be premature to delve too deeply into them at this stage. Instead only one aspect of molluscan distribution will be examined, that dealing with co-occurrence between selected taxa of molluscs and hominoids. For this particular study, efforts are confined to easily recognizable gastropod forms in order to minimize identification errors. Forms which are ubiquitous in their habitat tolerances, such as *Homorus,* are excluded from this study since they occur at virtually every locality. They will, of course, be examined in future research programs.

Verdcourt (1963) has suggested that many of the Miocene gastropods are so similar to the modern ones that they may have lived in rather similar habitats. For the examples of taxa used in this study a paraphase of Verdcourt yields the information shown in Table 3. (1963, pp. 32-ff).

Co-occurrence indices for hominoid/gastropod pairs have been calculated for West Kenya localities (Table 4). These indices are a rough guide to the frequency with which two taxa occur together. Table 4, for example, shows that *Proconsul africanus* occurs rarely with the snail *Lanistes* (index of 15%), but more commonly with *Maizania* (73.9%). *Micropithecus clarki,* in contrast, occurs only with the *Gonaxis* group and *Primigulella* (67% and 63% indices, respectively) and has never been recorded with *Maizania, Ligatella,* or *Lanistes.* The fact that co-occurrence is repeated over a number of localities indicates that accidental or *postmortem* effects cannot wholly account for these patterns. All Set I and II localities were counted in calculating the co-occurrence indices.

The Maboko and Fort Ternan gastropod assemblages are either deficient in or have few individuals of the snail genera listed in Table 4. This in itself is of some interest since their absence or paucity need not be merely due to selective preservation.

The gastropods from the Maboko Formation consist of abundant *Limicolaria, Cerastua,* and *Edouardia,* with fewer *Burtoa, Trochonanina,* and *Homorus* and traces of *Gonaxis protocavalii, Pila ovata,* and *Gulella leakeyi.* According to Verdcourt (1963), this sort of assemblage would be indicative of dry woodland (26 in. ≈ 660 mm mean annual rainfall). The mammalian fauna and paleopedology do not contradict this suggestion, and a wooded but relatively dry habitat for Maboko is accepted here as being representative of the conditions during deposition. The sediments accumulated as part of a floodplain sequence and contain many of the sedimentary and pedological facies usually found in floodplains. The main mammalian assemblages derive from overbank clays,

Table 3. Verdcourt's Data on Habitat Preference of Some Gastropoda

Taxon	Habitat	Climate
Lanistes	Rivers, small lakes, swamps	In dry areas
Pila	Common in lakes and rivers of Nile system	—
Maizania	Drier forests	600–1140 mm rain
Ligatella	Savannah and bushland	Up to 1020 mm rain
Gonaxis group	Rain forest	1270–1520 mm rain
Primigulella	Rain forest	1270–1780 mm rain

Table 4. Co-occurrence Indices: Lower Miocene Hominoidea and Gastropoda

	Bivalves	*Lanistes*	*Pila*	*Ligatella*	*Maizania*	*Gonaxis* group	*Primigulella*
P. africanus	38.2	15.8	12.9	59.3	73.9	42.3	27.9
P. nyanzae	21.6	56.8	45.3	58.9	76	42.9	16.4
P. major	0	0	0	22.8	27.1	77.1	76.4
R. gordoni	8.9	8.3	0	68.6	78	61.3	19.1
D. macinnesi	50	45.7	45.2	64.1	58.3	47.5	14.4
D. songhorensis	0	0	0	12.6	7.7	76.4	90
L. legetet	0	17.1	13.4	32	42	60	42
M. clarki	0	0	0	0	0	67	63
Taxon A	0	17.1	13.4	66	79	30	30

silts, and calcretes, but channel conglomerates are rich in reworked and rolled fossil remains. It is perhaps significant that fossil grass is preserved at Majiwa, the earliest local record of the group. A poorly developed swamp facies yields crocodiles, turtles, and aquatic snails (?*Lanistes* and *Pila*).

The Fort Ternan snail assemblage differs from the Maboko suite in the dominance of *Homorus* over all other genera. *Gulella leakeyi* is poorly represented, as are *Burtoa* and helicoid snails (cf. *Tayloria*). In addition, *Maizania* and *Helicarion* are rare at Fort Ternan. This assemblage would be most likely to occur in woodland to dry forest with a mean annual rainfall between 26 and 60 in. (660 –1520 mm) according to Verdcourt (1963). The suggested environment is more wooded and wetter than is felt to be the case at Maboko. The Fort Ternan faunule accumulated subaerially but was apparently influenced by surface water activity. No aquatic elements have been found at the locality. (Reports of crocodiles are based in all cases on carnivore lower canines, incompletely formed, and the supposed bivalves are either algal or of abiogenic origin.) The mammalian fauna is compatible with the environment suggested by the gastropods (Gentry, 1970; Andrews *et al.*, 1979).

Woodland environments at Rusinga are suggested by some of the gastropod species as well as by the dung beetles (Coleoptera) studied by Paulian (1976). This evidence is at variance with some previous ideas based on incomplete studies of fossil plants (Chesters, 1957; Andrews and Van Couvering, 1975).

Discussion

Of the many hypotheses and their variations dealing with early developments in human evolution, there is one in particular which stands out. While the fossil record cannot provide definitive proof of this or any other hypothesis, it nevertheless supports it to a certain extent. This hypothesis is based on the assumption that the primitive condition among hominoids was an adaptation to life in forested environments, while the more advanced or derived condition is represented by an adaptation to life in open woodland and grassland.

A glimpse at the only fossil record complete enough to throw light on the matter is preserved in East Africa and reveals in this one case that a progression through time from primitive forest-adapted hominoids toward more advanced savannah-adapted populations did take place within a relatively small geographic area. Furthermore, in intermediate periods and environmental settings, an intermediate condition in hominoid evolution is suggested by the fossil record. In the absence of comparable information elsewhere, especially as regards a complete range of environmental samples through time, an opinion concerning the details of hominoid history is unable to be completely formulated. The sample of late Neogene forested environments is nil, likewise for the samples of lower and middle Miocene grassland ecosystems. There are therefore tremendous gaps in the sample, of such a nature that with present knowledge, an adequate test for the hypothesis cannot be provided. Nevertheless, a pragmatist might argue that the chances that the only available sample is wildly in error are minimal.

Briefly, the evidence available in Kenya is as follows. In the lower Miocene, a large number of localities sample forested to woodland environments of several types. All these samples are categorized by diverse primate faunas covering a wide spectrum of body sizes and adaptations. By middle Miocene time, the indications are that woodland and bushland had become established in the same general areas, a change accompanied by a great reduction in the diversity of hominoid taxa, as well as a reduction in their population density (as fossils) relative to other mammals (micromammals excepted). Finally, after a long period for which there is little evidence, the fossil record preserves a large number of Plio-Pleistocene savannah-oriented assemblages. In these the diversity of Hominoidea is small, although cercopithecoids appear to be more diverse than at earlier points in time. The hominoid samples from these three ecosystems do indeed show a relative progression through time from "primitive" forest-adapted "apes" to an "advanced" open-country adaptation displayed by australopithecines and humans. But so does a completely modern sample based on extant Great Apes and humans. The only thing missing from the modern sample is an intermediate form of hominoid living in woodland to bushland, comparable to the middle and upper Miocene ramapithecines. This feature alone lends credence to the hypothesis and imparts to it a time perspective. But in the presence of large gaps in the information, the precise timing of events is difficult to obtain. One way of improving the outlook is to concentrate our research efforts on middle and upper Miocene deposits, preferably in East Africa, so that geographic factors may be kept to a minimum.

The preliminary results presented here reveal the extent to which the knowledge concerning the temporal and environmental contexts of Miocene hominoids has progressed. The sequence of localities has only recently been satisfactorily determined. The environments of deposition of many localities has been reviewed, indicating that early reconstructions, based essentially on

Kent's (1944) Miocene Lake hypothesis, are incorrect. Instead, a wide range of depositional types is preserved of which the lacustrine facies is but an insignificant proportion. The distribution of mammalian and other taxa within the various depositional types reveals such a consistency of pattern over many individual fossil concentrations that co-occurrences due to *post mortem* chance alone are considered to be insignificant. Considerations of entropy suggest that the major factor in the origin of the consistency of these distribution patterns is that the animals actually lived within the bounds of the areas undergoing each particular kind of deposition. The West Kenya sample of some 100 separate fossil concentrations shows remarkably few examples of "mixing" of faunal elements of different types. There are, of course, several wide-ranging taxa found ubiquitously in many sites, but in general, fossil assemblages from foodplain deposits are readily distinguishable from faunules derived from subaerial settings. Similarly, swamp assemblages differ widely in character from both of these.

Finally, the study of terrestrial gastropods found at the same localities as hominoids allows a reconstruction of some of the past environmental conditions. These results are based primarily on a knowledge of modern gastropod assemblages collected from a number of habitats in Kenya allied to the observation (Verdcourt, 1963) that the Miocene species are so similar in shell shape to modern snails that they probably survived under comparable conditions. Ideally there would be more reliance on the overall aspect of controlled collections of gastropod assemblages than on a few "*indicator*" taxa, but many of the old collections do not possess adequate provenience data. The raw material has, in effect, been placed out of focus, and until replacement collections are made, only second-rate material will be available for interpretation. Nevertheless, some feeling of viability and importance for the assemblages can be obtained, a sense given confidence by the repeatability of the results. Repetition of assemblages observed in a number of recent controlled collections from several localities is comparable to that observed in the case of depositional environments. Co-occurences between selected taxa of snails and hominoids are remarkable for their consistency. Four types of lower Miocene snail assemblages occur consistently at many localities with four assemblages of hominoids (as well as with other mammals). Examples of hominoid taxa occurring with the "wrong" snail assemblages are statistically insignificant. These taxa comprise the "trace" elements in Table 5, likewise for "misplaced" gastropods relative to hominoid faunules. These assemblages are therefore considered to be relatively uncontaminated biocoenoses, despite some evidence of *postmortem* disturbance.

In essence these results modify those of Andrews (1978) and Andrews *et al.* (1979). who have successively proposed that *Dendropithecus macinnesi, Limnopithecus legetet, Proconsul major,* and *Rangwapithecus gordoni* lived in forest, while *Proconsul africanus, P. nyanzae,* and *Ramapithecus wickeri* lived outside forest. Finally, they postulated that *Australopithecus* and early *Homo* species lived in grassland to woodland–bushland settings.

Recent taxonomic revisions based on greatly improved samples of homi-

Table 5.   Suggested Ecological Preferences of Hominoidea in Western Kenya based on Associated Gastropod Assemblages

	Dominant spp.	Common spp.	Rare spp.
A. Wet Rain forest (1270–1780 mm) *Primigulella, Gonaxis* group Koru, Legetet, Chamtwara Subaerial	*L. legetet,* *M. clarki*	*P. major,* *D. songhorensis,* *D. macinnesi*	*P. africanus*
B. Drier forest (up to 1020 mm) *Primigulella, Ligatella* Mteitei Valley, Songhor Subaerial	Taxon, A, *R. gordoni*	*P. major,* *D. songhorensis*	*R. vancouveringi,* *D. macinnesi,* *?L. legetet,* *P. africanus,* *?P. nyanzae*
C. Dry forest (600–1140 mm) *Ligatella, Maizania* Hiwegi, Kiahera Floodplain and channel	*D. macinnesi,* *P. africanus,* *P. nyanzae*	—	*L. legetet,* *R. vancouveringi,* *R. gordoni*
D. Woodland to dry forest (660–1520 mm) *Maizania, Helicarion, Burtoa* Fort Ternan Subaerial	*R. wickeri*	*Pliopithecus* sp.	*Proconsul* sp. or spp.
E. Dry woodland (up to 660 mm) *Limicolaria, Burtoa, Cerastua* Maboko, Majiwa, Kaloma Floodplain	Cercopithecoidea	Ramapithecine small and large spp.	Small-bodied ape
F. Swamp *Lanistes, Pila,* bivalves Gumba, Karungu, Chianda, Ombo	—	—	*P. nyanzae, D. macinnesi*

noids from the Koru area have led to a reassessment of the diversity of lower Miocene hominoids, in particular the smaller species, and to a revised conception of the distribution of each species. The latter is based mainly on a reshuffle of material between taxa as identifications of specimens have changed since the work of Andrews (1978).

According to Harrison (1981), the hypodigm of *D. macinnesi* has had to be split into two, with the subspecies *D. m. songhorensis* differing from the type subspecies at least at the specific level. Much of the hypodigm of *L. legetet* has now been shown to represent a different taxon (Taxon A), a feature which was not observable until the amplified Koru sample was made available. Finally, *Micropithecus clarki* is now known in large numbers from selected sites in

West Kenya. The position of the large hominoids has not changed greatly since Andrews (1978), although the new Koru sample has amplified our ideas on the distribution of certain species.

The suggested environmental preferences of the lower Miocene hominoids is summarized in Table 6.

The earliest known large sample of monkeys (from Maboko) is from a wet floodplain setting. Although no gastropods occur in direct association with this assemblage, the snails from the nearby sites of Majiwa and Kaloma, higher in the succession, are of open-woodland, low-rainfall affinities. This is the earliest open-country, low-rainfall community found in West Kenya, and it is perhaps significant that it yields also the earliest known ramapithecines.

Although recent studies have placed the sequence of events in West Kenya on a firmer foundation, no major revision of previously published orders has been necessitated. Recently completed and future radiometric dating programs will refine placement of sites on the temporal scale in West Kenya and may result in movement of certain sites up or down the scale, but no inversions should occur, since the order is based largely on superpositional knowledge allied with biostratigraphic data.

One result of the recent mapping in the Tinderet area has been the identification of two slightly different habitats which occurred in close proximity over a substantial period of time, during which the constituent faunas of each remained relatively constant, with little if any discernible mixing of faunas. All the localities in the Koru, Legetet, and Chamtwara Formations contain wet forest assemblages of snails. The associated mammalian faunas remain alike throughout the sequence. In the Mteitei Valley—Songhor area just 11 km to the northwest, in deposits mapped as lateral equivalents of the Koru and Chamtwara Formations, there occur gastropod assemblages with dry forest constituents (*Ligatella*, rare *Maizania*) among others. The hominoid assemblage at Songhor is markedly different from its stratigraphic equivalent at Chamtwara, in the presence of *R. gordoni* and Taxon A, both unknown at

Table 6. Suggested Environmental Preferences of Lower and Middle Miocene Hominoids in West Kenya

L. legetet	Wet rain forest in subaerial settings
M. clarki	Wet rain forest in subaerial settings
P. major	Wet to drier forest in subaerial conditions
D. songhorensis	Wet to drier forest in subaerial conditions
R. gordoni	Drier forest in subaerial conditions
D. macinnesi	Dry forest to woodland in floodplain settings
P. africanus	Dry forest to woodland in floodplain settings
P. nyanzae	Dry forest to woodland in floodplain settings
R. vancouveringi	Too few specimens
Ramapithecinae	Woodland to dry forest in subaerial and floodplain settings (small locality sample)
Pliopithecus sp.	?Woodland to dry forest (small locality sample)
Taxon A	Dry forest in subaerial settings

Chamtwara. Conversely, *L. legetet* and *M. clarki* occur at Chamtwara in large numbers, but are not yet reliably reported from Songhor (Harrison, 1981). The only taxa relatively common in both areas are *P. major* and *D. songhorensis*, with rarer *D. macinnesi*. The little that is known about Mteitei Valley conforms with Songhor, although the primate sample is too poor for the purposes of conviction.

The Rusinga (Hiwegi) hominoid assemblage, which on balance is drawn from drier and more open forest to woodland (Paulian, 1976) than appears to be the case at Songhor and Koru, is dominated by species of greater body weight than are the two latter faunules. The most common Rusinga "small-bodied ape," *D. macinnesi*, is larger than any of the small-bodied apes of the Songhor–Koru area. It should be noted, however, that the largest lower Miocene hominoid species of all, *P. major*, occurs in the company of a diverse fauna of small-bodied species. There are surely some interesting ecological factors to be explored here.

Summary and Conclusions

The fossil record of the Nyanza Rift Valley and Baringo in Kenya permits examination of the hypothesis that the evolution of the Hominidae was in response to the occupation of nonforested environments by hominoids some time during the latter part of the lower Miocene (ca. 17–16 m.y.a.). Most fossil hominoids found in East Africa immediately prior to this period are associated with faunal and floral remains suggestive of wet to dry forested environments, with only limited woodland components present. In contrast, most sites in the middle Miocene strata of Kenya contain fossils suggestive of more open vegetation types and drier climatic conditions. Therefore, in West Kenya, there was an opening up of the countryside accompanied by a reduction in mean annual rainfall at the end of the lower Miocene.

Almost all the hominoids recovered from sites in West Kenya older than 17 million years have "thin enameled" molars. Many of the sites younger than 17 million years contain hominoids with "thick" molar enamel, but also present are some that have "thin" enamel.

From these general observations, it appears that an adaptive radiation of hominoids in which molar enamel thickness increased (in conjunction with other changes in morphology) began at approximately the same time that major changes in the environment occurred. Therefore, it is postulated that the adaptive radiation of hominoids was in some way a response to the environmental changes. In deposits that accumulated under more closed and wetter conditions discussed in this chapter, there is no (or little) evidence of hominoids involved in this particular adaptive radiation (i.e., increase of enamel thickness and allied changes). The sample, both of sites (60+) and specimens (916+), is impressive, suggesting that the absence of thick enamel is not likely to be due to vagaries of the fossil record. In deposits that accumulated under drier, more open conditions, thick enamel occurs, although there is a far more incomplete fossil record to examine (10+ sites and 69+ hominoids, of which 21 are thick-enameled). In West Kenya, the available evidence strongly supports the view that "thick enamel" in hominoids and open environments certainly coincide. Hominoids with thin molar enamel are found in sites of open as well as closed environmental types. They are, however, considerably more rare in the latter sites than they are in strata that accumulated in or near forests. In the former, hominoids seldom comprise more than 1% of macromammalian specimens, while in the latter they often comprise up to 30% of the recovered specimens.

In West Kenya, in a restricted area (80 km × 30 km), the fossil record indicates that the regional environment which, in the lower Miocene was predominately forested, changed until by the middle Miocene it was generally wooded and drier than it had previously been. Whether hominoids with "thick molar enamel" evolved autochthonously as a result of these environmental changes or whether they translocated into the area from elsewhere as conditions became suitable for their particular requirements cannot as yet be resolved. The understanding of the continental fossil record is still too poor to indicate which model (environmental change or immigration) is correct, but the sites at Buluk (Harris and Watkins, 1974) and elsewhere in North Kenya (Moruorot, Loperot, Losidok) may yield crucial evidence, especially if their placement in the lower Miocene is sustained. If these sites, some of which contain ramapithecines, are indeed of lower Miocene age (17–17.5 m.y.a. or older), then the West Kenyan ramapithecines may realistically be viewed as immigrants, and their arrival in West Kenya as coincident with the development of the appropriate niches for them. It should be stressed, however, that the sample of very early ramapithecines (Maboko, Majiwa, Kaloma, Moruorot) suggests that a morphological shift away from their putative ancestors (*Proconsul, Rangwapithecus,* or some similar form of hominoid) had only just taken place. Under these circumstances, it is suggested that the adaptive radiation of which *Ramapithecus* was a part would be unlikely to have begun any earlier than 17.5 m.y.a.

These, however, are merely hypotheses which require much further examination. The fossil record has, in effect, given only a glimpse of the possibilities. Certainly, a better understanding of the paleoenvironments, the hominoids, and the sequence and detailed timing of events during the lower and middle Miocene is needed in order to reconstruct adequately the evolution of the Hominidae.

ACKNOWLEDGMENTS

I wish to thank the office of the President, the Attorney General's Chambers, and the Director of the Kenya National Museum for permission to study in Kenya. This research was financed by the Boise Fund, the L. S. B. Leakey Foundation. and the Foundation for Research into the Origins of

Man. I wish particularly to thank N. Brooks of Koru for much help and encouragement in the field. Many people have helped at one time or another in the field and in the laboratory, and to them I extend my thanks. T. Harrison allowed me access to his preliminary findings in the revision of small-bodied apes currently being undertaken. Without this information the results of this chapter would have been different, especially with regard to the meaning of the Koru and Songhor assemblages. Final thanks go to Kiptalam Cheboi, my field assistant at Koru and Baringo.

References

Andrews, P. J. 1978. A revision of the Miocene Hominoidea of East Africa. *Bull. Br. Mus. (Nat. Hist.) Geol.* **30**(2):85–224.

Andrews, P. J., and Evans, E. N. 1979. The environment of *Ramapithecus* in Africa. *Paleobiology* **5**:22–30.

Andrews, P. J., and Van Couvering, J. A. 1975. Palaeoenvironments in the East African Miocene, in: Approaches to Primate Paleobiology (F. S. Szalay, ed.). *Contrib. Primatol.* **5**:62–103.

Andrews, P. J., Lord, J., and Evans, E. N. 1979. Patterns of ecological diversity in fossil and modern mammalian faunas *Biol. J. Linn. Soc.* **11**:177–205.

Bishop, W. W., Chapman, G. R., Hill, A. P., and Miller, J. A. 1971. Succession of Cainozoic vertebrate assemblages from the northern Kenya Rift Valley. *Nature (Lond.)* **233**:389–394.

Chesters, K. I. M., 1957 The Miocene flora of Rusinga Island, Lake Victoria, Kenya. *Palaeontographica B* **101**:30–67.

Gentry, A. 1970. The Bovidae (Mammalia) of the Fort Ternan fossil fauna, in: *Fossil Vertebrates of Africa*, Volume 2 (L. S. B. Leakey and R. J. G. Savage, eds.), pp. 243–323, Academic, London.

Harris, J., and Watkins, R. 1974. New early Miocene vertebrate locality near Lake Rudolf, Kenya. *Nature (Lond.)* **252**:576–577.

Harrison, T. 1981. New finds of small apes from the Miocene locality at Koru in Kenya. *J. Hum. Evol.* **10**:129–137.

Hay, R., and Reeder, R. J. 1978. Calcretes of Olduvai Gorge and the Ndolanya Beds of northern Tanzania. *Sedimentology* **25**:649–673.

Hopwood, A. T. 1948. Discussion, in: *Abstracts of the Proceedings of the Geological Society of London* **1445**(6 Dec. 1948):12.

Kent, P. E. 1944. The Miocene beds of Kavirondo, Kenya. *Q. J. Geol. Soc. Lond.* **100**:85–118.

King, B. C., and Chapman, G. R. 1972. Volcanism of the Kenya Rift Valley. *Phil. Trans. R. Soc. Lond. A* **271**:185–208.

Leeder, M. R., 1975. Pedogenic carbonates and flood sediment accretion rates: A quantitative model for alluvial arid-zone lithofacies. *Geol. Mag.* **112**(3):257–270.

Oswald, F., 1914. The Miocene Beds of the Victoria Nyanza and the geology of the country between the lake and the Kisii Highlands. *Q. J. Geol. Soc. Lond.* **70**:128–162.

Paulian, R., 1976. Three fossil dung beetles from the Miocene of Kenya. *J. E. Afr. Nat. Hist. Soc. Nat. Mus.* **31**(158):1–4.

Pickford, M. 1981. Preliminary Miocene mammalian biostratigraphy for Western Kenya. *J. Hum. Evol.* **10**:73–97.

Pickford, M., and Andrews, P. J. 1981. The Tinderet Miocene sequence in Kenya. *J. Hum. Evol.* **10**:11–33.

Pilbeam, D., Barry, J., Mayer, G., Shah, I., Pickford, M., Bishop, W. W., Thomas, H., and Jacobs, L., 1977. Geology and palaeontology of Neogene strata of Pakistan. *Nature (Lond.)* **270**:684–689.

Shackleton, R. M., 1951. A contribution to the geology of the Kavirondo Rift Valley. *Q. J. Geol. Soc. Lond.* **101:**345–383.

Van Couvering, J. A. 1972. *Geology of Rusinga Island and Correlation of the Kenya Mid-Tertiary Fauna,* Ph.D. Dissertation, Cambridge University.

Van Couvering, J. A., and Van Couvering, J. H., 1976. Early Miocene mammal fossils from East Africa: Aspects of geology, faunistics and paleo-ecology, in: *Human Origins: Louis Leakey and the East African Evidence* (G. L. Isaac and E. R. McCown, eds.), pp. 155–207, Benjamin, Menlo Park, California.

Verdcourt, B., 1963. The Miocene non-marine Mollusca of Rusinga Island, Lake Victoria and other localities in Kenya. *Palaeontographica A* **121:**1–37.

Whitworth, T., 1953. A contribution to the geology of Rusinga Island, Kenya. *Q. J. Geol. Soc. Lond.* **109:**75–96.

The Natural History of Sivapithecus

17

P. J. ANDREWS

Introduction

In the study of hominoid evolution much of the attention over the last few years has centered on the genus *Ramapithecus*. This fossil ape has been claimed in the past to be a hominid ancestor (Lewis, 1934; Simons and Pilbeam, 1965), but more recently it has been shown to be less hominid-like and more ape-like than previously proposed (Andrews and Walker, 1976). All such discussion, however, is to a large extent hypothetical because the evidence on which it is based is so limited. Even after a decade of intensive efforts to find additional *Ramapithecus* material from Pakistan, Turkey, Greece, and Kenya, many advances in assessing and interpreting hominoid evolution have come about through more critical evaluation of the known fossils rather than through the discovery of more complete material.

Almost as a by-product of this search for *Ramapithecus* has been the discovery of important new material of the genus *Sivapithecus* in association with it. From preliminary investigations of this new material, it is beginning to appear that *Sivapithecus* has a number of very interesting characters. Not least of these is a close resemblance to extant apes in cranial and postcranial morphology and, in particular, great similarity to the orangutan in facial and dental morphology. But perhaps the most interesting character of the species of *Sivapithecus* is that they are invariably found associated with woodland rather than tropical forest habitats, that is, in ecological conditions that would generally be considered marginal for living primates today. This prompts the

P. J. ANDREWS • Department of Paleontology, British Museum (Natural History), Cromwell Road, London SW7 5BD, England.

question of what kind of animals were *Sivapithecus* species, and this chapter is a preliminary attempt to answer that question.

Four sources of evidence will be drawn on to describe the natural history of *Sivapithecus* in as much detail as is possible for a fossil. First, the distribution of *Sivapithecus* species will be discussed in relation to their habitat associations, and some of the implications on adaptive strategies arising from this will be discussed as well. Second, estimates on body size and sexual dimorphism will be attempted, and the relevance of these to social structure will be discussed. Third, the morphological evidence on dietary adaptation will be described and the possible diet of *Sivapithecus* interpreted. Fourth, the morphological evidence in regard to locomotor adaptations will be mentioned briefly, and the adaptive strategy of the animal discussed in terms of body size. Finally, the adaptive strategy of *Sivapithecus* will be related to what is known of its phylogenetic relationships.

Distribution and Ecology

The distribution of *Sivapithecus* species is difficult to determine because there is no consensus on the taxonomic status of what loosely may be called the sivapithecines. *Sivapithecus indicus* from India and Pakistan is well established in the literature, and the name has been used in this chapter to refer exclusively to the samples from India and Pakistan. But perhaps some of the more recently discovered material may also be assigned to this taxon. *Sivapithecus* has been described from Lufeng in China (Xu and Lu, 1979) and, if true, this extends the range of *Sivapithecus* east from the Indian subcontinent. To the west there are no certainly known specimens of *S. indicus*. There is a canine from Can Llobateres in Spain that may represent *S. indicus*, but while it is far too large to belong with any of the other material from Spain, it is difficult to make conclusions based on a single isolated tooth. Some of the material from Hungary referred to "*Bodvapithecus altipalatus*" (Kretzoi, 1975) also appears to be related to *Sivapithecus*. In particular, it seems to have some resemblance to the *Sivapithecus* from China, but only preliminary descriptions of all this material have been published. Finally, there are the specimens of *Sivapithecus* from Greece and Turkey, which have been combined as a separate species, *S. meteai* (Andrews and Tekkaya, 1980), although new and undescribed material from Pakistan invalidates some of the distinctions drawn between *S. meteai* and *S. indicus*. However, disregarding species distinctions, the genus *Sivapithecus* seems to have had a distribution across Europe and Asia from Greece in the west to China in the east, with the largest collections coming from Greece/Turkey, India/Pakistan, and China.

Sivapithecus has also been described from Africa. Le Gros Clark and Leakey (1951) described a species of *Sivapithecus*, *S. africanus*, based on a maxilla and two isolated teeth. This was later transferred by Leakey to his new genus *Kenyapithecus* (Leakey, 1967) and subsequently referred to *Proconsul* species through lack of evidence (Pilbeam, 1969; Andrews, 1978), although this last course of action has been criticized by Madden (1980). Greenfield (1979) reverts to Leakey's position in advocating a relationship between this taxon and the material that Leakey (1962) described as *Kenyapithecus wickeri,* but Greenfield joins them into one species which, by priority, must be *Sivapithecus* or *Kenyapithecus africanus.* Whatever the outcome of this taxonomic wrangle is, it seems evident that there is a *Sivapithecus*-like primate in Africa. Therefore, the distribution of this genus can loosely be considered to have encompassed East Africa as well as Europe and Asia.

With so wide a distribution it must be assumed that the different species of *Sivapithecus* occupied a varied range of habitats. There is strong evidence for the paleoecological association of some of the *Sivapithecus* samples, and some unusual findings emerge. The best evidence comes from the recent publication of a flora from Lufeng based on pollen analysis (Sun and Wu, 1980), which indicates a vegetation association dominated by temperate forms. Three floras from three separate levels are described. The lowest unit in the sequence is a black clay and contains many temperate forms that are typical of temperate broad-leaved deciduous woodland or forest, but it also contains cycads and ferns, suggesting warmer and wetter conditions. However, there is no evidence for the sclerophyllous vegetation that Axelrod (1975) indicated would have been present in this area in the middle Miocene. The overlying bed is the lignite that contains the vertebrate fossils (including *Sivapithecus*). This bed has a mixture of temperate and a few tropical forms, mainly the former, while the tropical forms are in some cases taxa associated with tropical mountains. Conditions would have been drier than in the underlying layer and almost certainly more seasonal. It is likely that the vegetation was a form of deciduous woodland with some evergreen sclerophyllous components. This is the vegetation association that borders on the sclerophyllous vegetation of the Mediterranean region today (Axelrod, 1975). The final stage in the sequence is a mainly coniferous association, suggesting cooler conditions succeeding the warm temperate phase. In summation, there is a sequence, therefore, of woodland floral association ranging from probably subtropical woodland in the lowest level to cool, temperate, coniferous woodland in the upper level. The level in between, in which *Sivapithecus* is associated, has a flora that is intermediate between the other two beds and is a form of warm, temperate, deciduous woodland.

This is an extraordinary vegetation association in which to find fossil primates. It suggests that they could survive in conditions to which few living primates are adapted, for very few primates are known today from such habitats. The nearest equivalent would be among the Old World macaques, for example, the North African Barbary macaque or the Japanese macaque, both of which can survive in seasonal temperate climates. They both achieve this with the aid of a number of physiological adaptations, but partly also through increasing variability of diet and through changes in social structure (Crook and Gartlan, 1979). Climatic seasonality imposes food shortages, which result in high seasonal mortality, while during the favorable seasons food is superabundant and allows vigorous seasonal breeding (Crook and Gartlan, 1979, p. 367). While social groups in such a situation may be larger than in more favorable conditions, allowing exploitation of abundant food sources, in extreme cases at the limits of their range group size may decrease, as in *Macaca fuscata* in northern Japan. Crook and Gartlan (1979, p. 368) conclude that this system of group structure is the result of terrestrial adaptations in open conditions favoring increase in group size and cohesion within groups, with an increase in intrasexual selection between males, leading to marked group structuring and to the development of protective roles by males. The association of *Sivapithecus* with a vegetation type such as that established for Lufeng must be considered evidence for its occupation of seasonal habitats. Therefore, by analogy with extant primates, it must be considered likely that *Sivapithecus* developed some of these aspects of group structure. (This will be considered further in the next section in the discussion of sexual dimorphism in *Sivapithecus*.)

The paleoecological evidence from the other *Sivapithecus* localities is unfortunately not as good as the Chinese evidence. There is a flora from Rudabánya in Hungary, where a *Sivapithecus*-like primate is known ("*Bodvapithecus altipalatus*"), but the flora comes from different levels from the vertebrate fauna and from a different part of the Rudabánya Quarry (Kretzoi *et al.*, 1976). The floras vary from subtropical to temperate woodland. Grasses are present throughout, but the species lists are dominated by trees, many of which are deciduous broad-leaved species or conifers, but which do include some sclerophylls. In addition to the latter, there are also a few tropical to subtropical forest species. These do not appear to be common, and the majority of forms are mainly warm temperate. There is no evidence in the published floras for grass to forest zonation such as suggested by Kretzoi (1975), although there is evidence for temperature fluctuations of the kind described for the Lufeng profile, resulting in a similar range of associations, namely from cool, coniferous woodland to probably evergreen subtropical woodland. With which flora the primates may have been associated is impossible to say on present evidence, and the fauna that is associated with these primates does not offer much help, for the dominant animals appear to be pigs. It does not seem possible at this time to come to any conclusions concerning Rudabánya paleoecology except to acknowledge that it appears to have been some form of woodland.

No floras are known from the localities where *Sivapithecus indicus* and *S. meteai* are best known. Moreover, the faunas associated with the latter species do not provide a clear indication of the paleoecology at either the Greek locality (de Bonis *et al.*, 1974) or the Turkish locality (Ozansoy, 1965). Both faunas appear to be strongly biased, although the nature of the bias in each case is not known. Bovids are abundant, but restricted in diversity in the Macedonian locality, while they are both abundant and diverse in the middle Sinap of Turkey. Small mammals are rare at both sites, and *Hipparion*, although present at both, is also poorly represented. No indicators of closed forest are present, so it is likely that a more open form of vegetation was present. But what that may have been cannot be reconstructed on present evidence.

The fauna associated with *S. indicus* is much better known, although until recently there has not been good evidence on association and even now the faunas from Pilbeam's Pakistan collections have only been provisionally listed (Pilbeam *et al.*, 1979). Ecological diversity spectra have been prepared for the Zone 6 fauna of Pilbeam's collections (see Fig. 1) following the method given in Andrews *et al.* (1979). One departure from this method is in the size analysis, which has been modified by omitting the carnivores. The reason for this omission is that carnivores are at a higher tropic level than the plant-eaters and hence reflect habitat at second hand. In modern habitats, the modified size spectra show *decrease* in frequency from small to large mammals for tropical forest faunas when the carnivores are omitted, while in grassland habitats the frequency *increases* from small to large size except for smallest size category. Woodland and bushland faunas are intermediate, with two peaks for small and large mammals, but with lower numbers of the largest size category. This is clearly the pattern for the Siwalik fauna. Statistical tests have not been performed, because the lists in Pilbeam *et al.* (1979) are neither complete nor definitive, but the pattern for the size analysis of the Siwalik Zone 6 fauna appears very different from the patterns for modern forest and grassland faunas and is remarkably similar to the woodland–bushland pattern. The other analyses show the same result, with high proportions of artiodactyls and carnivores in the taxonomic analysis, a high proportion of large terrestrial mammals in the locomotor analysis, and a low proportion of frugivores and a high proporation of herbivores in the feeding analysis [see Andrews *et al.* (1979) for comparative data on modern habitats]. Based on this evidence, the Zone 6 fauna of the Siwaliks of Pakistan, which includes *Sivapithecus indicus,* appears to be a woodland fauna. Methods of analysis available at present do not indicate what form of woodland it may have been.

For comparison with the Siwalik fauna two other faunas are also shown in Fig. 1. These are the faunas from Paşalar, in Turkey, where *Sivapithecus darwini* has been reported (Andrews and Tobien, 1977), and from Fort Ternan, in Kenya, where *Ramapithecus wickeri* is represented (Leakey, 1962). These will not be examined here in detail, but they both show general similarity to the Siwalik pattern, that is, to the woodland–bushland pattern of modern faunas (Andrews *et al.*, 1979). The fauna from Paşalar shows some evidence of more open woodland conditions than the Siwalik fauna in the higher proportions of Carnivora, of mammals in the largest size category (over 100 kg), and of large terrestrial mammals, while the Fort Ternan fauna may indicate a more closed form of woodland, based on the low proportions of large mammals over 100 kg and in the high proportion of frugivores relative to other feeding classes.

The evidence from the association of *Sivapithecus* species with paleo-ecological indicators, whether of flora or fauna, appears to be consistent, always accepting the wide margin of error inherent in such analyses. Tropical forest setting is never indicated, and tropical to subtropical woodland (adapted for a moderately seasonal climate and distinguished on the basis of a single tree canopy dominated by relatively few species and with grasses often common in the ground vegetation) may be the preferred habitat, as at Fort Ternan and the Siwaliks. There is only slight evidence for the association of *Sivapithecus* species with sclerophyllous vegetation, which may have bordered on the subtropical woodlands along their northern boundary (Axelrod, 1975),

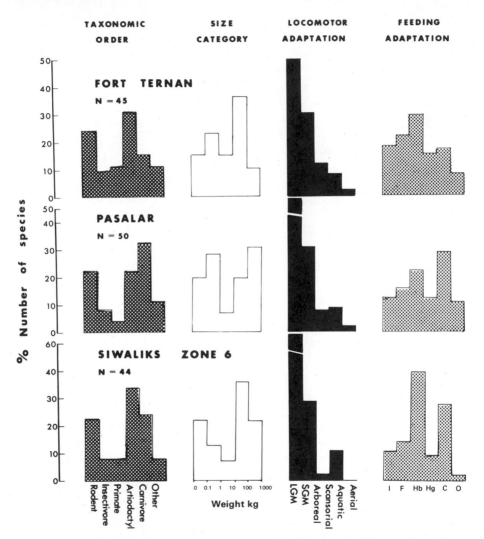

Fig. 1. Ecological diversity histograms for three middle Miocene fossil faunas: Fort Ternan, Kenya, whole fauna with 45 species; Paşalar, Turkey, fauna from single level, 50 species (Andrews and Tobien, 1977); Siwaliks, Pakistan, Zone 6 fauna (Pilbeam *et al.*, 1979), 44 species. The vertical scale shows the proportions of the faunas for four categories, taxonomic, by order; size, in kg (see text); locomotor adaptive zones, with LGM signifying large, fully terrestrial mammals and SGM signifying small, scrambling mammals that live both on the ground and in the bush canopy and/or fallen trees; and feeding classes, with I, insectivores; F, frugivores; Hb, herbivorous browsers; Hg, herbivorous grazers; C, carnivores and O, omnivores. Bats are excluded from the size analysis (see text).

but there is good evidence again for their association with the warm, broad-leaved, deciduous woodlands to the north of the sclerophyll belt. In terms of physical structure there is a certain degree of similarity between tropical and warm, temperate, deciduous woodlands, as both are adapted for moderately seasonal climates, and the problems that they pose for primates are similar, both in terms of the seasonal variation in food and the availability of continuous tree cover for species with an arboreal heritage.

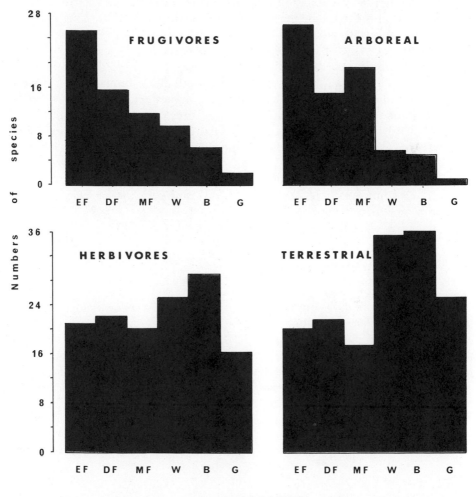

Fig. 2. Distribution of frugivorous, herbivorous, arboreal, and terrestrial mammals in six tropical African habitats. The vertical scale shows the numbers of species. The habitat types are, from left to right: EF, evergreen tropical forest; DF, semideciduous tropical forest; MF, montane tropical forest; W, tropical deciduous woodland; B, tropical bushland; G, tropical grassland. The species counts are the means for 22 faunas derived from the six habitat types: the evergreen forest count is based on the mean from three forest faunas, semideciduous forest on the mean from five faunas, montane forest on three, woodland on five, bushland on three, and grassland on one. Data and references on habitats are from Andrews et al. (1979).

Seasonality results in periodic times of shortage of specialist foods such as fruit, and the number of species of frugivorous mammals in woodland habitats is therefore lower than in tropical forest (Fig. 2). The number of species of herbivorous mammals does not vary as greatly, because even in the driest (or coldest) part of the year some vegetative matter, which is much more abundant in the first place, remains. Those woodland species that retain a partially frugivorous diet are therefore dependent on alternative sources of food for large parts of the year.

Regarding the availability of tree cover, the restriction of the tree canopy to a single level means that inevitably it will be discontinuous. This makes it difficult for an arboreal species to remain all the time in the trees, with the resulting loss of some of the advantages of arboreality, for example, protection against predators. As would be expected, therefore, the numbers of species of arboreal mammals are lower in woodland habitats than in forest, while the numbers of terrestrial mammals are higher (Fig. 2). The association of *Sivapithecus* with woodland habitats, therefore, is strong presumptive evidence that it was far from being an arboreal frugivore, like the forest-living chimpanzee or orangutan of today. Some degree of terrestriality is indicated for *Sivapithecus* by this ecological association and perhaps a generalized diet containing some fruit but tending toward a moderately omnivorous range of food. These, however, are only indications, and for further evidence it is necessary to look at the sivapithecines themselves.

Body Size and Dimorphism

Estimates of body size of *Sivapithecus* species have been made based on molar length. Several recent studies have commented on the accuracy of this approach (Gingerich, 1977; Fleagle, 1978; Kay and Simons, 1980) and have provided regression formulas based on samples of varying sizes from living primates. Gingerich (1977) lists two regression formulas relating body weight to mesiodistal length of M_2, first for just seven hominoid primates and second for a more representative series of 38 species of monkeys and apes. These two regression equations given by Gingerich are:

$$\log_{10}(\text{body weight}) = 2.989 \log_{10}(M_2 \text{ length}) + 1.462 \quad (r = 0.942) \quad (1)$$

$$\log_{10}(\text{body weight}) = 3.289 \log_{10}(M_2 \text{ length}) + 1.095 \quad (r = 0.963) \quad (2)$$

Both these equations are based on data originally compiled by Kay (1975), and more recently Kay (in Kay and Simons, 1980) has published more comprehensive data based on 106 species of primates and yielding a third regression equation:

$$\log_{10}(\text{body weight}) = 2.86 \log_{10}(M_2 \text{ length}) + 1.37 \quad (r = 0.949) \quad (3)$$

This gives body weight predictions that are considerably lower than the two equations given by Gingerich (about 60% lower), but since it is based on the greatest range of material, it will be used here as the primary source for predicting body weights of *Sivapithecus* species. These weights have a considerable built in error even when assessing species of known weight, and great caution is needed in interpreting estimated weights of fossil species. With a correlation coefficient of 0.949 for Kay and Simons (1980) data, 90% of the variance is accounted for, but there is still no way of testing the relevance of the regression equation to any one species of unknown weight. Also, the weights used here must be considered as no more than best estimates.

Using formula (3) from Kay and Simons (1980), the following body weights are obtained:

1. *Sivapithecus indicus:* 46.1 kg (95% confidence limits 30.0–65.5 kg).
2. *Sivapithecus meteai:* 64.0 kg (one specimen).
3. *"Ouranopithecus macedoniensis":* 65.1 kg (range for five specimens of 46.3–80.1 kg, and 95% confidence limits 34.4–109.7 kg).

Weights for the smaller *Sivapithecus* species have not been calculated, as these are of uncertain taxonomic status. The larger species shown here have predicted body weights within the range of variation known for the orangutan and are at the top of the chimpanzee range and bottom of the gorilla range. Despite the inaccuracies inherent in regression predictions, it can reasonably be concluded that these fossil species were rather large, at least the size of male chimpanzees, and within the overall range of orangutans.

It has recently been suggested (D. R. Pilbeam, personal communication) that *Sivapithecus* and *Ramapithecus* both had rather large teeth for their body size. If this were the case, they would diverge from the mean regression lines based on less megadont primate species and have body weights lower than those indicated by their molar teeth. With this possibility in mind, it might be better to use the minimum estimate of body size at the 95% level of probability, giving estimated body weights for *S. indicus* of 30 kg and for *"O. macedoniensis"* of 34 kg. These figures correspond very closely to Pilbeam's estimate of 32 kg based on other cranial and postcranial measurements.

The confidence limits given above for the ranges in body weight are dependent on the standard deviation of the length of M_2 and do not reflect true weight variation or sexual dimorphism. It has been claimed (Pilbeam and Zwell, 1972) that variation in continuous variables such as molar dimensions has no relation with sexual dimorphism, but that variables that are discontinuous or show bimodel distribution can be used as a measure of dimorphism. This does not explain the origin of the dimorphism, however, and merely distinguishes between dimorphic and nondimorphic characters. Pilbeam and Zwell use lower canine length as an example of a dimorphic character, and they show values for the coefficient of variation (CV) for pooled male/female samples ranging from 14.5 to 28.0 in the most dimorphic species of living primates (baboons, macaques, gorillas, and orangutans) down to 5.1–11.9 for less dimorphic species (gibbons to chimpanzees). If these are compared with the lower canine coefficients for variation for *Sivapithecus* species, it would appear that they fall into the higher category, with values of

13.6 for *Sivapithecus indicus* and 17.2 for the Greek and Turkish specimens of *Sivapithecus meteai* (Andrews and Tekkaya, 1980). They are most closely comparable in canine dimorphism with the gorilla (CV = 18.3) and orangutan (CV = 14.5), and these have female/male body weight proportions of 55–60%. This is consistent with the ranges of body weight predicted from the M_2 regressions and reinforces the similarity of *Sivapithecus* species to these extant Great Apes.

Sexual dimorphism was first shown by Darwin (1871) to be the manifestation of intrasexual selection. Darwin related sexual selection to social structure, and although he had very little evidence of animal social groups on which to base his conclusions, his theory proved consistent with the evidence once it became available. In animals forming monogamous groups the sexual relationships balance, both within and between sexes, on a one-to-one basis, and as a result the intrasexual (male/male and female/female) and intersexual (male/female) selective forces balance each other on a one-to-one basis as well, with the result that males and females are not distinguished from each other by size or secondary sexual characteristics. Similarly, males do not vary greatly from the other males, nor females from other females. In contrast to this, in polygynous group structure there is an imbalance in both intra- and intersexual selection, intrasexual selection giving rise to sexual dimorphism as a result of competition between the males for access to females, and intersexual selection giving rise to role differentiation between the sexes and the development of secondary sexual characteristics (R. Short, personal communication). Turning this argument around, it may be possible to predict something of the social structure of fossil animals based on their degree of sexual dimorphism, inferring monogamous social groups in a nondimorphic species and polygyny in dimorphic species. This has been attempted recently by Fleagle *et al.* (1980), who showed that the early catarrhine primate *Aegyptopithecus* was sexually dimorphic and hence may have lived in polygynous social groups. Similar degrees of sexual dimorphism are present also in *Sivapithecus*, as just demonstrated for canine length, and it can be suggested for these fossil apes that they also had polygynous social groups like the living Great Apes but unlike the monogamous gibbons. This is consistent with the evidence of the previous section, which showed that the habitat with which *Sivapithecus* species appear to be associated was a seasonal one, leading to large polygynous groups and high rates of sexual selection, and hence high rates of sexual dimorphism.

Diet

It has been shown in previous sections that *Sivapithecus* species are associated with a habitat type that implies some degree of seasonality in climate. Food sources such as fruit or insects would not be available year-round in even moderately seasonal habitats, and this suggests, therefore, that the diet of *Sivapithecus* species could not have been entirely insectivorous or frugivorous. By analogy with primates living today in seasonal habitats, it appears likely that their diet could have been a highly variable one. The evidence for diet in *Sivapithecus* species will now be examined by looking at the proportions of the anterior teeth, at molar and premolar morphology, at tooth wear, and at body size.

The incisors in both *Sivapithecus indicus* and *S. meteai* appear to be large, and, in particular, the upper central incisor is a broad spatulate tooth very much larger than the lateral incisor (Andrews and Tekkaya, 1980). It has been shown (Hylander, 1975) that the width of the maxillary incisor row relative to body weight distinguishes between frugivorous primate species, which have relatively large incisors, and folivorous species, which have relatively small incisors. The widths of the maxillary incisor row for the Turkish (Andrews and Tekkaya, 1980) and Greek (de Bonis and Melentis, 1977) specimens of *Sivapithecus* are 35.6 and 34.8 mm, respectively, and when these are compared with Hylander's data, using the estimated body weights given in the previous section, both *Sivapithecus* specimens fall into an intermediate position between frugivores and folivores. Hylander's analysis distinguishes between the more frugivorous diet of the chimpanzee, with its large incisors, and the more herbivorous diet of the gorilla, with its small incisors. But the orangutan, despite its apparently greatly enlarged incisors, is intermediate and in this respect is the living species that is most similar to *Sivapithecus*. This is not all that helpful, as the diet of orangutans presents some ambiguities which will be discussed later.

Another marked dental similarity between *Sivapithecus* and *Pongo*, and one that also lacks clear dietary significance, is the relative size of the two upper incisors. The central incisor in orangutans is very much larger than the lateral, up to one and one-half to two times the size, whereas in most other primates there is less of a size discrepancy. The I^1/I^2 length proportions in orangutans thus varies from 150% to over 200%, while in *Sivapithecus meteai* it ranges from 179% in the Greek material to 211% in the Turkish specimen. Percentage figures for *S. indicus* are not available.

It was shown earlier that *Sivapithecus* species have high degrees of sexual dimorphism in the canines. Despite this, however, the canine size is small overall when compared to the postcanine teeth. The canine to first molar proportions are low (Table 1) and are closer to the proportions seen in female chimpanzees and gorillas than in males. Sample sizes are so small that this conclusion is of doubtful significance, although in the cases of the mandibular proportions, the extremes of the size ranges appear to be represented in the *Sivapithecus* samples. It may be, therefore, that the canine in *Sivapithecus* is small relative to the postcanine teeth, and this might be related to diet in the following way. Lucas (in preparation) has shown that much of the variability of the canine might be explained by the height of the temporomandibular joint relative to the tooth row length. The higher the joint above the occlusal plane, the greater will be the posterior displacement of the front of the lower jaw relative to the upper during jaw movement (Herring, 1972) and therefore the smaller the canines will have to be. If the canines were too large, they

Table 1.   Canine Proportions in *Sivapithecus*

	N	Upper C/M1[a] %	*N*	Lower C/M1[a] %
S. meteai	2	95–122	5	48–95
S. indicus	3	63–114	2	91–105
Chimpanzee[b]	Female	78–108	—	79–115
	Male	114–210	—	112–184
Gorilla[b]	Female	66–113	—	55–92
	Male	111–186	—	92–172

[a]The C/M1 values are expressed as percentages of the cross-sectional area of the canine (length × breadth) divided by the area of the first lower molar, × 100.
[b]The data for chimpanzees and gorillas are taken from Greenfield (1972).

would prevent the jaws opening and closing. Raising the height of the temporo-mandibular joint increases the compressive forces during chewing (Maynard Smith and Savage, 1959; Herring, 1972), and the most extreme example of this in primates is to be seen in some of the fossil and extant species of the gelada, *Theropithecus* (Jolly, 1972). In these species, the higher is the ascending ramus, the smaller are the canines, and it may be that a similar morphological response was taking place in *Sivapithecus* which may have been adaptive for strong compressive forces during chewing.

The possibility of *Sivapithecus* having had strong compressive forces acting during chewing is certainly consistent with the molar morphology. Work in progress on enamel thickness (L. Martin, personal communication) shows that *Sivapithecus* had extremely thick and functionally strong enamel, and if this is combined with the low, rounded cusps and flattened occlusal relief of the molars, it appears that the teeth are well adapted for resisting very powerful compressive forces. (The body of the mandible is similarly robust). Shearing blades as defined by Kay (1977a,b) are hardly developed on the molars, and so a high fiber content in the diet appears unlikely. It seems equally unlikely that the diet of *Sivapithecus* contained high proportions of ductile tissue, because the occlusal basins, especially the trigon and talonid basins of the upper and lower molars, are relatively restricted in area: mammals adapted for a mainly fruit diet, in which the plant tissues are soft and ductile, normally have large and often concave occlusal basins with small cusps, as, for example, in the fruit bats, and the large cusps, flat occlusal surfaces, and restricted occlusal basin in *Sivapithecus* suggest a rather different diet. What this may have been is not immediately obvious.

Some general guidelines may be available from consideration of the body weight of *Sivapithecus*. Body weight is related to diet in several ways (Jarman, 1974; Gaulin, 1979). The first and most obvious is that larger animals need more food than smaller animals, but since their metabolic rate is lower, they may be able to subsist on lower quality foods. In other words, they may go in for quantity rather than quality. In addition to this, however, high-quality food sources, such as insects, leaf buds, fruit, and seeds, are limited in supply and moreover often have a discontinuous distribution, so that the effort needed by a large animal to locate the food may outweigh its dietary value. Finally, there is also the consideration, important in other groups of animals, such as bovids (Jarman, 1974), but less important in primates, that larger animals can eat more bulky food items and that smaller items, no matter how nutritious, are harder for large animals to locate and ingest. Some of these differences are immediately obvious in the primates, since there is a tendency for smaller species to be insectivorous and the larger ones to be herbivorous. In the herbivorous class the small species tend to be frugivorous and the larger ones tend to be folivorous (Kay, 1978; Kay and Simons, 1980).

The larger *Sivapithecus* species fall toward the top of the frugivore range based on the analysis of extant primates provided by Kay and Simons (1980). Their analysis, however, is restricted to primates and therefore makes necessary the assumption that a fossil primate being compared with it had a similar range of adaptations to the living primates. But I would suggest that the possibility must also be considered that the diet of *Sivapithecus* species might

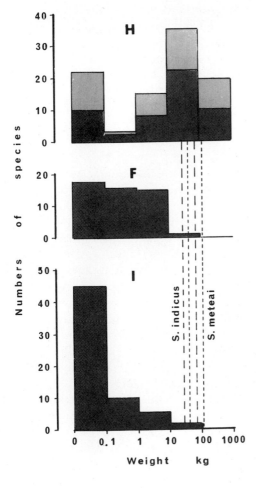

Fig. 3. Histograms of dietary classes against body weight ($\log_{10}$) for 254 mammalian species from tropical Africa (excluding bats). The three categories shown represent combinations of feeding classes: I, totally insectivorous mammals and those with mixed insectivorous and herbivorous diets; F, mammals with frugivorous and mixed frugivorous–herbivorous diets; H, mammals that are browsers, both folivorous and herbivorous (lower unit), and grazers, in which grass is a major dietary element (upper unit). The dashed lines show the 95% confidence limits of the predicted weight ranges of *Sivapithecus indicus* and *Sivapithecus meteai*.

be unlike that of any of the extant primates because the morphology and proportions of their teeth have no modern counterpart. I have therefore extended the analysis to encompass the known dietary range for all mammals in one geographical area, and the results of the analysis of 254 mammalian species in the tropical region of Africa are shown in Fig. 3. This list is not exhaustive, but it contains all the mammals known from 23 collecting localities representing a wide range of habitat type [for locality details and references see Andrews *et al.* (1979)]. Over the lower weight ranges, frugivores and insectivores overlap (Fig. 3), with higher numbers of insectivores; this is similar to the results of Kay and Simons (1980). Toward the higher weight ranges, however, these dietary types diminish sharply, while the herbivorous mammals, which include the folivores of Kay and Simons (1980), increase, and it is at just this level that the predicted weight ranges place the *Sivapithecus* species. Even if one assumes a wider margin for error for these weight predictions, one finds that they are still well above the level at which most mammals have had to change from essentially specialist high-nutrition foods like insects and fruit to less nutritious but much more abundant plant foliage. On this evidence, therefore, and in view of the ambiguities in the evidence from tooth morphology and proportions, it seems more likely that the larger *Sivapithecus* species were adapted for a herbivorous diet that may have contained only a limited amount of fruit. This conclusion is consistent with the evidence presented earlier on the distribution and ecology of *Sivapithecus* species.

Locomotion and Adaptive Zone

Evidence on the locomotor adaptations of *Sivapithecus* species is scanty. There are only fragmentary fossil remains, and these for the most part have not yet been fully studied. Such conclusions as can be drawn are the result of body size considerations similar to those relating to diet.

Postcranial remains that may probably be assigned to *Sivapithecus indicus* are known from Pilbeam's collections from Pakistan (Pilbeam *et al.*, 1977). Several femoral fragments and part of the distal articular surface of a humerus may belong to *S. indicus* on the basis of size. The humerus is similar in some respects to the more complete distal humerus from Rudabánya in Hungary (Kretzoi, 1975). They are similar to the equivalent bones in the living Great Apes, although since these show some variety in locomotor patterns, no definite conclusions can be drawn from this similarity. They show evidence for stability of the elbow joint in compression, together with full extension of the elbow joint as is present in the extant apes.

Within broad limits there is a relationship between body size and terrestriality. Kay and Simons (1980) showed that within the Old World monkeys the terrestrial species were generally larger than arboreal species, and this division was even clearer when comparisons were made between pairs of related species, of which the arboreal one was smaller and the terrestrial one was larger. The factors causing this relationship are not clear, for there would not appear to be any dietary constraint on increased body size in an arboreal species. The abundance of leaves as a food source ought not to impose a limit on the body size of a folivorous mammal, because leaves are as widespread a food source in evergreen forest as grass and herbs are in open country environments. Kay and Simons found, for instance, that even within the leaf-eating colobine monkeys, the arboreal species were smaller than the terrestrial, despite their adaptation to a high-quantity and low-quality diet comparable with the diet of large-bodied unselective browsers in the antelopes (Jarman, 1974). The limit to body size in arboreal species might therefore be the physical limitation imposed by the strength of the trees that form the substrate on which they must move around (or conversely, the increase in size of nonar-

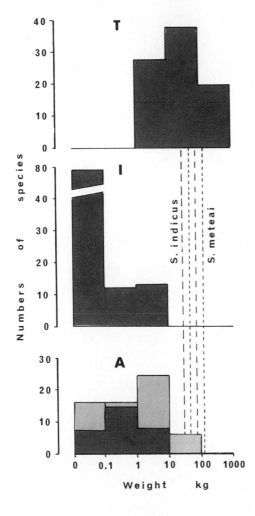

Fig. 4. Histograms of adaptive zones against body weight ($\log_{10}$) for 254 mammalian species from tropical Africa (excluding bats). The zone classes are: A, arboreal, including scansorial mammals, which make up the lower unit of the histogram; I, intermediate lower canopy and terrestrial small mammals; and T, larger terrestrial mammals. The dashed lines show the 95% confidence limits of *Sivapithecus indicus* and *Sivapithecus meteai*. Further details are in the text.

boreal species may be related to other factors, such as increased predation pressure).

This appears to be the main limiting factor within the mammals generally. The same fauna of 254 species of mammals described in the previous section (see Fig. 3) was divided also into four classes corresponding to four adaptive zones related to degrees of arboreality and terrestriality. These divisions are not absolute, and species that pass from one to another have been counted for both. The four classes consist of: arboreal mammals, which spend much of their time in the upper tree canopies; scansorial mammals, which are partly canopy dwellers but depend more on the large branch niche consisting of the tree trunks and main branches; small ground mammals, which live both on the ground and in the bush canopy and on fallen trees; and finally, large ground mammals, which are completely terrestrial (although even in this category there is the odd exception, like tree-climbing otters and goats). The distribution of these classes is shown in Fig. 4, and it may be seen that there is a marked distinction between the sizes of the more arboreal mammals, which are smaller, and the more terrestrial mammals, which are larger. The size ranges of *Sivapithecus* species are shown on Fig. 4, and they fall at or beyond the limits of arboreal mammals in Africa today but in the middle of the terrestrial range of variation. It appears likely, therefore, that *Sivapithecus* species were terrestrial, at least in part, although it is not possible, for lack of evidence, to say anything about their method of locomotion on the ground. This result also is consistent with the ecological evidence presented earlier, where it was predicted that in the habitat indicated for *Sivapithecus* some degree of terrestriality would have been necessary because of discontinuities in the woodland tree canopy.

Phylogeny

The status of *Sivapithecus* is in some doubt at present, and much remains to be done to evaluate its relationships with other taxa and the relationships of its contained species. The evidence presented here on ecology, body size, diet, and terrestriality may contribute to a better understanding of these relationships, but these will be discussed, insofar as they are known at present, from the opposite point of view to see if they provide any further evidence on *Sivapithecus* natural history.

Two species of *Sivapithecus* have been discussed in detail in this chapter, *S. indicus* and *S. meteai*, because these are the two best known. They share a number of characters, which may briefly be listed as follows:

1. Face with large, flaring zygomatic process; narrow interorbital septum; high orbits set well above the top of the nasal aperture; short, broad nasal aperture.

2. Dentition with broad I^1; small I^2—very much smaller than I^1; massive

lower incisors, subequal in size; low-crowned, robust canines; robust P_3; large, flat-crowned molars with thick enamel.

3. Postcrania: humerus with large trochlea on distal articular surface; well-developed lateral ridge on trochlea; deep olecranon fossa; ulna with short olecranon process.

These characters are all derived with respect to the primitive conditions of the catarrhine ancestral morphotypes (Delson and Andrews, 1975) and are also derived with respect to the primitive hominoid condition as inferred from the ancestral condition of the sister groups (a) gibbons (Hylobatidae) and (b) Great Apes and humans (Hominidae).

The other fossil species that are referred to *Sivapithecus* are not as well known and share only a few of the characters listed above. *Sivapithecus darwini* (Abel, 1902; Lewis, 1937; Andrews and Tobien, 1977) is known only from the dentition and shares only the last three dental characters on this list, while its incisors and lower molar morphology are sufficiently distinct to justify its specific status. *Sivapithecus africanus* (Le Gros Clark and Leakey, 1951) is even less well known and shares only the last of the listed dental characters, while it has distinctive characters in the great enlargement of the premolars and narrow palate. *Ramapithecus* species have also been suggested to be congeneric with *Sivapithecus* (Greenfield, 1979; also see Kay and Simons, this volume, Chapter 23), and it is clear that they share at least as many characters with *S. indicus/S. meteai* as do *S. darwini* and *S. africanus*. Moreover, they lack the distinctive species-specific characters that are present in *S. darwini* and *S. africanus* and differ from *S. indicus* and *S. meteai* mainly in being smaller. It is clear from this discussion that a revision of the genus *Sivapithecus* is long overdue, but more specimens are needed for the less well-known groups so that relationships can be determined on the basis of similarity in morphology and not on the basis of absence of differences in poor specimens.

The problem of the species diversity of *Sivapithecus* can be left for the moment and the question of the phylogenetic affinities of the genus considered. It has been suggested that the facial morphology and dentition of *S. meteai* has many points of similarity with the orangutan, including such features as the narrow interorbital septum, the deep zygomatic region with strong lateral flare, the broad nose, and the broad I^1 that is very much larger than I^2 (Andrews and Tekkaya, 1980). These characters are all considered derived relative to the primitive hominoid condition (Delson and Andrews, 1975; Andrews and Tekkaya, 1980; McHenry *et al.*, 1980), and since they are shared exclusively with the orangutan, there is a good case for making *Sivapithecus* and *Pongo* sister groups, that is, two groups that are more closely related to each other than either is to anything else. In phylogenetic terms, *Sivapithecus* may therefore be the closest known relative to the orangutan (Fig. 5), closer than the living hominoids, and since it is earlier in time, it may be considered to be broadly ancestral to the orangutan, although this makes many assumptions as to whether, for instance, the orangutan lineage was distinct by the time of *S. meteai*, that is, about 11 million years ago. Regardless

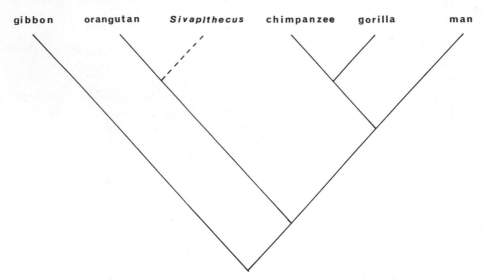

Fig. 5. Cladogram showing the relationships of the Great Apes and humans with the position of *Sivapithecus* referring specifically to the species pair *S. indicus* and *S. meteai*. Other species that have been attributed to *Sivapithecus* (*S. darwini* from Turkey, *S. yunnanensis* from China) or that may be closely related to *Sivapithecus* ("*Bodvapithecus altipalatus*" from Hungary, "*Austriacopithecus*" from Czechoslovakia) cannot be included until their relationships with *Sivapithecus indicus* have been established.

of this, however, its possible relationship with the orangutan suggests that we should look at this living species to see if it provides any further information on the natural history of the fossil *Sivapithecus* species.

The orangutan is generally considered to be an arboreal frugivore (MacKinnon, 1971; Rijksen, 1978), very different from the adaptive niche indicated here for *Sivapithecus*. It has been postulated, however, that because the orangutan has a well-developed repertoire of terrestrial locomotion it may have been more terrestrial in the past (MacKinnon, 1971), and recently the anatomical evidence for such a supposition has been reviewed by Smith and Pilbeam (1980). The teeth and jaws of the orangutan are not typical of arboreal frugivores, and the high mandibular condyle on the mandible and massive zygomatic of the maxilla and the great thickness and wrinkling of the enamel of its molar teeth (Gantt, 1977) suggest adaptation to greater stresses than are normally imposed by a diet of fruit (Herring, 1972; Lucas, 1981). Smith and Pilbeam (1980) draw parallels between the orangutan and the terrestrial monkeys that have a more omnivorous diet, and they suggest that the orangutan may also have been more terrestrial and omnivorous in the past. If this is the case, it is consistent with the morphological evidence suggesting closeness of relationship between *Sivapithecus* and *Pongo*. On the other hand, if the orangutan has changed so drastically in its comparatively recent past, its present adaptations do not provide a good model for its fossil relatives and certainly not for *Sivapithecus*.

Summary and Conclusions

Sivapithecus is widely distributed in Eurasia and Africa, and its preferred habitat appears to be woodland as opposed to tropical or subtropical forest. At least some of the evidence points to deciduous woodland (in China and Hungary), while for other areas the woodland type is not distinguished. Primates living in such habitats today are at least partly terrestrial because the tree canopy is not sufficiently continuous to allow progression from tree to tree, and so it is concluded on this evidence that some degree of terrestriality is indicated for *Sivapithecus*. In addition, a moderate degree of seasonality is implied in woodland habitats, especially if it is deciduous, and this suggests that fruit-eating could not be the major dietary adaptation of *Sivapithecus*, because fruit would not be available for much of the year.

The body weights of *Sivapithecus indicus* and *S. meteai*, from India/Pakistan and Greece/Turkey, respectively, have been estimated by regression on $\log_{10}(M_2$ length) to be 30–68 kg for *S. indicus* and 34–110 kg for *S. meteai*. The lower end of this range may provide the best estimate if it is proven that these species had megadont dentitions. The inferred dietary adaptation of *Sivapithecus*, based on the dietary/weight spectrum for living primates, would be frugivorous (Kay and Simons, 1980). The possibility exists, however, that there is no living counterpart to *Sivapithecus* among primates, and an extension of the analysis to include other mammals (254 species of African mammals) shows that the large *Sivapithecus* species are at or beyond the weight range of mammalian frugivores and are well into the herbivorous category. This is consistent with the conclusions of paleoecological association, that is, that *Sivapithecus* species were not frugivorous.

The jaws and dentition of *Sivapithecus* species do not provide any clear indication of diet at present levels of analysis. Hylander's (1975) data indicate they are not frugivores, on the basis of incisor size, and Kay's (1975) data suggest that they are not folivores. The low-cusped and thick-enamel type of dentition that is present in *Sivapithecus* and *Ramapithecus* has been claimed by Jolly (1970) to be functionally adapted for crushing small, hard objects, and he gave grass seeds as a possible example. The best modern analog for this kind of dietary adaptation, and the one that Jolly used, is the gelada baboon, but this primate has deeply crenulated teeth which produce an efficient grinding surface after only a slight degree of wear. The flattened crowns of *Sivapithecus* molars, on the contrary, show little relief, and since the dentine is not exposed until the teeth are heavily worn, the occlusal surfaces remain smooth and poorly adapted for grinding.

More recently, Hatley and Kappelman (1981) have suggested that the similarities between the thick-enameled hominoids such as *Sivapithecus* and other thick-enameled mammals such as bears and pigs may be related to similarities in diet. Bears and pigs are dietary opportunists, living for the most part in seasonal environments, and they rely to a considerable extent on underground food sources from geophytes during times of food shortage.

Geophytic plants with underground plant storage parts are an especially valuable food source, both because they are not available to browsing and grazing herbivores, so that animals exploiting underground food sources can coexist with the specialist herbivores like the ungulates, and because, as agents of storage and vegetative propagation, they are most abundant in seasonal habitats and during the times of greatest shortage of above-ground plant parts. Bears and pigs have both developed digging adaptations, and Hatley and Kappelman (1981) conclude that Plio-Pleistocene hominids may have paralleled them through the development of tools for digging in tropical African savannas. The most widespread distribution of geophytes, however, is not in tropical savannas but in warm, temperate (Amaryllidaceae, Iridaceae) or temperate (Liliaceae) evergreen and deciduous woodlands (Hutchinson, 1967). It is interesting to speculate that the newly emergent thick-enameled hominoids like *Sivapithecus,* expanding into less favorable woodland habitats of this type, perhaps as a result of competition with the cercopithecoid monkeys (Andrews, 1981), may have started to exploit this very abundant food source in seasonal woodland habitats.

In terms of locomotion, an association also exists between body weight and degree of terrestriality in primates (Kay and Simons, 1980), and this was further emphasized by the enlarged sample of 254 mammalian species used here. *Sivapithecus* is grouped on the basis of body size with mainly terrestrial species, and this is consistent with the terrestrial indications for paleoecological analyses. It is also consistent with the scanty evidence from the postcranial remains thought to represent *Sivapithecus,* but these are presently under study and should provide more information when the studies are completed.

The large body size of *Sivapithecus* species is linked with a high degree of sexual dimorphism. This implies a similarly high degree of sexual selection for these hominoids and therefore they may have had a polygynous group structure. The same conclusion is reached by analogy with living primates that are at least partly terrestrial, with the additional indication of large group sizes. This is also indicated by the association of *Sivapithecus* species with moderately seasonal woodland habitats of a kind that would be considered marginal for living primates, for primates in this situation usually form large aggregations of animals. They also are characterized by variable diets, becoming to a large extent omnivorous, and it seems likely that in the absence of positive indications of frugivory or folivory in *Sivapithecus,* and its dental similarities to mammalian omnivores such as pigs and bears, *Sivapithecus* species were omnivores, eating a wide variety of vegetable foods.

In evolutionary perspective, hominoids have been shown to have been primitively forest-adapted (Andrews, 1982) and to have retained a dentition that in most respects does not differ from the primitive catarrhine pattern (Delson and Andrews, 1975). Kay (1977b) has shown that this pattern is usually associated with a frugivorous diet, which may also therefore be considered primitive for hominoids. Postcranially, the hominoids have been shown to be adapted for below-branch posture (Ripley, 1979), and this is also considered primitive for both catarrhines and haplorhines. In all these aspects, therefore, the hominoids appear to have retained primitive adaptations, whereas the ancestral cercopithecoid morphotype seems to have been characterized by the derived condition: the specialized bilophodont dentition (Delson, 1975), diet (Kay, 1978), habitat preference (Andrews, 1982), and posture (Aiello, 1981). This view is also supported by Fleagle and Kay (this volume, Chapter 7). The success of the monkeys relative to the apes has been attributed to changing environment during the Miocene leading to greater niche availability for above-branch feeders (Ripley, 1979), cooling climate in the Miocene leading to reduction in fruit availability (Temerin and Cant, 1983), and to the evolution of differing feeding strategies that enabled monkeys to compete successfully with apes for the same food source (Andrews, 1981). Some of the living apes have converged on the monkeys in some respects, for example, the folivorous adaptations in gorillas (Kay, 1977a); and others have evolved highly specialized locomotor systems, as, for instance, in the gibbons (Fleagle, 1976). At least one group of apes appears to have moved out of the arboreal forest niche and developed some degree of terrestriality in woodland habitats, and that group contains the *Sivapithecus* species discussed here. It may also have contained the orangutan, based on the evidence that it used to be more terrestrial and that it covered a much wider range than at present, extending into the southern Chinese grasslands (Smith and Pilbeam, 1980), and it may be that the shared derived characters in *Sivapithecus* and *Pongo* are the product of this shared habitat.

In conclusion, therefore, it appears that *Sivapithecus* species were large hominoids the size of the orangutan, were adapted to woodland conditions, ranging from tropical to warm temperate, were at least partly terrestrial, and were neither frugivores nor folivores but something in between, eating a wide variety of other food. They appear also to be the closest known relative of the orangutan.

ACKNOWLEDGMENTS

I am very grateful to Wendy Bosley, Steven Dreyer, Libby Evans, Terry Harrison, Lawrence Martin, and Chris Stringer for comments on this chapter.

References

Abel, D. 1902. Zwei neue Menschenaffen aus den Leithakalkbildungen des Wiener Beckens. *Sitzungber. Akad. Wiss. Wien Math.-Naturwiss. Kl.* **111:**1171–1207.
Aiello, L. 1981. Locomotion in the Miocene Hominoidea, in: *Aspects of Human Evolution* (C. B. Stringer, ed.), pp. 63–97, Taylor and Francis, London.
Andrews, P. 1978. A revision of the Miocene Hominoidea of East Africa. *Bull. Br. Mus. (Nat. Hist.) Geol.* **30:**85–224.

Andrews, P. 1981. Species diversity and diet in monkeys and apes during the Miocene, in: *Aspects of Human Evolution* (C. B. Stringer, ed.), pp. 25–41, Taylor and Francis, London.

Andrews, P. 1982. Ecological polarity in primate evolution *Zool. J. Linn. Soc.* **74:**233–244.

Andrews, P., and Tekkaya, I. 1980. A revision of the Turkish Miocene hominoid *Sivapithecus meteai. Palaeontology* **23:**85–95.

Andrews, P., and Tobien, H. 1977. New Miocene locality in Turkey with evidence on the origin of *Ramapithecus* and *Sivapithecus. Nature (Lond.)* **268:**699–701.

Andrews, P., and Walker, A. 1976. The primate and other fauna from Fort Ternan, Kenya, in: *Human Origins: Louis Leakey and the East African Evidence* (G. L. Isaac and E. R. McCown, eds.), pp. 279–304, Benjamin, Menlo Park, California.

Andrews, P., Lord, J. M., and Evans, E. M. N. 1979. Patterns of ecological diversity in fossil and modern mammalian faunas. *Biol. J. Linn. Soc.* **11:**177–205.

Axelrod, D. I. 1975. Evolution and biogeography of Madrean–Tethyan sclerophyll vegetation. *Ann. Mo. Bot. Gard.* **62:**280–334.

Crook, J. H., and Gartlan, J. S. 1979. Evolution of primate societies, in: *Primate Ecology* (R. W. Sussman, ed.), pp. 363–373, Wiley, New York.

Darwin, C. 1871. *The Descent of Man and Selection in Relation to Sex.* John Murray, London.

De Bonis, L., and Melentis, J. 1977. Un nouveau genre de Primate hominoide dans le Vallesian (Miocene supérieur) de Macédoine. *C. R. Acad. Sci. Paris D* **284:**1393–1396.

De Bonis, L., Bouvrain, G., Geraads, D., and Melentis, J. 1974. Première découverte d'un Primate hominoide dans le Miocène supérieur de Macédoine (Grèce). *C. R. Acad. Sci. Paris D* **278:**3063–3066.

Delson, E. 1975. Evolutionary history of the Cercopithecidae, in: Approaches to Primate Paleobiology (F. S. Szalay, ed.). *Contrib. Primatol.* **5:**167–217.

Delson, E., and Andrews, P. 1975. Evolution and interrelationships of the catarrhine primates. in: *Phylogeny of the Primates: An Interdisciplinary Approach.* (W. P. Luckett and F. S. Szalay, eds.), pp. 405–446, Plenum, New York.

Fleagle, J. G. 1976. Locomotion and posture of the Malayan Siamang and implications for hominoid evolution. *Folia Primatol.* **26:**245–269.

Fleagle, J. G. 1978. Size distribution of living and fossil primate faunas. *Paleobiology* **4:**67–76.

Fleagle, J. G., Kay, R. F., and Simons, E. L. 1980. Sexual dimorphism in early anthropoids. *Nature (Lond.)* **287:**328–330.

Gantt, D. G. 1977. *Enamel of Primate Teeth: Its Thickness and Structure with Reference to Functional and Phyletic Implications,* Ph.D. Dissertation, Washington University.

Gaulin, S. J. C. 1979. A Jarman/Bell model of primate feeding niches. *J. Hum. Evol.* **7:**1–20.

Gingerich, P. D. 1977. Correlation of tooth size and body size in living hominoid primates, with a note on relative brain size in *Aegyptopithecus* and *Proconsul. Am. J. Phys. Anthropol.* **47:**395–398.

Greenfield, L. O. 1972. Sexual dimorphism in *Dryopithecus africanus. Primates* **13:** 395–410.

Greenfield, L. O. 1979. On the adaptive pattern of "*Ramapithecus.*". *Am. J. Phys. Anthropol.* **50:**527–548.

Hatley, T., and Kappelman, J. 1981. Bears, pigs and Plio-Pleistocene hominids: A case for the exploitation of belowground food resources. *Hum. Ecol.* **8:**371–387.

Herring, S. W. 1972. The role of canine morphology in the evolutionary divergence of pigs and peccaries. *J. Mammal.* **53:**500–512.

Hutchinson, J. 1967. *The Genera of Flowering Plants. II Dicotyledones,* The Clarendon Press, Oxford.

Hylander, W. 1975. Incisor size and diet in anthropoids with special reference to Cercopithecidae. *Science* **189:**1095–1098.

Jarman, P. J. 1974. The social organization of antelope in relation to their ecology. *Behaviour* **48:**215–266.

Jolly, C. J. 1970. The seed-eaters: A new model of hominid differentiation based on a baboon analogy. *Man* **5:**5–28.

Jolly, C. J. 1972. The classification and natural history of *Theropithecus (Simopithecus)* (Andrews 1916) baboons of the African Plio-Pleistocene. *Bull. Br. Mus. (Nat. Hist.) Geol.* **22:**1–122.

Kay, R. F. 1975. The functional adaptations of primate molar teeth. *Am. J. Phys. Anthropol.* **43:**195–216.

Kay, R. F. 1977a. The evolution of molar occlusion in the Cercopithecidae and early Catarrhines. *Am. J. Phys. Anthropol.* **46:**327–352.

Kay, R. F. 1977b. Diet of early Miocene African hominoids. *Nature (Lond.)* **268:**628–630.

Kay, R. F. 1978. Molar structure and diet in extant Cercopithecidae, in: *Development, Function and Evolution of Teeth* (P. M. Butler and K. A. Joysey, eds.), pp. 309–339, Academic, London.

Kay, R. F., and Simons, E. L. 1980. The ecology of Oligocene African Anthropoidea. *Int. J. Primatol.* **1:**21–37.

Kretzoi, M. 1975. New ramapithecines and *Pliopithecus* from the Lower Pliocene of Rudabanya in north-eastern Hungary. *Nature (Lond.)* **257:**578–581.

Kretzoi, M., Krolopp, E., Lovinz, H., and Palfalry, I. 1976. Floren- und Faunenfunde der altpannonischen Prähominden-Fauna von Rudabanya und ihre stratigraphische Bedeutung. *M. Áll. Földtani. Intézet. Évi. Jelentése.* **1974:**365–394.

Leakey, L. S. B. 1962. A new Lower Pliocene fossil primate from Kenya. *Ann. Mag. Nat. Hist.* **4:**689–696.

Leakey, L. S. B. 1967. An early Miocene member of Hominidae. *Nature (Lond.)* **213:**155–163.

Le Gros Clark, W. E., and Leakey, L. S. B. 1951. The Miocene Hominoidea of East Africa. *Fossil Mammals of Africa* (Br. Mus. Nat. Hist.) **1:**1–117.

Lewis, G. E. 1934. Preliminary notice of new man-like apes from India. *Am. J. Sci.* **27:**161–179.

Lewis, G. E. 1937. Taxonomic syllabus of Siwalik fossil anthropoids. *Am. J. Sci.* **34:**139–147.

Lucas, P. W. 1982 (in preparation) An analysis of canine size and jaw shape in higher primates.

MacKinnon, J. 1971. The Orang-utan in Sabah today. *Oryx* **11:**141–191.

Madden, C. T. 1980. East African *Sivapithecus* should not be identified as *Proconsul nyanzae*. *Primates* **21:**133–135.

Maynard Smith, J., and Savage, R. J. G. 1959. The mechanics of mammalian jaws. *School Sci. Rev.* **40:**389–401.

McHenry, H. M., Andrews, P., and Corruccini, R. S. 1980. Miocene hominoid palatofacial morphology. *Folia Primatol.* **33:**241–252.

Ozansoy, F. 1965. Etude des gisements continentaux et des mammiferes du Cénozoique de Turquie. *Mem. Soc. Geol. Fr.* **44:**5–89.

Pilbeam, D. R. 1969. Tertiary Pongidae of East Africa: Evolutionary relationships and taxonomy. *Bull. Peabody Mus. Nat. Hist.* (Yale Univ.) **31:**1–185.

Pilbeam, D. R., and Zwell, M. 1972. The single species hypothesis, sexual dimorphism, and variability in early hominids. *Yearb. Phys. Anthropol.* **16:**69–79.

Pilbeam, D. R., Meyer, G. E., Badgley, C., Rose, M. D., Pickford, M. H. L., Behrensmeyer, A. K., and Shah, S. M. I. 1977. New hominoid primates from the Siwaliks of Pakistan and their bearing on hominoid evolution. *Nature (Lond.)* **270:**689–695.

Pilbeam, D. R., Behrensmeyer, A. K., Barry, J. C. and Shah, S. M. I. 1979. Miocene sediments and faunas of Pakistan. *Postilla* (Peabody Mus. Nat. Hist., Yale Univ.) **179:**1–45.

Rijksen, H. 1978. *A Field Study on Sumatran Orang-utans (Pongo pygmaeus Abelii Lesson 1827)*, H. Veenman and Zonen, Wageningen.

Ripley, S. 1979. Environmental grain, niche diversification, and positional behaviour in Neogene primates: An evolutionary hypothesis, in: *Environment Behaviour and Morphology: Dynamic Interaction in Primates* (M. E. Morbeck, H. Preuschoft, and N. Gonsberg, eds.), pp. 37–74, Gustav Fischer, Stuttgart.

Simons, E. L., and Pilbeam, D. R. 1965. Preliminary revision of the Dryopithecinae (Pongidae, Anthropoidea). *Folia Primatol.* **3:**81–152.

Smith, R. J., and Pilbeam, D. R. 1980. Evolution of the orang-utan. *Nature (Lond.)* **284:**447–448.

Sun, X. and Wu, Y. 1980. Palaeoenvironment during the time of *Ramapithecus lufengensis*. *Vertebr. Palasiat.* **18:**255.

Temerin, L. A., and Cant, J. G. H. 1983. The evolutionary divergence of Old World monkeys and apes. *Am. Nat.* (in press).

Xu, Q., and Lu, Q. 1979. The mandibles of *Ramapithecus* and *Sivapithecus* from Lufeng, Yunnan. *Vertebr. Palasiat.* **17:**1–13.

Facts and Fallacies Concerning Miocene Ape Habitats

18

A. KORTLANDT

The Problem

In recent years a controversy has arisen concerning the East African land-scape and vegetation in the time range from about 23 to about 16 million years ago (m.y.a.), i.e., in the early Miocene. Until a few years ago there was no such difference of opinion. It was generally accepted that the fossil ape hab-itats were characterized by a mosaic landscape consisting of woodlands and savannas interlaced with broad riverine forests and valley forests, alternating with rain forests and mountain forests on the slopes of ridges and mountains, and with grass plains on periodically inundated grounds. This concept origi-nated primarily with Leakey (1955, 1969) and Le Gros Clark and Leakey (1951). Supporting evidence came from Bishop (1963, 1968), Butzer (1978), Chaney (1933), Chesters (1957), Hamilton (quoted in Walker, 1969), Mac-Innes (1953), Omaston (quoted in Bishop, 1968), Verdcourt (1963, 1972), and Whitworth (1953/54, 1958). This was the kind of landscape that is still widespread over enormous parts of Africa, in those areas between the rain forest belt and the dry bushland belt where the natural vegetation has not yet been too degraded by wood-cutting, agriculture, overgrazing, and bushfires resulting from human activity.

The proposed range and diversity of Miocene habitats, according to this concept, fitted in very well with the anatomy of the dryopithecines, which

A. KORTLANDT • Vakgroep Psychologie en Ethologie der Dieren, Universiteit van Amster-dam, Nieuwe Achtergracht 127, 1018 WS Amsterdam, The Netherlands.

were characterized, among other things, as "active, running and leaping creatures and in no way specialized for a purely arboreal habitat." Moreover, "the type of hind limb which they evidently possessed provides a suitable antecedent for the subsequent evolutionary development along divergent lines of the brachiating specializations of the modern large apes on the one hand, and of the type of limb structure required for the erect posture of the Hominidae on the other" (Le Gros Clark and Leakey 1951, pp. 113–114). Thus this concept of a semiarboreal, semiterrestrial taxon of ancestral apes, as formulated in the 1950s, combined a "missing link" type of primate between apes and hominids with a "missing link" type of habitat between forests and grass plains.

The concept also fitted in very well with the subsequent evidence indicating the occurrence of savanna-adapted behavior patterns among chimpanzees and gorillas. That is, behavioral data suggested that "the protohominid ancestors of the contemporary chimpanzees and gorillas were originally 'eurytopic' species, i.e. inhabitants not only of forests, but also of savannas, park-like landscapes, and other semi-open and diversified habitats" (Kortlandt and Kooij, 1963, p. 68). This was called the dehumanization hypothesis of African ape evolution (Kortlandt, 1959; Kortlandt and Kooij, 1963). Thus convergent evidence from both anatomy and behavior resulted in a consistent picture of the dryopithecine ancestors. A unified theory of ape and human evolution had come within reach.

At that time, the chimpanzee was still believed to be an (almost) exclusively forest-dwelling ape that moved about in the canopy mainly by brachiating. I had not yet done my chimpanzee habitat field survey in which I found that about 40% of the natural geographical range of these apes was situated in the savanna, woodland, and dry forest belt of Africa, whereas only about 60% was situated in the rain forest belt. We now know that these apes are ecologically and behaviorally much more "human" and less "dehumanized" than we could know at that time. Since then the hypothesis has been accordingly revised (Kortlandt, 1965, 1968, 1972, in preparation; Kortlandt and van Zon, 1969).

In recent years, Andrews and Van Couvering (1975), Andrews and Walker (1976), Pilbeam (1979), Van Couvering and Van Couvering (1976), and Walker (1969) have proposed a rather different picture of East African early Miocene ape habitats. In their words, "The [equatorial] lowland forest belt may have extended eastwards as far as the [then] continental divide and possibly to the East African coast" (Andrews and Van Couvering, 1975, p. 98 and Fig. 2); "Western Kenya . . . was under evergreen forest during the preceding [early] Miocene" (Andrews and Walker, 1976, p. 302); "Geological, paleobotanical, and paleontological evidence suggests that habitats were mainly forested" (Pilbeam, 1979, p. 343); "There may have . . . existed a continuous forest cover from Ethiopia to the eastern part of the equatorial lowland forest" (Van Couvering and Van Couvering, 1976, p. 163); and "Large tracts of unbroken rain forest covered most of East Africa in early Miocene times" (Walker, 1969, p. 593). These authors present a quite new and rather surprising view. The implication would appear to be that the Miocene apes should be considered to have been much more forest-dwelling (and probably also much more treetop-dwelling) than has been believed since Le Gros Clark and Leakey (1951).

It is a rather odd situation. On the one hand, all field workers who are currently studying the ecology and behavior of the Great Apes depict an image of them that is steadily becoming more and more human-like. On the other hand, a new generation of paleoscientists depicts an image of the ancestral taxon in which the humanoid behavioral elements have been erased.* In that image the early Miocene apes are, in their ecological adaptations, no longer precursors of humans. They have lost their hominid predestination. The emergence of *Ramapithecus* now becomes, so to speak, a creationist event. In any case, these new ideas are rather disastrous to what I called above a unified theory of ape and human evolution.

It seems to me, however, that the new view is a fallacy. While the evidence presented in the 15 classical papers quoted above is too substantial to be easily discarded, one wonders what the roots of the reinterpretation by Andrews, Van Couvering, and Walker might have been. In the following section I shall review and analyze their arguments. Pilbeam's views will not be discussed, because he advanced no evidence and simply referred to Andrews and Van Couvering (1975).

A Critical Review

Topography, Geology, and Climate

Africa is generally believed to have been, in early Miocene times, a continent characterized by a low relief. During the very long preceding time span of tectonic quiescence, which lasted some 70 million years (Sowerbutts, 1972), the preexisting high surfaces and mountain chains had largely been weathered, eroded, and denuded. The resulting topography was predominantly a gently undulating landscape, with here and there some remnants of older plateaus and mountain ranges. However, a process of uplift and volcanism had already begun, which would eventually result in the formation of the Ethiopian Mountains and the Great Rift Valley system that runs from Lebanon through the Red Sea to Mozambique and Zimbabwe (e.g., Baker *et al.*, 1972; Holmes, 1966; King, 1978; Mohr, 1971; Shackleton, 1978; and map in Kortlandt, 1972). According to Andrews and Van Couvering (1975, p. 72), this early Miocene relief (i.e., before the major uplift of the Rift highlands) "would result in an eastward extension of the high year-round rainfall zone

---

*By "humanoid behavior" I designate those natural behavior patterns among nonhominids that result from parallel or convergent evolution and are functionally identical or equivalent to those behavior patterns that, in a more advanced form, characterize the hominid lineage of specialization.

that today characterizes the Congo Basin, to cover most of Uganda and western Kenya and Tanzania."

It is obvious that the Rift mountain chains do contribute to the relative aridity of East Africa compared with the equatorial rainfall belt in West and Central Africa. The fact has been generally recognized for many decades. I have elaborated on its role in hominid evolution (Kortlandt, 1968, 1972) and proposed ideas partly similar to those of Andrews and Van Couvering (1975). However, the influence of the Rift mountains cannot easily be quantified in meteorological terms. Moreover, greater desiccating influences must have been exerted already during the early Miocene by the gradual closure of the Tethys Sea, by the uplift of the Ethiopian Plateau and the Taurus–Himalayan mountain range, and by the jet stream from the Philippines to Senegal that resulted from this topography. Among other things, the Ethiopian basalts had already reached a thickness of 2000–2500 m about 25 m.y.a. The formation of large volcanoes in Kenya and Uganda had started, and the rift-forming processes in southern East Africa were in full swing in the beginning of the Miocene (references above). At any rate, from a climatological viewpoint, the most decisive elements of the continental and intercontinental topography were to a large extent already present (e.g. Axelrod and Raven, 1978; Butzer, 1978; Dewey *et al.*, 1973; Flohn 1964; Frakes and Kemp, 1972; Kennett *et al.*, 1974; Kortlandt, 1972; Mohr, 1971; van Zinderen Bakker, 1977/78). In this constellation the subsequent rise of the Rift mountain ridges was only an additional factor. Before the major uplift, therefore, the climate of eastern Africa must already have been substantially drier than that of the equatorial belt in central and western Africa at that time. This is supported by a comparison of the Miocene floral data from East Africa (see section on Vegetation) with those from Cameroon (Salard-Cheboldaeff, 1977).

Furthermore, Andrews and Van Couvering (1975, pp. 76–77) mentioned the presence of old erosional remnant surfaces no less than 1500 m high in the Cherangani and Chemerongi area at that time. Brock and MacDonald (1969) have described a "fossilized" escarpment of more than 600 m high which showed lateritic weathering and was covered by lavas and tuffs from Mount Elgon dating from 24–22 m.y.a. A subsequent tectonic movement in the early phase of volcanic activity increased this relief by another 300 m. We may be certain that similar relief features were present elsewhere in East Africa because the remnants of the Jurassic and Cretaceous erosion surfaces still exist at many places even today (Bishop and Trendall, 1966/67; Cahen, 1954; Gautier, 1965*a,b*; King, 1967; Saggerson and Baker, 1965). Admittedly, the overall altitude of eastern and southern Africa was lower than it is nowadays, but the amount of relief would have been at least as much as (or even more than) in many other geomorphologically old, but nevertheless still quite rugged landscapes in Africa in our time (de Swardt, 1964), e.g. in western Guinea (my observations). Apparently Andrews and Van Couvering were not sufficiently aware that such a rugged topography would have resulted in large rain-shadow areas, locally quite diversified climates, and forest–woodland–savanna mosaic landscapes similar to those which can be observed today. In short, their climatic argument based upon topography gives an oversimplified picture and suggests a much more humid overall climate during the early Miocene in East Africa than is actually justified by the facts.

Much the same emerges from the geological data. Many early Miocene beds contain conglomerates and coarse gravels in several places and in several horizons, beginning already 24 m.y.a., i.e., before the Oligocene–Miocene boundary (e.g. Van Couvering and Van Couvering, 1976, Fig. 3). This indicates that vigorous tectonic movements and erosional rejuvenations were occurring again and again. That is, the supposed monotonous East African peneplain had been broken up repeatedly, both before and during the early Miocene, and had given way to a lively, rugged, and diversified type of landscape. Tectonic studies suggest that this process may have started already in the Eocene (e.g. Baker *et al.,* 1972).

As to the climate, the presence of laterites at several sites indicates that there have been periods and areas characterized by a marked dry season, both in the early Miocene and in earlier times. The Gumba Beds at Rusinga, whose age is somewhere in the 18 million year range, contain red earths, concretionary clay pellets, and other indications of a locally and/or temporarily semi-arid climate, with a short, seasonal, or irregular rainfall, i.e., a *sahélien* or dry-*soudanien* type of climate.* The same is indicated by the presence of ephemeral lakes (Whitworth, 1953/54). The inference of the existence of dunes by Butzer (1978) is perhaps controversial, but the remaining evidence is convincing enough. Such deposits could never have been formed in a forest type of climate, not even on a local scale when the forest had been destroyed by whatever unusual event, because in such a climate the soil would immediately have been overgrown by secondary vegetation, except on flood plains. Thus, again, the geological data show that there was no continuous forest cover over most of East Africa.

Morphology of Vegetation

The new interpretation of the East African early Miocene landscape as an (almost) continuous rain forest is especially surprising because the botanical data indicating less humid and more open types of habitat are clear enough. The only practical difficulty is that the collected material has been published in a preliminary and rather incomplete form. However, even the incompleteness of the data does not give much leeway for subjective interpretations. I shall now review and interpret the published data.

According to Bishop (1963, p. 260) and Chesters (1957, p. 31), thorny wood was "common" at both Napak and Rusinga, and even "especially com-

*The French technical terms are more precisely defined than are the English ones. The distinction between the *sahélien* belt and the *soudanien* belt in Africa, together situated between the Sahara and the rain forest belt, is primarily based upon a difference in flora and vegetation (see page 470). The terms are also used to indicate climatic zones because the ecological adaptations of plants are highly dependent on climate.

mon" at Rusinga. Since thorns can serve as a protection against browsers, it would follow that browsing mammals should have been equally "common," *casu quo* "especially common," because they tend to spare the thorny plant species. As a matter of fact, thorny wood is generally rare in tropical rain forests and browsers are equally rare there, apparently because nearly all the biomass production takes place at treetop level, beyond reach of the browsers. However, in open sites within the rain forest, for example, along water courses, and particularly in swampy areas where hoofed animals come to drink and wallow, and also in places where elephants have seriously damaged the forest, thorny plants and trees are sometimes more common. There are even places where they predominate. In open, less humid types of forest, and in open woodlands and bushlands, where more biomass is produced at the lower levels, thorny plants and trees occur more abundantly, again apparently as a result of selection pressures by browsers. In *sahélien* areas they are the dominant vegetation. Thus the "common" presence of thorns at Napak and Rusinga suggests not only the frequent appearance of browsers, but also a partly open arboreal vegetation and a seasonal climate. The qualifications "common" and "especially common" are quantitively inadequate for a final assessment, but I would estimate that, in present-day Africa, in areas without bushfires due to human activity but with many browsers, vegetation zones with an annual rainfall between 500 and 1000 mm (in West Africa up to 1300 mm) might represent suitable models for the amounts of thorns as suggested in the papers by Bishop and Chesters. Such zones are Belts Nos. 17, 18, and 20 on the AETFAT (1959) Vegetation Map of Africa, i.e., the *soudano-sahélien*, the central *soudanien*, and the Zambesian belt.*

According to Chesters (1957) climbers and lianas were abundant at Rusinga. Her list of fossil plant species included at least seven (probably ten) tree species and at least five (probably seven) climbers. When the provisionally identified families were added the proportion became probably 22 tree spe-

---

*The *sahélien* may be defined for our purposes as a vegetation south of the Sahara consisting of wooded steppes, low-grass savannas, and bushlands, with a more or less open, or very open, arboreal stratum composed mainly of deciduous, fine-leaved, and quite thorny *Acacia* and *Commiphora* species, not higher than 7–12 m as a rule, but often much lower. The rainfall is less than about 500–600 mm (in West Africa 600–750 mm) per year, and the dry season lasts longer than 7 months on the average, with a rainfall of less than 30 mm per month. In the last several decades the natural vegetation has been largely destroyed by overgrazing, overbrowsing, and wood-cutting.

The *soudanien* may be defined as the belt originally covered by dry, predominantly deciduous forests, with a somewhat open stand of trees that generally reached heights of 12–15 m, sometimes 20 m, locally even more, and with a dense undergrowth that consisted of partly evergreen grasses, herbs, shrubs, and creepers. The annual rainfall amounts to less than 1300–1500 mm, in West Africa less than 1700 mm, and the dry season lasts at least 3–4 months as a rule, in West Africa 5 months or more. The natural vegetation in this belt has been almost entirely destroyed by wood-cutting, agriculture, and bushfires of human origin, but a few relict forests survive, mainly at sacred places, graveyards, etc. The present vegetation consists of dry, wooded savannas and woodlands, mainly with a dense growth of tall grasses that are kindled and burnt almost every year as a result of human activity.

The Zambesian may be defined for our purposes as the equivalent on the southern hemisphere of the *soudanien*. For more elaborate definitions see Schnell (1976).

cies (small and large) to 11 climbers. From this and other data she concluded that her sample represented largely the vegetation of gallery forests and forest margins because climbers need much light and grow from clearings and forest fringes into the tall trees. This would imply a mosaic type of vegetation as conceived by Leakey and others (see first section). (The material probably came entirely from the Hiwegi Formation, which is about 18 million years old; M. Pickford, personal communication).

I would like to broaden this assessment by concluding that the abundance of climbers indicates a so-called secondarized type of humid forest.* Such a secondarization could have been caused either by the action of watercourses (as implied by Chesters), or by the action of elephants and other destructive herbivores, or by a combination of both (see Fig. 1A and 1B). Elephants are as a rule not very destructive in the equatorial, evergreen type of rain forest (Curry-Lindahl, 1974; Guillaumet, 1967; Kortlandt, 1972, 1974; Letouzey, 1968; Merz, 1977). They can, however, be very devastating in semideciduous and riverine forests (e.g. Buechner and Dawkins, 1961; Eggeling, 1947; Jackson, 1956; Kortlandt, 1972; Laws *et al.*, 1975; Mitchell, 1960; Wing and Buss, 1970), and even more so in *soudanien, sahélien,* and Zambesian vegetations (Fig. 2) (e.g., Barnes, 1980; Caughley, 1976; Cobb, 1976; Croze, 1974; Harrington and Ross, 1974; Harris and Fowler, 1975; Kortlandt, 1976; Laws *et al.*, 1975; Ross *et al.*, 1976). In a forest habitat, this destruction can increase plant biomass production. The relative abundance of both thorny wood and climbing plants at Rusinga may, therefore, perhaps be interpreted as an indication of heavy herbivore pressure on a semideciduous or *soudanien* vegetation. This could have resulted in a mosaic or parkland type of forest, shrub, and savanna landscape. Such an interpretation is supported by Chesters' statement that a few tree species in her sample "are said to grow in more open situations."

Some colleagues have difficulty in believing that, *under natural conditions,* herbivores can indeed inflict as much damage to the vegetation as I have attributed to them. Forestry experts, as a rule, are familiar only with forest reserves where the number of wild game is severely restricted by hunting and/or poaching. Nature conservationists, on the other hand, while familiar with the enormous damage which elephants can cause, tend to attribute this predominantly to an overpopulation of animals that have been driven away elsewhere and have fled to a national park. The abnormal extra damage caused by these immigrants tends to obscure the normal influence of herbivores in a state of ecological equilibrium. Those who feel doubts about my assessment should visit, for instance, the northern part of Manyara National Park in Tanzania, or the centuries-old wild parts of the New Forest near

---

*A forest unaffected by human activity is called a "primary forest." When a forest has been cut and is later allowed to regrow (e.g., in rotational farming), one gets a "secondary forest." When, however, in a primary forest only some of the trees have been destroyed (e.g., in timber exploitation) one gets a "secondarized forest" when the felled trees have been replaced by the spontaneous growth of new trees. In the course of this regeneration process, climbers get their chance to grow from the clearings into the trees that have not been felled. This is one of the so-called "edge effects" in a forest vegetation.

Fig. 1. Two characteristic views of a semideciduous monsoon rain forest in a national park in Uganda, showing the thinning out of the arboreal vegetation caused by browsing, debarking, breaking, and uprooting of trees, mainly by elephants. [Rabongo Forest, a *Cynometra–Holoptelea* forest in Kabalega National Park, 1964; photographs copyright A. Kortlandt.]

Fig. 2. A group of elephants in Tsavo National Park destroying a large baobab tree and eating its core. The branches of the two acacia trees have been torn off by elephants and the trees will eventually die from debarking. [Photograph 1975, copyright A. Kortlandt.]

Southampton in England, where horses, cattle, and deer are allowed to browse and graze. Here they can observe the damage herbivores can cause.

Floristic Composition

Chesters did not give an ecological specification of the sample she described, except for stating that the fruit of the Apocynaceae are adapted for water dispersal. Nevertheless, certain preliminary conclusions can be drawn from her list of species, genera, and provisionally identified families. (1) Some of these taxa occur, in our time, in various types of habitat, both humid and dry. They do not, of course, give any indications of climate and type of vegetation. (2) Some other taxa occur both in continuous rain forests and in riverine forests that intersect woodlands and savannas. They are, therefore, not very indicative either. (3) Among the arboreal taxa, *Entandrophragma*, *Antrocaryon*, *Celtis*, and the Lauraceae live mainly in continuous or extensive tracts of evergreen, semideciduous, and secondarized types of rain forest, but *Entandrophragma* has been found only on Mfwanganu [not on Rusinga Island itself (Chesters, 1957, p. 34)]. (4) The climbers on the list belong mainly to the small species that are characteristic of secondarized forest vegetation. (5) Among the remaining taxa, *Zizyphus*, *Berchemia*, *Odina* (*Lannea*), *Schrebera-Schreberoides*, and Capparidaceae occur predominantly or exclusively in dry forests, woodlands, and savannas, i.e., in the *soudanien* and *sahélien* vegetation belt (Dale and Greenway, 1961; Hutchinson and Dalziel, 1954–1972; J. J. F.

E. de Wilde, personal communication). (6) Trees are, generally speaking, quite conservative from an evolutionary viewpoint. Thus we may conclude that, in the Miocene Rusinga vegetation, genera adapted in our time for surviving a long dry season probably predominated somewhat over genera characteristic of the evergreen rain forest.

When one considers all the evidence together, i.e., the thorns, the lianas, and the floral list, Chesters' account suggests a *soudano-guinéen* and partly *soudanien* mosaic or parkland type of woodland and more or less open forest, presumably shaped by herbivore activity, and interspersed with luxurious riverine forests. Such a vegetation might also include some true rain forest (either locally, due to edaphic and soil conditions, or temporarily, as a result of climatic fluctuations), as well as some more or less arid types of woodland (either locally or temporarily). The overall impression indicates a rainfall somewhere in the 1000–1500 mm range, correlated with Belts Nos. 8 (in part), 16, and 17 (in part) on the AETFAT Vegetation Map of Africa (1959) at present, defined, respectively, as "Forest–Savanna Mosaic," "Woodlands, Savannas (and Steppes), undifferentiated—relatively moist types," and "Woodlands, Savannas (and Steppes), northern areas: with abundant *Isoberlinia doka* and *I. dalziellii*." Thus the floral composition suggests perhaps a slightly more humid habitat than what has been concluded above from the thorny plants. This is not surprising, because, on the one hand, a waterside community would have harbored a relatively more humid flora, while on the other hand, the damage by herbivores would have caused an increased abundance of thorns. An estimate of 800–1200 mm of annual rainfall may, therefore, be the best guess.

Similar conclusions can be drawn from the data presented by Chaney (1933) and Omaston (in Bishop, 1968, p. 30). From the character and size of the leaves, Chaney estimated for the vegetation at the Bugishu site at Mount Elgon an annual rainfall "much lower than characterizes the tropical rain forest of Africa" and "not much less than 40 inches," which is somewhat less than nowadays in that area and which would be correlated with a "savanna or woodland." Phillips [quoted by Chaney (1933)] estimates "25 to 50 inches per annum—probably under conditions of a definite dry and a definite rainy season," resulting in "Open Woodland and Woodland." This means that this vegetation would be comparable to the southern edge of Belt 20 on the AETFAT map, defined as "Woodlands, Savannas (and Steppes), Undifferentiated: relatively dry types," i.e. the *soudano-sahélien* belt. (When drawing such comparisons one should, of course, look in the field for a vegetation which has not been too degraded by human activity.) Omaston (quoted in Bishop, 1968) concludes for Napak "a tropical vegetation of trees and shrubs," without broad-leaved monocotyledons, ferns, savanna grasses, or microphylls, with a climate "not very moist," but "not very dry" either, i.e., certainly not a rain forest, but rather like the type of *soudano-guinéen* orchard landscape (with moderate browsing and grazing influences) described by Kortlandt (1968, p. 60, photo 1). Similar data from Ethiopia have been summarized by Axelrod and Raven (1978, pp. 107–108).

The extensive evidence reviewed above has largely been insufficiently evaluated by Andrews and Van Couvering (1975), Van Couvering and Van Couvering (1976), and Walker (1969). For instance, the presence of only one mahogany (*Entandrophragma*) species at the adjacent site of Mfwanganu, and its absence in the main (Hiwegi) deposits of Rusinga, was all too readily interpreted as "abundance" of this genus and as an "indication of extensive tracts of evergreen forest" (Andrews and Van Couvering, 1975, p. 78). In fact, *Entandrophragma* spp. occur not only in evergreen forests but also in semideciduous forests, and they tend to occur especially in the middle part of the succession, i.e., in what is often called old mature secondary forest (T. J. Synnott, personal communication). This again would fit in with the suggestion that large herbivores had opened up and thinned out the forest zone. Equally surprising is that Andrews and Van Couvering (1975, p. 79) have derived a list of "most common taxa" from Chesters' floral list. The absentees are two genera, two families, and eight provisionally identified families. They include one family which belongs predominantly in woodlands, bushlands, and savannas (Capparidaceae).

Andrews and Van Couvering also blurred the ecological distinctions between a dry deciduous forest (*forêt sèche dense*), a dry evergreen forest (*muhulu*, *Baikiaea* forest, etc.), a riverine forest, a montane forest, a subequatorial semideciduous rain forest, an equatorial evergreen rain forest, and a forest in general. In addition, their paper is marred by the remark: "Chesters, apparently in an attempt to rationalize the presence of fossil forest trees with the idea that the dominant habitat was savanna, suggested that the countryside was open with forest present along the water courses" (p. 63). In fact, Chesters gave a different version: "The abundance of climbers in association with the evidence of their occurrence given above might be regarded as support for the theory advanced by the faunal experts, on purely faunal evidence, that the Rusinga deposits represent, in part, a gallery forest type of vegetation" (Chesters, 1957, p. 33).

Finally, Walker (1969) ignored the presence of grass horizons and the alkaline herb cf. *Juncellus laevigatus* at Bukwa when he inferred from Hamilton's data (Walker, 1969, p. 593) "evergreen rain forests, gallery forests and freshwater lakes." Moreover, a gallery forest in an evergreen rain forest is a contradiction in terms.

The above conclusions are still rather tentative because too little is known about the precise origin of the collected material. The Rusinga and Mfangano deposits cover a time span of some 4 million years, during which time substantial changes of climate may have occurred (Pickford, 1981). Moreover, it is generally assumed that plant fossils from several sites have been mixed up in the material made available to Chesters (Andrews, personal communication). It is far from certain, therefore, whether these fossils are indeed representative of the dryopithecine habitats. This is, however, partly compensated by the wide diversity of environments in which the bones and teeth of these apes have been found (Pickford, 1981, and this volume, Chapter 16). One can only hope that the mixing up of the plant fossils has resulted in a sample that represents a good average. For the rest one should take the data for what they are: the only ones available on this issue.

Unfortunately, both Chaney and Chesters have published only preliminary and incomplete reports on the early Miocene plant material. According
to Andrews (personal communication), the thorns from Rusinga belong to the
cone-shaped type which one commonly finds on certain rain forest trees,
whereas the needle-shaped and recurved types characteristic of *Acacia* savannas are absent. Actually, these cone-shaped thorns are also found on many
soudanien trees and on species that occur in both belts (e.g., on the well-known
cotton tree or kapok tree, *Ceiba pentandra*). Thus, the character of the thorns
would hardly contradict my opinion stated above. Furthermore, according to
Andrews (personal communication), thorny wood appears to be less common
than is suggested in Chesters' description. In his opinion it is possible that in
the original collection an oversampling of thorny wood took place, reflecting
the personal interests of the collector. In any case, a new, thorough inspection
of the material would be highly desirable. Fortunately, Dr. M. E. Collinson has
made a beginning with this work.

It is regrettable that in the several decades since this material was collected no paleobotanist has had a second look at it. The fossil vegetations are
of crucial importance for an understanding of the ecological conditions which
eventually led to the emergence of humans. It is strange that this subject has
been neglected for such a long time.

Taphonomy

It is a common cliché that "the apes" do not fossilize as a rule because they
inhabit forests, and forest soils are acid and destroy the skeleton. The matter
is, of course, not as simple as that. In a forest environment, fossilization can
take place in rivers and lakes, in caves, or under volcanic ashes. A more
complicated situation arises in mixed forest and nonforest habitats. A typical
African mosaic landscape usually consists of woodlands, savannas, and sometimes bushlands on the plateaus and watersheds, while the valleys, ravines,
and slopes are covered by a forest vegetation. Periodically inundated valley
floors may be covered by either swamp forest or open grassland, according to
local edaphic conditions. An important factor in shaping such a mosaic vegetation is that most hoofed mammals dislike foraging on slopes. Another factor
is that the water table in the bottom of the valley supports a lusher vegetation.

Thus the sites where sedimentation takes place, and hence where fossilization of dying animals is possible, are often fringed by forest. Consequently
the forest-adapted animal species will often be over-represented in the fossil
record of a former mosaic landscape. Andrews, Van Couvering, and Walker
have disregarded this aspect. Another factor is that primates often spend the
night in riverine forest trees and consequently will often die in a forest environment, even if they are full-time savanna-dwellers during the day. It is
quite possible, therefore, that primates whose fossils have been found in (almost) exclusively forest faunal assemblages were during their lifetimes exclusively savanna foragers. This would apply, e.g., to our present-day baboons

in much of their geographical range. Again, this aspect has not been considered by these authors.

Nonprimate Vertebrates

With regard to the nonprimate mammals and the gastropods, one gets the impression that Andrews, Van Couvering, and Walker unintentionally played down the evidence for local or temporary nonforest habitats. In the literature, the fossil springhare *Megapedetes* has always been considered as an indicator of the existence of dry, open habitats and sandy soils in the East African early Miocene. Andrews and Van Couvering (1975, pp. 81, 84), instead of drawing the same conclusion, point to the occurrence of a convergent form of jumping and burrowing large rodent in what they call the "forests" of western Madagascar, without any further specification. They disregard the fact that the occurrence of an animal in a *dry* forest with a 7-month dry season on a subcontinent without felids and canids does not imply that a more or less similar animal could have lived in an equatorial rain forest on a continent where felids, canids, and hyaenodonts did exist. Other examples of biased reasoning could be added. Another indication is that the presence of flamingoes at Rusinga (Harrison and Walker, 1976) is not mentioned by Andrews, Van Couvering, and Walker. These birds prefer saline and alkaline lakes.

Of course, this comment does *not* mean that there would have been *no* forest at Rusinga. In fact, in a faunal analysis of four sites at Rusinga, by means of Shotwell's (1958) method, Andrews and Van Couvering (1975) produced convincing evidence that three of these sites represented some kind of forest, whereas the fourth was probably an open area (Andrews and Van Couvering, 1975, Fig. 3 and pp. 88–89). One of the forest sites (KF) was apparently either a forest at a streamside (Andrews and Van Couvering, 1975) or a swamp forest (alternative possibility, A. K.), as is indicated by the high numbers of crocodiles and water turtles, and by the high numbers of remaining specimens per minimum number of fossilized individuals. This site contained *Dendropithecus macinnesi*, an ape with semisuspensory adaptations. Another forest site (KG) was probably further away from the waterside, but not too far away from it, as it held some scanty crocodile remnants. Both *Dendropithecus macinnesi* and *Proconsul africanus* were found here. The third forest site (KB) may have been still further away from the water, because aquatic animals were absent. There were no ape remnants. The fourth site (KH) yielded a rather intriguing spectrum. Typical forest animals were absent, crocodiles were numerous, the springhare *Megapedetes* was present, suggesting an open vegetation with a sandy soil, and *Proconsul africanus* was present. The number of specimens per minimum number of individuals was quite low, which suggests that running water had washed away most of the bones and teeth. The site might, therefore, have been a levee in a flood plain where seasonal inundation prevented forest growth.

The numbers of fossils from these four sites are too few to allow any definitive conclusions to be drawn. However, if we extrapolate beyond what is statistically allowed, the data would suggest that the semisuspensory climber *Dendropithecus* lived in riverine and swamp forests, while the semi-ground-walker *Proconsul* appears to have lived in both riverine forests and open habitats, perhaps with scattered trees and bushes. At any rate there is no evidence against L. S. B. Leakey's concept of a mosaic landscape. Moreover, the presence of the saltatorial lagomorph *Kenyalagomys* at sites KF and KG suggests that there were clearings in the forest (see also MacInnes, 1953). Much more and new evidence on the same theme is given by Pickford (1981, and this volume, Chapter 16) and by Pickford and Andrews (1981).

Gastropods

The gastropods are important climatic and ecological indicators. The enormous amount of data on these invertebrates produced by Verdcourt (1963, 1972) has been streamlined by Andrews and Van Couvering, but in so doing they have underemphasized those aspects of the data indicating relatively drier types of climate and relatively open types of habitat. The reader should compare pp. 32–34 in Verdcourt (1963) and pp. 308–348 in Verdcourt (1972) with pp. 82–83 in Andrews and Van Couvering (1975). The following review may suffice:

1. Verdcourt (1963) stated that, at present, *Lanistes carinatus* "favours rivers, small lakes and swamps in rather dry areas," and that the associations in which it was found at seven fossil sites indicate "swampy lakes (7), evergreen gallery forest (1), evergreen forest (1), dry forest (1), savanna (1), and bushland (2)," with rainfall figures mostly "? under 35 in." (~900 mm). This was reduced by Andrews and Van Couvering (1975) to a characterization "rivers and swamps."

2. Verdcourt's extensive discussion concluding that the fossil *Maizania* species indicate definitely dry types of rain forest, relict forests, and riverine forests, with an annual rainfall of 25–45 in. (635–1140 mm), was condensed to "evergreen forests," which in current botanical parlance means a *forêt ombrophile (forêt sempervirente)*, i.e., with more than 1500 or 1600 mm rainfall per year (Devred, 1958; Lebrun and Gilbert, 1954; Letouzey, 1958/59; Mangenot, 1955; Schnell, 1952). (Admittedly, Verdcourt may have caused some confusion by using the term "dry evergreen forest" for *both* so-called dry evergreen forests in the southern hemisphere and semideciduous forests in both hemispheres.)

3. Andrews and Van Couvering did quote from Verdcourt that the nearest relative of *Cerastua miocenicus* lives at present in Somaliland, but did not mention "in areas of bushland and woodland with up to 26 inches [660 mm] of rain per year."

4. The fossil occurrence of *Homorus* subgenus *Subulona*, at present "mostly restricted to rain forest and very numerous in the equatorial forest region of

East and West Africa," was rendered incorrectly as "wet evergreen forest," i.e., excluding the semideciduous type of rain forest.

5. *Thapsia's* habitat "in moist wooded places, particularly abundant in rain forest" was condensed to "widespread in forest."

6. The occurrence of *Trochonanina* was paraphrased by Verdcourt as "in riverine forest along gorges surrounded by dry country, coastal bushland, in dry savanna country as well as in drier parts of rain forests. They are not common in very wet ancient types of rain forest." Andrews and Van Couvering's summary, "evergreen forest and bush," appears, however, to include wet types of rain forest and to exclude dry savannas.

7. *Gonaxis* subgenus *Marconia* lives "in rain forest but also in drier upland forest and in coastal bushland," with an annual rainfall of ~1000 mm or more. This was summarized as "wet evergreen forest," i.e., excluding semideciduous lowland forest, drier upland forest, dry evergreen forest, and coastal bushland in the 1000–1500 or 1600 mm rainfall zone.

8. Similarly, the genus *Edentulina,* which occurs not only in "true rain forest usually in areas receiving more than 40 inches [~1000 mm] of rain per year" (Verdcourt 1963), i.e., including both semideciduous and [apparently] dry evergreen forests, but also in "coastal forest and bushland" (Verdcourt, 1972), is stated to occur only in "wet lowland evergreen forest."

An interesting feature of Verdcourt's (1963) list is that, among the fossil land mollusc fauna, the number of species that are characteristic of bushland, thicket, and arid habitats constitute 15% of the number of forest species, while among the living fauna they constitute 18% (Verdcourt 1972). The difference is statistically not significant, due to the small number of fossils, but the figures do imply that the mollusc faunal list does not give any positive indications of a moister climate in the early Miocene.

Andrews and Van Couvering (1975) also have disregarded Verdcourt's (1963) conclusion that the East African early Miocene gastropod fauna is related mainly to the present-day submontane and montane gastropods of East Africa, in a lesser degree to the recent coastal species, and hardly at all to the West African fauna. This fact obviously did not fit in with their idea of a continuous rain forest from West to East Africa. Similarly, Walker (1969) did not take into account the presence of *Tayloria* at Bukwa. Again, one gets the impression that these authors may have reasoned from a bias.

The most striking indication emerging from Verdcourt's data is the apparently wide diversity of ecological conditions in the East African Miocene. Rainfall estimates based upon gastropod adaptations vary from under 35 in. (<900 mm, *soudano-sahélien,* Belt 20 on the AETFAT map, *miombo* and *Commiphora* woodlands South of the equator) to 50–70 and 40–80 in. (1300–1800 and 1000–2000 mm, Belts 16, 8, and 7, including *forêts semidécidues* and *forêts ombrophiles*). One can hardly assume that the climatic fluctuations actually were that large in relatively short periods during the African Miocene. Moreover, drought-adapted and rain-forest-adapted gastropods often occur at the same site. The only plausible explanation is that the *apparent* climatic diversity resulted largely from local ecological diversity, i.e., from a mosaic or parkland vegetation.* If the ecological diversity is real, it raises the question of how such areas of *apparently* dry vegetation could have developed in regions where the climate was humid enough to establish at least some rain forest and to create a niche for molluscs adapted to year-round rainfall. This question becomes even more intriguing when we look at the other faunal material (e.g., spring-hare together with bunodont browsers).

Here again, an answer to the problem lies in taking into account the enormous destruction of vegetation that can be caused by herbivores, especially the elephants and other large pachyderms. While the large game that occur in forest reserves are usually hunted to such an extent that they no longer cause much damage, in African national parks and game reserves one can still watch how the "bulldozer" herbivores can transform a woodland almost into a desert, and semideciduous forest into a disrupted patchwork of "broken forest" alternating with tall grass savanna. The ensuing differences in microclimate are enormous: the maximum air temperatures at 1½ m above ground level can vary on sunny days from site to site by as much as 5–10°C. Such contrasts in vegetation and microclimate might largely explain the ecological diversity of gastropods and mammals in Miocene Africa. These contrasts might even have been enlarged by an undulating or rugged topography. Moreover, in those days there were more and larger species of big herbivores than occur at present.

Van Couvering and Van Couvering (1976, pp. 183–184) had a problem concerning Napak and Songhor: "These deposits were formed in an alkaline environment although the fauna is obviously derived, at least in part, from a forest environment in which the soils would have been acid. This apparent paradox has not yet been fully resolved." Actually, in Manyara National Park, a series of springs emanating from the Rift escarpment has created a luxurious edaphic forest within a mile of the alkaline lake. Similar situations might have existed during the Miocene in rain shadow zones behind mountains and ridges. To quote Charles Lyell: "The present is the key to the past."

Primates

It has been a traditional belief that "the apes" climb and swing from bough to bough in the canopy of the moist evergreen rain forest, whereas man "shall inherit the Earth." These ideas are centuries old. Camper (1782) argued on both anatomical and philosophical–theological grounds that the orangutan could not stand upright; in the 1950s and 1960s many anatomists continued to consider "the apes" as brachiators [e.g., Le Gros Clark (1959),

---

*According to Andrews (personal communication), however, the wide diversity resulted probably from the mixing up of fossils from different fossiliferous units. Pickford (this volume, Chapter 16), on the other hand, concludes from Verdcourt's data that, in general, it seems the climate was getting drier in West Kenya during the early and middle Miocene, but adds that the climate and vegetation may have been as varied in the lower miocene as it is now. Fortunately, W. Martin is now working on this material, so that we may hope that this issue will be clarified in due course.

still being reprinted unchanged], and Goodall (1965, p. 426) still declared "the closed rain forest" to be "the normal habitat" for the chimpanzee. The results of my chimpanzee field work since 1960 have shown that this image is untrue (e.g., Kortlandt, 1962, 1965, 1967, 1968, 1972, 1974, 1975a; Kortlandt and Kooij, 1963). While about 60% of the natural geographical range of the chimpanzee falls in the evergreen, semideciduous, and mountain rain forest belts, 40% falls in the woodland, savanna, and dry forest belts (Kortlandt, 1965, 1968, 1972; Kortlandt and van Zon, 1969). Chimpanzees may walk for several miles through entirely treeless grass steppes from one narrow gallery forest strip to the next (de Bournonville, 1967; Izawa, 1970; Kano, 1971, 1972; Kortlandt, 1965, 1968; Suzuki, 1969). Photographs and tv films have shown the everyday life of chimpanzees, as well as their nests for sleeping, in open woodlands and in savannas where trees are scarce. In western Tanzania south of the Malagarasi River, within the chimpanzee range, the surface covered by forest varies from area to area from 1.1% to 10.0%. In the Masito area the apes migrate during the dry season over distances of up to 15 km from the humid valleys to the dry *Brachystegia bussei* woodlands on the escarpments, because too little food remains in the narrow riverine forest strips (Kano, 1971, 1972). Nevertheless, Andrews and Van Couvering (1975, p. 80) still wrote: "One of the most characteristically forest-living groups of mammals of the present day are the higher primates, the pongids and hylobatids in particular." Van Couvering and Van Couvering (1976, p. 177) persisted: "Pongids . . . are, in our opinion, excellent indicators of nearby forest." Opinions are sometimes stronger than widely published facts.

Another aspect has recently emerged from research by Kay (1977). He made an attempt to infer the choice of food among six early Miocene ape species from East Africa belonging to *Limnopithecus, Proconsul,* and *Dendropithecus.* He compared measurements of the lower second molar teeth of the fossil apes with those of 11 recent ape species from Africa and Asia and five selected monkey species from Africa and South America. This was done by means of principal components analysis. The results suggest that the scores on the second axis of analysis (Kay, 1977, Fig. 1) give a good indication of the specialization of the various species in the series from primarily fruit-eaters (including most recent apes) to primarily leaf-eaters (gorilla and siamang among the apes, howler monkey and colobus monkey among the monkeys). None of the Miocene apes fall within the range of scores for the recent leaf-eaters. Thus, apparently they were fruit-eaters.

However, surprisingly, their range of scatter on the second axis in the diagram is twice as wide as that of the recent fruit-eaters (the orangutan, two chimpanzee species, five gibbon species, and three monkey species). The range of scatter of the species and individuals *from Songhor* is the same as that of the recent fruit-eaters, but the species and individuals *from Rusinga* fall *beyond* the fruit-eating range, on the opposite side from the leaf-eaters. That is, they appear to represent some kind of "super fruit-eaters." Kay did not try to explain this phenomenon. A clue might possibly emerge from the fact that the orangutan, the pygmy chimpanzee, and the spider monkey fall not very far away from the leaf-eating end of the fruit-eating spectrum, while the common chimpanzee and the capuchin monkey fall at the opposite end: The common chimpanzee is the only semi-savanna-dweller and the most carnivorous one among the apes, and the capuchin monkey appears to be the most parkland-adapted and most insectivorous/carnivorous species among the New World monkeys. From this viewpoint one may tentatively suggest that the Rusinga apes were perhaps more savanna-adapted and/or more carnivorous than our present-day chimpanzees and capuchin monkeys [compare with the data in Kortlandt and Kooij (1963) and Teleki (1973, 1975)]. The issue deserves further study, but for now one can say that the adaptive zones of these Miocene apes appear to have been much wider than those of extant fruit-eating apes.

Yet another aspect is suggested by an anatomical comparison. According to Napier and Davis (1959), the arm bones of *Proconsul africanus* resemble particularly those of the langurs (*Presbytis*) of southern Asia. Such comparisons are, of course, always controversial, and this inference has been contested (e.g., Conroy and Fleagle, 1972; Lewis, 1972; Preuschoft, 1973; Schön and Ziemer, 1973; Zwell and Conroy, 1973; but see Corruccini *et al.*, 1975, 1976). However, if there is some core of truth in it, this could mean that the locomotor patterns and the habitats of the langurs *might* perhaps to some extent be regarded as a model of what the locomotor patterns and the habitats of the dryopithecines could have been. Or rather, in an even more cautious formulation, it would mean that we have no anatomical reasons to reject the possibility that the dryopithecines moved about and lived in habitats similar to those of the langurs.

The langurs are both ground-walkers and branch-walkers. The amount of time they spend on the ground varies locally and seasonally from 1% to 80% of the daylight hours. They live in an array of habitats, varying from very humid tropical rain forests to very hot semidesertic scrub steppes (with occasional temperatures up to 48°C and annual rainfall figures as low as 280 mm on the average), and they also occur in deciduous oak forests, coniferous forests, and alpine shrubs in the Himalayas (at altitudes up to 4300 m and at winter temperatures of −3°C). They regularly pass through wide open plains and have been seen to cross snow fields with 30 cm deep snow [references in Oppenheimer (1977)]. It is also known that at least some European dryopithecines lived in oak–chestnut–pine forests, as mentioned already by Lartet in 1856. They lived in a type of vegetation that at present still survives locally in the southern Appalachian Mountains. Furthermore, among present-day monkeys, not only the langurs but also the baboons, macaques, guenons, and patas have forms adapted to live in semidesertic environments. Other monkey species live in cold-temperate climates, in winters with up to 4 months of frost and snow, and with average January temperature of −4°C (Izawa and Nishida, 1963). There are no obvious physiological or ecological reasons why the dryopithecines should not have been able to do the same. One can therefore be reasonably sure that the dryopithecines (*s.l.*) did indeed live in an array of habitats as wide as that seen among the modern monkeys: where there are niches available, animals will evolve to fill them.

A Wider View

Habitats and Adaptations

From the critical reevalution in the preceding section, in combination with other data, a fairly consistent picture of African ape habitats emerges. I shall now outline this picture in terms of ecological adaptations and in a more general framework.

The earliest known apes (i.e., primates currently classified as Hominoidea) have been found in middle and upper Oligocene delta deposits in Egypt, dating from about 30 m.y.a. It has been generally assumed that these apes lived in tall gallery forests along rivers that meandered through woodlands and grass plains (Bowen and Vondra, 1974; Simons, 1967, 1968). However, a comprehensive analysis of the geological and palaeontological data shows that this site had a *sahélien* climate and that the fossilized trees were probably driftwood from elsewhere (Kortlandt, 1980c). Thus the earliest known apes must apparently have been semiterrestrial woodland-, bushland-, and savanna-dwellers, and were adapted to survive prolonged annual droughts.

Undoubtedly, these apes were already foraging to a large extent on fruit. Many African trees, both inside and outside the rain forest belt, bear fruits that are well adapted to dispersal by primates. Several of these adaptations date from Oligocene and even from Eocene times. However, there are also good reasons to infer that many Oligocene and early Miocene primates were also semicarnivorous. Among other things, the number of primate species among the total mammal fauna was about equal to that of the then carnivores, the Creodonta, i.e., much higher than the proportion of primates to carnivores in comparable habitats during middle Miocene, Pliocene, Pleistocene, and Recent times (see also Andrews *et al.*, 1979; Kortlandt, 1980b, 1980c; Kortlandt and Kooij, 1963). Thus the adaptive zone of the Oligocene and early Miocene higher primates may have been as wide as the *combined* ecological niches of the present-day apes and monkeys plus a part of the niche of the genets and possibly a part of the niche of the civets in Africa. That is, the taxon has been from the onset a highly eurytopic, versatile, and probably omnivorous group of mammals.

This again implies that the East African early Miocene habitats analyzed in this chapter should not be considered as indicative of all higher primates in those times. On the contrary, one can be sure that the rain forest belt in western and central Africa was inhabited by a large variety of specifically forest-adapted primates, including prosimians, cercopithecoids, and hominoids, and possibly even including one or more brachiating or semibrachiating forms, e.g., the as yet unknown ancestors of *Oreopithecus*. Again, we can be equally sure that even quite dry, *sahélien* types of vegetation also harbored various primates, exactly as they do today, including such semicursorial forms as the present-day patas monkey. Where food could be reached by climbing trees, shrubs, and bushes, there must have been primates to utilize it.

Among the apes, at least two lineages (*Limnopithecus–Pliopithecus* and *Proconsul–Dryopithecus*) managed to cross from Africa to Eurasia, probably via Arabia, about 16 million years ago.* Van Couvering and Van Couvering (1976) have constructed an elaborate theory in order to explain how such alleged rain-forest-dwelling creatures could have migrated through such an arid region: They postulated an evergreen forest bridge all the way from Kenya through Ethiopia, the Red Sea, Jordan, and Syria to Turkey! In fact it is known that all this area was definitely arid, and locally even desert-like, in the entire early and middle Miocene (e.g. Axelrod and Raven, 1978; Butzer and Hansen, 1968; Frakes and Kemp, 1972; Hamilton *et al.*, 1978; Kedves, 1971; Kortlandt, 1972, 1974, 1980c; Louvet, 1971; Powers *et al.*, 1966; Said, 1962; Sen and Thomas, 1979; Thomas *et al.*, 1978; Whiteman, 1971). Nevertheless, two primitive species of dryopithecine lived in eastern Arabia about 15–17 m.y.a. (Andrews *et al.*, 1978). It may be significant that exactly such primitive forms could survive under such adverse climatic conditions so far away from the rain forest zone.

It may be equally significant that *Pliopithecus* arrived in Austria and France at a time when the climate was becoming drier and when there was a subtropical and relatively dry type of forest, as indicated by the plant fossils, while about one-quarter or one-fifth of the fauna consisted of savanna elements (Ginsburg, 1968; Rabeder, 1978; Thenius, 1955; Zapfe, 1960). This was the upper part of the European Mammal Age NM 5, either just below or just above the Burdigalian–Langhian boundary (in Mediterranean stratigraphy), i.e., the Karpatian–Badenian boundary (in Central Paratethys stratigraphy), estimated as 16.5 m.y.a. (Ginsburg, 1975a,b; Ginsburg and Mein, 1980; Mein, 1980; Rabeder, 1978; Rögl and Müller, 1978; Steiniger *et al.*, 1976). Thus the arrival of the first apes occurred 1 or 2 million years after the appearance of the elephants, which occurred in NM 4a when the climate was more humid and the savanna element was less (Mein, 1980; Thenius, 1955).†

It may also be significant that *Dryopithecus* arrived in Austria and France yet another 1 or 2 million years later, in the late part of NM 6, when the climate was becoming still drier and the vegetation was developing toward a true mosaic landscape, probably with both savannas and riverine forests, and with a still larger savanna fauna (Ginsburg, 1968; Mein, 1980; Rabeder, 1978; Thenius, 1955). This is supported by paleobotanical data also indicating that, in the Vienna Basin at least, the dessication of the climate culminated in Mammal Age NM 7, i.e., the "Sarmatian" in Central Paratethys terminology, which was thereafter followed by the "Pannonian," a more humid and somewhat colder interval, characterized by the predominance of a deciduous vegetation. The dryopithecines survived all these events. There is even a "Sarmatian" site in Vienna where the plant fossils indicate a valley covered entirely with macchia, savanna, and bush vegetation interspersed with riverine forest,

---

*The ancestor of *Oreopithecus* was probably a third migrant, but we have no indications of its time of crossing.

†Glaessner (1932/33) mentioned already that the alleged find of *Pliopithecus* in the Burdigalian Sables d'Orléans was due to an incorrect determination. Yet the error continues to wander around in the literature to this day.

but without any indication of leafy forest (Laubwald) on nonriverine soil (Berger and Zabusch, 1954). The period of drought is also confirmed by the occurrence of lightning-kindled forest fires (Abel, 1927).

An even greater delay may have occurred between the first appearance of African faunal elements and the arrival of the first apes in Central Asia: the earliest representatives of the *Dryopithecus–Sivapithecus* lineage are known from the Chinji Formation about 12 or 13 m.y.a., while they have not been reported from the underlying Kamlial Formation (Pilbeam *et al.*, 1977). Thus it appears that the arrival of *Pliopithecus* in Europe coincided with a phase of beginning desiccation, and the arrival of *Dryopithecus–Sivapithecus* in both Europe and Asia with a phase of increasing desiccation. Should one assume, therefore, that the European faunal assemblage was insufficiently preadapted to survive long, dry seasons, and consequently offered a chance to the African invaders? Or were the Miocene apes so much adapted to mosaic habitats that they could not disperse to central and western Europe as long as their route was blocked by continuous forests in the preceding humid phase? Were they ecologically dependent, to some extent, on the presence of forest edges? Did they have to wait until the elephants had thinned out and broken up the once-continuous forests? I proposed the idea a decade ago (Kortlandt, 1972). It is known, after all, that the optimum habitats of chimpanzees (which are better climbers than were the dryopithecines) are the "broken forests" created by elephants and/or human action in the semideciduous zone, whereas the intact semideciduous forests harbor fewer of these apes, and the evergreen climax forests in some areas even none at all. While no one seems to have taken the suggestion seriously, it may be worthwhile to reconsider it.

Incidentally, it is currently assumed that the Miocene land connection between Africa and Eurasia was established about 17–18 m.y.a. However, the canid *Hecubides*, the felid *Metailurus* (=*Afrosmilus*), the viverrid *Kichechia*, and a mustelid, which are all of Eurasian origin, have been found at African sites dated 20 m.y.a. (Koru) and/or 19 m.y.a. (Songhor) (Pickford, 1981; Pickford and Andrews, 1981; Savage, 1978). Thus the time lag in the date of arrival is even greater than the one derived above from the first appearance of the elephants in Europe. This reinforces the argument that the continuous forest cover in the Eurasian Aquitanian and Burdigalian acted as a barrier against immigration of East African apes.

It would be tempting here to continue surveying Eurasian Miocene ape habitats. However, I postpone this for a future publication (Kortlandt, in preparation), except for the methodological discussion of the faunal analysis presented in the discussion section. Neither shall I discuss the influence of the middle and late Miocene climatic changes. Suffice it here to state that the Eurasian early and middle Miocene habitats of the *Pliopithecus* and *Dryopithecus–Sivapithecus* lineages show a wide variety, ranging from subtropical swamp forests (resembling parts of the Everglades in Florida) to moist, warm–temperate, oak–chestnut–pine forests (probably interspersed with glades), to open woodlands in a dry, karstic landscape to savanna-woodlands with riverine forests in the African style. This confirms the general picture of two highly versatile and highly adaptive taxa that could survive for no less than 10 and 15 million years, respectively, from Bukwa and Koru in East Africa to Rudabánya in Hungary, and from Muhuroni in East Africa to the Siwaliks in India and Pakistan. Judging from the taxonomic confusion, one can only conclude that, in the meantime, there were apparently no radical changes in their overall anatomy and dental adaptations. (Quite a long time span in mammalian evolution!). The main change perhaps was the thickening of the dental enamel in the *Proconsul–Dryopithecus–Sivapithecus* lineages. This may possibly be explained by the greater abundance of nuts in proportion to soft fruit on Eurasian trees. If so, it would *not* indicate a fundamental change in the ecological habits of this taxon. (Langurs, after all, also feed on acorns.)

The Disappearance of the Miocene Apes

One wonders what caused the disappearance of these highly successful lineages, in Africa 14 m.y.a., in Europe 10 m.y.a., and in Asia 8 m.y.a. Many authors have proposed that climatic changes, *casu quo* climate-induced changes in vegetation, were responsible. I cannot find any evidence supporting this suggestion, except on a regional scale. The disappearance of the fossil apes came much earlier in Africa than in Eurasia, even though the amount of climatic change was much less in Africa. Vegetational changes occurred on the three continents, but these consisted mainly of geographical shifts of the vegetation belts. The apes would have followed the moving belts. There were no large-scale floral extinctions in the middle and late Miocene. Nor did the forests and the mosaic habitats as such vanish from any one of these continents. Thus the apes should have survived. The cooling and drying of the climate cannot in itself explain the events either. Chimpanzees occur at altitudes of up to 3000 m and gorillas up to ~3500 m. As already mentioned, some monkey species can survive in habitats varying from semidesert to high-altitude mountain shrubs, and Japanese macaques can withstand 4 months of frost and snow. So why should the dryopithecines not have been able to do the same? The absence of their fossils north of the line from the Loire River to Mainz to Opole (in southern Poland) can easily be explained by the general scarcity of fossilization conditions for mammals in that area, but it certainly constitutes no proof that the Miocene apes could not adapt to cold-temperate climates and vegetation. (See also section on Future Research).

Several authors have suggested that the radiation of the monkeys caused the extinction of the apes (except for the modern survivors), due to ecological competition. The most elaborate interpretation of this idea has been proposed by Andrews (1981). The suggestion seems plausible at first sight. A major difficulty, however, is that in Africa the decline of the apes appears to have begun already around the end of the early Miocene, while the main radiation of the monkeys (so far as we know from their fossils) seems to have begun in the latest Miocene. Similarly, in Eurasia, the main dispersal of the monkeys appears to have occurred several million years after the apes had disappeared. During the intervening time span, judging from the fossils, monkeys were still so rare and so little differentiated that it seems impossible that these humble creatures could ever have threatened the apes (Delson, 1975; Szalay and Delson, 1979).

Perhaps a better suggestion is that, first, in Africa, the incipient hominids successfully competed against the other apes in those areas where our museum fossil collections have been gathered, i.e., in East Africa (Kortlandt, 1972, 1975b, 1980a). In West and Central Africa, after all, the chimpanzee and the gorilla have survived to date, notwithstanding the presence of an abundance of monkeys. Second, in Eurasia, it may have been primarily the radiation of the squirrels and the bears that led to the extinction of the apes. These three groups are all ground-walkers as well as tree-climbers (even bears often forage quite high in the trees), all three feed (among other things) on berries, fruits, and nuts, all three are excellent manipulators, and bears use their hands in fighting in a manner that would be effective against apes. In short, there are many parallels and convergences among them which might have led to severe competition. The squirrels, however, evolved the behavioral advantage of hoarding, the bears evolved the physiological advantage of fat-storing, and both evolved the extra advantage of hibernation during cold spells. In a cooling climate this may have been decisive. It would explain the slow radiation and the scarcity of monkeys in southern and central Europe and in the Siwaliks. Furthermore, since the true bears (Ursinae) never managed to cross the Sahara, their absence from tropical Africa may perhaps explain the survival of the chimpanzee, the gorilla, and the (incipient) hominids (Kortlandt, 1968, 1972). Or perhaps conversely, south of the Sahara, the presence of large apes and hominids who did not need to waste time and effort in storing energy because they were better arboreal foragers in a richer environment, and who, moreover, could use tools to crack nuts, and weapons to defend themselves, may have made penetration by the bears impossible.

The Ecosystem

What I have been advocating since 1972 is an ecosystem approach to the problems of ape and human evolution, an approach shown also by some earlier writers. The basic idea is that it is always the ecosystem as a whole that shapes the flora and fauna that together constitute the ecosystem. Any speculation about evolutionary processes in animals should begin with thorough knowledge of the ecosystem in which these processes took place. Furthermore, any attempt at a functional interpretation of teeth, bones, brain size, etc. should begin with a definition of their survival value in that particular ecosystem.

Seen from this perspective, an ecosystem is an assemblage of life within an area, which is separated by a floral and faunal barrier from another ecosystem. The barriers isolate breeding populations, split evolutionary lineages, and consequently determine the ecological niches to which the species within the ecosystem will eventually adapt—or become extinct. The same applies, of course, to the subsystems that can be distinguished within a larger ecosystem. In each system or subsystem, natural selection will eventually result in an ecological balance between all the adaptive types of plant and animal that can be distinguished: trees, shrubs, herbs and grasses; predators, prey and scavengers; flying animals, tree-climbers, ground-walkers and burrowers; large, medium-sized, small and very small animals; diurnal and nocturnal creatures; and so on. To illustrate this point, the following examples may be helpful, even if described in a rather oversimplified manner.

In Australia one finds among the marsupials the approximate equivalents of rats, mice, rabbits, moles, squirrels, hares, insectivores, hedgehogs, weasels, cats, wolves, otters, bears, ungulates, and, last but not least, primates. The kangaroos and wallabies take the place of hoofed mammals. There has even been a marsupial "hippo" (*Diprotodon*) and a marsupial "lion" or "tiger" (*Thylacoleo*), both probably exterminated by humans in fairly recent times. Similarly, before the formation of the Panama isthmus and the subsequent large-scale extinction of the original fauna, South America had a faunal assemblage greatly resembling the African and Eurasian ones, including a complete spectrum of carnivorous animals and an equally complete spectrum of hoofed mammals, with "antelopes," "horses," "buffalos," "rhinos," "hippos," and "elephants"—in short, everything that a tourist expects to see in an African or Indian National Park. A comparison of the spectrum of the New World primates with that of their Old World relatives also yields many inspiring resemblances. (Admittedly, there are no predominantly ground-walking primates in South America, but they may have been wiped out when the Panama land bridge was formed: about one-half of the ground-dwelling families among the South American mammals shared the same fate.)

The manner of ecosystem analysis exemplified above is picturesque and often instructive, but it cannot be regarded as very scientific. It might be termed euromorphic, asiomorphic, and afromorphic labeling of a species list. Nevertheless, such labeling has one advantage: it makes one aware of certain basic processes that shape the faunal diversity within an ecosystem. The result of these processes is that two ecosystems which have different evolutionary origins, but where the overall ecological conditions are more or less the same, will tend to evolve marked similarities in their faunal diversity, even while the taxonomic composition of their faunas will be quite different.

Research into mammalian paleoecology has not yet developed very far above the level of what may be called "recentomorphic" labeling of fossil species lists. Furthermore, one can still all too often hear such pronouncements as "These teeth are brachyodont, thus the animal was a forest-dweller," or "Hypsodont teeth indicate a grass steppe." Within such a restricted frame of reference there is no place for broken forests, parklands, woodlands, savannas, bushlands, and all the infinite variety of natural landscapes on this planet. Moreover, many animals adapt their food choice to the vegetation. Elephants and buffalos, for instance, are browsers in leafy vegetation, but grazers in grassland. As to the choice of habitat, animals do not always follow the textbooks. Baboons, for instance, are supposed to be savanna, woodland, and bushland specialists, but they also live in forests that have been secondarized by elephants or human activity (Rowell, 1966; my observations in Zaïre). Similar behavior has been reported for vervet monkeys (Kavanagh, 1980). Conversely, I have seen forest-dwellers such as colobus and mitis monkeys traveling (though not foraging) in a grass savanna with some bushes and in a papaya plantation respectively. Mammal palaeontologists should be aware of all these intricacies.

Analysis of the Ecosystems

"La critique est aisée, mais l'art est difficile." Of course we need a faunal inventory and some kind of ecological labeling of the listed species in order to get a first impression of how a fossil ecosystem might have looked. More difficult to answer is the question of what should be done next.

Recently Andrews *et al.* (1979) have proposed a new and quite original approach to this problem. To summarize briefly, they do not look for fossil species that are indicative of a certain type of habitat according to the modern equivalent species, but instead they compare the composition and diversity of the *entire* mammal assemblage from a fossil site with the present composition and diversity of an *array* of living mammal assemblages in *various* habitats within the same geographical and climatic zone. For this purpose the fossil and living mammal fauna of East Africa is broken down multidimensionally into the following categories: (1) Six taxonomic groups (Rodentia, Insectivora, Primates, Artiodactyla, Carnivora, and other), (2) five body size groups (ranging from <1 kg to >180 kg), (3) six locomotor categories [large terrestrial, small and more or less terrestrial, arboreal, scansorial (i.e., clawed climbers), aquatic, and aerial], and (4) six types of feeding adaptation (insectivores, frugivores, browsers, grazers, carnivores, and omnivores). Thus 23 categories under four headings are distinguished.* The numbers of species in each of these categories are expressed as percentages of the total mammal fauna of the locality. These analyses of the composition and diversity of the fauna from five fossil sites (in current chronological order: Karungu, Songhor, Rusinga, Fort Ternan, and Olduvai—see Fig. 3) are then compared with similar analyses of 23 recent assemblages, mainly from places in eastern Africa, whose faunal inventory is well known from the literature and whose vegetation is briefly characterized by Andrews *et al.* (1979). An accompanying series of histograms gives a clear picture of the faunal composition and diversity of each locality (Figs. 4–7). The procedure and the final results are carefully controlled by means of statistical techniques. [In a subsequent, short paper, Andrews and Nesbit Evans (1979) produced faunal histograms of Songhor, Fort Ternan, and Olduvai which were based on a somewhat different sample selection. Since the results were consistent with those obtained by Andrews *et al.* (1979), I will discuss only the more extensive paper.]

My only major objection is that the characterizations of the vegetation of

---

*When reading the paper one should be aware that "insectivores" sometimes stands for Insectivora and sometimes for insect-eaters. Similarly, "carnivores" may mean either Carnivora or meat-eaters.

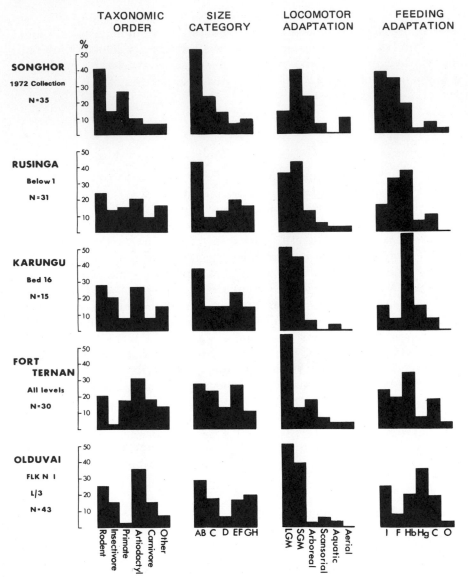

Fig. 3. Ecological diversity histograms for five fossil localities. The site details are given on the left, with *N* signifying the number of species available in the sample. The vertical scale represents the percentage number of species of the faunas for four categories: taxonomic (by order), size (in kg), locomotor adaptive zones (with LGM signifying large, fully terrestrial mammals and SGM signifying small, scrambling mammals that live both on the ground and in the bush canopy), and feeding adaptations (with I indicating insectivores; F, frugivores; Hb, herbivorous browsers; Hg, herbivorous grazers; C, carnivores; and O, omnivores. These same categories are used in Figs. 4–7. [Figure redrawn from Andrews *et al.* (1979) by courtesy of the authors and the *Biological Journal of the Linnean Society*.]

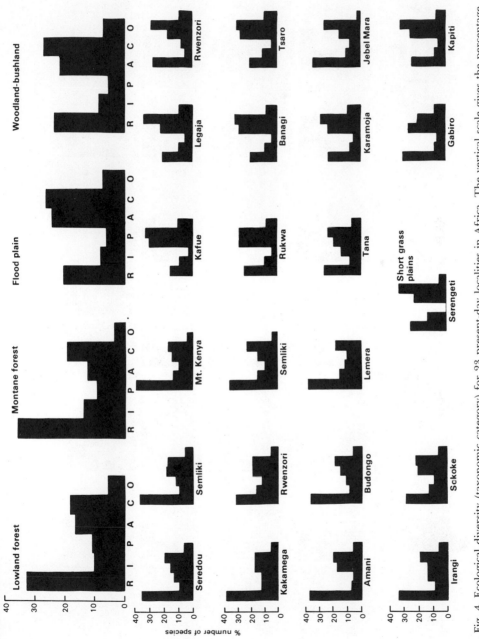

Fig. 4. Ecological diversity (taxonomic category) for 23 present-day localities in Africa. The vertical scale gives the percentage number of species in the taxonomic division based on Orders. Bats are excluded. R, Rodentia; I, Insectivora; P, Primates; A, Artiodactyla; C, Carnivora; O, others. [Figure redrawn from Andrews *et al.* (1979) by courtesy of the authors and the *Biological Journal of the Linnean Society.*]

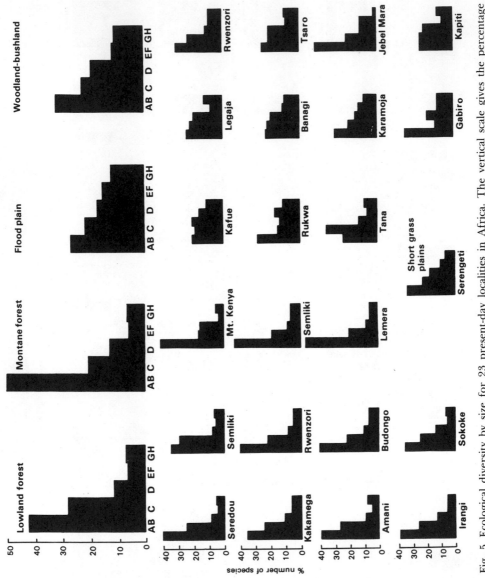

Fig. 5. Ecological diversity by size for 23 present-day localities in Africa. The vertical scale gives the percentage numbers of species within the size categories, which are divided as follows: AB, less than 1 kg; C, 1–10 kg; D, 10–45 kg; EF, 45–180 kg; GH, more than 180 kg. [Figure redrawn from Andrews et al. (1979) by courtesy of the authors and the Biological Journal of the Linnean Society.]

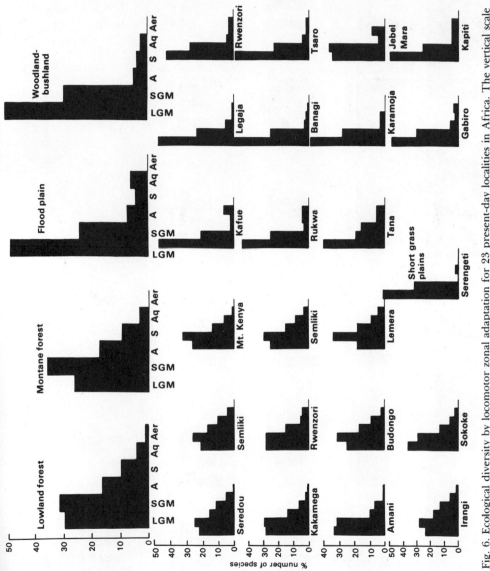

Fig. 6. Ecological diversity by locomotor zonal adaptation for 23 present-day localities in Africa. The vertical scale gives the percentage numbers of species in the categories of locomotor adaptation. SGM and LGM defined as in legend to Fig. 3. A, arboreal; S, scansorial; Aq, Aquatic; Aer, aerial. [Figure redrawn from Andrews *et al.* (1979) by courtesy of the authors and the *Biological Journal of the Linnean Society*.]

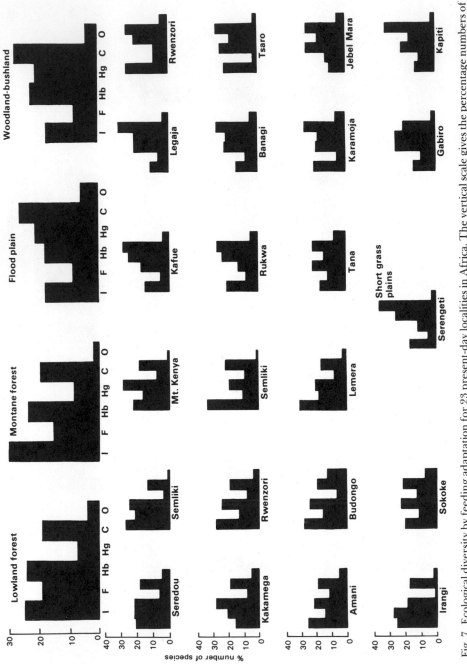

Fig. 7. Ecological diversity by feeding adaptation for 23 present-day localities in Africa. The vertical scale gives the percentage numbers of species in the categories of feeding adaptation, which are divided as follows: I, insectivores; F, frugivores; Hg, herbivorous browsers; Hg, herbivorous grazers; C, carnivores; O, omnivores. [Figure redrawn from Andrews *et al.* (1979) by courtesy of the authors and the *Biological Journal of the Linnean Society*.]

the 23 recent sites were much too summarily made: an average of less than eight words per site cannot be called a habitat description. Even the rainfall figures were not mentioned. Fortunately I know several of these localities and many comparable sites nearby, but for the general reader the information is inadequate. Some of the characterizations are incorrect or incomplete; for instance, the Kakamega forest is not an "impoverished Congo type." These technical points do not, however, affect the overall conclusions.

The results obtained by Andrews *et al.* (1979) clearly demonstrate that the Songhor fossils represent a lowland and/or mountain forest fauna. This is fully in line with the findings of earlier paleontologists. Songhor was situated on a large volcano. Such a relief causes much rainfall and consequently the slopes would have been covered (mainly) by forest, even if the wider environment had a drier vegetation. The data do not, however, exclude the possibility that there were open glades and areas of low vegetation within the forest, as is often found on active volcanoes, due to young lava currents. I shall return to this point.

The Karungu fauna [currently dated 22–23 m.y.a., but redated 18 m.y.a. by Pickford (1981)] according to its taxonomic composition "most closely resembles that of the woodland–bushland faunas." The Rusinga assemblage (below Marker Bed I, 18 m.y.a.) taxonomically "resembles forest communities, but its additional similarities to the flood plain communities suggests that the habitat may have been mixed." The Fort Ternan fauna (13–14 m.y.a.) "appears to be mixed" but shows "the strongest (taxonomic) affinities with the woodland–bushland communities" (Pickford, 1981, pp. 188–189). The diversity of body size at both Karungu and Rusinga "could indicate that the faunas are either mixed or heavily biased," while Fort Ternan in this respect resembles modern "bushland communities . . . particularly close to that of the Kafue flood plain fauna" (Pickford, 1981, p. 191). As to locomotor diversity, the Karungu assemblage "is significantly correlated with the woodland–bushland faunas," while the Rusinga fauna "resembles the montane communities most closely." Fort Ternan "has an exaggerated bushland pattern because of its large number of terrestrial species," but shows at the same time a "relatively high proportion of arboreal mammals due to the primates" (Pickford, 1981, pp. 192–193).

From all this one gets the impression that these three fossil sites, over a time span of 5–10 million years, yielded mixed faunas, which indicates mosaic landscapes consisting of both forest and woodland–bushland vegetation, but with relatively sparse savanna and open grassland compared with the vegetation of both Olduvai and present-day nonforest communities in East Africa. These data again largely support the classical concept of East African Miocene habitats as outlined by L. S. B. Leakey and many others, *contra* the newer ideas proposed by Andrews, Van Couvering, and Walker. They also suggest that— if Karungu is 22 million years old, and if there was any significant trend of climatic change at all in this region—the time span from 22 to 18 m.y.a. was perhaps a period of slight increase of arboreal vegetation, and the time span from 18 to 14 m.y.a. a period of equally slight decrease of arboreal vegetation. These small differences could, however, also have been caused by local differences in relief and soil condition, similar to what one can observe in Africa today. The real climatic change apparently came later, between Fort Ternan and Olduvai Bed I times, when a grassland type of fauna gained the ascendancy.

With regard to the feeding adaptations among the analyzed fauna, Andrews *et al.* (1979) are somewhat reticent in their conclusions. The Karungu pattern "is unlike that of any modern community" (p. 195). Its prominent feature is a record high percentage (60%) of browsers combined with an almost record low percentage of frugivores. My guess is that this represents an analogy similar to either a present-day very dense secondary vegetation growing on abandoned farmland in the rain forest belt, or a young *forêt sèche dense* that has been growing for several years on a site protected from bushfires in the *soudanien* belt. Zoologists cannot make faunal inventories of such habitats. In the same category belong the impenetrable *forêts claires à brousse de Marantacées* in West and Central Africa, which presumably result from forest destruction followed by bush encroachment, in combination with the action of elephants, buffalos, and pygmy hippos (Guillaumet, 1967; Letouzey, 1968). More or less comparable dense bush occurs locally on young lavas on Mount Cameroon and on the Nyamulagira–Nyiragongo volcanoes. Karungu may, therefore, represent a regenerating vegetation after a volcanic eruption.

Rusinga (below Marker Bed I) is interpreted by Andrews *et al.* (1979) as "similar to the forest pattern," in spite of a low percentage of insect-eaters, because the browsers are much more common than the grazers. The Fort Ternan pattern, however, which shows high percentages of both insect-eaters and frugivores, and also many more browsers than grazers, is nevertheless described as being "closest to that of the woodland–bushland communities." The argument based upon the feeding adaptations appears to be inconsistently applied here. Maybe Andrews *et al.* (1979) were still under the influence of Andrews and Van Couvering (1975), Andrews and Walker (1976), and other authors who overemphasized the differences between Rusinga and Fort Ternan.

In fact, a closer look at the histograms (Fig. 3) shows that at *all four* Miocene sites (i.e., including Songhor) the percentage of grazers is almost exactly equal to one-fifth of the percentage of browsers in the total mammal fauna, irrespective of the fluctuating percentage of browsers plus grazers taken together. This is quite surprising. It is often assumed that this time span of 5–10 million years was a period of gradual, world-wide climatic cooling and drying out (e.g. Chaney, 1940; Schwarzbach, 1961; Woldstedt, 1954), which resulted in a shrinking of forests, an extension of grasslands, and, consequently, a partial replacement of browsing species by grazing species. It now appears from the faunal percentages that, during this time span in East Africa, *the proportion and diversity of bush vegetation relative to grass vegetation was remarkably constant,* irrespective of whether the bush and grass plants grew in a predominantly forest habitat or in open woodlands with scattered trees. Fort Ternan may still have been somewhat drier than the early Miocene sites on the average, but the difference has probably been greatly exaggerated (see also Pickford, 1981). Savannas in a strict sense, i.e., large plains of wooded grassland, apparently did not yet exist. They emerged between Fort Ternan and Olduvai times. This finding agrees with recent deep-sea drilling results showing virtually constant ocean water temperatures from the earliest Oligocene until the early middle Miocene (about 14 m.y.a.), after which a steady cooling (although with oscillations) set in (Shackleton and Kennett, 1976).

Discussion

Leakey's Reconstruction

In summary, the data emerging from the new method of faunal analysis of Andrews *et al.* (1979) largely support the classical picture of early and early–middle Miocene Africa as a mosaic landscape. However, the *type* of mosaic vegetation implied by these new findings is somewhat different from that which L. S. B. Leakey and others had in mind.

A revealing picture of the classical image is J. H. Matternes' drawing of a Miocene landscape, which was prepared in consultation with L. S. B. Leakey and J. R. Napier and published in Leakey's (1969) book *Animals of East Africa*. It was reprinted opposite the title page of the Van Couvering and Van Couvering (1976) paper in which Leakey's views were rejected. The illustration shows a wide-ranging grassland speckled with isolated small *Acacia* trees and some mini-bushes. The grassland is interfingered by many narrow forest-and-bush gallery strips along rivulets. The grassland, the small trees, and the sharply defined boundaries of the gallery vegetation indicate annual bush-fires—a condition that was not possible in the Miocene, because thunder-storms would only kindle such fires now and then over the course of many years. Moreover, the numerous gallery strips suggest an amount of pluvial or edaphic humidity irreconcilable with such wide, open grassland. (Periodically flooded grassland has no gallery forest.) On the other hand, the portrayal of the forest-and-bush as a "broken forest," with many gaps covered by second-ary growth and with a quite irregular canopy, is an excellent rendering of a gallery forest under heavy herbivore pressure, especially from elephants, but without an overpopulation of browsers. Obviously the artist saw these things in nature, but could not break away from his recentomorphic idea that wide, open grasslands must have been abundant. Much the same criticism applies to his depiction of the dryopithecines as tailless and to his inclusion in the picture of a dryopithecine making a pathetic attempt at brachiation. True, "the pres-ent is the key to the past," but one should add: *"mutatis mutandis"*!

Vegetation Dynamics

In order to get a more realistic idea of Miocene habitats it is worthwhile considering the other end of the spectrum. Recently Douglas-Hamilton (1980) published a fascinating aerial photograph of a herd of 160 elephants browsing in ⅔ hectare of *Capparis–Euphorbia* bushland a few meters high, in the Virunga National Park in a rather dry part of Zaïre. In the course of decades of trampling, the elephants had worn a dense network of paths, each about as wide as an elephant, so that the vegetation consisted of innumerable little clumps of bush, each about 4–6 m wide. During the rainy season one may expect that these paths will be overgrown with grass, so that a bush-land–grassland mosaic will develop in which each compartment is only some tens of square meters in size. Observations like these, amplified by long-term studies, may provide a clue to the problem of why the proportion of browsing to grazing species remained constant throughout the entire early and ear-ly–middle Miocene: the answer may be found in the dynamic interactive processes between vegetation and herbivores.

Bushland often provides an ecological basis for spontaneous reforesta-tion; first because certain types of bush protect both the soil and the saplings from damage by fires, and second because they often protect the young trees from browsers. In the case of the aforementioned national park in Zaïre, no forest vegetation could develop because elephant pressure was too heavy, due to the suppression of poaching and the fact that large sabretooth cats are extinct. In other cases, however, one can recognize the basic elements of a long-term cyclical process:

grassland→bushland→tall forest→dying forest→grassland

Such cyclical processes can, under certain ecological conditions, result from the dynamic interactive processes in a predator–prey relationship. In the present context the herbivores are the predators and the plants the prey. The result can be an irregularly mottled vegetation reflecting at different sites all the successive phases of the cyclical process. This is also a mosaic of a sort, though different from the one depicted in the Leakey-inspired picture. These small-scale mosaic patterns may develop on stretches of homogeneous land within the larger-scale mosaic that results from differences in relief, soil, water-table, etc. They contribute, therefore, to the diversity of the flora and fauna.

There are also cyclical interactive processes that cause very large-scale mosaics. Some animal species may have malfunctioning population regulatory mechanisms, or perhaps their natural predators have become extinct. Conse-quently their population density gradually builds up, until famine ensues, followed by a population crash, a large-scale migration, or both. Locusts, lemmings, and elephants provide examples, but there are many more, es-pecially among the insects. (Since the advent of agriculture, humans should be added, of course.) The devastating effects of such events are widely known. They can cause mosaic effects of an almost subcontinental extent within an otherwise homogeneous vegetation belt.

In short, cyclical interactive processes leading to mosaic patterns within an ecosystem are as inherent to life as waves are to the interaction of wind and water (e.g., Miles, 1979). The scale of these patterns may vary from a few m² to millions of km², but they are all superimposed upon one another. Paleontologists will rarely be able to recognize such patterns directly in the fossil record. Nevertheless we can be quite sure that they have existed, and sometimes we can infer them indirectly from the mixed or "blurred" appearance of the fossil assemblage. A case in point are the Miocene ape habitats in west and central Europe. Another case are Zones 5–6 in the Siwaliks, dated about 10 m.y.a. and characterized by an abundance of fossil apes (Badgley and Behrensmeyer, 1980).

These and other data show that the current dichotomy of forest *vs.* savanna, with woodland as an intermediate form, is long past its usefulness. It should be replaced by the trichotomy of forest–bushland–grassland, with wooded bushland, wooded grassland (=woodland), and bushy grassland as intermediate forms. Such a tripartition will greatly improve our understanding of paleoenvironments.

Abel's Reconstruction

Abel more than anyone else shaped and propagated the classical (and still more or less current) ideas about early and middle Miocene paleoenvironments in central and southern Europe (e.g., Abel, 1927). In his view the general character of the European "Vindobonian" mammal fauna was approximately similar to that in Indonesia at present ("Der allgemeine Charakter der Säugetierwelt des europäischen Miozäns ist zwar nicht in den Einzelheiten, aber doch in den grossen Zügen derselbe wie jener der heute noch im Gebiete des indomalaiischen Archipels lebenden Säugetierfauna"; Abel, 1927, p. 209). His artistic reconstruction of a swamp forest in the Vienna Basin about 15 m.y.a. resembled nothing so much as a supertropical rain forest, with a tree 35 m high—taller than any swampy soil could ever carry, even in the tropics! (Abel, 1927, Fig. 208). He called the pliopithecines "long-armed gibbons" ("langarmigen Gibbons"; Abel, 1927, p. 256). He had, however, a problem with the floristic composition of the vegetation, which appeared to be definitely moist, warm–temperate, and partly subtropical, but almost devoid of any really tropical elements. Conifers predominated, leafy trees were also abundant, palms were rare, and the overall character was reminiscent of the floristically rich natural forests in humid areas and swamps in the southern parts of the United States. In this picture there was no place, of course, for glades, woodlands, and similar open vegetation.

More Recent Views

Later authors have corrected this reconstruction. Among others, Ginsburg and Tassy (1977), Hünermann (1969), Kretzoi (1975), Tobien (1956), and Zapfe (1960) found evidence of the mixed occurrence of forest and grassland faunas at certain sites, or the simultaneous occurrence of forest, woodland, and grassland faunas at sites near to each other, or a combination of both phenomena. That is, the clues indicated both small-scale mosaic and large-scale mosaic landscapes. Unfortunately, they did not sufficiently distinguish between bushland-adapted and grassland-adapted mammal species.

Recently Andrews (1980) applied the new technique of ecological analysis to the faunas of the "Helvetian" and "Badenian" *Pliopithecus* sites of Sansan, La Grive, and Neudorf-Spalte (NM lower 6–lower 7, 16–14 m.y.a.). The faunal composition and diversity of all three sites show the closest affinities to those of Rusinga, and to a lesser extent to those of Karungu (at least according to my visual impression of the histograms—no statistical analysis was added). Thus they represent some form of mixed forest–woodland–bushland habitat, rather than closed forests as was thought in most of the previous literature. The percentage of grazers is one-third of the percentage of browsers. This, again, is rather surprising. Every self-sustaining forest and woodland contains, of course, an amount of bushy undergrowth that includes the youngest generation of trees. In this case there was apparently also quite an amount of grassland, possibly because the canopy was far from closed while the young undergrowth was consumed by the browsers. It could also have been a park landscape with open glades. One wonders, of course, how such a vegetation could have developed. According to foresters, the natural climax vegetation on good soil in a moist, warm–temperate or subtropical climate consists of closed forest, except for some grassland on periodically inundated grounds along rivulets.

A Modern Analog

A solution to the problem may be found in the vegetation of the New Forest near Southampton in England. In some parts of this royal forest, according to historical records and tree-ring counts, no significant wood-cutting or tree-planting seems to have taken place for centuries, at some sites perhaps even since William the Conqueror, about 900 years ago. The original purpose was to protect the deer, but cattle and swine also roamed around. Fallen trees were left to rot and still are at many sites. Fallow deer and domesticated, but semi-wild-living ponies and cattle are still allowed to roam freely, and to browse and graze as they like, but serious overpopulation is prevented by commercial exploitation. The old, wild parts of the forest can therefore be regarded to some extent as a model of an ecosystem where plants, herbivores, and (human) predators are, on the average, more or less in dynamic equilibrium, albeit with large fluctuations resulting from events in human history (Tubbs, 1968). The effects of browsing and grazing can be studied by inspecting the ends of twigs, the length of grasses, the condition of saplings, the shape of shrubs, the traces of debarking, the botanical succession, the distribution of the age classes, and the composition of the various plant communities at different sites. In some places a fence allows a comparison with herbivore-free plots. To recognize the dynamic processes a keen pair of eyes suffices. Tree-ring counts in combination with historical data show that temporary browsing pressures of 0.28–0.37 "animal feeding units" or more per

Fig. 8. Characteristic views of two sites in an English royal forest. These areas have been protected from wood-cutting and interference by forestry for a very long time, supposedly even since medieval times. However, deer, cattle, and ponies have been allowed to roam in them freely. Browsing, debarking, and grazing have brought about an irregularly mottled mosaic landscape consisting of both closed and semi-open forests, thorn and fern bushes, and glades of various sizes. [Denny Wood and Black Bush in the New Forest near Southampton, England, a forest that has a beech–oak (*Fagus–Quercus*) natural climax vegetation, 1980 and 1967; photographs copyright A. Kortlandt and Meridian Airmaps Ltd.].

acre virtually suppressed forest regeneration* (Peterken and Tubbs, 1965; Tubbs, 1968).

The most prominent feature of this wild forest is its irregular, patchy physiognomy. Sites with a dense canopy of fully grown oaks and beeches, and consequently with very little undergrowth, alternate with predominantly younger wood and more undergrowth, with dense fern bushes where hardly a tree grows, with minimeadows of classroom size, and with larger glades on waterlogged ground (Fig. 8A). Aerial photographs show a quite irregular, patchy pattern (Fig. 8B). Visual inspection of the canopy from the ground reveals that in some places a monkey or an ape would easily find its way by climbing and leaping from branch to branch and from tree to tree; in other places it would have to make large detours on its route and it could then be trapped by a pine-marten; and at still other places it would have to cross on the ground. There are also sites with mainly old, dying trees, due to the last severe summer drought.

Much of this diversity is triggered and maintained by the action of herbivores. Defecation places in meadows are avoided by grazers. In them develop bramble bushes and other thorny thickets whose protection permits their expansion and later colonization by birches and other fast-growing trees. These are, in their turn, succeeded by the slower growing trees, oaks and beeches, which form a dense canopy and thus suppress much of the undergrowth. When they die, either grasses or fern bushes may get a chance. The latter are less browsed upon, because of their content of toxic chemicals, but will be suppressed by a new arboreal cycle. The herbivores also contribute to the irregular pattern by browsing and grazing, so to speak, mosaic-wise: they move in herds and concentrate their foraging at places where the palatability seems to be just a little bit better than elsewhere. A fascinating place for field work by a primate paleoecologist! Such a study will be facilitated by the literature on the food choice and feeding strategies of European herbivores (e.g., Borowski and Kossak, 1972; Borowski et al., 1967; Dzieciolowski et al., 1975; Hofmann and Stewart, 1972; Kossak, 1976; Ruckebusch and Thivenx, 1980; van de Veen, 1979). Again, the present is the key to the past. However, one must not forget the dictum *mutatis mutandis:* the largest and most destructive game, the proboscids, rhinos, chalicotheres, and pigs, have still to be inserted into the picture. Present-day Africa and Asia provide the clues to their impact. Data are found in many papers in the *East African Wildlife Journal,* now called the *African Journal of Ecology.*

Future Research

The new method of analysis of fossil faunas, as applied by Andrews *et al.* (1979), is a quite important step forward in the development of paleoecology. True, the assessment of the ecological categories according to which the species are classified still involves a subjective element, but this is partly compen-

*One head of cattle = one unit, one deer = three units, one pony = five units of browsing damage.

sated by the comparative procedure. To some extent the subjective element can be reduced by objective methods of assessment. For instance, the food adaptations of the teeth can be evaluated according to the principles outlined by Kay (1977), Kay and Hylander (1978), and Lucas (1979).

A technical refinement aimed at an improved "visibility" of the results would be to make use of the so-called concentration index, i.e., an index figure often used in geographical statistics. It is defined as the proportion of a category of people within the total population of an area divided by the proportion of the same category of people within the total population of a larger comprehensive area, e.g., a nation. The advantage of this method is that, especially in histograms, the regional differences can be identified much more easily. Moreover, when applied to fossil material, the differences in fossil abundance among the 23 ecological categories would be reduced to comparable levels. The inconsistencies in the interpretation of the Karungu, Rusinga, and Fort Ternan material would then have occurred less easily. A disadvantage, however, is that the figures for the four main categories would be unequally weighted. Both methods of analysis should, therefore, be used side by side.

Faunal lists and analyses of the faunal composition and diversity are excellent tools of paleoecology, but they are not enough. They are insufficient even to determine the geo-ecological boundaries of a fossil species. For example, the site list in Szalay and Delson (1979) and the maps in Collier (1979), Kowalski and Zapfe (1974), and Papp and Thenius (1959) show that no specimens of *Dryopithecus* and *Pliopithecus* have been found north of the line from the Loire River via Mainz to Opole.* South of this line, in central Europe, in a belt that ranges from France through Switzerland, Germany, and Austria to Hungary and Poland, 30–35 Miocene ape sites are known. Thus it would seem that the line from the Loire to Opole constitutes the northern ecological limit of the Miocene apes. However, a different picture emerges when we look at Papp's and Thenius' map of "important Tertiary mammal sites" in central Europe (excluding France). It shows two important fossiliferous belts: one in and around the Alps and the Carpathians, and the other north of the line from Eppelsheim (between Mainz and Mannheim) to Krakow (on the Vistula). Between these two belts is a virtual vacuum about 150–200 km wide. The northern belt can be divided by a line from Cologne via Göttingen to Wroclaw (Breslau) into a northern and a southern half. Now the interesting point is that the northern half consists of 23 mapped Tertiary sites without including an ape site, and the southern half 27 mapped sites including three Miocene ape sites (11%), while the Alpine and Carpathian fossil zone contains 130 sites but has only seven ape sites (5%). When those sites from Szalay and Delson's (1979) list that can be traced in the *Times Atlas* are added, the figures become 15% and 11%, respectively. These figures are statistically too insignificant to allow a conclusion to be drawn concerning the northern limit for the Miocene apes in central Europe. (They are also geologically inadequate

---

*In Szalay and Delson's world list of fossil primate localities the presumed age of each site and the continent where it is located are stated, but its geographical position is not indicated. Several of these sites cannot be traced on current atlases.

because the Tertiary is not subdivided in Papp's and Thenius' map, but that is not the point I am making here.)

I have been unable to trace more satisfactory data. Sadly, paleontologists have the habit of publishing, as a rule, only the location and the species list of the site, without stating the *numbers* of specimens for each taxon. If one had these numbers, then it would be possible to make an estimate of the statistical chance that a taxon could be absent due to the small number of specimens collected at the site. Such an estimate can be based upon the numbers collected at other, comparable sites. It is ironic that these numbers are not available. Thousands of geologists have dedicated their lives to extracting millions of fossils from the earth, and to labeling, meticulously describing, and taxonomically assigning all these millions, but few of them have done any "bookkeeping" about all this work, and still fewer published their counts. Yet counting the numbers and stating them in the species list of the site would have been the easiest part of the whole job.

Another important category of figures that are rarely reported are the minimum numbers of individuals of mammals. If other researchers availed themselves of these data, together with the numbers of specimens, one could apply Shotwell's (1958) method of analysis of mammal communities. [This technique has been used by Andrews and Van Couvering (1975) in their study of the four selected sites discussed above.] The basic principle is that the skeletons of those animals that died at or near to the site of fossilization may be expected to be represented much more completely in the fossil record than those animals that died further away from it. One can raise many objections, of course, but nevertheless this method can facilitate our understanding of the composition of mixed and mosaic assemblages, e.g., at the East African sites. Furthermore, with knowledge of the numbers of specimens as well as the minimum numbers of individuals, one would then be better equipped to test various hypotheses. Among other things, it would be possible to estimate the amount of food competition among apes, monkeys, bears, and squirrels. This would still be rather far away from the ecologist's ultimate goal, which is drawing a chart of the food chain and the energy flow, but it would represent a major step forward toward the realization of this goal.

Lack of (potentially available) data also hampers an evaluation of the method of analysis of Andrews *et al.* (1979). First, one would like to know to what extent the calculated diversity patterns might have been deformed by the accidental absence of species belonging to any particular category. Because of the lack of the numbers of specimens, no such statistical check is possible. Second, a site with, for instance, many *Hipparion* but few deer specimens is ecologically quite different from a site with few deer but many *Hipparion,* even if the species diversity pattern happens to be approximately the same. Without the numbers, the difference remains hidden. Thus it is quite feasible that some of my critical remarks will turn out to be unjustified when one day the numbers stored in the British Museum (Natural History) are counted. It is, however, equally feasible that some of my conclusions may be corroborated when the numbers become available. Much useful work could be done here.

There is still much more hidden in museums. I have already mentioned that all the material on the Miocene flora of East Africa has so far only been published in a preliminary form. The same applies to the gastropods. Material also exists on birds, but so far only some flamingo fossils have been published. Finally, nothing has been done on the magnificently preserved collection of insects (Leakey, 1952). In the near future this material will be elaborated on by M. P. Clifton. Since insects are excellent habitat indicators and, moreover, since they are evolutionarily very conservative, a study of them might solve several problems treated in the present paper. One occasionally wonders why so often so much effort and money are directed toward finding still more fossils, while so much relevant material is left unexplored in the vaults of museums.

Final Comments

I would have preferred not to have written this chapter in its present form, for several reasons: First, it is unpleasant to criticize esteemed colleagues, especially when there is a friend among them. Second, the critical tone may take attention away from the more positive elements. Third, readers of scientific books do not like to have their trust in such books shattered by too much criticism.

Let me, therefore, say only this. Those who work hard and produce much data, but do not try to penetrate into the underlying truth, will rarely be criticized. Those, on the other hand, who honestly search for the truth, do relevant research, and publish significant results will inevitably make errors and will receive much criticism. The difference is similar to that between Linnaeus and Darwin. Both kinds of scientists are needed, of course. Criticism, therefore, is not necessarily a form of blame, though it occasionally can be.

Still, a problem remains. Even open-minded scientists, familiar with both the fossil evidence and the recent fauna and flora of Africa, have made serious misjudgments in trying to outline the past in terms of the present. Often these scientists unintentionally ignored, misquoted, or distorted the evidence available to them, so that a misinterpretation became inevitable. The result was that the habitats of the Miocene apes were usually depicted as being more forested, *casu quo* more wooded, than was justified by the facts. Thus two questions arise: (1) Why do people depict ape habitats as being more forested, or more wooded, than they are nowadays, and were in the past? (2) How do such errors come to be generally accepted?

The answer, I think, is: because since time immemorial we have been indoctrinated. From Camper and Tulp, through Darwin and Haeckel, to Le Gros Clark, Yerkes, and Washburn—and also from kindergarten through high school to university—we have all been taught that the living apes are specialized as canopy-dwellers and arm-swingers of the evergreen rain forest. Admittedly, the gorilla was known to be a quadrupedal ground-walker. Never mind, he was declared to be a "modified brachiator." Both Nissen (1931) and

Gromier (1952) had extensively described chimpanzee wild life in the savannas of Guinea. Never mind again, Dekeyser (1955, p. 157) informed us that chimpanzees are forest animals which sometimes visit gallery forests and wooded savannas near forests ("sont des animaux forestiers, mais ils fréquentent parfois les galeries et les savanes boisées préforestières"). Malbrant and Maclatchy (1949) wrote that the preferred habitat was the mature secondary forest ("haute forêt secondaire"), and that they had only once seen a chimpanzee in a savanna, and that was walking from one gallery forest to another one (p. 59). Even Goodall (1965, p. 426) continued to maintain that "the closed rain forest" was "the normal habitat" for these apes. Napier and Davis (1959, p. 63) called them "structurally brachiators and capable of this form of locomotion." Finally, the same authors concluded that the "upper limb" of *Proconsul africanus* "foreshadows the extreme degree of mobility necessary for brachiation of the pongid type" (Napier and Davis, 1959, p. 66). Such terms as "upper limb" (as opposed to "forelimb") and "foreshadows" (as opposed to "might possibly evolve") illustrate how subtly this kind of indoctrination can sometimes operate.

This arboreal bias has much deeper roots than a scientific disagreement about habitats and locomotion patterns. In fact, modern ape research in the fields of ecology, ethology, psychology, sign language, and sociology has been a perpetual battle against the bulwarks of preconceived ideas and *Weltanschauung*. All these controversies show how deeply rooted is our cultural aversion to considering anything human-like among the apes.

Similar prejudices, though less strong, are also found among anatomists, zoologists, and paleoscientists. It is well known how vehemently Charles Darwin was attacked and ridiculed. Eugene Dubois' *Pithecanthropus* was so fiercely contested that he almost became mentally disturbed. Raymond Dart originally could not believe in his find and named it *Australopithecus,* rather than *Australoanthropus.* G. G. Simpson continued to classify the "ape-men" among the Pongidae even in the late 1940s, when their fossil remains had become available by the dozen. Taxonomists continue to lump the African apes with the orangutan in the Pongidae, in spite of all the biochemical evidence, and in spite of the entirely different motor patterns and climbing adaptations of these two taxa. Similarly, the dryopithecines have been consigned to the rain forest by unintentionally ignoring, misquoting, and distorting the plain facts. Thus a special seat in the universe is still being reserved for the human species. Apparently, even among dedicated and honest scientists, some peculiar psychological mechanisms ["mindscapes" (Maruyama, 1980)] are at work here.

Even such a simple formulation as the "dehumanization hypothesis of African ape evolution" shocked many colleagues and met with serious resistance. Yet it was just a renaming of the well-known phenomenon called character displacement by taxonomists, and which in this case was applied to ape evolution. The crux of the matter apparently is that humans do not like to admit the presence of anything humanoid outside the human race. Theologians, philosophers, behaviorists, and ethologists alike have stated either that animals have no soul and no mind, or at least that their mental processes are unknowable and therefore scientifically nonexistent. The separation is then made between the apes and humans. Apes can, therefore, be put behind bars in zoos, or used for medical experiments, with moral impunity. In such a cultural setting the humanoid aspects of ape behavior and ecology were not taken (and could not be taken) seriously, whatever lip service was paid to evolutionary biology.

One wonders what is behind all this. A possible (though quite tentative) explanation may perhaps be found in the principle of competitive exclusion. Observe in the wild how chimpanzees and baboons mostly avoid one another, but occasionally threaten, fight, and even kill one another. Or just go to the local zoo and watch the aggressive behavior between non-conspecific primates in adjoining cages. Also watch the behavior of uninhibited human beings in front of the chimpanzee cage. The analogy (or should I say homology?) is obvious, though scientifically unproven. The phenomenon is reminiscent of high-intensity racist responses. I have seen similar facial and bodily expressions in mobbing episodes during the war, when Nazis dealt with Jews. One can also occasionally observe such expressions among bullying schoolchildren. One gets the impression that very, very ancient behavioral archetypes are at work in such cases.

Comparable behavior occurs among the carnivores. The relationship between cat and dog is proverbial. In national parks in Africa one can observe similar incidents between non-conspecific beasts of prey. Thus, mobbing is a basic mechanism in the process of competitive exclusion among ground-dwelling and semi-ground-dwelling species that depend on relatively scarce food resources. (The exclusively arboreal, canopy-dwelling primate species are much more tolerant toward one another in zoos, and may travel in multi-species parties in the wild.) One may expect, therefore, that our ancestors, ever since ramapithecine times, have chased all the other apes and monkeys high up into the trees whenever they came across them, exactly as dogs do with cats. Accepting this inference by analogy as a working hypothesis, one may extrapolate from it by speculating that an instinctive perceptual archetype has evolved identifying the competing primates as "should-be-in-the-trees" creatures. Psychologists and ethologists may wish to try to test this suggestion.

At any rate, both paleoprimatologists and paleoanthropologists should scrutinize the data more scrupulously than even the best among them have often done. They should also be more aware of stereotypes that all too often lead scientists along false tracks. Furthermore, they should become much more precise in their interpretations of past habitats by incorporating all the data that modern climatology, ecology, zoology, and botany can provide. And finally, they should widen their field of view and consider the newest trends, not only in comparative anatomy and taxonomy, but also in the sciences that study life when it is still alive.

Modern ecological, behavioral, and social research on primates, both in the wild and in captivity, has shown that our relatives are much more human-like than all the Horatios of science had ever dreamt. At the same time the living primates constitute the final end-products of an infinite series of evolutionary events that shaped each of them in its own niche in a wide variety of continuously changing ecosystems. That is, all of them carry with them the hidden traces of their paleohistory. Unraveling these traces is a matter of dialog between the paleosciences and the sciences that study the Recent. In the first part of this chapter I criticized a new generation of paleoscientists who depicted an image of the hominoids in which the humanoid elements had been erased, ecologically speaking. In that image the dryopithecines were no longer a prelude to what followed. The evolution of the hominids had become a creationist event. I hope my criticism will eventually contribute to a reemergence of a unified theory of ape and human evolution.

ACKNOWLEDGMENTS

I am indebted to the following people, who kindly provided me with advice and information, or assisted in other ways: P. Andrews, I. R. Ball, H. de Bruijn, E. Delson, J. B. Gillett, I. Isheim, W. Kerkhof, M. Kretzoi, R. Letouzey, J. D. Paterson, M. Pickford, T. Synnott, J. C. Thackray, E. Thenius, H. Tobien, C. R. Tubbs, H. E. van de Veen, B. Verdcourt, J. Verschuren, and H. Zapfe.

References

Abel, O. 1927. *Lebensbilder aus der Tierwelt der Vorzeit*, 2nd ed., Fischer, Jena. 714 pp.

AETFAT 1959. *Vegetation Map of Africa—Carte de la végétation de l'Afrique*, Oxford University Press, Oxford. 24 pp.

Andrews, P. 1980. Ecological adaptations of the smaller fossil apes. *Z. Morphol. Anthropol.* **71:**164–173.

Andrews, P. 1981. Species diversity and diet in monkeys and apes during the Miocene, in: *Aspects of Human Evolution* (C. B. Stringer, ed.), pp. 25–61, Taylor, London.

Andrews, P., and Nesbit Evans, E. 1979. The environment of *Ramapithecus* in Africa. *Paleobiology* **5:**22–30.

Andrews, P., and Van Couvering, J. A. H. 1975. Palaeoenvironments in the East African Miocene, in: Approaches to Primate Paleobiology (F. S. Szalay, ed.). *Contrib. Primatol.* **5:**62–103.

Andrews, P., and Walker, A. 1976. The primate and other fauna from Fort Ternan, Kenya, in: *Human Origins: Louis Leakey and the East African Evidence* (G. L. Isaac and E. R. McCown, eds.), pp. 278–304, Benjamin, Menlo Park, California.

Andrews, P., Hamilton, W. R., and Whybrow, P. J. 1978. Dryopithecines from the Miocene of Saudi Arabia. *Nature (Lond.)* **274:**249–250.

Andrews, P., Lord, J. M., and Nesbit Evans, E. M. 1979. Patterns of ecological diversity in fossil and modern mammalian faunas. *Biol. J. Linn. Soc.* **11:**177–205.

Axelrod, D. I., and Raven, P. H. 1978. Late Cretaceous and Tertiary vegetation history of Africa, in: Biogeography and Ecology of Southern Africa (M. J. A. Werger, ed.). *Monogr. Biol.* **31:**77–130.

Badgley, C., and Behrensmeyer, A. K. 1980. Paleoecology of Middle Siwalik sediments and faunas, northern Pakistan. *Palaeogeogr. Palaeoclimatol. Palaeoecol.* **30:**133–155.

Baker, B. H., Mohr, P. A., and Williams, L. A. J. 1972. Geology of the eastern rift system of Africa. *Geol. Soc. Am. Spec. Pap.* **136**:1–67.

Barnes, R. F. W. 1980. The decline of the baobab tree in Ruaha National Park, Tanzania. *Afr. J. Ecol.* **18**:243–252.

Berger, W., and Zabusch, F. 1953. Die obermiozäne (sarmatische) Flora der Türkenschanze in Wien. *Neues Jahrb. Geol. Palaeontol., Abh.* **98**:226–276.

Bishop, W. W. 1963. The Later Tertiary and Pleistocene in eastern equatorial Africa, in: *African Ecology and Human Evolution* (F. C. Howell and F. Bourlière, eds.), pp. 246–275, Aldine, Chicago, Illinois.

Bishop, W. W. 1968. The evolution of fossil environments in East Africa. *Trans. Leicester Lit. Phil. Soc.* **62**:22–44.

Bishop, W. W., and Trendall, A. F. 1966/67. Erosion-surfaces, tectonics and volcanic activity in Uganda. *Q. J. Geol. Soc. Lond.* **122**:385–420.

Borowski, S., and Kossak, S. 1972. The natural food preferences of the European bison in seasons free of snow cover. *Acta Theriol.* **17**:151–169.

Borowski, S., Krasiński, Z., and Milkowski, L. 1967. Food and role of the European bison in forest ecosystems. *Acta Theriol.* **12**:367–376.

Bowen, B. E., and Vondra, C. F. 1974. Paleoenvironmental interpretations of the Oligocene Gabal el Qatrani Formation, Fayum depression, Egypt. *Ann. Geol. Survey Egypt* **4**:115–137.

Brock, P. W. G., and MacDonald, R. 1969. Geological environment of the Bukwa mammalian fossil locality, eastern Uganda. *Nature (Lond.)* **223**:593–596.

Buechner, H. K., and Dawkins, H. C. 1961. Vegetation change induced by elephants and fire in Murchison Falls National Park, Uganda. *Ecology* **42**:752–766.

Butzer, K. W. 1978. Geoecological perspectives on early hominid evolution, in: *Early Hominids of Africa* (C. Jolly, ed.), pp. 191–217, Duckworth, London.

Butzer, K. W., and Hansen, C. L. 1968. *Desert and River in Nubia*, University of Wisconsin Press, Madison. 562 pp.

Cahen, L. 1954. *Géologie du Congo Belge*, Vaillant-Carmanne, Liège. 577 pp.

Camper, P. 1782. *Natuurkundige verhandelingen over den Orang Outan; en eenige andere aap-soorten*, Meijer & Warnars, Amsterdam, 120 pp.

Caughley, G. 1976. The elephant problem—An alternative hypothesis. *E. Afr. Wildl. J.* **14**:265–283.

Chaney, R. W. 1933. A Tertiary flora from Uganda. *J. Geol.* **41**:702–709.

Chaney, R. W. 1940. Tertiary forests and continental history. *Bull. Geol. Soc. Am.* **51**:469–488.

Chesters, K. I. M. 1957. The Miocene flora of Rusinga Island, Lake Victoria, Kenya. *Palaeontographica B* **101**:30–71.

Cobb, S. 1976. *The Distribution and Abundance of the Large Herbivore Community of Tsavo National Park, Kenya*, Ph. D. Dissertation, Oxford University. 208 pp.

Collier, A. 1979. Notes sur le Pliopithèque des Faluns de Touraine. *Bull. Soc. Géol. Touraine* **1**:19–24. .

Conroy, G. C., and Fleagle, J. G. 1972. Locomotor behaviour in living and fossil pongids. *Nature (Lond.)* **237**:103–104.

Corruccini, R. S., Ciochon, R. L., and McHenry, H. M. 1975. Osteometric shape relationships in the wrist joint of some anthropoids. *Folia Primatol.* **24**:250–274.

Corruccini, R. S., Ciochon, R. L., and McHenry, H. M. 1976. The postcranium of Miocene hominoids: Were dryopithecines merely "dental apes"? *Primates* **17**:205–233.

Croze, H. 1974. The Seronera bull problem. *E. Afr. Wildl. J.* **12**:1–27, 29–47.

Curry-Lindahl, K. 1974. In: CA book review: New perspectives on ape and human evolution. *Curr. Anthropol.* **15**:432–433.

Dale, I. R., and Greenway, P. J. 1961, *Kenya Trees and Shrubs*, Buchanan, Nairobi. 654 pp.

De Bournonville, D. 1967. Contribution à l'étude du Chimpanzé en République de Guinée. *Bull. Inst. Fond. Afr. Noire* **29**(A):1188–1269.

Dekeyser, P. L. 1955. *Les Mammifères de l'Afrique noire française*, IFAN, Dakar. 426 pp.

Delson, E. 1975. Evolutionary history of the Cercopithecidae, in: Approaches to Primate Paleobiology (F. S. Szalay, ed.) *Contrib. Primatol.* **5**:167–217.

De Swardt, A. M. J. 1964. Lateritisation and landscape development in parts of equatorial Africa. *Z. Geomorphol., N.F.* **8**:313–333.

Devred, R. 1958. La végétation forestière du Congo belge et du Ruanda-Urundi. *Bull. Soc. R. For. Belg.* **65**:409–468.

Dewey, J. F., Pitman, W. C., Ryan, W. B. F., and Bonnin, J. 1973. Plate tectonics and the evolution of the Alpine system. *Geol. Soc. Am. Bull.* **84**:3137–3180.

Douglas-Hamilton, O. 1980. Africa's elephants. Can they survive? *Nat. Geogr. Mag.* **158**:568–603.

Dzieciolowski, R., Kossak, S., Borowski, S., and Morow, K. 1975. Diets of big herbivorous mammals. *Pol. Ecol. Stud.* **1**:33–50.

Eggeling, W. J. 1947. Observations on the ecology of the Budongo rain forest, Uganda. *J. Ecol.* **34**:20–67.

Flohn, H. 1964. Über die Ursachen der Aridität Nordost-Afrikas. *Würzburg. Geogr. Arbeiten* **12**:21–37.

Frakes, L. A., and Kemp, E. M. 1972. Influence of continental positions on Early Tertiary climates. *Nature (Lond.)* **240**:97–100.

Gautier, A. 1965a. *Geological Investigation in the Sinda-Mohari (Ituri, NE-Congo)*, Rijksuniversiteit, Gent. 161 pp.

Gautier, A. 1965b. Relative dating of peneplains and sediments in the Lake Albert Rift area. *Am. J. Sci.* **263**:537–547.

Ginsburg, L. 1968. L'évolution du climat au cours du Miocène, en France. *Bull. Assoc. Nat. Orléanais Loire, N.S.* **41**:3–13.

Ginsburg, L. 1975a. Une échelle stratigraphique continentale pour l'Europe occidentale et un nouvel étage: l'Orléanien. *Nat. Orléanais, Ser. 3* **18**:3–11.

Ginsburg, L. 1975b. Le Pliopithèque des faluns Helvétiens de la Touraine et de l'Anjou, in: Évolution des Vertebrés—Problèmes Actuels de Paléontologie. *Colloq. Int. Cent. Nat. Rech. Sci.* **218**:877–885.

Ginsburg, L., and Mein, P. 1980. *Crouzelia rhodanica*, nouvelle espèce de primate catarhinien, et essai sur la position systématique des Pliopithecidae. *Bull. Mus. Natn. Hist. Nat. Sér. 4* **2**(C):57–85.

Ginsburg, L., and Tassy, P. 1977. Les nouveaux gisements à mastodontes du Vindobonien moyen de Simorre (Gers). *C. R. Somm. Soc. Géol. Fr.* **1977**:24–26.

Glaessner, M. F. 1932/33. Neue Zähne von Menschenaffen aus dem Miozän des Wiener Beckens. *Ann. Naturhist. Mus. Wien.* **46**:15–27.

Goodall, J. 1965. Chimpanzees of the Gombe Stream Reserve, in: *Primate Behavior* (I. DeVore, ed.), pp. 425–473, Holt, Rinehart & Winston, New York.

Gromier, E. 1952. *La Vie des Anthropoïdes*, Amiot-Dumont, Paris. 190 pp.

Guillaumet, J.-L. 1967. *Recherches sur la végétation et la flore de la région du Bas-Cavally (Côte-d'Ivoire)*, ORSTOM, Paris, 247 pp.

Hamilton, W. R., Whybrow, P. J., and McClure, H. A. 1978. Fauna of fossil mammals from the Miocene of Saudi Arabia. *Nature (Lond.)* **274**:248.

Harrington, G. N., and Ross, I. C. 1974. The savanna ecology of Kidepo Valley National Park I. The effects of burning and browsing on the vegetation. *E. Afr. Wildl. J.* **12**:93–105.

Harris, L. D., and Fowler, N. K. 1975. Ecosystems analysis and simulation of the Mkomazi Reserve, Tanzania. *E. Afr. Wildl. J.* **13**:325–346.

Harrison, C. J. O., and Walker, C. A. 1976. Cranial material of Oligocene and Miocene flamingos: With a description of a new species from Africa. *Bull. Br. Mus. (Nat. Hist.) Geol.* **27**:305–314.

Hofmann, R. R., and Stewart, D. R. M. 1972. Grazer or browser: A classification based on the stomach-structure and feeding habits of East African ruminants. *Mammalia* **36**:227–240.

Holmes, A. 1966. *Principles of Physical Geology*, Nelson, London. 1288 pp.

Hünermann, K. A. 1969. Über den Leitwert der Suidae im europäischen Neogen. *Eclogae Geol. Helv.* **62**:715–730.

Hutchinson, J., and Dalziel, J. M. 1954/58/63/68/72. *Flora of West Tropical Africa*, 2nd ed. Crown Agents for Overseas Governments and Administration, London.

Institut des Parcs Nationaux du Congo Belge 1947. *Parc National Albert. Territoires Biogéographiques—Formations Végétales*, Bruxelles. 1 map.

Izawa, K. 1970. Unit groups of chimpanzees and their nomadism in the savanna woodland. *Primates* **11**:1–46.

Izawa, K., and Nishida, T. 1963. Monkeys living in the northern limits of their distribution. *Primates* **4**:67–88.

Jackson, J. K. 1956. The vegetation of the Imatong Mountains, Sudan. *J. Ecol.* **44**:341–374.

Kano, T. 1971. The chimpanzee of Filabanga, Western Tanzania. *Primates* **12**:229–246.

Kano, T. 1972. Distribution and adaptation of the chimpanzee on the eastern shore of Lake Tanganyika. *Kyoto Univ. Afr. Studies* **7**:37–129.

Kavanagh, M. 1980. Invasion of the forest by an African savannah monkey: Behavioural adaptations. *Behaviour* **73**:238–260.

Kay, R. F. 1977. Diets of early Miocene African hominoids. *Nature (Lond.)* **268**:628–630.

Kay, R. F., and Hylander, W. L. 1978. The dental structure of mammalian folivores with special reference to Primates and Phalangeroidea (Marsupialia), in: *The Ecology of Arboreal Folivores* (G. G. Montgomery, ed.), pp. 173–191, Smithsonian Institution Press, Washington, D.C.

Kedves, M. 1971. Présence de types sporomorphes importants dans les sédiments préquaternaires égyptiens. *Acta Bot. Acad. Sci. Hung.* **17**:371–378.

Kennett, J. P., Houtz, R. E., Andrews, P. B., Edwards, A. R., Gostin, V. A., Hajos, M., Hampton, M. A., Jenkins, D. G., Margolis, S. V., Ovenshine, A. T., and Perch-Nielsen, K. 1974. Development of the Circum-Antarctic Current. *Science* **186**:144–147.

King, B. C. 1978. Structural and volcanic evolution of the Gregory Rift Valley, in: *Geological Background to Fossil Man* (W. W. Bishop, ed.), pp. 29–54, Scottish Academic Press, Edinburgh.

King, L. C. 1967. *The Morphology of the Earth* 2nd ed., Oliver and Boyd, Edinburgh. 699 pp.

Kortlandt, A. 1959. *Tussen mens en dier,* Wolters, Groningen. 27 pp.

Kortlandt, A. 1962. Chimpanzees in the wild. *Sci. Am.* **206**(5):128–138.

Kortlandt, A. 1965. *Some Results of a Pilot Study on Chimpanzee Ecology,* Zoologisch Laboratorium, Amsterdam. 59 pp.

Kortlandt, A. 1967. Experimentation with chimpanzees in the wild, in: *Neue Ergebnisse der Primatologie—Progress in Primatology* (D. Starck, R. Schneider, and H.-J. Kuhn, eds.), pp. 208–224, Stuttgart, Fischer.

Kortlandt, A. 1968. Handgebrauch bei freilebenden Schimpansen, in: *Handgebrauch und Verständigung bei Affen und Frühmenschen* (B. Rensch, ed.), pp. 59–102, Huber, Bern.

Kortlandt, A. 1972. *New perspectives on Ape and Human Evolution,* Stichting voor Psychobiologie, Amsterdam. 100 pp.

Kortlandt, A. 1974. New Perspectives on ape and human evolution. A CA* book review. *Curr. Anthropol.* **15**:427–448.

Kortlandt, A. 1975*a*. Ecology and paleoecology of ape locomotion, in: *Proceedings from the Symposia of the Fifth Congress of the International Primatological Society 1974,* pp. 361–364, Japan Science Press, Tokyo.

Kortlandt, A. 1975*b*. Reply. In: On new perspectives on ape and human evolution. *Curr. Anthropol.* **16**:647–651.

Kortlandt, A. 1976. Tree destruction by elephants in Tsavo National Park and the role of man in African ecosystems. *Neth. J. Zool.* **26**:449–451.

Kortlandt, A. 1980*a*. The ecosystem in which the incipient hominines could have evolved, in: *Proceedings of the 8th Panafrican Congress of Prehistory and Quaternary Studies* (R. E. Leakey and B. A. Ogot, eds.), pp. 133–136, The International Louis Leakey Memorial Institute for African Prehistory, Nairobi.

Kortlandt, A. 1980*b*. How might early hominids have defended themselves against large predators and food competitors? *J. Hum. Evol.* **9**:79–112.

Kortlandt, A. 1980*c*. The Fayum primate forest: Did it exist? *J. Hum. Evol.* **9**:277–297.

Kortlandt, A., and Kooij, M. 1963. Protohominid behaviour in primates. *Symp. Zool. Soc. Lond.* **10**:61–88.

Kortlandt, A., and van Zon, J. C. J. 1969. The present state of research on the dehumanization hypothesis of African ape evolution, in: *Proceedings of the Second International Congress of Primatology Atlanta 1968,* Volume 3, pp. 10–13, Karger, Basel.

Kossak, S. 1976. The complex character of the food preferences of Cervidae and phytocenosis structure. *Acta Theriol.* **21:**359–373.

Kowalski, K., and Zapfe, H. 1974. *Pliopithecus antiquus* (Blainville, 1839) (*Primates, Mammalia*) from the Miocene of Przeworno in Silesia (Poland). *Acta Zool. Cracov.* **19:**19–30.

Kretzoi, M. 1975. New ramapithecines and *Pliopithecus* from the Lower Pliocene of Rudabánya in north-eastern Hungary. *Nature (Lond.)* **257:**578–581.

Lartet, E. 1856. Note sur un grand signe fossile qui se rattache au groupe des singes supérieurs. *C. R. Acad. Sci.* **43:**219–223.

Laws, R. M., Parker, I. S. C., and Johnstone, R. C. B. 1975. *Elephants and Their Habitats,* Clarendon, Oxford. 376 pp.

Leakey, L. S. B. 1952. Lower Miocene invertebrates from Kenya. *Nature (Lond.)* **169:**624–625.

Leakey, L. S. B. 1955. The environment of the Kenya Lower Miocene apes, in: *Congrès Panafricain de Préhistore: Actes de la 11e Session, Algiers* (L. Balout, ed.), pp. 323–324, Arts et Métiers graphiques, Paris.

Leakey, L. S. B. 1969. *Animals of East Africa,* National Geographic Society, Washington. 199 pp.

Lebrun, J., and Gilbert, G. 1954. *Une classification écologique des forêts du Congo,* INEAC, Sér. Scient. 63. 90 pp.

Le Gros Clark, W. E. 1959. *The Antecedents of Man.* University Press, Edinburgh. 374 pp.

Le Gros Clark, W. E., and Leakey, L. S. B. 1951. The Miocene Hominoidea of East Africa. *Fossil Mammals of Africa* (Br. Mus. Nat. Hist.) **1:**1–117.

Letouzey, R. 1958/59. Phytogéographie camerounaise, in: *Atlas du Cameroun,* IRCAM, Yaoundé.

Letouzey, R. 1968. *Etude phytogéographique du Cameroun,* Lechevalier, Paris. 511 pp.

Lewis, O. J. 1972. Osteological features characterizing the wrists of monkeys and apes, with a reconsideration of this region in *Dryopithecus (Proconsul) africanus. Am. J. Phys. Anthropol.* **36:**45–58.

Louvet, P. 1971. *Sur l'évolution des flores tertiaires de l'Afrique nord-équatoriale,* Thèse, Université de Paris VI, Paris.

Lucas, P. W. 1979. The dental–dietary adaptations of mammals. *Neues Jahrb. Geol. Palaeontol. Monatsh.* **1979:**486–512.

MacInnes, D. G. 1953. The Miocene and Pleistocene Lagomorpha of East Africa. *Fossil Mammals of Africa* (Br. Mus. Nat. Hist.) **6:**1–30.

Malbrant, R., and Maclatchy, A. 1949. *Faune de l'Equateur Africain Français 2, Mammifères,* Lechevalier, Paris, 323 pp.

Mangenot, G. 1955. Etude sur les forêts des plaines et plateaux de la Côte-d'Ivoire. *Etud. Eburnéennes* **4:**5–61.

Maruyama, M. 1980. Mindscapes and science theories. *Curr. Anthropol.* **21:**589–599.

Mein, P. 1980. Biozonation du Néogène méditerranéen à partir des mammifères (photocopy, personal communication).

Merz, G. 1977. *Untersuchungen über Ernährungsbiologie und Habitatspräferenzen des afrikanischen Waldelefanten, Loxodonta africana cyclotis, Matschie, 1900,* Zoologisches Institut, Heidelberg. 58 pp.

Miles, J. 1979. *Vegetation Dynamics,* Chapman, London. 80 pp.

Mitchell, B. L. 1960. Ecological aspects of game control measures in African wilderness and forested areas. *Kirkia* **1:**120–128.

Mohr, P. A. 1971. Outline tectonics of Ethiopia, in: *Tectonique de l'Afrique—Tectonics of Africa,* pp. 447–458, UNESCO, Paris.

Napier, J. R., and Davis, P. R. 1959. The fore-limb skeleton and associated remains of *Proconsul africanus. Fossil Mammals of Africa* (Br. Mus. Nat. Hist.) **16:**1–69.

Nissen, H. W. 1931. A field study of the chimpanzee. *Comp. Psychol. Monogr.* **8,** 1 (Ser. 36):1–122.

Oppenheimer, J. R. 1977. *Presbytis entellus,* the hanuman langur, in: *Primate Conservation* (Prince Rainier III of Monaco and G. H. Bourne, eds.), pp. 469–512, Academic, New York.

Papp, A., and Thenius, E. 1959. Tertiär II, Wirbeltierfaunen, in: *Handbuch der Stratigraphischen Geologie* (Fr. Lotze, ed.), pp. 1–328, Enke, Stuttgart.

Peterken, G. F., and Tubbs, C. R. 1965. Woodland regeneration in the New Forest, Hampshire, since 1650. *J. Appl. Ecol.* **2:**159–170.

Pickford, M. 1981. Preliminary Miocene mammalian stratigraphy for western Kenya. *J. Hum. Evol.* **10:**73–97.

Pickford, M., and Andrews, P. 1981. The Tinderet Miocene sequence in Kenya. *J. Hum. Evol.* **10:**11–33.

Pilbeam, D. 1979. Recent finds and interpretations of Miocene hominoids. *Annu. Rev. Anthropol.* **8:**333–352.

Pilbeam, D., Barry, J., Meyer, G. E., Shah, S. M. I., Pickford, M. H. L., Bishop, W. W., Thomas, H., and Jacobs, L. L. 1977. Geology and palaeontology on Neogene strata of Pakistan. *Nature (Lond.)* **270:**684–689.

Powers, R. W., Ramirez, L. F., Redmond, C. D., and Elberg, E. L. 1966. *Geology of the Arabian Peninsula* (Geological Survey Professional Paper 560-D.) U.S. Government Printing Office, Washington.

Preuschoft, H. 1973. Body posture and locomotion in some East African Miocene Dryopithecinae, in: *Human Evolution* (M. H. Day, ed.), pp. 13–46, Symposium of the Society for the Study of Human Biology, Volume 11, Taylor, London.

Rabeder, G. 1978. Die Säugetiere des Badenien, in: *Chronostratigraphie und Neostratotypen, Miozän der zentralen Paratethys, 6, M4, Badenien (Moravien, Wielicien, Kosovian)* (J. Senes and E. Brestenska, eds.), pp. 467–480, VEDA, Bratislava.

Robyns, W. 1948. *Les territoires biogéographiques du Parc National Albert*, Institut des Parc Nationaux du Congo Belge, Bruxelles. 51 pp.

Rögl, F., and Müller, C. 1978. Middle Miocene salinity crisis and paleogeography of the Paratethys (middle and eastern Europe), in: *Initial Reports of the Deep Sea Drilling Project*, Volume 42, pp. 985–990, U.S. Government Printing Office, Washington, D. C.

Ross, I. C., Field, C. R., and Harrington, G. N. 1976. The savanna ecology of Kidepo Valley National Park, Uganda. III. Animal populations and park management recommendations. *E. Afr. Wildl. J.* **14:**35–48.

Rowell, T. E. 1966. Forest living baboons in Uganda. *J. Zool. Lond.* **149:**344–364.

Ruckebusch, Y., and Thivend, P. 1980. *Digestive Physiology and Metabolism in Ruminants*, M.T.P. Press, Lancaster, England.

Saggerson, E. P., and Baker, B. H. 1965. Post-Jurassic erosion-surfaces in eastern Kenya and their deformation in relation to rift structure. *Q. J. Geol. Soc. Lond.* **121:**51–72.

Said, R. 1962. *The Geology of Egypt*, Elsevier, Amsterdam, 377 pp.

Salard-Cheboldaeff, M. 1977. *Paléopalynologie du bassin sédimentaire littoral du Cameroun dans ses rapports avec la stratigraphie et la paléoécologie*, Thèse, Université de Paris VI.

Savage, R. J. G. 1978. Carnivora, in: *Evolution of African Mammals* (V. J. Maglio and H. B. S. Cooke, eds.), pp. 249–267, Harvard University Press, Cambridge.

Schnell, R. 1952. Contribution à une étude phytosociologique et phytogéographique de l'Afrique occidentale: les groupements et les unités géobotaniques de la région guinéenne. *Mém. Inst. Fr. Afr. Noire* (Dakar) **18:**41–235.

Schnell, R. 1976. *Introduction à la phytogéographie des pays tropicaux. 3: La flore et la végétation de l'Afrique tropicale*, 1re partie, Gauthiers-Villars, Paris. 459 pp.

Schön, M. A., and Ziemer, L. K. 1973. Wrist mechanism and locomotor behavior of *Dryopithecus (Proconsul) africanus. Folia Primatol.* **20:**1–11.

Schwarzbach, M. 1961. The climatic history of Europe and North America, in: *Descriptive Palaeoclimatology* (A. E. M. Nairn, ed.), pp. 255–291, Interscience, New York.

Sen, S., and Thomas, H. 1979. Découverte de rongeurs dans le Miocène moyen de la Formation Hofuf (Province du Hasa, Arabie Saoudite). *C. R. Somm. Soc. Géol. Fr.* **1:**34–37.

Shackleton, N. J., and Kennett, J. P. 1976. Paleotemperature history of the Cenozoic and the initiation of Antarctic glaciation: Oxygen and carbon isotope analyses in DSDP sites 277, 279, and 281. *Initial Report Deep Sea Drilling Project* **17:**743–755.

Shackleton, R. M. 1978. Structural development of the East African Rift System, in: *Geological Background to Fossil Man* (W. W. Bishop, ed.), pp. 19–28, Scottish Academic Press, Edinburgh.

Shotwell, J. A. 1958. Inter-community relationships in Hemphillian (mid-Pliocene) mammals. *Ecology* **39:**271–282.

Simons, E. L. 1967. The earliest apes. *Sci. Am.* **217**(6):28–35.

Simons, E. L. 1968. Early Cenozoic mammalian faunas. Fayum Province, Egypt. Part I. African Oligocene mammals: Introduction, history of study, and faunal succession. *Bull. Peabody Mus. Nat. Hist.* (Yale Univ.) **28**:1–105.

Sowerbutts, W. T. C. 1972. Rifting in eastern Africa and the fragmentation of Gondwanaland. *Nature (Lond.)* **235**:435–436.

Steiniger, F., Rögl, F., and Martini, E. 1976. Current Oligocene/Miocene biostratigraphic concept of the Central Paratethys (Middle Europe). *Newsl. Stratigr.* **4**:174–202.

Suzuki, A. 1969. An ecological study of chimpanzees in a savanna woodland. *Primates* **10**:103–148.

Szalay, F. S., and Delson, E. 1979. *Evolutionary History of the Primates,* Academic, New York, 580 pp.

Teleki, G. 1973. *The Predatory Behavior of Wild Chimpanzees,* Bucknell University Press, Lewisburg. 232 pp.

Teleki, G. 1975. Primate subsistence patterns: Collector-predators and gatherer-hunters. *J. Hum. Evol.* **4**:125–184.

Thenius, E. 1955. Zur Entwicklung der jungtertiären Säugetierfaunas des Wiener Beckens. *Paläontol. Z.* **29**:21–26.

Thomas, H., Taquet, P., Ligabue, G., and Del'Agnola, C. 1978. Découverte d'un gisement de Vertébrés dans les dépôts continentaux du Miocène moyen du Hasa (Arabie saoudite). *C. R. Somm. Soc. Géol. Fr.* **2**:69–72.

Tobien, H. 1956. Zur Ökologie der jungtertiären Säugetiere vom Höwenegg/Hegau and zur Biostratigraphie der europäischen Hipparion-Fauna. *Schr. Ver. Gesch. Naturgesch. Baar* **24**:208–223.

Tubbs, C. R. 1968. *The New Forest: An Ecological History,* David and Charles, Newton Abbot. 248 pp.

Van Couvering, J. A. H., and Van Couvering, J. A. 1976. Early Miocene mammal fossils from East Africa, in: *Human Origins: Louis Leakey and the East African Evidence* (G. L. Isaac and E. R. McCown, eds.), pp. 155–207, Benjamin, Menlo Park, California.

Van de Veen, H. E. 1979. *Food Selection and Habitat Use in the Red Deer (Cervus elaphus L.),* Ph.D. Dissertation, Rijksuniversiteit, Groningen. 263 pp.

Van Zinderen Bakker, E. M. 1977/78. Late-Mesozoic and Tertiary palaeoenvironments of the Sahara region, in: *Antarctic Glacial History and World Palaeoenvironments* (E. M. van Zinderen Bakker, ed.), pp. 129–135, Balkema, Rotterdam.

Verdcourt, B. 1963. The Miocene non-marine Mollusca of Rusinga Island, Lake Victoria and other localities in Kenya. *Palaeontographica A* **121**:1–37.

Verdcourt, B. 1972. The zoogeography of the non-marine Mollusca of East Africa. *J. Conchol.* **27**:291–348.

Walker, A. 1969. Fossil mammal locality on Mount Elgon, eastern Uganda. *Nature (Lond.)* **223**:591–593.

Whiteman, A. J. 1971. *The Geology of the Sudan Republic,* Clarendon, Oxford, 290 pp.

Whitworth, T. 1953/54. A contribution to the geology of Rusinga Island, Kenya. *Q. J. Geol. Soc. Lond.* **109**:75–96.

Whitworth, T. 1958. Miocene ruminants of East Africa. *Fossil Mammals of Africa* (Br. Mus. Nat. Hist.) **15**:1–48.

Wing, L. D., and Buss, I. O. 1970. Elephants and forests. *Wildl. Monogr.* **19**:1–92.

Woldstedt, P. 1954. *Das Eiszeitalter,* 1. *Die allgemeinen Erscheinungen des Eiszeitalters,* Enke, Stuttgart. 374 pp.

Zapfe, H. 1960. Die primatenfunde aus der Miozänen spaltenfüllung von Neudorf an der March (Děvinská Nová Ves), Tschechoslowakei. *Schweiz. Palaeontol. Abh.* **78**:1–293.

Zwell, M., and Conroy, G. C. 1973. Multivariate analysis of the *Dryopithecus africanus* forelimb. *Nature (Lond.)* **244**:373–375.

Descriptive Analyses of Siwalik Miocene Hominoids

VI

The Significance of Hitherto Undescribed Miocene Hominoids from the Siwaliks of Pakistan in the Senckenberg Museum, Frankfurt

19

G. H. R. VON KOENIGSWALD

Introduction

With a team from the Geology Department of the University of Utrecht, as well as assistance from Tasir Hussain of Lahore and a grant allocated by the Werner-Reimers Foundation, several weeks were spent in Pakistan during the winter season of 1964–1965. A good faunal assemblage, now deposited in the Geological Institute of the University of Utrecht, was collected. Included in this faunal assemblage were a number of Miocene primate fossils which to date have yet to be presented to the scientific community. Therefore, it is the intent of this chapter to describe the hominoid primates recovered and to discuss their significance to the taxonomic and phylogenetic status of *Ramapithecus*.

G. H. R. VON KOENIGSWALD • Section of Paleoanthropology, Senckenberg Museum, 6000 Frankfurt 1, West Germany. Professor von Koenigswald died on July 10, 1982.

Fossil remains of early hominoids have been recovered from the Indian Siwaliks for more than a century (Lydekker, 1879). A canine, mentioned by Falconer (1868) and classified as a fossil orangutan, has never been formally described. Since these early finds, and in spite of many efforts undertaken by individual scientists and team expeditions, subsequent recoveries have been meager. Isolated teeth and mandibles from the Siwaliks are not too rare, but no complete cranial material and very little postcranial remains have been recovered. One explanation for this paucity may be apparent from the following scenario. The rivers of tropical India, especially during the Chinji, were filled by crocodiles. These carnivores consumed most parts of the hominoid carcasses floating down river before they were deposited in the sediments of river bottoms. Therefore, little deposition of hominoid skeletal remains occurred, and as a result, little material can be recovered today.

With little hominoid material available, interpretation of both taxonomic and phylogenetic information is very tentative. When dentition constitutes the majority of the recovered material, another problem arises. Because the teeth of apes and humans are similar in pattern and show only restricted morphological variation, it is not wise to rely exclusively on dentition. Sexual dimorphism within a given species is great and must certainly be considered in the assessment and interpretation of fossil hominoids. Based upon this, I do not trust the statistics relating to dentition very much.

The Taxonomic Problem

The number of hominoid species described from the Siwaliks is approximately 20, and the number of genera is eight. This is entirely too many taxonomic distinctions and indicates that many of the names are certainly *nomina vana*. Conversely, the conviction that too many species were being distinguished led to lumping, which also causes problems in the interpretation of hominoid evolution. Presently, only four taxa from the Siwaliks are generally recognized [taxa established by Simons and Pilbeam (1965) and Szalay and Delson (1979), respectively]:

Dryopithecus indicus	*Sivapithecus indicus*
Dryopithecus sivalensis	*Sivapithecus sivalensis*
(*Dryopithecus indicus*)	*Gigantopithecus giganteus*
Ramapithecus punjabicus	*Ramapithecus punjabicus*

In a recent publication, Greenfield (1979) changed several taxonomic names: *Ramapithecus* to "*Sivapithecus brevirostris*" and "*Kenyapithecus*" (often included in *Ramapithecus*) to *Sivapithecus africanus*. In my opinion, this inflation of "*Sivapithecus*" does not reflect reality. It is hoped that this taxonomic problem will be resolved via more fossil discoveries, thorough reinvestigations, and scientific consensus.

The Siwalik Fauna and Environmental Reconstruction

The Siwalik fauna is very diverse. By compiling the data Colbert (1935) obtained from three different beds, the following correlations result:

Chinji beds	85 species
Nagri beds	42 species
Dhok Pathan beds	102 species

Species in common:

Chinji and Nagri	17
Nagri and Dhok Pathan	9
Chinji and Dhok Pathan	7

Colbert (1935) mentioned 229 species, but only seven of these are common in all three horizons. It is unlikely that *Ramapithecus* or *Sivapithecus* is among them because primates, certainly evident from the studies of extant forms, react very quickly to climate or faunal changes. Such changes can easily be studied in the Siwaliks. In both the sedimentation and fauna of the Lower Siwaliks, the change from humid tropical forest to more open country and finally to a dry savanna is recorded. The basal layers of the Lower Siwaliks, the Chinji beds, are deeply colored red by laterite, an indication of a wet tropical climate. In the younger Nagri beds, the soil is more brown, although a red soil component is still evident, and in the Dhok Pathans, the soil is gray; both of these beds indicate past drier environmental conditions. An environmental change to an open savanna, together with a new biotope, may have aided in the development of bipedal locomotion in an originally forest-dwelling primate, hence the beginning of hominization.

Hipparion, the most important index fossil for the "classical" Pliocene, is rarely recovered from the Chinji beds but certainly is abundant in its upper levels. With faunal remnants from older periods, the Chinji can be compared to the Vallesian of Europe. The Nagri fauna can be correlated with the European Turolian. The fauna changed, especially with the loss of many pig-like Artiodactyla and the expansion in the number of *Hipparion*. An absolute date for the first appearance in Europe of *Hipparion*, migrating from the east, is 12 m.y.a. (Pilbeam and Boyer, 1971). A new date for *Ramapithecus* from the Potwar Plateau Siwaliks, based on paleomagnetics, is 8 million years (Tauxe, 1979). This would correspond to the Samos horizon of the upper Turolian.

Description of the Siwalik Primates

Chinji.   A single right lower molar was brought to the expedition by a native collector. The exact provenience is unknown, but according to circumstances, the horizon must be Chinji, perhaps from the upper level.

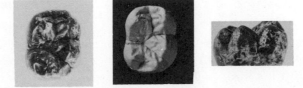

Fig. 1. *Chinjipithecus atavus,* von Koenigswald, 1981, right second lower molar. The figure in the middle is a cast to better illustrate the cusp pattern. Recovered from the Chinji beds near Chinji, Pakistan (× c.1.4).

It is a well-preserved elongated molar (probably an M_2) with the roots not yet developed (Fig. 1). There is discoloration caused by grass roots, and the molar can best be studied in the cast. The *Dryopithecus* pattern is well developed; fovea anterior and posterior are closed by crests. Coarse wrinkles cover the cusps. The broad field of the trigonid section with detailed morphology seems to really distinguish the molar from all *Sivapithecus* specimens (Fig. 1). This specimen has recently been described in a separate paper (see, von Koenigswald, 1981) as *Chinjipithecus atavus,* a possible ancestor of the Nagri taxon, *Gigantopithecus giganteus.** The measurements for the M_2 are as follows: length, 13.1 mm; breadth, 10.5 mm; index, 80.15 mm.

Nagri. During two short visits to the type locality of Nagri in January 1964 and 1965, a number of primate teeth were collected, apparently belonging to a large species of *Sivapithecus.* The teeth are from three separate individuals. Two are represented by a single tooth each: Individual II, an upper first molar (see Fig. 2), and Individual III, a lower third molar. Individual I, after analysis and assemblage, has 11 teeth belonging to both the upper and lower jaws.

The maxillary dentition of Individual I is composed of M^3 to P^3 from the right (Fig. 3), and P^4 and the outer walls of M^1 to M^3 from the left side. The teeth are badly worn and no specific characters can be recognized. The occlusal surface is completely smooth, without any traces of wrinkles (Fig. 3). A well-preserved palate of the same genus, with canines intact and the teeth well worn but in a better state of preservation, has recently been recovered from the Siwaliks of the Potwar Plateau (Pilbeam *et al.,* 1977, 1980). Pilbeam provides measurements of this specimen (designated GSP 9977), which are presented in Table 1 along with those of Individual I and Pilgrim's type specimen

---

*A careful study of the isolated molar from Chinji, illustrated above, has revealed a close resemblance in the trigonid to the lower molar described by Pilgrim from the Nagri beds as "*Dryopithecus giganteus*" [=*Gigantopithecus bilaspurensis* Simons and Chopra (1969) = *Gigantopithecus giganteus,* new combination after Szalay and Delson (1979)]. I have arrived at the conclusion that this find most probably indicates a "normal-sized" ancestor of the *Gigantopithecus* lineage. Such a form could be expected from the Lower Siwaliks. I have named this new form *Chinjipithecus atavus* von Koenigswald, 1981. An analogous comparable increase in size from the Chinji level on to higher levels in the Siwaliks can also be observed in other genera, most notably among the Suidae (*Tetraconodon–Sivachoerus*).

Fig. 2. *Sivapithecus indicus,* right upper first molar, designated Individual II. Recovered from the type Nagri beds, Nagri, Pakistan (×c.2.1).

of "*Sivapithecus orientalis.*" The *Sivapithecus* maxillary dentition described in this chapter is notably the largest of these three specimens (see Table 1).

The mandibular dentition of Individual I consists of M_3 and M_2 from the left side and M_3, M_2, and the canine from the right. Professor Dehm of Munich recovered teeth from the same locale 8 years earlier. The canine of Individual I fitted his recovery, and it was determined that his recovered dentition was from the same jaw. (By exchange, he now has the right dentition and the left is in my possession.) Unfortunately, the molars of the mandibular dentition are badly worn and little detail is apparent. The third molar is very elongated, tapering toward the posterior end (Fig. 4). Measurements of the mandibular dentition of Individual I are compared with other *Sivapithecus* specimens in Table 1.

Fig. 3. *Sivapithecus* cf. *indicus,* M^3 to P^3 of right upper jaw, designated Individual I. From type Nagri beds (× c.1.5).

Table 1. Measurements (in mm) of *Sivapithecus* cf. *indicus* Teeth from Nagri-Level Beds in Pakistan and India[a]

	Individual I (type Nagri beds)	Individual II (type Nagri beds)	Individual III (type Nagri beds)	GSP 9977/9564 (Khaur Locality 260)	GSI D-196/197 (Haritalyangar)	(Bandal)
Upper dentition						
M^3 length	13.5			12.4	12.4	
breadth	17.5			13.4	13.2	
M^2 length	14.4			12.9	12.8	
breadth	17.6			14.0	14.2	
M^1 length	12.5	12.7		11.3	11.0	
breadth	16.7	15.1		13.3	12.0	
P^4 length	8.4			8.2	7.3	
breadth	15.0			12.2	12.0	
P^3 length	10.0			8.9	7.7	
breadth	14.3			10.4	—	
Lower dentition						
M_3 length	19.1		17.6	(15.5)	—	
breadth	15.6		15.1	—	—	
M_2 length	16.5			14.5	15.1	
breadth	13.7			12.7	13.7	
M_1 length	(15.8)			(13.0)	11.8	13.9
breadth	(13.2)			(12.0)	10.9	—
P_4 length	11.6			(9.7)	8.5	
breadth	12.3			(10.6)	11.2	

[a]Data for Individuals I, II, and III are derived from this study; data for GSP 9977/9564 are taken from Pilbeam *et al.* (1977, 1980); data for GSI D-196/197 are based on Pilgrim's (1927) type description of "*Sivapithecus orientalis*," and data for the Bandal specimen are based on a cast of the type specimen of "*Sivapithecus lewisi*" (Pandey and Sastri, 1968). Measurements in parentheses are estimates based on very worn or fragmentary teeth.

Fig. 4. *Sivapithecus* cf. *indicus,* M_3 to M_2 of right lower jaw, designated Individual I (same as Fig. 3). From type Nagri beds. Colored to enhance morphology ($\times$ c.1.3).

The teeth of Individuals I, II, and III certainly belong to *Sivapithecus* and are among the largest on record. They will here be included in *Sivapithecus indicus,* but a new species may be warranted. Professor Dehm has another, much more complete and better preserved mandible from the Nagri; therefore, the taxonomic question may be better assessed and answered by him. The jaw in his possession has a much shorter M_3 than our Individual I jaw, but the isolated third molar I recovered from the Nagri(Individual III), which is unfortunately badly weathered, is quite similar to the M_3 of the jaw in Prof. Dehm's possession.

The isolated upper molar, Individual II (Fig. 2), is a little smaller and not as worn as the Individual I maxillary molars. The pattern is very simple. There is a well-marked fovea anterior and no trace of any wrinkles.

It has recently been suggested that *Sivapithecus* may be closely related to the orangutan (Andrews and Tekkaya, 1980). Orangutan dentition is adapted to soft food; therefore, the dental surface is covered with fine wrinkles. The smooth surface of *Sivapithecus* dentition may indicate a diet of coarser food, which reflects a different habitat.

Dhok Pathan. The expedition discovered in the Dhok Pathans near Khaur, on the right bank of the Soan River, the youngest representatives of hominoids yet recovered from the Siwaliks. The three specimens, two isolated teeth and a fragment of a mandible, indicate three separate individuals, all belonging to a small primate species. From the same locale, the crown of an unworn hypsodont molar from *Hipparion* (approximately 73 mm in height) was recovered; this indicates a high geological level, therefore a younger date for the hominoid remains.

The mandibular fragment (Fig. 6) represents a portion of the jaw from the left side of the posterior border of the canine to the posterior wall of the first molar. The lower border of the mandible is missing. The dentition is badly weathered. The anterior premolar is missing, and of the posterior premolar, only fragments of the posterior half have been preserved. The first molar is rectangular in shape. In spite of poor preservation, and discoloration

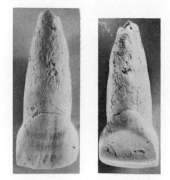

Fig. 5. cf. *Ramapithecus*, central upper incisor. Recovered from Dhok Pathan beds near Khaur, Pakistan. Anterior view is on left side and posterior view is on right. Colored to enhance morphology (× c.2.1).

attributable to grass roots, the *Dryopithecus* pattern can be seen. The hypoconulid is in the middle of the posterior border. The body of the mandible is narrow: the height below the first molar, reconstructed, is about 23 mm. The measurements for M_1 and P_4 are as follows: M_1, length, 9.7 mm and breadth, 8.9 mm; P_4 length, 6.9 mm and breadth, 8.0 mm.

The upper molar (Fig. 7) of the two isolated teeth recovered, an M^1, is well worn. The posterior–lingual corner is missing. The dentine is exposed on the protocone and hypocone and the trigonid crest is interrupted. The tooth is very small (length, 9.2 mm; breadth, 10.4 mm) and has the same dimensions as the first molar in Lewis' (1934) type specimen, "*Ramapithecus brevirostris*" (9.2 mm and 10.5 mm).

The upper central incisor (Fig. 5) is well preserved, being just worn at the edge. Some wrinkles are indicated, but there is no evidence of a cingulum. Measurements for this incisor are: breadth, 8.0 mm; length, 6.1 mm; crown height, 7.3 mm; and length of root, 12.7 mm.

The material from Khaur is too fragmentary to permit a definite conclusion regarding its taxonomic position. Upon close examination, the isolated

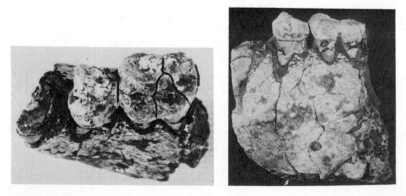

Fig. 6. cf. *Ramapithecus*, fragment of lower jaw, left side with P_4 and M_1. From Dhok Pathan beds, same locality as in Fig. 5. Occlusal view is on left side (× c.2.0) and buccal view is on right (× c.1.5).

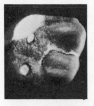

Fig. 7. cf. *Ramapithecus,* upper left first molar. From Dhok Pathan beds, same locality as in Figs. 5 and 6 (× c.1.8).

upper molar gives the impression of being on the "hominid" side. Regarding the mandibular fragment, the depth of its body is very similar to GSI D-298, the *Ramapithecus* cf. *brevirostris* jaw of Gregory *et al.* (1938), both of which have a low position of the foramen mentale. The GSI D-298 mandible [also designated YPM 13870; see Simons (1964, p. 141)] has caused many difficulties in taxonomic classification, and there is still no conclusive designation (however, see Kay and Simons, this volume, Chapter 23). Simons and Pilbeam (1965, p. 123) have listed it under *Dryopithecus laietanus* (Villalta and Crusafont, 1944). But this is a wholly European taxon and for geographical and faunal reasons is not represented in the Siwaliks. Also, the suggested presence of the Indian *Ramapithecus* in Southern Germany (Villalta and Crusafont, 1944, p. 136) is based on an incorrect interpretation of a single molar. This tooth, which comes from Melchingen (Branco, 1898, plate I, Fig. 2), is heavily worn, but an unusual and rare ridge connecting the metacone and the hypocone is still recognizable. The same ridge can be observed in an unworn molar from Melchingen which certainly belongs to *Dryopithecus*. In addition, Szalay and Delson (1979) find it difficult to classify the GSI D-298 jaw, designating it as cf. *Sivapithecus sivalensis* ("*Ramapithecus* cf. *brevirostris*"). I am not at all sure that it is justifiable to refer the Dhok Pathan jaw fragment described in this chapter to either *D. laietanus* or to cf. *S. sivalensis* and will therefore at present classify this jaw as cf. *Ramapithecus*.

Conclusion

The whole question of the taxonomy of *Ramapithecus* is greatly confused. Szalay and Delson (1979) regard *Ramapithecus* as a hominid. Greenfield (1979, p. 544) arrives at the conclusion that "the morphological and adaptive patterns suggested by the known parts of '*Ramapithecus*' and *Sivapithecus* are virtually identical." Moreover, there seems to be a tendency to classify *Ramapithecus* together with *Sivapithecus, Gigantopithecus,* and other forms as Ramapithecidae, leaving the question of direct ancestry of hominids more or less open (Pilbeam and Jacobs, 1978). At present, I have to conclude that the new *Ramapithecus* material described in this chapter does not lend itself to any conclusive interpretations, other than descriptive ones, concerning the phyletic position and potential hominid status of *Ramapithecus*. An answer to this "*Ramapithecus* question" will have to await the recovery of more complete remains from well-documented stratigraphic units.

References

Andrews, P., and Tekkaya, I. 1980. A revision of the Turkish Miocene hominoid *Sivapithecus meteai*. *Palaeontology* **23**:85–95.

Branco, W. 1898. Die menschenähnlichen Zähne aus dem Bohnerz der schwäbischen Alb. *Jahrsh. Ver. Vaterl. Naturk. Württ.* **54**:1–144.

Colbert, E. H. 1935. Siwalik mammals in the American Museum of Natural History. *Trans. Am. Phil. Soc. N.S.* **26**:1–401.

Falconer, H. 1868. *Palaeontological Memoirs and Notes of the Late Hugh Falconer* (C. Murchison, ed.), Robert Hardwicke, London. Vol. 1, 590 pp., Vol. 2, 675 pp.

Greenfield, L. O. 1979. On the adaptive pattern of "*Ramapithecus.*" *Am. J. Phys. Anthropol.* **50**:527–548.

Gregory, W. K., Hellman, M., and Lewis, G. E. 1938. Fossil anthropoids of the Yale–Cambridge India Expedition of 1935. *Carnegie Inst. Wash. Publ.* **495**:1–27.

Lewis, G. E. 1934. Preliminary notice of new man-like apes from India. *Am. J. Sci.* **27**:161–181.

Lydekker, R. 1879. Further notices of Siwalik Mammalia. *Rec. Geol. Surv. India* **12**:33–53.

Pandey, J., and Sastri, V. V. 1968. On the new species of *Sivapithecus* from the Siwalik rocks of India. *J. Geol. Soc. India* **9**:206–211.

Pilbeam, D., and Boyer, S. J. 1971. Appearance of *Hipparion* in the Tertiary of the Siwalik Hills of North India, Kashmir and West Pakistan. *Nature (Lond.)* **229**:408–409.

Pilbeam, D., and Jacobs, L. L. 1978. Changing views of human origins. *Plateau* (Mus. N. Arizona) **51**:18–30.

Pilbeam, D., Meyer, G. E., Badgley, C., Rose, M. D., Pickford, M. H. L., Behrensmeyer, A. K., and Shah, S. M. I. 1977. New hominoid primates from the Siwaliks of Pakistan and their bearing on hominoid evolution. *Nature (Lond.)* **270**:689–695.

Pilbeam, D. R., Rose, M. D., Badgley, C., and Lipschutz, B. 1980. Miocene hominoids from Pakistan. *Postilla* (Peabody Mus. Nat. Hist., Yale Univ.) **181**:1–94.

Pilgrim, G. E. 1927. A *Sivapithecus* palate and other primate fossils from India. *Mem. Geol. Surv. India (Palaeontol. Ind.)* **14**:1–24.

Simons, E. L. 1964. On the mandible of *Ramapithecus*. *Proc. Natl. Acad. Sci. USA* **51**:528–535.

Simons, E. L. and Chopra, S. R. K. 1969. *Gigantopithecus* (Pongidae, Hominoidea) a new species from North India. *Postilla* (Peabody Mus. Nat. Hist., Yale Univ.) **138**:1–18.

Simons, E. L., and Pilbeam, D. R. 1965. Preliminary revision of the Dryopithecinae (Pongidae, Anthropoidea). *Folia Primatol.* **3**:81–152.

Szalay, F. S., and Delson, E. 1979. *Evolutionary History of the Primates*, Academic, New York. 580pp.

Tauxe, L. 1979. A new date for *Ramapithecus*. *Nature (Lond.)* **282**:399–401.

Villalta, J. F., and Crusafont, M. 1944. Dos nuevos antropomorfos del Mioceno español y su situación dentro de la moderna sistemática de los simidos. *Notas Commun. Inst. Geol. Min. Espana* **13**:91–139.

Von Koenigswald, G. H. R. 1981. A possible ancestral form of *Gigantopithecus* (Mammalia, Hominoidea) from the Chinji Layers of Pakistan. *J. Hum. Evol.* **10**:511–515.

Miocene Hominoid Primate Dental Remains from the Siwaliks of Pakistan

20

R. DEHM

Introduction

Two regions of the Old World show a thick, uninterrupted series of Miocene freshwater sediments with rich, successive mammalian faunas: the Upper Freshwater Molasse (Obere Süsswasser-Molasse) of subalpine southern Bavaria together with the neighboring subalpine parts of Austria, and the Siwalik series of Pakistan. In 1939, in order to compare the faunas and sections of both series, Dr. J. Schröder (now conservator at the BSPhG in Munich) and I undertook a 6-week visit to the classic Siwalik area of Chinji, Sethi Nagri Dhok, and Dhok Pathan (Schröder and Dehm, 1940). Unfortunately, war action in Munich destroyed most of the field notes and specimens in 1944. In 1955–56 Dr. Th. Zu Oettingen-Spielberg, Dr. H. Vidal, and I made a visit of 7 weeks to this area (Dehm *et al.*, 1958). More details on localities, sections, accompanying faunas, and on the preliminary taxonomic attributions will be given elsewhere.

At this point it is necessary to note that these collections do not contain remains of *Ramapithecus,* nor any postcranial remains of primates. The described specimens (all dental remains) are deposited in the Bayerische Staats-

R. DEHM • Institut für Paläontologie und historische Geologie, Universität München, 8000 München 2, West Germany.

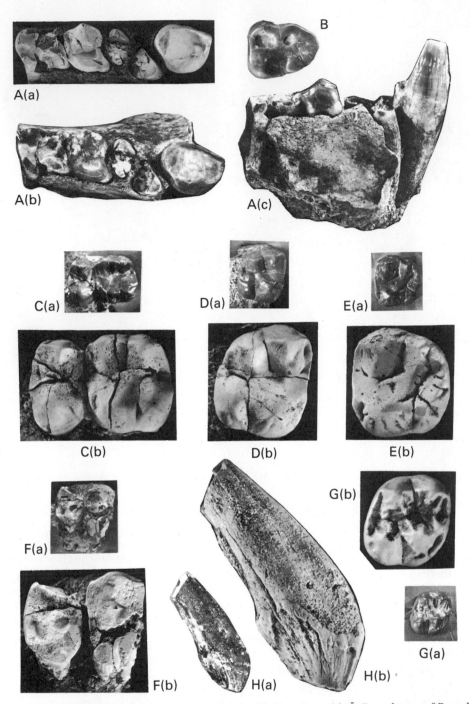

Fig. 1. (A) *Sivapithecus* cf. *indicus* Pilgrim; right mandibular ramus with $\bar{C}$, P_4 and parts of P_3 and M_1 (BSPhG 1956 II 39); Nagri Zone, Sethi Nagri Dhok; natural size; (a) occlusal view, colored to reveal morphology; (b) occlusal view; (c) lateral view. (B) *Sivapithecus* cf. *indicus* Pilgrim; right M_3 (BSPhG 1956 II 40); Nagri Zone, Sethi Nagri Dhok; natural size. (C) *Sivapithecus* cf. *orientalis* Pilgrim; left upper jaw fragment with P^4–M^1 (BSPhG 1939 X 2); Chinji Zone, upper part,

sammlung für Paläontologie und historische Geologie (BSPhG) in Munich. The locality names have been given to us by local guides, mostly from Chinji Village.

The measurements of the teeth are given as length (L) × breadth (B) = product, rounded off in square millimeters. This product adequately represents the efficiency of the dentition, especially its change through time. All measures are in millimeters.

Description of the Specimens

Chinji Zone

The middle and upper parts of the Chinji Zone mostly yielded only isolated primate teeth from thin, red and gray reworked layers of lime nodules in a sand and clay "conglomerate," together with jaw fragments and teeth of Selenodontia, teeth of Suidae, teeth and plates of crocodilians, fish vertebra, and isolated bones and bone fragments.

Locality Kundal Nala: 2.3 km SW of Chinji, Chinji Zone, middle part. Yielded a fresh enamel crown of a right M^3 partially filled with dentine (BSPhG 1939 X 1; see Fig. 1G); $L \times B = 10.8 \times 12.0 = 130$ mm^2. The four main cusps are well separated by deep furrows; there is a rich relief of secondary ridges, especially the fovea anterior, which is accompanied by a well-developed crista transversa anterior and by a mesial cingulum. These both join the mesial crest of the protocone at a well-defined point, from which a distinct cingulum runs to the lingual base of the protocone. The distal border is formed by the long distal crest of the hypocone. The fovea posterior has connection with a central furrow since the crista transversa posterior and the crista obliqua of the protocone do not reach the metacone. The rich details of the enamel and the size of this tooth indicate affinities with *Sivapithecus sivalensis.*

Locality Bhuriwala: 2.5 km SE of Chinji, Chinji Zone, middle part. Yielded a right C (BSPhG 1956 II 38; see Fig. 1H); base of the crown is 14.5 × 9.5 mm; length is estimated at 45. Top and root of tooth are broken; the mesiobuccal side is flattened by an occlusal facet. The surface shows vertical wrinkles; a buccal cingulum is very faint. The size corresponds with *Sivapithecus sivalensis.*

Parrewali, 7 km W of Chinji; (a) natural size; (b) ×2, colored. (D) Same individual as (C), right M^2; (a) natural size; (b) ×2, colored. (E) *Sivapithecus* cf. *indicus* Pilgrim; left M^3 (BSPhG 1939 X 3); Chinji Zone, upper part, NW Sosianwali, 3 km WSW of Chinji; (a) natural size; (b) ×2, colored. (F) *Sivapithecus* cf. *sivalensis* (Lydekker); fragment of right upper jaw with damaged P^3–P^4 (BSPhG 1956 II 37); Chinji Zone, middle part, Sosianwali, 2 km SW of Chinji; (a) natural size; (b) ×2, colored. (G) *Sivapithecus* cf. *sivalensis* (Lydekker); right M^3 (BSPhG 1939 X 1); Chinji Zone, middle part, Kundal Nala, 2.3 km SW of Chinji; (a) natural size; (b) ×2, colored. (H) *Sivapithecus* sp., size of *S. sivalensis* (Lydekker); right C (BSPhG 1956 II 38); Chinji Zone, middle part, Bhuriwala, 2.5 km SE of Chinji; (a) natural size; (b) ×2, colored.

Locality Sosianwali–Kundal Nala: 2 km SW of Chinji, Chinji Zone, middle part. Yielded a fragment of a right upper jaw with a badly damaged and worn P$^{3-4}$ (BSPhG 1956 II 37; see Fig. 1F); P^3: $L \times B = 6.5 \times$ —; P^4: $L \times B = 8.0 \times$ —. The damage to the tooth appears to have occurred during deposition, as chips of enamel and dentine are fixed in displaced positions. Both teeth have contact facets. The main cusp of the P^3 lies in the middle of the buccal border; the tooth has no mesiobuccal projection. The P^4 shows two points on its lingual half where dentine is exposed (each about 1 mm in diameter); they may correspond to the protocone and some kind of an unusual hypocone. The specimen is, of course, not suitable for a taxonomic determination; the slenderness of the P^3 seems unique; the size range fits into the group of *Sivapithecus sivalensis*.

Locality NW Sosianwali: 3 km WSW of Chinji, Chinji Zone, upper part. Yielded a slightly worn crown of a left M^3 (BSPhG 1939 X 3; see Fig. 1E); $L \times B = 12.3 \times 13.8 = 170$ mm^2. The mesial wall shows a contact facet (3 $\times$ 2 mm). The rounded quadrangular outline of the tooth is tapered distally. On the unworn enamel surface some wrinkling can be seen. The mesiobuccal ridge and the crista obliqua of the protocone are flattened and connect the protocone with the buccal cusps. The fovea anterior must have been very faint, whereas the fovea posterior is deep. On the mesiolingual side of the protocone there is a tiny furrow, too faint to call a Carabelli's furrow. The hypocone is not a well-separated cusp, as the whole distal border forms a dam which is interrupted by several notches. The size range may fit that of *Sivapithecus indicus*.

Locality Parrewali: 7 km W of Chinji, Chinji Zone, upper part. Yielded a left upper jaw fragment with P^4–M^1 and a right upper jaw fragment with M^2 (BSPhG 1939 X 2; see Figs. 1C, 1D); $L \times B$: P^4, 7.0 $\times$ 12.5 = 88 mm^2; M^1, 10.7 $\times$ 12.6 = 135 mm^2; M^2, 12.4 $\times$ 15.0 = 186 mm^2. On a steep slope both fragments had been washed out of a reworked layer; the red-brown color of the bone, the dark gray color of the enamel, and the moderate degree of wear—there is no dentine exposed—are so similar that both specimens most likely belong to one individual. The enamel surface is smooth, not wrinkled. The P^4 has a prominent, subequal protocone and paracone, which are joined by two not very strong crests; the fovea anterior and fovea posterior are shallow. The M^1 has a rounded quadrangular outline; the protocone is voluminous, the hypocone is much smaller; the crista obliqua connects the protocone with the metacone in an uninterrupted fashion. The mesiobuccal slope of the protocone ends in a very faint swelling which may be considered a protoconule; paracone and metacone are subequal, there is a tiny depression on the buccal slope of the hypocone. The outline of M^2 tapers distally; the protocone is very voluminous; on its mesiobuccal side a rather flat protoconule is raised in the middle of the mesial border of the tooth. The hypocone is separated from the protocone by a deep furrow. Paracone and metacone, both slightly smaller, are separated from each other and from the lingual cusps by clear grooves; only on the mesial side of the paracone is there a short fovea.

The two molars have nearly the same dimensions as those of *Sivapithecus orientalis* Pilgrim (Pilgrim 1927, pp 5, 16) from the Nagri Zone of Haritaly-angar, but their hypocones are considerably larger.

Nagri Zone

In the fossiliferous reworked layers of the Chinji Zone two or more remains of one individual could only be found on exceptional occasions; however, associated remains of a single individual were more common in the sandy marls of the Nagri type locality, the deep valley near Sethi Nagri Dhok. In 1939 a native peasant found the two mandibular rami (BSPhG 1939 X 4) illustrated in Fig. 2. These had lain in the marl at a "distance of about a handspan" from one another; both specimens still showing on their mesial surface a red to gray-green clay matrix. In 1955–56 we recovered a right lower jaw fragment of a primate (BSPhG 1956 II 39) and found by digging on the spot two isolated left teeth. In 1964–65 Prof. von Koenigswald visited the same spot and recovered a left primate jaw fragment and two isolated right teeth; he later discovered that our specimens of 1955–56 and his specimens of 1964–65 fitted together. Therefore it seems that in the Nagri sediments of the type locality parts of one individual are sometimes not too widely scattered.

Locality Nagri Type Site: Valley near Sethi Nagri Dhok, Nagri Zone. Yielded a powerfully constructed mandible consisting of incomplete right and left rami (BSPhG 1939 X 4, see Fig. 2). The right ramus of the mandible (Figs. 2A, 3) contains the $\bar{\text{C}}$, P_3, distal half of M_1, M_2, and the sockets of P_4 and M_3. The symphysis, incisors, and most of the ascending ramus are lacking. Two features characterize this robust mandible: its transverse thickness below M_3 and M_2, each about 35 mm (see Fig. 3), and its increasing depth from 35 mm below M_3 and M_2 to 55 mm below $P_3/\bar{\text{C}}$. On the external surface a distinct foramen mentale with a 2.5 mm diameter opens 25 mm below the sockets of P_4. The ascending ramus begins below M_1 with a remarkable swelling; between this swelling and the molars there extends a rather wide, flat depression. On the internal surface, behind the symphyseal breach, there is a groove which marks the location of the "simian shelf."

The $\bar{\text{C}}$ (base of the crown is 17.7 × 13.1 mm, height of the crown is 24 mm, total height is 60 mm) has lost parts of the root by breakage and part of the apex by wear; the occlusal facet blunts the uppermost part of the distolingual crest. The surface of the enamel exhibits vertical wrinkling. The mesiobuccal side is well rounded; three crests run down from the top, a mesial, a distal, and a distolingual one. Between the mesial and the distal crest the crown is well rounded, but between the distal and the distolingual crest it is nearly flat. A cingulum begins at the base of the mesial crest, surrounding the distal part, and ends in a faint bud where the distal and distolingual crests unite.

The P_3 (L maximum × B transverse = 17.4 × 9.9 mm) shows wide mesiolingual wear on the protoconid (dentine is exposed: 4 × 3 mm), on the

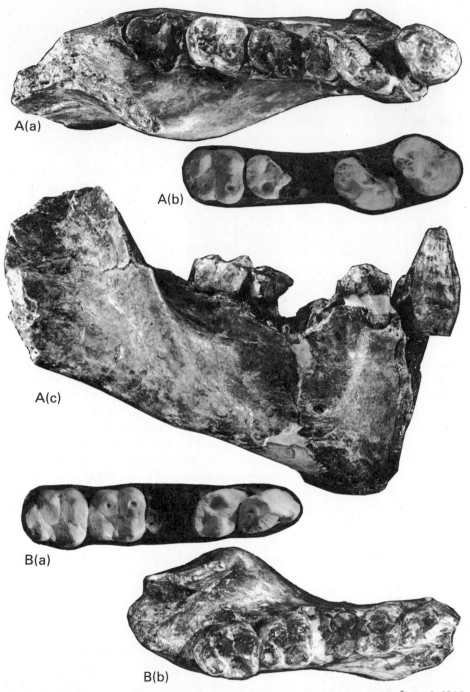

Fig. 2. *Sivapithecus* n. sp.? cf. *aiyengari* Prasad. (A) Right mandibular ramus with $\bar{C}$, P_3, half M_1 and M_2 (BSPhG 1939 X 4); Nagri Zone, Sethi Nagri Dhok; all natural size; (a) occlusal view; (b) dentition only, colored to reveal morphology; (c) lateral view. (B) Fragment of a left mandibular ramus with P_3, P_4, M_2, M_3 (BSPhG 1939 × 4); Nagri Zone, Sethi Nagri Dhok; natural size; (a) occlusal view, dentition only, colored; (b) occlusal view of entire specimen.

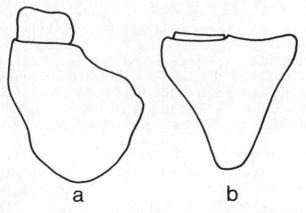

Fig. 3. *Sivapithecus* n. sp.? cf. *aiyengari* Prasad; right ramus of the mandible BSPhG 1939 X 4 from the Nagri zone, Sethi Nagri Dhok. Vertical sections made through the mandible: (a) across midline of M_2 and (b) across midline of M_3. Figure is natural size.

metaconid (1×1.4 mm), and on the distal crest of the protoconid, which is also a distal contact facet. The outline of the tooth is quadrangular, with a greater length lingually than buccally. The protoconid must have been very prominent, whereas the metaconid is rather faint, yet still distinct. A mesial crest runs down from the protoconid and surrounds, along with a short cingulum, a small mesiolingual depression. Distally from the metaconid a shallow groove opens distolingually. The enamel is, where unaffected, wrinkled.

The M_1 ($L \times B = {\sim}12.5 \times 13 = {\sim}163$ mm^2) is only represented by its distal half; on the hypoconid, dentine is exposed (3 mm in diameter) by occlusal wear, and the entoconid and hypoconulid are flattened. A distal contact facet shortens the tooth a little.

On the M_2 ($L \times B = 15.2 \times 14.6 = 222$ mm^2) occlusal wear has exposed dentine on the protoconid (2 mm in diameter) and hypoconid (1 mm in diameter) and has flattened the hypoconulid. The mesoconid, still rather high, and the entoconid are sharpened by mesio- and distolingual crests; there are contact facets. The outline is well rounded and quadrangular. The most voluminous cusp is the metaconid, then the protoconid, entoconid, hypoconid, and hypoconulid follow; the furrows between these cusps are mostly smooth.

The M_3 is broken away; the root sockets show that the tooth must have been shortened distally; its outline may have been more triangular than quadrangular, and its length not greater than the breadth.

The left ramus of the mandible (see Fig. 2B) comprises the corpus with P_{3-4} and M_{2-3}; it has lost the symphyseal region mesial to the P_3, the ascending ramus, most of the lower border, and the M_1. The ramus is strikingly thick and similar to the right-side ramus.

The left P_3 (L maximum $\times B$ transverse $= 17.0 \times 9.8$ mm) equals the right P_3, but occlusal wear has united the protoconid and metaconid in a dentine field of 3.5×2.2 mm in size.

The P_4 ($L \times B = 12.2 \times 13.0 = 159$ mm^2) has been worn down, especially on the lingual half, where on the site of the protoconid the dentine is exposed (4 × 5 mm); mesial and distal walls of the P_4 are smoothed by contact facets. The outline of the tooth is trapezoidal and the mesiobuccal corner forms a rounded projection. The metaconid is cone-like and sharpened by a distolingual ridge; it is separated from the distal border of the tooth by a furrow.

The left M_2 ($L \times B = 14.9 \times 14.5 = 216$ mm^2) equals the right one in most details; it differs only in size and proportions, not much, but recognizably so.

The M_3 ($L \times B = 16.0 \times 15.4 = 246$ mm^2) shows occlusal wear on the buccal side, but no exposed dentine; the mesial wall touches the distal wall of the M_2. The outline of the tooth narrows distally; therefore the tooth is rather short. The metaconid is the highest cusp; then follow the entoconid, protoconid, hypoconulid, and hypoconid. Concerning the areas that they occupy, the protoconid is first in size, followed by the metaconid, hypoconid, entoconid, and hypoconulid. The surface still shows some traces of wrinkling. The fovea anterior is weak and the fovea posterior is missing. Similar to the M_2, the metaconid is sharpened by a distinct mesial ridge and by a less clear distobuccal one. The hypoconulid forms the distal end of the ridge, and is situated in the midline, equally far from the buccal and lingual border. The furrow between the metaconid and entoconid is deep and marked by a tiny point at the distal foot of the metaconid; the other furrows are less clear.

The taxonomic assignment of this mandible must chiefly rest on the bulky form of its ramus—compared with the rami of other Siwalik primates (Simons, 1964, p. 532)—and on its robust dentition. These characteristics point to the group of larger Siwalik hominoids which Simons and Pilbeam (1965, pp. 125–126) unite under "*Dryopithecus indicus* (Pilgrim)," and among this group to two Nagri species from the Haritalyangar area, namely *lewisi* Pandey and Sastri and *aiyengari* Prasad (see Fig. 4). But I hesitate to identify this mandible with one of these two species, as its premolars are decidedly bigger and its robustness stouter.

The Nagri type locality also yielded a fragment of a right mandibular ramus with worn P_4–M_1 and sockets of P_3 (BSPhG 1956 II 39; see Fig. 1A). The teeth only reveal a few details. The $\bar{C}$ (crown base 14.0 × 11.0 mm, total height estimated 53 mm) shows in distal aspect a broad plane of occlusal wear (8 × 15 mm); the enamel is vertically wrinkled and the mesiolingual base bears a faint cingulum which joins with a mesial crest. On the P_4 ($L \times B = 11.0 \times 12.3 = 134$ mm^2) dentine is exposed at the protoconid (6 × 3 mm) and mesial and distal contact facets have shortened the tooth. Its outline forms a trapezoid with a well-rounded mesiobuccal projection; the metaconid is high, its lingual side sharpened by wear, and distal to the metaconid lies a small groove. On the M_1 no details are in evidence, due to nearly complete occlusal wear.

A right M_3, found by Prof. von Koenigswald (BSPhG 1956 II 40; see Fig. 1B), has also been recovered. Von Koenigswald and I are in agreement that this worn M_3, which was collected on the same spot as the above-mentioned mandibular fragments, could belong to the same individual. But in spite of

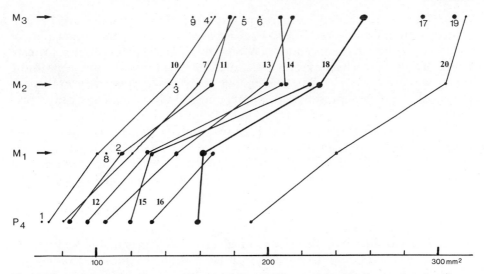

Fig. 4. Size comparisons of some *Sivapithecus* species discussed in this chapter based on lower teeth of a stratified series of Siwalik primates [not including *Hylopithecus, Palaeopithecus*(?), *Ramapithecus,* or *Sugrivapithecus*]. Length × breadth product in mm^2. (1–10) Lower Chinji–Lower Nagri primates: Lower Chinji: 1, *pilgrimi* type (AMNH 19411). Chinji: 2, *chinjiensis* (GSI D-181); 3, *chinjiensis* type (GSI D-179); 4, *chinjiensis* (GSI D-180); 5, *sivalensis* (YPM 13832); 6, *indicus* (GSI D-176). Upper Chinji: 7, *indicus* (GSI D-177). Uppermost Chinji: 8, *cautleyi* (YPM 13813); 9, cf. *darwini* (Mus. Jammu). Lower Nagri: 10, *cautleyi* type (AMNH 19412). (11–19) Nagri primates: 11, *frickae* type (AMNH 19413); 12, *himalayensis* type (GSI D-197); 13, *indicus* (YPM 13828); 14, *indicus* (GSI 18040); 15, *aiyengari* type (GSI 18039); 16, *lewisi* type (ONGC/V/790); 17, *giganteus* type (GSI D-175); 18,? cf. *aiyengari* (BSPhG 1939 X 4), M_1 dimensions estimated; 19, cf. *indicus* (BSPhG 1956 II 40). (20) Dhok Pathan primates: 20, *bilaspurensis* type (CYP 359/68).

this, something is missing: the color of the enamel is gray instead of light brown; the dentine is lilac-colored instead of yellowish, and, perhaps the most important, the size of the M_3 seems considerably too large to go with the jaw fragments. Therefore I wish to keep this M_3 separate for the time being. At any rate, all these specimens belong to a large primate morph of the group *Sivapithecus indicus*. This M_3 ($L \times B = 19.3 \times 15.5 = 299$ mm^2) is flattened by occlusal wear with dentine exposed on the protoconid and hypoconid, and the mesial border has a deep contact facet. The outline of the tooth gradually shortens to a well-rounded distal end. In spite of wear, the metaconid is still prominent. The size of this M_3 surpasses nearly all other Siwalik M_3's and even comes near to the M_3 size of the Dhok Pathan species, *G. bilaspurensis* (see Fig. 4).

Significance of the Specimens

In the Kamlial Zone and in the lower part of the Chinji Zone we did not find primate remains. The middle part of the Chinji Zone yielded primate teeth at the localities of Kundal Nala, Bhuriwala, and Sosianwali–Kundal

Nala. The three isolated teeth recovered from these localities are best assigned to the middle-sized *Sivapithecus sivalensis* taxon. In the upper part of the Chinji Zone, at the localities NW Sosianwali and Parrewali, two primate forms are present which exceed the size of the former and come near to the larger species, *S. indicus*. The succeeding Nagri Zone yielded several specimens, all of which represent considerably bigger forms than any of the preceding ones.

This increase of size (see Fig. 4) from Chinji to Nagri continues into the Dhok Pathan, where *G. bilaspurensis* reaches very large size and marks a general trend in the evolution of this primate group. As to the change of Siwalik primates through the course of time, Gregory and Hellman (1926, p. 24) have stated that the specimens which they described as three new species, *pilgrimi, cautleyi,* and *frickae,* come from successive horizons and "exhibit as many successive stages of evolutionary advance."

As long as the small differences in age of local faunas do not seem essential, differences among specimens may be treated as examples of normal variability, meaning that there is no need to keep them taxonomically separated. In the Bavarian field, in the Upper Freshwater Molasse, mastodonts at first analysis seemed to belong to one species with a large amount of variability. However, as soon as the stratigraphic sequence of their localities had been considered, it became clear that an evolutionary increase in size had been hidden by their supposed variability (Dehm, 1951, pp. 142–145); *Dinotherium* shows a similar trend (Dehm, 1949, pp. 22–25). Therefore size and other differences among specimens from discernible horizons must not be definitely regarded as only examples of morphologic variability, even if they fall within the range of recent variability. Without a reasonable number of stratigraphically-controlled specimens which reveal the range of variability, it will be difficult to come to a final judgement. Elsewhere in the Siwalik series there are many signs of evolutionary processes within several different groups (Prasad, 1969, pp. 297–298), such as mastodonts, suids, selenodonts, and also in the rare *Dinotherium* (Dehm, 1963). A deeper knowledge of evolutionary processes concerning Siwalik mammals will certainly result in stratigraphic refinement of the various Miocene hominoid discoveries and possibly in a better understanding of their cladogenesis.

Acknowledgments

The field work of 1955–56 was accompanied by Dr. A. F. M. Mohsenul Haque, at that time Senior Geologist and Paleontologist of the Geological Survey of Pakistan in Quetta, and whose support is gratefully acknowledged. I also thank Prof. Dr. G. H. R. von Koenigswald, of Frankfurt/Main, for many valuable discussions concerning interpretations of our finds. The senior editor of this volume, R. L. Ciochon, of Charlotte, North Carolina, has kindly encouraged me to describe and briefly interpret the hominoid specimens collected in the Chinji and Nagri Zones of Pakistan.

References

Dehm, R. 1949. Das jüngere Tertiär in Südbayern als Lagerstätte von Säugetieren, besonders *Dinotherium*. *Neues Jahrb. Min. Abh.* **90:**1–30.

Dehm, R. 1951. Zur Gliederung der jungtertiären Molasse in Süd-deutschland nach Säugetieren. *Neues Jahrb. Geol. Palaeontol. Monatsber.* **1951:**140–152.

Dehm, R. 1963. Paläontologische und geologische Untersuchungen im Tertiär von Pakistan. 3. *Dinotherium* in der Chinji-Stufe der Unteren Siwalik-Schichten. *Abh. Bayer. Akad. Wiss. Math.-Naturw. Kl. N.F.* **114:**1–34.

Dehm, R., Prinzessin zu Oettingen-Spielberg, Th., and Vidal, H. 1958. Paläontologische und geologische Untersuchungen im Tertiär von Pakistan. 1. Die Münchener Forschungsreise nach Pakistan 1955–1956. *Abh. Bayer. Akad. Wiss. Math.-Naturw. Kl. N.F.* **90:**1–13.

Gregory, W. K., and Hellman, M. 1926. The dentition of *Dryopithecus* and the origin of man. *Anthropol. Pap. Am. Mus. Nat. Hist.* **28:**1–123.

Gregory, W. K., Hellman, M., and Lewis, G. E. 1938. Fossil anthropoids of the Yale–Cambridge India expedition of 1935. *Carnegie Inst. Wash. Publ.* **495:**1–27.

Lewis, G. E. 1937. Taxonomic syllabus of Siwalik fossil anthropoids. *Am. J. Sci.* **34:**139–147.

Pandey, J., and Sastri, V. V. 1968. On a new species of *Sivapithecus* from the Siwalik rocks of India. *J. Geol. Soc. India* **9:**206–211.

Pilbeam, D. R., Rose, M. D., Badgley, C., and Lipschutz, B. 1980. Miocene hominoids from Pakistan. *Postilla* (Peabody Mus. Nat. His., Yale Univ.) **181:**1–94.

Pilgrim, G. E. 1927. A *Sivapithecus* palate and other primate fossils from India. *Mem. Geol. Surv. India (Palaeontol. Ind.)* **14:**1–26.

Prasad, K. N. 1962. Fossil primates from the Siwalik beds near Haritalyangar, Himachal Pradesh, India. *J. Geol. Soc. India* **3:**86–96.

Prasad, K. N. 1969. Critical observations on the fossil anthropoids from the Siwalik System of India. *Folia Primatol.* **10:**288–317.

Schröder, J., and Dehm, R. 1940. Bericht über eine paläontologisch-geologische Reise nach Vorderindien und Australien. *Sitz. Ber. Bayer. Akad. Wiss. Math.-Naturw. Abt.* **1940:**167–173.

Simons, E. L. 1964. On the mandible of *Ramapithecus*. *Proc. Natl. Acad. Sci. USA* **51:**528–535.

Simons, E. L., and Chopra, S. R. K. 1969. *Gigantopithecus* (Pongidae, Hominoidea) a new species from North India. *Postilla* (Peabody Mus. Nat. Hist., Yale Univ.) **138:**1–18.

Simons, E. L., and Pilbeam, D. R. 1965. Preliminary revision of the Dryopithecinae (Pongidae, Anthropoidea). *Folia. Primatol.* **3:**81–152.

Significance of Recent Hominoid Discoveries from the Siwalik Hills of India

21

S. R. K. CHOPRA

Introduction

In earlier communications (Chopra, 1974, 1976, 1978) I have given a brief historical review of paleoprimatological studies in India and the scientific issues that engaged earlier researchers. It is clear that past research was largely exploratory in nature, and only recently have careful followup studies both in the field and the laboratory been attempted with a view to identifying the region's potential. Of interest are studies on the paleoecology of the Siwalik* Formations, on stratigraphic ranges and faunal assemblages for reconstructing paleoenvironments on the basis of lithology and paleocommunity structures, and, finally, on human origins (Vasishat *et al.*, 1978*a,b*; Gaur *et al.*, 1978*a,b*). As an integral part of the theme of a new interpretation of human ancestry, it is relevant to note the new lower primate finds [tree shrews (Chopra *et al.*, 1979; Chopra and Vasishat, 1979), adapids and lorisids (Chopra and Vasishat, 1980*a,b*)], which have provocative implications for the evolutionary diversification of primates on the Indian subcontinent. It would also be relevant to discuss Pleistocene correlations with climatic and human

*The author prefers the phonetic spelling *Sivalik*; however, the more common spelling *Siwalik* will be used.

S. R. K. CHOPRA • Department of Anthropology, Panjab University, Chandigarh-160014, India.

cultural levels in various parts of the subcontinent, to bring into sharper focus the possibility of the origin of the hominid family in India and the crucial need for new investigations and acquisition of future support for the extension of this work. However, as the title of this chapter suggests, it is intended here to discuss only the significance of new Miocene hominoids in the Siwalik Hills as providing new evidence for a reinterpretation of ape and human ancestry.

Historical Background

The descriptions and formulations regarding the phyletic relationships of the Siwalik primate fossil finds have many implications. These must be reviewed before the impact and nature of the newer finds can be understood. In all events, interpretations concerning early humans and their relatives should be built up from immediate observations as well as antecedent knowledge and reason.

The fossil hominoids of the Siwaliks were first brought into focus by Lydekker (1879, 1884), who described a partial palate assigned to *Palaeopithecus sivalensis,* which was later renamed *Sivapithecus* by Lewis (1937) and *Dryopithecus* by Simons and Pilbeam (1965). Subsequent to Lydekker's contributions, Pilgrim (1910, 1915, 1927) described and proposed a whole array of new species of fossil hominoids: *Sivapithecus indicus, S. middlemissi, S. orientalis, S. himalayensis; Dryopithecus punjabicus, D. chinjiensis, D. giganteus; Palaeosimia rugosidens; Paleopithecus* (?) *sylvaticus; Hylopithecus hysudricus.* In 1923 Barnum Brown, of the American Museum of Natural History, recovered three dryopithecine fossils, which were later described by Brown *et al.* (1924) as belonging to three different species: *Dryopithecus pilgrimi, D. cautleyi, D. ? frickae.* Gregory and Hellman (1926) published an exhaustive monograph detailing a comparative study of the dryopithecine dentition. Following these earlier contributions the Yale–North India expedition of 1931–1933 and the Yale–Cambridge expedition of 1935 added further to the collections of fossil primates from the Siwaliks. Lewis (1934) and Gregory *et al.* (1938) communicated the scientific results of these expeditions.

Lewis (1934) also proposed a number of new taxa: *Sugrivapithecus salmontanus; Dryopithecus cautleyi, D. sivalensis; Bramapithecus thorpei; Adaetontherium incognitum.* These taxa were based on his recovery of new fossil hominoid material and a reanalysis of existing material in museums. Colbert (1935) published a detailed memoir on the Siwalik mammals, which remains the most thorough description of the Siwalik collection housed in the American Museum of Natural History. It also included a systematic account of the various described genera and species of the primates. In a subsequent publication Lewis (1937) proposed a reduction, by synonymy, of over 20 species and half as many genera of the Siwalik hominoids then reported. This reduced the number of taxa to only four genera and ten species: *Bramapithecus*

(B. punjabicus, B. sivalensis, B. thorpei); Ramapithecus (R. brevirostris); Sivapithecus (S. darwini, S. giganteus, S. indicus, S. sivalensis), and Sugrivapithecus (S. salomontanus, S. gregoryi). Wadia and Aiyengar (1938) provided a list, with brief notes, of fossil primate collections of the Geological Survey of India which were earlier recovered from the Tertiary deposits of the Indian subcontinent. According to them approximately 82 distinct primates arranged into 11 genera and 21 species had been recovered. Gregory et al. (1938) described the fossil primate collections of the Yale–Cambridge India Expedition as well as a small part of the primate teeth collections in the Geological Survey of India and an isolated third molar loaned by the Prince of Wales College Museum, Jammu. On the basis of Wadia and Aiyengar's (1938) extensive studies on the structural variations of the shape and size of the mandible and particularly of the teeth, it was suggested that the Indian dryopithecines were more advanced than their European counterparts (Gregory et al., 1938). Recently, Prasad (1962, 1964, 1968, 1969a,b) recovered and described a number of fossil hominoids from the Nagri beds of Haritalyangar in the Bilaspur district of Himachal Pradesh: Sivapithecus (S. aiyengari, S. indicus, and S. sivalensis); Sugrivapithecus (S. gregoryi and S. salmontanus); Dryopithecus punjabicus.

Simons and Pilbeam (1965) included all the fossil hominoids from the Siwaliks into two genera, Dryopithecus (Family: Pongidae) and Ramapithecus (Family: Hominidae). The Indian material included in the genus Dryopithecus was assigned to three species: D. laietanus. D. sivalensis, and D. indicus. The other Siwalik hominoid material [Dryopithecus punjabicus; Ramapithecus brevirostris (YPM 13799), Bramapithecus thorpei, B. ? sivalensis] is included in the genus Ramapithecus represented by a single species, Ramapithecus punjabicus. This taxonomic revision proposed by Simons and Pilbeam (1965) seeks to limit the conception of generic diversity among the Mio-Pliocene dryopithecine fauna. While the proposed classification is instructive insofar as it points to a considerable amount of synonymy among the Siwalik primates, the conclusions, "cannot be regarded as final, for recovery of more material may well alter the taxonomy" (Simons and Pilbeam, 1965). Prasad (1969a) suggested a generic distinction for Sivapithecus on the basis of detailed morphological, biometric, and x-ray studies of the Sivapithecus material and included Ramapithecus cf. brevirostris (GSI D-298) under what he termed "Sivapithecus chinjiensis (Pilgrim), a pygmy sized species of the Sivapithecus from the Siwaliks hitherto unrecognised." This mandible (D-298), Ramapithecus cf. brevirostris, was earlier assigned to Dryopithecus laietanus by Simons and Pilbeam (1965). Prasad (1969a) does not consider Ramapithecus to be monotypic as Simons and Pilbeam (1965) had suggested but recognizes two species: R. punjabicus Pilgrim and R. brevirostris Lewis. According to him, R. punjabicus was "a short faced form [and] smaller in size than R. brevirostris."

In 1968 a new species of Sivapithecus, S. lewisi, based on a broken right ramus from the Nagri beds in Kangra was described by Pandey and Sastri (1968). Gupta (1969) reported on additional dryopithecine material, which he assigned to Sivapithecus indicus. The material consisted of a fragment of a mandibular ramus with M_1 and P_4 and an isolated premolar. These fossils were recovered from the same site in Kangra as the one Pandey and Sastri (1968) had earlier reported.

In addition to this, Sahni *et al.* (1974) reported on a left P_3 of *Dryopithecus*, which he assigned to the subgenus *Sivapithecus*. This find was recovered from a site 3 km northeast of Nathupani in the district of Garhwal in Uttar Pradesh. Dutta *et al.* (1976) described a left M^2 and a right M^3 of *Ramapithecus*, including a right and two left canines, left P^4, left M^2–M^3, and a left M_2 assigned to *Dryopithecus sivalensis* from Ramnagar (Jammu and Kashmir).

New Siwalik Fossil Hominoids

The bulk of the new primate fossil collections have come from the Nagri and Dhok Pathan Formations of the Middle Siwaliks, while there are a few specimens in the collection, currently under study, recently been recovered from the Lower Siwalik strata (Chinji Formation).

Pliopithecus krishnaii (Chopra, 1978; Chopra and Kaul, 1979)

Pleistocene gibbons (*Hylobates*) are known from sites in Szechuan and Kwangsi in China and from the lower to middle Pleistocene beds on Central Java (Hill, 1972). Other evidence of early hylobatids comes from the Miocene of Europe and East Africa, represented, respectively, by *Pliopithecus* and *Dendropithecus* (=*Limnopithecus macinnesi*). The majority of paleoprimatologists generally consider these two genera to constitute, or to be close relatives of, the Miocene ancestors of the recent Hylobatidae (Tuttle, 1972). To date, no remains of gibbons dating earlier than the Pleistocene have been recovered from Asia. As von Koenigswald (1968) observes, the problem of whether the Hylobatidae ever occurred outside Asia has not been settled satisfactorily, as all African and European fossils related to the ancestry of gibbons have been disputed (Hill, 1972). Von Koenigswald (1968) questions the inclusion of *Pliopithecus* and Ferembach (1958) contested the inclusion of *Dendropithecus* in the Hylobatidae. On the other hand, Groves (1972), following Remane (1965), is inclined to put them in a separate family, Pliopithecidae, which, according to him, would contain the common ancestors of both the Cercopithecoidea and Hominoidea. He further suggests that the small Siwalik dryopithecine, *Sivapithecus chinjiensis* (Prasad, 1969b), could be the possible ancestor of gibbons. These diverse views notwithstanding, the consensus of opinion favors the inclusion of *Pliopithecus* and *Dendropithecus* in Hylobatidae on the basis of comparative anatomy, morphology, and paleozoogeography (Tuttle, 1972).

Recognizing that the evidence from Mio-Pliocene deposits of Asia is crucial in the consideration of gibbon ancestry, an isolated upper molar of a hominoid recently recovered from the Middle Siwalik deposits of north-

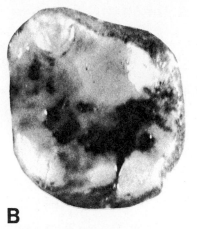

A   B

Fig. 1. Occlusal views of (A) the type (PUA 950–69) of *Pliopithecus krishnaii*, a left M³, and (B) *Hylobates lar carpenteri*, also a left M³ for comparison. Teeth are oriented so that lingual (internal) surfaces are facing down the page (× c.7.5).

western India assumes considerable significance, as it displays definite hylobatid affinities in its size and morphology (Chopra, 1978), and provides a possible link between the Miocene forms of Europe and Africa on the one hand and the Pleistocene–Recent forms of Southeast Asia on the other.

The hominoid tooth (PUA 950-69) is a left M³ (Fig. 1A). It was recovered from the Nagri beds exposed near Haritalyangar at a site which has yielded dryopithecine dental remains (Chopra and Kaul, 1975) in addition to those of lorisids. In its size (A–P length 5.70 mm; B–L breadth 7.05 mm) and morphology the tooth comes well within the range of fossil and living gibbons (see Figs. 1A, 1B). The three cusps of the trigon are well developed. The two buccal cusps are placed higher than the lingual one. The protocone is somewhat rounded and descends less steeply into the central basin than do the buccal cusps. The protocone is placed markedly distal to the paracone by a moderately developed christa obliqua. The lingual cingulum begins from the mesial aspect halfway from the protocone. It surrounds the base of the protocone and ascends distally to form the hypocone. The lingual cingulum in the fossil tooth is only moderately developed. The degree of its expression and its gradual reduction have been considered as a progressive feature of hominoid, including hylobatid, evolution (Frisch, 1973). There is a very clear contact facet. The slight tooth wear is suggestive of the fact that it belonged to a young animal which had probably just erupted its teeth.

In general, the Haritalyangar molar resembles the African and European forms, such as *Dendropithecus* and *Pliopithecus,* in having four cusps and a lingual cingulum. It differs from them in its small absolute size and a relatively weaker development of the cingulum. This tooth is relatively broad compared to those of the extant gibbon (Fig. 1B) and the Miocene hylobatid from Napak IV in Uganda described by Fleagle (1975), and has lingually and buccally expanded faces, so that the sides of the tooth have steeper slopes.

The African *Dendropithecus* (=*Limnopithecus macinnesi*) lived in ecological conditions corresponding to areas of primary rain forests, whereas the European *Pliopithecus* represents a type of adaptation to a wide range of habitats, including open woodlands and dry forests. The Indian form was, however, adapted to *subtropical* forest conditions which characterized the middle to late Miocene in the Siwaliks. During the successive periods of time in the Pliocene and the Plesitocene when open lands appeared and the areas became less forested, these forms must have been in southeast Asia and south China, where some post-Miocene hylobatid dental material has been reported from the Pleistocene deposits.

The present find permits its placement in the gibbon family and has been referred to a new species of *Pliopithecus, P. krishnaii.**

Sivasimia chinjiensis (new genus and new species; Chopra, this publication)

Type: PUA 187-76, isolated left M^1 (see Fig. 2).

Type locality: 1 km northeast of the Ramnagar Rest House, Jammu and Kashmir, India.

Hypodigm: Type only.

Diagnosis: *Sivasimia chinjiensis* is significantly different from all other Chinji-level hominoids in its possession of strongly wrinkled enamel on its occlusal surface. In this feature and in other aspects of its cusp morphology it bears a striking resemblance to the Asian orangutan *Pongo* (=*Simia* Linnaeus, 1758, in part) *pygmaeus*.

Etymology: Generic name is a combination based on the Hindu god *Siva*, who represents the principle of destruction in the Trimurti, and the suppressed nomen "*Simia*," which was the first proposed generic nomen for the orangutan. Species name is after the Chinji Stage of the Lower Siwaliks, where the type specimen, PUA 187-76, was found.

Discussion: While exploring the fossiliferous localities in and around Ramnagar (Jammu and Kashmir) during the field season of 1976 and 1977, the Panjab University crew recovered an isolated first upper left molar (?) of an orangutan-like primate (Fig. 2) from a site about 1 km northeast of the Ramnagar Rest House. This find (PUA 187-76) was recovered from the Chinji beds of the Siwaliks with a suggested date of upper middle Miocene to upper Miocene in age and is the first of its kind ever reported from anywhere in the world. While the morphology of the tooth and its comparison with other fossil and living hominoid dentitions are currently under study, a preliminary examination reveals that in its features, it shows greater essential similarities to the orangutan of southeast Asia than to any other anthropoid apes, either living or extinct.

---

*This species has recently been placed in a new genus, *Krishnapithecus*, by L. Ginsburg and P. Mein (for discussion see Ciochon, this volume, Chapter 30).

Fig. 2. Occlusal view of the type (PUA 187–76) of *Sivasimia chinjiensis* (new genus and new species, Chopra, this publication), a left M^1. Tooth is oriented so that distal (posterior) surface is facing down ($\times$ c.4.0).

Only the crown of the tooth is preserved, which is in an excellent state of preservation except that it is slightly chipped on the posterior side. In its size (A–P length $\sim$ 12.5 mm; B–L breadth 12.0 mm) and shape it falls well within the range of the Pleistocene and living orangutans. On its occlusal surface there are four well-developed cusps. The lingual cingulam is barely visible. A clear contact facet is present on the anterior aspect. The occlusal surface is markedly wrinkled. The tooth is also characterized by the presence of discernible secondary wrinkles which form the buccal half of the posterior wall dividing the anterior fovea, which, according to von Koenigswald (1952), are generally well developed in orangutans.

Hooijer (1948) undertook the study of prehistoric and recent teeth of orangutans in the collections that Dubois made from the islands of Sumatra and Java, with a view to establishing, "whether the species *had* undergone a certain amount of differentiation in the course of time." The earlier fossil orangutans are known from the Pleistocene of Java (*Pongo pygmaens hopius*) and from Southern China (*P. pygmaeus weidenreichi*). While both the Javanese and the Chinese forms were found in or near tropical Pleistocene caves, the Indian form was probably adapted to generally woodland-forested conditions. During successive periods in the Pliocene and Pleistocene, when open lands appeared and areas became less forested, these forms, like the ancestral gibbons (*P. krishnaii*) of India, migrated to southeast Asia and China. The present find has been tentatively assigned to a new genus and species, *Sivasimia chinjiensis*.

New Remains of Previously Known Siwalik Hominoids

Additional dryopithecine material, consisting of a fragment of a left mandible, an isolated left lower first molar, and a crown of a lower last molar, was recovered from the Nagri beds of Haritalayangar at the same site that had yielded the *Pliopithecus krishnaii* specimen.

Sivapithecus sivalensis

A mandibular fragment (PUA 1047-69) consisting of P_3, P_4, and broken I_1, I_2, and C (see Fig. 3), plus an isolated left M_1 (PUA 760-69), have been described and assigned to *Sivapithecus* (=*Dryopithecus*, in part)* *sivalensis* (Chopra and Kaul, 1975).

In its height the symphysis of *S. sivalensis* approximates both *Dryopithecus fontani* and *Proconsul nyanzae*, but is not as massive. The slenderness of the Indian species is also shown in its symphyseal cross section. The symphyseal cross section in this specimen also exposes the root of the central incisor, which is slightly inclined. The vertical axis of the symphysis in *D. fontani* and *P. nyanzae* shows a distinct slope, which is not so well marked in the Indian specimen. The mandibular fragment is shallow, as is indicated by its depth, which is 33.5 and 32.2 mm at the levels of P_3 and P_4, respectively. The depth of the mental foramen below P_3 is 20.6 mm and that below P_4 is 19.0 mm. The mental foramen below P_3 is distinct, but the one below P_4 is smaller and less marked. The distance between the two is about 3 mm. P_3 is rectangular and has a sectorial facet for the maxillary canine. It shows the beginnings of the metaconid. P_4 is slightly laterally compressed and indicates the presence of the hypoconid. A slight diastema is present between the canine and P_3. The premolars have distinct anterior and posterior roots.

The incisors and the canine are partly broken, and therefore no reliable measurements could be taken. In their length and breadth dimensions the P_3 and P_4 of the new find approximate the figures recorded for two other lower jaws, GSI 18069 and AMNH 19412, assigned to *S. sivalensis*. Among the living anthropoid apes these dimensions best approximate those recorded for *Pan troglodytes* and also fall within the ranges recorded for female *Pongo*.

An isolated left M_1 designated PUA 760-69 (see Fig. 4) is slightly smaller in dental dimensions when compared with known specimens of *S. sivalensis*, *S. indicus*, *D. fontani*, and the living Great Apes. The lingual cusps are placed at a higher level than the buccal cusps, which show marked perforations in the enamel. The buccal cusps are somewhat rounded, with primary ridges and wrinkles present on the occlusal surface. The hypoconulid is small and centrally placed and is less visible from the buccal side. The external cingulum is vestigial or absent.

From the foregoing observations it is clear that the mandibular fragment

---

*In this chapter I depart from my earlier writings (e.g., Chopra and Kaul, 1975; Chopra, 1978) in the usage of the generic nomen *Sivapithecus* Lewis. I now recognize that generic distinctions should be made among the thick-enameled large fossil hominoids of Asia (=*Sivapithecus*, in part), the thin-enameled medium-to-large fossil hominoids of Europe (=*Dryopithecus*, in part), and the early Miocene ancestral hominoids of East Africa (=*Proconsul*, in part). Formerly, these three genera were grouped by Simons and Pilbeam (1965) into one pan-Afro-Euro-Asiatic genus, *Dryopithecus* Lartet. Recently a number of authors (e.g., Pilbeam, 1979) have reformulated this interpretation in light of the many recent fossil hominoid discoveries. In my opinion, the evidence now supports a tripartite generic distinction as I have outlined here.

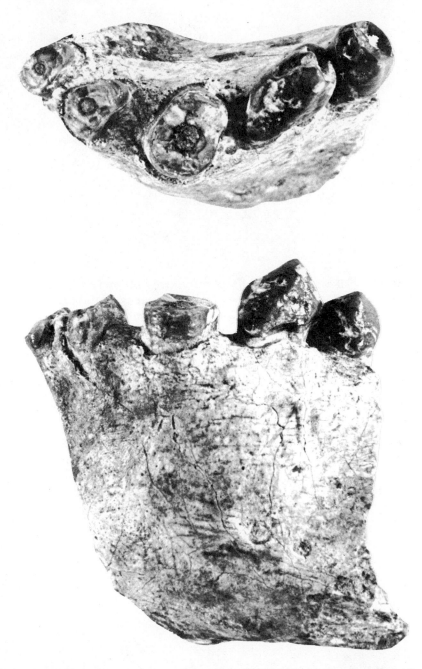

Fig. 3. Lateral view (bottom) and occlusal view (top) of PUA 1047–69, a left mandibular fragment containing P_3, P_4 and partial $\bar{C}$, I_2, and I_1, which has been assigned to *Sivapithecus sivalensis* (× c.2.5).

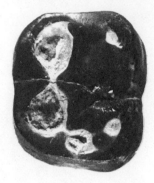

Fig. 4. Occlusal view of PUA 760-69, a left M$_1$ assigned to *Sivapithecus sivalensis*. Tooth is oriented so that distal (posterior) surface is facing down ($\times$ c.4.0).

and the isolated M$_1$ are referable to the genus *Sivapithecus* and show close structural and metrical similarities to the species *S. sivalensis*.

Sivapithecus indicus

A further isolated molar (PUA 1052-69) represents merely the crown of a lower third molar (Fig. 5). While in its overall dimensions the present specimen is somewhat intermediate between *S. indicus* and *R. punjabicus*, in its length and breadth dimensions it approximates closely the figures for male orangutans and also falls within the ranges for female orangutans and female gorillas. The molar crown is rectangular and shows slight tapering on the buccal side at the level of the hypoconulid.

The metaconid is the largest and highest of all the cusps present in the specimen. There is no indication of a cingular connection. The furrows between the buccal cusps are deeper and run down to the base of the crown. In these and other features of its morphology, including the enamel foldings, it resembles more or less the description of the crown of the third molar re-

Fig. 5. Occlusal view of PUA 1052–69, a crown of a left M$_3$ assigned to *Sivapithecus indicus*. Tooth is oriented so that distal (posterior) surface is facing down ($\times$ c.4.0).

covered from Alipur (GSI D-175), which was later assigned to *S. indicus*. Apart from the precise relationship of this form to fossil and living Great Apes, it is reasonable to infer that this specimen belongs to a comparatively large ape which in its dental morphology approximates *S. indicus*.

Past Significant Discoveries of Siwalik Hominoids

Gigantopithecus bilaspurensis (Chopra, 1968a,b; Simons and Chopra, 1969a,b)

This find (CYP 359/68) preserved both horizontal rami complete to points about 2 cm behind the M_3's and is joined at the symphysis (Fig. 6). Only the incisors and the left P_4 are missing. The locality from which it was re-

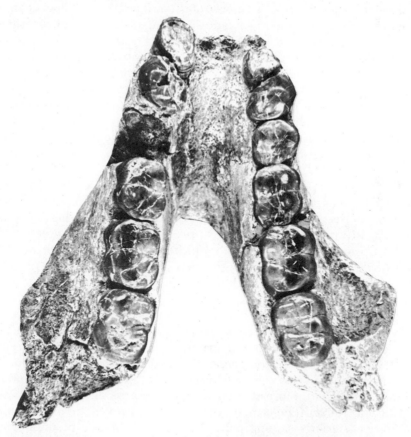

Fig. 6. Occlusal view of the type (CYP 359/68) of *Gigantopithecus bilaspurensis*, a nearly complete lower jaw (× c.0.85).

covered by a peasant is situated in the upper levels of the Dhok Pathan horizon of the Siwaliks, which gives the find an early Pliocene age.

This specimen is the most completely preserved, and perhaps the only known, mandible of a hominoid primate from early Pliocene deposits of Eurasia or Africa. All other jaws of apes or hominoids in the Old World appear to be associated with faunas which are either much older or much younger. This find therefore gains importance as probably a unique documentation of Hominoidea in the early Pliocene. Chopra (1968a,b) first announced and identified this important find as constituting a new species of *Gigantopithecus*, namely *Gigantopithecus bilaspurensis*. Subsequently Simons and Chopra (1969a,b) and Chopra (1974) published more scientific details of the find. This fossil has also been a topic of discussion among, e.g., Jolly (1970), Groves (1970), Simons and Ettel (1970), Pilbeam (1970), Frayer (1973), Robinson and Steudel (1973), and Corruccini (1975).

The jaw is slightly smaller in absolute size than the smallest known *Gigantopithecus* found in the Pleistocene deposits of China. Its large molars constitute almost the same jaw length as in apes, but the horizontal rami of the mandible in *Gigantopithecus* are more robust and relatively deeper than in any of the living apes. The dental arcade shows marked posterior divergence, a primitive character which it shares with the Miocene apes, as also with the Fayum Oligocene hominoid *Aegyptopithecus*. The space between the lower canines in all these early forms ranges from about 50% to 65% of the breadth between M_2's and M_3's. The intercanine width ranges between 70% and 110% of the width between the M_2's in *Pan, Pongo,* and *Gorilla*. From present evidence it would seem likely that in those species of the *Sivapithecus* that gave rise to the living Great Apes there has been a relative increase in the size of the symphyseal region and front teeth, while the forerunners of Hominidae (and *Gigantopithecus*?) experienced a relative reduction of the same area.

The degree of posterior divergence has been expressed as the percentage ratio of the breadth between the insides of the canines to that between the insides of the second molars. In *Homo* and *Australopithecus* this percentage is below 50%. In modern pongids the approximate range is between 70% and 110%. In primitive apes like the dryopithecines the percentage varies from 55% to 65%, showing marked posteriorly divergent rami. The percentages for *Gigantopithecus* are between those for *Australopithecus* and primitive hominoids. These are 54.8%, 55.5%, and 44.6% in the four known mandibles.

The incisors of *G. bilaspurensis* are missing, but the greatly reduced anterior region in the mandible shows crowding of the front teeth (see Fig. 6), unlike the condition found in modern apes, where the incisors are relatively broad. The incisors in *Gigantopithecus* formed a compressed anterior-facing arc, as is evidenced from the wear facet on the canine formed by interstitial attrition, and from the breadth across the incisor teeth (Woo, 1962; Simons and Chopra, 1969b). The canines, though partly broken, do give evidence of wear in a posteriorly sloping plane. Relative to the size of premolars and molars the canine teeth are reduced and should have been overlapped very little when in occlusion with the upper canines.

The premolars are relatively large. The metaconid is well developed and gives the premolar a bicuspid look. The crenulations or ridges on the metaconid are barely noticeable and there is no wear facet on the anterolateral base of P_3 as is usually found in the Great Apes. The trigonid in P_4 is extremely low and the talonid is greatly reduced.

There is a progressive increase in the overall size of molars from M_1 to M_3. The occlusal surfaces are worn flat and the cusps lack prominent ridges or crenulations. The molars in *G. bilaspurensis* are brachydont, unlike those of *G. blacki*, which show incipient hypsodonty.

In the context of the phyletic position of *Gigantopithecus*, this genus shows greater affinity with the pongids than with the hominids. It is, however, clear that the fossil jaw of *G. bilaspurensis* shows some hominid characteristics in its anterior region, as is evidenced by a marked reduction of the front teeth, relatively small canine, and the tooth wear. A particularly distinctive feature of the molars of *G. bilaspurensis* is the extraordinary flatness of their occlusal surfaces. The teeth lack the deeply incised crenulations and polycuspidation of molars and premolars seen in the newly erupted teeth of *G. blacki*. In *G. bilaspurensis* all the molar cusps are delineated on the occlusal face by shallow grooves, but even before wear the apices of the cusps could not have risen to an extent any way similar to cusp height in most apes (except possibly in some *Pongo*), because the enamel has only been perforated by wear on the protoconid and hypoconid of both M_1's. Such hominid-like features are related to a change in masticatory function and could be considered parallelisms, with implications that hominid-like tendencies were already underway some 5–10 m.y.a. in the late Miocene and early Pliocene. It would then appear that *Gigantopithecus* represents a divergent branch of the Pongidae with distinctive dental specializations which perhaps fitted it for foraging in open country. It may have been derived from earlier, larger, forest-dwelling dryopithecine forms such as *S. indicus*.

If the dimensions of the jaw are any indication of the body size, this giant ape might have weighed and attained a size equivalent to that of a present-day gorilla. An average gorilla weighs about 400 lb, and, when standing bipedally, is approximately 6 ft tall. The horizon that yielded this find is characterized by the presence of antelopes and primitive elephants. This would be suggestive of open woodland conditions, which is unlike the present-day rain-forest habitat of the gorilla.

Ramapithecus brevirostris and punjabicus

Among the fossil hominoids of the Siwaliks, *Ramapithecus* alone has been considered a probable Miocene hominid. Lewis (1934) reported principally on this new genus and proposed two species of this taxon. The nomen *Ramapithecus brevirostris* was assigned to the type consisting of a ". . . right maxilla and premaxilla with M^2, M^1, P^4, P^3, the alveolus of the canine, the root of I^2 and the alveolus of I^1" (YPM 13799), found ¼ mile east of

Chakrana, which is situated 4 miles east of Haritalyangar village, Himachal Pradesh, India. The other species, *Ramapithecus hariensis*, was based on a type-specimen consisting of a fragment of the right maxilla containing the first and second molars (YPM 13807*). This specimen was found ¼ mile east of Haritalyangar village.

The geologic and taxonomic position of *Ramapithecus* has been a subject of much discussion. Lewis (1934) first assigned *Ramapithecus brevirostris* (YPM 13799) to the ". . . Pliocene, either Latest Middle Siwalik (Dhok Pathan) or the basal Upper Siwalik (Tatrot)" horizon and *Ramapithecus hariensis* (YPM 13807) to ". . . the lower Middle Siwalik Nagri" horizon. In a later publication, however, Lewis (1937) fixed the provenance of *Ramapithecus brevirostris* (YPM 13799) in the Nagri Formations representing "the Lower Middle Siwaliks" (Pliocene) at Haritalyangar. Hooijer and Colbert (1951) refer to *Ramapithecus* as occurring in the Tatrot horizon. However, Prasad (1964) and Simons and Pilbeam (1965) assigned *Ramapithecus* to Nagri beds on the basis of detailed studies, including analysis of the matrix.

Simons and Pilbeam (1965) proposed a preliminary revision of the dryopithecines and included *Ramapithecus brevirostris* (YPM 13799) in the family Hominidae and *Ramapithecus hariensis* [YPM 13807 (=MCZ 8336)] in the family Pongidae. *Ramapithecus* was judged to be represented by a single species, *Ramapithecus punjabicus*, which included such other fossil material as *Dryopithecus punjabicus*, "*Bramapithecus*," and the YPM 13799 specimen of *Ramapithecus brevirostris* (Simons and Pilbeam, 1965). As has been pointed out, Prasad (1969a) does not consider *Ramapithecus* to be monotypic, but recognizes at least two species, *R. punjabicus* and *R. brevirostris*. In a more recent paper, Prasad (1978) states that "there are at least three recognizable species of *Ramapithecus* of which two are discernable." According to Prasad, *Ramapithecus punjabicus* (Pilgrim, 1910) differs from *Ramapithecus brevirostris* (Lewis, 1934) ". . . in the shallower robust mandible with a more complex pattern of tooth crenulations with a reduced cingulum. A progressive short faced hominid with molars showing low cusps."

Tattersall (1969a) suggested that the Nagri environment in which *Ramapithecus* lived consisted of tropical forests which were interspersed with broad rivers and woodland savanna. From the environmental picture reconstructed by Tattersall (1969a) and Prasad (1971), one would expect *Ramapithecus* to possess bipedal locomotion usually typical for such forms in an open country. However, no postcranial skeletal parts of *Ramapithecus* have been recovered that could indicate its locomotor adaptation. Simons and Pilbeam (1965) had earlier speculated that *Ramapithecus* could have been an "ad hoc tool user," and Pilbeam (1968) has suggested that *Ramapithecus* should be classified as a hominid even if it should be shown not to possess a habitual bipedal locomotor pattern!

Pilbeam (1969) described a damaged mandibular fragment (BMNH M 15423) in the collections of the British Museum of Natural History which he claimed was very similar to *Ramapithecus* and had not been hitherto identified as such. This contention was effectively countered by Leakey (1970), who held that Simons and Pilbeam (1965) had originally assigned this specimen, which came from the Upper Chinji beds at Domeli, to *Dryopithecus laietanus*. Conroy and Pilbeam (1973) assigned GSI D-199 and BMNH M 15423 to *Ramapithecus* and argued for their placement in the family Hominidae. Earlier Jolly (1970) had identified an adaptive "T" complex in the dental characteristics, including cheek teeth crowding, extensive interstitial wear, a steep molar gradient, and broad molars with thick enamel and a thick mandibular body. These criteria were used in the interpretation of the ecology of both *Ramapithecus* and *Gigantopithecus*. However, Wolpoff (1971) and Greenfield (1974) have critically examined these criteria in determining the hominoid status of fossils and consider the presence or absence of these features as ". . . neither unique to hominids nor taxonomically significant," but largely dependent upon ". . . age and/or sex of the individual, regardless of familial association." Some of these critical considerations notwithstanding, *Ramapithecus* is very possibly a hominid as far as can be judged from the hominid-like characteristics of its facial and dental morphology.

A review of the Siwalik hominoid material including the new finds and their distribution and ecological locations on the Indian subcontinent shows that the bulk of this material comes from the Mio-Pliocene beds in the Potwar Plateau of Pakistan, and in the vicinity of Ramnagar and Haritalyangar in India. Krynine (1937) wrote extensively on the paleoecology of the deposits during Chinji, Nagri, and Dhok Pathan times based on lithological analysis of the Siwalik group. Tattersall (1969*b*, pp. 821–822) summarizes Krynine's conclusions and states:

> The Chinji deposits are primarily composed of fine-grained red clastics, while the rest of the section consists of medium-grained clastics of the channel type. Intraformational conglomerates indicate river-channel boundaries and there is evidence of postdepositional weathering penecontemporaneous with deposition. . . . these sediments indicate a humid tropical climate with heavy seasonal rainfall resulting in short destructive floods; and shifting, wide rivers flowing through a heavily-forested, flat country with numerous water bodies and a relatively small number of open clearings. During Nagri time the medium-grained predominated over the fine-grained clastics. Rainfall was probably more seasonal than during Chinji time, but a forested environment prevailed though perhaps more broken than during the Chinji. Only with Dhok Pathan is the lithological and faunal evidence indicative of the formation of the open prairie and, at the top of this zone an arid environment.

The Chinji, Nagri, and Dhok Pathan faunas, comprising rodents, carnivores, ungulates, and proboscideans in addition to primates, is broadly indicative of a mixed-forest environment with open grasslands and scattered trees interspersed with meandering rivers and streams. My own observations, based on fossil floral remains, including leaf impressions of dicotyledonous plants and fossilized wood samples from the Nahans and the Nagris, suggest a thick floral cover during these times. The leaf impressions exhibit entire margins with characteristic drip points, indicating general tropical climatic conditions with adequate rainfall.

Conclusion

While engaged in paleoanthropological research in the Siwaliks the Panjab University team searched for early human relics without any positive evidence of the exact geological period, including the Pleistocene, which marked the advent of humans in the fauna of the Indian subcontinent. The numerous facts which have been brought to our notice regarding the discovery of *Ramapithecus* in the Miocene and the various stone implements in the Pleistocene in association with animal remains constitute important evidence, but are not yet conclusive. In reflecting upon the past conditions during the Mio-Pliocene, especially concerning the habitable areas of the Indian subcontinent, and assuming the rapid succession of births and deaths, one is easily led to believe that large areas would contain animal life in the succeeding Pleistocene period. It is, however, rather puzzling that remains of generations of animals, including humans, appear to have become exceedingly scarce during this epoch.

Since the Pleistocene, while the general configuration of the subcontinent has remained more or less unchanged down to the present time, the deposits and geological formations in which one could most advantageously study the earliest traces of early humans are, in the greatest degree, to be found in the northern river terraces, alluvial fans and cones, fluviatile accumulations and laterite formations of Peninsular India, at the raised beaches and strand lines along the sea coast, and in sand sheets and fossil desert dunes in Rajasthan. It is hoped that these will yield satisfactory evidence regarding the appearance of *Homo* in India. Coeval with the glacial periods of the northern part, which at the present time exhibits a similarity to the Extra-Peninsular region, early humans may have already inhabited the Peninsular region. It is also likely that during the gradual emergence of glacial conditions in the northern zone *Homo* may have followed the rivers southward toward areas of warmer and more temperate climate. While no final evidence has yet been adduced in favor of the early appearance of *Homo* in India, I strongly suspect that such evidence will be discovered, just as it already has regarding the ancestry of Asian apes.

References

Brown, B., Gregory, W. K., and Hellman, M. 1924. On three incomplete anthropoid jaws from the Siwaliks, India. *Am. Mus. Novit.* **130**:1–9.

Chopra, S. R. K. 1968*a*. The early man in Sivaliks, I. *Akashvani,* **33**(45):5.

Chopra, S. R. K., 1968*b*. The early man in Sivaliks, II. *Akashvani,* **33**(47):9.

Chopra, S. R. K. 1974. Palaeoprimatological studies in India with special reference to recent finds in the Sivaliks. Presidential Address, in: *Proceedings of LXVIth Session, Indian Science Congress,* Part II, pp. 1–16.

Chopra, S. R. K. 1976. Primatological studies in India, in: *Anthropology in India* (H. K. Rakshit, ed.), pp. 124–143, Anthropological Survey of India, Calcutta.

Chopra, S. R. K. 1978. New fossil evidence on the evolution of Hominoidea in the Sivaliks and its bearing on the problem of the evolution of early man in India. *J. Hum. Evol.* **7**:3–9.

Chopra, S. R. K., and Kaul, S. 1975. New fossil *Dryopithecus* material from the Nagri beds at Haritalyangar (H.P.), in: *Contemporary Primatology* (S. Kondo, M. Kawai, and A. Ehara, eds.), pp. 2–11, Karger, Basel.

Chopra, S. R. K., and Kaul, S. 1979. A new species of *Pliopithecus* from the Indian Sivaliks. *J. Hum. Evol.* **8**:475–477.

Chopra, S. R. K., and Vasishat, R. N. 1979. Sivalik fossil tree shrews from Haritalyangar, India. *Nature (Lond.)* **281**:214–215.

Chopra, S. R. K., and Vasishat, R. N. 1980a. Premiere indication de la presence dans le Mio-Pliocene des Siwaliks de l'Inde d'un Primate adapide, *Indoadapis shivaii*, nov. gen., nov. sp. *C. R. Acad. Sci. Paris D* **290**:511–513.

Chopra, S. R. K., and Vasishat, R. N. 1980b. A new Mio-Pliocene *Indraloris* (Primate) material with comments on the taxonomic status of *Sivanasua* (carnivore) from the Sivaliks of the Indian sub-continent, *J. Hum. Evol.* **9**:129–132.

Chopra, S. R. K., Kaul, S., and Vasishat, R. N. 1979. Miocene tree shrews from the Indian Sivaliks. *Nature (Lond.)* **281**:213–214.

Colbert, E. H. 1935. Siwalik mammals in the American Museum of Natural History. *Trans. Am. Phil. Soc. N.S.* **26**:1–401.

Conroy, G. C., and Pilbeam, D. R. 1973. *Ramapithecus:* A review of its hominid status, in: *Paleoanthropology, Morphology and Paleoecology* (R. H. Tuttle, ed.), pp. 59–86, Mouton, The Hague.

Corruccini, R. S. 1975. Multivariate analysis of *Gigantopithecus* mandibles. *Am. J. Phy. Anthropol.* **42**:167–170.

Dutta, A. K., Basu, P. K., and Sastry, M. V. A. 1976. On the new finds of hominids and additional finds of pongids from the Siwaliks of Ramnagar area, Udampur District, J & K state. *Ind. J. Earth Sci.* **3**(2):234–235.

Ferembach, D., 1958. Les Limnopithèques du Kenya. *Ann. Paleontol.* **44**:149–249.

Fleagle, J. G. 1975. A small gibbon-like hominoid from the Miocene of Uganda. *Folia Primatol.* **24**:1–15.

Frayer, D. W., 1973. *Gigantopithecus* and its relationship to *Australopithecus. Am. J. Phys. Anthropol.* **39**:413–426.

Frisch, J. 1973. The hylobatid dentition. *Gibbon and Siamang* **2**:55–95.

Gaur, R., Vasishat, R. N., and Chopra, S. R. K. 1978a. Upper Sivalik vertebrate communities of the Indian subcontinent. *Recent Res. Geol.* **5**:206–218.

Gaur, R., Sablok, A. K., Chopra, S. R. K., and Suneja, I. J. 1978b. A palaeoenvironmental study of the Sivalik Formations in parts of Bilaspur District (Himachal Pradesh). *Man and Environ.* **2**:69–71.

Greenfield, L. O. 1974. Taxonomic re-assessment of two *Ramapithecus* specimens. *Folia Primatol.* **22**:97–115.

Gregory, W. K., and Hellman, M. 1926. The dentition of *Dryopithecus* and the origin of man. *Anthropol. Pap. Am. Mus. Nat. Hist.* **28**:1–123.

Gregory, W. K., Hellman, M., and Lewis, G. E. 1938. Fossil anthropoids of the Yale–Cambridge India expedition of 1935. *Carnegie Inst. Wash. Publ.* **495**:1–27.

Groves, C. P. 1970. *Gigantopithecus* and the mountain gorilla. *Nature (Lond.)* **226**:973–974.

Groves, C. P. 1972. Systematics and phylogeny of gibbons. *Gibbon and Siamang* **1**:1–89.

Gupta, V. J. 1969. Fossil primates from the Lower Sivaliks of Kangra District, H.P. *Research Bull. (N.S.) Panjab Univ. India* **20**:577–578.

Hill, W. C. O. 1972. *Evolutionary Biology of the Primates,* Academic, London, 233 pp.

Hooijer, D. A. 1948. Prehistoric teeth of man and of the orang-utan from Central Sumatra, with notes on the fossil orang-Utan from Java and Southern China. *Zool. Meded. Rijks Mus. Nat. Hist.* **29**:175–301.

Hooijer, D. A., and Colbert, E. H. 1951. A note on the Plio-Pleistocene boundary in the Siwalik-Series of India and Java. *Am. J. Sci.* **249**:533–538.

Jolly, C. J. 1970. The seed-eaters: A new model of hominid differentiation based on baboon analogy. *Man* **5**:5–26.

Krynine, P. D. 1937. Petrography and genesis of the Siwalik series. *Am. J. Sci.* **34**:422–446.

Leakey, L. S. B. 1970. "Newly" recognised mandible of *Ramapithecus*. *Nature (Lond.)* **225**:199–200.

Lewis, G. E. 1934. Preliminary notice of new man-like apes from India. *Am. J. Sci.* **27**:161–179.

Lewis, G. E. 1937. Taxonomic syllabus of Siwalik fossil anthropoids. *Am. J. Sci.* **34**:139–147.

Lydekker, R. 1879. Notices of Siwalik mammals. *Rec. Geol. Surv. India* **11**:64–85.

Lydekker, R. 1884. Rodents, ruminants and synopsis of Mammalia. *Mem. Geol. Surv. India (Palaeontol. Ind.) Ser. X* **3**:105–134.

Pandey, J., and Sastri, V. V. 1968. On a new species of *Sivapithecus* from the Siwalik rocks of India. *J. Geol. Soc. India* **9**(2):206–211.

Pilbeam, D. R. 1968. The earliest hominids. *Nature (Lond.)* **219**:1335–1338.

Pilbeam, D. R. 1969. Newly recognised mandible of *Ramapithecus*. *Nature (Lond.)* **222**:1093–1094.

Pilbeam, D. R. 1970. *Gigantopithecus* and the origins of the Hominidae. *Nature (Lond.)* **225**:516–519.

Pilbeam, D. R. 1979. Recent finds and interpretations of Miocene hominoids. *Annu. Rev. Anthropol.* **8**:333–352.

Pilgrim, G. E. 1910. Notices of a new mammalian genera and species from the tertiaries of India. *Rec. Geol. Surv. India* **40**:63–71.

Pilgrim, G. E. 1915. New Siwalik primates and their bearing on the question of the evolution of man and the Anthropoidea. *Rec. Geol. Surv. India* **45**:1–74.

Pilgrim, G. E. 1927. A *Sivapithecus* palate and other primate fossils from India. *Mem. Geol. Surv. India (Palaeontol. Ind.)* **14**:1–26.

Prasad, K. N., 1962. Fossil primates from the Siwalik beds near Haritalyangar, Himachal Pradesh, India. *J. Geol. Soc. India* **3**:86–96.

Prasad, K. N. 1964. Upper Miocene anthropoids from the Siwalik beds of Haritalyangar, Himachal Pradesh, India. *Palaeontology* **7**:124–134.

Prasad, K. N. 1968. The vertebrate fauna from the Siwalik beds of Haritalyangar, Himachal Pradesh, India. *Palaeontol. Ind.* **39**:1–55.

Prasad, K. N. 1969a. Fossil anthropoids from the Siwalik system of India, in: *Proceedings Second International Congress Primatology, Atlanta, GA. 1968,* Vol. 2, pp. 131–134, Karger, Basel.

Prasad, K. N. 1969b. Critical observations on the fossil anthropoids from the Siwalik system of India. *Folia Primatol.* **10**:288–317.

Prasad, K. N. 1971. Ecology of the fossil Hominoidea from the Siwaliks of India. *Nature (Lond.)* **232**:413–414.

Prasad, K. N. 1978. Observations on the genus *Ramapithecus,* in *Recent Advances in Primatology,* Volume 3, *Evolution* (D. J. Chivers and K. A. Joysey, eds.), pp. 495–497, Academic, New York.

Remane, A. 1965. Die Geschichte der Menschenaffen, in: *Menschliche Abstammungslehre* (G. Heberer, ed.), pp. 249–309, Fischer, Stuttgart.

Robinson, J. T., and Steudel, K. 1973. Multivariate discriminate analysis of dental data bearing on early hominid affinities. *J. Hum. Evol.* **2**:509–528.

Sahni, A., Kumar, V., and Srivastava, V. C. 1974. *Dryopithecus* (Subgenus: *Sivapithecus*) and associated vertebrates from the Lower Siwaliks of Uttar Pradesh. *Bull. Ind. Geol. Assoc.* **7**(1):54.

Simons, E. L., and Chopra, S. R. K. 1969a. A preliminary announcement of a new *Gigantopithecus* species from India, in: *Proceedings Second International Congress Primatology, Atlanta, GA. 1968,* Vol. 2, pp. 135–142, Karger, Basel.

Simons, E. L., and Chopra, S. R. K. 1969b. *Gigantopithecus* (Pongidae, Hominoidea). A new species from North India. *Postilla* (Peabody Mus. Nat. Hist., Yale Univ.) **138**:1–18.

Simons, E. L., and Ettel, P. C. 1970. *Gigantopithecus. Sci. Am.* **222**(1):76–85.

Simons, E. L., and Pilbeam, D. R. 1965. Preliminary revision of the Dryopithecinae (Pongidae, Anthropoidea). *Folia Primatol.* **3**:81–152.

Tattersall, I. 1969a. Ecology of north Indian *Ramapithecus*. *Nature (Lond.)* **221**:451–452.

Tattersall, I. 1969b. More on the ecology of *Ramapithecus*. *Nature (Lond.)* **224**:821–822.

Tuttle, R. H. 1972. Functional and evolutionary biology of hylobatid hands and feet. *Gibbon and Siamang* **1**:136–206.

Von Koenigswald, G. H. R. 1952. *Gigantopithecus blacki* von Koenigswald, a giant fossil hominoid from the Pleistocene of southern China. *Anthropol. Pap. Am. Mus. Nat. Hist.* **43**(4):291–326.

Von Koenigswald, G. H. R. 1968. The phylogenetical position of the Hylobatidae, in: *Taxonomy and Phylogeny of the Old World Primates with Special Reference to the Origin of Man* (A. B. Chiarelli, ed.), pp. 271–276, Rosenberg and Sellier, Torino.

Vasishat, R. N. 1979. *Contributions to the Tertiary Mammals from Haritalyangar, Himachal Pradesh, India, and Their Evolutionary Significance in the Context of Palaeoprimatological Studies*, Ph.D. Dissertation, Panjab University, Chandigarh.

Vasishat, R. N., Gaur, R., and Chopra, S. R. K. 1978*a*. Community structure of Middle Sivalik vertebrates from Haritalyangar (H.P.), India. *Palaeogeogr. Palaeoclimatol. Palaeoecol.* **23**:131–140.

Vasishat, R. N., Gauer, R., and Chopra, S. R. K. 1978*b*. Geology, fauna and palaeoenvironments of Lower Sivalik deposits around Ramnagar, India. *Nature (Lond.)* **275**:736–737.

Wadia, D. N., and Aiyengar, N. K. N. 1938. Fossil anthropoids from India: A list of the fossil material hitherto discovered from the Tertiary deposits of India. *Rec. Geol. Surv. India* **72**:467–494.

Wolpoff, M. H. 1971. Interstitial wear. *Am. J. Phys. Anthropol.* **34**:205–228.

Woo, J. K. 1962. The mandibles and dentition of *Gigantopithecus*. *Palaeontol. Sin.* **11**:1–94.

Historical Notes on the Geology, Dating and Systematics of the Miocene Hominoids of India

22

K. N. PRASAD

Introduction

Recent discoveries of fossil hominoids mainly from Spain, the Middle East, East Africa, India, and China have added considerably to our understanding of problems concerning the radiation and adaptation of the Hominoidea in the Mio-Pliocene. It is generally agreed that the origin of both the Pongidae and Hominidae can be identified with some species of the subfamily Dryopithecinae of the Miocene. During the last two decades, significant discoveries of much interesting material have given scope for a critical analysis of various factors concerning human origins and hominoid evolution. The main intent of this review is to piece together the relevant material recovered from India and from different segments of the Tertiary in the Euro-Asiatic region and to make observations on its morphology and distribution. During the research for this survey, the author had the occasion to study the originals housed in the various museums in U.S. and Great Britain.

Among the more important hominoids, *Ramapithecus* from the Siwaliks

K. N. PRASAD • Geological Survey of India, 4-3-542, Bogulkunta, Tilak Road, Alladin Building, Hyderabad-500001, India.

has received considerable attention, as potentially the earliest hominid yet recorded, with a date of about 8–10 m.y.a. The actual status of *Ramapithecus* has not yet been defined. For example, the characteristic dryopithecine "Y5" molar pattern has been modified considerably, with a splaying apart of the cusps resulting in broader areas for a particular type of dietary habit, mainly graminivorous. There was a well-marked dental and facial change, with shortening of mandible, which reduced the diastema space. There was also a marked absence of cingular extension and a molarization of the premolars. The vertical emplacement of the incisors and possible parabolic contour of maxilla are also notable features. Thus, in *Ramapithecus,* such as maxillary fragment GSI 18064, the upper molars reveal a morphology of broad, squarish cusps. This cusp morphology, together with other features mentioned above, suggests a potential phylogenetic relationship with the Plio-Pleistocene hominids of Africa.

Geological Setting and Background

A threefold division for the Siwaliks was proposed by Pilgrim (1934, 1940) mainly on the basis of the associated fauna, since a classification based on lithology alone was found unreliable (see Table 1). Geological unconformities within the System, except between Kamlial and Chinji, do not exist. Due to tectonogenic movements, the Upper Siwaliks were laid down on the Middle Siwaliks after a period of folding, uplift, and denudation. Studies have shown that the Boulder Conglomerates contain relics of Pleistocene Early Man (*Homo* sp.), whose appearance in India roughly coincided with the disappearance of the rich Siwalik fauna (1.8 m.y.a.) during a phase of climatic cooling. The maximum development of the fossil hominoid apes of India occurred earlier, during the Chinji–Nagri stages of the Siwalik System. Figure 1 illustrates all these hominoid sites. The possible reason for the absence of any hominoids in the early Tertiary sediments of India, which are fairly well known, has been discussed by Wadia and Aiyengar (1938).

Table 1. Siwalik Succession and Classical Mammal Age Correlations

Upper Siwaliks (1880–2400 m)	**Boulder Conglomerate:** Sands, gravels, and conglomerates; no correlation **Pinjor Beds:** Sandstones, mudstones, and grits; Villafranchian **Tatrot Beds:** Soft sandstones, mudstones, and conglomerates; Ruscinian
Middle Siwaliks (1800–2100 m)	**Dhok Pathan Beds:** Brown sandstones, orange mudstones, shales; L. Turolian **Nagri Beds:** Hard gray sandstones and shales; E. Turolian
Lower Siwaliks (1500–2000 m)	**Chinji Beds:** Bright red shales and purple shales; Vallesian **Kamlial Beds:** Gray fine sandstones and dark purple clays; E. Astaracian

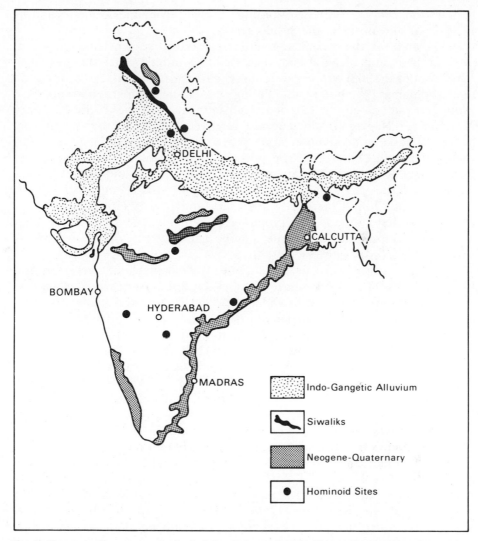

Fig. 1. Neogene–Quaternary geological formations of India illustrating location of hominoid sites. Though there are a number of hominoid-bearing sites in India spanning the time range from upper Miocene through upper Pleistocene, only three sites in the sub-Himalayan foothills (Siwaliks) are known to yield Miocene hominoids.

Fossil hominoid primates in the Siwaliks are known from three important sites: (1) the much dissected and eroded sedimentary belt of the Lower and Middle Siwaliks of the Potwar Synclinal Basin in Pakistan, (2) the area around Ramnagar, Jammu, and Kashmir, and (3) the Haritalyangar area of Bilaspur District, Himachal Pradesh. A few other localities, for example, the Narmada Alluvium, the Pranhita Godavari Basin, the Karnul Caves, and Gokak Beds, where human implements have been recorded, await detailed exploration. The age of these localities ranges from the lower to middle Pleistocene and postdate the Siwalik strata.

The bulk of the Siwalik hominoid material is known from the Chinji–Nagri beds, and I have already discussed this fauna in detail (Prasad, 1964, 1968). Simons (1968) expressed doubts regarding the actual age of the sediments containing the hominoids. Though the clastic sediments have not been radiometrically dated, the faunal assemblage generally does reveal an upper Miocene age. Nagri fauna in general is relatively less abundant in proportion to Chinji fauna, but contains a larger percentage of hominoid material. Matthew (1929) and Colbert (1935) proposed that the Lower Siwalik fauna is of Pontian age. Recent work by Simons and Pilbeam (1965) and Prasad (1971*a*) has revealed that the Vallesian fauna of northeast Spain compares favorably with those of Chinji–Nagri. Le Gros Clark and Leakey (1951) and Leakey (1962) recorded a number of dryopithecine remains from East Africa. They indicate a Miocene age and some may be equated with the Lower Siwalik formations of India with reservation.

Recent application of new techniques on paleomagnetic reversals and magnetostratigraphic studies have considerably enlarged geochronologic correlation methods and have provided new means for elucidating problems connected with age determination of various stratotype sections. Stratigraphic inquiry aided by paleomagnetic dating may provide a basis for analyzing critically the various stratotypes from which the hominoid remains have been recovered. Some of the areas where such studies could be conducted lie in the Siwalik belt around the Haritalyangar–Hamipur area of Himachal Pradesh (Figs. 2 and 3), and the Ramnagar section in Jammu and Kashmir (Figs. 4 and 5).

The Haritalyangar Area.   The Siwalik formations at Haritalyangar first drew the attention of G. Pilgrim about 1910. Somewhat prior to this date Vinayak Rao had collected fauna from this area and had examined its geological strata (see Figs. 2 and 3). On faunal and lithological grounds Pilgrim (1913) assigned the beds at Haritalyangar to the Nagri level of the Middle Siwaliks. Over the years great significance has been attached to the beds at Haritalyangar because of the occurrence of fossil hominoid material first recovered by officers of the Geological Survey of India in the early part of this century and by the Yale (and Cambridge) North India Expeditions of the 1930s. Both the paratype maxilla of *R. punjabicus* and the type maxilla of *R. brevirostris* were recovered from Haritalyangar.

During four field seasons in 1951, 1954, 1962, and 1967, I made extensive faunal collections and mapped localities in and around the Bilaspur-Haritalyangar area (Fig. 2). The results of these investigations are summarized in Prasad (1970). More than 65 species of mammals representing 20 families have now been recovered from the Haritalyangar area. This analysis reconfirmed the Nagri level affinities of the Haritalyangar vertebrates and suggests an age of about 8 m.y.a. for the deposits.

In the late 1960s Elwyn Simons of Yale teamed up with S. R. K. Chopra of Panjab University for the most recent expedition to the Haritalyangar area.

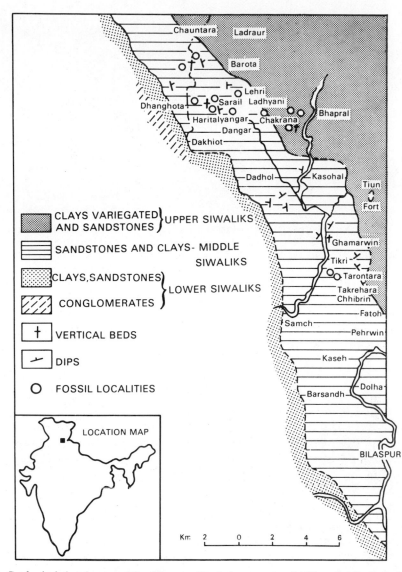

Fig. 2. Geological sketch map of the Bilaspur–Haritalyangar area in Himachal Pradesh. The site of Haritalyangar, situated on the Nagri Formations of late Miocene age (~8 m.y.a.) is the type locality of *Ramapithecus*.

Fig. 3. View of Nagri Formation near Haritalyangar, Himachal Pradesh, India. Mudstones between the sandstone ridges are source of the *Ramapithecus* remains. At the left extremity below the topmost *cuesta* scarp, one crude pebble tool (or eolith?) was recovered by the author from the same litho-unit where *Ramapithecus* and *Sivapithecus* had earlier been found.

Some of the field crew of this expedition included S. Kaul, now at Panjab University, T. Bown, now at the U.S. Geological Survey in Colorado, K. Rose, now at Johns Hopkins University in Baltimore, and J. Fleagle, now at SUNY in Stony Brook. This field crew was the first to attempt large-scale quarrying at the site. Perhaps the most notable discoveries of this collecting effort at this important site are the type and to date only specimen of *Gigantopithecus bilaspurensis* (Simons and Chopra, 1969) and several dozen jaws of as yet undescribed *Indraloris* specimens.

The Ramnagar Area. The first recorded hominoid recovered from the Ramnagar area of Jammu and Kashmir was discovered by C. S. Middlemiss in the 1920s and described by Pilgrim (1927). Based on its associated mammalian fauna Pilgrim (1927) judged the specimen, which he named *Sivapithecus middlemissi,* to be from beds equivalent to those of the Chinji. Recent geological mapping in the area around Ramnagar village by teams from the Geological Survey of India in Calcutta (see Fig. 4) has shown that Chinji beds, Nagri beds, and Dhok Pathan beds are all represented here.

The Chinji–Nagri litho-units at Ramnagar show a NNE–SSW trend forming a syncline structure which abuts against the Dalser Fault and the Taktal Thrust to the northeast (Fig. 4). The Chinji's are here represented by variegated, compact mudstones and coarse-grained hard sandstones with intraformational conglomerates. The Nagri units are characterized by bright yellow mudstone alterations in the lower units and sand/mudstone alterations in the upper part. The mudstone to sand ratio is much lower than in the Chinji units.

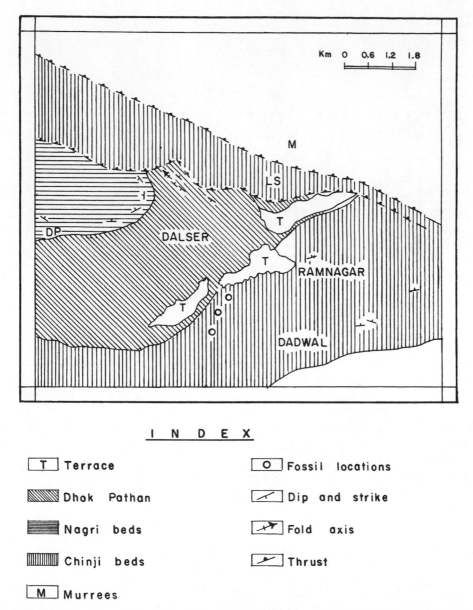

INDEX

T	Terrace	O	Fossil locations
	Dhok Pathan		Dip and strike
	Nagri beds		Fold axis
	Chinji beds		Thrust
M	Murrees		

Fig. 4. Geological map of the hominoid-bearing Ramnagar area in Jammu and Kashmir, India.

Fig. 5. View of the Chinji Formation (Lower Siwaliks), showing layered sequence of pink mudstones and gray sandstones near Ramnagar, Jammu and Kashmir, India. The mudstones in the foreground are known to have yielded remains of *Ramapithecus* and *Sivapithecus*, though in significantly lower numbers than the Haritalyangar Nagri beds. *Hipparion* remains appear to be totally absent from the Ramnagar Chinji beds.

The upper Chinji and lower Nagri levels at Ramnagar yield the bulk of the fossil material (Fig. 5). Thanks to the collecting efforts of A. K. Dutta and P. K. Basu of the Geological Survey in Calcutta, the primate remains now include *Sivapithecus, Ramapithecus,* and *Indraloris.* Associated with these hominoid remains were cricetids, *Aceratherium, Macrotherium, Listriodon, Conohyus,* and *Sus.* Tragulids are represented by *Dorcabune* and *Dorcatherium.* The giraffid *Giraffokeryx* and the bovids *Protragocerus, Gazella,* and *Miotragocerus* can also be listed. Reptilian remains such as *Crocodilus, Gavialis* and *Chelonia* complete the faunal picture. It is most interesting to note that no remains of *Hipparion* have yet been recorded at Ramnagar.

Review of Hominoid Material from India

Fossil hominoids from the Siwaliks were first investigated by R. Lydekker during 1879–1885. Though the samples he analyzed were in a good state of preservation, their exact provenience had yet to be fully established. Therefore, their stratigraphic value in charting hominoid evolution in the Siwaliks was greatly diminished. The partial palate of *Palaeopithecus sivalensis* (GSI D-1) received considerable attention when first described (Lydekker, 1879), but its provenience was not established. Pilgrim (1915) assigned the specimen to the

Dhok Pathan, the latest Miocene horizon, where to date only cercopithecoids have been recorded. Hooijer (1951) broke up the sample of *P. sivalensis*, assigning the molars to *Sivapithecus indicus* and the premolar to *Pongo* (?). I feel that GSI D-1 is probably best assigned to *Sivapithecus sivalensis* and that its actual provenience in the Siwaliks is probably the Nagri beds.

One of the most often discussed isolated teeth from the Siwaliks is an upper right third molar from the Chinji horizon, which was originally described as *Palaeosimia rugosidens* by Pilgrim (1915). This tooth has highly crenulated enamel and does share features in common with the upper molars of the extant Great Ape, *Pongo*. In my opinion, it also recalls certain features found in the East African hominoid material described by Le Gros Clark and Leakey (1951). It is interesting to note here that both Lewis (1937) and Simons and Pilbeam (1965) referred this specimen to *S. sivalensis*. Whether or not the *Palaeosimia* specimen, designated GSI D-188, shares any special affinity with the Asian orangutan cannot be decided on the basis of a single tooth. However, it is worth noting that other isolated molars also showing crenulated enamel have recently been described from Chinji level beds in India and Pakistan. Chopra (this volume, Chapter 21) describes a new taxon from Ramnagar which he feels shares a strong affinity with the Asian orangutan, *Pongo*. In a similar fashion, von Koenigswald (1981, and this volume, Chapter 19) named a new taxon from the Chinji beds of Pakistan which he feels may potentially prove to be the ancestor of *Gigantopithecus bilaspurensis*. The names of both these new taxa will probably not stand the test of time, but their presence does indicate that some potentially exciting new specimens remain to be recovered from the Chinji levels of the Lower Siwaliks.

During 1922 the American Museum of Natural History fossil collector Barnum Brown undertook an expedition to the Siwaliks. Though he collected a large number of fossil vertebrate specimens, his team only managed to secure three dryopithecine lower jaw fragments. These lower jaws were tersely described by Brown *et al.* (1924) as three new species of *Dryopithecus, D. pilgrimi, D. cautleyi,* and *D. frickae.* Lewis (1937) synonymized the former two species with *S. sivalensis* and the latter with *S. indicus.* I am of the opinion that all three species may represent but one genus and species, namely *Sivapithecus indicus* (Table 2).

Pilgrim (1927) also proposed a number of new hominoid taxa, which were all later synonymized with existing taxa (see Table 2). For example, Pilgrim (1927) assigned a left mandibular ramus (GSI D-197) to *S. himalayensis*, which Lewis (1937) referred to *S. indicus*. Likewise, a maxilla from Haritalyangar preserving the canine through M^3 (GSI D-196), which Pilgrim named *S. orientalis*, was later included in *S. indicus* by Lewis (1937). From the Chinji level beds of Ramnagar a lower jaw assigned to *S. middlemissi* by Pilgrim (1927) is also referrable to *S. indicus*. This specimen (GSI D-198) has a well-developed basal cingulum, a rare feature in Indian dryopithecines. Finally, a partial M_3 in a mandibular fragment (GSI D-200) was named *Hylopithecus hysudricus* by Pilgrim (1927). It was recovered from the Nagri horizon(?) near Haritalyangar and was considered a potential early hylobatid by Pilgrim, al-

Table 2. Dryopithecine Taxa Discussed in this Chapter and Their Probable Synonyms

Date	Previous terminology	Suggested terminology
1879	*Palaeopithecus sivalensis* Lydekker Type: palate, GSI D-1	*Sivapithecus sivalensis*
1910	*Sivapithecus indicus* Pilgrim Type: lower molar, GSI D-176	*Sivapithecus indicus*
1910	*Dryopithecus punjabicus* Pilgrim Type: lower jaws, GSI D-118, D-119	*Ramapithecus punjabicus*
1915	*Dryopithecus chinjiensis* Pilgrim Type: lower molar, GSI D-179	*Sivapithecus chinjiensis*
1915	*Dryopithecus chinjiensis* Pilgrim Paratype: lower molar, GSI D-180	*Sivapithecus sivalensis*
1915	*Dryopithecus giganteus* Pilgrim Type: lower molar, GSI D-175	*Gigantopithecus giganteus*
1915	*Palaeosimia rugosidens* Pilgrim Type: upper molar, GSI D-188	*Sivapithecus sivalensis?*
1924	*Dryopithecus pilgrimi* Brown, Gregory, & Hellman Type: lower jaw, AMNH 19411	*Sivapithecus indicus*
1924	*Dryopithecus cautleyi* Brown, Gregory, & Hellman Type: lower jaw, AMNH 19412	*Sivapithecus indicus*
1924	*Dryopithecus frickae* Brown, Gregory, & Hellman Type: lower jaw, AMNH 19413	*Sivapithecus indicus*
1927	*Sivapithecus himalayensis* Pilgrim Type: lower jaw, GSI D-197	*Sivapithecus indicus*
1927	*Sivapithecus orientalis* Pilgrim Type: palate, GSI D-196	*Sivapithecus indicus*
1927	*Sivapithecus middlemissi* Pilgrim Type: lower jaw, GSI D-198	*Sivapithecus indicus*
1927	*Palaeopithecus sylvaticus* Pilgrim Type: lower jaw, GSI D-199	*Sivapithecus sivalensis*
1934	*Ramapithecus brevirostris* Lewis Type: right maxilla, YPM 13799	*Ramapithecus punjabicus*
1934	*Sugrivapithecus salmontanus* Lewis Type: lower jaw, YPM 13811	*Sivapithecus* sp. indet.
1934	*Dryopithecus sivalensis* Lewis Type: lower jaw, YPM 13806	*Ramapithecus punjabicus*
1934	*Bramapithecus thorpei* Type: lower jaw, YPM 13814	*Ramapithecus punjabicus*
1936	*Sugrivapithecus gregoryi* Lewis Type: lower jaw, YPM 13825	*Sivapithecus* sp. indet.
1957	*Dryopithecus keiyuanensis* Woo Type: five lower molars	*Sivapithecus sivalensis*
1962	*Sivapithecus aiyengari* Prasad Type: lower jaw, GSI 18039	*Sivapithecus indicus*
1962	*Kenyapithecus wickeri*, Leakey Type: left maxilla, KNM-FT 46	*Kenyapithecus africanus*
1968	*Sivapithecus lewisi* Pandey & Sastri Type: lower jaw, ONGC/V/790	*Sivapithecus indicus*
1969	*Gigantopithecus bilaspurensis* Simons & Chopra Type: lower jaw, CYP 359/68	*Gigantopithecus giganteus*
1974	*Dryopithecus* sp. Sahni, Kumar, & Srivastava Referred specimen: lower premolar	*Sivapithecus indicus*

though probably not "... on the direct line of ancestry of the Gibbon ..." (Pilgrim, 1927, p. 12). On the basis of size alone Simons and Pilbeam (1965, p. 90) considered GSI D-200 a non-dryopithecine and went on to conclude "... that it is, or is closely related to, *Pliopithecus*." Recently Chopra and Kaul (1979) have announced the recovery of a left M^3 from Quarry D at Haritalyangar which they have assigned to a new species of *Pliopithecus*, *P. krishnaii* (see also Chopra, this volume, Chapter 21). This new specimen, designated PUA 950-69, may prove to represent the same primate taxon as GSI D-200, especially since Chopra and Kaul (1979) conclude that PUA 950-69 also appears to be specially related to the Hylobatidae.

Pilgrim (1910, 1915, 1927) described a number of important specimens from the Siwaliks. Of these, two mandibles (GSI D-118, GSI D-119) are now assignable to *Ramapithecus punjabicus*, potentially the earliest known hominid. Lewis (1934) created the genus *Ramapithecus* based on a right maxilla (YPM 13799) recovered near the Haritalyangar scarp in the early 1930s. Its identification as a genus separate from *Sivapithecus* has stood the test of time, but has recently come under intense scrutiny (Greenfield, 1979, and this volume, Chapter 27; Kay and Simons, this volume, Chapter 23). The Yale (and Cambridge) North India expeditions of the 1930s also recovered a number of other specimens from the area around Haritalyangar which were assigned a variety of new names, such as *Bramapithecus* and *Sugrivapithecus* (Lewis, 1934). *Bramapithecus* has since been reasonably included under *Ramapithecus* (see Table 2) but the status of *Sugrivapithecus* is still uncertain. In my opinion, it appears to show some similarities with the Vallesian *Hispanopithecus* material from Spain. Szalay and Delson (1979), in their treatment of the dryopithecine material from the Miocene of Eurasia, refer *Sugrivapithecus*, *Sivapithecus*, and *Gigantopithecus* to the new tribe Sugrivapithecini. They also question whether *Sugrivapithecus* should be maintained as a separate genus. I feel that *Sugrivapithecus* differs from *Sivapithecus* in a number of important features. For example, the upper molar cusp morphology is less complicated and the molars are longer than they are broad, unlike *Sivapithecus*, which has squarish molars. Furthermore, the crown area is considerably smaller, the cingulum is either absent or vestigial, and the symphysis is elongated like the Spanish Vallesian material. If these features do not warrant separate generic status for *Sugrivapithecus*, they might at least be used to argue for a separate specific status from all other *Sivapithecus* species.

One of the more interesting specimens described by Pilgrim (1915) is GSI D-175, a lower molar recovered from the Nagri beds of Alipur. He assigned this specimen to *Dryopithecus giganteus*, based on the fact that it was the largest molar tooth yet recovered. Lewis (1937) synonymized this specimen with *S. indicus*. Recently Szalay and Delson (1979) have suggested that it is very much like the molars of *Gigantopithecus bilaspurensis* [CYP 359/68 (Simons and Chopra, 1969)]. Based on this fact they have proposed a new combination of names, sinking the species *bilaspurensis* and assigning GSI D-175 and CYP 359/68 to *Gigantopithecus giganteus*. In my opinion, this represents a very sound decision.

The author (Prasad, 1954, 1962, 1964, 1970, 1971*b*, 1974, 1975, 1978) has carried out intensive research on anthropoid material recovered from the Haritalyangar area, Himachal Pradesh (Fig. 2). One of the most interesting finds is a large mandible (GSI 18039) with crowns of P3–M3, alveolus of the canine, and part of the symphysis. Extreme wear in the holotype indicates an aged individual. X-ray photographs reveal the absence of a diastema between the third premolar and the canine. The left canine has been lost and the alveolus is filled with cancellous bone. I assigned the mandible to a new species *S. aiyengari* on morphologic and metric grounds (Prasad, 1962). Somewhat later Simons and Pilbeam (1965) advocated its retention under *Dryopithecus indicus* (Pilgrim). Its jaw proportions and molar pattern in some ways recall those of *Proconsul major* from East Africa. In addition to this mandible, three other finds have also been recorded by Prasad (1962). Two isolated first molars (GSI 18041, 18042) have been assigned to *S. sivalensis.* The only upper third molar of *Ramapithecus punjabicus* (GSI 18068) recovered by the author is from the same Nagri horizon. In addition to the holotype of *Ramapithecus brevirostris* (YPM 13799), another maxilla (GSI 18064) has also been collected from these Miocene beds. This maxilla needs restudy.

A mandible with P3–M3 from Nagri level strata, near Bandel, Himachal Pradesh, collected by the Oil and Natural Gas Commission, Dehra Dun, India, (ONGC), is one of the latest hominoid specimens to come to light. It has been assigned to *Sivapithecus* by the author. Three more finds by the ONGC need study. The recovery of a lower third premolar from Garhwal by Sahni *et al.* (1974) indicates an extension of the Siwalik range about 400 km eastward. The premolar was assigned to *Sivapithecus indicus.* S. R. K. Chopra and his associates from the Anthropology Department, Chandigarh have carried out intensive exploration of fossiliferous sites in the Siwaliks. Chopra (1974) and Chopra and Kaul (1975) reviewed and described new material, including the first record of *Gigantopithecus.*

Remarks on the Classification of the Siwalik Dryopithecines

The Siwalik deposits contain both the short-jawed as well as the elongated varieties of dryopithecines. Complete jaws, where the lower border is preserved, indicate that three species of *Sivapithecus* can be justified. The relatively short-jawed forms with the typical dryopithecine pattern of molars are assignable to *Sivapithecus sivalensis.* The robust, relatively elongated forms with traces of a cingulum in the second and third molars belong to *Sivapithecus indicus.* The third variety, with massive, robust, deep mandibles with high mandibular and symphyseal slopes at P3 and M3, respectively, can be assigned to *S. aiyengari.* However, for lack of additional preserved material, *S. aiyengari* is here provisionally retained under *S. indicus* (Table 2). The fourth variety includes pygmy-sized forms with smaller, delicate jaws and narrow cheek teeth; these are assigned to *Sivapithecus chinjiensis,* a nomen originally proposed by Pilgrim (1915).

Many of the new species of *Dryopithecus* and *Sivapithecus* described by Pilgrim (1915, 1927) from the Siwaliks have mostly been invalidated, while some of them have been included under new taxa (Table 2). In many cases, pathological deformation of various components of the dentition, differences in weathering, and lack of associated or ankylosed upper and lower dentitions have made the correlation of isolated specimens difficult. Lack of precise locality data on the various hominoid finds and inability to examine the actual specimens (by vertebrate paleontologists in general and paleoanthropologists in particular) are some of the reasons that have contributed to a lack of understanding of their taxonomy. There seems to be no consensus regarding the status of various finds of certain dryopithecines. Gregory *et al.* (1938) clarified the taxonomy of the dryopithecines and lucidly discussed the character differences between *Sivapithecus* and *Dryopithecus*. Le Gros Clark and Leakey (1951) justified the species separation of *S. africanus* from the Indian species based on certain features, such as the flatness of the palate; Pilgrim (1927) had earlier noted that it is high and arched in *S. orientalis*. The persistence of a trace of the anterointernal cingulum on the upper molars, the development of an internal cingulum on P4, and the slightly smaller width of the upper molars were also stressed as distinct features of *S. africanus* by Le Gros Clark and Leakey (1951). It is interesting to note that none of the upper molars of the Indian dryopithecines show any development of the cingulum, though it is sometimes faintly present in the lower molars.

Remarks on the Taxonomic Usage of "Sivapithecinae"

Prasad (1968) advocated the usage of the term "Sivapithecine" as a subfamilial designation for the group *Sivapithecus* and related forms that showed morphological similarity but had no defined status. Some earlier workers (e.g., Simons and Pilbeam, 1965) questioned the validity of adducing subfamilial status for the *Sivapithecus* group. From a detailed study of the various dental elements, both upper and lower, I tentatively conclude that the Miocene apes allocated to the *Dryopithecus* and *Proconsul* groups of Europe and Africa, respectively, were substantially different from the *Sivapithecus* group of Asia. The various morphological characters justifying this distinction have been dealt with elsewhere (Prasad, 1977). In this chapter I have used the term "dryopithecine" not so much as a taxonomic term but instead as a general descriptive term to characterize the Miocene hominoid apes.

Remarks on the Occurrence of Ramapithecus in East Africa

There has been frequent reference to the occurrence of *Ramapithecus* in the East African Miocene sediments. *Kenyapithecus wickeri* is now considered a synonym of *R. punjabicus* of the Siwaliks (Simons and Pilbeam, 1965). *Ken-*

yapithecus wickeri is of middle Miocene extraction, whereas *Ramapithecus* from the Siwaliks is late Miocene in age. The index fossil *Hipparion* is associated with hominoids of late Miocene age. The age of the first occurrence of *Hipparion* in Europe and Asia is generally misunderstood. During Dhok Pathan times (latest Miocene) the second upheaval in the Himalayas occurred and probably the Alpine Chain was also involved. A change in climate and fauna occurred from Chinji through Nagri and Dhok Pathan. Wet, humid, and forested conditions prevailed during Chinji, which were replaced by steppe and savanna during Nagri–Dhok Pathan. It is not clear whether such conditions existed in the East African Tertiary belt. Stratigraphically, *Kenyapithecus* can be assigned to a part of the Miocene much older than the *Ramapithecus* biostratigraphic unit of the Siwaliks. Apart from these, *Kenyapithecus* differs from *Ramapithecus* in the morphology of the symphysis, P3 shape, and in having a larger base for the cusps, which are more rounded. The molars are more progressive in *Ramapithecus*, with a larger occlusal surface. The molars in *Ramapithecus* are more square, with heavy transverse wear facets. The morphologic construction of the molars is therefore related to diet and adaptation, which presumably was controlled by the prevailing ecosystem in each of the two different geographical provinces. Therefore, in my opinion, the East African Fort Ternan "*Ramapithecus*" specimens, on the basis of morphological, temporal, and ecological factors, should not be placed in the same taxon as the Indian *Ramapithecus*. In other words, perhaps the taxon *Kenyapithecus* should be resurrected for those specimens currently classified as *Ramapithecus* in East Africa. The differences outlined above strongly warrant this suggestion. Similarly, some Chinese hominoid material which has at times been attributed to *Ramapithecus* (Woo, 1957; Chow, 1958) could best be assigned to *Sivapithecus sivalensis* on the nature of its cusp morphology, canines, and third molar shape. The final question that remains to be answered is whether the Indian specimens of *Ramapithecus* will continue to be regarded as a separate valid genus or will one day also be sunk back into *Sivapithecus*. That question is not yet fully answerable.

Concluding Remarks

In this chapter, I have tried to briefly summarize the history of Indian hominoid Siwalik discoveries and place them in perspective with regard to recent discoveries in India and elsewhere. The great number of new finds made in Europe, Africa, the Middle East, Pakistan, and China have in some ways overshadowed the recent finds in the Indian Siwaliks. This is to be expected, since many of the most productive areas in India were extensively explored early in the 20th century and the level of research underway here is far below that now undertaken outside of India. The importance of the hominoid samples from the Siwaliks of India should not be underestimated, however. Many of our original conceptions of hominoid evolution in the Miocene were based on the Indian discoveries. Also, many of the type specimens of hominoid taxa still in use today were recovered from the Indian Siwaliks.

Much new work remains to be done in India. Intensive collecting and quarrying efforts should now be undertaken at important type localities such as Ramnagar and Haritalyangar. It may also be possible to better date these localities through the use of paleomagnetic geochronological techniques. Perhaps the time is ripe for the advent of multidisciplinary collaborative research efforts in the Indian Siwaliks. Many critical questions regarding ape and human ancestry remain to be answered and intensive research in the historically significant and productive Indian Siwaliks may very well yield the answers for future generations of scientists.

References

Brown, B., Gregory, W. K., and Hellman, M. 1924. On three incomplete anthropoid jaws from the Siwaliks, India. *Am. Mus. Novit.* **130**:1–9.

Chopra, S. R. K. 1974. Palaeoprimatological studies in India with special reference to recent finds in the Siwaliks. Presidential Address, in: *Proceedings of LXVIth Session, Indian Science Congress,* Part II, pp. 1–16.

Chopra, S. R. K., and Kaul, S. 1975. New fossil *Dryopithecus* material from the Nagri beds at Haritalyangar (H.P.), in: *Contemporary Primatology* (S. Kondo, M. Kawai, and A. Ehara, eds.), pp. 2–11, Karger, Basel.

Chopra, S. R. K., and Kaul, S. 1979. A new species of *Pliopithecus* from the Indian Siwaliks. *J. Hum. Evol.* **8**:475–477.

Chow, M. C. 1958. Mammalian faunas and correlation of Tertiary and early Pleistocene of South China, *J. Palaeontol. Soc. Ind.* **3**:123–130.

Colbert, E. H. 1935. Siwalik mammals in the American Museum of Natural History. *Trans. Am. Phil. Soc. N.S.* **26**:1–401.

Greenfield, L. O. 1979. On the adaptive pattern of "*Ramapithecus.*" *Am. J. Phys. Anthropol.* **50**:527–548.

Gregory, W. K., Hellman, M., and Lewis, G. E. 1938. Fossil anthropoids of the Yale–Cambridge India Expedition of 1935. *Carnegie Inst. Wash. Publ.* **495**:1–27.

Hooijer, D. A. 1951. Questions relating to a new anthropoid ape from the Mio-Pliocene of the Siwaliks. *Am. J. Phys. Anthropol.* **9**:79–95.

Lewis, G. E. 1934. Preliminary notice of new man-like apes from India. *Am. J. Sci.* **27**:161–179.

Lewis, G. E. 1937. Taxonomic syllabus of Siwalik fossil anthropoids. *Am. J. Sci.* **34**:139–147.

Leakey, L. S. B. 1962. A new Lower Pliocene fossil primate from Kenya. *Ann. Mag. Nat. Hist.* **13**(4):689–696.

Le Gros Clark, W. E., and Leakey, L. S. B. 1951. The Miocene Hominoidea of East Africa. *Fossil Mammals of Africa* (Br. Mus. Nat. Hist.) **1**:1–117.

Lydekker, R. 1879. Further notices of Siwalik Mammalia. *Rec. Geol. Surv. India* **11**:64–85.

Matthew, W. D. 1929. Critical observation on Siwalik mammals. *Bull. Am. Mus. Nat. Hist.* **56**:437–560.

Pilgrim, G. E. 1910. Notices of new mammalian genera and species from the tertiaries of India. *Rec. Geol. Surv. India* **40**:63–71.

Pilgrim, G. E. 1913. The correlation of the Siwaliks with mammal horizons of Europe. *Rec. Geol. Surv. India* **43**:264–326.

Pilgrim, G. E. 1915. New Siwalik primates and their bearing on the question of the evolution of man and the Anthropoidea. *Rec. Geol. Surv. Ind.* **45**(1):1–74.

Pilgrim, G. E. 1927. A *Sivapithecus* palate and other primate fossils from India. *Mem. Geol. Surv. India (Palaeontol. Ind.)* **14**:1–26.

Pilgrim, G. E. 1934. Correlation of fossiliferous sections in the upper Cenozoic of India. *Am. Mus. Novit.* **704**:1–5.

Pilgrim, G. E. 1940. Middle Eocene mammals from North India. *Proc. Zool. Soc. Ind. B* **110**:127–152.

Prasad, K. N. 1954. A preliminary report on the Siwalik rocks at Haritalyangar and Ukhli Bilaspur, Dist. Punjab. Progress Report (unpublished).

Prasad, K. N. 1962. Fossil primates from Siwalik beds from Haritalyangar, H.P. *J. Geol. Soc. Ind.* **3**:86–96.

Prasad, K. N. 1964. Upper Miocene anthropoids from the Siwalik beds of Haritalyangar H.P. *Palaeontology* **7**(1):124–134.

Prasad, K. N., 1969. Fossil anthropoids from the Siwalik System of India, in: *Proceedings Second International Congress Primatology. Atlanta, GA. 1968*, Vol. 2, pp. 131–134, Karger, Basel.

Prasad, K. N. 1970. The vertebrate fauna from the Siwalik Beds of Haritalyangar, H.P. India. *Palaeontol. Ind.* **39**:1–56.

Prasad, K. N. 1971a. Observations on the Dryopithecines of India and Europe, in: *Proceedings Third International Congress Primatology. Zurich*, Vol. I, pp. 48–53, Karger, Basel.

Prasad, K. N. 1971b. Ecology of the fossil Hominoidea from the Siwaliks of India, *Nature (Lond.)* **232**:413–414.

Prasad, K. N. 1974. The hominid status of *Ramapithecus*. *J. Ind. Acad. Geosci.* **17**:77–80.

Prasad, K. N. 1975. Observations on the paleoecology of South Asian Tertiary primates, in: *Paleoanthropology, Morphology and Paleoecology* (R. Tuttle, ed.), pp. 21–30, Mouton, The Hague.

Prasad, K. N. 1977. Review of Miocene Anthropoidea from India and adjacent countries. *J. Palaeontol. Soc. Ind.* (Orlov Memorial Volume) **20**:382–390.

Prasad, K. N. 1978. Observations on the genus *Ramapithecus*, in: *Recent Advances in Primatology*, Volume 3, *Evolution* (D. J. Chivers and K. A. Joysey, eds.), pp. 495–497, Academic, New York.

Sahni, A., Kumar, V., and Srivastava, V. C. 1974. *Dryopithecus* (Subgenus: *Sivapithecus*) and associated vertebrates from the Lower Siwaliks of Uttar Pradesh, *Bull. Ind. Geol. Assoc.* **7**(1):54.

Simons, E. L. 1968. A source for dental comparison of *Ramapithecus* with *Australopithecus* and *Homo. S. Afr. J. Sci.* **64**(2):92–112.

Simons, E. L., and Chopra, S. R. K. 1969. *Gigantopithecus* (Pongidae, Hominoidea). A new species from North India. *Postilla* (Peabody Mus. Nat. Hist., Yale Univ.) **138**:1–18.

Simons, E. L., and Pilbeam, D. R. 1965. Preliminary revision of the Dryopithecinae (Pongidae, Anthropoidea). *Folia Primatol.* **3**:81–152.

Szalay, F. S., and Delson, E. 1979. *Evolutionary History of the Primates*, Academic, New York, 580 pp.

Von Koenigswald, G. H. R. 1981. A possible ancestral form of *Gigantopithecus* (Mammalia, Hominoidea) from the Chinji Layers of Pakistan. *J. Hum. Evol.* **10**:511–515.

Wadia, D. N., and Aiyengar, N. K. N. 1938. Fossil anthropoids of India. *Rec. Geol. Surv. Ind.* **72**(4):467–494.

Woo, J. K. 1957. *Dryopithecus* teeth from Keiyuan Yunnan. *Vertebr. Palasiat.* **1**:25–32.

Assessments of the Mio-Pliocene Evidence

VII

A Reassessment of the Relationship between Later Miocene and Subsequent Hominoidea

23

R. F. KAY AND E. L. SIMONS

Introduction

In consideration of the significance of middle and later Miocene Hominoidea to the origin of the taxonomic family of humans, Hominidae, there are three central problems. These are: (1) definition of the taxa concerned, (2) demonstration of significant anatomical (and thus phyletic) relationship of any of these later Miocene taxa to later hominids, and (3) establishment of a derivation that is zoogeographically sound in relation to a plausible theory of human descent.

Continent of Hominid Origin

There have been many recent discussions of fossils that have changed considerably the viewpoint of scientists on each of these three topics. Nevertheless, the bulk of new evidence found or published in the past 4 or 5 years pertaining to Miocene apes (and/or protohominids) comes from a few scat-

R. F. KAY • Department of Anatomy, Duke University Medical Center, Durham, North Carolina 27710.   E. L. SIMONS • Duke University Center for the Study of Primate Biology and History, Durham, North Carolina 27705.

tered areas in Eurasia. In spite of these additions, most scientists want more evidence about higher primate evolution from the Pliocene or later Miocene of Africa. This is because most students believe that the principal stage for origin and evolution of hominids was in Africa. The concept that this was the place of origin goes back to, and beyond, the following remark of Charles Darwin in *The Descent of Man* (1871):

> In each great region of the world the living mammals are closely related to the extinct species of the same region. It is therefore probable that Africa was formerly inhabited by extinct apes closely allied to the gorilla and chimpanzee; and as these two species are now man's nearest allies, it is somewhat more probable that our early progenitors lived on the African continent than elsewhere. But it is useless to speculate on this subject. . .

(Darwin, 1871, p. 199)

After the discovery of archaic human relatives in Java, reported by Dubois (1894), and later finds by others in Java and China (von Koenigswald, 1935, 1936; Weidenreich, 1937, 1943), the stage for human origins preferred by many scholars became Asia. Although these two choices, either an African or an Asian origin for hominids, have often been proposed, hardly anyone ever thought that Europe could have been the cradle of the taxonomic family of *Homo*. Until recently, relatively few large fossil hominoids have ever been found in Europe.

Charles Darwin is often cited as the earliest exponent of the view that the African apes provide anatomical evidence of a nearer relationship to humans than does the Asian great ape, *Pongo*, the orangutan. In this, however, Darwin was clearly influenced by the earlier findings of his colleague and friend, the anatomist T. H. Huxley. In *Man's Place in Nature,* originally published in 1863, Huxley (1897, p. 97) had already remarked: "It is quite certain that the Ape which most nearly approaches man, in the totality of its organization, is either the Chimpanzee or the Gorilla."

Apart from these anatomically based associations of the African apes and the evidence of Java "ape man" recovered in the 1890s, little further information about the phyletic affinity and geographic origin of humans was produced in the 19th century. By 1904, however, pioneering work on the chemistry of the blood of apes and humans had been published by Nuttall, working at Cambridge, England. From his early work, measuring the degree of cross-reaction in precipitation of the blood of related primates, humans seemed to stand closer to the African apes than to the Asian orangutan.

The Evidence from Fossils

Unfortunately, the fossil record of human forebears accumulated in the first three-quarters of this century does not reveal as clear a picture of the location or the exact line of hominid descent as one would wish. Certainly the now copious record of Plio-Pleistocene hominids from Africa confirms the central position of that continent as a stage for later hominid evolution. Between about 1.5 and 5.5 million years ago (m.y.a.), the only known fossils related to humans came from Africa and not from Eurasia. Perhaps the proper faunas to contain early hominids in the time range of 1.5–5.5 m.y.a. have yet to be found in Eurasia. Nevertheless, it is probable that Eurasia has been searched more extensively for fossils of that time period than has Africa, without turning up either *Australopithecus* or any *Homo* more primitive than *Homo erectus*.

The Eurasian and African fossil evidence of higher hominoids (apes and prehumans) in the period from 6 to 16 m.y.a. was for a long time either very scanty or difficult to interpret. From the 1930s to the present, large collections of fossil apes (17–23 million years old) were made in the early Miocene of Kenya. Yet African Miocene sites younger than this are few, and the large hominoid fossils from them are incomplete. Only three of these finds are more than fragments: (1) an associated incomplete maxillae of *Sivapithecus* (formerly *Ramapithecus*) from Fort Ternan, Kenya, at 14 million years; (2) a lower jaw of *Sivapithecus* from Maboko Island of similar age; and (3) a palate with teeth and partial face of ? *Proconsul major* from Moroto, Uganda, possibly as young as 16 million years. In Eurasia, the later Miocene fossil record of higher primates is quite different. Temporally the situation is almost reversed from that of Africa, for the scanty and very incomplete hominoid fossils come first, from the period 14–17 m.y.a. Then species of *Dryopithecus* appear in Austria and southern France, and a little later in Spain and southern Russia. After approximately 12 m.y.a. the only reasonably extensive samples of fossil hominoids from anywhere in the Eurasian Miocene are the Greek and the Indo-Pakistan higher primate fossils. These latter samples were recovered from the turn of the century onward for 30 years by Pilgrim and his co-workers at the Geological Survey in India. These collections were made in the Potwar Plateau and Salt Range areas of northeastern Pakistan and from Himachal Pradesh State in north India near Haritalyangar, 100 miles north of Delhi. Fossil hominoids came from sites that appeared to be widely scattered both geographically and temporally. After the period of Pilgrim's work, a few additional specimens reached the American Museum of Natural History and the British Museum (Natural History). Then in the 1930s, two Yale University north India expeditions greatly enlarged the number of fossil hominoids from north India and Pakistan while ape fossils were also found in Kashmir. Following World War II, further additions to the series of fossil hominoids from India were made during two expeditions to the Haritalyangar region in the late 1960s (see Simons and Chopra, 1969). Simons (1964) reemphasized that the Siwalik mammalian faunas had strong ties with those of the East African Miocene, that is, both held many genera in common. He showed that the Indian genus *"Ramapithecus,"* regarded in this chapter as *Sivapithecus,* occurs at Fort Ternan, Kenya. This genus had also been reported elsewhere in Africa by Le Gros Clark and Leakey (1950, 1951). The research of Gentry (1970) and Shipman (1977) on the Fort Ternan fauna has reconfirmed its ties with those of the Siwaliks, especially among carnivores, proboscideans, and artiodactyls.

Simons and Pilbeam (1965) published a review of the extinct Dryopithecinae in which they included *Dryopithecus* (with subgenera *Dryopithecus* in Europe, *Sivapithecus* mainly in Asia, and *Proconsul* in Africa) and *Gigantopithecus*. The genus *Ramapithecus* was also discussed by them, as a hominid, and its derivation from earlier forms of *Dryopithecus* was suggested. Perhaps the principal deficiency which their review exposed was that of a general inadequacy in dating the fossils concerned. Although the largest sample of fossil hominoids available to them was extensive, mainly from the Salt Ranges and Siwaliks of Pakistan and India, the best specimens came from scattered, poorly dated sites. It was not possible to group sets of these fossils in a sound time-successive sequence. Nevertheless, it seemed to Simons and Pilbeam (1965) that in the Indo–Pakistan Tertiary, several different species of hominoids occurred which to varying degrees resembled *Australopithecus,* a species which by then had become relatively well understood morphologically. Attempts to sort these Siwalik specimens statistically into groupings were made in the course of their research, but results were inconclusive. On morphological grounds some of the best specimens, such as the type maxilla of *Ramapithecus* (YPM 13799), were associated by them with later hominids. In the case of certain other species, such as *Dryopithecus indicus*, affinities with *Gigantopithecus* were suggested. *Gigantopithecus* itself has, in turn, frequently been ranked as a hominid. Other somewhat hominid-like finds, such as the type of "*Sugrivapithecus,*" were referred to an intermediate-sized Indian species *Dryopithecus sivalensis*. Because the majority of specimens were fragments (often only single teeth), most sites were poorly dated, and because the meaning of morphological similarities to (and differences from) other forms were hard to assess, defining and distinguishing these Miocene taxa were difficult. In fact, their 1965 publication was entitled a "preliminary" revision. At that time, no clear statistical groupings could be found; rather, the best specimens were morphologically gradational, with individual fossils at one extreme showing clusters of hominid features, while at the opposite extreme of the series stood specimens of mainly pongid character. No single find preserved even an associated upper and lower dentition. Thus the first two of the criteria for judgement we have mentioned (definition of taxa and demonstration of relationships) were confused by the lack of completeness or lack of dating of these fossils. The third factor, zoogeographic relevance of the Asian fossils, seemed less of a problem because contemporary hominoids from Africa were referrable to the same genus, *Sivapithecus,* and the Fort Ternan fauna resembled those approximately contemporary to it in Pakistan and India.

New Siwalik Hominoids

For the past 6 years, David Pilbeam of Yale University has directed a research project in the Potwar Plateau area of Pakistan in cooperation with the Pakistan Geological Survey (Pilbeam *et al.,* 1977*a,b;* 1980). This large-scale project has been oriented toward improving the understanding of the hominoid fossils recovered there together with other vertebrate and plant fossils and the dating of all these finds. Attention has been given to paleomagnetic, biostratigraphic, and taphonomic studies (Pilbeam *et al.*, 1979; Barry *et al.*, 1980).

As a result of this important program, the number of relatively complete Indo-Pakistan Miocene fossils has more than doubled. Equally important is the dating of the Pakistan sites and the somewhat unexpected demonstration that the bulk of the hominoids come from a limited time range. The Pakistan fossils are largely about 7 million years old and those found near Haritalyangar about 8 million years old (Pilbeam, personal communication). This new dating evidence means that, in consideration of the Indo-Pakistan Miocene hominoids, it is not possible to say that one is sampling a sequence with time-successive genera, nor is there even much likelihood of species successions. As stated earlier, previous attempts to sort Indo-Pakistan hominoids statistically were largely inconclusive, but now the way is cleared to reassess the problem statistically because the sample size is much larger and the temporal range much shorter than formerly thought.

Classification of the Siwalik Hominoids

The first major revision of the fossil apes of the Siwaliks by Simons and Pilbeam (1965) recognized three species of pongids, *Dryopithecus (D.) laietanus, D. (Sivapithecus) sivalensis,* and *D. (S.) indicus,* and one hominid, *Ramapithecus punjabicus.* Subsequent amendments to this scheme constituted both description of new taxa based on new specimens and some taxonomic juggling at the generic level. Since 1965 two additional Siwalik ape taxa have been described: *Gigantopithecus bilaspurensis* and *Sivapithecus lewisi.* Most authors now recognize the generic distinctiveness of *Sivapithecus* from *Dryopithecus.* One recent proposal is that *Ramapithecus* be considered as another distinct species of *Sivapithecus* (Greenfield, 1979). These developments and the accumulation of many more specimens have necessitated a further revision of the Siwalik material as a whole (Kay, 1982*a*). The revision is grounded on improved scientific understanding of variability within living species. For the first time, variation within the recognized extinct taxa can be shown to be similar in kind and degree to that seen in modern ape and monkey species.

A difficult problem which has always faced those who have studied the hominoid fossil record, in addition to the scarcity of the material, has been an insufficiency of published information about variability among extant hominoids and other anthropoids. Over the last 10 years, this information has accumulated to the point where useful generalizations can be made about the kinds and degrees of variability in living populations and applied, accordingly, to the assessment of fossils [for a review of this literature, see Gingerich and Schoeninger (1979)]. Among the principal findings of variability studies are the following: (1) Almost without exception, dental variability in living primates is lowest in the upper and lower first and second molars. (2) Sexual dimorphism in dental dimensions is greatest in the canines and anterior lower premolars. Such dimorphism often leads to bimodal size distributions in histograms of these latter teeth. Although occasionally significant differences may be detected between the mean molar dimensions of males and females of sexually dimorphic species, these differences are never of sufficient magnitude to produce bimodal distributions in histograms of molar dimensions. These findings have several important implications for studies of fossil hominoids. When coefficients of variation (CV) of upper and lower first and second molars exceed about 8.5 in a collection of fossils with similar morphology, it must be assumed that one is dealing with a heterogeneous sample of two or more species. Similarly, with sufficiently large samples, when length–breadth scatter plots or histograms of $\log_{10}$ (length $\times$ breadth) M_1 or M_2 are examined, the presence of two distinct clusters, or bimodal distributions, strongly indicates the presence of two species. Once decisions have been made about the number of species and the identity of individual specimens based on molar dimensions, high CVs and/or bimodal distributions in the canine–P_3 region are indicative of sexual dimorphism within a single species. Moreover, the reverse findings point toward low sexual dimorphism. In some cases, the failure to adhere strictly to these two procedures in a stepwise fashion has led to misleading conclusions with respect to *Sivapithecus* taxonomy at the species level.

The findings presented here concerning Siwalik hominoids are drawn from Kay (1982*a*). Figure 1 is a scatter plot of M_1 length and breadth and a histogram of $\log_{10}(M_1$ length $\times$ breadth) of Siwalik specimens. Included for comparison is a small sample of male and female chimpanzees. Most specimens fall within two discrete size clusters. Combined, the specimens in the two clusters have CVs of 11.12 and 12.26 for M_1 length and breadth, respectively. These CVs fall outside the 99% confidence intervals for CVs of M_1 length and breadth in any extant anthropoid species, but taken separately, the variability within each cluster is just what would be expected for a single anthropoid species. Taken in conjunction with similar clusterings and CVs for other upper and lower molar dimensions, the two species may be identified as *Sivapithecus sivalensis* and *Sivapithecus indicus*.

Variability within *S. sivalensis* is sufficient to include all the specimens formerly attributed by Simons and Pilbeam (1965) to *Ramapithecus punjabicus*. This latter taxon was formally sunk within *S. sivalensis* by Kay (1982*a*). *Sivapithecus lewisi*, alluded to above, falls within the hypodigm of *S. indicus*, as indicated by Szalay and Delson (1979). Two specimens fall well away from *S. sivalensis* and *S. indicus* in known dental dimensions. The largest species is *Gigantopithecus giganteus* [formerly *G. bilaspurensis;* see Szalay and Delson (1979)]. It includes only the type, an isolated M_2 or M_3 described by Pilgrim (1915), and a lower jaw (Simons and Chopra, 1969). The smallest species is represented by two mandibles: GSI D-298 from Kundal Nala (Gregory *et al.,* 1938), and BMNH M 15423 from Domeli (Pilbeam, 1969*a*). A maxilla, GSI D-185, from Haritalyangar also belongs to this species. Simons and Pilbeam

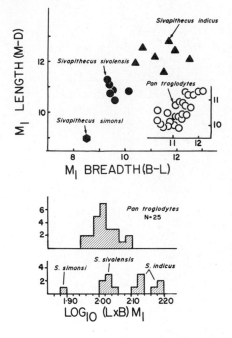

Fig. 1. Bivariate plot of M_1 dimensions and histograms of $\log_{10}$ (length × breadth M_1) for Indo-Pakistan ramapithecines (*Gigantopithecus* omitted). Shown in the inset at the same scale and in a separate histogram are length–breadth plots of a balanced-sex sample of *Pan troglodytes*. The smaller sample of *S. sivalensis* plus *S. indicus* exceeds the range of the *Pan* specimens.

(1965, p. 124) correctly observed that the two mandibles are very small and "cannot possibly pertain to *Ramapithecus*" (*Sivapithecus sivalensis* in this revision); accordingly, they assigned these specimens to *Dryopithecus laietanus*, a European species. However, *D. laietanus* has a much narrower P_4–M_2 than the Siwalik material. Furthermore, casts and excellent photographs of GSI D-298 (which is apparently lost) reveal that it had very thick enamel, like *Sivapithecus* and unlike *D. laietanus* (Fig. 2). The maxilla (Fig. 3) GSI D-185 has extremely narrow (buccal–lingually) P^{3-4}, as distinct from *Sivapithecus sivalensis,* and is the right size to go with GSI D-298. Kay (1982a) has named a new species, *Sivapithecus simonsi,* based on these three specimens. Further Pakistan material collected by von Koenigswald (this volume, Chapter 19) probably belongs to this small species, especially a mandibular fragment from Khaur (see Chapter 19, Fig. 6). Thus, only four hominoids come from the Siwaliks:

> *Sivapithecus indicus* Pilgrim, 1910
> *Sivapithecus sivalensis* (Lydekker, 1879)
> *Sivapithecus simonsi* Kay, 1982a
> *Gigantopithecus giganteus* (Pilgrim, 1915)

The anatomy of Siwalik *Sivapithecus* has been reviewed in detail by Simons (1972) and recently by Pilbeam *et al.* (1980). Both *Sivapithecus sivalensis* and *S. indicus* have reduced premaxillary prognathism meaning the premaxilla is short mid-sagittally, comparable to *Gorilla, Proconsul,* and *Australopithecus,* but much lower than either *Pan* or *Pongo.* In both species, the root of the zygomatic arch is situated high above the tooth row at M^1 or M^2 in a fashion

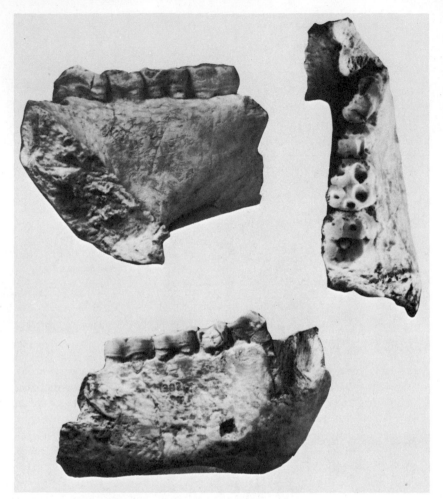

Fig. 2. *Sivapithecus simonsi*, Type, GSI D-298 (field no. 618), from Kundal Nala, Chinji (× c.1.2).

similar to modern pongids and the earliest *Australopithecus*. *Sivapithecus* species also resemble these latter taxa in having arched palates in the region of the molars. Pronounced canine fossae and well-developed canine jugae are always visible. The shape of the palate is ape-like and a considerable space intervenes between the root of the upper canine and I^2 in both species, as in pongids and the earliest *Australopithecus*.

The Siwalik sample of ramapithecines offers a unique opportunity to investigate sexual dimorphism. Canine sexual dimorphism, when expressed in the usual fashion as a ratio of male to female size, is difficult or impossible to gauge accurately in an extinct species because identification of specimens as to sex is most unreliable. The usual procedure is to allocate large-sized, large-canined specimens as males and small-sized, small-canined specimens as females. This procedure inevitably leads to an overestimation of the degree of

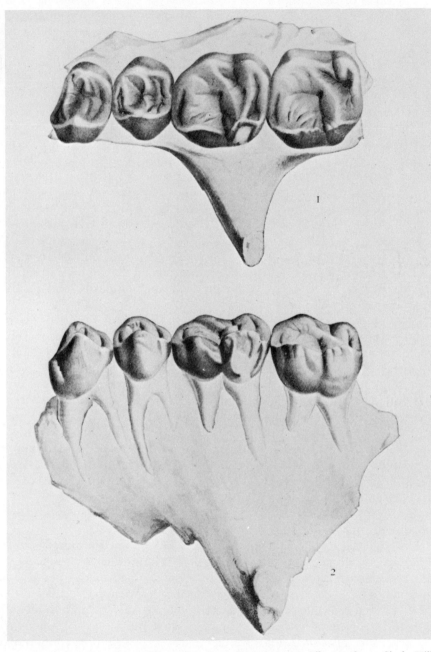

Fig. 3. cf. *Sivapithecus simonsi*, GSI D-185, from Haritalyangar, Bilaspur State, Simla Hills (×
c.2.5). [Drawing from Plate 3 of Pilgrim (1915).]

sexual difference because large females and small males are often misattributed. An alternative approach is to express the degree of dimorphism in the population in terms of the coefficient of variation in the population. In living species of known sex the sample CV is highly correlated with the ratio of male to female size in that character (Kay, 1982b). CVs of mandibular canine dimensions range between 14 and 19 in balanced-sex samples of the dimorphic species like chimpanzee, gorilla, and orangutan; the values for maxillary canines range between 15 and 21. CVs in *Hylobates,* a species with low canine dimorphism, range between 4 and 10 in upper and lower teeth.

The samples of canines definitely attributable to single Siwalik hominoid species are very small—between two and six per species—so that CVs cannot be calculated reliably. Although the large number of isolated maxillary canines of Siwalik hominoids cannot be assigned to a particular species, the overwhelming proportion must belong either to *Sivapithecus sivalensis* or to *S. indicus,* inasmuch as the other two species are extremely rare. Taken together, 21 maxillary canines of *Sivapithecus* have CVs for mesial–distal length of 10.90 and for buccal–lingual breadth of 14.37. *The CVs of this combined-species, combined-sex sample are lower than those of chimpanzees, the single least dimorphic extant pongid.* This strongly indicates that the canine sexual dimorphism in separate *Sivapithecus* species was very low by modern ape standards, and probably comparable to the values for *Hylobates.*

Low canine dimorphism is an important feature of similarity between *Sivapithecus* and *Australopithecus.* Notwithstanding recent statements by Johanson and White (1979), Johanson *et al.* (1982), and Brace and Ryan (1980) to the contrary, canine dimorphism in *Australopithecus* is quite low by modern pongid standards. CVs of the mesial–distal and buccal–lingual dimensions of upper and lower canines of Swartkrans and Sterkfontein *Australopithecus* never exceed 10.5 [data from Robinson (1956)]. The CV of canine root socket breadth for a combined sample of 43 australopithecine canines which come from a minimum of two species is 9.2 (Wolpoff, 1975). CVs of upper and lower canine dimensions of a combined sample of Laetolil and Hadar australopithecines do not exceed 11.3 except in mandibular canine breadth (Johanson and White, 1977). CV for the latter dimension is 16.81, but the sample size for this tooth is only five, compared with nine or more for the other dimensions, so the high CV may be unreliable.

Although canine sexual dimorphism is low in Siwalik *Sivapithecus,* the canines are larger than in australopithecines. The canines of *Sivapithecus indicus* are comparable in size to those in large-canined extant pongids and hylobatids. Those of *S. sivalensis* seem to be comparatively smaller (Kay, 1982a). The ratio of mean maxillary canine mesial–distal length to mean M^1 length for *Sivapithecus indicus* is comparable to those for *Pongo, Gorilla, Pan troglodytes,* and *Hylobates;* that for *Sivapithecus sivalensis* is comparatively smaller and comparable to *Pan paniscus.* The mandibular canines of *S. indicus* are smaller than in *Pan paniscus* but larger than in *Australopithecus.* The maxillary canines of Siwalik ramapithecines are remarkable for having their greatest occlusal dimension near the buccal–lingual axis, as in *Australopithecus*, but unlike living pongids.

The structure of P_3 in *Sivapithecus* and *Gigantopithecus* has been the focus of considerable attention *vis-à-vis* the hominid status of these taxa. Dryopithecines and extant pongids have P_3's with an oval occlusal outline and a prominent paracristid which wears against (or "hones") the postparacrista of the upper canine. P_3 is generally single-cusped, with a short talonid heel. A metaconid is generally absent; however, a small one is present in a low frequency of cases in *Gorilla* and more frequently in *Pan* (Huag, 1977; Remane, 1960) and *Pongo* (R. F. Kay, personal observation). The P_3 of *Australopithecus* is more rounded, with a greatly reduced paracristid, but the earliest fossils still retain a prominent buccal face which wore against a prominent canine postparacrista (Johanson and White, 1979). In the earliest *Australopithecus*, the metaconid is usually present or the protocristid presents an inflated appearance (Johanson and White, 1979).

The P_3 of Siwalik *Sivapithecus* retains the primitive ape-like oval occlusal outline and a prominent, generally ape-like paracristid in most cases. The metaconid is variable in occurrence and expression. It is absent in some *Sivapithecus sivalensis* (AMNH 19411), but in GSP 9563, its size surpasses the average and even the maximum of living apes and falls within the range of the earliest *Australopithecus*. Similarly, it is frequently very large in *S. indicus* (e.g., GSI 18039). This cusp is also extremely large in *Gigantopithecus giganteus*. Thus, a derived tendency seen in *Sivapithecus* and *Gigantopithecus* for the P_3 metaconid to reach large size is otherwise restricted to Pliocene–Recent hominids alone.

Molars of Siwalik *Sivapithecus* are typically low-crowned, with limited occlusal relief that often wears nearly smooth. Molar enamel is extremely thick: on the hypoconid it averages about 2.5 times thicker than that typically seen in extant catarrhines with comparably sized molars (Kay, 1981). Recently, attention has been drawn to the presence of thick molar enamel in orangutans as well (Andrews and Tekkaya, 1980; Kay, 1981). The former authors have suggested that thick enamel may be merely a primitive retention that is not indicative of an affinity between ramapithecines and *Australopithecus*. We do not subscribe to this possibility. *Gorilla* and chimpanzee still retain thin enamel, as do their early Miocene dryopithecine forebears and the extant hylobatids. Even though the molar enamel of orangutans is thicker than that of chimpanzees and gorillas, it remains on average about 60% thinner than that of ramapithecines. Thus, an evolutionary scenario placing ramapithecines in or near the ancestry of orangutans would require the retrogressive reduction of enamel thickness in the orangutan ancestry, a process that would have to be carried still further in ancestors of modern *Pan* and *Gorilla* in order to return to a condition where enamel is so thin as to be indistinguishable from that observed in early Miocene East African hominoids.

Two decades ago, Simons (1961) called attention to the extreme robusticity of the mandibles of *"Ramapithecus"* as a feature allying it to *Aus-*

tralopithecus. Research since then has confirmed this observation and extended it to ramapithecines in general. *Sivapithecus indicus, S. sivalensis,* and *Australopithecus* have broad or shallow mandibles, whereas early Miocene dryopithecines and extant pongids have comparatively narrower or thinner mandibles.

Given our unification within a single species of specimens previously allocated to *Ramapithecus punjabicus* and those assigned to *Sivapithecus sivalensis,* some of the gnathic features that were considered to separate the small and large Siwalik hominoids now break down. However, some, such as the configuration of the mandibular symphysis of *S. sivalensis* and *S. indicus,* remain somewhat different. Even so, the values for mandibular robusticity are essentially the same (Table 5). Another gnathic feature separates *S. sivalensis* from *S. indicus.* In the former species, as exemplified by GSP 4622/4857, the mandibular rami diverge posteriorly to a considerable degree; in the latter species (e.g., GSP 9977/01/05, 9564), the tooth rows diverge posteriorly only slightly. In this respect, *Gigantopithecus giganteus* resembles *S. indicus* in having subparallel tooth rows. The configurations of GSP 4622/4857 is nearly identical to that of *Australopithecus.* Parenthetically, "broadly spaced" (i.e., less parallel) tooth rows surely have nothing to do with a relatively enlarged cranial capacity or a broadened cranial base in Pliocene hominids, as suggested by Greenfield (1979).

Newly described material from Pakistan and Turkey indicates that the large species of *Sivapithecus* has large upper central incisors relative to the laterals. A recently described specimen of *Sivapithecus indicus* (GSP 15000) (Pilbeam, 1982) has an I^1/I^2 length ratio of 1.88. This value falls outside the upper range for *Pan* and *Gorilla,* but within that of *Pongo.* A similar high ratio in Turkish *Sivapithecus* led Andrews and Tekkaya (1980) to suggest this as a derived similarity linking *Sivapithecus* with *Pongo* phyletically. For two reasons we hold the alternative view that this feature of similarity may be an evolutionary parallelism and not indicative of a close relationship: the demonstrated common occurrence of parallelism in incisor size, and the fact that large incisors are not ubiquitous among *Sivapithecus* species.

Enlargement of the upper incisors of catarrhines has occurred independently in many Old World monkeys. Incisor enlargement is characteristic of some but not all species of the genera *Cercopithecus, Macaca, Presbytis,* and *Colobus;* it also characterizes some, but not all, members of the papionin tribe. In a similar fashion a large degree of variation occurs among closely related species of extant apes: *Hylobates* and *Pan* have relatively large upper incisors; *Symphalangus* and *Gorilla* have relatively small ones. Such variation appears to be related to the amount of incisal preparation of foods prior to mastication. Thus, an increased frequency and duration of incisal preparation causes increased amounts of attrition and abrasion of the anterior dentition. Enlarged incisors can be seen to represent an adaptive response for delaying dental obsolescence (Hylander, 1975). It seems that frequent shifts have occurred in relative incisor size in response to relatively minor changes in the mode of feeding. As a consequence of such fluctuations, not much phyletic weight should be placed on the common upper central incisor enlargement of *Pongo* and some *Sivapithecus*.

Although *Sivapithecus indicus* has large upper central incisors, the same is not true of *S. sivalensis*, indicating that incisor enlargement is a variable feature of low phyletic valence in ramapithecines. In addition to GSP 15000, which preserves the upper incisors in place, 13 other hominoid upper central incisors are to be found in Siwalik collections. These come in two sizes. Seven incisors were similar to GSP 15000 and may belong to *Sivapithecus indicus*. The ratio of I^1 length/M^1 length gives an estimate of their relative size. This ratio for GSP 15000 is 1.06. The ratio of means of eight *S. indicus* I^1's to the mean M^1 length for that species is about the same, 1.07. Six smaller isolated I^1's may be allocated to *S. sivalensis*. The ratio of mean I^1 length to mean M^1 length for *S. sivalensis* is 0.86. A similar low ratio of 0.93 is obtained for a specimen described by von Koenigswald (this volume, Chapter 19). (Interestingly, I^1/M^1 ratios for two *A. afarensis* specimens, A.L. 200-1a and L.H.-3, are 0.90 and 0.91 respectively.) The range in this ratio for species of *Sivapithecus* (1.06 to 0.86) is equivalent to that between chimp and gorilla (1.11 vs. 0.89). Given this variation within the genus, its value for phylogeny reconstruction is questionable.

Hominoids from China

In 1956 and 1957, hominoid teeth were recovered from the Hsiaolungtan lignite beds of Keiyuan County, Yunnan Province in south-central China. Woo (1957) described five lower cheek teeth as a new species, *Dryopithecus keiyuanensis*. Later another series of larger teeth, including P_3-M_3 were recovered here and attributed to the same species (Woo, 1958). Associated faunas indicated a late Miocene age. Simons and Pilbeam (1965) attributed the first set of specimens to *D. (Sivapithecus) sivalensis*, and the second to *D. (Sivapithecus) indicus*, a position which we follow here, although we give *Sivapithecus* generic rank.

Expeditions from 1975 through 1980 have been carried out by the Academia Sinica (IVPP) and the Yunnan Provincial Museum in the coal fields close to Lufeng in Yunnan Province. The material reported so far, two jaws, a portion of a skull, and more than 100 isolated teeth, is about 8 million years old (Xu *et al.*, 1978; Xu and Lu, 1979). These two jaws have been described as *Ramapithecus lufengensis* (PA 580) and *Sivapithecus yunnanensis* (PA 548). The partial skull found in 1979 was assigned to *S. yunnanensis*. PA 580 fits comfortably within *Sivapithecus sivalensis* in overall size and proportions and in the details of its dental morphology. PA 580 is also of special interest since it is the only published specimen of *Sivapithecus sivalensis* to preserve some information about relative incisor size. This specimen has a relative I_2 size comparable to that of *S. indicus*, and is, in proportion, very small by modern pongid standards (see Table 1). Although I_1's are not preserved in this specimen, the space available to them indicates that they were quite small, as in *S. indicus*. In

Table 1. Lower Incisor Proportions for Ramapithecines, *Australopithecus afarensis*, and Pongids[a]

	I_1L/M_1L	I_2L/M_1L	I_1L/I_2L
Gorilla gorilla (20)	0.47 (0.36–0.61)	0.56 (0.36–0.61)	0.84 (0.68–1.04)
Pan troglodytes (20)	0.79 (0.69–0.87)	0.79 (0.69–0.87)	0.89 (0.78–1.05)
Pongo pygmaeus (20)	0.64 (0.52–0.72)	0.69 (0.50–0.79)	0.93 (0.82–1.07)
Sivapithecus sivalensis			
PA 580[b]	—	0.56	—
S. indicus			
PA 548[b]	0.44	0.48	0.92
GSP 15000	0.48	0.55	0.86
RPl 54	0.43	0.51	0.80
RPl 85	—	—	0.83
S. sp. (Macedonia)			
RPl 55	0.45	0.54	0.83
RPl 75	0.42	0.50	0.84
Gracile *Australopithecus*			
STS 24[c]	0.48	0.56	0.86
STS 52b[c]	0.45	0.54	0.83
A.L. 400-1a[d]	0.42	0.52	0.84
L.H.-2[e]	0.57	—	—
Dryopithecus fontani[f]	0.52	0.55	0.94
Proconsul africanus[g]	0.46	0.52	0.89
Proconsul nyanzae[h]	0.50	0.56	0.88

[a]Figures in parentheses are sample sizes and ranges. All data for fossils based on originals except where noted. Data for extant pongid dental dimensions from Mahler (1973).
[b]Measurement of sharp cast.
[c]Data from Robinson (1956).
[d]Data from Johanson *et al.* (1982).
[e]Data from White (1977).
[f]Data from Greenfield (1979) and Mottl (1957) ($N = 1$–2).
[g]Data from Andrews (1978b) ($N = 2$–3).
[h]Data from Andrews (1978b) ($N = 1$–2).

PA 580 lower canine roots are long, but canine crowns are small and project very little above the level of the tooth row. In this specimen, as in other *S. sivalensis*, there is some lateral expansion of the P_3, implying a more pointed upper canine than usually occurs in *Australopithecus*. The long axis of P_3 in Lufeng PA 580 is situated more transversely and the P_3 metaconid is larger (distinctly two-cusped) than typical for modern apes.

In addition to the nearly complete Lufeng mandible referable to *S. sivalensis*, the much larger species, *Sivapithecus yunnanensis*, referable to *S. indicus*, already cited, is represented by many teeth, a mandible preserving all teeth except M_3's, and the most complete skull of an ape of this age found in Eurasia (Fig. 4). This material is very similar to known parts to *S. indicus* and probably can be referred to it.

The skull, figured by Xu and Lu (1980), is crushed nearly flat, so details

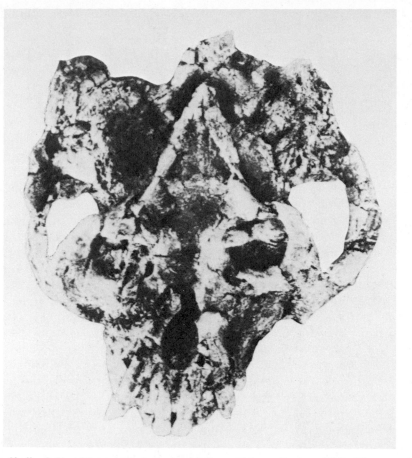

Fig. 4. Skull of *Sivapithecus indicus* recovered from Chinese lignites, 1979. [Photograph reproduced from Xu and Lu (1980).]

of its proportions are problematic. What we say about it is based on the published photograph and another kindly provided by Dr. Woo Ju Kang.

Comparable parts and proportions of the lower face of the Lufeng skull are reminiscent of those seen in a *S. indicus* specimen from Turkey and mentioned below. There are prominent canine fossae and canine jugae. There is little premaxillary prognathism. The upper central incisor is much bigger than the upper lateral one, and the zygomata are deep and flaring. The latter features, together with a very large temporal fossa (the space for the temporalis musculature), great postorbital constriction, and strongly developed temporalis muscle scars on the cranium all suggest very powerfully developed jaw musculature. Despite extreme crushing in this region, there can be little doubt that the interorbital space was comparable to that seen in species of *Pan* and much broader than typical of *Pongo*. On the other hand, there is not strong evidence for a heavy brow ridge midsaggitally, and the piriform aperture appears tall and narrow, features of resemblance to *Pongo,* although the latter feature could be the result of crushing.

Hominoids from Hungary

Kretzoi (1975) summarized recent discoveries of Miocene hominoids from Rudabánya in northeastern Hungary. The extensive fauna and flora from this site, particularly the presence of a primitive *Hipparion,* suggested to him an age just about synchronous with the invasion of *Hipparion* in Europe, or approximately 12 m.y.a. Kretzoi recognized three higher primate species at Rudabánya: a large species of *Pliopithecus,* and two larger hominoids which he called *Rudapithecus hungaricus* and *Bodvapithecus altipalatus. Bodvapithecus* is known from a nearly adult mandible, Rud 14, preserving all the lower teeth on one side or the other except M_3. The type, Rud 7, is a maxilla with P^3-M^1. These specimens resemble *Sivapithecus indicus* in size, dental proportions, and in having a deep palate and a zygomatic process rooted high above the tooth row. Interestingly, Rud 44, part of a frontal bone, preserving the dorsal portions of the orbits, appears to demonstrate that the larger Hungarian species had a broad interorbital region unlike Siwalik *Sivapithecus indicus.*

As for *Rudapithecus hungaricus,* Kretzoi argued for a close taxonomic association with *Ramapithecus,* although he gave some reasons for maintaining its generic distinctiveness. Simons (1976) also stressed the affinity of this material with *Ramapithecus,* a view later subscribed to by Andrews (1978a), Greenfield (1979), Pilbeam (1979), Szalay and Delson (1979), and Wolpoff (1980). However, after more extensive study such a view now appears unlikely. *Rudapithecus hungaricus* may even be conspecific with *Dryopithecus fontani* (as also suggested by Andrews, personal communication). Unlike any ramapithecine, this smaller Hungarian hominoid has very thin enamel, as indicated by the crater-like appearance of the dentinal exposures on the lateral cusps of the lower molars. Precisely the same thin-enameled configuration is seen also on the *Dryopithecus fontani* mandible from Lérida Province, Spain. Moreover, a mandible of *D. fontani* from St. Stefan, Austria, although somewhat older geologically, is very similar in size and dental proportions to *Rudapithecus hungaricus,* including having a small, non-projecting canine (Mottl, 1957). Furthermore, as Wolpoff (1980) notes, the lower jaws are rather gracile, not broadened transversely as is generally the case for *Sivapithecus.*

The importance of the *"Rudapithecus hungaricus"* material, then, is for what it can indicate about the anatomy of the upper jaw and teeth of *D. fontani,* and not, as previously published, about the structure of the palate and teeth of *Ramapithecus.* Upper jaws of *D. fontani* were hitherto unknown. Interestingly, this material illustrates that *D. fontani* was greatly advanced over the structure of early Miocene African *Proconsul* in having a deep palate and a high position for the zygomatic arch (see Kretzoi, 1975, Fig. 2).

Hominoids from Germany

Branco (1898) described a number of isolated upper and lower molars from the Swabian Jura which he assigned to *Dryopithecus fontani*. This material comes from a Pontian equivalent fauna with a probable age of about 10 million years. Later Schlosser (1901) set aside an M_3 as the type of *Neopithecus brancoi*. Simons and Pilbeam (1965) recognized an M^1 or M^2 from Melchingen as a possible specimen of *Ramapithecus*. Simons (1972) suggested the possibility that *"Neopithecus" brancoi* might be synonymous with *Dryopithecus laietanus* (Villalta and Crusafont, 1944) from Spain, a view adopted formally by Szalay and Delson (1979). In summary, the taxonomic arrangement which we follow is that three species may have been present in Germany at the relevant time: *Dryopithecus fontani, Dryopithecus laietanus,* and a small ramapithecine.

Among the German specimens, two examples, both from Melchingen, are regarded as possibly referable to *Sivapithecus sivalensis*. The two specimens shown in Branco's (1898) Plate I, Figs. 1, 6, and 7 and Plate II, Fig. 1 (reproduced in Fig. 5) appear to be a lower third molar and an upper first or second molar. The lower third molar has low rounded cusps with sloping sides, lacking any evidence of a cingulum. In size [13.1 mm length, 11.0 mm breadth (Branco, 1898)] it is larger than *D. fontani* and falls well within the range of variation of *S. sivalensis* from India and Pakistan. The upper molar has a similar constellation of features and in size [10.7 mm length, 11.3 mm breadth (Branco, 1898)] is very close to the means of those M^1 dimensions in *S. sivalensis*. Another specimen from Melchingen figured in Plate I, Fig. 2 by Branco (1898) is probably a deciduous molar and may also belong to this same species.

An analysis of the thickness of enamel on these specimens might shed additional light on their affinities. There is a unique opportunity to make such an assessment inasmuch as they are actually enamel caps alone without the dentinal tissues. Andrews (personal communication) has examined the material and suggests that the molar enamel may be quite thin. If so, this would favor placement of this material with *Dryopithecus*.

Hominoids from the Vienna Basin

From Neudorf an der March, CSSR, in the Vienna Basin come several isolated teeth of apes. Two of these were described by Abel (1902): a left dp^4 as *Griphopithecus suessi*, and a left M_3 as *Dryopithecus darwini*. Later, Glaessner (1931) described a left M^2 and Steininger (1967) described a right M_3 from the same locality. The specimens appear to be about 15 million years old (Steininger *et al.*, 1976). It is probable that all these specimens come from the same species. Abel (1902) divided them into two species only because he misidentified the tooth now known to be a dp^4 (see Abel, 1931).

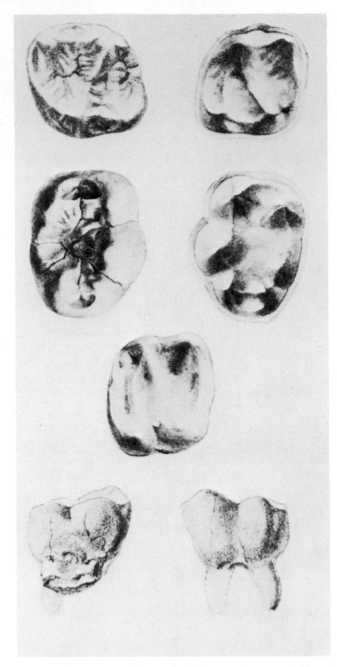

Fig. 5. Isolated possible ramapithecine teeth from the Melchingen Bohnerz, Germany. Top row, occlusal view (left) and underside view (right) of an upper molar cap in the Tubingen collection; second row, an occlusal view (left) and underside view (right) of a lower molar cap; bottom rows, three views (occlusal, lingual, and buccal) of a milk upper molar.

The M_3's from Neudorf an der March are of comparable size to those of Siwalik *S. sivalensis* and considerably larger than that tooth in *Dryopithecus fontani.* Although both are quite broad relative to length, they do not exceed the range of variation of *S. sivalensis* in this feature. Also like *Sivapithecus,* the molar cusps are low and rounded. Despite considerable occlusal wear on one specimen, the dentin is not exposed, suggesting that the enamel may be considerably thicker than in *Dryopithecus fontani.* One very distinctive morphological feature of these teeth which may justify their specific distinctiveness from *S. sivalensis* is the comparatively well developed buccal cingulum. This degree of cingulum development is not seen in any Siwalik ape.

Unfortunately, the name *Griphopithecus suessi* takes priority over *Dryopithecus darwini,* and if the Vienna basin species belongs to the same genus as the Siwalik apes, *Griphopithecus* Abel, 1902 has priority over *Sivapithecus* Pilgrim, 1910. Szalay and Delson (1979) argue that *Griphopithecus* should be suppressed due to its lack of usage except as a junior synonym of *Dryopithecus,* a view to which we subscribe. Even Abel abandoned the use of *Griphopithecus* soon after he coined it.

Hominoids from Greece

Beginning in 1973, a team of Greek and French paleontologists have uncovered an extensive series of jaws and teeth of apes from a site called the Rain Ravine about 4 km east of the village of Vathylakkos near Salonika, in Macedonian Greece. The fauna has been suggested as Vallesian, which would yield an age of 10–11 million years. Altogether, 12 specimens have two or more teeth. De Bonis and Melentis (de Bonis *et al.,* 1974; de Bonis and Melentis 1977, 1978, 1980) take the view that a single gorilla-sized species called *Ouranopithecus macedoniensis* is represented. They consider this species to be close to *Sivapithecus* and possibly near the ancestry of *Gigantopithecus.*

There is little question that the Macedonian material belongs with *Sivapithecus,* because of the following resemblances between the two. The lower molars are low-crowned, lack a buccal cingulum (as is often the case in *Sivapithecus*), and have rounded cusps with thick enamel. The mandibular body is broad and shallow (see Table 5). The upper canine is short and stout, with its greatest occlusal dimension in the buccal–lingual axis. There is a prominent canine fossa of the maxilla. The upper central incisor is much larger than the lateral and the lower incisors are small (compared to M_1 length).

That only a single species of *Sivapithecus* is represented, however, appears very doubtful. The coefficients of variation (CV) derived from dental dimensions on the original material are unusually high for a single species. We suspect that two species are present here. The statistics for the P_4 dimensions of "*Ouranopithecus*" compared with those of living apes (Table 2) illustrate our

Table 2.   Statistics of the Sample of Hominoid P_4s from Rain Ravine, Macedonia[a]

Tooth dimension	Sample size	CV	Range of CVs for five Extant Hominoids[b]
P_4 length (M–D)	7	9.23	5.2–7.3
P_4 breadth (B–L)	7	12.44	5.0–8.4

[a]Data from Kay (1982*b*).
[b]*Gorilla, Pan* (two species), *Pongo, Hylobates* (one species).

point. *Ouranopithecus* is much too variable to be a single species. If the accumulation of more material confirms our hypothesis that two species are represented, the type of *Ouranopithecus macedoniensis*, RPl 54, and the mandibles RPl 85 and RPl 197 belong to the smaller species. The unnamed larger species would include six mandibles (RPl 55, 56, 75, 76, 196, and 391), and three maxillae (RPl 199, 128, and 193).

The smaller species falls within the range of variation of *Sivapithecus indicus* in lower incisor and P_4–M_3 dimensions. The larger species is considerably larger than *S. indicus,* but smaller than *Gigantopithecus giganteus.* However, both species resemble *Sivapithecus sivalensis* in having very small canines, much smaller than known in *S. indicus.* Three specimens of the larger species have a mean lower canine-size ratio (mesial–distal lower canine length divided by mesial–distal M_1 length) of 0.95; a single specimen of the smaller species has a ratio of 0.82. Two Indian *S. sivalensis* have similar values (0.78 and 0.95), but the canines of two *S. indicus* specimens are much larger (1.15 and 1.20).

Interestingly, the relative small size of the canines of both large and small Macedonian species also resembles *Gigantopithecus giganteus.* The ratio of C̄ length/M_1 length for the sole specimen of *G. giganteus* is 0.97. On the other hand, *Gigantopithecus giganteus* has an extremely well-developed P_3 metaconid, also frequently seen in *S. indicus* and *S. sivalensis,* but this cusp is completely lacking from P_3's of the Macedonian material.

The taxonomic placement of this material *vis à vis* Siwalik and Turkish material (see below) is unclear. The smaller of the two species is *S. indicus*-sized but with comparatively smaller canines. It could be regarded either as a distinct species, *S. macedoniensis,* or as female *Sivapithecus indicus.* The larger species is distinct from material recovered elsewhere from the Siwaliks, Turkey, or Africa.

Elsewhere in Greece a lower jaw of a ramapithecine comes from Pyrgos near Tour la Reine on the outskirts of Athens. This battered specimen was found in 1944 but not described until later by von Koenigswald (1972) as *Graecopithecus freybergi.* The site is dated as Pontian, between 8 and 12 m.y.a. Very little remains of this jaw that would be diagnostic at the species level, but its great robusticity and the overall proportions of the teeth or tooth roots indicate ramapithecine affinities.

Hominoids from Turkey

Ramapithecines are known from three Turkish localities. From Paşalar in northwest Anatolia come about 100 isolated teeth of at least 20 individuals, perhaps of two species, recovered from deposits dated by faunal correlation at between 15 and 16.5 m.y.a. (Andrews and Tobien, 1977). Andrews (personal communication) now suggests a younger 13–15 m.y. age. From the Çandir Formation northeast of Ankara in Ankara Province comes a mandible, the type of *Sivapithecus alpani* of "Tortonian" age (Tekkaya, 1974), which could place it as old as 15 m.y.a. If *Hipparion* is actually present, as indicated tentatively by Tekkaya (1974), a probable age of 12 m.y.a. or less would be more likely. Finally, from the Middle Sinap series exposed northwest of Ankara in central Antolia come two specimens of middle Vallesian age, about 11 m.y.a. Mandibular fragments with teeth, the type of *Ankarapithecus meteai* (Ozansoy, 1957), come from the lower level of the Middle Sinap series, while a palate and lower face, referred to hereafter as the Mt. Sinap face, described recently by Andrews and Tekkaya (1980) is from the upper level.

The Paşalar material is not completely published, but from the preliminary note of Andrews and Tobien (1977) the following can be derived. Material of the larger species includes I^1, P^3–M^3, $\bar{C}$–M_3. The molars of this species are about *S. indicus*-sized and the upper incisors are *Proconsul major*-sized. If so, the upper central incisor of this species may be proportionately smaller than *S. indicus*. Unlike any Siwalik ape, the lower molars have a prominent buccal cingulum, reminiscent of the condition in *Proconsul* and *Sivapithecus darwini*. The molar enamel is quite thick, a feature of resemblance to all *Sivapithecus*. The smaller species from Paşalar is based on isolated upper and lower dp^4 and M^1–M^3. It resembles closely the larger species in having occasionally well-developed molar cingula, low, rounded molar cusps, and thick molar enamel.

Andrews and Tobien (1977) assigned the larger species to *Sivapithecus darwini* because they claimed that these Paşalar specimens were similar to the Vienna Basin species in M_3 size and because both have molars with a combination of low, rounded cusps, prominent cingula, and thick enamel. However, a bivariate plot of M_3 dimensions (Andrews and Tobien, 1977, Fig. 3) illustrates that these Paşalar specimens are *S. indicus*-sized, somewhat larger than *S. darwini*. The smaller specimens from Paşalar were attributed to the African taxon *Ramapithecus wickeri* on the basis of M_3 size, but with recognition that the presence of a cingulum is unlike any "*Ramapithecus.*" Actually, given that Andrews and Tobien (1977) recognize the distinctiveness of *S. darwini* from Siwalik *Sivapithecus* based on the presence of a well-developed molar cingulum in the former, it is difficult to understand how they can argue that the smaller species is conspecific with "*Ramapithecus wickeri.*" The smaller Pasalar species resembles the larger in having a well-developed cingulum, a feature never found in "*Ramapithecus wickeri*" (=*Sivapithecus africanus*). More study and better material will be needed to clarify the taxonomic placement of Paşalar material.

From the Çandir Formation in Anatolia comes a mandible with a complete corpus including the symphysis, but lacking the rami. Left $P_3–M_3$ and right $P_4–M_3$ are preserved. This specimen, the type of *Sivapithecus alpani*, was found in 1973. Based on the contained faunas, Sickenberg and Tobien (1971) placed the Çandir find in the Tortonian. From a later paper, it is apparent that an age of between 12.5 and 16 m.y.a. was intended (Steininger *et al.*, 1976). However, Tekkaya (1974) reports the possible presence of *Hipparion*, suggesting a somewhat younger age, perhaps less than 12 m.y.a. Andrews (personal communication) suggests that this *Hipparion* material is from upper Miocene deposits resting unconformably on the middle Miocene deposits from which the Çandir mandible is derived.

The morphology of the Çandir mandible is similar to that of other small *Sivapithecus*. Mandibles are very shallow and broad in cross section across the molar region, resembling all *Sivapithecus*. The mandibular symphysis is robust, with well-developed superior and inferior transverse tori. The plane of the symphysis is erect and the inferior transverse torus goes to the distal edge of P_4, both of which are resemblances to *S. sivalensis*. In contrast, the symphysis of *S. africanus* is more horizontal and extends further back (Andrews, 1971). It must be remembered, however, that both these features are highly variable in living apes. The Çandir mandible is very short anteriorly: the exterior surface of the mandible turns in sharply toward the midline just in front of the P_3. This and the small space between the P_3's suggest that the canines and incisors were small. P_3 is elongate but with a small metaconid. The molars apparently had thick enamel.

In overall size of the cheek teeth, *S. alpani* is quite a bit smaller than typically seen in Siwalik or African *Sivapithecus*. The distinctiveness of this species from either *S. africanus* or *S. sivalensis* is open to question on the present evidence. We place it with the latter tentatively.

Two specimens of hominoids come from the Middle Sinap series, northwest of Ankara, Turkey. The type was described under the name *Ankarapithecus meteai* by Ozansoy (1957, 1965). It is a lower jaw with left $P_4–M_3$ associated with the symphysis and right corpus containing the left $\bar{C}–P_3$ and right $I_2–\bar{C}$. A complete palate with all the teeth described recently by Andrews and Tekkaya (1980) comes from a higher stratigraphic level, but the two fit together and presumably belong to the same species. The published dimensions of the lower molars and the palatal teeth are all within the range of Siwalik *Sivapithecus indicus* (unfortunately, dimensions of other mandibular teeth are unavailable). The proportions of the palate and face closely resemble comparable parts of *S. indicus* specimens GSP 9977 and GSP 15000 recently recovered from Pakistan. Thus, it is likely that this Middle Sinap hominoid is *S. indicus*. The presence of *S. indicus* in Turkey further supports the possibility that the smaller Macedonian *Sivapithecus* also belongs to this species. As mentioned above, maxillary and mandibular material from Macedonia cannot be distinguished morphologically or metrically from Siwalik *S. indicus* except on the basis of canine and P_3 size. These teeth are proportionately smaller in the Macedonian sample, as the data in Table 3 show.

Table 3. Upper and Lower Canine Size[a]

		C̲ Length/M¹ length	C̄ Length/M₁ length
Gorilla gorilla[b]	Males (10)	1.40 (1.19–1.60)	1.18 (1.02–1.46)
(lowland)	Females (10)	1.04 (0.95–1.10)	0.89 (0.79–0.96)
Pan troglodytes	Males (10)	1.39 (1.17–1.66)	1.21 (1.02–1.42)
	Females (10)	1.14 (1.02–1.24)	1.03 (0.92–1.16)
Pan paniscus[c]	Males (17)	1.32 (1.17–1.66)	—
	Females (19)	1.05 (0.89–1.22)	—
Pongo pygmaeus	Males (10)	1.49 (1.39–1.58)	1.21 (1.13–1.35)
(Borneo)	Females (10)	1.13 (0.95–1.26)	0.99 (0.80–1.13)
S. sivalensis	YPM 13811	—	0.77
	AMNH 19412	—	0.96
	GSI D-1[d]	1.22	—
	GSI D-299/300[d]	1.04	—
	PA 580[d]	—	1.00
S. indicus	RPl 128	1.06	—
	RPl 54	—	0.82
	MTA 2125[e]	1.26	
	YPM 13828/		
	GSI D-189/190	—	1.14
	GSP 15000	1.36	1.15
	GSP 9977/01/05	1.25	—
	GSI D-196	1.32	—
	PA 548[d]	—	0.98
	GSP 11704 (?S. i.)	1.31	—
S. africanus	KNM-FT 46[f]	0.90	—
S. species			
(big Macedonian specimens)	RPl 55	—	0.95
	RPl 75	—	1.03
	RPl 56	—	0.88
Gigantopithecus			
giganteus	CYP 359/68[d]	—	0.98
Dryopithecus			
fontani	Rud 2[f]	—	1.04
	Rud 17[f]	—	0.92
	St. Stefan[g]	—	1.01
Gracile			
Australopithecus	STS 52a/b[h]	0.81	0.77
	TM 1512[h]	0.78	—
	A.L. 199[i]	0.86	—
	A.L. 200[i]	0.89	—
	L.H.-6 (Loc. 7)[j]	0.85	—
	A.L. 128-23[i]	—	0.79
	A.L. 198-1[i]	—	0.87
	A.L. 400-1a[i]	—	0.70
Proconsul africanus		1.24	1.11
(means)[k]			
Proconsul nyanzae		1.33	1.30
(means)[k]			
Proconsul major	UMP 62-11[k]	1.59	—

(continued)

Table 3 (*Continued*)

[a]Canine length refers to the maximum dimension in the occlusal plane. Data from personal observations of originals except as indicated.
[b]Data on extant apes except *Pan paniscus* from Mahler (1973).
[c]Data by courtesy of Brian Shea.
[d]Measured on high-resolution plaster or epoxy cast.
[e]Data from Andrews and Tekkaya (1980).
[f]Data from Greenfield (1977).
[g]Data from Mottl (1957).
[h]Data from Robinson (1956).
[i]Data from Johanson *et al.* (1982).
[j]Data from White (1977).
[k]Data from Andrews (1978b).

Comparative data for canines of living apes show that the range of variation in the fossils would not be surprising even within a single sex sample of a living hominoid. The only fact that precludes this interpretation is that all those specimens representing *Sivapithecus indicus* males come from the Siwaliks and Turkey and the females come from Greece. However, given that the samples are very small (a total of seven specimens) and if we assume males and females are preserved in equal abundance, this sort of distribution will occur at least 25% of the time.

The Mt. Sinap material is important for the details it reveals about the structure of the palate and face of *Sivapithecus indicus*. Andrews and Tekkaya (1980) offer a detailed description of the Mt. Sinap palate and face, concluding that it is within the range of modern Great Apes and unlike *Proconsul* in overall proportions. Andrews and Tekkaya (1980) also note that in addition to many generally pongid-like features, the deep and widely flaring zygomatic process, marked alveolar prognatism, short upper face, and narrow interobital distance are shared only with the orangutan among hominoids. Additionally they argue that relatively large I^1 compared with I^2 and large, squared molars also are exclusively orangutan-like features. Unfortunately, they do not provide any comparative data for living pongids in support of these points, and our study of pongid dental, palatal, and facial anatomy gives somewhat different conclusions. We will discuss these disagreements in detail when we consider the question of ramapithecine relationships. However, generally speaking, our uncertainties about Andrews and Tekkaya's conclusions come from four sources:

1. One of us (ELS) who has studied the original material believes that in some places diagenic distortion to the bones of the face has not been accounted for adequately by Andrews and Tekkaya. The biorbital dimension for the Mt. Sinap face has been given as relatively very narrow, but study of the original specimen indicates that the bones of the upper face have been bent plastically and dislocated by a crack between two halves of the face which are glued together somewhat out of orientation. These distortions can be seen clearly in Andrews and Tekkaya (1980), Figs. 1a and 1c. In both figures, the right frontal process of the maxilla actually crosses the mid-sagittal plane of the cranium; no space is left for the nasal bones. Thus, the reconstructed biorbital dimension must

have been greater than that given by Andrews and Tekkaya (1980). Inasmuch as narrowness here would be an interesting similarity to *Pongo,* the inadequacies of this specimen are regrettable.

2. There has been an unfortunate tendency to underestimate the variability in facial shape among the extant pongids with which the Mt. Sinap face was compared, particularly when overall size is accounted for. For example, Andrews and Tekkaya note that the estimated nasal height (nasopinale– nasion, with nasion reconstructed) is a minimum of 66.5 mm. This they correctly note is below the range of variation in *Gorilla,* above that for *Pan,* and between male and female *Pongo.* They go on from this to conclude that a short upper face, as indicated by this dimension, corrected for overall size (size correctors not given), is one of the features "shared only with the orangutans among living hominoids" (Andrews and Tekkaya, 1980, p. 94). Actually, a relative height of this magnitude is not unusual in gorillas as well as orangutans, as the data in Table 5 show.

In Table 5, nasal height as defined in Table 4 (both tables appear in the section, Possible Phyletic Affinities of the Ramapithecinae), has been expressed as a ratio with two dimensions given by Andrews and Tekkaya (1980), palate breadth at M^2 and M^{1-3} length. Although the relative nasal height of MTA 2125 is closest to the mean of *Pongo,* we find *Gorilla gorilla* females that have a shorter relative nasal height in a randomly selected sample of six full-adult specimens. Thus, a nasal height as short as that seen in the Mt. Sinap face appears to be common among female gorillas and is not a unique feature shared only with orangutans.

3. There is an evident failure by Andrews and Tekkaya (1980) to consider any characters with distributions that do not support the conclusion that *Sivapithecus "meteai"* is related to orangutans. For example, the mandibular corpus of the type specimen of this animal is shallow and broad in the molar region, corresponding to Siwalik *Sivapithecus,* but not to be seen even as an extreme variant in any living pongid species. Robust mandibular corpora occur elsewhere only in later hominids and appear to represent a shared derived character linking the two groups. Failure to consider this character and others mentioned below leads to an unbalanced view of the probable affinities of *Sivapithecus.*

4. We have some difficulty understanding the conclusion of Andrews and Tekkaya (1980) that the Mt. Sinap specimen exhibits a relatively great amount of premaxillary prognathism, as expressed by the dimension alveolare–nasospinale. A great deal of prognathism would indicate a resemblance to both *Pongo* and *Pan.* The value given for this dimension is 19.5 mm, which is absolutely smaller than the values provided by Pilbeam (1969b) for a large sample of gorillas and chimpanzees. Our own data for gorilla, chimpanzee and orangutan in Table 5 clearly show that the premaxillary prognathism in MTA 2125 was both relatively and absolutely small compared with living pongids. Actually MTA 2125 more closely resembles early hominids in having a short premaxilla.

In summary, published data do not support Andrews and Tekkaya's (1980) conclusion that the Mt. Sinap face is cloest morphologically to *Pongo* among living pongids. We feel that the Mt. Sinap face was pongid-like but not specifically *Pongo*-like.

Hominoids from Kenya

Only a few specimens, all from the Kenya Rift Valley, document the existence of ramapithecines in the African Miocene. These specimens come from Fort Ternan, Maboko, Majiwa, and Kaloma, of the middle Miocene, all older than about 13 million years. A single hominoid upper molar with thick enamel from the Ngorora Formation, in the Baringo Basin, Kenya, is all that is known for the 9–12 m.y.a. interval. A lower molar from the Lukeino Formation in the Baringo Basin has an estimated age of about 7 million years. Finally, a partial right mandibular fragment containing M_1 and the roots of M_2 and M_3 from Lothagam is between 5.5 and 6.0 million years old.

The middle Miocene site at Fort Ternan was first excavated in 1961 by L. S. B. Leakey. The age of the sediments from which the fossils come is bracketed between 12.5 and 14.0 million years (Bishop *et al.*, 1969). Four specimens were described by Leakey as *Kenyapithecus wickeri* (Leakey, 1962). These are a partial left maxillary fragment containing P^3 roots and P^4–M^2 with an associated maxillary canine, a partial right maxilla with M^{1-2} and the roots of M^3 and an isolated right M^2. Additional fragments described in recent years include a left mandibular specimen with two premolars (probably belonging to the same individual as the left maxilla), a right mandibular fragment with P_4–M_1, an isolated right lower canine, and an unerupted left upper canine still in its crypt. Andrews and Walker (1976) provide detailed descriptions of all these fragments.

One species of ramapithecine among other apes is represented at Fort Ternan. In most details of the known morphology, this animal closely resembles the Indian *Sivapithecus* (including "*Ramapithecus*"). Maxillary fragments indicate a highly arched palate. The zygomatic process of the maxilla swings laterally from the body at the level of M^1, and a canine fossa is well developed. The mandible is shallow and broad. The mandibular symphysis has a well-developed inferior transverse torus which reaches back to the front of M_1. Molars are low-crowned with thick enamel; the lower molars lack a buccal cingulum. P_3 is a narrow, compressed tooth with its longest axis set at an acute angle to the mesial-distal plane. A small P_3 metaconid is present. A well-developed P_3 hone is present for the upper canine. Upper and lower canines are small and the lower incisors must have been quite small also, as judged by the preserved portions of their root sockets. The reconstructed shape of the lower dental arcade, as first proposed by Walker and Andrews (1973), has been demonstrated to be wrong by Simons and Pilbeam (1978), who point to several unlikely features, such as tooth rows which actually converge posteriorly to reach their narrowest point at P_4 before diverging again posteriorly. Clearly the tooth rows are set too closely together in this reconstruction.

Independent confirmation of this comes from a recently described specimen, KNM-MJ 5, of roughly similar age from the Maboko Formation (Pickford, 1982; see below). In this latter specimen, which preserves both corpora and the symphyseal region, the ratio of M_{1-3} length to inter-M_2 breadth is about 1.0, as judged from photographs. Walker and Andrews' (1973) reconstruction gives a much higher ratio, greater than 1.4, and therefore leaves too narrow a space for the tongue.

Based on their mandibular reconstruction and the available maxillary materials, Walker and Andrews (1973) also provide a reconstruction of palate shape based on the shape of the mandibular arch. (Neither of the maxillary fragments reaches the midline.) We do not accept this reconstruction, for two reasons. First, it is based on a mandibular corpus shape which is not plausible for reasons given above and by Simons and Pilbeam (1978). Second, even if the mandibular shape were correctly estimated, it is well known that palatal arch form cannot be reconstituted accurately from mandibular arch shape alone. For example, M^2's of cebids are anywhere from 6% to 25% further apart than are M_2's, and the ratio of the width between the lower and upper dental arches can vary widely in a single animal at different points along the tooth row (Zingeser, 1976).

In overall size of the postcanine dentition, there is a remarkable similarity between the Fort Ternan material and the Siwalik species *Sivapithecus sivalensis* (here including material previously assigned to *Ramapithecus punjabicus*), although specimens of the latter are up to 5 million years younger in age. However, the maxillary canine assigned to Fort Ternan *Sivapithecus*, although quite small, is buccal–lingually compressed, as is typical for most apes, but unlike Indian *Sivapithecus*, where the long axis of the canine in the occlusal plane is in a buccal–lingual orientation.

Other middle Miocene ramapithecine material comes from unknown localities and from Maboko Island and nearby Majiwa and Kaloma. The known localities are all from the Maboko Formation, capped by a phonolite dated at 12.3 ± 0.2 and 13.1 ± 0.6 m.y.a. (Bishop *et al.*, 1969).

Le Gros Clark and Leakey(1951) described *Sivapithecus africanus* based on a maxilla with three teeth and two other isolated teeth. The maxilla, first thought to have come from Rusinga Island (early Miocene), seems not to have come from there but rather to have come from Maboko Island (Andrews and Mollison, 1979). One isolated tooth is from an unknown locality and another comes from Maboko Island. Recently, Pickford (1982) has described six new specimens from Majiwa and Kaloma: a right M^{2-3} (KNM-MJ 1), a mandible with left P_{3-4}, and M_{2-3} and right I_{1-2} and P_4-M_3 (KNM-MJ 5), as well as isolated right P_4, right C, right M^1, and right I_1, the latter perhaps belonging with the mandible. He also refers a maxillary canine from Maboko Island, KNM-MB 70, to the same species.

All the specimens are of comparable size, consistent with belonging to the same species. The palate of the type specimen of *Sivapithecus africanus* is slightly less arched than that of the Fort Ternan species, and the zygomatic process of the maxilla arises closer to the alveolar border (Pilbeam, 1969*b*).

Both of these are primitive characters reminiscent of early Miocene apes and unlike the deeply arched palate and high alveolar process typical of *Sivapithecus*. The preserved portions of the upper canine socket suggest a large tooth (Simons and Pilbeam, 1965). The premolars and molars of the type are simple in construction and resemble *Sivapithecus* in lacking a lingual cingulum.

Specimens which come definitely from the Maboko Formation resemble dentally both the Fort Ternan material and the type of *Sivapithecus africanus*. The molars were low-crowned with thick enamel and without prominent cingula. P_3 was apparently anteroposteriorly elongate and lacked a metaconid (Pickford, 1982). Unassociated maxillary canines are small, as are the mandibular canines of the only known lower jaws, as judged by their preserved roots. The lower incisors are also small in proportion to the molars, comparable in size to those of *Proconsul africanus*.

The Kaloma mandible resembles other *Sivapithecus* and differs from *Proconsul* in having a well-developed inferior transverse torus. The inferior border of the mandibular symphysis extends posteriorly in a fashion similar to that seen in Fort Ternan *Sivapithecus*. The mandibular corpora diverge only slightly posteriorly, and the tooth rows were apparently more nearly parallel than in *Sivapithecus sivalensis* (as judged from GSP 4857) (Pickford, 1982). The jaws were shallow and broad under the molars, like other *Sivapithecus*.

The generic distinction of these Fort Ternan and Maboko Formation species from *Sivapithecus* cannot be sustained on present evidence. Furthermore, all the specimens are of similar size and could belong to the same species. The age disparity of 5–7 million years between this sample and the bulk of the Siwalik material, as well as the geographic separation, and the slightly more primitive nature of the upper canine structure, all suggest a species different from *S. sivalensis*. For this, *Sivapithecus africanus* is the name which has priority.

A left M^2 from the Ngorora Formation which has an age of about 10 million years, and a lower left molar from Lukeino which is about 7 million years old are informative only in that they confirm the presence of a thick-enameled African species from the time between the middle Miocene thick-enameled *Sivapithecus* of Fort Ternan and the Maboko Formation and the earliest *Australopithecus*.

Revision of the Ramapithecinae

To summarize all the available evidence, we list below the taxa of Miocene ramapithecines we recognize and the names of several taxa in use over the past ten years which we believe to be invalid junior synonyms:

Genus   *Sivapithecus* Pilgrim, 1910.
        Includes: *Ramapithecus, Ouranopithecus, Graecopithecus, Kenyapithecus, Bodvapithecus, Ankarapithecus*

Sivapithecus indicus Pilgrim, 1910.
>   Includes: *S. lewisi, S. yunnanensis, Bodvapithecus altipalatus, Ouranopithecus macedoniensis* (part?), *Graecopithecus freybergi, Ankarapithecus meteai*
>   Distribution: China, India, Pakistan, Turkey, Greece, Hungary

Sivapithecus sivalensis (Lydekker, 1879).
>   Includes: *Ramapithecus lufengensis, Sivapithecus alpani,* not *Rudapithecus hungaricus*
>   Distribution: China, India, Pakistan, Turkey, not Hungary

Sivapithecus africanus (Le Gros Clark and Leakey, 1951).
>   Includes: *Kenyapithecus* (= *Ramapithecus*) *wickeri*
>   Distribution: Kenya

Sivapithecus darwini (Abel, 1902)
>   Distribution: Czechoslovakia, ?Turkey

Sivapithecus simonsi (Kay, 1982a)
>   Distribution: Pakistan, India

Sivapithecus, possible unnamed species A
>   Distribution: Pakistan, Turkey

Sivapithecus, unnamed species B
>   Includes: part of *Ouranopithecus macedoniensis* material
>   Distribution: Greece

?Sivapithecus sp. indet.
>   Distribution: Germany

Genus   *Gigantopithecus* von Koenigswald 1935
>   Includes: *Indopithecus*

Gigantopithecus giganteus (Pilgrim, 1915)
>   Includes: *Gigantopithecus bilaspurensis*
>   Distribution: India, Pakistan

Possible Phyletic Affinities of the Ramapithecinae

How are the ramapithecines related to the living Great Apes and hominids? It is generally accepted that there are three natural groups of extant large hominoids: hominids, including *Homo* and its Plio-Pleistocene relative *Australopithecus;* African Great Apes (*Gorilla* and *Pan*); and the Asian orangutan, *Pongo.* It is well known how these three living groups are related to one another. Albumin and transferrin immunology, immunodiffusion, DNA annealing, and amino acid analyses all indicate and confirm that chimpanzee, gorilla, and human share a substantially more recent common ancestor than any do with orangutans. If so, then ramapithecines can be related to these groups in five possible ways:

1. Ramapithecines may be in or near the line of ancestry leading to humans, the African apes, and *Pongo,* with all these forms passing through a

ramapithecine-like stage in their evolution. By this scheme, ramapithecines are the cladistic sister group of all later forms, or different species with ramapithecine anatomy gave rise to Asian and African apes and humans.

2. Ramapithecines *as a group* are specifically related to orangutans alone and not related to either hominids or African Great Apes.

3. Ramapithecines *as a group* are specifically related to both African Great Apes and Humans, but not orangutans.

4. Ramapithecines are related to living African Great Apes alone.

5. Ramapithecines are related to hominids alone.

Proposition number one has gained popularity recently, particularly among those who favor a recent time of divergence between the modern groups on the basis of molecular clock evidence. The most thorough treatment of the morphological evidence for this proposition is by Greenfield (1980) (see also Wolpoff, this volume, Chapter 25). Other proponents of this view include Pilbeam *et al.* (1977*b*) and Zihlman *et al.* (1978). The second view for ramapithecine relationships has been championed recently by Andrews (Andrews and Tekkaya, 1980, Andrews, this volume, Chapter 17; Ward and Pilbeam, this volume, Chapter 8). They point to certain possible shared derived features of the palate, face, and dentition linking *Sivapithecus* and *Pongo*. We know of no current proponents of views 3 or 4. Proposition 5, that ramapithecines are specially related to *Australopithecus* and *Homo* alone and not to the extant Great Apes of Africa and Asia, is the view that has been advocated by Simons and many others.

Many of the claims supporting relationships among living apes, and practically all of those supporting similar claims for extinct taxa, are based on cranial and dental characters. This will continue to be the case as long as fossil postcranial remains are so scarce. To assess these claims, we measured a series of skulls of male and female extant Great Apes closely approximating the cranial dimensions provided in various publications on early Miocene to early Pliocene hominoids. The dimensions and their size correctors are described in Table 4. Table 5 summarizes the data for living and fossil species. We will consider, in turn, the evidence for three proposals: that *Sivapithecus* is (1) specially related to *Pongo,* (2) in or near the ancestry both of modern Great Apes and humans, or (3) in or near the ancestry of humans alone.

Evidence for a Sivapithecus–Orangutan Relationship

The clearest case for a link between *Sivapithecus* and *Pongo* has been made by Andrews (Andrews and Tekkaya, 1980; Andrews, this volume, Chapter 17). The principal evidence claimed in support of this case is found in the structure of the face of *Sivapithecus* "*meteai,*" MTA 2125, from the Middle Sinap series in Turkey of middle Vallesian age. This specimen, which we have assigned here to *Sivapithecus indicus*, was described by Andrews and Tekkaya (1980) and further reported on by McHenry *et al.* (1980). According

Table 4. Description of Characters of the Jaws and Face Mentioned in the Text and Measured in Table 5

1. Zygomatic depth: Depth of the root of the zygomatic process from the anteromedial-most extent of the scar of the superficial masseter muscle to the closest point on the ventral orbital rim; this is expressed as $100 \times$ zygomatic depth divided by external palate breadth at M^2.
2. Zygomatic height: Point of anteromedial-most extent of the superficial masseter origin to the cemento-enamel junction of the closest maxillary tooth, usually M^1 or M^2; this is expressed as $100 \times$ zygomatic height divided by M^{1-3}
3. Position of zygomatic root: The position of the anteromedial-most origin of superficial masseter projected perpendicularly onto the maxillary tooth row with the cemento-enamel junctions of P^3 and M^2 set at horizontal
4. Zygomatic flare: The maximum bizygomatic width expressed as $100 \times$ bizygomatic breadth divided by external palate breadth at M^2
5. Nasal aperture shape: $100 \times$ nasal aperture height (nasospinale to rhinion) divided by maximum nasal aperture breadth
6. Nasal aperture position: The position of rhinion *vis à vis* the ventral oribital rim, viewed from the P^3–M^2 horizontal
7. Premaxillary prognathism: $100 \times$ nasospinale–alveolare divided by external palate breadth at M^2
8. Nasal height: $100 \times$ nasospinale–nasion divided by external palate breadth at M^2
9. Interorbital distance: $100 \times$ distance between the left and right maxillofrontale (the points of intersection of the anterior lacrimal crest, or the crest prolonged, with the frontomaxillary suture) divided by external palate breadth at M^2
10. Palate breadth: $100 \times$ distance between lingual canine margins divided by distance between the most lingual points on left and right M^2's
11. Mandibular corpus shape: $100 \times$ jaw depth at M_{1-2} divided by jaw breadth at that point taken in the buccal–lingual axis
12. Mandibular arcade shape: $100 \times$ breadth between lingual aspects of the left and right lower canines divided by breadth between lingual aspects of the left and right P_4's

to Andrews and Tekkaya, *Sivapithecus "meteai"* shares with the orangutan the following uniquely shared derived similarities:

(1) a deep zygomatic process, by which is meant that the facial buttress of the zygomatic arch is broad;
(2) flaring zygomatic arches, as expressed by a relatively large bizygomatic dimension;
(3) high alveolar or premaxillary prognathism;
(4) a short upper face (relatively low nasal height);
(5) a narrow interorbital distance;
(6) large, squared upper molars;
(7) a large I^1;
(8) I^1 relatively much larger than I^2.

To these, Andrews (this volume, Chapter 17) adds a ninth character of *Sivapithecus* claimed to be shared uniquely with orangutans, i.e., a nasal aperture that is short and broad. The claim for orangutan affinities of *Sivapithecus* cannot be sustained on the basis of these nine characters or definitively on other characters of which we are aware.

Table 5. Cranial and Mandibular Proportions of Living and Fossil Hominoids[a]

Taxon	Zygomatic depth	Zygomatic height	Zygomatic flare	Nasal aperture shape	Nasal aperture position	Palate breadth	Mandible corpus shape	Mandibular arcade shape	Premaxillary prognathism	Nasal height	Interorbital distance
Pongo pygmaeus (N = 20)	59 (48–71)	80 (65–98)	215 (185–245)	137 (115–171)	Male: 50% below orbit. Female: 100% near orbit	100 (87–109)	204 (165–235)	77 (61–93)	53 (39–65)	Males 104 (96–112) Females 95 (88–105)	18 (13–26)
Gorilla gorilla (N = 13)	56 (45–66)	99 (77–130)	227 (198–248)	100 (77–134)	Well below orbit	109 (90–133)	179 (150–204)	81 (69–102)	46 (38–54)	Males 134 (118–143) Females 118 (105–126)	26 (17–32)
Pan troglodytes (N = 12)	45 (39–54)	92 (80–102)	209 (195–230)	107 (86–120)	At or near orbit	104 (96–116)	183 (165–199)	85 (77–93)	51 (44–65)	Males 93 (87–99) Females 87 (77–97)	28 (22–36)
Early gracile *Australopithecus*	41[b]	—	200[b]	96[b]	At or near orbit[b]	74[n]	149 (121–179)[p]	53, 71[v]	—	70[b]	26[b]
Gigantopithecus giganteus	—	—	—	—	—	—	156[q]	70[q]	—	—	—
Sivapithecus indicus	64[c]	60[d]	251[c]	106[c]	Well below orbit	e.103[o]	154[r]	82[w]	34[z]	107[c]	?219[aa]
Sivapithecus sp.	—	—	—	105[h]	—	99[h]	167[h]	56[x]	—	—	—
Sivapithecus sivalensis	—	—	—	—	—	—	152[s]	77[x]	—	—	Broad
Procsul africanus	—	35[e]	—	164[i]	At or near orbit[l]	80[i]	176[i]	68[e]	—	—	Broad
Procsul nyanzae	—	45[f]	—	e.155[j]	Well below orbit[u]	102[j]	189[u]	—	—	—	Broad
Procsul major	—	38[g]	—	e.119[k]	Well below orbit[m]	109[k]	186[f]	76[y]	26[m]	157[m]	Broad

[a] See Table 4 for an explanation of the indices.
[b] Type of "*Plesianthropus transvaalensis*", measurements from cast.
[c] MTA 2125 (Turkey), data from Andrews and Tekkaya (1980).
[d] GSP 9977 (Pakistan).
[e] N = 2, data from Andrews (1978b).
[f] N = 3, data from Andrews (1978b).
[g] UMP 62-11, data from Andrews (1978b).
[h] RPI 128 (Macedonia), data from cast.
[i] BMNH M 32363, data from Whybrow and Andrews (1978).
[j] BMNH M 16647, data from Whybrow and Andrews (1978).
[k] UMP 62-11, data from Whybrow and Andrews (1978).
[l] BMNH M 32363, personal observations.
[m] UMP 62-11, data from Pilbeam (1969b).
[n] From casts of "*P. transvaalensis*" and A.L. 200-1a.

[o] Estimate from cast of MTA 2125.
[p] *A. afarensis*, N = 6, data from White and Johanson (1982).
[q] CYP 359/68.
[r] Siwalik sample, N = 7.
[s] Siwalik sample, N = 6.
[t] N = 7, data from Andrews (1978b).
[u] N = 4, data from Andrews (1978b).
[v] Data for A.L. 288-1, and A.L. 400-1a, from White and Johanson (1982).
[w] GSP 9564.
[x] GSP 4857/4622 (60), AMNH 19411 (81).
[y] KNM-SO 404, data from Andrews (1978b).
[z] Mean of two specimens: GSP 9977, MTA 2125; data for the latter from Andrews (1978b).
[aa] Interorbital breadth estimated from MTA 2125 may be incorrect; see discussion in text.

The claim that *Sivapithecus* has premaxillary prognathism to the same degree as *Pongo* cannot be supported by the available evidence. Even if it were supported, this claim ignores the fact that *Pongo* has evolved a comparable degree of prognathism in parallel with *Pan*. The values for the index of premaxillary prognathism described in Table 4 and given in Table 5 show, based on Andrews and Tekkaya's (1980) published measurements and the same dimensions on Macedonian specimens, that *Sivapithecus indicus* has a very low degree of prognathism, comparable to or lower than *Gorilla* or gracile *Australopithecus* and much smaller than that seen in either *Pongo* or *Pan*. As demonstrated by the very low values for this index in *Proconsul* [based on data from Andrews (1978b)] and slightly larger values in *Gorilla, Australopithecus,* and *Sivapithecus,* this probably represents the primitive condition for apes. *Pongo* and *Pan* apparently have increased prognathism in parallel.

There is one character of *Sivapithecus* that is claimed by Andrews (this volume, Chapter 17) to exhibit shared derived similarity with orangutans but is actually not shared uniquely between them. Contrary to Andrews' claim, the short, broad nasal aperture typical of *Sivapithecus* is not normally seen in *Pongo.* Actually, this short, broad nasal opening could be construed to be a shared derived feature linking *Sivapithecus* with *Australopithecus* and the African Great Apes. In contrast, *Pongo* actually has a tall or narrow nasal aperture, similar to the probably primitive conditions of *Proconsul* (see Table 5 for data).

Another supposed shared derived character claimed to weld a special *Pongo–Sivapithecus* link is more plausibly a primitive retention and not indicative of such a special relationship. Andrews claims that *Sivapithecus indicus* has a relatively large upper central incisor, but I^1 size, expressed as a function of molar size, does not differ greatly from the range of values in *Proconsul,* primitive *Australopithecus* or *Gorilla.* The average value for *S. indicus* is much smaller than the average for either *Pongo* or *Pan.* Actually, this distribution indicates also that I^1 size increase has occurred in parallel in *Pongo* and *Pan* and may be linked to the relatively high degree of premaxillary prognathism observed in both the latter taxa. What little preliminary evidence we have for *Sivapithecus sivalensis* suggests that it may have had relatively somewhat smaller upper central incisors than did *S. indicus.* Thus, neither *Sivapithecus* species for which fossil evidence is available exhibits an I^1 that is enlarged relative to cheek teeth.

As Andrews and Tekkaya (1980) note, the ratio of I^1/I^2 length is quite high in *Sivapithecus indicus,* as in *Pongo,* compared to the means for other living or fossil apes (Table 6). Superficially, this appears to be a shared derived resemblance between the two. However, *Pongo* has such a high ratio because of a relatively very large I^1 and average-sized I^2, both compared to M^1. In *Sivapithecus indicus,* on the other hand, I^1 is similar in relative size to most living and fossil apes, but I^2 is very small. Thus, a similar I^1/I^2 ratio for *Sivapithecus indicus* and *Pongo pygmaeus* is produced in a different way and seems an obvious example of parallel evolution. It should also be mentioned that South African gracile *Australopithecus* have I^1/I^2 ratios similar to those of *Pongo* but again, as with *Sivapithecus* (and in contrast to *Pongo*) a high value for this ratio is accounted for by the comparatively small I^2. *Australopithecus africanus* have very small I^2s, comparable in relative size to those of *Gorilla* (Table 6). Thus, a high I^1/I^2 ratio, if it proves to be found generally among *Sivapithecus*, is a possible shared derived similarity with *Australopithecus*.

Four possibly derived characters of the facial anatomy of *Sivapithecus indicus* bear some similarity to *Pongo* but among all extant apes and *Australopithecus* are not exclusively shared with the Asian Great Ape (Table 5). The deep zygomatic seen in *Sivapithecus* and *Pongo* is found also in *Gorilla*. Flaring zygomatic arches, seen in *Sivapithecus* and *Pongo*, flare even more widely in *Gorilla*. A short upper face characterizes both *Sivapithecus* and *Pongo* but is present in *Pan* as well. Large, squared upper molars are seen not only in *Sivapithecus* and *Pongo*, but *Gorilla* and *Australopithecus* as well. We are not impressed by the argument for a *Sivapithecus–Pongo* link bolstered by these characters, given the demonstrated high degree of parallelism among living pongids.

The last linking character proposed by Andrews and Tekkaya (1980)

Table 6.   Upper Incisor Proportions Based on Mesial–Distal Dimensions[a]

	I^1/M^1	I^2/M^1	I^1/I^2
Gorilla gorilla (20)	0.88 (0.69–1.00)	0.63 (0.50–0.74)	1.40 (1.20–1.59)
Pan troglodytes (20)	1.11 (0.83–1.26)	0.85 (0.70–0.98)	1.31 (1.15–1.64)
Pongo pygmaeus (20)	1.10 (0.88–1.22)	0.72 (0.59–0.85)	1.52 (1.26–1.73)
Sivapithecus indicus			
RPl 128	0.88	0.56	1.56
MTA 2125[b]	0.98	0.47	2.11
GSP 15000	1.06	0.56	1.88
Dryopithecus fontani			
Rud 15[c]	1.02	0.61	1.66
Rud 12[c]	0.79	—	—
Gracile *Australopithecus*			
STS 52a[d]	0.77	0.59	1.31
STS 69/70[e]	0.76	0.51	1.50
TM 1512[d]	—	0.48	—
A.L. 200-1a[f]	0.89	0.60	1.49
Mean for South African			
graciles[g]	—	—	1.57
Proconsul africanus ($N = 2$)[h]	0.95	0.65	1.45
Proconsul nyanzae ($N = 1-2$)[h]	1.01	0.65	1.45
Proconsul major			
UMP 62-11[h]	0.92	0.73	1.26

[a]All data for fossils based on originals except as noted. All data for extant pongids from Mahler (1973). Kay (1982a) mistakenly listed those same ratios as being based on buccal–lingual dimensions.
[b]Data from Andrews and Tekkaya (1980).
[c]Data from Greenfield (1977).
[d]Data from Robinson (1956).
[e]Data from Grine (1981).
[f]Data from Johanson *et al.* (1982).
[g]Data from Tobias (1980).
[h]Data from Andrews (1978b).

between *Sivapithecus* and *Pongo* is the narrow interorbital space typical of *Pongo* and supposed to exist in the Turkish *Sivapithecus* face. We have outlined previously in this chapter why we think that *postmortem* distortion in the Mt. Sinap face is responsible for giving the impression that this animal had an unusually narrow interorbital dimension. It should be noted further that the interorbital region is much broader in two other specimens of *S. indicus* discussed previously (see sections on Chinese hominoids and Hungarian hominoids) although it is narrow and orangutan-like in a newly described specimen from Pakistan (Pilbeam and Smith, 1981: Pilbeam, 1982). In any event, the known ranges of variation for this width indicated in Table 5 weaken conclusions based on this feature in one or a few specimens. Furthermore, many specimens of African apes have relative interorbital breadths which fall below the mean for *Pongo* in this dimension. Kay (1982*a*) figures an adult specimen of *Pan paniscus* with an extremely narrow interorbital space.

Pilbeam (1982; Pilbeam and Smith, 1981) has recently described new cranial fragments of *Sivapithecus indicus* from Pakistan (catalogued as GSP 15000). He is very cautious in assessing possible similarities between this new specimen and *Pongo*, but Andrews and his coworkers have been less so. In addition to repeating some of the resemblances mentioned above between this new facial fragment and *Pongo*, several new characters are listed by Andrews (Andrews, 1982; Andrews and Cronin, 1982) as uniquely shared-derived features of *Sivapithecus* and *Pongo* therefore indicating a close relationship between them. But on careful examination and comparison most of these characters do not prove to be unique to just these two taxa; they are found as well in at least one other hominoid species:

(1) The incisive foramen is small in *Sivapithecus* and *Pongo*, but it is also small in *Pan*, some *Australopithecus*, and *Homo sapiens* (Kay, personal observation; Ward *et al.*, 1983).
(2) The floor of the nasal cavity is smooth and unstepped in *Sivapithecus* and *Pongo*, but it also has this configuration in some *Australopithecus*, in *Homo sapiens* and occasionally in *Pan* (Kay, personal observation; Ward *et al.*, 1983).
(3) *Sivapithecus* and *Pongo* have a relatively large zygomatic foramen (or foramina) set above the lower rim of the orbit, but *Pan* frequently exhibits equally large, similarly positioned foramina (Kay, personal observation, see USNM 176240).
(4) The palatine foramina of *Sivapithecus* and *Pongo* are said to be narrow and slitlike, but a similar configuration is seen frequently in other Great Apes and *Australopithecus*.
(5) The zygomatic bone is flattened, flaring and anteriorly facing in *Sivapithecus* and *Pongo*, but this is equally so in *Australopithecus*.

Another character listed by Andrews, the poor development of the supra-orbital torus of *Sivapithecus* (at least in GSP 15000) and *Pongo*, is more likely a primitive character than a shared-derived one because *Proconsul africanus* and *Pliopithecus* have a similar condition.

On the other hand, as noted by Andrews, assuming that distortion from diagenesis is not a problem, GSP 15000 has tall, narrow orbits, a striking unique resemblance to *Pongo*.

To summarize, there are a number of probably derived resemblances shared between *Pongo* and some *Sivapithecus* but all except orbit shape are seen as well among other living apes reducing their value as strong indicators of a close phyletic link of *Sivapithecus* to *Pongo*. Furthermore, as will be reviewed below numerous other shared-derived features link *Sivapithecus* with *Australopithecus* making the *Sivapithecus-Pongo* relationship appear even less likely.

Evidence that Sivapithecus May Be a Common Ancestor of Extant Great Apes and Homo

Thus far, Greenfield (1980) has been the only author to have seriously considered the morphological case that *Sivapithecus* might be ancestral to, or the sister group of, all later Great Apes and hominids. Wolpoff (this volume, Chapter 25) favors this argument, but has added no additional information to bolster it. His analysis reestablishes the position of Delson and Andrews (1975), among others, that *Sivapithecus* has advanced to an extant pongid "grade" of organization. This information is sufficient to establish that *Sivapithecus* should be united with Great Apes and hominids as a clade, but for *Sivapithecus* to be a sister group of all later taxa in this clade, it must be shown also (1) that hominids and extant Great Apes share a set of derived features for which *Sivapithecus* remains primitive, and (2) that *Sivapithecus* shares no derived features with any individual member of the hominid–extant ape clade. Greenfield has failed to satisfy either of these criteria. As will be shown, a great number of unique shared derived features actually link *Sivapithecus* with early hominids.

Is Sivapithecus a Hominid?

A large number of anatomical characteristics have been advanced to support a relationship between some or all *Sivapithecus* (as presently understood to include "*Ramapithecus*") and *Australopithecus*. These include:

(1) parabolic (not U-shaped) dental arcade
(2) an arched palate
(3) a canine fossa
(4) small upper and lower incisors relative to cheek tooth size
(5) small upper and lower canines relative to cheek tooth size
(6) a low degree of premaxillary prognathism
(7) zygomatic set well forward
(8) zygomatic deep and flaring
(9) vertically oriented canine roots

(10) long axis of the upper canine cross-section set more or less buccolingually
(11) cheek teeth with low relief
(12) thick molar enamel
(13) extreme molar wear gradient, with much interstitial wear
(14) lower molar buccal cingula poorly developed or absent
(15) mandibular corpora shallow, broad
(16) P_3 with metaconid
(17) P_3 occlusal long axis set relatively transverse to long axis of tooth row
(18) lower dental arcade V-shaped
(19) widely spread molar cusps with correspondingly wide foveae.

A review of these and other characters of *Sivapithecus* and other living and extinct apes (Tables 5 and 6) shows that many are actually retained primitive features from a *Proconsul*-like ancestor and many others are features of the last common ancestor of living Great Apes and hominids. But when these are removed, a number of characters remain which can only be seen as unique shared derived features linking *Sivapithecus* with *Australopithecus* and confirming that the two form a separate clade from other hominoids.

Sivapithecus shares with *Proconsul* a number of primitive features which also survive in *Australopithecus*. (1) As noted by one of the authors (ELS), the lower tooth rows diverge posteriorly in a "V-shaped" configuration in *Sivapithecus, Proconsul,* and other primitive apes (Table 5), whereas in extant Great Apes the tooth rows are more nearly parallel. This correlates well with (2) lower incisor size. *Sivapithecus* and *Proconsul,* as well as *Dryopithecus fontani* and practically all other Miocene apes, have relatively very small lower incisors, much smaller than seen in any living ape, even *Gorilla,* and similar to *Australopithecus* (Table 6). Perhaps V-shaped tooth rows simply reflect the presence of small incisors. (3) I^1 is relatively about the same size as that of *Proconsul* species and *Australopithecus* (see above discussion *vis à vis* I^1 size in *Pongo*). Apparently functionally correlated with this is (4) the low degree of premaxillary prognathism in these forms. *Pongo* and *Pan,* with large upper incisors, have much more premaxillary prognathism (Table 6). (5) The proportions of the palate of *Sivapithecus indicus* are reminiscent of *Proconsul major* and modern pongids. Only in *Australopithecus* does the palate narrow anteriorly (e.g., the intercanine dimension of the palate is proportionately reduced; Table 5). (6) The position of the root of the zygomatic arches above M^1 or M^2 in *Sivapithecus indicus* and *S. sivalensis* is similar to that in *Proconsul* and early *Australopithecus*. (In advanced *Australopithecus,* like *A. boisei,* the zygomatics are much further forward.)

In several ways *Sivapithecus* is advanced over *Proconsul* and has reached the pongid "grade" of organization cranially but has not advanced further in the direction of *Australopithecus:* (1) The root of the zygomatic arch is raised well above the level of the occlusal plane (Table 5), and (2) the palate is deeply arched (Greenfield, 1979); these resemblances it shares with the Great Apes and *Australopithecus*. In contrast, *Proconsul* species have very low zygomata and shallow palates. (3) The zygomata of *Sivapithecus* are very deep (Table 5), resembling *Gorilla* and *Pongo,* and probably primitive for the Great Ape clade. (4) The cheek teeth of *Sivapithecus* and *Dryopithecus* are greatly advanced over those of *Proconsul* and resemble extant pongids and hominids by having greatly reduced molar cingula and relatively limited occlusal relief.

Overlying this impressive number of primitive and ape-like characteristics are a series of uniquely hominid shared derived characters which indicate and uphold the long-held majority opinion of paleontologists that *Sivapithecus* species are phyletically closer to Plio-Pleistocene hominids and *Homo* than to Asian or African Great Apes. This list of characters relate mainly to the teeth and lower jaws.

Sivapithecus species have very thick jaws (Table 5). In fact, the *mean* values for our index of jaw depth in *Sivapithecus indicus* and *S. sivalensis* are below the total range of values of this index for all but one of the 45 specimens of extant Great Apes we examined and all other published Oligocene through Miocene ape material. Conversely, the shallow-jawed condition is typical of the earliest *Australopithecus* and its Plio-Pleistocene descendants.

A second unique shared derived feature linking *Sivapithecus* species with *Australopithecus* is the great enamel thickness on the cheek teeth. Kay (1981) has shown that the enamel on the hypoconid of *Sivapithecus indicus* and *S. sivalensis* averages 1.81 times thicker than that seen in *Pongo,* 2.50 times thicker than in *Pan,* and 3.16 times thicker than in *Gorilla.* A comparable degree of enamel thickness is seen in Plio-Pleistocene hominids and *Homo* alone among other living and fossil hominoids and cercopithecoids [Gantt's (1977) data agree with this]. Although orangutan molar enamel is somewhat thicker than typical for hominoids, a comparable degree of thickening is present in several cercopithecoids, including some *Macaca sylvanus,* some *Cercopithecus mona denti,* and average *Cercocebus* species. As a correlate of the very thick enamel of *Sivapithecus,* the lower molar occlusal surfaces wear practically flat before the enamel becomes perforated.

In Turkish, Chinese, and Siwalik *Sivapithecus indicus* and *S. sivalensis,* as well as in *Gigantopithecus,* the P_3 metaconid tends to be enlarged. The presence of an enlarged P_3 metaconid is not a unique occurrence among extant Great Apes. We observed it as a very small but discrete entity in 40% of cases in *Pan* and *Pongo,* though not in *Gorilla.* However, in no instance have we observed such well-developed P_3 metaconids in living apes as are seen in some *Sivapithecus.* This sort of P_3 metaconid development is seen elsewhere alone among Pliocene–Recent hominids. It is interesting to observe in *Sivapithecus* such well-developed P_3 metaconids on teeth which still maintain a large contact facet for the upper canine, indicating that the cutting function of the canine is still maintained.

So far, the similarities of *Sivapithecus* and *Australopithecus* have been restricted to the cheek teeth and jaws. In fact, except for slightly better developed P_3 honing, and slightly less well-developed P_3 metaconids, it is virtually impossible to distinguish some *Sivapithecus* postcanines from those of *Australopithecus.* We doubt that an expert could wholly separate specimens of the two taxa on the basis of these parts. This conclusion takes on added significance when it is demonstrated, as we have done, that much of the similarity is in shared derived features. Similarly, it would be difficult to distinguish the maxillary or mandibular incisors of *Sivapithecus* from early *Australopithecus* in size or proportions, although these similarities are probably of a shared primitive character.

The main way in which *Sivapithecus* jaws and teeth may be distinguished from *Australopithecus* is in the size and morphology of the upper and lower canines. The upper canines of *Sivapithecus*, represented by more than 20 specimens, retain a primitive hominoid cutting postparacrista, which wears on P_3, and a prominent mesial groove which does not extend onto the canine root. However, as distinct from all extant pongids, the canine of *Sivapithecus* is oval in occlusal cross section, with its long axis being buccal–lingual, or at an oblique angle to a line drawn between the mesial groove and the distalmost point of the postparacrista. This condition is not seen elsewhere among living or fossil pongids. We have examined more than 150 maxillary canines of male and female *Pan* (two species), *Pongo*, and *Gorilla;* the long axis of the maxillary canine is *always* in the mesial–distal orientation. We suspect that this derived difference in *Sivapithecus* is due to the reduction of the length of the postparacrista in this animal compared to extant pongids. The canines of early *Australopithecus* from Afar and Laetoli resemble those of *Sivapithecus* in this respect (White, 1977, 1980; Johanson and White, 1979; Johanson *et al.*, 1982), as do South African *A. africanus* (Robinson, 1956). We regard this as an important unique shared derived similarity between *Sivapithecus* and *Australopithecus*.

The canines of *Sivapithecus* are quite small by modern pongid standards, although still much larger than *Australopithecus*. The relative size ranges of both upper and lower canines of *Sivapithecus* species are about the same as those for females of *Pan paniscus*, *Pan troglodytes*, *Pongo pygmaeus*, and *Gorilla gorilla* (Table 3). No single specimen of any *Sivapithecus* falls at or above the male *means* for upper or lower relative canine size of any species of extant pongid. The available sample of canines associated with molars (which allows an estimate of relative size) is nine for upper and ten for lower canines. The probability of sampling one male of one of these species with a canine at or above the average in relative size is extremely high. That being the case, the absence of specimens with canines at or above the average for male extant pongids is strong evidence that *Sivapithecus* males have relatively small canines. Reference to individual values in Table 3 gives a hint that the canines of *Sivapithecus sivalensis* were smaller than those of *Sivapithecus indicus*.

The problem of determining the amount of sexual dimorphism in canine size for *Sivapithecus* species is confounded by the presence of more than one taxon at most sites, the scarcity of specimens with canines which can be assigned to a taxon, and the difficulty of determining the sex of specimens once they are allocated to species. Kay (1982a) has outlined a technique, also described above, based on high correlation among living species of male : female canine-size ratios and the coefficient of variation of a mixed-sex sample. A high coefficient of variation in canine dimensions implies a high male to female canine-size ratio even when the sex of any specimen is unknown.

The coefficients of variation for the dimensions of a sample of 21 maxillary canines from all Siwalik localities almost certainly representing both sexes of two species, *Sivapithecus indicus* and *Sivapithecus sivalensis* (see Fig. 1), is lower than that of any single extant ape species. Therefore, although only a handful of individual specimens can be assigned to a single taxon, let alone a particular sex, the canine sexual dimorphism of *Sivapithecus* must be much lower than that of any living pongid species. A low degree of canine sexual dimorphism in *Sivapithecus* is a shared derived feature with *Australopithecus*. Johanson and White (1979) contend that *A. afarensis* exhibits considerable sexual dimorphism, but they have not specified whether this refers to canine size, body size, or both. It is well known that these two measures of dimorphism often differ (Fleagle *et al.*, 1980). If all the specimens referred to *A. afarensis* belong to a single species, which is yet to be adequately demonstrated, the CVs for the canine dimensions are below 11.0. This is lower than any species of pongid, comparable to what we find in *Sivapithecus*, but still high by the standards of other *Australopithecus* species or *Homo*.

What Evidence Exists for the Ancestry of the Great Apes?

If our assessment that 15 million year-old ramapithecines are hominids is correct, and if Huxley's (1863) dictum that among apes the human lineage is closest to that of the African species, it follows that the ancestors of *Pongo* and the African apes should have split into separate lineages by about 15 m.y.a. Is there any evidence documenting the African Great Ape or *Pongo* lineages in the later Miocene? We would hazard the guess that African Great Apes and *Pongo* derive from stocks of thin-enameled apes represented perhaps co-laterally by western and northern European later Miocene *Dryopithecus*. In support of this, *Dryopithecus* of Europe, as far as meagre evidence goes, appears to be at a similar grade of postcranial, dental, and gnathic structure relative to ramapithecines and living Great Apes.

The skeleton of European *Dryopithecus* is poorly known, but a few advances toward the condition of modern apes are apparent (see Morbeck, this volume, Chapter 14). A 15 million year-old humeral fragment from Saint Gaudens, France, probably of *Dryopithecus fontani*, compares favorably with *Pan paniscus* (Pilbeam and Simons, 1971). An ulna, possibly of *Dryopithecus*, from the Vienna Basin from about 16 m.y.a. exhibits marked reduction of the olecranon process (Zapfe, 1960), a derived feature of modern apes and hominids. On the other hand, a femur of *Dryopithecus* from Eppelsheim, Germany, collected more than 150 years ago, most closely resembles those of colobines and hylobatids and is more primitive than those of modern pongids (Le Gros Clark and Leakey, 1951).

No cranial material of European *Dryopithecus* has yet been identified with certainty. If our suggestion is correct that the Hungarian fossils described as *Rudapithecus* are referrable to *Dryopithecus*, then the genus resembles modern pongids (and early hominids) in having the root of the zygomatic high above the tooth row, and an arched palate. These are shared derived features of all later Miocene–Recent pongids, but lacking in known early Miocene apes. Dentally, *Dryopithecus* from Europe resembles early hominids and extant pongids in having reduced molar cingula. Early Miocene *Proconsul* have well-developed molar cingula, a primitive character.

In short, European *Dryopithecus* exhibit several shared derived similarities with hominids and modern pongids, not seen in early Miocene apes. Any specific shared derived resemblance to a particular modern pongid is not yet established. Unfortunately, African fossils of 10–15 million year-old *Dryopithecus* are even rarer than those from Europe. About all that can be said with certainty is that large-bodied apes with thin enamel coexisted with ramapithecines in Africa at this time.

We visualize the following scenario consistent with the available evidence. On the one hand, ramapithecines may be considered descendants of a form structurally like *Dryopithecus fontani*, specialized dentally and gnathically in the hominid direction. On the other hand, Great Apes also presumably passed through a stage in their evolution where they resembled *Dryopithecus fontani*. However, as yet, we can identify no shared derived characters linking European *Dryopithecus* with either *Pongo* or extant African apes. It may even be that known members of this group will prove to be sterile side branches of the extant pongid lineages.

Concluding Remarks

One of the novel features of the present interpretation is our conclusion that two widely recognized Siwalik species, *Sivapithecus sivalensis* and *Ramapithecus punjabicus*, are actually the same. Such a suggestion has not been made heretofore. In terms of taxonomy, the name *Sivapithecus* Pilgrim 1910 has priority over *Ramapithecus* Lewis, 1934, and the species *S. sivalensis* (Lydekker, 1879) takes priority over *R. punjabicus* (Pilgrim, 1910). Therefore the combination *Sivapithecus sivalensis* (Lydekker, 1879) is the correct nomen for this species. In the event this animal proves to be distinct generically from *S. indicus*, the new combination *Ramapithecus sivalensis* should be used.

Whether *Ramapithecus*, *Sivapithecus*, and *Gigantopithecus* should all be considered as separate genera or the former two be lumped is a matter of taste. Some tendency is perceived recently toward what we view as excessive splitting at the generic level. At one extreme, Pilbeam (1979, Fig. 2 and text) recognizes six genera of ramapithecines and two of contemporary thin-enameled apes. Perhaps we have erred in the opposite end of the spectrum by recognizing just two genera of ramapithecines (*Sivapithecus* and *Gigantopithecus*) and only one of contemporary large-bodied, thin-enameled apes

(*Dryopithecus*). Certainly there are some recognizable differences between *S. sivalensis* and *S. indicus,* such as mandibular shape and perhaps incisor and canine size. But recent finds have largely bridged the morphological gap which was formerly thought to separate the two. For example, both animals are now known to have short faces, thick-enameled, flat-wearing molar crowns, and transversely thick mandibular corpora. Earlier, these features were thought to have characterized *Ramapithecus* alone. Because of the extremely fragmentary fossil record and variability in the known material, we have had to use statistical arguments to demonstrate the existence of several contemporary sympatric species of *Sivapithecus;* many of the less complete specimens are difficult to assign to a particular species. As a rule of thumb, when the species affinity even of relatively well-preserved material are difficult to recognize, a generic distinction cannot be warranted.

In assessing the phylogenetic position of Ramapithecinae (*Sivapithecus* and *Gigantopithecus*), we believe that only shared derived characters can indicate a close relationship between taxa. By this analytical strategy, first we identified preserved anatomical characters which show some variation (exhibit several different "states") among Miocene to Recent Great Apes. Once this was done, we determined which character states were likely primitive for the group and which represent derived conditions. Ideally, any time two taxa share a derived state, they should be more closely related to one another than either is to the other taxa being considered. But what has become apparent to us is that considerable parallel evolution among Miocene to Recent apes frequently obscures the true phylogenetic picture. Thus, several conflicting views of phylogeny may be supported by shared derived characters. Evolutionary changes in the degree of premaxillary prognathism provide a case of such parallel evolution. This evidence, taken by itself, might lead to erroneous conclusions as to ape phylogeny. All early and middle Miocene apes for which this character is known have primitively low premaxillary prognathism, as defined in Table 4. *Gorilla* and *Australopithecus* are known from abundant evidence to be more closely related to one another and to *Pan* than any is to *Pongo*. Inasmuch as *Gorilla* and *Australopithecus* have low premaxillary prognathism, a primitive retention, the last common ancestor of these taxa and *Pan* also probably had a short premaxilla. Therefore, the high premaxillary prognathism shared by *Pan* and *Pongo* must have evolved in parallel. Thus, an uncritical look at this single feature out of context could lead to the erroneous conclusion that *Pan* and *Pongo* were closely related.

Given that a certain amount of parallelism is inevitable in all comparisons of this kind, we analyzed as many characters as possible and assumed that the phylogeny which is most likely has the fewest parallelisms and, where the parallelisms are found, they occur in characters showing a high frequency of similar parallelisms in other primate groups (e.g., the characters in question are assumed to have a low "phyletic valence").

Hewing to the above principles, we find that among the possible phylogenies considered we feel the one in which ramapithecines and *Australopithecus* are linked is considerably more parsimonious than any other. The following uniquely shared derived characters link ramapithecines and early *Australopithecus:* (1) shallow, broad mandibular corpora, robust mandibular symphyses, (2) low-cusped molars with extremely thick enamel, (3) somewhat reduced canine size, (4) somewhat reduced canine sexual dimorphism in *Sivapithecus* [both characters 3 and 4 foreshadowing the condition of *Australopithecus*], (5) buccal-lingually broad, mesial–distally short upper canines, and (6) a tendency toward great enlargement of the P_3 metaconid. Additionally, *Sivapithecus* exhibits no "crossing specializations" (with the possible exception of the unexpected narrow interorbital distance in one species) which would rule it out of the ancestry of *Australopithecus*. Furthermore, the genus is well situated geographically and temporally to be ancestral to *Australopithecus*. Simply stated, ramapithecines are ideally suited to be the ancestors of *Australopithecus* and *Homo*.

Accumulated evidence of the last century indicates that the ancestors of human beings arose by gradual divergence and developed distinctively human adaptations rather slowly. Fragmentary evidence of the anatomy of bones of the arm and foot (Pilbeam *et al.,* 1980) indicates that ramapithecines were mainly arboreal creatures, similar to living chimpanzees and orangutans although some of the larger species, such as *Gigantopithecus giganteus,* would surely have been somewhat terrestrial and gorilla-like. The brain case was small, like extant apes, as judged from the great postorbital constriction in the skull of Chinese *Sivapithecus*. In these ways, and in terms of facial structure, ramapithecines, although not precisely like any living Great Ape, were at an ape-like grade of anatomical organization.

Dentally, ramapithecines were much like later hominids in some ways and show the initiation of other hominid trends. The canines of ramapithecines show some size reduction and reduced sexual dimorphism compared to early Miocene *Proconsul* and modern apes, foreshadowing the condition of early *Australopithecus*. Reduced canine dimorphism may be indicative of a social structure centering on the pair bond by analogy with living gibbons (Kay 1982*b*). The cheek teeth of some ramapithecines are virtually indistinguishable from those of early *Australopithecus*. The extremely thick enamel and low, relatively featureless molar crowns of both are indicative of a similar feeding regime, which probably included the exploitation of very hard objects such as forest nuts and large, tough-skinned berries (Kay, 1981). Similarly, ramapithecines resemble *Australopithecus* in having shallow, broad mandibular corpora.

In a sense, then, ramapithecines are dental hominids. If the available material is correctly interpreted, the hominid lineage was initiated in the middle Miocene by a group that was cranially and postcranially like extant Great Apes, but with hominid-like dental adaptations, indicating the initiation of a nut-cracking adaptation, and, more problematically, pair bonding. By Pliocene times, hominids had advanced further and bipedality was achieved. The late Pliocene saw the initiation of tool manufacture and brain enlargement. Just as we see no major adaptive discontinuity between the earliest phyletic hominids and apes, we expect that late Miocene hominids, as they become known, will show a gradual transition dentally between ramapithecines and hominines (*Australopithecus* and *Homo*). The transition to bipedality may have been more abrupt: only the recovery of fossils between 7 and 4 million years old will resolve this problem.

We believe available evidence is strong enough to indicate that ramapithecines are broadly ancestral to Pliocene–Recent *Australopithecus* and *Homo*, but not to any living ape. Cladistically speaking, ramapithecines are the "sister group" of Pliocene–Recent hominids. But in terms of their overall grade of organization, ramapithecines were probably quite ape-like. So should ramapithecines be placed taxonomically in the family of humans (Hominidae)? A satisfactory solution of this problem is elusive, perhaps impossible, because it relies as much on one's taxonomic philosophy as it does on the evidence of relationship. Those who would relegate ramapithecines to a paraphyletic pongid "waste basket" might feel that ramapithecines, even if they are the sister group of true hominids, have not crossed a gradistic Rubicon into hominid-dom. Others would argue that if ramapithecines belong to a hominid, not an extant Great Ape, clade, this alone is enough to admit them to hominid status.

The principal difficulty with a gradistic definition of the family is that no two specialists can agree on where to set the boundary. *Homo* is derived with respect to extant Great Apes in a variety of features which could be preserved in the fossil record—bipedality, enlarged brains, relatively small canines, reduced canine sexual dimorphism, and thick molar enamel are a few of these features. Unfortunately for a gradistic concept, evidence is accumulating that these features were not acquired simultaneously as part of a common adaptive package; that is, for example, brain enlargement occurred long after bipedal locomotion had evolved. Because of this, ultimately, as the fossil record improves, a gradistic definition will have to depend on a single characteristic to define the hominid-grade boundary. The alternative to this, which we lean toward, is to define Hominidae in the cladistic sense as being all creatures specially related to *Homo* and *Australopithecus* going back to the split between these forms and the ancestry of any of the extant apes.

References

Abel, O. 1902. Zwei neue Menschenaffen aus den Leithakbildungen des Weiner Beckens. *Sitzungber. Akad. Wiss. Wien Math.–Naturwiss. Kl.* **111**:1171–1208.

Abel, O. 1931. *Die Stellung des Menschen in Rahmen der Wirbeltiere*, Gustav Fischer, Jena. 398 pp.

Andrews, P. 1971. *Ramapithecus wickeri* mandible from Fort Ternan, Kenya. *Nature (Lond.)* **231**:192–194.

Andrews, P. 1978a. Taxonomy and relationships of fossil apes, in: *Recent Advances in Primatology*, Volume 3, *Evolution* (D. J. Chivers and K. A. Joysey, eds.), pp. 43–56, Academic, London.

Andrews, P. J. 1978b. A revision of the Miocene Hominoidea of East Africa. *Bull. Br. Mus. (Nat. Hist.) Geol.* **30**:85–224.

Andrews, P. 1982. Hominoid evolution. *Nature*, **295**:185–186.

Andrews, P., and Cronin, J. 1982. The relationships of *Sivapithecus* and *Ramapithecus* and the evolution of the orang-utan. *Nature (Lond.)* **297**:541–546.

Andrews, P. J., and Molleson, T. I. 1979. The provenance of *Sivapithecus africanus. Bull. Br. Mus. (Nat. Hist.) Geol.* **32:**19–23.

Andrews, P., and Tekkaya, I. 1980. A revision of the Turkish Miocene hominoid *Sivapithecus meteai. Palaeontology* **23:**85–95.

Andrews, P., and Tobien, H. 1977. New Miocene locality in Turkey with evidence on the origin of *Ramapithecus* and *Sivapithecus. Nature (Lond.)* **268:**699–701.

Andrews, P., and Walker, A. 1976. The primate and other fauna from Fort Ternan, Kenya, in: *Human Origins: Louis Leakey and the East African Evidence* (G. L. Isaac and E. R. McCown, eds.), pp. 279–304, Benjamin, Menlo Park, California.

Barry, J. C., Behrensmeyer, A. K., and Monaghan, M. 1980. A geologic and biostratigraphic framework for Miocene sediments near Khaur village, Northern Pakistan. *Postilla* (Peabody Mus. Nat. Hist., Yale Univ.) **183:**1–19.

Bishop, W. W., Fitch, F. J., and Miller, J. A. 1969. New potassium–argon age determinations relevant to the Miocene faunal and volcanic sequence in East Africa. *Am. J. Sci.* **267:**669–699.

Brace, C. L., and Ryan, A. S. 1980. Sexual dimorphism and human tooth size differences. *J. Hum. Evol.* **9:**417–436.

Branco, W. 1898. Die menschenähnlichen Zähne aus dem Bohnerz der schwäbischen Alb. *Jahrsh. Ver. Vaterl. Naturk. Württ.* **54:**1–144.

Darwin, C. 1871. *The Descent of Man, and Selection in Relation to Sex,* Vol. 1, John Murray, London. 423 pp.

De Bonis, L., and Melentis, J. 1977. Les primates hominoides du Vallésien de Macédoine (Gréce). Étude de la mâchoire inférieure. *Géobios* **10:**849–885.

De Bonis, L., and Melentis, J. 1978. Les primates hominoides du Miocéne Supérieur de Macédoine. Étude de la mâchoire supérieure. *Ann. Paleontol. Vertebr.* **64:**185–202.

De Bonis, L., and Melentis, J. 1980. Nouvelles remarques sur l'anatomie d'un primate hominoide du Miocéne: *Ouranopithecus macedoniensis.* Implications sur la phylogenie des Hominidés. *C. R. Acad. Sci. Paris* **290:**755–758.

De Bonis, M. L., Bouvrain, G., Geraads, D., and Melentis, J. 1974. Première découverte d'un primate hominoide dans le Miocéne supérieur de Macédoine (Gréce). *C. R. Acad. Sci. Paris* **278:**3063–3066.

Delson, E., and Andrews, P. 1975. Evolution and interrelationships of the catarrhine primates, in: *Phylogeny of the Primates* (W. P. Luckett and F. S. Szalay, eds.), pp. 405–446, Plenum, New York.

Dubois, E. 1894. *Pithecanthropus erectus. Eine menschenahnliche Ubergangsform aus Java,* Landersdruckerei, Batavia.

Fleagle, J. G., Kay, R. F., and Simons, E. L. 1980. Sexual dimorphism in early anthropoids. *Nature (Lond.)* **287:**328–330.

Gantt, D. G. 1977. *Enamel of Primate Teeth,* Ph.D. Dissertation, Washington University, St. Louis. 403 pp.

Gentry, A. W. 1970. The Bovidae (Mammalia) of the Fort Ternan fossil fauna, in: *Fossil Vertebrates of Africa,* Volume 2 (L.S.B. Leakey and R.J.G. Savage, eds.), pp. 243–323, Academic, London.

Gingerich, P. D., and Schoeninger, M. 1979. Patterns of tooth size variability in the dentitions of Primates. *Am. J. Phys. Anthropol.* **51:**457–466.

Glaessner, M. F. 1931. Neue Zähne von Menschenaffen aus dem Miozän des Weiner Beckens. *Ann. Naturhist. Mus. Wien* **46:**15–27.

Greenfield, L. O. 1977. *Ramapithecus and Early Hominid Origins,* Ph.D. Dissertation, University of Michigan.

Greenfield, L. O. 1979. On the adaptive pattern of "*Ramapithecus*". *Am. J. Phys. Anthropol.* **50:**527–548.

Greenfield, L. O. 1980. A late divergence hypothesis. *Am. J. Phys. Anthropol.* **52:**351–365.

Gregory, W. K., Hellman, M., and Lewis, G. E. 1938. Fossil anthropoids of the Yale–Cambridge India expedition of 1935. *Carnegie Inst. Wash. Publ.* **495:**1–27.

Grine, F. E. 1981. A new composite juvenile specimen of *Australopithecus africanus* (Mammalia, Primates) from Member 4, Sterkfontein Formation, Transvaal. *Ann. S. Afr. Mus.* **84:**169–201.

Huag, J. 1977. *A Study of Qualitative Morphological Variation in the Dentition of Gorilla gorilla gorilla*, Ph.D. Dissertation, University of Colorado.

Huxley, T. H. 1863. *Evidence as to Man's Place in Nature*, Williams and Norgate, London. 159 pp. (1897 reprinting by D. Appelton, New York cited herein).

Hylander, W. L. 1975. Incisor size and diet in anthropoids with special reference to Cercopithecidae. *Science* **189:**1095–1098.

Johanson, D. C., and White, T. D. 1979. A systematic assessment of early African hominids. *Science* **202:**321–330.

Johanson, D. C., White, T. D., and Coppens, Y. 1982. Dental remains from the Hadar Formation, Ethiopia: 1974–1977 Collections. *Am. J. Phys. Anthropol.* **57:**605–630.

Kay, R. F. 1981. The nut-crackers—A new theory of the adaptations of the Ramapithecinae. *Am. J. Phys. Anthropol.* **55:**141–151.

Kay, R. F. 1982a. *Sivapithecus simonsi,* A new species of Miocene hominoid with comments on the phylogenetic status of the Ramapithecinae. *Int. J. Primatol.* **3:**113–174.

Kay, R. F. 1982b. Sexual dimorphism in Ramapithecinae. *Proc. Natl. Acad. Sci. USA* **79:**209–212.

Kretzoi, M. 1975. New ramapithecines and *Pliopithecus* from the lower Pliocene of Rudabánya in north-eastern Hungary. *Nature (Lond.)* **257:**578–581.

Le Gros Clark, W. E., and Leakey, L. S. B. 1950. Diagnosis of East African Miocene Hominoidea. *Q. J. Geol. Soc. Lond.* **105:**260–262.

Le Gros Clark, W. E., and Leakey, L. S. B. 1951. The Miocene Hominoidea of East Africa. *Fossil Mammals of Africa* (Br. Mus. Nat. Hist.) **1:**1–117.

Leakey, L. S. B. 1962. A new lower Pliocene fossil from Kenya. *Ann. Mag. Nat. Hist.* **13:**689–696.

Lewis, G. E. 1934. Preliminary notice of new man-like apes from India. *Am. J. Sci.* **27:**161–181.

Lydekker, R. 1879. Further notices of Siwalik Mammalia. *Rec. Geol. Surv. India* **12:**33–54.

Mahler, P. E. 1973. *Metric Variation in the Pongid Dentition,* Ph.D. Dissertation, University of Michigan. 467 pp.

McHenry, H. M., Andrews, P., and Corruccini, R. S. 1980. Miocene hominoid palatofacial morphology. *Folia Primatol.* **33:**241–252.

Mottl, von M. 1957. Bericht über die neuen Menschenaffenfunde aus Österreich, von St. Stefan im Lavanttal, Kärnten. *Naturwiss. Beitr. Heimatk. Kaerntens* **67:**39–84.

Ozansoy, F. 1957. Faunes des mammifères du Tertiare de Turquie et leurs révisions stratigraphiques. *Bull. Min. Res. Expl. Inst. Turkey* **49:**29–48.

Ozansoy, F. 1965. Étude des gisements continentaux et des mammifères du Cénozoique de Turquie. *Mem. Soc. Geol. Fr.* **44:**5–89.

Pickford, M. 1982. New higher primate fossils from the middle Miocene deposits at Majiwa and Kaloma, Western Kenya. *Am. J. Phys. Anthropol.* **58:**1–19.

Pilbeam, D. 1969a. Newly recognized mandible of *Ramapithecus. Nature (Lond.)* **222:**1093–1094.

Pilbeam, D. R. 1969b. Tertiary Pongidae of East Africa. *Bull. Peabody Mus. Nat. Hist.* (Yale Univ.) **31:**1–185.

Pilbeam, D. 1979. Recent finds and interpretations of Miocene hominoids. *Annu. Rev. Anthropol.* **8:**333–352.

Pilbeam, D. R. 1982. *Nature (Lond.)* **295:**232–234.

Pilbeam, D., and Simons, E. L. 1971. Humerus of *Dryopithecus* from St. Gaudens, France. *Nature (Lond.)* **229:**408–409.

Pilbeam, D. R. and Smith, R. 1981. New skull remains of *Sivapithecus* from Pakistan. *Mem. Geol. Surv. Pakistan* **II:**1–13.

Pilbeam, D. R., Berry, J., Meyer, G. E., Shah, S. M. I., Pickford, M. H. L., Bishop, W. W., Thomas, H., and Jacobs, L. L. 1977a. Geology and paleontology of Neogene strata of Pakistan. *Nature (Lond.)* **270:**684–689.

Pilbeam, D., Meyer, G. E., Badgley, C., Rose, M. D., Pickford, M. H. L., Behrensmeyer, A. K., and Shah, S. M. I. 1977b. New hominoid primates and their bearing on hominoid evolution. *Nature (Lond.)* **270:**689–695.

Pilbeam, D. R., Behrensmeyer, A. K., Barry, J. C., and Shah, S. M. I. 1979. Miocene sediments and faunas of Pakistan. *Postilla* (Peabody Mus. Nat. Hist., Yale Univ.) **179:**1–145.

Pilbeam, D. R., Rose, M. D., Badgley, C., and Lipschutz, B. 1980. Miocene hominoids from Pakistan. *Postilla* (Peabody Mus. Nat. Hist., Yale Univ.) **181:**1–94.

Pilgrim, G. E. 1910. Notice of new mammalian genera and species from the tertiaries of India. *Rec. Geol. Surv. India* **50**:63–71.

Pilgrim, G. E. 1915. New Siwalik primates and their bearing on the question of the evolution of man and the Anthropoidea. *Rec. Geol. Surv. India* **45**:1–74.

Remane, A. 1960. Zähne und Gebiss. *Primatologia* **3**:637–846.

Robinson, J. T. 1956. The genera and species of the Australopithecinae. *Am. J. Phys. Anthropol.* **12**:181–200.

Schlosser, M. 1901. Die menschenahnlichen Zähne aus dem Bohnerz der Schwabischen Alb. *Zool. Anz.* **24**:261–271.

Shipman, P. L. 1977. *Paleoecology, Taphonomic History, and Population Dynamics of the Vertebrate Fossil Assemblage from the Middle Miocene Deposits Exposed at Fort Ternan, Kenya*, Ph.D. Dissertation, New York University. 395 pp.

Sickenberg, O., and Tobien, H. 1971. New Neogene and lower Quaternary vertebrate faunas in Turkey. *Newsl. Stratigr.* **1**:51–61.

Simons, E. L. 1961. The phyletic position of *Ramapithecus*. *Postilla* (Peabody Mus. Nat. Hist., Yale Univ.) **57**:1–57.

Simons, E. L. 1964. On the mandible of *Ramapithecus*. *Proc. Natl. Acad. Sci. USA* **51**:528–535.

Simons, E. L. 1972. *Primate Evolution: An Introduction to Man's Place in Nature*, Macmillan, New York. 322 pp.

Simons, E. L. 1976. The nature of the transition in the dental mechanism from pongids to hominids. *J. Hum. Evol.* **5**:511–528.

Simons, E. L., and Chopra, S. R. K. 1969. *Gigantopithecus* (Pongidae, Hominoidea). A new species from North India. *Postilla* (Peabody Mus. Nat. Hist., Yale Univ.) **138**:1–18.

Simons, E. L., and Pilbeam, D. R. 1965. Preliminary revision of the Dryopithecinae (Pongidae, Anthropoidea). *Folia Primatol.* **3**:81–152.

Simons, E. L., and Pilbeam, D. R. 1978. *Ramapithecus*, in: *Evolution of African Mammals* (V. J. Maglio and H. B. S. Cooke, eds.), pp. 147–153, Harvard University Press, Cambridge.

Steininger, von F. 1967. Ein weiterer Zahn von *Dryopithecus (Dry.) fontani darwini* Abel, 1902 (Mammalia, Pongidae) aus dem Miozän des Weiner Beckens. *Folia Primatol.* **7**:243–275.

Steininger, F., Rögl, F., and Martini, E. 1976. Current Oligocene/Miocene biostratigraphic concept of the Central Parathethys (Middle Europe). *Newsl. Stratigr.* **4**:174–202.

Szalay, F. S., and Delson, E. 1979. *Evolutionary History of the Primates*, Academic, New York. 580 pp.

Tekkaya, I. 1974. A new species of Tortonian Anthropoid (Primates, Mammalia) from Anatolia. *Bull. Min. Res. Expl. Inst. Turkey* **83**:148–165.

Tobias, P. V. 1980. "*Australopithecus afarensis*" and *A. africanus:* Critique and an alternative hypothesis. *Palaeontol. Afr.* **23**:1–17.

Villalta, J. F., and Crusafont, M. 1944. Dos nuevos antropomorfos del Mioceno español y su situación dentro de la moderna sistemática de los simidos. *Notas Comun. Inst. Geol. Min. Espana* **13**:91–139.

Von Koenigswald, G. H. R. 1935. Eine fossile Säugetierfauna mit *Simia*, aus Südchina. *Proc. K. Ned. Akad. Wet. Amsterdam* **38**:872–879.

Von Koenigswald, G. H. R. 1936. Erste mitteilungen über einen fossilen Hominiden aus dem Altpleistocän Ostjavas. *Proc. K. Ned. Akad. Wet. Amsterdam* **39**:1000–1009.

Von Koenigswald, G. H. R. 1972. Ein unterkiefer eines fossilen hominoiden aus dem Unterpliozän Griechenlands. *Proc. K. Ned. Akad. Wet. Amsterdam B* **75**:385–394.

Walker, A., and Andrews, P. J. 1973. Reconstruction of the dental arcades of *Ramapithecus wickeri*. *Nature (Lond.)* **244**:313–314.

Ward, S. C., Kimbel, W. H., and Pilbeam, D. 1983. Subnasal alveolar morphology and the systematic position of *Sivapithecus*. *Am. J. Phys. Anthropol.* **61**(1).

Weidenreich, F. 1937. The dentition of *Sinanthropus pekinensis:* A comparative odontography of the hominids. *Palaeontol. Sin. D* **1**:1–180.

Weidenreich, F. 1943. The skull of *Sinanthropus pekinensis;* A comparative study on a primitive hominid skull. *Palaeontol. Sin. D* **10**:1–484.

White, T. D. 1977. New fossil hominids from Laetoli, Tanzania. *Am. J. Phys. Anthropol.* **46**:197–230.

White, T. D. 1980. Additional fossil hominids from Laetoli, Tanzania: 1976–1979 specimens. *Am. J. Phys. Anthropol.* **53:**487–504.

White, T. D., and Johanson, D. C. 1982. Pliocene hominid mandibles from the Hadar Formation, Ethiopia: 1974–1977 collections. *Am. J. Phys. Anthropol.* **57:**501–544.

Whybrow, P. J., and Andrews, P. 1978. Restoration of the holotype of *Proconsul nyanzae. Folia Primatol.* **30:**115–125.

Wolpoff, M. H. 1975. Sexual dimorphism in the Australopithecines, in: *Paleoanthropology* (R. H. Tuttle, ed.), pp. 245–284, Mouton, The Hague.

Wolpoff, M. H. 1980. *Paleoanthropology,* Knopf, New York, 379 pp.

Woo, J. K. 1957. *Dryopithecus* teeth from Keiyuan, Yunnan province. *Vertebr. Palasiat.* **1:**25–32.

Woo, J. K. 1958. New materials of *Dryopithecus* from Keiyuan, Yunnan. *Vertebr. Palasiat.* **2:**38–43.

Xu, Q., and Lu, Q. 1979. The mandibles of *Ramapithecus* and *Sivapithecus* from Lufeng. Yunnan. *Vertebr. Palasiat.* **17:**1–13.

Xu, Q., and Lu, Q. 1980. The Lufeng ape skull and its significance. *China Reconstructs* **29:**56–57.

Xu, Q., Lu, Q., Pan, J., Chi, K., Zhang, C., and Zheng, L. 1978. On the fossil mandible of *Ramapithecus lufengensis. Kexue Tongbao* **23:**554–556.

Zapfe, H. 1960. Die primatenfunde aus der Miozänen spaltenfülung von Neudorf an der March (Dĕvinská Nová Ves), Tschechoslowakei. Mit anhang: Der primatenfund aus dem Miozän von Klein Hadersdorf in Niederösterreich. *Schweiz. Palaeontol. Abh.* **78:**1–293.

Zihlman, A. L., Cronin, J. E., Cramer, D. L., and Sarich, V. M. 1978. Pygmy chimpanzee as a possible prototype for the common ancestor of humans, chimpanzees, and gorillas. *Nature (Lond.)* **275:**744–746.

Zingeser, M. R. 1976. Arch form, tooth size, and occlusomandibular kinesis in the Ceboidea. *Am. J. Phys. Anthropol.* **45:** 317–330.

Phyletic Relationships of Miocene Hominoids and Higher Primate Classification

24

L. DE BONIS

Introduction

The superfamily Hominoidea (Gray, 1825), in its classical form, includes the apes (chimpanzee, gorilla, orangutan, and gibbon) and humans. The study of the links between humans and those creatures sometimes considered the "inferior kin" to humans has long fascinated naturalists, zoologists, and paleontologists. The spread of Darwin's ideas about the evolution of species and the origins of the human race marked the beginning of a new phase in these studies. Their purpose was no longer limited to a static examination of the morphological characters and their translation into a classification, whether or not hierarchical, but to search for an ancestor sometimes called the "missing link," which would allow a bridging of the gap between humans and apes.

The first fossil hominoid, discovered by Lartet (1837), was related to the gibbons. Gervais (1849) called it *Pliopithecus*. Gibbons along with siamangs constitute the family Hylobatidae. This group occupies a special position within the Hominoidea, and is outside the scope of this investigation. In 1856, only three years before the publication of *The Origin of Species,* the discovery of

L. DE BONIS • Laboratoire de Paléontologie des Vertébrés et Paléontologie Humaine, Université de Poitiers, Faculté des Sciences, 86022 Poitiers, France.

another fossil ape came just in time to illustrate the controversies surrounding the problem of human origins. In a clay pit in southwestern France near Saint Gaudens, the geologist Fontan unearthed a jaw which Lartet (1856) named *Dryopithecus fontani*. A few years later, following other finds at the same locality, Gaudry resumed the study of these fossils. One of the main advocates of the new ideas concerning evolution in France, he undertook the search for traces of hominization in the Saint Gaudens primates. In 1890 he published a study presenting his conclusions. For Gaudry (1890), speech was the test of hominization and the *Dryopithecus* mandible, which allowed limited space to the tongue, showed that this animal was a long way from articulated speech. The gap between the apes, whether living or fossil, and humans remained wide and the link between these groups remained to be found.

In subsequent years the discoveries continued in several other areas. The fossil primates recovered were often accompanied by a certain amount of taxonomic inflation. Toward the close of the last century the Siwaliks Hills, on the southern foothills of the Himalayas, yielded a series of specimens described by various names, most notably *Sivapithecus* and *Ramapithecus*. In East Africa many primate fossils were unearthed in the early Miocene layers throughout the 1930s. After extensive examination most were classified in the genus *Proconsul*. In China, a very large primate, *Gigantopithecus*, was identified from teeth purchased in apothecaries. In most of these cases, close links to humans were considered at one time or another, only to later be rejected. Therefore, in spite of the large number of taxa in existence, about 20 years ago no Miocene primate was considered as directly related to the Plio-Pleistocene hominines. All of these fossil forms were placed in the family Pongidae alongside the living apes (Fig. 1). The morphology of the lower premolar P_3 played a large role in determining this taxonomic placement. This tooth is unicuspid and has a more or less cutting (shearing) edge in apes, while it is bicuspidate in humans.

In the mid-1960s Simons and Pilbeam (1965) reevaluated fossil hominoids and simplified the taxonomic imbroglio, perhaps to excess. Following Simons' opinion (1961, 1964), they classified *Ramapithecus* as a hominid. Since then an intensification of research has produced many new discoveries. The most important occurred in the Siwalik Hills of India and Pakistan, in East Africa (particularly Fort Ternan), in Asia Minor (Çandir and Paşalar), in Greece, in Spain (Catalonia), in Hungary (Rudabánya), and in China (Lufeng). Part of this new material is still under study, and it would be difficult to make a fully synthetic statement regarding all the new discoveries. Nevertheless it appears necessary to use the first published results and also the new data from comparative anatomy and molecular biology in order to critically examine the question of hominoid phyletic relationships.

The approach that I have used, wherever possible, is the cladistic method. The principle of this method is founded on the statement that two monophyletic taxa A and B are more closely related to each other than to another monophyletic taxon C if the former share a common ancestor which is not an ancestor of C. In this case the two taxa A and B form a monophyletic group.

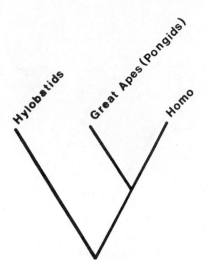

Fig. 1. Cladogram of relationships within the Hominoidea most commonly proposed in recent years. Note that the hylobatids, sometimes called hyperbrachiators, branched out first from the common hominoid stem.

However, it is not necessary to know the ancestor in order to define this monophyletic group. In most cases the two taxa share derived characters which are not shared by C. That is, there is one (or more) synapomorphy between A and B (Hennig, 1966). A and B are called sister groups, but can be distinguished from one another since each of them has its own derived characters (autapomorphy). The search for sister groups to build a phylogeny is generally easier when dealing with living forms than with fossils. However, it is the method best adapted for constructing phylogenies and then testing and translating them into a classification (Delson, 1977). For this reason I have attempted to use the cladistic method in spite of the difficulties resulting from the fragmentary nature of these fossils.

One of the purposes of this chapter is to bring together data from a wide variety of sources that have a potential bearing on the cladogenesis of the Hominoidea. In this regard it is necessary to discuss the numerous biochemical methods used in comparative biology for taxonomic and phylogenetic analyses. These methods are relatively varied and include electrophoresis, immunological distances (microcomplement fixation or immunodiffusion), comparison of amino acid sequences, and examination of karyotypes by banding (Goodman and Tashian, 1976; Sarich, 1968; Goodman and Moore, 1971; Dutrillaux, 1979; see also Goodman *et al.*, this volume, Chapter 3; Cronin, this volume, Chapter 5; and Mai, this volume, Chapter 4). The concordance of results from these studies is remarkable, in that one nearly always finds the same relative disposition within the living Primates. With regard to the hominoids, the genus *Pongo* occupies a phylogenetic position further apart (a cladistic out-group) from the other large hominoid taxa (gorilla, chimpanzee, human) than any of these latter taxa do when compared among themselves. Such results raise a definitive question regarding the family Pongidae, which now appears to be nothing more than a paraphyletic grouping. On the other hand, based on the molecular biological data, human, gorilla, and chimpanzee

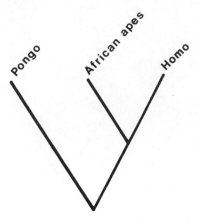

Fig. 2. Cladogram based on molecular biological data. The biological distances between the Great Apes and humans (*Homo*) indicate that the *Pongo* lineage branched out before the divergence between *Homo* and the African apes.

would seem to constitute a monophyletic grouping like the one now consisting of the two large African apes (Fig. 2). A major purpose of this chapter is to attempt a synthesis of these results and recent paleontological discoveries.

The Miocene Open-Environment-Dwelling Primates

During the Miocene, hominoid primates are present on all three continents of the Old World: Africa, Europe, and Asia. The oldest are found in the lower Miocene of Kenya and have been dated about 22 m.y.a. Until the Turolian the documentation is relatively good, but starting about 8 m.y.a. there is a noticeable gap until the appearance of the australopithecines from Hadar and Laetoli around 3.5 m.y.a. (Johanson and White, 1979). Between 8 and 3.5 m.y.a. the presence of hominoid primates is known only from a few remains that are difficult to analyze. For this time period (between 22 and 8 m.y.a.) paleoanthropologists face two types of problems. In the first place, it is necessary to search for the ancestral forms, or more precisely for the sister groups which mark the history and phylogeny of the various lineages. starting backward from the modern groups. Second, it is important to establish the most probable date for the divergence between the groups being considered, or, at least, the time period from which one first notes that such groups are separated. These facts will permit construction of cladograms, showing phylogeny, and the establishment of a classification scheme for all Recent and extinct Hominoidea.

Differentiation of the Hominids

The oldest known hominoids, either in the Fayum (Oligocene) or in East Africa (early Miocene), were forest-dwelling animals. The study of their environment through sedimentology and paleobotany and the study of their vertebrate faunas point to a humid, tropical forest setting (Andrews and Van Couvering, 1975). On the other hand, the Plio-Pleistocene hominines lived in a more open environment, and their adapatation to more abrasive food is shown by the considerable thickness of the enamel of the jugal teeth (Kay, 1981). The molars also show a characteristic wear gradient: the dentine appears on M_2 only after substantial wear of M_1 and on M_3 only after strong abrasion of M_2. The molar enamel of forest-dwelling hominoids is thin and as a result the dentine is exposed very early on all the molars. The first hominoids whose teeth show a substantial thickening of the enamel are known from middle Miocene levels (about 14 or 15 m.y.a.) in Asia Minor, Europe, and Africa and have been placed in the two genera, *Sivapithecus* and *Ramapithecus* (Pilbeam *et al.*, 1977). During the upper Miocene a whole series of forms occurred and have been classified as *Bodvapithecus, Rudapithecus, Ouranopithecus, Ankarapithecus,* and *Graecopithecus* (Kretzoi, 1975; de Bonis and Melentis, 1977a; Ozansoy, 1970; von Koenigswald, 1972). Finally, *Gigantopithecus,* a large Plio-Pleistocene hominoid, must be added to this list. The similarities between the dentitions of these taxa constitute a derived character which may be explained either by convergence or by synapomorphy. Until proven otherwise, it is wise to adhere to the latter, more parsimonious, hypothesis (see also Gantt, this volume, Chapter 10). The postcranial skeleton of all these primates is not well known, but it is probable that none of them had yet achieved bipedalism. Based on current evidence, it appears probable that this first occurred with *Australopithecus afarensis* in the Pliocene (Johanson and White, 1979). Bipedalism then can be considered as the basic synapomorphy of the hominines. The other primates mentioned above then constitute the sister group of the hominines, plesiomorphous for this feature, although they show some autapomorphies (Fig. 3). They can be attributed to the group ramapithecines. If one considers that the first ramapithecines appeared about 15 m.y.a., one must admit that the divergence between them and the large African apes is prior to that date.

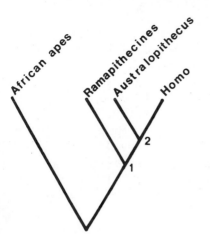

Fig. 3. Cladogram showing relationships between the thin-enameled African apes and some other thick enameled hominoid primates. (1) Appearance of thick enamel. (2) Appearance of bipedalism.

Relationships among the Ramapithecines

The Third Lower Premolar

The first problem in determining the phyletic relationships among the ramapithecines concerns the shape and function of the third lower premolar. Since the discovery of *Australopithecus afarensis* in the Hadar area, the presence of a single cusp on P$_3$ can no longer be considered as a valid criterion for excluding a fossil from the human lineage. The mandible A.L. 288, from the famous "Lucy" skeleton, and other Hadar specimens have a single cusped premolar with a very simple occlusal pattern. The single cusp is located on the external, longitudinal ridge (eocrest), and the secondary ridge (epicrest), which is more or less perpendicular to the first one, begins from its top. This pattern may be considered as primitive (Vandebroek, 1969) and more archaic than the two-tubercle condition which frequently occurs in the more recent hominines. There is an extremely similar pattern in one of the oldest known hominoids, *Propliopithecus haeckeli* of the Fayum Oligocene (Kalin, 1961). If the anterior premolar of "Lucy" represents a symplesiomorphism with *P. haeckeli* in this respect, then the known Miocene ramapithecines show derived characters in the morphology of this tooth. The epicrest in ramapithecines is generally at a sharp angle to the eocrest, the anteroexternal wall is high with a well-developed honing facet, and there is a lateral flattening of the whole tooth. It is therefore difficult to consider them as ancestors of the recent hominines. But the structure of "Lucy's" premolar may actually be derived over the condition seen in the known Miocene ramapithecines. It could then be considered a return or reversal to the primitive pattern first evident in *Propliopithecus*. The small or medium-sized ramapithecines whose premolars are not too hypertrophied (*Ramapithecus* or *Sivapithecus*) can then be considered as belonging to the sister group of australopithecines. Nevertheless, this hypothesis is less parsimonious than the previous one, since it implies important but somewhat contradictory transformations.

The Origin of Gigantopithecus

The genus *Gigantopithecus* is represented by two species. One, *G. bilaspurensis,* can be dated approximately from the Pliocene (Simons and Chopra, 1969). The other one, *G. blacki,* is known from the Pleistocene of China. It is possible that the differences between the two species may justify the creation of a distinct genus. But they share several derived features which amply justify grouping them together. These characters are: very large size; massive mandibles; small incisors and canines; presence of an internal cusp on P$_3$; lengthened molars; horizontally worn areas on the jugal teeth, particularly P$_3$; very thick enamel.

Among the Miocene hominoids, *Sivapithecus,* which is sometimes considered as a possible ancestor of *Gigantopithecus* (Pilbeam, 1970), seems too specialized due to its lower canine, which is mediolaterally flattened and well developed, and by its lengthened P_3 with an epicrest at sharp angle to the eocrest and a well-developed honing complex. On the other hand, *Ouranopithecus*, another Miocene hominoid, shows several primitive characters (globular P_3 without developed honing complex, noncutting lower canine) which if associated with other features could all be considered synapomorphous for *Gigantopithecus* (de Bonis and Melentis, 1977*b*, 1978, 1980). These are: very large size; secondary cusps on M_3; and horizontally worn areas on jugal teeth, particularly on P_3.

It is easy to imagine going from *Ouranopithecus* to *G. bilaspurensis* through reduction of anterior dentition, development of a second cusp on P_3, and an increase in the size and the massive aspect of the jaw. In both species a masticatory system is found in which horizontal movements predominate, resulting in the jaws functioning as grindstones. Also, the wear pattern of the upper incisors differs little between *Ouranopithecus* and *G. blacki* (the upper incisors of *G. bilaspurensis* are not known). This could also be an apomorphous character common to this group of primates. Therefore *Ouranopithecus* may be considered the best candidate at present for the ancestor of *Gigantopithecus*. Therefore *Ouranopithecus* and *Gigantopithecus* would be sister groups (Fig. 4).

Phyletic and Systematic Position of Ramapithecus

Created by Lewis (1934) for a maxillary fragment (YPM 13799) from the upper Miocene beds of the Siwaliks, *Ramapithecus* is often considered as the

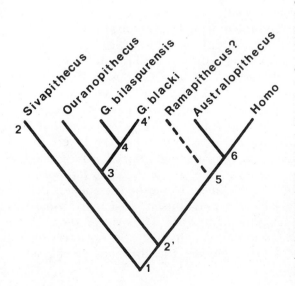

Fig. 4. Cladogram of relationships within the subfamily Homininae. (1) Appearance of thick enamel. (2) Increase of the honing complex between upper canine and lower P_3. (2′) Alternative proposal if a well-developed honing complex is primitive for Homininae; node 2′ would then indicate a decrease of the honing complex (it is a less parsimonious hypothesis). (3) Large size, accessory basal cusps on I^1, accessory cusps on M_3. (4) Very large size, small incisors and canines, well-developed medial cusp on P_3. (4′) huge size, very small incisors and canines, medial cusp on P_3 as high as the principal one, accessory cusps on molar teeth, high-crowned jugal teeth. (5) ?Hominine characters (shape of the palate, shape and robustness of the mandible, etc.; see text). An alternative proposal (see text) is to sink *Ramapithecus* into *Sivapithecus*. (6) Appearance of bipedalism.

forerunner of the Plio-Pleistocene hominines. Since its creation, views of *Ramapithecus* have changed somewhat. Some specimens placed by Lewis in *Ramapithecus* have been reevaluated and withdrawn from this genus, while new specimens have been allocated to it. In recent publications (Simons, 1961, 1964, 1976) the main criteria used to define *Ramapithecus* are as follows: parabolic palate; australopithecine-like symphysis; robustness of the mandible; thickness of the jugal tooth enamel; size and shape of the anterior teeth; shape of the third lower premolar; and small size.

Shape of the Palate.   The parabolic shape of the human palate sets it apart distinctly from that of the Great Apes. The shapes of the *Ramapithecus* palates reconstructed from the Yale maxillary fragment (*R. "brevirostris"*) and from the Fort Ternan (*R. wickeri*) specimens have yielded contradictory results (Simons, 1961; Walker and Andrews, 1973; Genet-Varcin, 1969). A maxilla from Rudabánya in Hungary (Rud 12, paratype of *Rudapithecus hungaricus* Kretzoi) displays a rather complete dental arcade (Kretzoi, 1975). *Rudapithecus* has been placed in the genus *Ramapithecus* by many specialists. The shape of the dental arch of this specimen is not truly parabolic, but rather the rows of jugal teeth diverge slightly toward the back, and the front portion is more or less rounded. In this respect, this arch is intermediate between the condition in humans and that found in the Great Apes. However, there is another primate in the Rudabánya sample, *Bodvapithecus altipalatus* Kretzoi, which differs from *Rudapithecus* by several features (larger size, relatively larger, more robust canines), which are possibly due to a sexual variation within the same species. If this were the case, the male dental arch, enlarged in front by the size of the canines, would be more U-shaped. In the extant apes the female dental arch is always more rounded than is that of the male. Finally, there is a great variation of palate shape in the australopithecines, which are without question hominines. It seems it may not be possible to reach any phyletic conclusion with regard to palate shape.

Symphysis and Shape of the Mandible.   In the list of characters that aids in distinguishing *Ramapithecus* from *Sivapithecus,* the degree of strength of the horizontal part of the mandible and the morphology of the symphysis are the most important. This matter was discussed in a recent article (de Bonis and Melentis, 1980), and the conclusions are as follows. A high index of robustness in the horizontal ramus would seem to characterize *Ramapithecus,* but this index is subject to sexual variations in *Ouranopithecus.* These variations are very similar to those separating *Ramapithecus* and *Sivapithecus.* With regard to the morphology of the symphysis, it is notable that male *Ouranopithecus* have a longer alveolar planum than the female specimens, while the latter show a better marked upper transverse torus and genioglossal pit. These are the same anatomical features found respectively in *Sivapithecus* and *Ramapithecus.*

Thickness of Tooth Enamel.   The thickness of the tooth enamel is one of the most salient characters of *Ramapithecus* and by itself sets it apart from fossil (*Proconsul, Dryopithecus*) as well from the living apes. The wear gradient of the molars, which allows the dentine to show through on M_2 and obviously on M_3 only after rather heavy wear of M_1, expresses this thickness of enamel in external morphology with the aid of a late eruption sequence of the posterior molars. In this respect it would seem that Frayer (1976) is making an incorrect assumption when he compares this type of wear to that seen in the chimpanzee. This interpretation could be explained by the fact that this author worked only from casts, for on original specimens, the thinness of the chimpanzee's molar enamel when considered jointly with the relatively fast eruption sequence leads to the precocious appearance of dentine at the level of most cusps before the first molar is too heavily worn. The morphology of the molar tubercles, pyramidal and little raised in relation to the total height of the crowns, also constitutes a good criterion for distinction. Nevertheless, in the Miocene, other primates with jugal teeth having a thick enamel covering, particularly *Sivapithecus* and *Ouranopithecus,* show a wear gradient very similar to that of *Ramapithecus.* The molar cusps are also relatively low and pyramidal in shape. However, on some specimens attributed to *Sivapithecus* (for instance, GSI 18040, YPM 13828, and GSI D-197) the wear facets of the relatively young molars meet together at sharp angle, like diamond facets, which implies a rather special type of occlusion. But the thickness of enamel constitutes a character which links the three genera and draws them nearer to the Plio-Pleistocene hominines as well as to *Gigantopithecus.*

Anterior Teeth. Incisors and canines are also very important in interpreting *Ramapithecus.* Unfortunately, most specimens assigned to this genus are without anterior dentition, leading to much speculation on the size and shape of those teeth. Once again it is the Rudabánya maxilla (Rud 12, *Rudapithecus*) which allows an *in situ* examination. Its central incisor, which is spatulate, is small in relation to the jugal teeth, and the lateral incisor is very reduced. But relative sizes very similar to these can be found in other Miocene hominoids, such as *Ouranopithecus.* The Rudabánya canine is barely above the occlusal level of the other teeth. The canine attributed to the Forst Ternan maxilla (KNM-FT 45) is also small (Andrews and Walker, 1976) and judging from the alveolus of YPM 13799, it could be the same in the case of the Haritalyangar upper jaw. Is this enough to characterize a separate genus? That would depend on the sexual variation in *Ramapithecus.* In the hominoids, living or fossils, the difference in size between male and female canines can be considerable (gorilla) or relatively small (human). If *Ramapithecus* resembled the first case, it could be considered the female representative of forms with better developed canines found at the same localities (*Sivapithecus* or *Bodvapithecus*). On both morphological and functional levels, the posterior side of the Fort Ternan upper canine has a long vertical wear facet which probably resulted from friction against the anteroexternal face of P_3, although the specimen is still a relatively young individual. In *Ouranopithecus* the upper canine shows a shorter vertical facet, but also an extensive basal facet. The Fort Ternan canine, therefore, seems to have a greater emphasis on shearing, a feature which does not support a phyletic position close to the Plio-Pleistocene hominines.

Premolar Morphology. The premolars of *Ramapithecus* are short relative to the length of the jugal tooth row, but this character is also found in other thick-enameled Miocene primates. There is also a large variation in premolar proportions in extant apes such as the gorilla or orangutan. The females have generally shorter premolars than do the males. The shape of the third lower premolar is very important for understanding the phyletic position of this genus. Among the living hominoids the problem is easy. Humans have a bicuspidate third premolar, more or less molarized, emphasizing more crushing than shearing. The apes have, in contrast, a single-cusp, lengthened, and sectoral third lower premolar. Is this simple criterion, presence or absence of the second cusp on P_3, adequate to distinguish a fossil hominine from a fossil ape? The solution is not evident, because it is possible to find a small accessory cusp on the epicrest of the chimpanzee; the tooth nevertheless emphasizes a shearing function. On the other hand, the australopithecines from Hadar, whose skeleton is fully hominine, may have a single-cuspidate P_3. Also, *Gigantopithecus* and *Oreopithecus* have a double-cuspidate P_3 (Simons and Ettel, 1970; Hürzeler, 1951), the second cusp appearing in both these taxa independently. The lack of the second cusp, as the presence of a small one on some specimens, may not have great phylogenetic significance. The morphology of this tooth, known now in several fossil specimens (Fort Ternan, Çandir, Paşalar), is more interesting than the simple account of cusps. There is a wear facet on the anteroexternal base of the crown, the honing facet, due to the friction of movement against the upper canine. This pattern is a result of a $\underline{C}–P_3$ shearing complex. The whole structure of the P_3, with the epicrest at sharp angle to the eocrest, projects out in the same way. Taking into account sexual dimorphic variation (as in the gorilla), it is possible to interpret any specimens called *Sivapithecus* and having a well-developed cutting $\underline{C}–P_3$ complex as males of the so-called *Ramapithecus* genus.

Size. The small size of *Ramapithecus* was an important factor in the original diagnosis of this genus, with Lewis (1934) noting "small and delicate jaws and teeth." This criterion is still often used in recent articles (e.g., Simons, 1977). Undertaking a detailed biometric study is difficult due to the small number of specimens. Nevertheless it is possible to take a look at the variation in the fossils allocated to this genus. I have chosen to examine M_2 and M_3, which are the most common teeth of *Ramapithecus* found. For both *Ramapithecus* and *Sivapithecus* only the measurements of the Siwalik material from Nagri level have been used, to avoid a sample which would be too heterogeneous. For comparison I have used two samples of gorilla and chimpanzee whose sexes were known. Sexual variation is marked in the gorilla sample, males being much larger than females. For chimpanzee, the variation is smaller and the sexual differences are not readily apparent on the graphs. Among the fossil groups, *Ouranopithecus* provides a good comparative sample from the upper Miocene. First, for M_2 (Fig. 5) the variation of the set *Ramapithecus* + *Sivapithecus* is larger than that seen in the gorilla. Taking an account the smaller mean size of this set, one may consider it to be too heterogeneous. But a sharp separation between the specimens called *Ramapithecus* and those called *Sivapithecus* is not apparent. There are several solutions. One alternative is that the variation of M_2 of *Ramapithecus* is small (as in chimpanzee), without

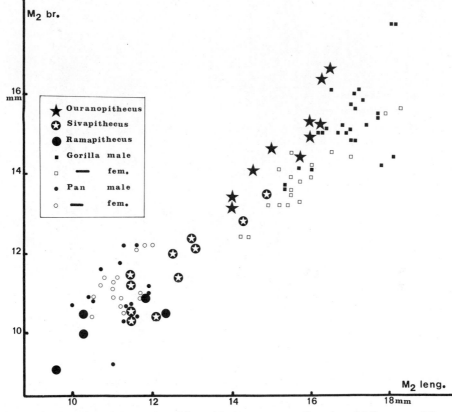

Fig. 5. Bivariate plot of the M_2 length/breadth measurements for selected Miocene and Recent hominoid primates. See text.

marked sexual differences, and the amplitude of the variation overlaps with that of the *Sivapithecus* females. A second alternative is that the smaller *Sivapithecus* group are males of *Ramapithecus* and the larger group belong to another species (*S. indicus?*). Finally, it is also possible that three species exist in the studied sample (Pilbeam *et al.*, 1977).

The problem is not much different for M_3 (Fig. 6). Except for an isolated tooth (GSI D-175) whose systematic assessment is unclear, the specimens plot pretty well into two sets. But in both of these sets the variation is less than that seen in the gorilla set, the *Ouranopithecus* set, or even the chimpanzee set. If there are really two different taxa (*Ramapithecus* and *Sivapithecus*), it must be that the variation within each of them is very small. If not, then the variation for the group as a whole is larger than in the chimpanzee but smaller than in gorilla. In this latter case it may be a sex-linked variation.

The hominoid fossils from new localities in Pakistan are still under study by D. Pilbeam (see Pilbeam *et al.*, 1980), but in every quarry the teeth show a variation ranging from small specimens to large ones, with the same problems as expressed above. It is the same for the locality of Paşalar in Turkey (An-

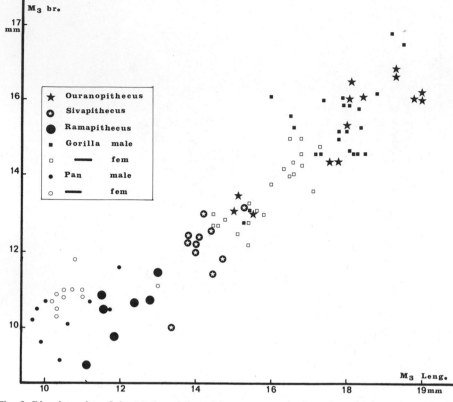

Fig. 6. Bivariate plot of the M_3 length/breadth measurements for selected Miocene and Recent hominoid primates. See text.

drews and Tekkaya, 1976; Andrews and Tobien, 1977). The ubiquitous presence in most of the *Ramapithecus* localities of another apparent taxon made up only of larger individuals with larger canines, such as *Bodvapithecus* (Rudabánya) or *Sivapithecus* (Siwaliks, Paşalar, Lufeng, etc.), leaves strong grounds for doubt about the validity of the genus *Ramapithecus*.

Forest-Dwelling Miocene Hominoids

Excluding the ramapithecines and, of course, the hylobatids, the remaining Miocene hominoids constitute a relatively homogeneous set which is often referred to as the dryopithecines. The dryopithecines are known principally from East Africa and western Europe. They were forest-dwelling primates. The East African early Miocene fossils have been recovered with a tropical forest fauna. Fossil seeds and fruits also indicate the same kind of environment. In western Europe, *Dryopithecus fontani* was found in the Saint Gaudens clay pit with the tragulid *Dorcatherium*, the cervid *Dicrocerus*, and the proboscidean *Deinotherium*. These three genera indicate that a forest was present in the middle Miocene, which is indirectly confirmed by the lack of open-environment animals. In Spain, a genus very similar to *Dryopithecus,* which is called *Hispanopithecus,* was found in late Astaracian and Vallesian localities (Crusafont, 1958; Crusafont and Hürzeler, 1961, 1969; Crusafont and Golpe, 1973). The fauna also indicates a wet tropical forest environment (tragulids, cervids, and large suids). The Astaracian fauna has been often compared, from an ecological point of view, to the recent Indo-Malaysian fauna (Crusafont, 1954; Crusafont and Golpe, 1971).

The dryopithecines have jugal teeth with thin enamel. The wear pattern is very characteristic. The enamel disappears very quickly on the top of the molar cusps and the dentine appears on all the molars. This kind of dentition, adapted to a diet with soft food (fruits, leaves, young plants), is typical of a forest environment. The canines are robust and the third lower premolar is high and flattened, with a large vertical honing facet.

The Genus Proconsul

It is possible to divide the dryopithecines into separate groups using both geographical and stratigraphical data. The first group is known from the early and middle Miocene deposits of Uganda and especially Kenya. This includes the genera *Proconsul, Limnopithecus, Rangwapithecus,* and *Micropithecus.* These genera differ from each other primarily by differences in the size and proportions of teeth. In all of them the jugal teeth, particularly the molars, have a thick cingulum on the lingual side of the upper teeth and on the labial side of lower ones (Andrews, 1978). The limb bones are known in *Proconsul,* but their analysis seems to have given rise to contradictory results. The first important work (Napier and Davis, 1959) concluded that the upper limb bones of *P. africanus* show many primitive (monkey-like) characters together with others which indicated an adaptation for a brachiating mode of life. Some primitive characters are, for example, the low brachial index and the lack of the elongation of the forelimb as a whole and the forearm in particular which are characteristic of modern aboreal brachiators. Napier and Davis (1959) also noted the shortness of the hand and concluded that *Proconsul* had a behavior closest to that of arboreal monkeys. This opinion fits well with what is known about the paleoenvironment in East Africa during the lower Miocene (Andrews and Van Couvering, 1975). Zwell and Conroy (1973), studying the forelimb of catarrhines by multivariate statistical analysis, argue that *Proconsul* was not a monkey-like quadruped and that the forelimb was much more similar to African apes than to monkeys. Nevertheless, for them, it is impossible to say whether or not *Proconsul* was a knuckle-walker. In their analysis *Proconsul* is plotted with African apes but it is also very near the monkeys. Because of the lack of data on *Homo* and *Pongo,* the significance of their study is not readily apparent. If some variables, such as radial neck length, separate quadrupeds from brachiators, others, such as trochlear projection, may indicate knuckle-walking or could be a hominoid derived character without any special locomotor significance.

Lewis (1972) claims that the wrist articulation of *Proconsul,* far from being of quadrupedal type, bears all the hallmarks of a meniscus-containing joint possessing the basic and unique attributes found only in extant Hominoidea; for him, brachiation appears to have been an early feature of hominoid evolution. But some would argue that the gibbon line and, to a lesser extent, the orangutan line diverged before the achievement of the *Proconsul* grade of structure.

Other recent studies provide a different view of the locomotor behavior and, inferentially, the phyletic position of *Proconsul.* Schön and Ziemer (1973) note, after a study of the anthropoid wrist, that the hand of *Proconsul* was not dorsoflexed and that its locomotion was probably palmigrade, indicating strong resemblances to New World monkey arboreal quadrupeds. Corruccini *et al.* (1975) demonstrate in a morphometric analysis that the wrist joint of *P. africanus* is completely monkey-like in nearly every essential anatomical detail. O'Connor (1976), studying the distal ends of the radius and ulna and some carpals, judged that, clearly, *Proconsul* was a quadruped in very much the same way that modern cercopithecoids are quadrupeds and that it would be premature to claim that the fossil wrist was more like modern cercopithecoid wrists than like some modern ceboid wrists. Morbeck (1975, 1976), in an analysis of the articular surfaces of the forelimb of *Proconsul,* reaches the conclusion that its movement capabilities can be reconstructed, and when compared to recent Primates, it was suggestive of palmigrade quadrupedalism. The knuckle-walking adaptations (modifications of the dorsal aspect of the radiocarpal joint) are totally lacking in *Proconsul.* McHenry and Corruccini (1975) show by a principal component analysis of distal humerus that the first principal component separates monkeys and hylobatids from other apes and hominids. *Proconsul* is intermediate, but nearest to the former. The second principal component separates all the apes and some early hominids from monkeys and modern humans. On the other hand, the femur of *Proconsul* (McHenry and Corruccini, 1976) is similar to the femur of cercopithecoids and not to the femur of living hominoids, which must have evolved after the early Miocene. The conclusion of all these workers is that *Proconsul* was probably an arboreal quadruped and was not a brachiator nor a knuckle-walker. But it shared some derived characters with Recent hominoids and, regarding its behavior, it was probably intermediate between monkeys and apes, particularly African apes.

Dryopithecus and the Origin of Pan

A second group of Miocene hominoids, more recent than the former, is made up of the European genus *Dryopithecus* and the closely related genus *Hispanopithecus.* When compared to *Proconsul* and allied genera, they exhibit several derived characters. They have a faint cingulum, which is weaker on both upper molars and the lower molars when compared with the cingulum of *Proconsul*. *Hispanopithecus* has relatively narrow and lengthened lower molars, whereas *Dryopithecus* has a more squared M_1 and M_2 and also a proportionally shorter M_3 (Table 1). These two features, a decrease in cingulum development and shortening of the molar row, are derived characters shared by *Dryopithecus* and *Pan*. The latter has several autapomorphies, such as the large size of its incisors, a more shortened M_3, the almost complete disappearance of the molar cingulum, and some details of the molar cusp pattern. But the two genera may be considered as sister groups. Little is known about the limb bones of *Dryopithecus*. Following Pilbeam and Simons (1971), the only known humerus and ulna of the European dryopithecines are very similar to that of *Pan paniscus*, the bonobo or pygmy chimpanzee (for an alternative view, see Morbeck, this volume, Chapter 14). Now, *P. paniscus* is a knuckle-walker and *Dryopithecus* could also have been a knuckle-walker. If Pilbeam and Simons are right, it is an additional derived character which is lacking in *Proconsul* and which adds another link between *Dryopithecus* and *Pan*. Together these two genera and probably *Hispanopithecus* constitute the sister group of *Proconsul*. The latter has plesiomorphous characters for both its dentition and limb bones (Fig. 7).

Proconsul and Gorilla

In *Proconsul* and *Gorilla* the same relation can be found between apomorphous and plesiomorphous characters as exists between *Proconsul* and the group *Dryopithecus–Hispanopithecus–Pan*. Pilbeam (1969) originally claimed that there was a direct line between *P. major*, the largest species of *Proconsul*, and *Gorilla*, with the small species *P. africanus* and *P. nyanzae* closely related to the ancestry of *Pan*. At first glance this hypothesis seems to be logical, but there are no special synapomorphies shared by *P. major* and *Gorilla* except for large size. Moreover, in light of more recent studies, *Proconsul* was an arboreal palmigrade quadruped, not a knuckle-walker. Of course, *Gorilla* and *Pan* are both knuckle-walkers. Following this hypothesis it would be necessary to admit that knuckle-walking appeared independently in both gorilla and chimpanzee lineages. It is more parsimonious to claim that knuckle-walking is a derived character shared by both African apes. In this case *Proconsul* and allied genera are the sister group of both African apes and *Dryopithecus*. The separation of the *Gorilla* line must have occurred before the separation of *Pan*, because *Dryopithecus* shares derived characters (molar proportions) with chimpanzee that are not shared with gorilla. It is difficult to find the placement of *Hispanopithecus*, whose limb bones are unknown. Its branching may have occurred before or after the gorilla's separation (Fig. 8). The above hypothesis would become testable by a better knowledge of the limb bones, particularly those of the wrist, the metacarpus, and the fingers of *Dryopithecus*. If I am correct, *Dryopithecus* would be a knuckle-walker. If one takes into account the fossil evidence, the divergence between *Gorilla* and *Pan*

Table 1. Lower Molar Length/Breath (L/B) Indices for Selected Miocene Hominoid Samples and for Sex-Sorted Samples of *Pan* and *Gorilla*

	R. gordoni	*L. legetet*	*P. africanus*	*P. nyanzae*	*P. major*	*D. fontani*	*P. troglodytes* males	*P. troglodytes* females	*Gorilla* males	*Gorilla* females
L/B M$_1$	130	126	115	118	115	100–105	109	109	120	116
L/B M$_2$	125	116	119	114	113	101–104	103	104	114	112
L/B M$_3$	139	123	126	122	128	113	101	101	114	112
LM$_3$/LM$_1$	140	118	130	137	141	118	96	95	104	107
LM$_3$/LM$_2$	114	107	113	116	146	103	93	92	95	98

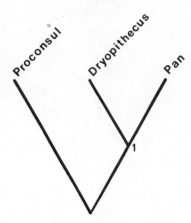

Fig. 7. Alternative cladogram of some Miocene to Recent hominoid primates. (1) Decrease in cingulum development; change in proportions of lower molar shape.

occurred before *Dryopithecus* and after *Proconsul,* somewhere between 22 and 14 m.y.a. This implies that the lineage which will give rise to Ramapithecini and Hominini branched out before gorilla, that is, about 20 m.y.a. or earlier, although thick-enameled hominoid primates are not known from localities much older than 15 m.y.a.

A recent theory (Zilhman *et al.,* 1978) claims that the bonobo (*Pan paniscus*) is a kind of archetype for the ancestor of living hominines and the African Great Apes. This hypothesis emphasizes cranial resemblances between living forms and takes into account the molecular clock data. The molecular data give distances between humans and living apes. Some data show that chimpanzees are closer to humans than to gorillas. Other data show chimpanzee and gorilla forming a monophyletic group which is the sister group of humans. Brachiation and knuckle-walking characterize the locomotion of chimpanzee and gorilla. In Zilhman *et al.*'s (1978) hypothesis that the bonobo is a prototype human ancestor there would have to be a phase of brachiation and knuckle-walking in the evolution of humans without any fossil evidence

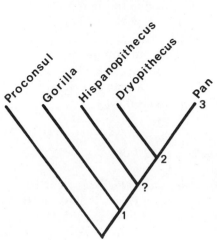

Fig. 8. Cladogram of the forest-dwelling hominoid primates. (1) Appearance of knuckle-walking, beginning of a decrease in cingulum development. (2) Further decrease in cingulum development, change in molar tooth proportions. (3) Disappearance of cingulum, decrease in length of M_3, change in molar tooth shape (*Dryopithecus* pattern is more or less lost).

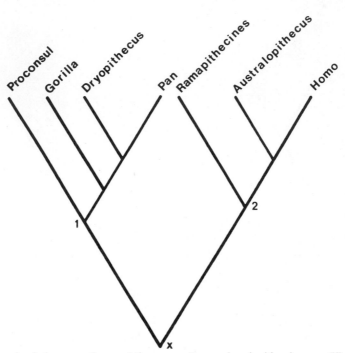

Fig. 9. Schematic cladogram of some Miocene to Recent hominoid primates. (X) Divergence point based on molecular data. (1) Beginning of the change in postcranial morphology (Zwell and Conroy, 1973; Lewis, 1972). (2) Appearance of thick enamel. For further nodes see legends to previous figures.

to support this contention. The hypothesis does not take in account the striking likeness between the teeth of some thick-enamel Miocene primates and those of early australopithecines. Finally, it would be necessary to believe in a tortuous evolutionary pathway, for example, the molar teeth of Miocene primates would become bonobo-like by decreasing in length, then *Australopithecus*-like by increasing, and finally human-like by a new decreasing.

I think the relationships between hominoids are better expressed by the cladogram of Fig. 9. The two groups, African apes and hominines, are linked by a node which represents a theoretical common ancestor whose behavior was that of an arboreal palmigrade quadruped very close to *Proconsul*. It was a dental ape with thin enamel and probably a nonsectorial P_3.

The Problem of Pongo and the Brachiation Question

All studies based on molecular biology show that *Pongo* is clearly separated from the group composed of *Pan, Gorilla,* and *Homo*. Following what I have written above on hominoid phylogeny, one would surmise that the divergence is relatively old, probably older than 20 m.y.a., somewhat before the

Proconsul step. *Pongo* is a brachiator, but not a knuckle-walker. Tuttle and Basmajian (1974) claim it is a palmigrade or fist-walker. On the other hand, the tooth enamel is thicker than that of African apes but thinner than in ramapithecines and hominines. The enamel shows a very characteristic wrinkling. Only specimens from the Asian Pleistocene may be allotted to the orangutan lineage, but they are so similar to the living orangutan that they do not contribute to an understanding of orangutan phylogeny. Other specimens found in the late Miocene of China would perhaps be potential orangutan ancestors. Two mandibles were described under the names *Sivapithecus yunnanensis* and *Ramapithecus lufengensis* (Xu and Lu, 1979). The two mandibles differ in general size and by the shape and size of canines and third lower premolars. But since they otherwise share similarities in incisor shape and molar cusp patterns, perhaps they actually belong to the same species. Both have wrinkled enamel on molar teeth as in *Pongo*. However, both also have very long and narrow incisors and *S. yunnanensis* (male?) has a very long, sharp, and narrow canine, which seem to be autapomorphous characters and differ significantly from orangutan canine and incisor morphology.

From the above discussions regarding the ancestry of both the Greater and Lesser Apes one important fact has emerged that now seems to have a high degree of probability. If paleoanthropologists accept the dental evidence presented in this chapter, then they will have to accept the fact that brachiation appears to have evolved three times in hominoid history. It would have evolved first in the hylobatids, second in *Pongo,* and last in the African apes, where it was subsequently modified into a knuckle-walking mode of locomotion.

Conclusion

The relationships that have been examined in this chapter may be expressed in terms of classification. In constructing a classification one encounters two kinds of problems. The first one, which is directly linked to the phylogenetic analysis, is to find the relative distances between the taxa. The method I have used is based on cladistic phylogeny—taxa are linked together according to the shared derived characters they possess. The second one is the problem of the rank of the taxonomic units, particularly the higher taxonomic units, and their boundaries. Whatever the method used (Simpsonian estimate of adaptational change, sequencing, recency of common ancestor, etc.), there is a measure of subjectivity, or rather, some high-handed decision-making. For example, for the hominid family, the boundary depends at once on a combination of scientific reasons and personal convictions in which ethics, philosophy, and religion are not always lacking. The species *Homo sapiens* has at times been raised to an ordinal rank or even a phylum rank! On the other hand, the specialty of the scientist may also have an important bearing on the choice. Morphologists, biochemists, and ethologists may all disagree on this problem. I tend to give great importance to cladistic relationships and I attempt to take into account evidence from comparative anatomy, molecular biology, and paleontology. I have shown that it is possible to separate hominoid primates into monophyletic groups or probable monophyletic groups. One of them is characterized by bipedalism, which is a derived character shared by *Australopithecus* and *Homo.* Together, they constitute the tribe Hominini. The term australopithecine, which represents a single genus, perhaps paraphyletic in origin, is not really needed as a systematic term. The tribe Hominini is in the hominid family, but this family also contains the other thick-enameled primates that I choose to call Ramapithecini (I also include *Gigantopithecus* in this tribe). The two tribes mentioned above make up the subfamily Homininae. It is difficult to precisely state the position of *Ramapithecus* in this classification. The status of this genus is uncertain; it probably should be included within the genus *Sivapithecus.* On the other hand, if it turns out to be a valid taxon and if it shares some derived characters with *Australopithecus* and *Homo,* it must be put in the same tribe. But since nothing about its affinities has yet been proven, I choose to place *Ramapithecus,* with grave doubt as to the validity of the genus, with the other Miocene thick-enameled primates.

For characterizing the African apes, the most important feature is knuckle-walking. Nevertheless, on the basis of dental characters, *Dryopithecus* seems to be close to *Pan* and it should be placed in the same group, the Dryopithecini. The same is true for the Spanish Miocene ape *Hispanopithecus.* Away from this group one must place the East African Miocene apes. These primates, which I call the Proconsulini, are plesiomorphous for a number of characters, but they share a few derived features in the postcranial skeleton and possibly in the $C–P_3$ honing complex. Should dryopithecines and proconsulines be placed in a separate family (Dryopithecidae) or included together in the hominid family? On the basis of results from molecular biological studies of the living apes and humans, I choose the second solution. The isolated placement of the orangutan, which is based more on molecular biological studies than on paleontological evidence, presents the possibility of assigning this taxon to the family Pongidae. The third family of hominoids, which was the first to branch off from the common ancestral trunk, is the hylobatid family. Documenting the fossil record of this family was not attempted in this chapter.

For the Miocene and more recent hominoid genera, excluding the hylobatids, I offer the following classification:

Hominoidea Gray, 1825
  Hylobatidae Gray, 1870
  Pongidae Elliot, 1870
    Pongo Lacepede, 1799
  Hominidae Gray, 1825
    Homininae Gray, 1825
      Hominini Gray, 1825
        Homo Linnaeus, 1758
        Australopithecus Dart, 1925

Ramapithecini Simonetta, 1957
  Sivapithecus Pilgrim, 1910
  ?Ramapithecus Lewis, 1934
  ?Bodvapithecus Kretzoi, 1975
  ?Rudapithecus Kretzoi, 1975
  ?Graecopithecus von Koenigswald, 1972
  Ouranopithecus de Bonis and Melentis, 1977
  Gigantopithecus von Koenigswald, 1935
Dryopithecinae Gregory and Hellman, 1939
Dryopithecini Gregory and Hellman, 1939
  Dryopithecus Lartet, 1856
  Hispanopithecus Villalta and Crusafont, 1944
  Pan Oken, 1816
  Gorilla Geoffroy, 1852
Proconsulini Leakey, 1963
  Proconsul Hopwood, 1933
  Limnopithecus Hopwood, 1933
  Rangwapithecus Andrews, 1974
  Micropithecus Fleagle and Simons, 1978

References*

Andrews, P. 1974. New species of *Dryopithecus* from Kenya. *Nature (Lond.)* **249**:188–190.
Andrews, P. 1978. A revision of the Miocene Hominoidea of East Africa. *Bull. Br. Mus. (Nat. Hist.) (Geol.)* **30**(2):85–244.
Andrews, P., and Tekkaya, I. 1976. *Ramapithecus* in Kenya and Turkey, in: *Les Plus Anciens Hominides* (P. V. Tobias and Y. Coppens, eds.), pp. 7–25, Colloque VI, IX Union Internationale des Sciences Prehistoriques et Protohistoriques, Nice. CNRS, Paris.
Andrews, P., and Tobien, H. 1977. New Miocene locality in Turkey with evidence on the origin of *Ramapithecus* and *Sivapithecus*. *Nature (Lond.)* **268**:699–701.
Andrews, P., and Van Couvering, J. A. H. 1975. Paleoenvironments in the East African Miocene, in: Approaches to Primate Biology (F. S. Szalay, ed.). *Contrib. Primatol.* **5**:62–102.
Andrews, P., and Walker, A. 1976. The primate and other fauna from Fort Ternan, Kenya, in: *Human Origins: Louis Leakey and the East African Evidence* (G. L. Isaac and E. R. McCown, eds.), pp. 279–304, Benjamin, Menlo Park, California.
Boule, M. 1915. Les singes fossiles de l'Inde. *Anthropologie (Paris)* **26**:397–410.
Burtschak-Abramovitch, N., and Gabachvili, E. 1950. Découverte d'un Anthropoïde fossile en Géorgie. *Priroda (Mosc.)* **9**:70–72 (in Russian).
Conroy, G., and Pilbeam, D. 1975. *Ramapithecus*: A review of its hominid status, in: *Palaeoanthropology, Morphology and Palaeoecology* (R. Tuttle, ed.), pp. 59–86, Mouton, The Hague.
Corruccini, R. S., Ciochon, R. L., and McHenry, H. M. 1975. Osteometric shape relationships in the wrist joint of some anthropoids. *Folia Primatol.* **24**:250–274.
Crusafont, M. 1954. Quelques considérations biologiques sur le Miocène espagnol. *Ann. Paleontol.* **40**:97–103.
Crusafont, M. 1958. Nuevo hallazgo del Pongido vallesiense *Hispanopithecus. Boll. Informativo Activ. Europeas Paleontol. Verteb.* **13–14**:37–43.

*Several references have been added which provide background information to the contents of this chapter; they are not cited in the text.

Crusafont, M., and Golpe, J. M. 1971. Biozonation des mammiferes Neogenes d'Espagne. *Bur. Recher. Geol. et Min., Lyon* **78:**121–129.

Crusafont, M., and Golpe, J. M. 1973. New Pongids form the Miocene of Valles Penedes Basin (Catalonia, Spain). *J. Hum. Evol.* **2:**17–23.

Crusafont, M., and Hürzeler, J. 1961. Les Pongidés fossiles d'Espagne. *C. R. Acad. Sci. Paris* **254:**582–584.

Crusafont, M., and Hürzeler, J. 1969. Catalago comentado de los Pongidos fosiles de Espana. *Acta Geol. Hisp.* **4**(2):44–48.

De Bonis, L., and Melentis, J. 1975. Première découverte de Muridés (Mammalia, Rodentia) dans le Miocène de la région de Thessalonique. Précisions sur l'âge géologique des Dryopithécinés de Macédoine. *C. R. Acad. Sci. Paris D* **280:**1233–1236.

De Bonis, L., and Melentis, J. 1976. Les Dryopithécinés de Macédoine (Grèce): Leur place dans l'évolution des Primates hominoïdes du Miocène, in: *Les Plus Anciens Hominides* (P. V. Tobias and Y. Coppens, eds.), pp. 26–38, Colloque VI, IX Union Internationale des Sciences Prehistoriques et Protohistoriques, Nice. CNRS, Paris.

De Bonis, L., and Melentis, J. 1977a. Un nouveau genre de Primate hominoïde dans le Vallésien (Miocène supérieur) de Macédoine. *C. R. Acad. Sci. Paris D* **284:**1393–1396.

De Bonis, L., and Melentis, J. 1977b. Les Primates hominoïdes du Vallésien de Macédoine (Grèce). Etude de la machiore inférieure. *Géobios* **10:**849–885.

De Bonis, L., and Melentis, J. 1978. Les Primates hominoïdes du Vallésien de Macédoine. Etude de la machiore supérieure. *Ann. Paleontol. Vertebr.* **64:**185–202.

De Bonis, L., and Melentis, J. 1980. Nouvelles remarques sur l'anatomie d'un Primate hominoïde du Miocène: *Ouranopithecus macedoniensis*. *C. R. Acad. Sci. Paris D* **286:**755–758.

De Bonis, L., Bouvrain, G., Keraudren, B., and Melentis, J. 1973. Premiers résultats des fouilles récentes en Grèce septentrionale (Macédoine). *C. R. Acad. Sci. Paris D* **277:**1431–1434.

De Bonis, L., Bouvrain, G., Geraads, D., and Melentis, J. 1974. Première découverte d'un Primate hominoïde dans le Miocène supérieur de Macédoine (Grèce). *C. R. Acad. Sci. Paris D* **278:**3063–3066.

De Bonis, L., Bouvrain, G., and Melentis, J. 1975. Nouveaux restes de Primates hominoïdes dans le Vallésien de Macédoine (Grèce). *C. R. Acad. Sci. D Paris* **182:**379–382.

Delson, E. 1977. Catarrhine phylogeny and classification: Principles, methods and comments. *J. Hum. Evol.* **6:**433–459.

Deperet, C. 1911. Sur la découverte d'un grand singe anthropïde du genre *Dryopithecus* dans le Miocène moyen de la Grive Saint Alban, Isère. *C. R. Acad. Sci. Paris* **53:**32–35.

Dutrillaux, B. 1979. Chromosomal evolution in Primates: Tentative phylogeny from *Microcebus murinus* (Prosimian) to man. *Hum. Genet.* **48:**251–314.

Dutta, A. K., Basu, R. F., and Sastry, M. V. A. 1976. On the new finds of hominids and additional finds of pongids from the Siwaliks. *Ind. J. Earth Sci.* **3:**234–235.

Fleagle, J., and Simons, E. 1978. *Micropithecus clarki*, a small ape from the Miocene of Uganda. *Am. J. Phys. Anthropol.* **49:**427–440.

Frayer, D. W. 1976. A reappraisal of *Ramapithecus*. *Yearb. Phys. Anthropol.* **18:**19–30.

Gaudry, A. 1890. Le Dryopithèque. *Mem. Soc. Geol. Fr.* **1:**5–11.

Genet-Varcin, E. 1969. *A la Recherche du Primate Ancêtre de l'Homme*, Boubée, Paris. 336 pp.

Gervais, P. 1849. Note sur une nouvelle espece de Singe fossile. *C. R. Acad. Sci. Paris* **28:**699–700.

Goodman, M., and Moore, G. W. 1971. Immunodiffusion systematics of the primates. I. The Catarrhini. *Syst. Zool.* **20:**19–62.

Goodman. M., and Tashian, R. E., Eds. 1976. *Molecular Anthropology*, Plenum, New York. 466 pp.

Gray, J. E. 1825. Outline of an attempt at the disposition of the Mammalia into tribes and families with a list of the genera apparently pertaining to each tribe. *Ann. Phil. N.S.* **10:**377–344.

Greenfield, L. O. 1973. Note on the placement of the most complete "*Kenyapithecus africanus*" mandible. *Folia Primatol.* **20:**274–279.

Gregory, W. K., and Hellman, M. 1939. The dentition of the extinct South African man-ape *Australopithecus (Plesianthropus) transvaalensis* Broom. A comparative and phylogenetic study. *Ann. Transvaal. Mus.* **19:**339–372.

Gupta, V. J. 1969. Fossil primates from lower Siwaliks. *Research Bull. (N. S.) Panjab Univ. India* **20:**577–578.

Harle, E. 1898. Une mâchoire de Dryopithèque. *Bull. Soc. Geol. Fr. Ser. 3* **26:**377–383.

Harle, E. 1899. Nouvelles pièces de Dryopithèque et quelques coquilles de Saint-Gaudens (Haute-Garonne). *Bull. Soc. Geol. Fr. Ser. 3* **27:**304–310.

Hennig, W. 1966. *Phylogenetic Systematics*, University of Illinois Press, Urbana.

Hopwood, A. T. 1933. Miocene Primates from Kenya. *J. Linn. Soc. Lond. Zool.* **38:**437–464.

Hürzeler, J. 1951. Contribution à l'étude de la dentition de lait d'*Oreopithecus bambolii*. *Eclogae Geol. Helv.* **44**(2):404–411.

Hürzeler, J. 1954. Contribution à l'odontologie et à la phylogénèse du genre *Pliopithecus* Gervais. *Ann. Paleontol.* **40:**5–63.

Johanson, D., and White, T. 1979. A systematic assessment of early African hominids. *Science* **203:**321–330.

Kalin, J. 1961. Sur les Primates de l'Oligocène inférieur d'Egypte. *Ann. Paleontol.* **47:**3–48.

Kay, R. F. 1981. The nut-crackers—A new theory of the adaptations of the Ramapithecinae. *Am. J. Phys. Anthropol.* **55:**141–151.

Kretzoi, M. 1975. New ramapithecines and *Pliopithecus* from the Lower Pliocene of Rudabánya in north eastern Hungary. *Nature (Lond.)* **257:**578–581.

Lartet, E. 1837. Note sur les ossements fossiles des terrains tertiaires de Simmorre, de Sansan, etc., dans le département du Gers, et sur la découverte récente d'une mâchoire de Singe fossile. *C. R. Acad. Sci. Paris* **4:**85–93.

Lartet, E. 1856. Note sur un grand singe fossile qui se rattache au groupe des singes supérieurs. *C. R. Acad. Sci. Paris* **43:**219–223.

Leakey, L. S. B. 1967. An early Miocene member of Hominidae. *Nature (Lond.)* **213:**155–163.

Leakey, L. S. B. 1968. Lower dentition of *Kenyapithecus africanus*. *Nature (Lond.)* **217:**827–830.

Lewis, G. E. 1934. Preliminary notice of new man-like apes from India. *Am. J. Sci.* **27:**161–179.

Lewis, G. E. 1936. A new species of *Sugrivapithecus*. *Am. J. Sci.* **31:**450–452.

Lewis, O. J. 1972. Osteological features characterising the wrists of monkeys and apes, with a reconsideration of this region in *Dryopithecus (Proconsul) africanus*. *Am. J. Phys. Anthropol.* **36:**45–58.

Lydekker, R. 1879. Further notices of Siwalik Mammalia. *Rec. Geol. Surv. India* **12:**33–52.

McHenry, H. M., and Corruccini, R. 1975. Distal humerus in hominoid evolution. *Folia Primatol.* **23**(3):227–244.

McHenry, H. M., and Corruccini, R. 1976. Affinities of Tertiary hominoid femora. *Folia Primatol.* **26**(2):139–150.

Morbeck, M. E. 1975. *Dryopithecus africanus* forelimb. *J. Hum. Evol.* **4:**39–46.

Morbeck, M. E. 1976. Problems in reconstruction of fossil anatomy and locomotor behavior: The *Dryopithecus* elbow complex. *J. Hum. Evol.* **5:**223–233.

Napier, J. R., and Davis, D. R. 1959. The forelimb skeleton and associated remains of *Proconsul africanus*. *Fossil Mammals of Africa* (Br. Mus. Nat. Hist.) **16:**1–69.

O'Connor, B. L. 1976. *Dryopithecus (Proconsul) africanus:* Quadruped or non-quadruped? *J. Hum. Evol.* **5:**279–283.

Ozansoy, F. 1965. Etude des gisements continentaux et des Mammifères du Cénozoïque de Turquie. *Mem. Soc. Geol. Fr.* **44:**5–89.

Ozansoy, F. 1970. Insani karabterli Turkiye pliosenfosil Ponjide si *Ankarapithecus meteai*. *Belleten (Ankara)* **34:**1–16.

Pilbeam, D. 1969. Tertiary Pongidae of East Africa: Evolutionary relationships and taxonomy. *Bull. Peabody Mus. Nat. Hist.* (Yale Univ.) **31:**1–185.

Pilbeam, D. 1970. *Gigantopithecus* and the origins of the Hominidae. *Nature (Lond.)* **225:**516–519.

Pilbeam, D., and Simons, E. L. 1971. Humerus of *Dryopithecus* from Saint-Gaudens (France). *Nature (Lond.)* **229:**407–409.

Pilbeam, D., Meyer, G. E., Badgley, C., Rose, M. D., Pickford, M. H. L., Behrensmeyer, A. K., and Shah, S. M. I. 1977. New hominoid primates from the Siwaliks of Pakistan and their bearing on hominoid evolution. *Nature (Lond.)* **270:**689–695.

Pilbeam, D. R., Rose, M. D., Badgley, C., and Lipschutz, B. 1980. Miocene hominoids from Pakistan. *Postilla* (Peabody Mus. Nat. Hist., Yale Univ.) **181:**1–94.

Pilgrim, G. E. 1910. Notices of new mammalian genera and species from the tertiaries of India. *Rec. Geol. Surv. India* **40:**63–71.

Pilgrim, G. E. 1915. New Siwalik primates and their bearing on the question of the evolution of man and the Anthropoidea. *Rec. Geol. Surv. India* **45**:1–74.

Prasad, K. N. 1962. Fossil primates from the Siwalik beds near Haritalyangar, Himachal Pradesh, India. *J. Geol. Soc. India* **3**:86–96.

Sarich, V. 1968. The origin of the hominids: An immunological approach, in: *Perspectives on Human Evolution*, Volume 1 (S. L. Washburn and P. C. Jay, eds.), pp. 94–121, Holt, Rinehart and Winston, New York.

Schön, M. A., and Ziemer, L. K. 1973. Wrist mechanism and locomotion behavior of *Dryopithecus (Proconsul) africanus*. *Folia Primatol.* **20**:1–11.

Simonetta, A. 1957. Catalogo e sinonimia annotata degli ominoidi fossili ed attuali (1758–1955). *Atti Mem. Soc. Toscana Sci. Nat. Ser. Ae, B* **63–64**:53–112.

Simons, E. 1961. The phyletic position of *Ramapithecus*. *Postilla* (Peabody Mus. Nat. Hist., Yale Univ) **57**:1–9.

Simons, E. L. 1964. On the mandible of *Ramapithecus*. *Proc. Natl. Acad. Sci. USA* **51**:528–535.

Simons, E. 1976. Relationships between *Dryopithecus, Sivapithecus, Ramapithecus* and their bearing on hominid origins, in: *Les Plus Anciens Hominides* (P. V. Tobias and Y. Coppens, eds.), pp. 60–65, Colloque VI, IX Union Internationale des Sciences Prehistoriques et Protohistoriques, Nice. CNRS, Paris.

Simons, E. L. 1977. *Ramapithecus*. *Sci. Am.* **236**(5):28–35.

Simons, E. L., and Chopra, S. R. K. 1969. *Gigantopithecus* (Pongidae, Hominoidea) a new species from north India. *Postilla* (Peabody Mus. Nat. Hist., Yale Univ.) **138**:1–18.

Simons, E. L., and Ettel, P. C. 1970. *Gigantopithecus*. *Sci. Am.* **222**(1):76–85.

Simons, E. L., and Pilbeam, D. 1965. Preliminary revision of the Dryopithecinae (Pongidae, Anthropoidea). *Folia Primatol.* **3**:81–152.

Simons, E. L., and Pilbeam, D. 1971. A gorilla-sized ape from the Miocene of India. *Science* **173**:23–27.

Steininger, F. 1967. Ein weiterer Zahn von *Dryopithecus (Dry.) fontani darwini* Abel 1902 (Mammalia, Pongidae) aus dem Miozän des Wiener Beckens. *Folia Primatol.* **7**:243–275.

Tekkaya, I. 1974. A new species of Tortonian anthropoid (Primate, Mammalia) from Anatolia. *Bull. Min. Res. Expl. Inst. Turkey* **83**:148–165.

Tuttle, R. H., and Basmajian, J. V. 1974. Electromyography of forearm musculature in *Gorilla* and problems related to knuckle-walking, in: *Primate Locomotion* (F. A. Jenkins, ed.), pp. 143–169, Academic, New York.

Vandebroek, G. 1969. *Evolution des Vertébrés*, Masson, Paris. 583 pp.

Villalta, J. F., and Crusafont, M. 1941. Hallazgo del *Dryopithecus fontani* Lartet en el Vindoboniense de la cuenca del Valles-Penedes. *Bol. Inst. Geol. Min. Espana* **55**:131–142.

Villalta, J. F., and Crusafont, M. 1944. Dos nuevos antropomorfos del Mioceno español y su situación dentro de la moderna systemática de los simidos. *Notas Comun. Inst. Geol. Min. Espana* **13**:91–139.

Vogel, C. 1975. Remarks on the reconstruction of the dental arcade of *Ramapithecus*, in: *Palaeoanthropology, Morphology and Palaeoecology* (R. Tuttle, ed.), pp. 87–98, Mouton, The Hague..

Von Koenigswald, G. H. R. 1935. Eine fossile Säugetierfauna mit *Simia*, aus Südchina. *Proc. K. Ned. Akad. Wet. Amsterdam* **38**:872–879.

Von Koenigswald, G. H. R. 1949. Bemerkungen zur *"Dryopithecus" giganteus* Pilgrim. *Ec. Geol. Helvet.* **42**:515–519.

Von Koenigswald, G. H. R. 1972. Ein unterkiefer eines fossilen hominoiden aus dem Unterplioczän Griechlands. *Proc. K. Ned. Akad. Wet. Amsterdam B* **75**:385–394.

Von Koenigswald, G. H. R. 1973. The position of *Proconsul* among the Pongidae. *Symposia Fourth International Congress of Primatology*, Vol. 3, *Craniofacial Biology of Primates*, pp. 148–153, Karger, Basel.

Walker, A., and Andrews, P. 1973. Reconstruction of the dental arcades of *Ramapithecus wickeri*. *Nature (Lond.)* **244**:313–314.

Washburn, S., and Ciochon, R. 1974. Canine teeth: Notes on controversies in the study of human evolution. *Am. Anthropol.* **76**:765–784.

Woodward, A. S. 1914. *Dryopithecus fontani*. *Q. J. Geol. Soc. Lond.* **19**:316–320.

Woo, J. K. 1957. *Dryopithecus* teeth from Keiyuan province. *Vertebr. Palasiat.* **1**:25–29.

Woo, J. K. 1958. New materials of *Dryopithecus* from Keiyuan, Yunnan. *Vertebr. Palasiat.* **2**:38–42.

Xu, Q., and Lu, Q. 1979. The mandibles of *Ramapithecus and Sivapithecus* from Lufeng, Yunnan. *Vertebr. Palasiat.* **17**:1–13.

Zihlman, A., Cronin, J., Cramer, D., and Sarich, V. M. 1978. Pygmy chimpanzee as a possible prototype for the common ancestor of humans, chimpanzees and gorillas. *Nature (Lond.)* **275**:744–746.

Zwell, M., and Conroy, G. C. 1973. Multivariate analysis of the *Dryopithecus africanus* forelimb. *Nature (Lond.)* **244**:373–375.

Ramapithecus and Human Origins

25

An Anthropologist's Perspective of Changing Interpretations

M. H. WOLPOFF

Introduction

According to the Baconian view of how science proceeds, the changing interpretations of *Ramapithecus* indicated in this volume (and elsewhere) reflect a shifting data base. Presumably, as more data have been recovered, the hypotheses regarding their interpretation have correspondingly shifted to adequately encompass a more complete (and presumably more revealing) data set. Because it has been contended at various times that *Ramapithecus* is the earliest hominid,* these changing interpretations have in turn resulted in different theories of hominid origins.

In this chapter, I argue that this does not describe what actually happened. Instead, it has been shifting hypotheses that have altered the in-

---

*By "hominid" I specifically mean those taxa on the lineage leading to *Homo sapiens,* and any collateral sidebranches of this lineage, after the divergence of this lineage with the one leading to the African apes.

M. H. WOLPOFF • Department of Anthropology, University of Michigan, Ann Arbor, Michigan 48109.

terpretation of the data base. The shifting hypotheses have been about hominid origins and not about *Ramapithecus* at all, because by in large the
importance of *Ramapithecus* has historically been in the claim that it represents
the earliest hominid. This argument is very Popperian, in that it regards the
interpretation of data as a consequence of theoretical framework and the
function of new data recovery mainly as a potential refutation of current
hypotheses (Wolpoff, 1976*b*, 1978). Unlike Popper (1957), or at least the
interpretations that have been given to his works (Halstead, 1980; but see
Popper, 1980), I regard evolutionary theory as a *scientific* theory, and specific
evolutionary hypotheses such as those concerning hominid origins as *scientific*
hypotheses because they *are* potentially refutable. The changing interpretations of *Ramapithecus* reflect just such a series of refutations.

Thus, whether or not *Ramapithecus* is a hominid, a hominid ancestor, or a
collateral sidebranch to the hominids, it is my contention that a complex
interplay of theory, analysis and interpretation, and accumulating discoveries
has wedded the interpretation of *Ramapithecus* with the problem of human
origins.

Ramapithecus and Darwin's Theory

The Darwinian theory of hominid origins has remained very powerful,
and in one way or another it has influenced virtually every attempt to hypothesize about this event. Darwin posited what we would call a positive feedback
relationship between what he viewed as the four critical elements that distinguish humans from the African apes: bipedalism, tool use, canine reduction, and the expansion of the brain. He hypothesized a fundamental adaptive shift associated with the origin of the human line, in which an arboreal
adaptation was exchanged for a terrestrial one and a primarily frugivorous
diet was replaced by one emphasizing meat obtained by hunting. The tools
attained their importance through their use in hunting and in defense. Bipedalism evolved as an adaptation for freeing the hands during locomotion so
that tools and weapons could be carried and used easily. Canines reduced as
tools replaced their functions in cutting, slashing, and social displays. Lastly,
expanding brain size resulted from selection for more complex cooperative
behavior and language, both of which were viewed as critical to the adaptations discussed above.

The initial interpretation of *Ramapithecus* as a hominid ancestor was completely within this Darwinian framework, just as the initial interpretations of
Australopithecus africanus (Dart, 1925) and *Gigantopithecus blacki* (Woo, 1962)
were. Thus, the characteristics Lewis (1934) emphasized in his hominid interpretation of the *Ramapithecus* maxillary remains were the parabolic dental
arcade, small canines that are transversely expanded, lack of a functional
diastema, and a small degree of maxillary prognathism. Because a small canine, and features related to it, represented the only characteristics in a max-

illa that can be related to the Darwinian theory, focus was brought on the interpretation of their morphology.

When Simons (1961) resurrected the *Ramapithecus* argument, he emphasized the same canine-related features, even removing one mandible from the taxon that did not fit the functional model because of its sectorial P_3. In a subsequent publication (Simons, 1964) assigning a number of mandibles to the taxon, one of the primary criteria he used was an inward turning of the corpus at the M_1 position, indicating a parabolic arcade and a small snout. These interpretations were fully consistent with the canine reduction aspect of Darwin's model.

Similarly, L. S. B. Leakey's (1962) publication of the *Kenyapithecus* maxilla emphasized the small canine seemingly associated with it (the tooth was found several feet away), although his discussion indicated that the tooth was probably used in cutting. While *Kenyapithecus* was not formally regarded as a hominid in this publication, its detailed resemblance to the *Ramapithecus* palates was noted.

In a later publication, L. S. B. Leakey (1967) formally allocated *Kenyapithecus* to the hominids, emphasizing the vertically short canine crown and roots, the small, shovel-shaped incisor crowns, and (what he termed) the arcuate shape of the dental arcade. Moreover, like Dart, he sought evidence for the behavioral implications of the reduced canine, ultimately claiming to have found artificially smashed bone at Fort Ternan (L. S. B. Leakey, 1968) as an indication that tools were being used.

The interpretive framework for these materials was probably most explicitly stated in a paper by Pilbeam (1966). *Ramapithecus* was claimed to be "completely hominid in known parts"; the hominid characteristics were found mainly in the features associated with canine reduction described above. Moreover, Pilbeam argued that since the small canines were ineffective in agonistic behavior and group defense, "presumably, weapon use was established by this time." Finally, because "food must have been *prepared* for chewing by non-dental means; hands were probably used extensively and perhaps tools as well. . . . The evidence, admittedly circumstantial at present, suggests a primate perhaps already bipedal and fully terrestrial" (p. 3).

With reduced canines, tool use, and (provisionally) bipedalism included in the *Ramapithecus* paradigm, there remained only increased intelligence to complete the predictions of Darwin's hypothesis. Indirect arguments suggesting improved intellectual capacities for the species were presented by Simons (1972) with an analysis of differential molar wear. He claimed that *Ramapithecus* differed from contemporary dryopithecines in showing a greater difference in wear between the adjacent molars (Steeper wear gradient). The steeper gradient was interpreted to indicate a longer period of time between successive molar eruptions, and consequently delayed maturation of *Ramapithecus* offspring. The maturational delay was presumably in order to learn more complex behaviors during childhood.

In sum, *Ramapithecus* was considered a hominid because the known remains fit the applicable aspects of Darwin's model. Thus, it was possible to use the model to speculate or interpret within the framework of interrelations that it provided. The arguments and interpretations indicating tool use, bipedalism, and more complex behavior clearly followed.

The Baconian interpretation would suggest that Darwinian-based arguments were ultimately dismissed because new data suggested a new theory about the phylogenetic status of *Ramapithecus*. Indeed, the period of the earlier 1970s when this framework was effectively questioned was also a period of intensive fossil discovery. However, for several reasons, I do not believe that the two phenomena are causally related.

The fit of the *Ramapithecus* data to the Darwinian model involved much more interpretation than actual analysis. The canine, for instance, was regarded as small and reduced long before a canine was found (the Fort Ternan specimen), and even when a canine was recovered and showed the morphology of a honing tooth, its "small size and reduction" were still emphasized. It did not require new materials to question these interpretations. Following the original paper by Lewis (1934), Hrdlička (1935) systematically rejected the morphological arguments supporting the hominid interpretation, and other workers showed that the relative sizes of the canines and the other anterior teeth do not differ from some of the African pongids (Wolpoff, 1971*a;* Yulish, 1970). The reconstruction of the dental arcade as parabolic was questioned (Frayer, 1976) and ultimately shown to be incorrect (Vogel, 1975; Greenfield, 1978). The evidence for a steep molar wear gradient in *Ramapithecus* was questioned (Greenfield, 1974), and in any event the presence of steep gradients in primate species without delayed maturation had already been shown (Mann, 1968; Wolpoff, 1971*b*). The questions described above [as well as others from this period (von Koenigswald, 1972; Robinson, 1967)] were not raised because new materials were discovered. Their basis was in the original specimens and their interpretations. However, historically it was too early for criticism to be effective and these objections were not widely noted. First, the Darwinian hypothesis of *hominid* origins had to be replaced by another.

Ramapithecus and Jolly's Theory

Publication of Jolly's (1970) "seed eaters" hypothesis received widespread attention (Jolly, 1973) and was generally (although not universally) well accepted among paleoanthropologists. Although Robinson (1963) had earlier argued that the masticatory apparatus of the *"Paranthropus"* remains from South Africa indicated a vegetarian adaptation in what he regarded as a little modified descendent of the original australopithecine stock, it was Jolly who publicized a formal evolutionary hypothesis indicating mechanisms that connected hominid *origins* with a diet-oriented masticatory shift and the behaviors associated with it.

Indeed, while Jolly focused on *"Paranthropus"* (and O.H. 5 from East

Africa) in support of his ideas, it soon became clear that the earlier *Australopithecus africanus* evidenced the same masticatory adaptation (Wolpoff, 1973). The model was immediately applied to the interpretation of *Gigantopithecus bilaspurensis* (Simons and Ettel, 1970), which was then still considered an "aberrant ape" (Pilbeam, 1970). The application to *Ramapithecus* was somewhat slower.

Jolly's model linked canine reduction, upright posture and finally bipedalism, the development of thumb opposability, the appearance of language, the development of single male groups, and the evolution of a masticatory apparatus adapted for powerful grinding and crushing, to a dietary and behavioral adaptation emphasizing the exploitation of seeds and other small objects. Jolly did not deal with increases in brain size (or more generally, behavioral complexity) since, by the end of the 1960s, it was evident that Pliocene *Australopithecus* had an essentially ape-sized brain. Moreover, Jolly did not accept the arguments for delayed maturation (Mann, 1968) or for neural reorganization (Holloway, 1966), so that in essence there was nothing to explain. For the above reasons, and since the early hominid dietary adaptation involved plant foods rather than meat (as Darwin had suggested), neither culture nor tool use played a role in the hominid origins model (much to the apparent relief of some of the more paleontologically oriented paleoanthropologists).

Interpretation of a terrestrial adaptation based on small-object feeding and a powerful-masticatory-apparatus found its way into the *Ramapithecus* discussions as new materials were discovered and older specimens reanalyzed. This interpretation was clear in the reconstruction and analysis of the Fort Ternan specimen (Andrews and Walker, 1976) and helped mitigate the effects of the narrow, parallel-sided dental arcade and associated mandible with a sectorial P_3 on the hominid interpretation of the specimen [although this process actually took some time, during which the premolar in question was "semi-sectorial" or "incipiently bicuspid" (Simons, 1976)].

By the earlier 1970s, the molars of already known *Ramapithecus* specimens were being reexamined and evidence was found that was interpreted to show powerful mastication in this form (Simons and Pilbeam, 1972), including thick enamel, interproximal attrition, and a steep molar wear gradient (the alternative to the delayed maturation interpretation). Other features now brought into focus were the flat, deep face, vertical incisors, and heavily buttressed mandible (especially at the symphysis).

Newly discovered specimens were interpreted in a framework that itself was changing (Wolpoff, 1975). Description of the Çandir mandible (Andrews and Tekkaya, 1976), for instance, emphasized the buttressing of the mandibular corpus and symphysis, and the shortening of the anterior face, while the shape of the dental arcade confirmed the Fort Ternan reconstruction. However, the sectorial P_3 was still being regarded as "molarized" (Simons, 1976). While the Pyrgos mandible was the only one known of the Greek specimens, it was described as extraordinarily australopithecine-like (Walker, 1976) with an "arcuate" mandibular arcade (Simons, 1978), although the (remains of) large molars and a thick mandibular body were also given some attention. The numerous more recent discoveries of much more complete Greek specimens have been interpreted somewhat differently (de Bonis and Melentis, 1977, 1978).

The Hungarian finds from Rudabánya (Kretzoi, 1975) were interpreted as *Ramapithecus*-like forms, although not actually allocated to *Ramapithecus*. Indeed, Kretzoi (1976) found greater similarities between *Rudapithecus* and early *Homo* than between *Rudapithecus* and *Australopithecus*. Simons (1976) disagreed with both points, allocating this form to *Ramapithecus* and noting the dental and gnathic adaptations for powerful mastication. At the same time, he compared the canine form and wear on the Rud 12 maxilla with the Hadar australopithecine palate A.L. 200, as well as with *Gigantopithecus*. The ground was shifting for the functional interpretation of the *Ramapithecus* canine; it no longer was incisiform [as it was in the middle of the decade (Conroy and Pilbeam, 1975)], but with the recovery of the Hadar australopithecines it could still be related to the hominids. In a similar manner, once analysis of australopithecine dental arcades showed that they were not parabolic (Genet-Varcin, 1969), this condition was no longer found in *Ramapithecus* (Simons, 1977; Simons and Pilbeam, 1978).

With the addition of the Çandir and *Rudapithecus* specimens to the *Ramapithecus* sample, it became generally accepted that *Ramapithecus* combined a short face, thick molar enamel, a nonparabolic dental arcade (with varying degrees of posterior divergence), a relatively thick corpus and symphysis, and widely divergent zygomatic processes in a pattern that so clearly indicated powerful mastication (Wolpoff, 1974; Hylander, 1979) that at least one author described the complex as most resembling a miniature hyper-robust australopithecine (Walker, 1976). At the same time, the canine function was seen to be pongid-like (although also like the earliest hominids). With the acceptance of Jolly's model, some of the criticisms of one decade were incorporated into the interpretations of the next.

Thus, by the closing years of the last decade, a firm case was being made that *Ramapithecus* showed the dental and gnathic adaptations of a powerful masticator, and could be considered a hominid because it fit Jolly's model of hominid origins (Simons and Pilbeam, 1978). These characteristics (in contrast to those emphasized by Darwin) related to an adaptive shift involving terrestrial small-object feeding in semi-open or open ecozones (Pilbeam, 1979). This interpretation took most of the decade to fully develop because the anterior teeth were never really deemphasized until the discoveries at Rudabánya and Hadar allowed a rather different comparison to sustain the hominid interpretation.

In all, this was truly a case in which (to paraphrase Samuel Butler) the foundations were changed while the superstructure remained the same. The focus shifted from the front to the back of the jaw, and *Ramapithecus* remained a hominid.

Although Jolly's hypothesis was about hominid origins, and the examples it drew upon were Pliocene *hominids*, it was the application of this model for interpreting *Ramapithecus* that carried the seeds of its destruction. Three factors combined to set the stage for questioning whether the seed eaters hypothesis can account for *hominid* origins. The first of these involves the widespread occurrence of the powerful masticatory complex, both within the primates and beyond them. Long before the hypothesis was published, the dental and gnathic characteristics associated with powerful grinding and crushing were recognized in a species whose diet was clearly not small objects. This complex, (perhaps mistakenly) referred to as the "T complex" because of Jolly's analogy using *Theropithecus,* was earlier described in the giant panda (Sicher, 1944) in a comparison with bears that parallels Jolly's comparison of *Theropithecus* and *Papio.* Indeed, in his discussion of this bamboo-eating species, Davis (1964) found the closest analogy to "*Paranthropus robustus.*" A similar morphological complex characterizes some of the ceboids (Kinzey, 1974). The point is that in living forms, the dental/gnathic complex described by Jolly does not necessarily indicate small-object feeding, let alone a terrestrial adaptation. It does not necessarily lead to hominization. Consequently, the same argument must apply to the interpretation of *Ramapithecus* as well as of other fossil primates. Thus, for instance, White (1975) indicates the possibility of a giant panda-like diet for *Gigantopithecus* and suggests that the pandas may have replaced this primate. Kay (1981) describes alternative dietary interpretations for the *Ramapithecus* specimens themselves (nut-eating) that involve neither a terrestrial adaptation nor any aspects of hominization hypothesized by Jolly. These arguments tend to remove some of the cause and effect aspects of the seed eaters hypothesis by showing that a morphological adaptation to powerful grinding and crushing does not require a diet of seeds or grains. Moreover, there is an ample variety of other difficult-to-masticate food sources that early or pre-hominids might have utilized (Coursey, 1973; Wolpoff, 1973; Kay, 1981); selecting for the same morphological complex.

Second, the fact is that Jolly's hypothesis has always been weak in its explanation of how other basal hominid features might have followed from small-object feeding. For instance, whatever the validity of the argument that small canines remove the restriction of canine interlock and allow free lateral movement of the jaws, the argument can only account for canine reduction; it does not account for the change in canine form and function in the hominids. Bipedalism, too, has never been adequately explained by this hypothesis and attempts to do so (Wrangham, 1980) have been less than convincing. Finally, not all workers are as willing as Jolly to regard culture, tool-making, and expanded behavioral complexity as hominid attributes that evolved after hominids originated. It is not clear that these lack all causal relation to hominid origins. Thus, while small-object feeding could account for some of the features found in the earliest Pliocene hominids, this dietary behavior does not necessarily account for others.

Third, the focus on this dental/gnathic complex resulted in renewed support for the earlier claims that *Ramapithecus* was similar, or identical, to the other Asian hominoids (Frayer, 1976, 1978; Greenfield, 1974, 1975, 1977, 1979) because many of these earlier claims were based on the same diet-related characteristics, but these features were not regarded as especially important under the Darwinian hypothesis. Of these hominoids, the special similarities (if not identity) of *Ramapithecus* and *Sivapithecus* (Andrews and Tekkaya, 1980; Greenfield, 1977, 1980; Frayer, 1978; Pilbeam, 1979), the combined mandibular morphology of both taxa in one species of *Ouranopithecus* (de Bonis and Meletis, 1980), and the likelihood that these hominoids were markedly dimorphic (Frayer, 1976; Greenfield, 1979; Pilbeam, 1979; Wolpoff, 1980) suggest that at the very least the paradigm for the genus *Ramapithecus* must be expanded to include more specimens and a wider range of variation. The features that attained importance in the newer interpretation, which was a consequence of the framework provided by the seed eaters hypothesis, were widespread, and when sexual dimorphism was taken into account, it became far from clear how many species were actually involved in what has increasingly come to be regarded as a single adaptive radiation (Wolpoff, 1980) which might be referred to by the nontaxonomic term "ramapithecine" (see also Kay and Simons, this volume, Chapter 23). By many accounts, the ramapithecines would include specimens allocated to *Ramapithecus, Kenyapithecus, Sivapithecus, Ankarapithecus, Rudapithecus, Bodvapithecus, Ouranopithecus, "Hemianthropus,"* and *Gigantopithecus* (Pilbeam, 1979; Wolpoff, 1980).

An adaptive radiation can result from the appearance of a new adaptation, with consequences that allow a previously unutilized set of niches to be entered (Simpson, 1953). Rapid speciation almost invariably follows (Stanley, 1979) and the resulting taxonomic group soon becomes highly diversified. In the ramapithecines, the common element in the remains now known for the group is the masticatory apparatus, adapted for a diet requiring powerful or prolonged grinding and crushing. It is likely that this reflects the novel adaptation that was the basis of the subsequent radiation. The exploitation of otherwise unusable dietary resources would allow adaptation to new ecozones because this dental/gnathic adaptation is not an adaptive specialization; it acts to expand the range of usable resources.

Apart from the masticatory complex, other common elements are difficult to identify, since few skeletal parts besides jaws and teeth have been found. Moreover, features unique to the ramapithecines cannot always be clearly distinguished from shared primitive features (such as the retention of marked sexual dimorphism), since little is known of the ancestral condition. Recent evidence suggests that the ramapithecines evolved from a *Proconsul* or *Proconsul*-like form of approximate *P. nyanzae* size (Pickford, 1982).

Temporal and geographic considerations alone suggest the existence of a fairly large number of ramapithecine species. Of greater importance is the number of lineages in the adaptive radiation. Although workers such as Pilbeam (1979) have indicated five or more, the fact is that no ramapithecine-bearing locality has provided evidence of more than two contemporary lineages, and in many cases the data can be interpreted to show only one. This could be a consequence of the generalizing aspects of the masticatory morphology underlying the radiation. Instead of more finely subdividing the new niche, competition between the emerging ramapithecines seems to have promoted their rapid spread. The resulting pattern emphasizes allopatric more than sympatric species proliferation in a manner similar to the distribution of baboons and baboon-like forms such as *Theropithecus.* Unlike those cercopithecoids, however, the extent of allopatric species proliferation may have been markedly greater.

There is evidence of variation within the ramapithecine adaptive radiation. This tends to be obscured by two circumstances. First, gross size would appear to be the most dramatic variant between the recognized ramapithecine forms (molars, for instance, range from smaller than *Homo sapiens* to *Gigantopithecus* size). The problem this creates is one of separating variation due to scaling from that resulting from other causes. Even still, the adaptive importance of size variation over this range should not be understated. Second, because jaws and teeth are in the most usual fossil remains, most comparisons are limited to the very adaptive complex that forms the basis of the radiation, and consequently the one that would not be expected to show dramatic adaptive differences within it.

However, there are important variants that can be observed. For instance, enamel thickness differs continuously from the extremely thick condition in *Ouranopithecus, Gigantopithecus,* and *"Hemianthropus,"* to a thinner expression in the Rudabánya forms. Even at individual sites this feature varies considerably. Thus at Rudabánya the larger specimens (*Bodvapithecus*) have quite thick enamel, while the smaller ones range from this thickness to a thinner condition such as that seen at its extreme in Rud 12. [Kay and Simons (this volume, Chapter 23) probably had this specimen in mind as the basis for their claim that molar enamel at Rudabánya is thin, and consequently that *Rudapithecus* should be allocated to the European dryopithecines rather than to the ramapithecines. However, Rud 12 lies at the low end of a marked range of enamel thickness variation, even within the *Rudapithecus* remains]. Similarly, marked variation can be found in the deeply incised molar wrinkles that characterize some of the ramapithecine specimens. As in the case of enamel thickness, the variation is one of frequency between samples. Cingulum expression represents yet another varying feature. Variation also can be seen in the development of the P_3 metaconid, which ranges from complete absence to the full development of an equal-sized cusp. [Kay and Simons (this volume, Chapter 23) report that the *frequency* of P_3 metaconid enlargement observable in some of the ramapithecine specimens (excluding *Gigantopithecus*) is comparable to that observed in chimpanzees and orangutans (although not gorillas). However, they further claim that the *degree* of development in some of the ramapithecine specimens exceeds that which they have observed in these living apes. This forms part of their argument for the contention of a special relationship between the ramapithecines and the hominids. Yet, the fact is that this feature is quite variable in the radiation, ranging from the gorilla condition (no metaconid) to the *Homo*-like bicuspid form of the *Gigantopithecus* premolars, with cusps of equal size].

Few nondental features allow comparison between the ramapithecine forms. The distal humeri from Hungary and Pakistan differ notably; the Fort Ternan humerus could represent a third variant, but its association with the African ramapithecine is uncertain. On the other hand, the cranial remains from Hungary, Turkey, Pakistan, Greece, and China are surprisingly similar.

The relation of the adaptive radiation of ramapithecines to hominid origins will be discussed below. The acceptance of the notion of a ramapithecine adaptive radiation provides the third basis for questioning the seed eaters model as a hypothesis about *hominid* origins. This is because the ramapithecine adaptive radiation was highly successful in terms of geographic range and survivorship of the taxonomic group. It recently became evident that there were at least two nonhominid Pleistocene survivors.

Ramapithecus and the Late Divergence Hyopthesis

The idea of a fairly recent divergence between humans and apes is hardly new. Early contentions of a late divergence were influenced by the very short estimate of the earth's age that preceded a full understanding of radioactive decay. The late divergence hypothesis as presented by Greenfield (1980, and this volume, Chapter 27) specifically focuses on the divergence between humans and the *African* apes, emphasizing that there were two different divergence points in the evolution of the recent hominoids. This distinction is an important one for reasons that will be discussed below.

Late divergence between humans and African apes, and an earlier separation of *Pongo,* has been supported morphologically, genetically, biochemically, and most recently on the basis of the fossil record itself.

Morphologically, it has long been recognized that the chimpanzee is the most human-like of the pongids (Huxley, 1861, 1863; Gregory, 1930; Simpson, 1963; Washburn, 1968) and the idea of a very *Pan*-like ancestor for the hominid line has been maintained for decades (e.g., Schwalbe, 1923; Coolidge, 1933; Gregory, 1934; Weinert, 1944; Washburn, 1968; Zihlman *et al.,* 1978; Zihlman, 1979), although not necessarily like *Pan paniscus* (McHenry and Corruccini, 1981; Johnson, 1981). This special relationship is one important aspect of the late divergence hypothesis: the morphological data provide information about divergence sequence, although not absolute date.

These morphological comparisons are independently supported by genetic analysis. Chromosome banding studies show an especially close relation of *Homo, Pan,* and *Gorilla,* with *Pongo* somewhat more divergent (Miller, 1977; Yunis *et al.,* 1980). The comparison of protein coding sequences indicates an extraordinary genetic similarity between *Homo* and *Pan* (King and Wilson, 1975). These genetic comparisons reveal less difference between the two genera than is common between sibling species. Taken at face value, these data would tend to indicate recent divergence as well as extreme closeness of relationship for the African hominids, and a more distant relation and earlier divergence for *Pongo.*

Biochemical analysis also supports the particular closeness of relationship between *Homo, Gorilla,* and *Pan.* In a recent series of summaries for numerous genetic systems (Goodman, 1976; Dene *et al.,* 1976), it was found that immunodiffusion studies, nucleotide replacements, and the analysis of various proteins consistently show humans, chimpanzees, and gorillas to be more closely related to each other than any of these are to orangutans. Thus, the same divergence sequence is supported that a number of workers suggested earlier on the basis of morphology.

Biochemical studies, in addition, have been used to calibrate a divergence "clock" for this sequence (Sarich and Wilson, 1967; Sarich, 1974; Sarich and Cronin, 1976). This "clock" seems to provide direct evidence for a very late *Pan–Homo* divergence (estimates based on this procedure have ranged between 3.5 and 5.5 m.y.a.). However, the procedures that result in these estimates give other divergence times that are at significant variance with virtually any interpretation of the fossil record (Uzzell and Pilbeam, 1971; Jacobs and Pilbeam, 1980; Read and Lestrel, 1972; Radinsky, 1978; Walker, 1976, and references therein). Moreover, there have been an extraordinary number of criticisms of the molecular "clock" (Lovejoy *et al.,* 1972; Lovejoy and Meindl, 1973; Read, 1975; Read and Lestrel, 1970; Jukes and Holmquist, 1972; Corruccini *et al.,* 1980; Goodman, 1974; Fitch and Langley, 1976; Jukes, 1980). Probably the best way to summarize the very disparate points raised is that the "clock" simply *should not* work. This conclusion supports the paleontological analyses that claim the "clock" *does not* work when applied to divergence times over broad time spans. Consequently, although biochemical evidence seems to support a late *Pan–Homo* divergence, I believe this is a red herring, and that the molecular "clock" does not support any divergence time, just as other independent evidence for late *Pan–Homo* divergence does not support the molecular "clock." While specific divergence *dates* may be rejected, I do not believe it is possible to dismiss the implications of the biochemical evidence about divergence *sequence* (Greenfield, this volume, Chapter 27).

Finally, paleontological evidence also supports the contention of an especially close relation between the hominid line and the ancestors of the African apes. The relationship has long been recognized for *Australopithecus africanus* (Le Gros Clark, 1947), and the discovery of an even closer approach to the chimpanzee condition in *Australopithecus afarensis* (Johanson and White, 1979) was not surprising. While the Pliocene fossil evidence can provide no more than a minimum divergence date (4 m.y.a.), the number of primitive and specifically champanzee-like features in the known crania support the contention of a late divergence.

In sum, these data strongly support the notion that the divergence of the lineage leading to *Pongo* from the lineage leading to *Homo, Pan,* and *Gorilla* was earlier than the divergence of these three African hominoids from each other. The data further suggests that the later (African) divergence might have been fairly recent.

The effect of the late divergence hypothesis on the interpretation of the ramapithecines is a consequence of the earlier divergence proposed for *Pongo*. This became evident as the first two Pleistocene survivors of the ramapithecine adaptive radiation were recognized. One of these, *Gigantopithecus*, survived into the middle Pleistocene, where contemporaneity with hominids has been established (Hsu *et al.*, 1975). The other survivor is *Pongo*.

Once again, the assertion that *Pongo* evolved from a ramapithecine was not initially established from recently discovered data, although the interpretations and implications of this contention are recent. Gregory and Hellman (1926) noted a number of morphological similarities between *Sivapithecus* and *Pongo* molars and suggested an ancestral relationship. Of course, these authors did not regard *Sivapithecus* as a ramapithecine (*Ramapithecus* had yet to be recognized). The similarities to *Pongo* can now be shown to extend throughout the ramapithecine dental remains, and include the following variably expressed features: (1) enamel thickness; (2) deeply incised wrinkles that persist after crown morphology has been worn away (especially in the Lufeng ramapithecines and "*Hemianthropus*," but also sporadically in the remains from Pakistan and Hungary); (3) asymmetric heteromorphic lateral maxillary incisor size and form; (4) central maxillary incisors that change in angulation during life so that in younger individuals a lingual wear plane typically extends from the tip to the base; (5) the morphological details of the molars noted by Gregory and Hellman (1926).

Newly discovered ramapithecine cranial remains also support the hypothesis of a ramapithecine ancestry for *Pongo*. In particular, the face and partial cranium from Pakistan (GSP 15000) reveals a suite of extraordinarily *Pongo*-like details (Pilbeam, 1982; see also Ward and Pilbeam, this volume, Chapter 8). The lower face and palate from Turkey (MTA 2125) has been similarly described: "the closest comparisons in most cases were with the orang-utan" (Andrews and Tekkaya, 1980, p. 94). The upper face from Rudabánya (Rud 44) is also characterized by *Pongo*-like characteristics, including relatively tall orbits, a very wide outer-orbital area, and marked converging anterior temporal ridges. Finally, numerous *Pongo*-like features characterize the newly discovered Lufeng crania (Wu, 1981). As in the Rudabánya remains, the most marked resemblances are in the upper face and frontal, including the marked temporal ridges, the wide outer-orbital (zygofrontal) pillars, supraorbital form, and the shape of the orbits (Xu and Lu, 1980).

Interestingly, Kay and Simons (this volume, Chapter 23) argue against a special affinity to *Pongo* in the ramapithecine facial remains, although their discussion does not include the most complete of these, GSP 15000 and the Lufeng crania. Their contention is mainly based on the Turkish face, MTA 2125, and involves arguments that are less than convincing. For instance, in claiming that the specimen is not as prognathic as *Pongo*, their measure of maxillary prognathism is actually one of relative alveolar height (alveolar height/palate breadth at M2) and bears no relation to how *prognathic* the premaxilla might be, independent of its relative size. Similarly, they admit to the marked maxillary incisor heteromorphy in the specimen, but feel it does not align it with *Pongo*, because the heteromorphy in this living ape was presumably attained by a relatively

large I^1 (this was calculated relative to molar size, and consequently potentially confuses I^1 expansion with molar reduction). Besides positing an *ad hoc*
explanation for what appears to be a simple relationship, this argument ignores the important morphological similarities of the MTA 2125 (and other)
I^2 form to *Pongo*. I differ with these authors, and would argue that the central
and lower portions of the known ramapithecine faces do show specific resemblances to *Pongo*. Based on the faces from Rudabánya, Ravin, Turkey,
Pakistan, and Lufeng, these similarities would seem to include the incisor
heteromorphy (metric *and* morphological), the changing incisor wear plane,
the marked premaxillary prognathism as measured by the angulation of the
premaxilla, and the strong superior–medial angulation of the canine roots.
However, as discussed above, I believe that an even stronger resemblance
between these faces and *Pongo* is shown in the upper face and frontal as
represented at Rudabánya, Pakistan, and Lufeng.

Moreover, I would emphasize the conclusions of Kay and Simons' discussion somewhat differently. While these authors seek to disprove a ramapithecine ancestry for *Pongo*, and instead establish a special relation between the
ramapithecines and the hominids, I believe that what they have actually
shown is the potential common ancestry this group might have for all of the
living hominoid forms. To the features they discuss might be added the
variable characteristics of the ramapithecine dentitions described above, including cingulum development, molar wrinkling, enamel thickness, and P_3
metaconid expression. The fact is that ramapithecine variation in these facial
and dental features would potentially allow the ancestry of every living hominoid group to be found.

Yet, the fact remains that the most specific resemblances are to *Pongo*.
Only one ramapithecine lineage could be ancestral to living orangutans; on
the basis of the present evidence (and geographic proximity) this would almost certainly be an Asian lineage. However, if one of the ramapithecines is
ancestral to *Pongo*, the implications are far-reaching for the entire adaptive
radiation.

There are two sets of hypotheses about how the ramapithecines may be
related to hominids, depending on when the ramapithecines appeared relative to the split between the line leading to *Pongo* and the line leading to the
African hominoids, and subject to the constraints of the data discussed above.
First is the hypothesis that the split between the African hominoids and the
line leading to *Pongo* occurred *before* the ramapithecines evolved. Second is
that this split took place *within* the ramapithecines.

Under the first hypothesis, the ramapithecines would have to be associated with either the *Pongo* line or the African hominoid line. The evidence for
the ancestry of *Pongo* among the ramapithecines indicates that if this hypothesis were correct, no ramapithecines contributed to hominid ancestry, because
they were on the *Pongo* side of the split (see Andrews, this volume, Chapter
17). It would then follow that the numerous dental and gnathic similarities of
the earliest hominids (especially *A. afarensis*) to the ramapithecines would have
to be interpreted as parallel independent acquisitions. This is far from impos-

sible, since most of the similarities are in the dental/gnathic complex associated with powerful mastication, and, as noted above, this complex has appeared again and again in unrelated forms as a common response to a similar adaptation. Thus, this interpretation cannot be easily dismissed.

On the other hand, it was exactly this sort of evidence involving similarity in the dental/gnathic complex and common functional adaptation that led to the contention of a ramapithecine adaptive radiation. If the same criteria were applied to the jaws and teeth of *A. afarensis,* the functional interpretation of the morphology of this extraordinarily megadont early hominid form would suggest the same conclusion, mainly that *A. afarensis* is part of or closely related to the ramapithecine adaptive radiation. Indeed, this is how R. E. F. Leakey (1976) interpreted the Hadar female A.L. 288-1. Thus, I consider the first hypothesis to be the far more unlikely one (Wolpoff, 1981). Indeed, unless one is willing to postulate a fairly extensive amount of parallel or even convergent evolution in the hominoids, the hypothesis can probably be rejected.

The second hypothesis indicates some sort of a ramapithecine ancestry for hominids as well as for *Pongo.* If the split with *Pongo* took place *within* the ramapithecines, one could account for the fundamental aspects of both sets of interpretations that have developed in the last decade: the relationship of the ramapithecines to *Pongo* and the similarities of the ramapithecines to the earliest hominids. Indeed, an alternative way of stating this hypothesis is that the earliest hominids are part of the ramapithecine adaptive radiation.

Under this hypothesis there is a ramapithecine ancestry for the African apes [as discussed by Greenfield (1977)], but these forms diverged markedly from the ancestral condition because of their specific dietary adaptations, with effects on their dental morphology and function described by Kay (1975, 1981). At the same time, however, it would account for the specific resemblances of *Australopithecus* and *Pan* by presuming a late divergence between them.

This hypothesis thus necessarily posits a specific ramapithecine ancestor for the hominids, a species which is *also* ancestral to *Pan* and *Gorilla.* It is possible that one of the Eurasian ramapithecines represents this ancestor, but I suspect that the limited evidence now available does not allow a decision as to which (if any) of these lineages is the most likely (Wolpoff, 1980). Traditional comparisons have emphasized the hominid-like aspects of the smaller species (albeit as a hominid and not as a common ancestor of the hominids and African apes) such as *Ramapithecus* or *Rudapithecus.* But detailed dental morphology and size considerations could argue for one of the larger forms (*Sivapithecus, Ouranopithecus*). However, middle Miocene geography would suggest the possibility that none of the Eurasian forms represent this ancestor.

If an African ramapithecine is a more likely candidate for this ancestor, the situation is little improved. There are virtually no hominoid fossils known between the Fort Ternan ramapithecines and *A. afarensis;* Lukeino, Ngorora, and Lothagam samples could be interpreted as either ramapithecines or early hominids, although there are substantial similarities between Lothagam and the mandibles from Laetoli. It is possible that the Fort Ternan ramapithecine represents or is closely related to the latest common ancestor of the African hominoids. On the other hand, it may be too early. The fact is that the date of the split between the African and Eurasian forms is unknown, and this split could be later than the Fort Ternan remains. Thus, in my view there is no ramapithecine ancestor for the African hominoids that can be identified unambiguously at this time. While it is possible that one of the Eurasian forms is this ancestor, I believe it more likely that as the African fossil record spanning the later Miocene is better explored, a suitable ramapithecine lineage (perhaps beginning with Fort Ternan) will be found.

Such a lineage would be expected to retain the dental/gnathic complex associated with powerful or prolonged chewing. At the same time, one might expect the more elongated cranial form also emphasizing anterior dental loading that characterizes the African apes in contrast to the shortened, more vertically oriented forms of *Pongo* crania. If one described this hypothetical ramapithecine as a chimpanzee-like vault with a ramapithecine face and dentition (with the consequent molar-loading-related superstructures on the vault), this description is fairly close to the known cranial remains of *A. afarensis* (Johanson, 1980).

The locomotor complex of this ramapithecine might be expected to reflect an at least partial arboreal adaptation (climbing, arm hanging) without the brachiating specializations of the African apes, or their associated knuckle-walking quadrupedalism. This is suggested by the lumbar elongation of the australopithecines (STS 14 has six functional lumbars) combined with the arm-hanging abilities inherent in the human upper torso, and even more markedly expressed in *Australopithecus*.

In sum, I believe the evidence best supports the late divergence hypothesis presented by Greenfield (1980). I would suggest that an adaptation for foods requiring powerful or prolonged chewing arose among one of the proconsuls, and that because this provided the basis for utilizing a much wider range of dietary resources, a very successful adaptive radiation of hominoids resulted. With the geographic spread of the ramapithecines, the radiation was split into Eurasian and African branches (the *Pongo*–African hominoid split), and perhaps into European and Asian branches as well. I predict that the eventual discovery of more complete remains will demonstrate a much wider adaptive range within this radiation than the analysis of the dental/gnathic complex now indicates. It would appear that the western portion of the Eurasian ramapithecines became extinct, while there were at least two Pleistocene survivors to the east: *Gigantopithecus* and *Pongo* (the so-called "giant orang" teeth and the "*Hemianthropus*" specimens* from south China may represent a third Asian Pleistocene survivor).

*The confusion of "*Hemianthropus*" teeth with *Australopithecus* teeth (von Koenigswald, 1957) is a consequence of their similar size and shared primitive characteristics related to the basal hominoid masticatory adaptation. These teeth are actually worn versions of the thick-enameled dental remains that have been attributed to "giant orangs." Even in Asian middle Pleistocene deposits, it is often difficult to distinguish worn orangutan postcanine teeth from worn hominid teeth.

Relatively little is known of the African ramapithecine branch.* Genetic evidence relating *Pan* and *Homo,* and the morphological relations of *Pan* and *A. afarensis* indicate that during the late Miocene, one of the African ramapithecine lineages further split into lines leading to the adaptively specialized African apes, and a hominid line.

Ramapithecus and Hominid Origins

While I have argued that none of the ramapithecines are hominids, I propose that the earliest hominids were a special form of ramapithecine. I believe this is a distinction *with* a difference, because it brings a different focus to the problem of hominid origins than is usually applied.

The seed eaters hypothesis, or something like it, would seem to pertain to *hominoid* rather than to *hominid* origins. Yet, this does not mean that powerful mastication was *unimportant* in the process of hominid origins. This adaptive complex simply did not originate then. Similarly, the evidence for chimpanzee toolmaking that has accumulated over the past two decades increases the likelihood that the (African ramapithecine) common ancestor of hominids and the African pongids was a toolmaker.† If so, this does not mean toolmaking was unimportant in the events leading to and following hominid origins, but rather that it did not originate then.

---

*I predict that when found the late Miocene African ramapithecine will be identifiably similar and fairly closely related to the known Eurasian remains. Craniodentally it will resemble *Australopithecus afarensis,* although with much more projecting and sexually dimorphic canines. It will *not* resemble a chimpanzee to a significantly greater extent than *A. afarensis* already does, and it will especially not resemble a pygmy chimpanzee. It is likely that at the time of the human–African ape divergence, character displacement and other consequences of competition had initially greater effects on the apes than on the human line. Thus, I suggest that the last common ancestor of the African apes and humans was probably as different from living apes as it was from living humans. No living ape, even an enculturated one such as described by Kortlandt (1972), could form an adequate model for this ancestral form, any more than a living human group could.

†There has been a continued confusion between *toolmaking* and *stone toolmaking* over the last decade. Numerous assertions that stone tools do not predate 2.5 m.y.a. and the absence of any stone tools associated with *A. afarensis* have been taken to mean that toolmaking originated after hominid origins and thus did not play a role in the event (e.g., Lovejoy, 1981). Undoubtedly, stone toolmaking was a critical discovery in hominid evolution. Its development may well have been associated with the divergence of the two hominid lineages of the earliest Pleistocene and the marked and rapid development of features associated with hominization in one of them. Nonetheless, it is likely that the manufacture of tools made of perishable materials, and the use of unmodified stone, greatly preceded the development of recognizable lithics. The potential adaptive importance of nonlithic tools could hardly be overstated (Wolpoff, 1980). In view of the fact that this behavior is shared with *Pan* (Harding and Teleki, 1981), its origin likely predates the hominid–African pongid divergence, and changes in the importance of this behavior are a potentially critical aspect of hominid origins.

In fact, what seems to have uniquely developed in the hominids is pretty much the Darwinian complex that has been discussed for the last 100 years, but with two modifications. First, marked brain size increase is not a dramatic event directly associated with hominid origins. The evolution of modern brain size (and presumably the cultural/behavioral changes that came with it) has taken most of the Pleistocene. Evidently, only moderate or even minimal endocranial expansion can be associated with hominid origins. Holloway (1980) has published figures for Jerison's encephalization quotient (EQ) for a sample of *Pan*, where

$$EQ = \frac{\text{brain weight}}{0.12(\text{body weight})^{0.666}}$$

The midsex value for *Pan* is 2.96. Using the midsex cranial capacity for *A. africanus* which I have determined (443 cm^3), and an approximate midsex body weight average of 40 kg (Wolpoff, 1973) results in an EQ of 3.20. These data, with the evidence for neural reorganization (Holloway, 1966, 1976), suggest that while some brain-related changes might have been associated with hominid origins, their magnitude was clearly small compared with subsequent Pleistocene changes. The extent of actual change will be better estimated when data for the earlier species *A. afarensis* are available.

The second modification is contextual, and involves a distinction between origins and importance. As an example, Darwin emphasized the causal influence of hunting in his model of adaptive change in the earliest hominids. Applying Jolly's model to the earliest stages of hominoid evolution and taking the continued evolution of the powerful masticatory apparatus in subsequent hominid evolution (Wolpoff, 1980) into account, it appears likely that organized hunting did not play a preeminent role in the earliest stages of hominization. Once again, however, this does not imply that organized hunting played *no* role in hominid origins, especially since almost all of the elements of organized hominid hunting appear in chimpanzee behavior (Galdikas and Teleki, 1981). Like toolmaking, this behavior was probably characteristic of the ancestral African ramapithecine form.

Of the elements discussed by Darwin, it is possible that only bipedalism actually originated at the time when the hominids became a distinct lineage. Toolmaking and the beginnings of organized hunting may have preceded this event, while significant expansion of the brain and functional change in the canine probably followed it. Similarly, the development of a powerful masticatory apparatus seems to have preceded the event.

Thus, the ramapithecine ancestor of the hominids and the African apes had already undergone a number of changes that are generally regarded as hominization. Shared characteristics of the living hominoids combined with the paleontological evidence discussed here indicate that this ramapithecine form was behaviorally complex, a tool user, and a rudimentary tool maker, an omnivore utilizing a wide variety of dietary resources ranging from foods difficult to masticate to protein obtained through organized hunting and systematically shared by at least part of the social group, an incipient biped (the behavior was possible but not morphologically efficient), and just possibly a more complex communicator than generally thought, utilizing an albeit limited but symbol-based open communications system. Hominid origins would seem to have involved as much a shift in the importance of these characteristics as the origin of uniquely hominid ones.

In sum, I propose that the continued reanalysis of *Ramapithecus* has ultimately affected the acceptable model of hominid origins, just as the reverse has been the case. In my view this continued reanalysis has resulted in the contention that the ramapithecines were not hominids, but rather that the earliest hominid was a ramapithecine. Many of the interpretations and implications of Jolly's model have been focused on the problem of hominoid, rather than hominid origins. Finally, this reanalysis, along with the last decade's advances in pongid behavioral studies and fossil hominid recovery, renews the focus on a modified Darwinian model for the origin of our lineage.

An Ecological Model of Hominid Origins

In my view, there clearly were changing interrelated ecological adaptations associated with hominid origins. These changes are best discussed in the context of the fact that the effects on the African *pongid** lineage (or lineages) were at least as great as on the hominids. It is critical to remember that *hominid* origins are also *African ape* origins.

The actual speciation event separating hominid and (African) ape lineages will probably always remain unknown, since it need not be directly associated with any of the adaptive changes that followed. The genetic isolation and ultimate speciation of ramapithecine populations was not necessarily a consequence of adaptive divergence (although this is always a possibility). However, following this event, the adaptive shifts in the two lineages could be interpreted as the result of subsequent competition between them. As I reconstruct the process, apes presumably reduced competition through dietary (and eventually dental/gnathic) specialization and locomotor changes (true brachiation, knuckle-walking), allowing an effective woodland/forest adaptation. Precluded from these ecozones by competition, the hominid adaptation was to more open regions. Building on their ramapithecine inheritance, a combination of powerful masticatory apparatus, the probably rapid development of efficient bipedalism, the use of rudimentary tools and weapons (digging sticks, clubs), and a series of social changes possibly related to the recog-

---

*If this model is correct, "pongid" is no longer an appropriate name for this African group, since Pongidae is named after *Pongo* and it is my contention that *Pongo, Pan,* and *Gorilla* no longer form a natural group by themselves (they would only if *Homo* were included). Their relationship is one of *grade,* and they can only be referenced together by a nontaxonomic term such as the "Great Apes."

nition of kinship relations as a basis for social behavior (Wolpoff, 1980) allowed a wide range of foods that were difficult to gather and difficult to masticate to help form an effective adaptation to a unique open-country niche.† One would suspect that dietary items included seeds and grains, nuts, roots, and hunted, gathered, and scavenged protein during the dry season (Coursey, 1973; Peters, 1979). Wet season food resources cover a much wider potential range and do not necessarily involve difficulties in obtaining or masticating.

Judging from its expression in *A. afarensis*, bipedalism would appear to have been an early critical aspect of the developing hominid adaptation. The main advantages of bipedal locomotion—freeing the hands, carrying, and long-distance, energy-efficient stride—probably all played a role in what was very likely a rapid locomotor shift. Arguments about whether carrying babies was more important than carrying clubs (i.e., Lovejoy, 1981) miss the entire point of this locomotor change; a *group* of early hominids could carry *all* of the items that have been deemed important in the various arguments about the origin of bipedalism.

The *A. afarensis* dentitions show that unlike bipedalism, the change in the canine cutting complex was more gradual. Various individuals in this earliest hominid species show canine and premolar wear that indicates a polyfunctional range, including chimpanzee-like honing (White, 1981), occlusal chiseling (Wolpoff, 1979; Wolpoff and Russell, 1981), and flat grinding (Taieb *et al.*, 1975). There is a corresponding polymorphism in the form of the P_3, ranging from single-cusped forms (sectorial) to a bicuspid morphology with equal-sized cusps (Johanson, 1980) and including all of the variants in between. Thus, while the functional change in the cutting complex may have

---

†This suggested model for divergence makes no assumptions about the niche of the ancestral ramapithecine species. This group may have been primarily arboreal or largely terrestrial prior to the speciation event. I believe that actual finds, ecological associations, and analysis of the locomotor skeleton (when discovered) will ultimately bring the needed data to resolve this question. However, given the consequences of competition following speciation, one credible hypothesis is that the initial niche was a broad one, involving dense to fairly open parklands with some utilization (perhaps seasonal) of even more densely forested localities. Besides best fitting a dietary regime indicated by the masticatory complex, such a hypothesis would allow one to view the initial results of competition as dividing the niche occupied by the parent ramapithecine species into two narrower and less overlapping adaptive zones. These would be the more open parklands grading onto savanna (hominids) and the denser woodlands grading into forest (African apes). The effects of subsequent competition on the differing adaptations outlined by this initial division would presumably be continued niche divergence that proceeded until the adaptive zones were sufficiently separate to significantly reduce competition over limiting resources. This model ties the appearance of a terrestrial, open grassland adaptation in the hominids to a *combination* of competitive exclusion *and* opportunism allowed by the expanded dietary resource base available to a hominoid species with both a powerful masticatory apparatus and rudimentary tool and weapon use. The model implies that there is no necessary link between the specific development of a powerful masticatory apparatus and any terrestrial adaptation (*contra* Pilbeam, 1979). Moreover, it suggests a ramapithecine adaptation to ecological circumstances that would support any of the several models of preadaptations for bipedalism that have been recently published (Tuttle, 1975; Post, 1980; Stern and Susman, 1981).

begun with hominid origins, it required continued selection to attain the modern condition in which the canine is morphologically and functionally incisor-like, and the P_3 is incorporated into the grinding dentition.

It is tempting to suggest that the graduality of this change reflects the gradualness with which tools replaced the cutting functions of the canine, but this argument is essentially circular and requires independent conformation. Moreover, I have hypothesized (Wolpoff, 1979) that the development of the bicuspid P_3 crown, with a ridge connecting the two cusps, may have provided a means of retaining a form of the cutting function of the canine while reducing its projection and overlap. This would presumably be an intermediate step in the process of functional change in the anterior cutting complex. Whatever the case, the association of this change with tool use remains an unsettled issue.

Finally, just as brain size expansion, associated with complex behavior and cooperation, were important in Darwin's model (we would term these developments "cultural" today), I believe that the origin of the cultural adaptation was probably central in this modification of it. Behavioral interpretations without obvious morphological correlates are very difficult to assess, and the relevant morphological data (relative brain size, endocast analysis) have not yet been published for the earliest hominid species. Nevertheless, the later species *A. africanus* shows evidence of both limited brain size expansion and neural reorganization, as well as delayed maturation, and it is surely short of a wild leap of faith to hypothesize that these correlates of cultural behavior had their origin in the social changes associated with the hominid adaptations to open country. Indeed, the specific elements suggested by Darwin, cooperation and language, may well have played a critical role in the group adaptations to this ecozone. Two decades of baboon studies have shown the importance of structured cooperative behavior in the savanna adaptations of this species, and the ramapithecine ancestor of the hominids probably brought a much more complex repertoire to the behavioral basis of the hominid adaptation, judging from the behavioral complexity of the African apes.

Whether or not culture (meaning structured learned behavior) actually originated with the hominids or developed as part of their successful open-country adaptation, its effects are demonstrable in the morphology of *A. africanus*. This early hominid species had already embarked on an evolutionary pathway that was and has remained unique.

Conclusions

I remarked at the beginning that the complex historical interplay between theories of hominid origins and the interpretations of *Ramapithecus* affected the development of both. In many respects, what I have described is a full circle in which modified versions of the original hominid origins theory and the original interpretation of *Ramapithecus* can be sustained, but not quite in the way they were first presented, and incorporating the bulk of the discoveries, interpretations, and criticisms that have appeared along the way. In the circumscribing of this circle, it is clear that the development of one could not have preceded without the development of the other, which is to say that there are neither factless theories nor theoryless facts.

If *Ramapithecus* itself was not a hominid, there is a great likelihood that the earliest hominid was a ramapithecine. If all of the details of Darwin's theory of hominid origins are not fully correct, virtually every one of them must still be accounted for by any current origins hypothesis. If Jolly's model cannot be applied to *hominid* origins, a modification of it may have critical importance in the interpretation of *hominoid* origins, and in any event the dental/gnathic complex he descibed for hominids was there at their beginning and played an important role in their earliest adaptive changes.

Of the participants in the intertwined developments of the last few decades, if it can be said that if none were completely right, it is also true that few were completely wrong, at least in the context of the model I have presented. Moreover, the interpretations that can be sustained in one form or another far outnumber those that must be completely rejected. In all, the historical development of human origins theories and ramapithecine interpretations presents a satisfying contrast to the Piltdown fiasco, and reflects the scientific aspect of paleoanthropological studies in a most positive manner.

ACKNOWLEDGMENTS

For permission to examine the specimens in their care, I am very grateful to P. Andrews, L. de Bonis, C. K. Brain, D. J. Johanson, G. H. R. von Koenigswald, M. Kretzoi, M. Leakey, R. E. F. Leakey, J. Melentis, M. Pickford, D. Pilbeam, Wu Ru-kang, P. V. Tobias, and A. C. Walker. I think D. W. Frayer, L. D. Greenfield, W. Jungers, R. Kay, F. B. Livingstone, M. Russell, and L. Shepartz for help in preparing the manuscript. This research was supported by NSF grant BNS 76-82729.

References

Andrews, P., and Tekkaya, I. 1980. A revision of the Turkish Miocene hominoid *Sivapithecus meteai*. *Palaeontology* **23**:85–95.

Andrews, P., and Walker, A. 1976. The primate and other fauna from Fort Ternan, Kenya, in: *Human Origins: Louis Leakey and the East African Evidence* (G. L. Isaac and E. R. McCown, eds.), pp. 279–304, Benjamin, Menlo Park, California.

Conroy, G. C., and Pilbeam, D. 1975. *Ramapithecus:* A review of its hominid status, in: *Paleoanthropology, Morphology and Paleoecology* (R. Tuttle, ed.), pp. 59–86, Mouton, The Hague.

Coolidge, H. J. 1933. *Pan paniscus:* Pygmy chimpanzee from south of the Congo river. *Am. J. Phys. Anthropol.* **18**:1–59.

Corruccini, R. S., Baba, M., Goodman, M., Ciochon, R. L., and Cronin, J. E. 1980. Non-linear macromolecular evolution and the molecular clock. *Evolution* **34:**1216–1219.

Coursey, D. G. 1973. Hominid evolution and hypogeous plant foods. *Man* **8:**634–635.

Dart, R. A. 1925. *Australopithecus africanus:* The man-ape of South Africa. *Nature (Lond.)* **115:**195–199.

Davis, D. D. 1964. The Giant Panda. A morphological study of evolutionary mechanisms. *Fieldiana: Zool. Mem.* **3:**1–339.

De Bonis, L., and Melentis, J. 1977. Les primates hominoides du Vallésien de Macédoine (Grèce). Étude de la machoire inférieure. *Géobios* **10:**849–885.

De Bonis, L., and Melentis, J. 1978. Les primates hominoides du Miocène de Macédoine. Étude de la machoire supérieur. *Ann. Paleontol. Vertebr.* **64:**185–202.

De Bonis, L., and Melentis, J. 1980. Nouvelles remarques sur l'anatomie d'un primate hominoide du Miocène: *Ouranopithecus macedoniensis.* Implications sur la phylogénie des Hominidés. *C. R. Acad. Sci. Paris D* **290:**755–758.

Dene, H. T., Goodman, M., and Prychodko, W. 1976. Immunodiffusion evidence on the phylogeny of the primates, in: *Molecular Anthropology* (M. Goodman and R. E. Tashian, eds.), pp. 171–195, Plenum, New York.

Fitch, W. M., and Langley, C. H. 1976. Evolutionary rates in proteins: Neutral mutations and the molecular clock, in: *Molecular Anthropology* (M. Goodman and R. E. Tashian, eds.), pp. 197–219, Plenum, New York.

Frayer, D. W. 1976. A reappraisal of *Ramapithecus. Yearb. Phys. Anthropol.* **18:**19–30.

Frayer, D. W. 1978. The taxonomic status of *Ramapithecus,* in: *Krapinski Pračovjek i Evolucija Hominida* (M. Malez, ed.), pp. 255–268, Jugoslavenska Akademija Znanosti i Umjetnosti, Zagreb.

Galdikas, B. M. F., and Teleki, G. 1981. Variations in subsistence activities of female and male pongids: New perspectives on the origins of hominid labor division. *Curr. Anthropol.* **22:**241–256.

Genet-Varcin, E. 1969. *A la Recherche du Primate Ancêtre de l'Homme,* Boubée, Paris. 336 pp.

Goodman, M. 1974. Biochemical evidence on hominid phylogeny. *Annu. Rev. Anthropol.* **3:**203–288.

Goodman, M. 1976. Toward a genealogical description of the primates, in: *Molecular Anthropology* (M. Goodman and R. E. Tashian, eds.), pp. 321–353, Plenum, New York.

Greenfield, L. O. 1974. Taxonomic reassessment of two *Ramapithecus* specimens. *Folia Primatol.* **22:**97–115.

Greenfield, L. O. 1975. A comment on relative molar breadth in *Ramapithecus. J. Hum. Evol.* **4:**267–273.

Greenfield, L. O. 1977. *Ramapithecus and Early Hominid Origins,* University Microfilms, Ann Arbor. 301 pp.

Greenfield, L. O. 1978. On the dental arcade reconstructions of *Ramapithecus. J. Hum. Evol.* **7:**345–359.

Greenfield, L. O. 1979. On the adaptive pattern of "*Ramapithecus.*" *Am. J. Phys. Anthropol.* **50:**527–548.

Greenfield, L. O. 1980. A late divergence hypothesis. *Am. J. Phys. Anthropol.* **52:**351–366.

Gregory, W. K. 1930. A critique of Professor Osborn's theory of human origin. *Am. J. Phys. Anthropol.* **14:**133–164.

Gregory, W. K. 1934. *Man's Place among the Anthropoids,* Clarendon, Oxford. 119 pp.

Gregory, W. K., and Hellman, M. 1926. The dentition of *Dryopithecus* and the origin of man. *Anthropol. Pap. Am. Mus. Nat. Hist.* **28:**1–123.

Halstead, B. 1980. Popper: Good philosophy, bad science? *New Sci.* **87:**215–217.

Harding, R. S., and Teleki, G., eds. 1981. *Omnivorous Primates: Gathering and Hunting in Human Evolution,* Columbia, New York. 673 pp.

Holloway, R. L. 1966. Cranial capacity, neural reorganization, and hominid evolution: A search for more suitable parameters. *Am. Anthropol.* **68:**103–121.

Holloway, R. L. 1976. Paleoneurological evidence for language origins, in: Origins and Evolution of Language and Speech. *Ann. N. Y. Acad. Sci.* **280:**330–348.

Holloway, R. L. 1980. Within-species brain-body weight variability: A reexamination of the Danish data and other primate species. *Am. J. Phys. Anthropol.* **53**:109–121.

Hrdlička, A. 1935. The Yale fossils of anthropoid apes. *Am. J. Sci.* **29**:34–40.

Hsu, C., Wang, L., and Han, K. 1975. Australopithecine teeth associated with *Gigantopithecus*. *Vertabr. Palasiat.* **13**:81–88.

Huxley, T. H. 1861. On the zoological relations of man with the lower animals. *Nat. Hist. Rev.* **1**:67–84.

Huxley, T. H. 1863. *Evidence as to Man's Place in Nature*, Williams and Norgate, London. 159 pp.

Hylander, W. L. 1979. The functional significance of primate mandibular form. *J. Morphol.* **160**:223–240.

Jacobs, L. L., and Pilbeam, D. R. 1980. Of mice and men: Fossil-based divergence dates and molecular "clocks." *J. Hum. Evol.* **9**:551–555.

Johanson, D. C. 1980. Early African hominid phylogenesis: A reevaluation, in: *Current Argument on Early Man* (L. K. Königsson, ed.), pp. 31–69, Pergamon, Oxford.

Johanson, D. C., and White, T. D. 1979. A systematic assessment of early African hominids. *Science* **203**:321–330.

Johnson, S. C. 1981. Bonobos: Generalized hominid prototypes or specialized insular dwarfs? *Curr. Anthropol.* **22**:363–375.

Jukes, T. H., and Holmquist, R. 1972. Evolutionary clock: Nonconstancy of rate in different species. *Science* **177**:531–532.

Jolly, C. 1970. The seed eaters: A new model of hominid differentiation based on a baboon analogy. *Man* **5**:5–26.

Jolly, C. J. 1973. Changing views of hominid origins. *Yearb. Phys. Anthropol.* **16**:1–17.

Jukes, T. H. 1980. Silent nucleotide substitutions and the molecular evolutionary clock. *Science* **210**:973–978.

Kay, R. F. 1975. The functional adaptations of primate molar teeth. *Am. J. Phys. Anthropol.* **43**:195–216.

Kay, R. F. 1981. The nut-crackers—A new theory of the adaptations of the Ramapithecinae. *Am. J. Phys. Anthropol.* **55**:141–151.

King, M. C., and Wilson, A. C. 1975. Evolution at two levels in humans and chimpanzees. *Science* **188**:107–116.

Kinzey, W. G. 1974. Ceboid models for the evolution of the hominoid dentition. *J. Hum. Evol.* **3**:193–203.

Kortlandt, A. 1972. *New Perspectives on Ape and Human Evolution*, Stichting voor Psychobiologie, Amsterdam. 100 pp.

Kretzoi, M. 1975. New ramapithecines and *Pliopithecus* from the lower Pliocene of Rudabánya in north-eastern Hungary. *Nature (Lond.)* **257**:578–581.

Kretzoi, M. 1976. Emberré Válás és az Australopithecinák. *Anthropol. Kozl.* **20**:3–11.

Leakey, L. S. B. 1962. A new lower Pliocene fossil primate from Kenya. *Ann. Mag. Nat. Hist.* **13**:689–696.

Leakey, L. S. B. 1967. An early Miocene member of Hominidae. *Nature (Lond.)* **213**:155–163.

Leakey, L. S. B. 1968. Bone smashing by late Miocene Hominidae. *Nature (Lond.)* **218**:528–530.

Leakey, R. E. F. 1976. Hominids in Africa. *Am. Sci.* **64**:174–178.

Le Gros Clark, W. E. 1957. Observations on the anatomy of the fossil Australopithecinae. *J. Anat.* **81**:300–333.

Lewis, G. E. 1934. Preliminary notice of new man-like apes from India. *Am. J. Sci.* **22**:161–181.

Lovejoy, C. O. 1981. The origin of man. *Science* **211**:341–350.

Lovejoy, C. O., and Meindl, R. S. 1973. Eurkayote mutation and the protein clock. *Yearb. Phys. Anthropol.* **16**:18–30.

Lovejoy, C. O., Burstein, A. H., and Heiple, K. G., 1972. Primate phylogeny and immunological distance. *Science* **176**:803–805.

Mann, A. 1968. *The Paleodemography of Australiopithecus*, University Microfilms, Ann Arbor. 171 pp.

McHenry, H. M., and Corruccini, R. S. 1981. *Pan paniscus* and human evolution. *Am. J. Phys. Anthropol.* **54**:355–367.

Miller, D. A. 1977. Evolution of primate chromosomes. *Science* **198**:1116–1124.

Peters, C. R. 1979. Toward an ecological model of African Plio-Pleistocene hominid adaptations. *Am. Anthropol.* **81**:261–278.

Pickford, M. 1982. New higher primate fossils from the middle Miocene deposits at Majiwa and Kaloma, Western Kenya. *Am. J. Phys. Anthropol.* **58**:1–19.

Pilbeam, D. R. 1966. Notes on *Ramapithecus*, the earliest known hominid, and *Dryopithecus. Am. J. Phys. Anthropol.* **25**:1–6.

Pilbeam, D. R. 1970. *Gigantopithecus* and the origins of the Hominidae. *Nature (Lond.)* **225**:516–519.

Pilbeam, D. R. 1979. Recent finds and interpretations of Miocene hominoids. *Annu. Rev. Anthropol.* **8**:333–352.

Pilbeam, D. R. 1980. Major trends in human evolution, in: *Current Argument on Early Man* (L. K. Königsson, ed.), pp. 261–285, Pergamon, Oxford.

Pilbeam, D. R. 1982. New hominoid skull material from the Miocene of Pakistan. *Nature (Lond.)* **295**:232–234.

Popper, K. R. 1957. *The Poverty of Historicism*, Harper, New York. 166 pp.

Popper, K. 1980. Letter to the Editor. *New Sci.* **87**:611.

Post, J. H. 1980. Origin of bipedalism. *Am. J. Phys. Anthropol.* **52**:175–190.

Radinsky, L. 1978. Do albumin clocks run on time? *Science* **200**:1182–1183.

Read, D. W. 1975. Primate phylogeny, neutral mutations, and "molecular clocks." *Syst. Zool.* **24**:209–221.

Read, D. W., and Lestrel, P. 1970. Hominid phylogeny and immunology: A critical approach. *Science* **168**:578–580.

Read, D. W., and Lestrel, P. 1972. Phyletic divergence dates of hominoid primates. *Evolution* **26**:669–670.

Robinson, J. T. 1963. Adaptive radiation in the australopithecines and the origin of man, in: *African Ecology and Human Evolution* (F. C. Howell and F. Bourlière, eds.), pp. 385–416. Aldine, Chicago.

Robinson, J. T. 1967. Variation and taxonomy of the early hominids, in: *Evolutionary Biology*, Volume 1 (T. Dobzhansky, M. K. Hecht, and W. C. Steere, eds.), pp. 69–100, Appleton-Century-Crofts, New York.

Sarich, V. 1974. Just how old is the hominid line? *Yearb. Phys. Anthropol.* **17**:98–112.

Sarich, V., and Cronin, J. E. 1976. Molecular systematics of the primates, in: *Molecular Anthropology* (M. Goodman and R. E. Tashian, eds.), pp. 141–170, Plenum, New York.

Sarich, V., and Wilson, A. C. 1967. Immunological time scale for hominid evolution. *Science* **158**:1200–1203.

Schwalbe, G. 1923. Die Abstammung des Menschen und die "altesten" Menschenformen, in: *Die Kultur der Gegenwart* (G. Schwalbe and E. Fischer, eds.), pp. 223–338, Teubner, Leipzig.

Sicher, H. 1944. Masticatory apparatus in the giant panda and the bears. *Field Mus. Nat. Hist. Zool. Ser.* **29**:61–73.

Simons, E. L. 1961. The phyletic position of *Ramapithecus. Postilla* (Peabody Mus. Nat. Hist., Yale Univ.) **57**:1–10.

Simons, E. L. 1964. On the mandible of *Ramapithecus. Proc. Natl. Acad. Sci. USA* **51**:528–535.

Simons, E. L. 1972. *Primate Evolution*, Macmillan, New York. 322 pp.

Simons, E. L. 1976. The nature of the transition in the dental mechanism from pongids to hominids. *J. Hum. Evol.* **5**:500–528.

Simons, E. L. 1977. *Ramapithecus. Sci. Am.* **236**(5):28–35.

Simons, E. L. 1978. Diversity among the early hominids: A vertebrate paleontologist's viewpoint, in: *Early Hominids of Africa* (C. Jolly, ed.), pp. 543–566, Duckworth, London.

Simons, E. L., and Ettel, P. C. 1970. *Gigantopithecus. Sci. Am.* **222**(1):77–86.

Simons, E. L., and Pilbeam, D. R. 1972. Hominoid paleoprimatology, in: *The Functional and Evolutionary Biology of Primates* (R. Tuttle, ed.), pp. 36–62, Aldine, Chicago.

Simons, E. L., and Pilbeam, D. R. 1978. *Ramapithecus* (Hominidae, Hominoidea), in: *Evolution of African Mammals* (V. J. Maglio and H. B. S. Cooke, eds.), pp. 147–153, Harvard University Press, Cambridge.

Simpson, G. G. 1953. *The Major Features of Evolution*, Columbia University Press, New York. 434 pp.

Simpson, G. G. 1963. The meaning of taxonomic statements, in: *Classification and Human Evolution* (S. L. Washburn, ed.), pp. 1–31, Aldine, Chicago.

Stanley, S. M. 1979. *Macroevolution*, Freeman, San Francisco.

Stern, J. T., and Susman, R. L. 1981. Electromyography of the gluteal muscles in *Hylobates, Pongo,* and *Pan:* Implications for the evolution of hominid bipedality. *Am. J. Phys. Anthropol.* **55:**153–66.

Taieb, M., Johanson, D. C., and Coppens, Y. 1975. Expédition internationale de l'Afar, Ethiopie (3e campagne 1974); découverte d'Hominidés plio-pleistocènes à Hadar. *C. R. Acad. Sci. Paris D* **281:**1297–1300.

Tuttle, R. H. 1975. Parallelism, brachiation, and hominid phylogeny, in: *Phylogeny of the Primates* (W. P. Luckett and F. S. Szalay, eds.), pp. 447–480, Plenum, New York.

Uzzell, T., and Pilbeam, D. 1971. Phyletic divergence rates of hominoid primates: A comparison of fossil and molecular data. *Evolution* **25:**615–635.

Vogel, C. 1975. Remarks on the reconstruction of the dental arcade of *Ramapithecus,* in: *Paleoanthropology, Morphology and Paleoecology* (R. Tuttle, ed.), pp. 87–98, Mouton, The Hague.

Von Koenigswald, G. H. R. 1957. Remarks on *Gigantopithecus* and other hominid remains from South China. *Proc. K. Ned. Akad. Wet. Amsterdam B* **60:**153–159.

Von Koenigswald, G. H. R. 1972. Was ist *Ramapithecus? Nat. Mus.* **102:**173–183.

Walker, A. C. 1976. Splitting times among hominoids deduced from the fossil record, in: *Molecular Anthropology* (M. Goodman and R. E. Tashian, eds.), pp. 63–77, Plenum, New York.

Washburn, S. L. 1968. *The Study of Human Evolution*, Oregon State System of Higher Education, Eugene, Oregon. 48 pp.

Weinert, H. 1944. *Ursprung der Menschenheit: Über den engeren Anschluss des Menschengeschlects and die Menschenaffen*, Ferdinand Enke, Stuttgart.

White, T. D. 1975. Geomorphology to paleoecology: *Gigantopithecus* reappraised. *J. Hum. Evol.* **4:**219–233.

White, T. D. 1981. Primitive hominid canine from Tanzania. *Science* **213:**348.

Wolpoff, M. H. 1971a. Metric trends in hominid dental evolution. *Case West. Res. Stud. Anthopol.* **2:**1–244.

Wolpoff, M. H. 1971b. Interstitial wear. *Am. J. Phys. Anthropol.* **34:**205–228.

Wolpoff, M. H. 1973. Posterior tooth size, body size, and diet in the South African gracile australopithecines. *Am. J. Phys. Anthropol.* **39:**375–394.

Wolpoff, M. H. 1974. The evidence for two australopithecine lineages in South Africa. *Yearb. Phys. Anthropol.* **17:**113–139.

Wolpoff, M. H. 1975. Comment on "*Ramapithecus* as a hominid," in: *Paleoanthropology, Morphology and Paleoecology* (R. H. Tuttle, ed.), pp. 174–176, Mouton, The Hague.

Wolpoff, M. H. 1976a. Some aspects of the evolution of early hominid sexual dimorphism. *Curr. Anthropol.* **17:**579–606.

Wolpoff, M. H. 1976b. Data and theory in paleoanthropological controversies. *Am. Anthropol.* **78:**94–96.

Wolpoff, M. H. 1978. Analogies and interpretation in paleoanthropology, in: *Early Hominids of Africa* (C. Jolly, eds.), pp. 461–503, Duckworth, London.

Wolpoff, M. H. 1979. Anterior dental cutting in the Laetolil hominids and the evolution of the bicuspid P_3. *Am. J. Phys. Anthropol.* **51:**233–234.

Wolpoff, M. H. 1980. *Paleoanthropology*, Knopf, New York. 379 pp.

Wolpoff, M. H. 1981. Comment on "Bonobos: Generalized hominid prototypes or specialized insular dwarfs?" *Curr. Anthropol.* **22:**370–371.

Wolpoff, M. H., and Russell, M. D. 1981. Anterior dental cutting at Laetolil. *Am. J. Phys. Anthropol.* **55:**223–224.

Woo, J.-K. 1962. The mandibles and dentition of *Gigantopithecus. Palaeontol. Sin., N.S.D* **11:**1–94.

Wrangham, R. W. 1980. Bipedal locomotion as a feeding adaptation in gelada baboons, and its implications for hominid evolution. *J. Hum. Evol.* **9:**329–332.

Wu, R.-K. (Woo, J.-K.) 1981. First skull of *Ramapithecus* found. *China Reconstructs* **30**(4):68–69.

Xu, Q. and Lu, Q. 1980. The Lufeng ape skull and its significance. *China Reconstructs* **29**(1):56–57.

Yulish, S. 1970. Anterior tooth reduction in *Ramapithecus*. *Primates* **11**:255–270.

Yunis, J. J., Sawyer, J. R., and Dunham, K. 1980. The striking resemblance of high-resolution G-banded chromosomes of man and chimpanzee. *Science* **208**:1145–1148.

Zihlman, A. 1979. Pygmy chimpanzee morphology and the interpretation of early hominids. *S. Afr. J. Sci.* **75**:165–167.

Zihlman, A. L., Cronin, J. E., Cramer, D. L., and Sarich, V. M. 1978. Pygmy chimpanzee as a possible prototype for the common ancestor of humans, chimpanzees, and gorillas. *Nature (Lond.)* **275**:744–746.

Ramapithecus and *Pan paniscus* 26
Significance for Human Origins

A. L. ZIHLMAN AND
J. M. LOWENSTEIN

Introduction

Attempts to explain human origins go back at least several thousand years, but only in the past 100 years or so have scientific methods begun to make headway against mythical and theological versions of creation. In the short time since Darwin's *The Origin of Species* provided a coherent framework for the evolution of life on earth, not only has the data base of fossil discoveries increased enormously, but rapid advances in atomic and molecular techniques have provided, even within the last decade, significant new information on the timing of human evolution and the genetic relations of humans to other primates. Traditional paleoanthropology has been slow to admit the existence and validity of the new information, with the result that many current textbooks in anthropology and much of the current literature fail to present views of human evolution that are compatible with the most objective available evidence—namely, the implications of the molecular data showing the very close relationship of humans to African apes.

It is difficult to explain why so many physical anthropologists, relying as they do on methods like the decay of isotopes for the timing of important

A. L. ZIHLMAN • Department of Anthropology, University of California, Santa Cruz, California 95064.   J. M. LOWENSTEIN • Department of Medicine (122 MR2), University of California, San Francisco, California 94143.

evolutionary events, are able to ignore or minimize a large and growing corpus of evidence from molecular biology on the timing of events in primate evolution, particularly the recent divergence of humans, chimpanzees, and gorillas. Much of this curious "blindness" to basic facts is due to a widely held belief that the Miocene hominoid *Ramapithecus* was a hominid, a belief inconsistent with the molecular data. An alternative proposal, consistent with and growing out of the molecular data, is that the living pygmy chimpanzee, *Pan paniscus*, represents a prototype for an African hominoid and hominid precursor (Zihlman *et al.*, 1978).

The hominid status of *Ramapithecus* is being reevaluated on the basis of new evidence of nonhominid features of jaws and teeth (Pilbeam *et al.*, 1977) and of discoveries of chimpanzee-like hominid fossils from East Africa at 3.5 m.y.a. (Johanson and White, 1979). The molecular data are absent in these and many other discussions on hominid origins, as though the morphological findings alone have led to changes in viewpoint. The "pygmy chimpanzee model" has not received a great deal of explicit support, though it remains the only well-defined precursor to *Australopithecus* other than a model based on *Ramapithecus*. These contrasting views of hominid ancestry make an interesting case history regarding hidden assumptions and what kinds of information are considered "facts" in the search for hominid ancestors. Here we briefly review the historical perspective on the application of molecular data to the proposed phyletic affinities of *Ramapithecus*, and then evaluate the pygmy chimpanzee model in light of current information and ideas.

Historical Background

For most anthropologists during the past two decades, the status of *Ramapithecus* as the earliest human ancestor has been an established fact. Following Lewis' work in the 1930s, in 1961 Elwyn Simons elevated *Ramapithecus* to the hominid grade on the basis of small canine teeth and a parabolic jaw reconstruction (G. E. Lewis, 1934, 1937; Simons, 1961). A progenitor of Miocene age was expected: *Ramapithecus* was between 10 and 14 million years old, dentally more human than ape, and of Asian or cosmopolitan distribution. Acceptance was rapid and nearly universal, though there were conspicuous exceptions, such as Washburn (1963), who thought that modern humans and African apes were so similar anatomically and *Australopithecus* so ape-like that an early separation of human and ape need not be postulated.

In the year after Simons published "The phyletic position of *Ramapithecus*," Morris Goodman in 1962 presented to a Wenner-Gren Symposium his early immunological data on the similarity of humans, gorillas, and chimpanzees. Goodman's immunological cross-reactions showed human, gorilla, and chimpanzee sera so nearly identical that initially he proposed that all three species should be included in the family Hominidae, leaving the farther removed gibbon and orangutan in the family Pongidae (Goodman, 1963). These results strongly implied a much more recent divergence of these three species than the 14–20 million years needed to make *Ramapithecus* human.

Goodman was challenged by a solid phalanx of anthropologists who asserted that an ape–human divergence less than 14 m.y.a. was in direct contradiction with the fossil record. Accepting these assertions as fact, Goodman felt he had to find a way of reconciling his results with the earlier divergence time, and so he hypothesized that a slowdown had taken place in molecular evolution among the primates, or at least among the hominoids, in comparison with other mammalian groups that had been studied (Goodman *et al.*, 1971; Goodman, 1976). His data on relationships among other primate groups—lemurs, lorises, Old and New World monkeys—fit reasonably well with prevailing opinions and no slowdown or speedup needed to be invoked.

In the 1960s Vincent Sarich began doing similar immunological research on the relations of primate albumins. Sarich, at the University of California, Berkeley, was trained in both anthropology (by S. L. Washburn) and biochemistry (by A. C. Wilson). In weighing the immunological versus the fossil evidence, Sarich (1968) and Sarich and Wilson (1967) concluded that the near identity of human, chimpanzee, and gorilla proteins was more impressive evidence of a recent divergence than the contrary evidence for a much earlier divergence embodied in a few sets of ramapithecine jaws and teeth, unaccompanied by skulls, limb bones, pelves, or other indications of brain size, locomotion, or other diagnostic human characters.

Sarich's and Goodman's research took place within the context of the extraordinary advances in molecular biology of the past three decades, which included the discovery of the DNA double helix by Watson and Crick (1953), as well as the discovery of methods for amino acid sequencing of proteins and base sequencing of DNA and RNA, and DNA hybridization; these all have brought forth the new science of molecular evolution. Margoliash (1963) showed that the cytochrome *c* of such diverse species as human, horse, and yeast were remarkably similar and that the countable differences corresponded well with the evolutionary divergence times of these groups estimated from the fossil record.

Since then, an enormous body of data has accumulated from many protein, DNA, and RNA sequences [summarized in Dayhoff (1978)] and there is increasing evidence that these molecules may be used as "molecular clocks" with sequence differences roughly proportional to divergence times (Wilson *et al.*, 1977). Changes in DNA bases (point mutations) do not, of course, occur with the regularity of a metronome, but have been shown to occur with statistical regularity (Jukes, 1980; Kimura, 1980; Miyata and Yasunaga, 1980). Rates of change may vary from one molecule to another (albumin is fast, cytochrome *c* is slow) or from one lineage to another (albumin evolves more slowly in birds than in mammals). Molecular clocks are rather like chronometers on 19th century sailing vessels: it was necessary to have on board several of them in order to obtain a sufficiently accurate average time for navigational purposes.

Goodman's proposed slowdown in primate molecular evolution was "explained" as being due to the longer generation time of the larger primates, on the assumption that mutations occur in connection with fertilization. This *ad hoc* explanation was enthusiastically accepted and is still cited by proponents of a Miocene hominid ancestor. Meanwhile, the "generation time" hypothesis has been tested by comparing long- and short-generation pairs, such as elephant and mouse, and human and lemur, and their proteins have been shown to have evolved at approximately the same rates despite tenfold or greater differences in generation time (Wilson *et al.*, 1977; Sarich and Cronin, 1977). Therefore, little evidence has been adduced to support the generation time hypothesis.

Sarich, Cronin, and Wilson have pointed out in numerous publications that the now massive body of molecular data from immunology, protein sequencing, and DNA hybridization and sequencing will not permit a divergence of humans and African apes earlier than about 5 m.y.a. *unless* molecules in these three species have behaved differently from those in the many hundreds of other species that have been studied.

For constructing phylogenies, the molecular data have two advantages over the morphological data. First, they are quantifiable and can be replicated in many laboratories, in contrast with the many conflicting phylogenies proposed by different paleoanthropologists. Second, the amount of immunological and sequence difference provides "molecular clocks" for estimating divergence times, whereas morphological change may occur slowly, as in frogs, or rapidly, as in primates. Yet rates of change in frog and human proteins such as albumin are quite comparable (Wilson *et al.*, 1977).

Molecular data are now being widely used for evaluating genetic and temporal relationships among bacteria, yeasts, fungi, invertebrates, fish, reptiles, birds, and mammals (Dayhoff, 1978). More than any other group of evolutionary scientists, anthropologists continue to resist the molecular data. A recent example of this resistence is a volume on the origin of New World monkeys in which a number of authors discuss the problem without acknowledging the molecular data showing relationships between New and Old World monkeys (Ciochon and Chiarelli, 1980).

Nevertheless, the biochemical evidence is having an impact. In recent years an increasing number of anthropologists have begun to question not only the hominid status of *Ramapithecus* but also the status of Oligocene primates as early apes. They tend to find new morphological bases for the new interpretations and either ignore or mention in passing the biochemical evidence which has long supported their new conclusions. Examples of this approach are: (1) the discovery that early hominid fossils such as those at Hadar and Laetoli are so "primitive" that "a late divergence must remain a possibility" (Johanson and White, 1979, p. 325); (2) the observation that ramapithecine characters, such as thick dental enamel, are not diagnostically hominid after all (Kay, 1981): (3) the new interpretation that Fayum Oligocene primates, rather than being ancestral hominoids, as previously maintained, may be ancestral catarrhines—giving rise to *both* Old World monkeys and apes (Kay *et al.*, 1981; see also Fleagle and Kay, this volume, Chapter 7). In our opinion, increasing numbers of morphologists are riding the molecular bandwagon without paying their fare—namely, the acknowledgment that their interpretations have been influenced, at least in part, by the increasing weight of the biochemical evidence.

The Oldest Hominid: 14 m.y.a. in Asia or 5 m.y.a. in Africa?

Scientific hypotheses make predictions from which competing theories may be verified or falsified. The contrast between the "*Ramapithecus*-as-hominid" and the "recent-divergence" hypotheses is particularly striking. If the latter is correct, one would expect to find the earliest humans in Africa, no older than 5 or 6 million years and closely resembling an African ape. If the first hypothesis is correct, then human ancestors as old as 14 million years in Asia, Africa, and Europe should be found, and hominids at least 3 million years old should be much more "human" than "ape."

Although there have been numerous new fossil finds in Asia, Africa, and Europe of between 8 and 14 m.y.a. (Pilbeam, 1979; Simons, 1981), none show evidence of bipedalism, a fundamentally hominid morphological characteristic. The oldest undisputed hominids, found in Africa, over 3 million years old, are bipedal, as demonstrated by a knee joint more human-like than ape-like, and by human-like footprints preserved in volcanic ash (Johanson and Taieb, 1976; Leakey, 1981). At present the fossil record does not disprove a recent divergence and does not support the "*Ramapithecus*-as-hominid" hypothesis.

The evidence for hominid status of *Ramapithecus* rested upon dental and gnathic features: a parabolic dental arch, small canines, and large molars with thick enamel (Simons, 1972). Eckhardt (1972) pointed out the pitfalls of reconstructing an entire dental arch with insufficient material. Walker and Andrews (1973) and Vogel (1975) presented evidence to challenge the parabolic arch reconstruction for *Ramapithecus*. Recent fossil discoveries have confirmed these "corrected" reconstructions; a complete mandible exhibits a V-shaped dental arch, unlike that of either living apes or humans (Pilbeam, 1978).

Both *Ramapithecus* and *Australopithecus* have large molars with thick enamel. This similarity was interpreted as supporting an ancestral relationship. Thickly enameled teeth were also presumed to correlate with tough or gritty terrestrial foods that required extensive chewing. Recently the function of thick enamel has been clarified by Kay (1981), who did a dental study of 37 anthropoid primate species. Living species with thick molar enamel include the capuchin (*Cebus*), a New World monkey, and orangutan (*Pongo*), an Asian ape, whose diets include hard nuts and seeds that require cracking in order to extract the edible kernels. Both of these arboreal species obtain their food in trees. Thus, thick enamel indicates neither a close genetic relationship between species nor a terrestrial adaptation.

It is notorious among paleontologists that dental characteristics may be misleading because of convergence between unrelated or distantly related species. Another recent example, disclosed by a new technique of fossil immunology, involved the Tasmanian wolf, a marsupial carnivore of Australia that has been extinct for only about 50 years (Lowenstein *et al.,* 1981). A cladistic analysis of 45 different tooth measurements of marsupial carnivores was interpreted as showing the Tasmanian wolf more closely related to the extinct South American *Borhyaena* than to any Australian group (Archer, 1976). But assay of the albumin from three different skin and muscle specimens of Tasmanian wolf in museums uniformly showed this protein to be very closely related to that of the two Australian species, *Dasyurus* and *Dasyuroides,* and very distant from that of living South American marsupials. Convergence in dental traits between the Tasmanian wolf and *Borhyaena* would explain these findings. This work further underscores the need for caution in drawing phylogenetic conclusions from dental similarities alone.

Given the problems involved in phylogenetic reconstruction based exclusively on paleontological evidence, it is interesting to see how anthropologists' views of *Ramapithecus* are changing (see Wolpoff, this volume, Chapter 25). Pilbeam (1978, 1979), once a firm supporter of *Ramapithecus* as a hominid, has abandoned this position and left open a number of possible pathways to the human line.

Greenfield (1980) proposes a "late divergence hypothesis" which is a compromise between the *Ramapithecus*-as-hominid hypothesis and the molecular phylogeny. In this scenario, presumably based almost entirely on shared dental and gnathic characteristics, *Ramapithecus* and *Sivapithecus* are considered synonymous, and the expanded *Sivapithecus* is proposed as ancestral to *Pongo, Pan, Gorilla,* and *Australopithecus.* The chimpanzee, gorilla, *Australopithecus* split is given as about 8 m.y.a., or about halfway between the molecular and the older paleontological dates for the origin of the hominid line (also see Greenfield, this volume, Chapter 27).

Simons (1977, 1981) continues to hold his position that the divergence of the hominid line occurred about 15 m.y.a. (see also Kay and Simons, this volume, Chapter 23). He maintains that most of the primate divergence times derived from molecular data, including human–African ape, human–Asian ape, ape–monkey, and Old World–New World monkeys, are only about half as old as they should be based on his assessment and view of the primate fossil record.

From the molecular viewpoint, the situation concerning relationships and divergences is consistent and clear. Humans, chimpanzees, and gorillas diverged from a common ancestor in Africa about 5 m.y.a. (possibly as much as 6 m.y.a.); orangutans diverged from the line heading to African apes 9 or 10 m.y.a. and gibbons about 11–12 m.y.a. (Sarich and Cronin, 1976). *Ramapithecus* is interpreted as an ape that lived earlier in Africa, later in Europe, and survived longest in Asia, from 14 to 8 m.y.a. at a time when the hominid line had not separated from that of the apes. What little postcranial fossil material there is appears to be ape-like rather than hominid. Only more fossil evidence, particularly of the locomotor system, will clarify the relationship of *Ramapithecus* to extinct and extant hominoids and hominids.

If the molecular data are correct, *Ramapithecus* should be genetically equidistant from humans and all living apes. A new method for studying fossil proteins makes it possible in principle to test this (Lowenstein, 1981). Ramapithecine proteins, if sufficient amounts have survived in the fossils, might be shown immunologically either to be more like human than like living hominoid proteins or equally distant from the proteins of both. Preliminary work suggests the latter result, but are not yet conclusive (Lowenstein, 1982).

The absence of fossil evidence between the time proposed for the divergence by molecular data—4–8 m.y.a.—means that hypotheses must be proposed to predict what kinds of hominoid fossils will be found in that time range. If, as we suggest here, *Ramapithecus* is not a hominid, what species would fit the anatomical and biochemical evidence and serve as an African hominoid and hominid precursor?

Testing the Chimpanzee Hypothesis

Long before abundant fossils of Miocene and Plio-Pleistocene age had been recovered, comparative anatomists looked to the African apes, particularly to the chimpanzee, to elucidate the transition from ape to human (Huxley, 1863; Gregory, 1916). Biochemistry reaffirms this chimpanzee, gorilla, human closeness and helps narrow the choices for a hominid precursor. So far, the Miocene fossil record provides no compelling hominid precursor.

A chimpanzee-like prototype derives logically from both molecular and anatomical similarities. Though the gorilla is equally close to humans biochemically (Sarich and Cronin, 1976), the chimpanzee is more similar to the earliest known African hominids. It is more "generalized" than the gorilla (Simpson, 1963), is smaller, less sexually dimorphic, and more omnivorous. In all these aspects, it is a more useful ancestral model for chimpanzees, for the terrestrial, omnivorous, minimally dimorphic humans, and for the relatively terrestrial, vegetarian, very dimorphic gorillas (Reynolds, 1967).

Before the close genetic affinity of humans to African apes had been demonstrated unequivocally, creators of models for human ancestors had much broader scope. Wood-Jones (1948) advocated the tarsier because of its long and heavy hindlimbs; Straus (1949) the cercopithecoid monkeys because of their longer hindlimb to forelimb ratio. Students of human evolution should welcome the results of the molecular data because they limit the choice of an ancestral morphological prototype to a form like the chimpanzee, as a number of comparative anatomists have long suggested (e.g., Washburn, 1951).

Pygmy Chimpanzees vs. Common Chimpanzees

There are two extant species of chimpanzees, *Pan troglodytes* and *Pan paniscus,* which are equally close biochemically to humans and diverged from each other between 2 and 3 m.y.a. (Cronin, 1977). Of the two, pygmy chimpanzees, or bonobos, may serve as a more specific model for understanding the precursor to humans, gorillas, and chimpanzees, given the three-way divergence based on the molecular data (Zihlman *et al.,* 1978).

Pygmy chimpanzees, named on the basis of their smaller skulls and dentition (Schwarz, 1929), are by no means "pygmy" when one compares the limb bones of the two species. Coolidge (1933) maintained that they are a separate, more generalized species than the common chimpanzee; he even decided that they might represent the common ancestor of chimpanzees and humans. Sarich (1967), integrating the molecular evidence with fossil evidence, also concluded that the common ancestor must resemble a small chimpanzee, like *P. paniscus.* Both Pilbeam (1972) and Simons (1972) mentioned pygmy chimpanzees as being closest to humans, but neither author elaborated on why this might be the case.

As behavioral observations began to be made on pygmy chimpanzees, it became apparent that they behaved differently from other chimpanzees (Tratz and Heck, 1954). A variety of systematic studies have begun, on sexual behavior and communication (Savage *et al.,* 1977; Savage-Rumbaugh and Wilkerson, 1978), ecology, social, and locomotor behavior in the wild (Horn, 1980; Kano, 1980; Kuroda, 1979, 1980; Susman, 1980). This information complements that on cranial and dental anatomy (Johanson, 1974; Cramer, 1977; Fenart and Deblock, 1973), so that the distinguishing characteristics of the two chimpanzees are beginning to be defined.

In setting up an anatomical model as a prototype, one makes a judgment concerning specialized versus generalized features. Generalized means "approaching the ancestral state." One should not discuss generalized and specialized traits abstractly; they must have a reference point. For our discussion, the Old World monkeys, or cercopithecoids, provide that reference point. These monkeys have relatively longer and heavier hindlimbs than forelimbs, resembling generalized mammals in this regard (such as the tree shrew). Mobile upper limbs which are long relative to trunk length or to lower limbs are an ape characteristic, and there is variation within the apes (Schultz, 1930, 1937). Gibbons and siamangs have the longest arms of all and are considered the most specialized for this characteristic (Gregory, 1916). In a single skeleton of *Pan paniscus* compared to Schultz's *P. troglodytes* sample (1929–30), Coolidge (1933) noted the shorter upper limbs and narrower chest of the former and thereby judged *P. paniscus* to be more generalized than *P. troglyodytes.*

A larger sample of *Pan paniscus* skeletons confirmed and elaborated Coolidge's initial findings: the two species of chimpanzee show statistical differences in their chest (scapular and clavicular measurements) and pelvic breadths (Zihlman and Cramer, 1978). Femur length is statistically longer in

P. paniscus than in female *P. troglodytes* of comparable body weight. To go beyond the skeletal system, Zihlman and Brunker (1979) have undertaken studies on the muscular system to determine tissue composition and relative weights of body segments using Grand's methods (1977). Preliminary results on two females, one of each species, of comparable body weight show that the pygmy chimpanzee has heavier lower limbs relative to body weight (24% vs. 18%) but the two species have similar upper limb weights (16%) (Zihlman, 1979a). The more massive lower limbs of pygmy chimpanzees are similar to those of Old World monkeys (Grand, 1977). Although *P. paniscus*, like all chimpanzees, is a quadrupedal knuckle-walker and an adept climber, behaviorally it appears to be bipedal in a greater variety of contexts and more frequently (Zihlman, 1980). Such behavior seems to correlate with its heavier lower limbs and lighter trunk.

With regard to sexual dimorphism, pygmy chimpanzees show a unique configuration of features which may or may not be generalized in comparison with common chimpanzees and humans. Cramer (1977) in a sample of 60 skulls demonstrated essentially no sexual dimorphism in cranial capacity (females average 348.5 cm^3, males 351.2 cm^3). There is some dimorphism in canine teeth, but little in incisors or posterior teeth (Almquist, 1974; Johanson, 1974); and little was observed in long bone lengths in a sample of 20 skeletons, though from a sample of body weights it appears that males may weigh 25% more than females (Cramer and Zihlman, 1978). There is a similar magnitude of difference in body weights of male and female humans and common chimpanzees, but in these species, there are also skeletal differences in long bone lengths, joint sizes, and cranial capacity (Zihlman, 1976).

There is a great deal of overlap between the two chimpanzee species in body weight, cranial capacity, skeletal dimensions, and dietary and social behavior. There are, however, differences in body build, degree of sexual dimorphism, social behavior, and diet. In these respects *Pan paniscus* is a suitable candidate to represent the common ancestor of African apes and hominids: (1) Its smaller body weight and heavier lower limb than in common chimpanzees is closer to early hominids and a bipedal adaptation; (2) sexual dimorphism is minimal and less than for the other hominoids, and we consider this a generalized character that can evolve rapidly through sexual selection; (3) *Pan paniscus* eats less fruit than the common species and more plant material, like the gorilla (Kano, 1979; Badrian *et al.*, 1980), and is therefore "intermediate" in diet. Pygmy chimpanzee social organization resembles that of the common chimpanzees in the variability of group size. But unlike them, pygmy chimpanzee groupings have larger temporary associations that are almost exclusively bisexual. Pygmy chimpanzees do not form all-male groups, nor do they move about alone as do common chimpanzees, but like gorillas, they spend their time in bisexual groups (Kuroda, 1979, 1980; Kano, 1980).

We argue that pygmy chimpanzees are more generalized than the common species and make a reasonable beginning point for the evolution of three very different adaptations: the slender, bipedal hominids, the large, dimorphic gorillas, and the moderately dimorphic, fruit-eating common chimpanzees. Such an argument implies that of these four species, pygmy chimpanzees have undergone the least amount of morphological change. Obviously, studies on distinctions among geographic races of *known* origin of *P. troglodytes* will clarify the range of variation in "common chimpanzees" and whether some races are more or less similar to *P. paniscus* (Groves, 1981).

Pygmy Chimpanzees and Early Hominids

The comparison of pygmy chimpanzees to early hominids helps to reconstruct the sequence of morphological changes that occurred in the transition from a quadrupedal, knuckle-walking, climbing ape to a structurally and behaviorally adapted bipedal hominid. The oldest hominids, now called *Australopithecus afarensis,* are defined by their hominid dentition and footprints, indicating that a bipedal locomotor pattern had evolved by 3.5 m.y.a. (Johanson and White, 1979). A cranial capacity of about 400 cm^3 (Kimbel, 1979) is still within the range of chimpanzees (both species may be as much as 420 cm^3). Some of these early hominid fossils, particularly A.L. 288 and STS 14, are similar in size to each other and to the average pygmy chimpanzee and present the possibility for comparing body builds of early hominid with ape. Figure 1, which is drawn to scale, compares a pygmy chimpanzee with a composite of African early hominid fossils. Table 1 compares known measurements of the fossils with those of pygmy chimpanzees and gives estimates where noted. The figure enables one to visualize the meaning of the measurements and morphological similarities and differences. Tobias (1981) emphasizes the similarities between *A. afarensis* and *A. africanus,* so that a composite of the two for comparison to *P. paniscus* is not misleading.

Note the similarities in the two species in overall size, stature (reconstructed from femur length), cranial capacity, joint sizes, and femoral length and the differences in shorter upper limb and innominate length in *Australopithecus* but greater iliac and sacral breadths. Measurements do not entirely reflect differences in morphology or orientation of the limbs, such as the long femoral neck, adducted femur, and convergent big toe in *Australopithecus* which are reflected in the figure.

With regard to relative weights, upper limbs constitute 16% of total body weight in *P. paniscus,* only 9.4% in modern humans, whereas the lower limb is 24% in *P. paniscus,* but 32% in humans. The upper limbs of *Australopithecus* must be lighter as well as shorter, probably about 12% of body weight and for the lower limbs about 28%. These estimates in the fossil would fit well with what is known of lengths: a ratio of humerus to femur is 98 in *P. paniscus,* 84 in *Australopithecus,* and 75 in modern humans.

This comparison suggests that if the ancestral form were similar to *P. paniscus* in body build, then during the shift from quadrupedal to bipedal locomotion one of the earliest changes might have been reduction in upper limb length and relative weight; this change would serve to lower the center of gravity, making upright posture more stable. Other early morphological changes might have included expansion of the ilium and sacrum in response

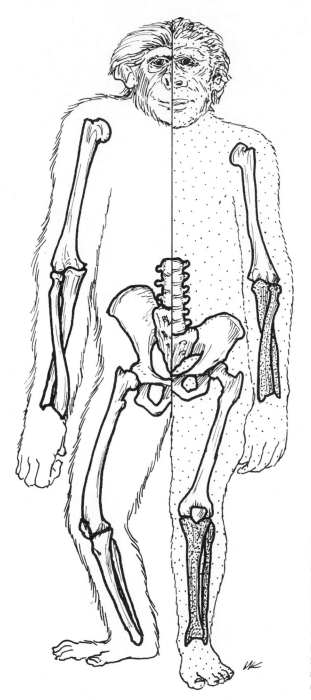

Fig. 1. Comparison of *Pan paniscus* with a reconstruction of *Australopithecus* using information from Hadar and Sterkfontein. Note similarities in overall size and in lengths of lower limbs. Note differences in the somewhat shorter upper limb in *Australopithecus* and in its modifications for bipedalism in the pelvis and foot. Drawn to scale and modified from Zihlman (1982).

Table 1.   Comparison of Measurements for *Pan paniscus* and *Australopithecus*[a]

| Measurement | *Pan paniscus* | | *Australopithecus* |
	Range	Mean	
Body weight, kg	25–48	35	25–50[b]
Cranial capacity, cm^3	260–420	350	390–440
Humerus length, mm	256–308	285	235 (AL 288)
Radius length, mm	235–284	262	192–216[c]
Ulna length, mm	250–299	274	210–226[c]
Pelvis			
Innominate length, mm	232–272	253	170 (STS 14)
Iliac breadth, mm	80–118	97	113 (STS 14)
Sacral breadth, mm	51–70	63	76 (STS 14)
Acetabulum diameter, mm	32–38	36	37 (STS 14)
Femur length, mm	275–316	293	280 (A.L. 288)
Tibia length, mm	225–271	242	233–250[c]
Ratio humerus/femur length	—	98	84 (A.L. 288)
Upper limb mass relative to			
body weight	—	16%	12%[b]
Lower limb mass[d]	—	24%	28%[b]

[a]Data from Zihlman (1979*b*), Zihlman and Cramer (1978), Cramer (1977), Johanson and Taieb (1976).
[b]Our estimates.
[c]Estimate of mean, not range.
[d]See text for discussion.

to increased stress of upright stance, resulting in a lumbar curve and vertical trunk, and reorientation of the femur, and therefore the entire lower limb, including the knee and ankle, under the body (Zihlman and Brunker, 1979).

Early hominid fossils vary in size, and this suggests body size variation among individuals. Much of this may be due to sexual dimorphism, perhaps marked dimorphism in *A. afarensis* (Johanson and White, 1979). The evidence for and extent of the dimorphism has not been documented, nor have appropriate comparisons been made with South African fossils, especially those from Makapan and Sterkfontein, which are closer in age. The Sterkfontein material indicates moderate sexual dimorphism (Zihlman, 1976).

If one accepts the principles of evolution, then *A. afarensis* was preceded by *some* kind of nonhominid—an ape, an African ape, a quadrupedal ape. The retention of ape-like, and particularly chimpanzee-like, characteristics has been noted for the australopithecine pelvis (Oxnard, 1975; Zihlman, 1978), hand (O. J. Lewis, 1977), foot (O. J. Lewis, 1980), and skull (Le Gros Clark, 1947). In a number of ways, the gap is indeed narrow between the morphology of early hominids and chimpanzees.

Is the Pygmy Chimpanzee Model Viable?

The pygmy chimpanzee model for hominid ancestry has been criticized on a number of grounds (Johnson, 1981, and comments). Some objections are: (1) *A. afarensis* has a dentition and masticatory apparatus so massive that it could not have evolved from a small-toothed form like *P. paniscus*. But this is contradicted by the claim that *A. afarensis* is ancestral to *Homo sapiens,* whose dentition and masticatory apparatus are smaller than those of the australopithecine ancestor. (2) *A. afarensis* is extremely sexually dimorphic compared to *P. paniscus* and could not have evolved from a less dimorphic species. However, sexual dimorphism is expressed in a variety of ways, depending upon adaptation and behavior, and morphological differences can evolve rapidly due to sexual selection. For example, the gorilla, with extreme dimorphism, is very closely related to the pygmy chimpanzee, with only slight sexual dimorphism. Also, to illustrate the rapidity with which dimorphism can evolve, note the patas monkey, which diverged from the vervet monkey about 2 m.y.a.; yet the former is the most dimorphic *Cercopithecus* species and the latter is only slightly dimorphic (Brunker, 1980). Because almost all the *Cercopithecus* species are not very sexually dimorphic, it seems likely that the ancestral species prior to the patas–vervet split was only slightly dimorphic as well. If a minimally dimorphic chimpanzee-like ape cannot be ancestral to *A. afarensis,* supposedly an extremely dimorphic species, then how can *A. afarensis* be ancestral to *Homo sapiens,* a minimally dimorphic species? (3) *Pan paniscus* is a rain-forest species restricted to an enclave in the Zaire River Basin, whereas the early hominid fossils are found in savanna environments. Because the savanna mosaic as we know it today became widespread only by the end of the Miocene, it would appear that at *some* period in evolutionary time, human ancestors were living in forests (Laporte and Zihlman, in preparation). Proponents of *Ramapithecus* place this period at 14 m.y.a. or earlier, but the molecular data place the transition nearer to 5 m.y.a., which correlates well with the time of widespread African savannas. The *common* ancestor of humans, chimpanzees, and gorillas must have been mainly forest-living and relatively arboreal unless one assumes that chimpanzees and gorillas descended from a terrestrial biped [argued by Wolpoff (1981)].

In response to these objections, we do not claim that *Pan paniscus* is an ideal prototype in every particular, but rather that, of the living hominoids, this species is probably most like the common ancestor of humans and African apes—more so than the common chimpanzee, gorilla, orangutan, or gibbon.

Most of those who reject the proposed model offer none of their own, but some do. McHenry (1981) (see also Corruccini and McHenry, 1979; McHenry and Corruccini, 1981) maintains that the pygmy chimpanzee is merely an allometric common chimpanzee—a position in which he is virtually alone—and so he concludes, not that the pygmy chimpanzee is not a valid model, but that a common chimpanzee would be just as valid. Tuttle proposed a gibbon-like progenitor in 1969 as preferable to Washburn's chimplike ancestor, on the grounds that gibbons are small and are sometimes bipedal in the trees (Tuttle, 1969, 1977, 1981). He assumes that bipedalism evolved fully *before* the ape ancestor left the trees. His response to the criticism that the gibbon is more distant genetically from humans, based on DNA and protein studies, than any other living ape, is this: "Peering into test tubes, even crystal ones, will not provide answers to questions about the morphological appearance and lifeways of pre- and protohominids" (Tuttle, 1977, p. 293). In other words, molecules are not facts which help us to make choices among proposed hominid ancestors.

Hominid Origins: Punctuation or Gradualism?

The proponents of punctuational versus gradual evolution have passively supported the hominid status of *Ramapithecus* by side-stepping the issue of human origins and the biochemical data (Stanley, 1979; Gould and Eldredge, 1977). One could use *Ramapithecus* to support punctuational human evolution in two ways: (1) to deduce that for 14 million years there was little change in the hominid line until the "punctuational event" of australopithecine origins, as implied by Kennedy (1980); or (2) to deduce that the punctuational event occurred 14 m.y.a., producing bipedal hominids who were around for millions of years prior to *Australopithecus,* so that by 3 m.y.a. merely "final touches" were being added to the hominid bipedal adaptation, as implied by Lovejoy (1973).

How can the *P. paniscus* versus *Ramapithecus* models be evaluated in terms of the punctuationalism versus gradualism controversy (Cronin *et al.,* 1981)? If one accepts a biochemical divergence date of 5 or 6 million years, the transition to bipedalism must have occurred relatively rapidly. Indeed, there are biomechanical reasons as well to argue for a quick transition, as it is difficult to imagine a condition halfway between four-legged and two-legged locomotion (Zihlman and Brunker, 1979). Human foot structure was long ago recognized as being very much like that of the African apes, and the Laetoli footprints are proof that there were bipedal hominids 3.5 million years ago. *Australopithecus afarensis* evinces many chimpanzee-like characters in its skull, body size, joints, ilium, and ischium, features similar to those observed in *A. afranicus* from Sterkfontein. Detailed comparison of *Australopithecus* with a pygmy chimpanzee (Fig. 1) shows the striking similarities and at the same time the pelvic–lower limb dissimilarities attributable to one being a biped, the other a quadruped. Nevertheless, the transformation of these features could easily be imagined to have taken place in 1 million years or less, the time between the proposed ape–human divergence and the earliest known bipedal hominids. We leave it to the contestants to discuss whether 1 million years is gradual or punctuational in the context of human evolution.

Pilbeam (1979) has raised the question of whether it is possible to say anything very sensible about past evolutionary stages. He takes the position that the present situation is very confused and that no phylogenetic certainty is to be found without much more fossil evidence (however, see Ward and Pilbeam, this volume, Chapter 8). This seems in part to be an over-reaction to the realization that his previous certainty about the hominid status of *Ramapithecus* was probably mistaken. No one can argue that paleoanthropologists are in possession of all the data they need, or that more fossils, especially in the time period 4–8 m.y.a., would not be critical to understanding human evolution. But there has been a lot of information gathered from the fossil record, from comparative anatomy, and from comparative biochemistry— enough at least to indicate that *Ramapithecus* was very likely an ape and that the earliest known hominids at 3.5 million years may have been only one step away from a small ape like the living *Pan paniscus*.

ACKNOWLEDGMENTS

We thank C. Borchert, L. Brunker, and J. Cronin for discussion and comments. We acknowledge research support from the Wenner-Gren Foundation for Anthropological Research, the Leakey Foundation, and the Faculty Research Committees at the University of California, Santa Cruz (ALZ) and the University of California, San Francisco (JML). To Wynn Kapit goes our appreciation for drawing the figure.

References

Almquist, A. J. 1974. Sexual differences in the anterior dentition in African primates. *Am. J. Phys. Anthropol.* **40:**359–369.

Archer, M. 1976. The dasyurid dentition and its relationship to that of didelphids, thylacinids, borhyaenids (Marsupicarnivora) and peramelids (Peramelina : Marsupialia). *Aust. J. Zool. (Suppl.)* **39:**1–34.

Badrian, N. L., Badrian, A. J., and Susman, R. L. 1980. Feeding ecology of *Pan paniscus* in Zaire. *Am. J. Phys. Anthropol.* **52:**202 (Abstract).

Brunker, L., 1980. A comparison of the locomotor adaptations of the vervet and patas monkeys, and implications for early hominid evolution. *Am. J. Phys. Anthropol.* **52:**208 (Abstract).

Ciochon, R. L., and Chiarelli, A. B., Eds. 1980. *Evolutionary Biology of the New World Monkeys and Continental Drift*, Plenum, New York, 528 pp.

Coolidge, H. J. 1933. *Pan paniscus*, pygmy chimpanzee from south of the Congo River. *Am. J. Phys. Anthropol.* **18:**1–59.

Corruccini, R. S., and McHenry, H. M. 1979. Morphological affinities of *Pan paniscus*. *Science* **204:**1341–1343.

Cramer, D. L. 1977. Craniofacial morphology of *Pan paniscus:* A morphometric and evolutionary appraisal. *Contrib. Primatol.* **10:**1–64.

Cramer, D. L., and Zihlman, A. L. 1978. Sexual dimorphism in the pygmy chimpanzee, *Pan paniscus*, in: *Recent Advances in Primatology*, Volume 3, *Evolution* (D. J. Chivers and K. A. Joysey, eds.), pp. 487–490, Academic, London.

Cronin, J. E. 1977. Pygmy chimpanzee (*Pan paniscus*) systematics. *Am. J. Phys. Anthropol.* **47:**125 (Abstract).

Cronin, J. E., Boaz, N. T., Stringer, C. B., and Rak, Y. 1981. Tempo and mode in hominid evolution. *Nature (Lond.)* **292:**113–122.

Dayhoff, M. O. 1978. *Atlas of Protein Sequence and Structure*, Volume 5, Supplement 3, National Biomedical Research Foundation, Washington, D.C.

Eckhardt, R. B. 1972. Population genetics and human origins. *Sci. Am.* **226**(1):94–103.

Fenart, R., and Deblock, R. 1973. *Pan paniscus* et *Pan troglodytes* craniometrie: Étude comparative et ontogénique selon les méthodes classiques et vestibulariers. Tome 1, *Ann. Mus. R. Afr. Cent., Tervuren, Ser. 8, Sci. Zool.* **204:**1–473.

Goodman, M. 1963. Man's place in the phylogeny of the primates as reflected in serum proteins, in: *Classification and Human Evolution* (S. L. Washburn, ed.), pp. 204–234, Aldine, Chicago.

Goodman, M. 1976. Toward a genealogical description of the primates, in: *Molecular Anthropology* (M. Goodman and R. E. Tashian, eds.), pp. 321–353, Plenum, New York.

Goodman, M., Barnabas, J., Matsudea, G., and Moore, G. W. 1971. Molecular evolution in the descent of man. *Nature (Lond.)* **233:**604–613.

Gould, S. J., and Eldredge, N. 1977. Punctuated equlibria: The tempo and mode of evolution reconsidered. *Paleobiology* **3:**115–151.

Grand, T. I. 1977. Body weight: Its relation to tissue composition, segment distribution and motor function. *Am. J. Phys. Anthropol.* **47:**211–239.

Gregory, W. K. 1916. Studies on the evolution of the primates. *Bull. Am. Mus. Nat. Hist.* **35:**239–355.

Greenfield, L. O. 1980. A late divergence hypothesis. *Am. J. Phys. Anthropol.* **52:**351–365.

Groves, C. P. 1981. Comment on "Bonobos: Generalized hominid prototypes or specialized insular dwarfs?" *Curr. Anthropol.* **22:**366.

Horn, A. D. 1980. Some observations on the ecology of the bonobo chimpanzee (*Pan paniscus* Schwarz 1929) near Lake Tumba, Zaire. *Folia Primatol.* **34:**145–169.

Huxley, T. H. 1863. *Evidence as to Man's Place in Nature*, Williams and Norgate, London. 159 pp.

Johanson, D. C. 1974. Some metric aspects of the permanent and deciduous dentition of the pygmy chimpanzee (*Pan paniscus*). *Am. J. Phys. Anthropol.* **41:**39–48.

Johanson, D. C., and Taieb, M. 1976. Plio-Pleistocene hominid discoveries in Hadar, Ethiopia. *Nature (Lond.)* **260:**293–297.

Johanson, D. C., and White, T. D. 1979. A systematic assessment of early African hominids. *Science* **203:**321–330.

Johnson, S. C. 1981. Bonobos: Generalized hominid prototypes or specialized insular dwarfs? *Curr. Anthropol.* **22:**363–375.

Jukes, T. H. 1980. Silent nucleotide substitution and the molecular evolutionary clock. *Science* **210:**973–978.

Kano, T. 1979. A pilot study on the ecology of pygmy chimpanzees, *Pan paniscus*, in: *The Great Apes* (D. A. Hamburg and E. R. McCown, eds.), pp. 122–135, Benjamin/Cummings, Menlo Park, California.

Kano, T. 1980. Social behavior of wild pygmy chimpanzees (*Pan paniscus*) of Wamba: A preliminary report. *J. Hum. Evol.* **9:**243–260.

Kay, R. F. 1981. The nut-crackers—A new theory of the adaptations of the Ramapithecinae. *Am. J. Phys. Anthropol.* **55:**141–151.

Kay, R. F., Fleagle, J. G., and Simons, E. L. 1981. A revision of the Oligocene apes of the Fayum Province, Egypt. *Am. J. Phys. Anthropol.* **55:**293–322.

Kennedy, G. E. 1980. *Paleoanthropology*, McGraw-Hill, New York. 479 pp.

Kimbel, W. H. 1979. Characteristics of fossil hominid cranial remains from Hadar, Ethiopia. *Am. J. Phys. Anthropol.* **50:**454 (Abstract).

Kimura, M. 1980. Was globin evolution very rapid at its early stages? *J. Mol. Evol.* **17:**110–113.

Kuroda, S. 1979. Grouping of the pygmy chimpanzees. *Primates* **20:**161–183.

Kuroda, S. 1980. Social behavior of the pygmy chimpanzees. *Primates* **21:**181–197.

Leakey, M. D. 1981. Tracks and tools. *Phil. Trans. R. Soc. Lond. B* **292:**95–102.

Le Gros Clark, W. E. 1947. Observations on the anatomy of the fossil Australopithecinae. *J. Anat.* **81:**300–334.

Lewis, G. E. 1934. Preliminary notice of new man-like apes from India. *Am. J. Sci.* **27:**161–181.

Lewis, G. E. 1937. Taxonomic syllabus of Siwalik fossil anthropoids. *Am. J. Sci.* **34:**139–147.

Lewis, O. J. 1977. Joint remodelling and the evolution of the human hand. *J. Anat.* **123:**157–201.

Lewis, O. J. 1980. The joints of the evolving foot. Part III. The fossil evidence. *J. Anat.* **131:**275–298.

Lovejoy, C. O. 1973. The gait of australopithecines. *Yearb. Phys. Anthropol.* **17:**147–161.

Lowenstein, J. M. 1981. Immunological reactions from fossil material. *Phil. Trans. R. Soc. Lond. B* **292:**143–149.

Lowenstein, J. M. 1982. Fossil proteins and evolutionary time, in: *Proceedings of the Pontifical Academy of Sciences,* The Vatican, Rome (in press).

Lowenstein, J. M., Sarich, V. M., and Richardson, B. J. 1981. Albumin systematics of the extinct mammoth and Tasmanian wolf. *Nature (Lond.)* **291**:409–411.

Margoliash, E. 1963. Primary structure and evolution of cytochrome C. *Proc. Natl. Acad. Sci. USA* **50**:672–679.

McHenry, H. M. 1981. Comment on "Bonobos: Generalized hominid prototypes or specialized insular dwarfs?" *Curr. Anthropol.* **22**:367.

McHenry, H. M., and Corruccini, R. S. 1981. *Pan paniscus* and human evolution. *Am. J. Phys. Anthropol.* **54**:355–367.

Miyata, T., and Yasunaga, T. 1980. Molecular evolution of mRNA: A method for estimating evolutionary rates of synonymous and amino acid substitutions from homologous nucleotide sequences and its application. *J. Mol. Evol.* **16**:23–36.

Oxnard, C. 1975. *Uniqueness and Diversity in Human Evolution: Morphometric Studies of Australopithecines,* University of Chicago Press, Chicago. 131 pp.

Pilbeam, D. R. 1972. *The Ascent of Man,* Macmillan, New York. 207 pp.

Pilbeam, D. R. 1978. Rearranging our family tree. *Hum. Nat.* **1**(6):38–45.

Pilbeam, D. R. 1979. Recent finds and interpretations of Miocene hominoids. *Annu. Rev. Anthropol.* **8**:333–353.

Pilbeam, D. R., Meyer, G. E., Badgley, C., Rose, M. D., Pickford, J. H. L., Behrensmeyer, A. K., and Shah, S. M. I., 1977. New hominoid primates from the Siwaliks of Pakistan and their bearing on hominoid evolution. *Nature* **270**:689–695.

Reynolds, V. 1967. *The Apes,* Dutton, New York. 296 pp.

Sarich, V. M. 1967. Man's Place in Nature, Presented at the American Anthropological Association Annual Meetings, Washington, D.C.

Sarich, V. M. 1968. The origin of the hominids: An immunological approach, in: *Perspectives on Human Evolution,* Volume 1 (S. L. Washburn and P. C. Jay, eds.), pp. 94–121, Holt, Rinehart and Winston, New York.

Sarich, V. M., and Cronin, J. E. 1976. Molecular systematics of the primates, in: *Molecular Anthropology* (M. Goodman and R. E. Tashian, eds.), pp. 141–170, Plenum, New York.

Sarich, V. M., and Cronin, J. E. 1977. Generation length and hominoid evolution. *Nature (Lond.)* **269**:354–355.

Sarich, V. M., and Wilson, A. C. 1967. Immunological time scale for hominid evolution. *Science* **158**:1200–1203.

Savage, E. S., Wilkerson, B. J., and Bakeman, R. 1977. Spontaneous gestural communication among conspecifics in the pygmy chimpanzee (*Pan paniscus*), in: *Progress in Ape Research* (G. H. Bourne, ed.), pp. 97–116, Academic, New York.

Savage-Rumbaugh, E. S., and Wilkerson, B. J. 1978. Socio-sexual behavior in *Pan paniscus* and *Pan troglodytes:* A comparative study. *J. Hum. Evol.* **7**:327–344.

Schwarz, E. 1929. Das vorkommen des schimpansen auf dem linken Kongo-Ufer. *Rev. Zool. Bot. Afr.* **16**:425–426.

Schultz, A. H. 1930. The skeleton of the trunk line and limbs of higher primates. *Hum. Biol.* **2**:303–439.

Schultz, A. H. 1937. Proportions, variability and asymmetries of the long bones of the limbs and the clavicles in man and apes. *Hum. Biol.* **9**:281–328.

Simons, E. L. 1961. The phyletic position of *Ramapithecus. Postilla* (Peabody Mus. Nat. Hist., Yale Univ.) **57**:1–9.

Simons, E. L. 1972. *Primate Evolution,* Macmillan, New York. 322 pp.

Simons, E. L. 1977. *Ramapithecus. Sci. Am.* **236**(5):28–35.

Simons, E. L. 1981. Man's immediate forerunners. *Phil. Trans. R. Soc. Lond. B* **292**:21–41.

Simpson, G. G. 1963. The meaning of taxonomic statements, in: *Classification and Human Evolution* (S. L. Washburn, ed.), pp. 1–31, Aldine, Chicago.

Stanley, S. M. 1977. *Macroevolution: Pattern and Process,* Freeman, San Francisco. 332 pp.

Straus, W. L. 1949. The riddle of man's ancestry. *Q. Rev. Biol.* **24**:200–223.

Susman, R. 1980. Acrobatic pygmy chimpanzees. *Nat. Hist.* **89:**33–39.

Tobias, P. V. 1981. The emergence of man in Africa and beyond. *Phil. Trans. R. Soc: Lond. B* **292:**43–56.

Tratz, E., and Heck, H. 1954. Der afrikanische anthropoide "Bonobo," eine neue menschenaffengattung. *Saugetierk. Mitt.* **2:**97–101.

Tuttle, R. H. 1969. Knuckle-walking and the problem of human origins. *Science* **166:**953–961.

Tuttle, R. H. 1977. Naturalistic positional behavior of apes and models of hominid evolution, 1929–1976, in: *Progress in Ape Research* (G. H. Bourne, ed.), pp. 277–293, Academic, New York.

Tuttle, R. H. 1981. Evolution of hominid bipedalism and prehensile capabilities. *Phil. Trans. R. Soc. Lond. B* **292:**89–94.

Vogel, C. 1975. Remarks on the reconstruction of the dental arcade of *Ramapithecus*, in: *Paleoanthropology, Morphology and Paleoecology* (R. H. Tuttle, ed.), pp. 87–98, Mouton, Paris.

Walker, A., and Andrews, P. 1973. Reconstruction of the dental arcades of *Ramapithecus wickeri*. *Nature (Lond.)* **244:**313–314.

Washburn, S. L. 1951. The analysis of primate evolution with particular reference to the origin of man. *Cold Spring Harbor Symp. Quant. Biol.* **15:**67–78.

Washburn, S. L. 1963. Behavior and human evolution, in: *Classification and Human Evolution* (S. L. Washburn, ed.), pp. 190–203, Aldine, Chicago.

Watson, J. D., and Crick, F. H. C. 1953. Molecular structure of nucleic acids. A structure for deoxyribose nucleic acid. *Nature (Lond.)* **171:**737–738.

Wilson, A. C., Carlson, S. S., and White, T. J. 1977. Biochemical evolution. *Annu. Rev. Biochem.* **46:**573–639.

Wolpoff, M. H. 1981. Comment on "Bonobos: Generalized hominid prototypes or specialized insular dwarfs? *Curr. Anthropol.* **22**(4):370–371.

Wood-Jones, F. 1948. *Hallmarks of Mankind*, Williams and Wilkins, Baltimore.

Zihlman, A. L. 1976. Sexual dimorphism and its behavioral implications in early hominids, in: *Les Plus Anciens Hominides* (P. V. Tobias and Y. Coppens, eds.), pp. 268–293, Colloque VI, IX Union Internationale des Sciences Prehistoriques et Protohistoriques, Nice. CNSR, Paris.

Zihlman, A. L. 1978. Interpretations of early hominid locomotion, in: *Early Hominids of Africa* (C. J. Jolly, ed.), pp. 361–377, Duckworth, London.

Zihlman, A. L. 1979*a*. Differences in body weight composition of pygmy and common chimpanzees. *Am. J. Phys. Anthropol.* **50:**496 (Abstract).

Zihlman, A. L. 1979*b*. Pygmy chimpanzees and early hominids, *S. Afr. J. Sci.* **75:**165–168.

Zihlman, A. L. 1980. Locomotor behavior in pygmy and common chimpanzees. *Am. J. Phys. Anthropol.* **52:**295 (Abstract).

Zihlman, A. L. 1982. *The Human Evolution Coloring Book,* Harper and Row, New York.

Zihlman, A. L., and Brunker, L. 1979. Hominid bipedalism: Then and now. *Yearb. Phys. Anthropol.* **22:**132–162.

Zihlman, A. L., and Cramer, D. L. 1978. Skeletal differences between pygmy (*Pan paniscus*) and common chimpanzees (*Pan troglodytes*). *Folia Primatol.* **29:**86–94.

Zihlman, A. L., Cronin, J. E., Cramer, D. L., and Sarich, V. M. 1978. Pygmy chimpanzees as a possible prototype for the common ancestor of humans, chimpanzees and gorillas. *Nature (Lond.)* **275:**744–746.

Toward the Resolution of Discrepancies between Phenetic and Paleontological Data Bearing on the Question of Human Origins

27

L. O. GREENFIELD

Introduction

Since the publication of comparative serological data and the protein clock (Sarich, 1968; Sarich and Wilson, 1967; Wilson and Sarich, 1969; Goodman, 1963, 1975, 1976), there has been a continuing controversy among serologists and paleoanthropologists over the timing of the divergence between human and Great Ape lineages. Based upon the high degree of similarity in the blood proteins and DNA of *Pan, Gorilla,* and *Homo* (King and Wilson, 1975), serologists have suggested a late divergence (5 m.y.a.), while most paleoanthropologists, citing middle Miocene *"Ramapithecus"* (Simons, 1961, 1964, 1968, 1972, 1976, 1977; Simons and Pilbeam, 1965, 1972; Tattersall, 1975; Conroy, 1972) or middle Miocene "thick-enameled" dryopithecines

L. O. GREENFIELD • Department of Anthropology, Temple University, Philadelphia, Pennsylvania 19122.

(Pilbeam *et al.,* 1977) as evidence of an independent human lineage, suggested an earlier divergence date of 15–20 m.y.a. (the early date possibilities are hereafter referred to as the early divergence hypotheses).

To resolve the discrepancies in divergence date estimates between the two schools of thought, the assumptions and available serological and fossil data were reanalyzed and the "late divergence hypothesis" supported (Greenfield, 1980). Largely it was the combination of serological data with a new interpretation of the paleontological evidence (Corruccini *et al.,* 1976; Greenfield, 1979) that led to formulation of the late divergence hypothesis. This hypothesis not only attempts to resolve the debate over divergence timing but also to define the phylogenetic history of the extant Great Apes and humans.

This chapter offers a brief comparison of early and late divergence hypotheses with a twofold purpose in mind: (1) to show that the late divergence hypothesis is, in light of all relevant evidence at hand, the *most* likely phylogenetic alternative (in part, this discussion will focus on the weaknesses of the competing early divergence hypotheses), and (2) to discuss early vs. late estimates in relation to their bearing on our view of the place of humans in nature.

Late vs. Early Divergence Hypotheses

Phenetics and Parsimony

A basic premise of the late divergence hypothesis is the phylogeny of the extant Great Apes and humans suggested by comparative studies of their sera and DNA (Fig. 1). From this work it can be concluded that there were *two* significant branching events.* The first branching point (marked 1) involved a split between the lineage leading to *Pongo* and the common ancestor of *Pan, Gorilla,* and *Homo;* the second branching point (marked 2) was the divergence of the *Pan, Gorilla,* and *Homo* lineages. In sum, comparative studies of characters of simple inheritance and DNA indicate that *Pan, Gorilla,* and *Homo* are more closely related to each other than any is to *Pongo.*

While this is not a startling revelation, paleontologists and paleoanthropologists have generally sought a single branching point based on, or at least reflected in, the traditional notion of "pongid" vs. "hominid." For example, the taxonomies or phylogenetic trees constructed by Gregory and Hellman (1926), Simons (1972), Pilbeam *et al.* (1977), Leakey and Lewin (1978), Wolpoff (1980), and Ciochon (1980) all deal with the divergence between humans and the extant Great Apes as a single branching point [for contrast with the two-divergence-point phylogeny and taxonomy see Goodman (1975)]. The reason this school has generally looked for a single branching point is probably because of greater emphasis and reliance on gross anatomy,

---

*Counting the (*Pan, Gorilla, Homo*) split as a single event.

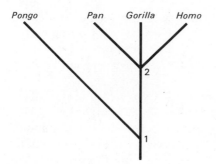

Fig. 1. Cladogram of the extant Great Apes and humans based on serological and biochemical studies. See text for further discussion.

behavior, ontogeny, etc., i.e., the characters of complex genetic inheritance. Their taxonomies or phylogenetic trees, reflecting the general lifestyles of the extant forms, suggest that *Pongo, Pan,* and *Gorilla* are more closely related to each other (although some polygenic characters sort these taxa in the same way that the serological and DNA data sort them) than any is to *Homo.* Hence, the simple taxonomic dichotomy (pongid and hominid) and a single branching point are supported.

A more convincing case, however, can be made for the correctness of the two-branching-point phylogeny because it is based on known genotypic change (in characters of simple inheritance) in the lineages since the times of divergence. The single-branching-point phylogeny is based only on genotypic change *estimated* from phenotypic change (in characters of complex inheritance) since divergence. Clearly, the best evidence available on this phylogenetic question is the data that measure aspects of the genotype more accurately, such as serological data and evidence from DNA hybridization experiments. The underlying sequences of organic bases in DNA are known for characters of simple inheritance and unknown for characters of complex inheritance. If the serological data are correct, it is possible to conclude that those who primarily use characters of complex inheritance have overestimated the genetic differences between *Homo* and *Pan/Gorilla* and underestimated the genetic differences between *Pan/Gorilla* and *Pongo.* The late divergence hypothesis proceeds on the premise that the serological data are the best phenetic evidence, conforming to the notions that *Pan, Gorilla,* and *Homo* are more closely related to each other than any is to *Pongo* and that two branching events must be accounted for.

These are important points, particularly the latter, because with two branching events, the early divergence hypotheses must cite a greater amount of independent acquisition to account for the evolution of many characters shared by the extant Great Apes and humans. This makes them, for reasons outlined below, less parsimonious and hence less likely alternatives.

If, as suggested by the early divergence hypotheses, the lineage leading to *Homo* separated from the lineages leading to *Pan* and *Gorilla* by 15–20 m.y.a. (branching point 2),* then the earlier divergence between the common ances-

*The evidence for this being *"Ramapithecus"* or thick-enameled dryopithecines.

tor of *Pan, Gorilla,* and *Homo* and the phyletic ancestor of *Pongo* (branching point 1) must have occurred perhaps 20–25 m.y.a. or earlier. Consequently, according to the early divergence hypotheses, the last common ancestor of all extant Great Apes and humans would have been a form predating early Miocene hominoids, such as *D. (Proconsul).**

It has been noted by a variety of workers (Le Gros Clark and Leakey, 1951; Napier and Davis, 1959; Robinson, 1952; Corruccini *et al.,* 1976; Greenfield, 1980) that *D. (Proconsul),* the best known early Miocene hominoid, lacks many dental, gnathic, cranial, postcranial, locomotor, and ontogenetic adaptations shared by the extant African Great Apes and *Australopithecus.* Consequently, as an early Miocene last common ancestor of just these three forms, it would necessarily have to be argued, within the context of an early divergence hypothesis, that numerous and unlikely independent acquisitions of the shared characters must have occurred. The earlier divergence (branching point 1) from a last common ancestor of *all* extant Great Apes and humans would then have involved a hominoid which was, in all likelihood, more "primitive" than *D. (Proconsul)* and perhaps intermediate in morphology between *D. (Proconsul)* and *Aegyptopithecus.* Derivation of all extant Great Apes and humans from this hypothetical late Oligocene form would have involved even greater (and less likely) independent acquisition of their shared characters (both in number and/or degree). Thus, with two branching points for which to account instead of one, the early divergence hypotheses become less parsimonious with respect to explaining the origin of characters shared by all extant Great Apes and humans.

Of the known Miocene hominoids, the genus *Sivapithecus* [which includes a number of junior synonyms listed in Greenfield (1980)] possessed the greatest number of characters shared by the extant Great Apes and *Australopithecus* (Greenfield, 1980). Their derivation from *Sivapithecus,* with respect to the shared characters, can be explained with far greater parsimony. It remains to be seen whether the less well known dryopithecine from Europe [*D. (Dryopithecus)*] is a more likely last common ancestor than *Sivapithecus.*

Thus, according to the late divergence hypothesis, *Sivapithecus* represents a structural grade from which it is most likely all extant Great Apes and humans could have evolved. Only this very general relationship, and not specific ancestor/descendant relationships, has been seriously suggested in the late divergence hypothesis. However, one variable which may hint at more specific relationships concerns the cingulum (Greenfield, 1980). Moreover, recently, Andrews and Tekkaya (1980) have noted that upper facial morphology may link one of the *Sivapithecus* species with *Pongo.*

In sum, the data on phenetics and parsimony suggest a two-branching-point phylogeny for the extant Great Apes and humans, with *Pan, Gorilla,* and *Homo* more closely related to each other, than any is to *Pongo,* and according-

---

*It should be noted here that other early Miocene hominoids are known, however, principally from dental remains. Thus there is still the question as to whether these forms were more or less "primitive" than *D. (Proconsul).*

ly, *Sivapithecus* is the most likely last common ancestor. The late divergence hypothesis is the only origin theory proposed thus far that is consistent with all of these conclusions.

Middle Miocene Hominoids, A. afarensis, and the Nature and Timing of the Divergences

Both early divergence hypotheses are based primarily on the claim that there are middle Miocene hominoids beginning the process of hominization and, for this reason, likely ancestors of *Australopithecus*. The most commonly cited phyletic ancestor, "*Ramapithecus*," was systematically compared to all relevant hominoid taxa (Greenfield, 1977, 1979) with special attention paid to those features purported to show its phyletic relationship to *Australopithecus*. That analysis showed "*Ramapithecus*" to be a smaller version of *Sivapithecus*, showing no *special* similarities to the earliest undoubted member of the human lineage, *A. afarensis*.

The major features which distinguish "*Ramapithecus*" from *Sivapithecus* are relative canine size and facial prognathism. The interspecific positive canine allometry and cheek tooth isometry found for primates in general (Wood, 1979; Gingerich, 1977; Kay, 1975) suggest that, all other factors being equal, smaller ape species like "*Ramapithecus*" should have smaller canines relative to cheek tooth size than larger ape species like *Sivapithecus* (these expectations are supported by the data). Similarly, Corruccini and Ciochon (1979) have found that large primates have disproportionately long faces (facial breadth is negatively allometric to facial length), and this may explain fully why *Sivapithecus* has greater facial prognathism than "*Ramapithecus*." Thus, the few real differences between the "*Ramapithecus*" species and other *Sivapithecus* species are most likely related to allometric scaling. Since the major differences are probably related to size, a generic distinction between "*Ramapithecus*" and *Sivapithecus* is unwarranted, just as it is unwarranted for the different sized *D. (Proconsul)* species.* This new interpretation of middle Miocene hominoids suggests that if there is a dryopithecine beginning the process of hominization in the middle Miocene (16–10 m.y.a.), it is apparently not "*Ramapithecus*" or *Sivapithecus*.

Recently, Pilbeam *et al.* (1977) proposed that a hominid/pongid split (again only one branching point is sought) is reflected in the appearance of "thick"- and "thin"-enameled middle Miocene dryopithecines. However, several points can be raised to argue against this scenario. First, the forms attributed to the "thick"-enameled group are characterized by a wide range in enamel thickness (Greenfield, 1980). Enamel thickness is a continuous variable and does not sort dryopithecines neatly into two discrete classes. Second, among extant Great Apes and humans, only *Pan* and *Gorilla* have thin enamel; *Pongo* and *Austalopithecus*/*Homo* have thick enamel (Molnar and Gantt,

---

***D (Proconsul)* species also show differences in relative canine size and snout length.

1977; Kay, 1981). It is therefore conceivable that *Pongo* is derived from a thicker enameled middle Miocene hominoid like *Sivapithecus* since there are also other resemblances (Greenfield, 1980; Gregory and Hellman, 1926; Andrews and Tekkaya, 1980) suggesting an ancestor/descendant relationship. Third, the major reason thick and thin enamel have been cited as marking a pivotal point of divergence between human and Great Ape lineages is because they are supposed to be related to the onset of a divergence in habitat and lifestyles. Smith and Pilbeam (1980) and Simons (1976) suggest thick enamel is indicative of more hominid-like ground feeding with associated high-attrition-rate diets, but thin enamel is related to arboreal feeding with associated low-attrition-rate diets (pongid-like). However, in a recent investigation of enamel thickness in extant primates, Kay (1981) has found that thick enamel is not associated with ground feeding: terrestrial *or* arboreal extant primates with thick enamel feed on hard nuts. This finding suggests that thick enamel is probably not an adaptation which delays the effects of attrition from gritty foods obtained in terrestrial feeding, but it is more likely an adaptation which resists occlusal stresses created by powerful vertical compression. In addition, Covert and Kay (1981) investigated dental microwear and concluded on the basis of observations on one thick-enameled "*Ramapithecus*" (*Sivapithecus*) molar that "this animal was not eating food containing exogenous grit or grasses or other plants containing phytoliths . . ." (p. 335). If similar observations are found for other *Sivapithecus* specimens, then it can be concluded that there is little evidence that these middle Miocene forms were ground feeding.

Also implicit in both early divergence hypotheses is the idea that the *earliest* members of the human lineage had, by 15 m.y.a., some distinguishing features or feature which marked the beginning of the adaptive divergence between humans and apes (one branching point). However, the present evidence, as discussed above and elsewhere (Greenfield, 1980), suggests other, more likely possibilities.

One possibility is that early divergences did occur (15–25 m.y.a., but the early–middle Miocene forms did not become distinctive enough to be identified with their respective extant descendants until long after the actual branchings occurred. In this scenario, however, many of the characters shared by extant Great Apes and *Australopithecus* must still be explained as independent acquisitions.

A second possibility is that the middle Miocene phyletic ancestors of the extant Great Apes and humans existed but are unknown as fossils. This negative view is viable if one believes that our phyletic ancestors have escaped collectors despite the now numerous dryopithecine samples from middle Miocene deposits all over the world.

A third possibility is that the divergences had not yet occurred by the middle Miocene, and that when they did occur (6–15 m.y.a.) they may or may not have immediately given rise to clearly distinguishable members of independent lineages leading to humans and the extant Great Apes. This is the central point of the late divergence hypothesis, and it is a statement concerned with both the timing and nature of the divergences (especially the latter one, branching point 2).

The later timing estimates for branching points are based on the magnitude of protein similarities among the extant forms, the morphology of *Sivapithecus* as it relates to the origin of characters shared by the extant Great Apes and *Australopithecus*, and the retention of numerous dryopithecine-like characters in the dentition and face of 3–3.8 million year-old *A. afarensis*, the earliest undoubted member of the human lineage.

While the measures of rates of protein evolution may not be precise (Radinsky, 1978) with "slow" and "fast" species recognized, the serological evidence favors later estimates of divergence. In other mammalian examples, where the differences in proteins and DNA are comparable to the differences between humans and extant African apes, very recent divergences took place. Many of the characters shared by the extant Great Apes and *Australopithecus* appear in dryopithecines after the start of the middle Miocene. Finally, *A. afarensis* is geologically young (Johanson and White, 1979) but exhibits, as does *Pongo*, numerous features reminiscent of *Sivapithecus* (Greenfield, 1980) in addition to a small brain on top of a bipedal locomotor system.

As to the nature of the branching events, especially branching point 2, several possibilities exist within the context of the late divergence hypothesis. If the divergence between the African Great Apes and humans occurred within the period from 15 to 8 m.y.a. (inclusive of the dated samples of *Sivapithecus*), then there is no evidence that the *earliest* members of the human lineage were making significant inroads toward hominization (judging from the mosaic of characters in *A. afarensis*, the most significant and documentable early hominid adaptation probably was the shift in the locomotor system). If branching point 2 occurred later, any scenario is possible due to the absence of relevant samples from the late Miocene and early Pliocene.

Summary

In sum, this brief comparison of early and late divergence hypotheses shows that the late divergence hypothesis is consistent with more of the data bearing on the question of human origins. The early divergence hypotheses have been based primarily on "*Ramapithecus*" or "thick"-enameled dryopithecines. "*Ramapithecus*" has now been shown to be dryopithecine-like and virtually indistinguishable from *Sivapithecus*. "Thick"-enameled dryopithecines could have given rise to *Pongo* as well as *Australopithecus*. The evidence for an independent lineage leading to *Homo* at the start of the middle Miocene is lacking. The late divergence hypothesis, setting the divergence between the lineage leading to *Homo* and the lineages leading to the extant African apes in the middle to late Miocene, is the simplest phylogenetic alternative at present and it should be the first one *tested*.

Early vs. Late Estimates in Relation to Our View of Humanity's Place in Nature

For those who have studied the evidence (fossils, comparative anatomy, serology, behavior, etc.), the evolutionary past and close relationship of humans to the extant Great Apes seem certain. It is the details of those phylogenetic events that are the focus of recent research into human origins. Perhaps the most commonly disputed detail in the phylogenetic history of the Great Apes and humans has been the date of separation of the human lineage from the lineages leading to the extant African apes (branching point 2). The question arises concerning the difference if the human lineage was independent 5–10 vs. 20 m.y.a.

In a recent article, Pfeiffer (1980) suggested (and not many have) why the date of divergence might be important. He stated that a late date would mean that "humans are not as different from other animals as one might like to suppose." However, this is not the case. The divergence dates are *not* measures of the differences between the extant descendants. A measure of the differences and similarities between extant humans and extant Great Apes is obtained by comparing their anatomy, biochemistry, behavior, etc. Therefore, the divergence dates are measures of the amounts of time it took for the differences to reach their present point. Whether the dates are early or late, the differences between living humans and other animals *remain the same*. The different dates of divergence in conjunction with estimates of differences between common ancestral and descendant (extant) forms yields little more than varying estimates of the average rates of evolutionary change for specific characters since divergence. Thus, it seems that the dates themselves may have very little impact on our view of humanity's place in nature.

References

Andrews, P., and Tekkaya, I. 1980. A revision of the Turkish Miocene hominoid *Sivapithecus meteai*. *Palaeontology* **23**:85–96.

Ciochon, R. L. 1980. Miocene ancestry. *L. S. B. Leakey Foundation News* **17**:3–4.

Conroy, G. C. 1972. Problems with the interpretation of *Ramapithecus:* With special reference to anterior tooth reduction. *Am. J. Phys. Anthropol.* **37**:41–48.

Corruccini, R. S., and Ciochon, R. L. 1979. Primate facial allometry and interpretations of australopithecine variation. *Nature* (Lond.) **281**:62–63.

Corruccini, R. S., Ciochon, R. L., and McHenry, H. M. 1976. The postcranium of Miocene hominoids: Were dryopithecines merely "dental apes"? *Primates* **17**:205–223.

Covert, H. H., and Kay, R. F. 1981. Dental microwear and diet: Implications for determining the feeding behaviors of extinct primates, with a comment on the dietary pattern of *Sivapithecus*. *Am. J. Phys. Anthropol.* **55**:331–336.

Gingerich, P. D. 1977. Correlation of tooth size and body size in living hominoid primates, with a note on relative brain size in *Aegyptopithecus* and *Proconsul*. *Am. J. Phys. Anthropol.* **47**:395–398.

Goodman, M. 1963. Man's place in the phylogeny of the primates as reflected in serum proteins, in: *Classification and Human Evolution* (S. L. Washburn, ed.), pp. 204–234, Aldine, Chicago.

Goodman, M. 1975. Protein sequence and immunological specificity, in: *Phylogeny of the Primates, A Multidisciplinary Approach* (W. P. Luckett and F. S. Szalay, eds.), pp. 219–248, New York.

Goodman, M. 1976. Protein sequences in phylogeny, in: *Molecular Evolution* (F. J. Ayala, ed.), pp. 141–159, Sinauer, Sunderland, Massachusetts.

Greenfield, L. O. 1977. *Ramapithecus and Early Hominid Origins,* University Microfilms, Ann Arbor, 301 pp.

Greenfield, L. O. 1979. On the adaptive pattern of *"Ramapithecus". Am. J. Phys. Anthropol.* **50:**527–548.

Greenfield, L. O. 1980. A late divergence hypothesis. *Am. J. Phys. Anthropol.* **52:**351–365.

Gregory, W. K., and Hellman, M. 1926. The dentition of *Dryopithecus* and the origin of man. *Anthropol. Pap. Am. Mus. Nat. Hist.* **28:**1–123.

Johanson, D. C., and White, T. 1979. A systematic assessment of early African hominids. *Science* **203:**321–330.

Kay, R. F. 1975. The functional adaptations of primate molar teeth. *Am. J. Phys. Anthropol.* **43:**195–216.

Kay, R. F. 1981. The nut-crackers—A new theory of the adaptations of the Ramapithecinae. *Am. J. Phys. Anthropol.* **55:**141–151.

King, M. C., and Wilson, A. C. 1975. Evolution at two levels in humans and chimpanzees. *Science* **188:**107–116.

Le Gros Clark, W. E., and Leakey, L. S. B., 1951. The Miocene Hominoidea of East Africa. *Fossil Mammals of Africa* (Br. Mus. Nat. Hist.) **1:**1–117.

Leakey, R. E. F., and Lewin, R. 1978. *Origins,* Dutton, New York, 264 pp.

Molnar, S., and Gantt, D. G. 1977. Functional implications of primate enamel thickness. *Am. J. Phys. Anthropol.* **46:**447–454.

Napier, J. R., and Davis, P. R. 1959. The forelimb skeleton and associated remains of *Proconsul africanus. Fossil Mammals of Africa* (Br. Mus. Nat. Hist.) **16:**1–69.

Pfeiffer, J. 1980. Current research casts new light on human origins. *Smithsonian* **11:**91–103.

Pilbeam, D. R., Meyer, G. E., Badgley, C., Rose, M. D., Pickford, M. H. L., Behrensmeyer, A. K., and Shah, S. M. I. 1977. New hominoid primates from the Siwaliks of Pakistan and their bearing on hominoid evolution. *Nature (Lond.)* **270:**689–695.

Radinsky, L. 1978. Do albumin clocks run on time? *Science* **200:**1182–1183.

Robinson, J. T. 1952. Note on the skull of *Proconsul africanus. Am. J. Phy. Anthropol.* **10:**7–12.

Sarich, V. M. 1968. The origin of the hominids: An immunological approach, in: *Perpectives on Human Evolution,* Volume 1 (S. L. Washburn and P. C. Jay, eds.), pp. 94–121, Holt, Rinehart, and Winston, New York.

Sarich, V. M., and Wilson, A. C. 1967. Immunological time scale for human evolution. *Science* **158:**1200–1203.

Simons, E. L. 1961. The phyletic position of *Ramapithecus. Postilla* (Peabody Mus. Nat. Hist., Yale Univ.) **57:**1–9.

Simons, E. L. 1964. On the mandible of *Ramapithecus. Proc. Natl. Acad. Sci. USA* **51:**528–535.

Simons, E. L. 1968. A source for dental comparison of *Ramapithecus* with *Australopithecus* and *Homo. S. Afr. J. Sci.* **64:**92–112.

Simons, E. L. 1972. *Primate Evolution,* Macmillan, New York. 312 pp.

Simons, E. L. 1976. The nature of the transition in the dental mechanism from pongids to hominids. *J. Hum. Evol.* **5:**511–528.

Simons, E. L. 1977. *Ramapithecus. Sci. Am.* **236**(5):28–35.

Simons, E. L., and Pilbeam, D. R. 1965. Preliminary revision of the Dryopithecinae (Pongidae, Anthropoidea). *Folia Primatol.* **3:**81–152.

Simons, E. L., and Pilbeam, D. R. 1972. Hominoid paleoprimatology, in: *The Functional and Evolutionary Biology of Primates* (R. Tuttle, ed.), pp. 36–62, Aldine Atherton, Chicago.

Smith, R. J., and Pilbeam, D. R. 1980. Evolution of the orang-utan. *Nature (Lond.)* **284:**447–448.

Tattersall, I. 1975. *The Evolutionary Significance of Ramapithecus,* Burgess, Minneapolis. 32 pp.

Wilson, A. C., and Sarich, V. M. 1969. A molecular time scale for human evolution. *Proc. Natl. Acad. Sci. USA* **63:**1088–1093.

Wolpoff, M. H. 1980. *Paleoanthropology,* Knopf, New York. 379 pp.

Wood, B. A. 1979. Tooth and body size allometric trends in modern primates and fossil hominids. *Am. J. Phys. Anthropol.* **50:**493.

Morphological Trends and Phylogenetic Relationships from Middle Miocene Hominoids to Late Pliocene Hominids

28

N. T. BOAZ

Introduction

It is perhaps too early to posit certain morphological continuities in the Neogene fossil record of Hominoidea since there are still so many hiatuses, both temporal and geographical. Nevertheless, the gaps are slowly closing and the datum points that can be established provide the basis for this chapter, the purpose of which is the comparison of Pliocene hominids and the late-to-middle Miocene hominoids which may have been their forebears.

Neogene Hominoidea

The first finds attributable to what would become the genus *Ramapithecus* Lewis, 1934 were described by Pilgrim (1910, 1915) based on specimens col-

N. T. BOAZ • Department of Anthropology, New York University, New York, New York 10003.

lected in the Siwalik Hills of India. Attention was drawn to the specimens, along with dryopithecines from other parts of the Old World, by W. K. Gregory (Gregory, 1920; Gregory and Hellman, 1926). Gregory maintained that the morphology of the dryopithecine dentition, especially the molar morphology, made the genus *Dryopithecus* a likely candidate for a forerunner of hominids. It was not until Simons and Pilbeam (1965) retained *Ramapithecus* as a valid, separate genus, while sinking many other taxa among Miocene Hominoidea, that particular attention was generally focused on *Ramapithecus*. The genus was considered until relatively recently to have possessed a parabolic dental arcade, relatively orthognathous facial profile, reduced canine size, and thickened molar occlusal enamel, all hominid-like characteristics. Specimens from sub-Saharan Africa, originally named *Kenyapithecus* (L. S. B. Leakey, 1962), were ascribed to *Ramapithecus* (Simons and Pilbeam, 1965) on the basis of the correspondence of the then known morphological traits of the Asian and African specimens. The sample of specimens from Asia has now been augmented by relatively complete specimens from Turkey, Pakistan, and China. These allow a much clearer picture of Asian *Ramapithecus*. The age relationships among these samples in both Eurasia and Africa have also been clarified. It is thus possible to assess much more clearly the morphological, temporal, and probable phylogenetic relationships among late Neogene hominoids.

In their revision of Miocene Hominoidea and in later papers, Simons and Pilbeam (1965; Pilbeam, 1969) suggested evolutionary relationships from early Miocene dryopithecines to modern pongids: *Dryopithecus* (*Proconsul*) *major* to *Gorilla; D. (P.) africanus* to *Pan;* and perhaps *Dryopithecus* (now *Sivapithecus*) *sivalensis* to *Pongo*. These hypothetical evolutionary relationships, based largely on aspects of size and dental morphology, when taken with the suggestion that *Ramapithecus* represented the basal hominid, implied an early Miocene or Oligocene divergence of the hominid and pongid lineages.

Several considerations argue against these phylogenetic relationships. Early Miocene dryopithecines do not share postcranial specializations characteristic of extant pongids (Napier and Davis, 1959; Morbeck, 1975; Corruccini *et al.,* 1976) and thus common pongid postcranial specializations would have to evolve in parallel in several pongid lineages. Similarly, metrical aspects of the palate and lower face of early Miocene hominoids are unlike those of extant Great Apes, based on canonical variates analysis (McHenry *et al.,* 1980). Any hypothetical morphological continuity from early Miocene forms to living Great Apes genera is difficult to demonstrate due to the lack of a fossil record, particularly in Africa, spanning the late Miocene to the Plio-Pleistocene. Recent fossil discoveries in Asia have confirmed that *Sivapithecus* (including "*Ramapithecus*") is anatomically similar to *Pongo* and is probably closely related to the ancestry of this genus (Andrews and Tekkaya, 1980; cf. Smith and Pilbeam, 1980). Furthermore, the geological ages of the hominoid-bearing deposits in the Potwar Plateau, and by extension in India and China, have been shown to be between 8 and 10.5 m.y.a., rather than about 12 m.y.a. as previously thought (Opdyke *et al.,* 1979; Tauxe, 1979). Finally, mac-

romolecular studies indicate much later (middle Miocene to Pliocene) hominid–pongid splits: approximately 8–10 m.y.a. for the *Pongo* split and 4–6
m.y.a. for the (*Homo, Pan, Gorilla*) split (Sarich and Cronin, 1976; Zihlman *et
al.*, 1978). These considerations have cast doubt on the theoretical position
that dryopithecines represent the direct forebears of extant hominoids.

Similarly, the evolutionary status of *"Ramapithecus"* as defined by Simons
and Pilbeam (1965) has been questioned on a number of grounds. Primary
among these have been the realization of close morphological similarity of
Asian samples of *"Ramapithecus"* and *Sivapithecus* to only one living hominoid
(*Pongo*), revised dating of these deposits, increased samples of Pliocene hominids which evince ape-like morphology (thus implying a relatively late divergence date from an ape-like ancestor), and macromolecular results, to which
allusion was made above. It also now appears to be likely that the inclusion of
the taxon *Kenyapithecus wickeri* (L. S. B. Leakey, 1962) within *"Ramapithecus"*
(Simons and Pilbeam, 1965) will prove unwarranted because of morphological differences which should become more demonstrable after study of increased samples of the former species. Based on morphological considerations, to be discussed later in this chapter, and geochronology, *"Ramapithecus"*
sensu stricto [now included within *Sivapithecus*, see Greenfield (1980) and Kay
and Simons (this volume, Chapter 23)] was likely already on a divergent
lineage leading to *Pongo*, whereas the earlier *Kenyapithecus* may, in fact, represent a close approach to the common ancestor of the Great Apes and humans,
or at least the African Hominoidea. Evidence on faunal provinces (Bernor,
1978, and this volume, Chapter 2) shows that late Miocene Indo-Pakistan was
characterized by a semi-isolated, woodland biome. This finding suggests that
South Asian and African hominoid populations were biogeographically and
probably phylogenetically separate by this time (8–10.5 m.y.a.).

Although the morphological distinctions between the Asian and African
samples referred to *Ramapithecus* have only recently been better documented,
the differences to be seen in the hypodigms of *Ramapithecus punjabicus* and
Kenyapithecus wickeri as originally erected caused Aguirre (1972, 1975) to argue for the retention of the latter taxon, *contra* Simons and Pilbeam (1965).
Aguirre particularly stressed the molar morphology, which in *Ramapithecus
punjabicus* shows cusps positioned toward the margins of "widely excavated"
crowns and steeply sloping cuspal ridges. *Kenyapithecus wickeri* possesses more
centrally located cusps on the crown, less steeply sloping sides to the molar
cusps, and molar crown occlusal outlines more rectangular, rather than squarish, as in *Ramapithecus*. Aguirre (1975, p. 102) concludes that the "paral-
lelogramic and bunodont crowns of *Kenyapithecus* . . . could more easily be
imagined as related to those of an ancestor of man." He noted that symphyseal and premaxillary morphology seemed to confirm this view.

Aguirre did not point out that in ways that *Ramapithecus* differs from
Kenyapithecus, it approximates *Pongo*. The molar crown morphology of *Pongo*
shows marginally positioned cusps, steep-sided valleys, and squared occlusal
outlines. This suggestion of a relationship between fossil hominoids from the
Siwaliks and the extant orangutan is not new. Lydekker (1886) described a

hominoid upper canine from the Siwaliks which he considered indistinguishable from the orangutan and which he ascribed to *Simia* (*Pongo*). Pilgrim (1915, p. 29) described and named a right upper third molar *Palaeosimia rugosidens* (GSI D-188). The species name refers to the crenulated molar occlusal surface, which along with the crown shape, closely resembles the orangutan. Remane (1922) reassigned a right lower third molar (GSI D-175) that had been assigned to *Dryopithecus giganteus* by Pilgrim (1915) to cf. *Pongo*. Hooijer (1951) discussed a right lower third premolar assigned by Pilgrim (1915) to *Sivapithecus* cf. *indicus* and by von Koenigswald (1950) to *Indopithecus giganteus* and concluded that "there is virtually nothing—no character at all—by which the present tooth can be distinguished from the P_3 of the common orangutan."

A recently discovered cranium from the Potwar Plateau of Pakistan (Pilbeam, 1982) shows closely set orbits, inflated and flaring zygomatic processes of the maxilla, strong I1–I2 heteromorphy, marked subnasal prognathism, and a relatively vertically oriented face. These are all characteristics which align the specimen, which has been assigned to *Sivapithecus*, with *Pongo*. If Greenfield (1979, 1980) and Kay and Simons (this volume, Chapter 23) are correct in their sinking of *Ramapithecus* into *Sivapithecus*, and this genus subsumes the vast panoply of taxonomic names that has been applied to Siwalik fossil hominoids, then there does not seem to be any strong evidence that this genus is closely related to the ancestry of hominids. Rather, there is mounting evidence that *Sivapithecus* is ancestral at least at a generic level to *Pongo*, a hypothesis that must be tested with fossil finds from the Asian late Miocene to Plio-Pleistocene horizons.

Andrews and Tekkaya (1980) described a hominoid lower face and palate from the Middle Sinap Series of Turkey (MTA 2125) of middle Vallesian age, approximately 10–11 m.y.a., and thus some 1–2 million years earlier than the bulk of the sample from the Siwaliks. Similarities in zygomatic morphology, alveolar prognathism, incisor heteromorphy, and squared molar occlusal outlines allied the specimen to *Pongo*. Considering the earlier age of this specimen, it is interesting that certain morphological features were cited by Andrews and Tekkaya (1980) as similarities with African apes: nasal form similar to *Pan*, and molar lengths, particularly a large M3, similar to *Gorilla*. Other, earlier hominoids from Turkey, such as the Çandir mandible (Tekkaya, 1974), show close similarities with African *Kenyapithecus wickeri*, to which this specimen has been referred. The Paşalar sample also contains teeth which have been referred to the Fort Ternan species, along with a larger form which has been referred to *Sivapithecus darwini* (Andrews and Tobien, 1977). Çandir is probably similar in age to Fort Ternan, while Paşalar is somewhat earlier, probably about 15 m.y.a. on faunal grounds (Andrews and Tobien, 1977).

Kenyapithecus wickeri from Fort Ternan has also been claimed to represent the species ancestral to later hominids (L. S. B. Leakey, 1962) on the basis of mandibular symphysis inclination, small superior torus, mandibular alveolar shape, and inferred small size of the upper canine. In fact, the morphology of the specimens referred to *Kenyapithecus wickeri* does not necessarily represent the derived hominid condition (Frisch, 1967), and may likely represent the primitive condition for hominoids. As such, *Kenyapithecus wickeri* is a possible candidate for the common ancestor for the Great Apes and hominids (Fig. 1). *Sivapithecus* and *"Ramapithecus"* in this view are already on or near divergent lineages leading to Asian *Pongo* and *Gigantopithecus*. Specimens from the Siwaliks, later in time than African and Turkish *Kenyapithecus*, would be expected to show morphology approximating the African hominoid or hominid condition, were they on the hominid lineage. Since the recently recovered cranial specimens seem to indicate orangutan-like morphology, this expectation is not borne out. One may thus hypothesize that the Siwalik hominoids, although probably near the branching point of African and Asian Hominoidea, have little direct bearing on the evolution of hominids, *Pan*, or *Gorilla*.

Pliocene Hominidae

The African record of fossil Hominoidea between about 14 m.y.a., after Fort Ternan, and 3.7 m.y.a., before Laetoli, is poor. Affinities of the available fossils are difficult to ascertain. The Ngorora upper molar, probably M^2

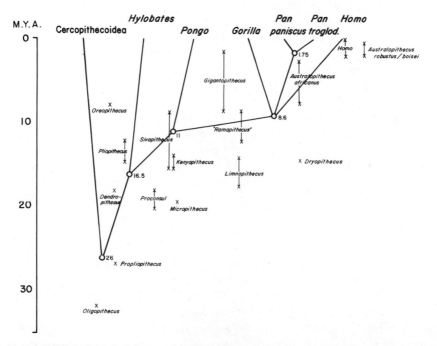

Fig. 1. A simplified view of phylogenetic relationships of Neogene hominoids, as discussed in the text. Usage of *Sivapithecus* follows Greenfield (1979, 1980) and includes *Ramapithecus*. *Kenyapithecus* refers to *K. wickeri* L. Leakey, 1962. *Australopithecus robustus* is used here as an inclusive term for robust australopithecines.

(KNM-BN 378), and Lukeino lower molar, probably M_2 (KNM-LU 335), have been ascribed to Hominidae gen. et sp. indet. (A) by Howell (1978a). McHenry and Corruccini (1980) concluded, on the basis of canonical variates analysis, based on cuspal proportions, that the Lukeino molar was most similar to *Pan troglodytes*. Although 50 specimens of *Pongo pygmaeus* were included in the study, the distance of this species to Lukeino was not reported. Unlike Howell (1978a, p. 159), who considered the Lukeino specimen to "deviate distinctively from known penecontemporaneous or antecedent pongids," McHenry and Corruccini (1980, p. 397) state that there are "no salient hominid apomorphies," although "there are other features of the fossil not measured by our study which differ from *Pan*."

The Lothagam mandible (KNM-LT 329), dated to 5.0–5.5 m.y.a. (Behrensmeyer, 1976), was also included in the study by McHenry and Corruccini (1980). They found the specimen to be midway between pongids and hominids along a multivariate axis with the interpreted distinctions of "square, relatively short molars with centrally situated hypoconulid, reduced trigonid breadth and protoconid and metaconid diameters, enlarged M_1 compared to M_2, shallow and thick mandibular corpus, expanded occlusal foveal length and crown breadth, large hypoconid, and high crown" (p. 398). These authors argue for classification of KNM-LU 335 as Hominoidea indet. and KNM-LT 329, on the basis of the relatively short occlusal length of its M_1, small M_1 entoconid, and shallow mandible, as Hominidae indet. These results, so far as they go, suggest divergence of hominids and African pongids between the dates of Lukeino (6.5–7.0 m.y.a.) and Lothagam (5.0–5.5 m.y.a.). Acceptance of such a late divergence date depends on the Ngorora and Lukeino specimens being viewed as generalized hominoids and possibly ancestral to both hominid and African pongid lineages, but not to hominids alone.

There is no fossil evidence of pongids in the African Plio-Pleistocene and the record of hominids, or at least hominid-like hominoids, is very sparse in the period postdating Lothagam and predating Laetoli. Boaz (1980) has reported a hominoid clavicle (26P4A) from the probably basal Pliocene (5.0 m.y.a.) Sahabi Formation of Libya. Morphology of the specimen hints at a shoulder girdle posteriorly positioned relative to the vertebral column, a characteristic possibly associated with bipedal posture. The Kanopoi humerus (KNM-KP 271), described and attributed to *A. africanus* by Patterson and Howells (1967), is judged to be about 4 million years old (Patterson, *et al.*, 1970). It has recently been restudied by Senut (1979), who suggests it is not australopithecine, but instead attributable to *Homo*. It is possible that the specimen, having been collected on the surface, may not be as old as other parts of the Kanopoi collection. A right temporal from Chemeron (KNM-BC 1), ascribed to Hominidae indet. by Tobias in Martyn (1967), and probably of similar age to Laetoli (Howell 1978a), is so incomplete as to be undiagnostic.

The Laetoli sample represents the earliest clear record of Hominidae, and the evolutionary origin of these australopithecines is a major unsolved question. These hominids were clearly fully bipedal (M. D. Leakey and Hay,

1979), and, as yet, the morphological and functionally intermediate stages through which hominids passed from a dryopithecine locomotor adaptation to bipedalism remain undocumented. Although somewhat earlier in time than Hadar and South African australopithecines, the Laetoli hominids are morphologically similar to both samples (see below). Johanson and White (1979) have treated the Laetoli and Hadar hominids as a group, termed *A. afarensis,* and separated them from South African *A. africanus.* The *A. afarensis* sample has been distinguished from *A. africanus* primarily on the basis of differences in the dentition mesial to the molars (Johanson *et al.,* 1978). Molar morphology has been more conservative throughout hominoid evolution in the Neogene.

A significant problem is assessing differences or similarities in the hypodigms of these two taxa is the small sample size of *A. africanus* anterior dentitions. There are cogent reasons for believing that the available sample of incisors and canines from South African gracile australopithecine sites do not represent the true range of variability in the population. These reasons will be discussed below.

Johanson and White (1979, p. 322) note the size difference to be seen between the upper central and the upper lateral incisors in *A. afarensis.* Means for mesiodistal diameters are 10.4 and 7.6 mm for I^1 and I^2, respectively. Only one specimen of *A. africanus* preserves these teeth, STS 52a.* The incisors are heteromorphic. I^1 measures mesiodistally 9.3 (R) and 9.5 mm (L), and I^2 6.8 (R) and 7.3 mm (L) (Robinson, 1956, pp. 24,28). In both the *A. afarensis* sample and in STS 52a, mesiodistal diameters, as good single measures of incisor size, are 25% larger for I^1 than for I^2.

Morphologically I^1 and I^2 are similar in *A. afarensis* and *A. africanus.* Both possess upper incisor heteromorphy, also characteristic of *Sivapithecus.* Lower central and lateral incisors are more similar in size in both *A. africanus* and *A. afarensis* samples. In both, the I_2 is slightly larger mesiodistally than I_1 and has a sloping distal crown. These teeth have not figured in distinguishing *A. africanus* from *A. afarensis.*

The canines have played an important part in assessments of *A. afarensis.* Metrically, the Hadar and Laetoli samples of upper canines show a wider size range up to larger size than the six Sterkfontein (two C's from one specimen, STS 52a) and one MLD upper canines. Mesiodistal diameters are between 8.9 and 11.6 mm for *A. afarensis* and 8.8 and 9.9 mm for *A. africanus,* and labiolingual diameters are 9.3–12.5 and ?7.3–9.9 mm, respectively. Robinson (1956) notes in several places (pp. 44, 48) that the dimensions of the available sample of lower canines of *A. africanus* are larger than the dimensions of the available sample of upper canines of this taxon. Robinson (1956, p. 44) concludes that "It would be most unusual if *Australopithecus (africanus)* had larger mandibular canines than maxillary ones and the conclusion is therefore

---

*TM 1512 without an I^1 possesses an *in situ,* quite worn I^2. Mesiodistal diameter is 5.8 mm (Robinson, 1956, p. 28). If this value were averaged in to the values for STS 52a, the incisor heteromorphy would be even more apparent.

drawn that the maxillary canines in the collection are either medium or small in size." Tobias (1980, p. 7) has drawn attention to the left lower canine, STW 21, which shows the crown asymmetry and strong lingual ridge supposedly diagnostic of *A. afarensis*. STS 3, a specimen originally described as an upper canine by Broom *et al.* (1950, p. 40), was rediagnosed a lower by Robinson (1956, p. 48), primarily because of its asymmetry, distal ridge, and enamel infolding. The specimen is also well worn, with a markedly distal sloping surface in the distal crown. Broom *et al.* (1950, p. 40) gave the dimensions of the tooth as 12.2 (labiolingual) × [11.3] (mesiodistal) mm, at the upper end or above the size range of Hadar and Laetoli lower (8.8–12.0, 7.9–11.7 mm) and upper (9.3–12.5, 8.9–11.6 mm) canines (Johanson and White, 1979, p. 322), while Robinson (1956, p. 48) gave its dimensions as 12.1 × 10.5 mm.

The primary question in regard to *A. afarensis* is not whether its represents a separate (though perhaps earlier and closely related) species distinct from *A. africanus,* but is the assessment of its phylogenetic position. Johanson and White (1979) have suggested that *A. afarensis* represents the ancestral population from which *Homo* sprang, and they have placed *A. africanus* on a corollary lineage leading to the robust australopithecines. Thus, there are two questions to be addressed: the *A. afarensis–Homo* relationship and *A. africanus–A. robustus/boisei* relationship. The argument for a special ancestor–descendant relationship between the populations represented by Laetoli and Hadar and the first species of *Homo* is tenuous. The two specimens of those ascribed to *A. afarensis* for which cranial capacities are estimated [A.L. 288-1, 450 cm^3; A.L. 166-9, 385 cm^3 (Kimbel, 1979)] are well below those for even early *Homo* and at the lower end of the known *A. africanus* range. As Johanson and colleagues have latterly emphasized [it must be remembered that Johanson and Taieb (1976) originally ascribed the A.L. 200-la palate, as well as other specimens, to *Homo*], the large canines, relatively large and mesiolingually reduced P_3, strong canine juga, prognathic lower face, upper incisor heteromorphy, reduced midfacial region, and small body size, all indicate a quite primitive hominid morphology, i.e., not trending in the direction of *Homo*. Nevertheless, the proportions of anterior to posterior dentition indicate a "harmonious" (Robinson, 1956) size relationship, similar in pattern to *Homo* and dissimilar to robust australopithecines. The Sangiran 4 and A.L. 200-la palates *are* similar in overall appearance. The morphologies of individual teeth are also similar in *A. afarensis* and *Homo*. One way to make sense of these observations is to hypothesize the Laetoli and Hadar populations as ancestral to *Homo,* but separated by some evolutionary distance from the first recorded instance of the latter.

Although it may not be relevant to discuss temporal relationships in constructing cladograms (e.g., Tattersall and Eldridge, 1977), in discussions of phylogeny, considerations of absolute and relative age are crucial. The Hadar sample has an upper age limit of 2.8 ± 0.08 m.y.a. (Walter and Aronson, 1982), revised from the original K–Ar date of 2.6 m.y.a. on the Bouroukie Tuff-2 [BKT-2 (Aronson *et al.*, 1977)], and a lower limit originally dated by K–Ar on the Kadada Moumou Basalt at 3.0 ± 0.2 m.y.a. (Aronson *et al.*, 1977) and now reportedly at 3.6 ± 0.15 m.y.a. (Walter and Aronson, 1982). Because of alteration, the age of the basalt is uncertain (Aronson and Taieb, 1981). The dating of the Laetoli sample is based on K–Ar dates of Tuff c and biotite near the top of the section of 3.59 m.y.a. stratigraphically closest to the hominid fossils, and a date of 3.77 m.y.a. on a tuff 50 m below Tuff c (M. D. Leakey *et al.*, 1976). The Hadar suids correspond to those of Omo Shungura lower Member B (Cooke, 1978) controlled by K–Ar and paleomagnetic dating [about 3.2 m.y.a. (Brown and Nash, 1976; Brown and Shuey, 1976)].

Attempts at geological (Partridge, 1973) and paleomagnetic dating (Brock *et al.*, 1977) of the South African caves have not proved satisfactory although great strides have been made recently in further refining the history of deposition and consequent naming of lithological members (Partridge, 1978, 1979; Butzer, 1976). Renewed studies of the faunal assemblages have enabled a clearer idea of relative biostratigraphic age of the South African cave sites to be obtained. Vrba (1981) estimates that Sterkfontein Member 4, from which the *A. africanus* samples derive, is between 2.5 and 3.0 m.y.a., based primarily on bovids. Harris and White (1979) estimate a similar age on the basis of suids. Makapansgat Member 3, the lithostratigraphic provenience of *A. africanus* at that site, is estimated to be somewhat older than Sterkfontain Member 4, perhaps 3.0–3.2 m.y.a. (Vrba, 1982). There is thus temporal overlap between Hadar, and Makapansgat and Sterkfontein, with the latter extending later in time. Laetoli may overlap Hadar, but it is also certainly earlier.

The earliest appearance of the genus *Homo* seems to be no earlier than 2.3 m.y.a., but the best-preserved specimens are later than this, 1.8–2.0 m.y.a. (Boaz and Howell, 1977; Tobias, 1978). The upper age limit of the Hadar sample is 2.9 m.y.a., but only A.L. 288-1 in the Hadar sample, deriving from the Denen Dora Member, may be this late in time. A hypothetical phylogenetic connection between *A. afarensis* and *Homo*, disregarding *A. africanus*, has the disadvantage of postulating a substantial temporal, as well as morphological gap. The gracile australopithecines from South Africa partially fill the lower portion of this time gap (up to about 2.5 m.y.a.). Although I believe that the eastern and southern gracile australopithecines, because of the morphological similarities discussed earlier, are both referrable to *A. africanus* (Boaz, 1979), some of the South African specimens show characteristics which may be interpreted as indicative of slightly more advanced morphology.

Primary among these are a decrease in the emphasis of the size of the anterior dentition and functionally related posterior fibers of *m. temporalis*, and perhaps an overall increase in body size. A greater degree of orthognathism and body size increase also characterizes the *Homo* lineage, and a more parsimonious hypothesis than that of Johanson and White (1979) is that South African *A. africanus* represents the populations descendent from earlier, slightly more primitive *A. africanus* and ancestral to early *Homo*. The morphology of the earliest *Homo* (see below) also argues for ancestry from *A. africanus*.

The second question to be considered in regard to the phylogeny of South African *A. africanus* is its proposed ancestral relationship to *A. robustus* or *boisei*. The robust australopithecines appear suddenly at about 2.0–2.1 m.y.a. with quite recognizable morphology in both South Africa and East Africa (Howell 1978*a,b*). The apparent suddenness of their appearance led Boaz (1977) to suggest a dispersal of these forms into East and South Africa from elsewhere, correlated with ecological changes at this time. Johanson and White (1979) suggest an *in situ* evolution of robust australopithecines. Although morphological comparison between available Kromdraai and Swartkrans hominid samples shows a range of variation possibly indicative of an evolutionary sequence (Rak, 1981; Cronin *et al.*, 1981, p. 120), this evolutionary sequence is not demonstrable through time. This is because all robust australopithecine samples must be considered more or less contemporaneous.

It has been recognized for some time that robust australopithecines must have evolved from a species with morphology similar to *A. africanus* (Tobias, 1967, pp. 240–244; *contra* Robinson, 1963). That the available sample of South African early australopithecines does not represent this evolutionary transition is indicated by the following considerations: differences between Hadar–Laetoli and South African *Australopithecus* samples (if not due to individual variability) can be understood as evolutionary change within the *Homo* lineage; the most "robust-like" of the South African sample, at Makapansgat (e.g., MLD 2 and MLD 37/38), occurs earlier than the less robust *A. africanus* (thus not supporting an evolutionary trend toward greater robusticity); and no specimens are known postdating the upper age limit of the South African sample that are clearly trending toward the robust australopithecine condition at the only site where this time period is represented, Omo Shungura Formations Members B (upper) through D. The origin of the robust australopithecines is clearly a major problem in early hominid studies, and more work in the period 2.0–3.0 m.y.a., and in other parts of Africa, is needed to resolve it. Evolution to larger body size occurred in both robust australopithecines and *Homo* lineages in parallel, but the former was characterized by increasing premolar and molar size and decreasing anterior dentition (related to masticatory grinding efficiency), whereas the latter emphasized cranial expansion while retaining a more generalized australopithecine dentition. For these reasons, suggestions that some (Aguirre, 1970; Wallace, 1975) or all (Johanson and White, 1979) South African early australopithecines and some East African specimens (Olson, 1981) represent ancestral robust australopithecines are difficult to accept on currently available evidence.

The earliest representatives of the genus *Homo*, such as KNM-ER 1470, KNM-ER 1813, and O.H. 24, show characteristics which ally them with *A. africanus*. In fact, all of these specimens have, in the recent past, been referred by some authors to *Australopithecus* [KNM-ER 1470 by Walker (1976); KNM-ER 1813 by R. E. F. Leakey (1974); and O.H. 24 by R. E. F. Leakey (1974)], although most authors consider them referrable to *Homo habilis* (Howell, 1978*a;* Tobias, 1978; Cronin *et al.*, 1981). The dentition of early *Homo* shows a reduction in size of the canine, a more squarish P_3, decrease in molar length relative to molar width, a frequent reduction in the amount of molar crenulation (and less frequent Carabelli complexes in the M^1 and protoconical cingula on lower molars), upper central incisors generally lacking labial longitudinal grooves, and somewhat less heteromorphic I^1 compared to I^2 than in *A. africanus*. The I^2–$\bar{C}$ and $\underline{C}$–P_3 diastemata are generally lacking. The cranial vault is generally more globular, with a relatively shorter length compared to height, than in *A. africanus*. Cranial capacity is usually, though not always, larger. The area of the temporal fossa is smaller, with a more anteriorly sloping root of the zygomatic process of the temporal. The facial skeleton in *Homo* shows raised margins of the nasal aperture and a more convex margin of the incisor region of the maxilla, causing a less flattened and straight naso-alveolar clivus. The canine jugae are also less pronounced in *Homo*. Postcranial remains ascribed to *Homo* indicate generally a larger body size than *A. africanus*.

The earliest fossils of *Homo* present a mosaic of characters recalling to varying degrees the australopithecine condition. KNM-ER 1470 possesses a relatively large cranial capacity while showing large canine juga, a relatively straight naso-alveolar clivus, and relatively long zygomatic arches subtending large temporal fossae. KNM-ER 1813 and O.H. 24, on the other hand, have small cranial capacities, within the *A. africanus* range, but show the midfacial, alveolar, and temporal morphology characteristic of *Homo*.

Conclusion

Figure 1 summarizes the phylogenetic conclusions at a generic level reached in this chapter. Research in some specific areas will be necessary to test the branching times and supposed phyletic ancestors in this scheme. Further work in the African middle to late Miocene on the nature of *Kenyapithecus wickeri*, especially its relationship with *Sivapithecus* (or *Ramapithecus*) *punjabicus*, is essential to test whether the former is a suitable common ancestor for both African and Asian Hominoidea. Much work is needed in Asia from Nagri times to the Pleistocene to elucidate the evolution of *Pongo* and *Gigantopithecus* and to determine what relations these genera have with *Sivapithecus*. Exploration is needed in Africa, particularly the central and western sections, for Miocene to Pleistocene deposits which might yield ancestors of the extant African pongids, documenting the phylogenetic relationships among *Pan*, *Gorilla*, and hominid lineages. Research in the period predating 4 m.y.a., especially in Africa, is essential to determining the evolutionary origin of early australopithecines and the relationship of these to the more derived robust australopithecines. Alternatively, this latter problem may be solved through concerted research effort in the period 3.0–2.0 m.y.a. in sub-Saharan Africa, if this was the site of robust australopithecine divergence, as some have suggested.

Although Fig. 1 shows closer agreement with phylogenies based on biomolecular comparisons of extant primates (Sarich and Cronin, 1976) than some other paleontologically based schemes (e.g., Simons and Pilbeam, 1965),

Table 1. Estimated Dates of Divergence for Hominoid Lineages from Biomolecular and Paleontological Data[a]

Taxa	Divergence dates (M.Y.A.)		
	Biomolecular[b]	Paleontological[c]	Average
Hylobates–Homo/Pan/Pongo	11–13	20–22	16.5
Pongo–Homo/Pan	9–10	10–15	11
Gorilla–Homo/Pan	4–5.5	10–15	8.8
Pan–Homo	4–5.5	10–15	8.6
Pan paniscus–P. troglodytes	1.5–2	?	1.75

[a]An average date of divergence of 26 m.y.a. for Hominoidea and Cercopithecoidea is obtained from estimates by Sarich and Cronin (1976) and Szalay and Delson (1979). Other averages are for illustrative purposes in comparing the differences between the two sets of interpretations (see Fig. 2).
[b]Sarich and Cronin (1976); Zihlman *et al.* (1978).
[c]Andrews (1978); Pilbeam *et al.* (1977); Szalay and Delson (1979); and this study.

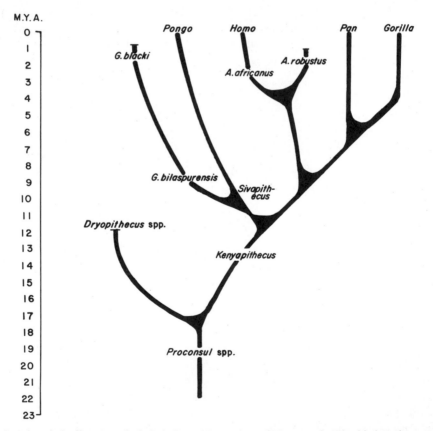

Fig. 2. A heuristic diagram of phyletic branching points of Neogene hominoids based on averaged paleontological and biomolecular estimates (see Table 1). Vertical lines and × symbols indicate known temporal ranges of fossil taxa.

there are still discrepancies. Using best-estimate ages derived from Andrews (1978, p. 210), Pilbeam *et al.* (1977), Szalay and Delson (1979), and this study, the probable divergence dates as determined from paleontological and geochronological data are listed in Table 1. Biomolecular ages for these same lineage divergences are derived from Sarich and Cronin (1976) and Zihlman *et al.* (1978), and are also presented in Table 1. On the assumptions that these two sets of data are measuring the same phenomena, and, to quote Jacobs and Pilbeam (1980, p. 552), that "both sets cannot be correct although either or both could be incorrect," the mean dates were averaged (Table 1, Fig. 2) as a heuristic exercise. The averaged dates are compared with known temporal spans of relevant fossil genera in Fig. 2. There is a relatively good fit with the available fossil record. A divergence data of hominids and African pongids of 8.6 m.y.a. is not inconsistent with the known Pliocene hominid record, but it is inconsistent with accepting "*Ramapithecus*" or *Kenyapithecus* as hominids. A divergence date for Asian and African pongids (and hominids) of 11 m.y.a. cannot be assessed by recourse to the largely nonexistent fossil record after this date, but it is consistent with revised dating of the Siwaliks. Occurrence of *Sivapithecus* species, if ancestral to *Pongo,* earlier than 11 m.y.a., as at Paşalar, would argue against this date. Similarly, although not discussed here, acceptance of *Dendropithecus* as ancestral to the hylobatid lineage would force an older revision of the 16.5 m.y.a. date for separation of the Great and Lesser Ape lineages.

Much further work in Neogene primate paleontology, geochronology, and biogeography will have to be carried out before a true rapprochement of fossil and biomolecular data and interpretations on hominoid evolution can be attained.

ACKNOWLEDGMENTS

The author thanks the American Museum of National History, British Museum (Natural History), University of California, Berkeley, Cleveland Museum of Natural History, Ethiopian National Museum, Musée de l'Homme, Paris, National Museums of Kenya, and Yale Peabody Museum for access to specimens examined in the course of this study, and a New York University Presidential Fellowship for release time and financial assistance.

References

Aguirre, E. 1970. Identificacion de "Paranthropus" en Makapansgat, in: *Crónica del XI Congreso Nacional de Arqueologia, Merida 1969,* pp. 98–124.
Aguirre, E. 1972. Les rapports phylétiques de *Ramapithecus* et *Kenyapithecus* et le origine des Hominidés. *Anthropologie (Paris)* **76:**501–523.

Aguirre, E. 1975. *Kenyapithecus* and *Ramapithecus*. in: *Paleoanthropology, Morphology and Paleoecology* (R. H. Tuttle, ed.), pp. 99–104, Mouton, The Hague.

Andrews, P. J. 1978. A revision of the Miocene Hominoidea of East Africa. *Bull. Br. Mus. (Nat. Hist.) Geol.* **30**(2):85–224.

Andrews, P. J., and Tekkaya, I. 1980. A revision of the Turkish Miocene hominoid *Sivapithecus meteai*. *Palaeontology* **23**:85–95.

Andrews, P. J., and Tobien, H. 1977. New Miocene locality in Turkey with evidence on the origin of *Ramapithecus* and *Sivapithecus*. *Nature (Lond.)* **268**:699–701.

Aronson, J. L., and Taieb, M. 1981. Geology and paleogeography of the Hadar hominid site, Ethiopia, in: *Hominid Sites: The Geological Settings* (G. Rapp, Jr. and C. F. Vondra, eds.), pp. 165–195, Westview, Boulder (AAAS Selected Symposium 63).

Aronson, J. L., Schmidt, T. J., Walter, R. C., Taieb, M., Tiercelin, J. J., Johanson, D. C., Naeser, C. W., and Nairn, A. E. M. 1977. New geochronologic and paleomagnetic data for the hominid-bearing Hadar Formation of Ethiopia. *Nature (Lond.)* **267**:323–327.

Behrensmeyer, A. K. 1976. Lothagam Hill, Kanapoi and Ekora: A general summary of stratigraphy and faunas, in: *Earliest Man and Environments in the Lake Rudolf Basin* (Y. Coppens, F. C. Howell, G. L. Isaac and R. E. F. Leakey, eds.), pp. 163–170, University of Chicago Press, Chicago.

Bernor, R. L. 1978. *The Mammalian Systematics, Biostratigraphy and Biochronology of Maragheh and Its Importance for Understanding Late Miocene Hominoid Zoogeography and Evolution*, Ph.D. Dissertation, University of California, Los Angeles. 314 pp.

Boaz, N. T. 1977. Paleoecology of early Hominidae in Africa. *Kroeber Anthropol. Soc. Pap.* **50**:37–62.

Boaz, N. T. 1979. Hominid evolution in eastern Africa during the Pliocene and early Pleistocene. *Annu. Rev. Anthropol.* **8**:71–85.

Boaz, N. T. 1980. A hominoid clavicle from the Mio-Pliocene of Sahabi, Libya. *Am. J. Phys. Anthropol.* **53**:49–54.

Boaz, N. T., and Howell, F. C. 1977. A gracile hominid cranium from upper Member G of the Shungura Formation, Ethiopia. *Am. J. Phys. Anthropol.* **46**:93–108.

Brock, A., McFadden, P. L., and Partridge, T. C. 1977. Preliminary palaeomagnetic results from Makapansgat and Swartkrans. *Nature (Lond.)* **266**:249–250.

Brown, F. H., and Nash, W. P., 1976. Radiometric dating and tuff mineralogy of Omo Group deposits, in: *Earliest Man and Environments in the Lake Rudolf Basin* (Y. Coppens, F. C. Howell, G. L. Isaac and R. E. F. Leakey, eds.), pp. 50–63, University of Chicago Press, Chicago.

Broom, R., Robinson, J. T., and Schepers, G. W. H. 1950. Further evidence of the structure of the Sterkfontein ape-man *Plesianthropus*. *Transvaal Mus. Mem.* **4**:1–117.

Brown, F. H., and Shuey, R. T. 1976. Magnetostratigraphy of the Shungura and Usno Formations, lower Omo Valley, Ethiopia, in: *Earliest Man and Environments in the Lake Rudolf Basin* (Y. Coppens, F. C. Howell, G. L. Isaac and R. E. F. Leakey, eds.), pp. 64–78, University of Chicago Press, Chicago.

Butzer, K. W. 1976. Lithostratigraphy of the Swartkrans Formation. *S. Afr. J. Sci.* **72**:136–141.

Cooke, H. B. S. 1978. Pliocene–Pleistocene Suidae from Hadar, Ethiopia. *Kirtlandia* **29**:1–63.

Corruccini, R. S., Ciochon, R. L., and McHenry, H. M. 1976. The postcranium of Miocene hominoids: Were dryopithecines merely "dental apes"? *Primates* **17**:205–223.

Cronin, J. E., Boaz, N. T., Stringer, C. B., and Rak, Y. 1981. Tempo and mode in hominid evolution. *Nature (Lond.)* **292**:113–122.

Frisch, J. E. 1967. Remarks on the phyletic position of *Kenyapithecus*. *Primates* **8**:121–126.

Greenfield, L. O. 1979. On the adaptive pattern of "*Ramapithecus*." *Am. J. Phys. Anthropol.* **50**:527–548.

Greenfield, L. O. 1980. A late divergence hypothesis. *Am. J. Phys. Anthropol.* **52**:351–365.

Gregory, W. K. 1920. The origin and evolution of the human dentition: A paleontological review. Part IV. *J. Dent. Res.* **2**:607–717.

Gregory, W. K., and Hellman, M. 1926. The dentition of *Dryopithecus* and the origins of man. *Anthropol. Pap. Am. Mus. Nat. Hist.* **28**:1–123.

Harris, J. M., and White, T. D. 1979. Evolution of the Plio-Pleistocene African Suidae. *Trans. Am. Phil. Soc.* **69:**1–128.

Hooijer, D. A. 1951. Questions relating to a new large anthropoid ape from the Mio-Pliocene of the Siwaliks. *Am. J. Phys. Anthropol.* **9:**79–94.

Howell, F. C. 1978*a*. Hominidae, in: *Evolution of African Mammals* (V. J. Maglio and H. B. S. Cooke, eds.), pp. 154–248, Harvard University Press, Cambridge.

Howell, F. C. 1978*b*. Overview of the Pliocene and earlier Pleistocene of the lower Omo basin, southern Ethiopia, in: *Early Hominids of Africa* (C. J. Jolly, ed.), pp. 85–130, St. Martin's, New York.

Jacobs, L. L., and Pilbeam, D. R. 1980. Of mice and men: Fossil-based divergence dates and molecular "clocks." *J. Hum. Evol.* **9:**551–555.

Johanson, D. C., and Taieb, M. 1976. Plio-Pleistocene hominid discoveries in Hadar, Ethiopia. *Nature (Lond.)* **260:**293–297.

Johanson, D. C., and White, T. D., 1979. A systematic assessment of early African hominids. *Science* **203:**321–330.

Johanson, D. C., White, T. D., and Coppens, Y. 1978. A new species of the genus *Australopithecus* (Primates: Hominidae) from the Pliocene of eastern Africa. *Kirtlandia* **28:**1–14.

Kimbel, W. H. 1979. Characteristics of fossil hominid cranial remains from Hadar, Ethiopia. Abstract. *Am. J. Phys. Anthropol.* **50:**454.

Leakey, L. S. B. 1962. A new lower Pliocene fossil primate from Kenya. *Ann. Mag. Nat. Hist.* **4:**689–696.

Leakey, M. D., and Hay, R. L. 1979. Pliocene footprints in the Laetolil Beds at Laetoli, northern Tanzania. *Nature (Lond.)* **278:**317–323.

Leakey, M. D., Hay, R. L., Curtis, G. H. Drake, R. E., Jackes, M. K., and White, T. D. 1976. Fossil hominids from the Laetolil Beds. *Nature (Lond.)* **262:**460–466.

Leakey, R. E. F. 1974. Further evidence of Lower Pleistocene hominids from East Rudolf, North Kenya. *Nature (Lond.)* **248:**653–656.

Lewis, G. E. 1934. Preliminary notice of new man-like apes from India. *Am. J. Sci.* **27:**161–181.

Lydekker, R. 1886. Siwalik Mammalia. Supplement 1. *Palaeontol. Ind.* **10**(4):1–18.

Martyn, J. E. 1967. Pleistocene deposits and new fossil localities in Kenya. *Nature (Lond.)* **215:**476–480.

McHenry, H. M., and Corruccini, R. S. 1980. Late Tertiary hominoids and human origins. *Nature (Lond.)* **285:**397–398.

McHenry, H. M., Andrews, P. and Corruccini, R. S. 1980 Miocene hominoid palatofacial morphology. *Folia Primatol.* **33:**241–252.

Morbeck, M. E. 1975. *Dryopithecus africanus* forelimb. *J. Hum. Evol.* **4:**39–46.

Napier, J. R., and Davis, P. R. 1959. The fore-limb skeleton and associated remains of *Proconsul africanus*. *Fossil Mammals of Africa* (Br. Mus. Nat. Hist.) **16:**1–69.

Olson, T. R. 1981. Basicranial morphology of the extant hominoids and Pliocene hominids: The new material from the Hadar Formation, Ethiopia and its significance in early human evolution and taxonomy, in: *Aspects of Human Evolution* (C. B. Stringer, ed.), pp. 99–128, Taylor and Francis, London.

Opdyke, N. D., Lindsay, E., Johnson, G. D., Johnson, N., Tahirkheli, R. A. K., and Mirza, M. A. 1979. Magnetic polarity stratigraphy and vertebrate paleontology of the Upper Siwaliks subgroup of Northern Pakistan. *Palaeogeogr. Palaeoclimatol. Palaeoecol.* **27:**1–34.

Partridge, T. C. 1973. Geomorphological dating of cave opening at Makapansgat, Sterkfontein, Swartkrans and Taung. *Nature (Lond.)* **246:**75–79.

Partridge, T. C. 1978. Re-appraisal of lithostratigraphy of Sterkfontein hominid site. *Nature (Lond.)* **275:**282–287.

Partridge, T. C. 1979. Re-appraisal of lithostratigraphy of Makapansgat Limeworks hominid site. *Nature (Lond.)* **279:**484–488.

Patterson, B., and Howells, W. W. 1967. Hominid humeral fragment from Early Pleistocene of northwestern Kenya. *Science* **156:**64–66.

Patterson, B., Behrensmeyer, A. K. and Sill, W. D., 1970. Geology and fauna from a new Pliocene locality in northwestern Kenya. *Nature (Lond.)* **226:**918–921.

Pilbeam, D. R. 1969. Tertiary Pongidae of East Africa: Evolutionary relationships and taxonomy. *Bull. Peabody Mus. Nat. Hist.* (Yale Univ.) **31**:1–185.

Pilbeam, D. 1982. New hominid skull material from the Miocene of Pakistan. *Nature (Lond.)* **295**:232–234.

Pilbeam, D. R., Meyer, G. E., Badgley, C., Rose, M. D., Pickford, M. H. L., Behrensmeyer, A. K. and Shah, S. M. I. 1977. New hominoid primates from the Siwaliks of Pakistan and their bearing on hominoid evolution. *Nature (Lond.)* **270**:689–695.

Pilgrim, G. E. 1910. Notices of new mammalian genera and species from the tertiaries of India. *Rec. Geol. Surv. India* **40**:63–71.

Pilgrim, G. E. 1915. New Siwalik primates and their bearing on the question of evolution of man and the Anthropoidea. *Rec. Geol. Surv. India* **45**:1–74.

Rak, Y. 1981. *The Morphology and Architecture of the Australopithecine Face*, Ph. D. Dissertation, Department of Anthropology, University of California, Berkeley.

Remane, A. 1922. Beitrage zur Morphologie des Anthropoidengebisses. *Arch. Naturgesch.* **A87**:1–179.

Robinson, J. T. 1956. The dentition of the Australopithecinae. *Transvaal Mus. Mem.* **9**:1–179

Robinson, J. T. 1963. Adaptive radiation of the australopithecines and the origin of man, in: *African Ecology and Human Evolution* (F. C. Howell and F. Bourliére, eds.), pp. 385–416, Aldine, Chicago.

Sarich, V. M., and Cronin, J. E. 1976. Molecular systematics of the primates, in: *Molecular Anthropology* (M. Goodman and R. Tashian, eds.), pp. 141–170, Plenum, New York.

Senut, B. 1979. Comparaison des hominidés de Gomboré 1B et de Kanapoi: Deux pièces du genre Homo? *Bull. Mem. Soc. Anthropol. Paris* **6**:111–117.

Simons, E. L., and Pilbeam, D. R. 1965. Preliminary revision of the Dryopithecinae (Pongidae, Anthropoidea). *Folia Primatol.* **3**:81–152.

Smith, R. J., and Pilbeam, D. R. 1980. Evolution of the orangutan. *Nature (Lond.)* **284**:447–448.

Szalay, F. S., and Delson, E. 1979. *Evolutionary History of the Primates*, Academic, New York.

Tattersall, I., and Eldredge, N. 1977. Fact, theory and fantasy in human paleontology. *Am. Sci.* **65**:204–211.

Tauxe, L. 1979. A new date for *Ramapithecus. Nature (Lond.)* **282**:399–401.

Tekkaya, I. 1974. A new species of Tortonian anthropoid (Primates, Mammalia) from Anatolia. *Bull. Min. Res. Exp. Inst. Turkey* **83**:148–165.

Tobias, P. V. 1967. *Olduvai Gorge,* Volume 2. *The Cranium and Maxillary Dentition of Australopithecus (Zinjanthropus) boisei,* Cambridge University Press, Cambridge. 264 pp.

Tobias, P. V. 1978. Position et rôle des australopithécinés dans la phylogenèse humaine, avec ètude particulière de *Homo habilis* et des théories controversées avancées à propos des premiers hominidés fossiles de Hadar et de Laetolil, in: *Les Origines Humaines et les Époques de l'Intelligence,* pp. 38–77, Foundation Singer-Polignac/Masson, Paris.

Tobias, P. V. 1980. "*Australopithecus afarensis*" and *A. africanus:* A critique and an alternative hypothesis. *Palaeontol. Afr.* **23**:1–17.

Von Koenigswald, G. H. R. 1950. Bemerkugen zu "Dryopithecus" giganteus Pilgrim. *Eclogae Geol. Helv.* **42**:515–519.

Vrba, E. S. 1982. Some remarks on the paleoecology associated with the Hominidae from Sterkfontein, Swartkrans and Kromdraai, in: *L'Environnement des Hominidés au Plio-Pleistocene,* Foundation Singer-Polignac, Paris (in press).

Walker, A. 1976. Remains attributable to *Australopithecus* in the East Rudolf succession, in: *Earliest Man and Environments in the Lake Rudolf Basin* (Y. Coppens, F. C. Howell, G. L., Isaac, R. E. F. Leakey, eds.), pp. 484–489, University of Chicago Press, Chicago.

Wallace, J. A. 1975. Dietary adaptations of *Australopithecus* and early *Homo,* in: *Paleoanthropology, Morphology and Paleoecology* (R. H. Tuttle, ed.), pp. 203–223, Mouton, The Hague.

Walter, R. C., and Aronson, J. L. 1982. Revisions of K/Ar ages for the Hadar hominid site, Ethiopia. *Nature (Lond.)* **296**:122–127.

Zihlman, A., Cronin, J. E., Cramer, D. L., and Sarich, V. M. 1978. Pygmy chimpanzee as a possible prototype for the common ancestor of humans, chimpanzees and gorillas. *Nature (Lond.)* **275**:744–746.

Australopithecus africanus   29

Its Phyletic Position
Reconsidered

T. D. WHITE, D. C. JOHANSON,
AND W. H. KIMBEL

In 1925 Dart proposed that his newly named genus *Australopithecus* was the ancestor of the genus *Homo* (Dart, 1925). Despite a recent rash of claims to the contrary based on misidentified fossils or erroneous dates (R. Leakey, 1970, 1973a, 1976a; Oxnard, 1975, 1979; M. Leakey, 1979), it has become increasingly evident that Dart was correct—*Australopithecus* was ancestral to *Homo*.

This chapter is reprinted with minor changes from an article by the same title which appeared in the *South African Journal of Science* **77**:445–470 (1981). In the process of standardizing the style of citation to conform with other chapters in this volume it was necessary to eliminate several lengthy text citations and references. The authors of this chapter and the editors of this volume both gratefully acknowledge the *South African Journal of Science* for permission to reprint this article.

T. D. WHITE • Department of Anthropology, University of California, Berkeley, California 94720.   D. C. JOHANSON • Institute of Human Origins, 2700 Bancroft Way, Berkeley, California 94704 and Laboratory of Physical Anthropology, Cleveland Museum of Natural History, Cleveland, Ohio 44106.   W. H. KIMBEL • Laboratory of Physical Anthropology, Cleveland Museum of Natural History, Cleveland, Ohio 44106 and Departments of Anthropology and Biology, Kent State University, Kent, Ohio 44240.

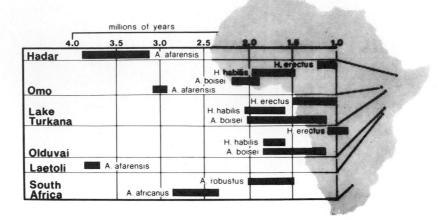

Fig. 1. Chronologic and geographic placement of key early hominid localities and taxa.

Introduction

In a reassessment of early African hominid systematics, Johanson and White (1979) proposed the Pliocene species *Australopithecus afarensis* (Johanson *et al.,* 1978) as a suitable common ancestor for later species of *Australopithecus* and *Homo*. Anatomical evidence placing *A. africanus* Dart, 1925 on the lineage leading to *A. robustus* (Broom), 1938 and *A. boisei* (Leakey), 1959* was cited: *Australopithecus africanus* was seen to lack ". . . elements in the suite of primitive characteristics described . . . for the Hadar and Laetolil hominids . . ." and ". . . to share several distinctive, derived characters with later robust australopithecines" (Johanson and White, 1979, p. 327). Thus far, no detailed comparison between *A. afarensis* and *A. africanus* has been published. Our reexamination of the South African fossil hominids yields further insight into the relationship between these taxa. Our purpose is to present an in-depth comparative analysis of their skeletal anatomy.

South African *Australopithecus* fossils are of fundamental importance in understanding human evolution. Fieldwork at the Transvaal sites of Sterkfontein, Swartkrans, Makapansgat, and Kromdraai (Tobias and Hughes, 1969; Tobias, 1973*a;* Hughes and Tobias, 1977; Brain, 1970; Brain *et al.,* 1974) is augmenting the already burgeoning hominid sample and adding much to our knowledge of the stratigraphy and natural history of the cave

---

*South African *A. robustus* and East African *A. boisei* are often considered together in this publication since the question of whether or not these taxa are conspecific remains unresolved. Where the South and East African "robust" *Australopithecus* species are contrasted or referred to individually, the generally accepted nomina are employed; otherwise they are referred to together as *A. robustus* + *A. boisei*.

deposits and their bone accumulations (Vrba, 1975, 1979; Brain, 1976*a,b;* Partridge, 1978, 1979; Butzer, 1974, 1976). Several detailed studies on the morphology of the hominids have been completed (Wallace, 1972; Sperber, 1973; Clarke, 1977*a;* White, 1977*a*). Biostratigraphic analyses have recently yielded an acceptable relative chronology for the fossils (White and Harris, 1977; Harris and White, 1979; Vrba, 1976, 1977) (Fig. 1). These new data combine with insights provided by *A. afarensis* to make our study appropriate.

The Two Taxa: A Historical Background

South African Discoveries

Dart's (1925) pronouncement that the Taung child represented a primitive ancestral hominid evoked immediate, sometimes vehement criticism from influential anatomists (Keith, 1925, 1931; Smith, 1925; Smith-Woodward, 1925). The skeptical reaction was not surprising considering the prevailing anthropological climate. The Neanderthal and Javanese discoveries had focused the search for early man on Europe and the Far East. Moreover, the fraudulent Piltdown find was still being hailed by many as the "missing link." Its combination of a human brain case and an ape's lower jaw precluded acceptance of Dart's Taung child as a human ancestor. Dart's announcement from remote South Africa claimed that a small-brained creature with human-like jaws and teeth represented an ancestral human stock. This claim stood in direct opposition to preconceived notions held by most students of human evolution.

Broom and Robinson's work at Kromdraai, Sterkfontein, and Swartkrans during the 1930s and 1940s combined with Dart's later efforts at Makapansgat to dramatically bolster the South African fossil hominid sample. The new fossils precipitated a series of detailed morphological analyses by these workers and others, including Gregory, Hellman, and Le Gros Clark (Broom, 1939; Gregory and Hellman, 1939; Broom and Schepers, 1946; Le Gros Clark, 1947, 1950, 1955; Broom *et al.*, 1950; Broom and Robinson, 1952; Robinson, 1956). These studies invariably concluded that in the skull, dentition, and postcranium the South African "ape-men" were indeed more human-like than ape-like. These results and the 1953 exposure of the Piltdown forgery (Weiner *et al.*, 1955) solidified the hominid status of *Australopithecus.*

Broom's (1950) taxonomic scheme proposed three genera and five species to accommodate the abundant South African "ape-men." Shortly after Broom's death Robinson simplified the taxonomy, proposing that hominids from Taung (*A. africanus*), Sterkfontein (*Plesianthropus transvaalensis*), and Makapansgat (*A. prometheus*) be subsumed under the first proposed nomen, *A. africanus*. Hominids from Kromdraai (*Paranthropus robustus*) and Swartkrans (*P. crassidens*) were combined in *P. robustus* (Robinson, 1954*a*).

Robinson provided a differential diagnosis of the two morphotypes which he distinguished at the generic level (Robinson, 1961, 1963). *Paranthropus* (*A. robustus*) he characterized as possessing molarized mandibular deciduous molars and permanent premolars, greatly expanded postcanine teeth relative to incisors and canines, a massive, heavily buttressed facial skeleton, and a deep mandible with a high, vertical ramus. Robinson contrasted this suite of features with what he perceived to be the lighter, smaller, more "gracile,"* or "hominine" morphology of *A. africanus*. Emphasizing differences between the two taxa, Robinson usually selected a few specimens to illustrate his conclusions.†

Building a behavioral and ecological model to explain the dichotomy in craniofacial form, Robinson invoked diet—*A. africanus* was described as a tool-making omnivore and *Paranthropus* as a cultureless herbivore. Stressing the less specialized morphology of the former, which he believed to have achieved the "hominine" grade, Robinson finally abandoned the nomen *Australopithecus* and transferred the entire *A. africanus* hypodigm to the genus *Homo* (Robinson, 1965).

Robinson's work, in retrospect, was both insightful and misleading. He accurately identified traits distinguishing the taxa, but was overly impressed by the superficial similarity between *Paranthropus* and the modern gorilla (Broom and Robinson, 1952; Robinson, 1961, 1972). Thus, he suggested that the specialized robust craniofacial morphology was primitive in hominids. More importantly, by stressing *differences* between the taxa, Robinson chose to ignore many detailed morphological and metrical *similarities* between *A. africanus* and *A. robustus*. The emphasis on sorting South African "ape-men" into "gracile" and "robust" categories overshadowed many robust features present in *A. africanus*.

This emphasis shifted in the 1960s as Tobias (1967) grappled with the Robinsonian dichotomy and concluded: ". . . the two australopithecine taxa are very much more closely related than has commonly been averred hitherto" (p. 231), and " . . . we cannot exclude the chance of crossing between *A. africanus* and members of the *A. boisei* to *A. robustus* line" (p. 244). This, of course, implied that the two taxa were conspecific. Later, Tobias even suggested that the holotype of *A. africanus* was actually a representative of *A. robustus* (Tobias, 1973*b*, 1978*a*, 1980). The realization that available site sam-

---

*The term "gracile" is a widely used misnomer for *A. africanus*, a hominid with cheek teeth almost twice as large as those of the chimpanzee (Wolpoff, 1973). Continued use of this term (Boaz and Howell, 1977; Zihlman, 1979) in East Africa has heightened confusion and clarified nothing. The present authors feel that there is no adequate substitute for the proper use of Linnean binomials in identifying species samples of fossil hominids and that the term "gracile" should no longer be used to characterize such species.

†Specimens SK 23 and 48 were most often used for *A. robustus* and STS 5 and 52b for *A. africanus*. One need only peruse the majority of textbook chapters on the South African early hominids to see that "consensus" opinion is largely based on these four specimens. In reference to the latter taxon, one worker even suggested that ". . . only one relatively complete adult cranium [from South Africa] exists (STS 5)" (Boaz, 1979, p. 77).

ples did not always segregate perfectly fostered a resurgence of the single-species hypothesis in the 1970s, a scheme that invoked the competitive exclusion principle in support of a unilineal hominid phylogeny (Brace, 1967; Wolpoff, 1968, 1970; Swedlund, 1974). While Robinson attributed differences between *A. africanus* and *A. robustus* to adaptive differentiation and ultimately cladogenesis, advocates of the single-species hypothesis ascribed the differences to ". . . a substantial allometric transformation. . ." (Brace, 1973, p. 35) and sexual dimorphism.

East African Discoveries

Mary Leakey's 1959 discovery of the 1.8 million year-old *"Zinjanthropus"* cranium (O.H. 5) at Olduvai Gorge (L. Leakey, 1959) in northern Tanzania established the potential for further discoveries in the East African Rift Valley. Anthropological attention then settled on East Africa as hominid-bearing sites were located near Lake Natron in Tanzania, the Omo Basin of southern Ethiopia, and Kanapoi, Lothagam, Baringo, and East Rudolf in northern Kenya.

The 1960s brought further discoveries at Olduvai (L. Leakey, 1960*a*, 1961; L. Leakey and M. Leakey, 1964). Hominid fossils contemporary with *"Zinjanthropus"* (e.g., O.H. 7, 13, 16) indicated the presence of a second hominid species, thereby confirming Broom and Robinson's discoveries at Swartkrans.* These smaller East African fossils were thought to differ from *Australopithecus* in a number of cranial and dental characters. Leakey *et al.* (1964) were inspired to create the new species *Homo habilis*. Initial scientific reaction was widespread and largely adverse (Robinson, 1965, 1966, 1967; Brace, 1973; Le Gros Clark, 1967). Critics were skeptical that *H. habilis* was anything besides a geographic variant of South African *A. africanus*. Others saw part of the paratype series as indistinguishable from *Homo erectus* (Brace *et al.*, 1973).† *Homo habilis* was only provisionally proposed and therefore never conformed to requirements of the International Code of Zoological Nomenclature. Many, therefore, consider it ". . . an empty taxon, inadequately proposed [which] should be formally sunk" (Brace *et al.*, 1973, p. 66).

Tobias valiantly continued to champion *H. habilis* despite these shortcomings and presented a revised phylogeny of early hominid evolution that incorporated the new species (Tobias, 1965, 1967). He commented that ". . . since they are contemporary with *H. habilis,* the australopithecine populations represented by the actual fossils recovered to date are clearly too late—and possibly slightly too specialized—to have been on the actual human line . . ." (To-

---

*Specimens recovered in 1949 at Swartkrans indicated a hominid species contemporary with *A. robustus*. Broom and Robinson (1949, 1950) named the taxon *Telanthropus capensis* and considered it to represent an independent evolutionary line ancestral to *Homo*.

†Louis Leakey (1966) himself offered his view 2 years after the initial announcement, calling O.H. 16 from Bed II a "protopithecanthropine."

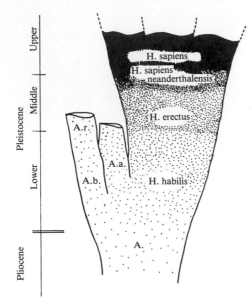

Fig. 2. A version of the phylogenetic tree proposed by Tobias (1965) to accommodate *Homo habilis:* "increasing intensity of shading represents increasing degrees of approach toward the structure and behavior of modern man. A, the hypothetical ancestral australopithecine; A.b., *Australopithecus* (*Zinjanthropus*) *boisei;* A.r., *Australopithecus robustus;* A.a., *Australopithecus africanus*" (p. 32).

bias, 1965, p. 32). In his accompanying phylogenetic diagram (Fig. 2), Tobias emphasized the close relationship between *A. africanus* and *A. robustus* and prophetically postulated a yet-to-be-recognized, undifferentiated Pliocene *Australopithecus* species.

The 1970s witnessed intensive paleoanthropological fieldwork in eastern Africa. Discoveries at Omo and Koobi Fora established a firm chronology for the appearance of East African "hyper-robust" *A. boisei* at approximately 2.2 m.y.a. (Howell and Coppens, 1976; Walker and Leakey, 1978; M. Leakey and R. Leakey, 1978). The announcement of KNM-ER 1470 (R. Leakey, 1973*b*), a large-brained hominid cranium from Koobi Fora, confirmed the claims of Leakey, Tobias, and Napier that *H. habilis* is a biologically valid taxon. Initially dated to about 3.0 m.y.a. (Fitch and Miller, 1970), the 1470 cranium profoundly altered perceptions of early hominid phylogeny. However, in light of detailed biostratigraphic (White and Harris, 1977; Harris and White, 1979; Cooke and Maglio, 1972; Shuey *et al.*, 1978; Brown *et al.*, 1978) and geochronologic (Curtis *et al.*, 1975; Cerling *et al.*, 1979; Drake *et al.*, 1980; Gleadow, 1980; McDougall *et al.*, 1980) studies it is now evident that claims of a mid-Pliocene age for KNM-ER 1470 were grossly exaggerated: the specimen is about as old as the holotype *H. habilis* specimen (O.H. 7) from Olduvai.

Hadar and Laetoli

Fossil hominids discovered in the mid-1970s at Hadar in Ethiopia and Laetoli in Tanzania have pushed knowledge of human evolution well into the Pliocene. Due to their antiquity, these fossils provide a test of Tobias' prediction about the ancestral Pliocene *Australopithecus.*

Fig. 3. A portion of the early hominid footprint trail in the Laetolil Beds of northern Tanzania. [Photo by Peter Jones.]

The site of Laetoli, situated about 50 km south of Olduvai, yielded hominid fossils as early as the 1930s (Kohl-Larsen, 1943; White, 1981). A team directed by Mary Leakey reinitiated exploration in 1974 and has recovered 22 hominid individuals consisting mostly of jaws and teeth (M. Leakey *et al.*, 1976; White, 1977*b*, 1980*a*). The Laetoli hominid fossils date to between 3.6 and 3.8 m.y.a. and are accompanied by a unique series of early hominid footprints (M. Leakey and Hay, 1979; White, 1980*b*; Day and Wickens, 1980) (Fig. 3).

In the northern reaches of the East African rift, the International Afar Research Expedition under the leadership of M. Taieb, D. C. Johanson, and Y. Coppens began a systematic exploration of Pliocene deposits at Hadar in 1973. Fossils representing at least 35 hominid individuals have been collected from the Hadar Formation. Dating between 3.0 and 4.0 m.y.a. (Aronson *et al.*, 1977; Walter, 1980; Walter and Aronson, 1982), these include a 40% complete skeleton (A.L. 288-1, "Lucy") and a unique set of over 200 hominid specimens from a single locality (a minimum of 13 individuals are represented at Afar Locality 333/333w).

The storehouse of African Plio-Pleistocene hominids has grown impressively since the 1924 Taung discovery. Many workers have presented systematic syntheses. Recognition that the Hadar and Laetoli fossils were temporally and morphologically distinct from other hominid forms prompted Johanson and White (1979) to critically review some of these earlier views of hominid phylogeny.*

Australopithecus afarensis: The Interpretive Background

Interpretations

The Hadar and Laetoli hominids received diverse taxonomic and phylogenetic treatment following their discovery. Some workers opted to view these finds as conspecific with *A. africanus* (Boaz, 1979; Anon, 1979*a,b;* Kennedy, 1980). Others have preferred a subspecific designation (Tobias, 1978*b,* 1979, 1980). The generic nomen *Praeanthropus* (originally coined for the Garusi maxillary fragment) has been applied to the Laetoli hominids (Clarke, 1977*a*). Still others have ascribed portions of the Laetoli and Hadar collections to *Homo* (M. Leakey, 1979; R. Leakey and Lewin, 1977; Olson, 1981), some of them specifically to *Homo erectus* (M. Leakey, personal communication in Oxnard, 1979). Other specimens from Hadar have been assigned to *Paranthropus*

---

*Provisional interpretations suggested the possibility of two or three hominid taxa at Hadar (Johanson and Taieb, 1976) and that the Laetoli sample represented the earliest evidence of the genus *Homo* (M. Leakey *et al.*, 1976). Recovery of additional specimens and detailed analysis of the Hadar and Laetoli fossils have shown these first provisional identifications to be incorrect.

(Olson, 1981) and it has even been suggested that "Lucy" represents a "late *Ramapithecus*" (R. Leakey, 1976*a*, p. 174; R. Leakey and Lewin, 1977).

Johanson and White (1979) argued that the known hominid specimens from Hadar and Laetoli could be accommodated in a single species. Since these creatures did not exhibit brain expansion normally associated with the genus *Homo*, yet were clearly bipedal, they were assigned to *Australopithecus*. A suite of primitive cranial, dental, and mandibular features were found to distinguish the Hadar plus Laetoli sample from other *Australopithecus* species. A new species name, *A. afarensis*, was therefore christened by Johanson *et al.* (1978). The species *A. afarensis* appears to have been ecologically diverse, geographically widespread, and temporally enduring.

While the specific Miocene ancestor of *A. afarensis* remains unidentified, middle Miocene species such as *Ouranopithecus macedoniensis* (de Bonis and Melentis, 1977*a*) are strikingly similar in some aspects of dentognathic anatomy (de Bonis *et al.*, 1981). Workers familiar with Miocene hominoid morphology have also drawn attention to resemblances between *A. afarensis* and the middle Miocene genera *Ramapithecus* and *Sivapithecus* (Greenfield, 1980).

Criticisms

The middle Pliocene *A. afarensis* provided a fresh perspective on early hominid phylogeny. The Johanson and White (1979) hypothesis of early hominid phylogeny is shown in Fig. 4. Both the naming of *A. afarensis* and the phylogenetic hypothesis have elicited criticism. Day *et al.* (1980) attacked the species on nomenclatural grounds and questioned the wisdom of selecting a

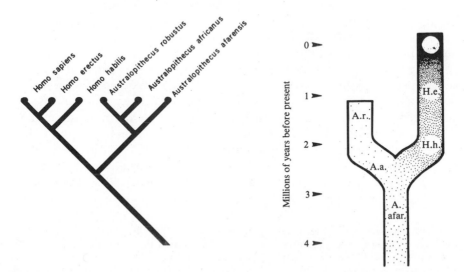

Fig. 4. Cladistic and evolutionary relationships of early hominid taxa postulated by Johanson and White (1979). [The original cladogram published in *Science* contained an error in its second branch.]

Laetoli mandible as the holotype for *A. afarensis*. Such criticisms remind us of Mayr's observations regarding nomenclature and biology (Mayr, 1969, p. 297). Nomenclatural procedures used in erecting *A. afarensis* were fully concordant with the rules outlined in the International Code of Zoological Nomenclature and did not depart from accepted taxonomic practice.*

Criticism was also aimed at the pooling of Hadar and Laetoli samples. This procedure was questioned on the basis of chronological and geographic separation of the two sites (M. Leakey, in Anon, 1979c). Objections concerning temporal separation of Hadar and Laetoli hominids are rendered academic by biostratigraphic studies (Harris and White, 1979; Cooke, 1978b) and the redating of the Hadar Formation basalt at about 3.7. m.y.a. (Walter, 1980; Walter and Aronson, 1982). The two sites are best considered essentially contemporaneous, pending further fieldwork. Geographic separation is not in itself a basis for sustaining taxonomic distinction between Hadar and Laetoli hominids. Mammalian species ranges of over 2000 miles are commonplace. For instance, it is universally accepted that Asian, European, and African Pleistocene hominids belong to a single, highly variable species, *H. erectus*, that ranges in time between 1.5 and 0.4 m.y.a.

We do not consider geographic and temporal factors as primary in taxonomic decisions unless a firm morphological basis is first demonstrated. Morphologically, the Hadar and Laetoli samples are essentially indistinguishable (Fig. 5). Some Laetoli dental metrics slightly exceed those of the Hadar sample (Tobias, 1978b),† but there is overwhelming overlap. In addition, the Laetoli dental sample is small enough to cast grave doubts on the significance of such differences. As Mayr observes, "The question is never whether or not the compared populations are completely identical. Population geneticists have demonstrated conclusively that no two natural populations in sexually reproducing animals are ever exactly alike. To find a statistically significant difference between several populations is, therefore, of only minor interest to the taxonomist; he takes it for granted" (Mayr, 1969, p. 187).

Concerns regarding the number of hominid taxa present at Hadar have been expressed. Size and shape differences between elements of the hominid sample are said to indicate species diversity (R. Leakey and Lewin, 1977; Olson, 1981). The large size variation typical of the Hadar hominids and of *A. afarensis* in general is not excessive when compared to that observed in living hominoid species (Fig. 6) (Johanson and White, 1979). Furthermore, equally large amounts of size variation, most of it due to sexual dimorphism, have long been known to characterize Asian (Weidenreich, 1936) and South African (Robinson, 1956; Brace, 1973; Wolpoff, 1974a) fossil hominids. Varia-

---

*The Laetoli specimen was chosen as the holotype because it was fully described and illustrated at the time the species was proposed. It therefore constituted the most thoroughly described holotype of any early hominid species named to that time.

†The metric analysis of Tobias (1978b) is rendered suspect because in at least one taxon (*H. habilis*) the number of individuals has been artificially inflated. Of six alleged individuals (*n* = 6 P4's, p. 70), three specimens were actually measured, but right *and* left teeth were entered into the calculations, a statistically invalid procedure.

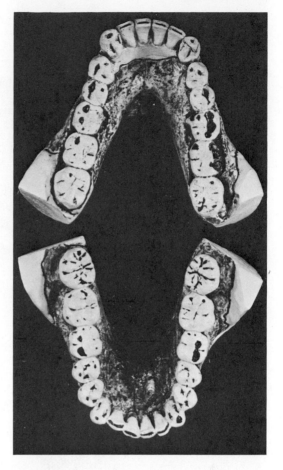

Fig. 5. A comparison of mandibles from Hadar (A.L. 400-1a, bottom) and Laetoli (L.H. 4 holotype, top). Minor reconstruction was required for the latter specimen. Note the near identity in mandibular and dental anatomy.

tion in *Australopithecus* was rediscovered by R. Leakey in East Africa during the 1970s: ". . . if such a range of dimorphism exists in East Africa, is it not possible that a similar range would exist in South Africa?" (R. Leakey, 1973*c*, p. 66). Such variation was forgotten as soon as *A. afarensis* was proposed to include both large and small individuals.

In 1953 Robinson published his objections to claims that the 1939 Garusi maxilla from Laetoli was distinct from *A. africanus* (*sensu stricto*). Robinson said that while the Garusi specimen ". . . cannot be distinguished satisfactorily from the latter [*A. africanus*], . . . it is clearly realized that with fuller material the two might not prove to be conspecific" (Robinson, 1955, p. 429). Robinson's remark mirrors that of Remane, who wrote: "There are needed further remains of this interesting form before its exact position may be determined" (Remane, 1954, p. 126).

We believe that the "further remains" alluded to by Remane and Robinson are those found at Hadar and Laetoli. In this light, the remainder of our presentation will address two specific considerations: the distinctiveness of *A.*

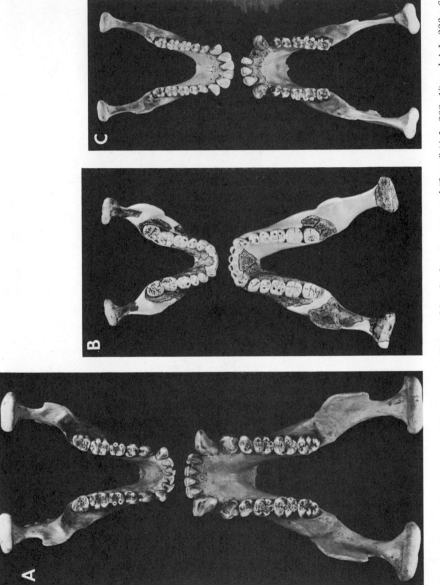

Fig. 6. Male and female hominoid jaws: sexual dimporphism in the reconstructed "Lucy" (A.L. 288–1i) and A.L. 333w-60 specimens (center) is not excessive when compared to that seen in the gorilla (left) or chimpanzee (right). Note the anterior dental arcade narrowing in the females.

afarensis from *A. africanus* and the placement of *A. africanus* in the *A. robustus* + *A. boisei* clade. First, however, we will place *A. africanus* in a geographic and temporal perspective.

Australopithecus africanus: The Taxon Circumscribed

South African A. africanus: The Hypodigm

The present authors follow the view that the hypodigm of *A. africanus* includes all known hominid fossils from Taung, Sterkfontein Member 4, and Makapansgat Members 3 and 4. We consider *A. robustus* to be a separate species thus far known from Kromdraai and Swartkrans Member 1. These hominid fossils are all recovered from cemented infillings of caves and fissures. Attempts to date the fossils have proceeded along diverse lines. Radiometric age determinations have proven entirely unsuccessful (MacDougall and Price, 1974). Geomorphological considerations were used by Partridge (1973) in an attempt to date the cave openings. His results generated excitement in at least one anthropologist (Tobias, 1973*b*, 1978*c*), but geologists remain unconvinced: ". . . variation observed in the parameters controlling the migration of nickpoints and valley flanks is so great that the geomorphologically derived dates cited by Partridge are meaningless" (Bishop, 1978, p. 262). Others have called Partridge's attempt at dating ". . . interesting but unacceptable" (Butzer, 1974, p. 416) or have held that ". . . Partridge's methods and basic assumptions are of doubtful scientific validity" (de Swardt, p. 683). We agree with Howell: "Geomorphological estimates employing nickpoint recession criteria are generally unreliable if not ill-founded" (Howell, 1978*a*, p. 159).

Recent paleomagnetic analysis of the cave sediments has suggested to McFadden *et al.* (1979) and Brock *et al.* (1977) that the age of the Makapansgat *A. africanus* fossils lies between 2.9 and 3.32 million years. This estimate is based on the upward sequence of reversed/normal/reversed/normal/reversed spanning Partridge's Members 1–4 (Partridge, 1979). Experience with paleomagnetic dating estimates at the sites of Koobi Fora and Hadar in eastern Africa has emphasized some hazards of matching isolated, local polarity patterns to the worldwide geomagnetic time scale (Walter, 1980; Hillhouse *et al.*, 1977). In cave situations like Makapansgat, variables in catchment area (size, shape, slope, substrate, precipitation, and vegetation) could combine with variation in the cave itself (size, shape, floor subsidence, bedrock type) to produce wildly fluctuating depositional rates (Brain, 1967, 1969). Partridge (1979) has documented major unconformities within the Makapansgat succession and has shown that deposition was complex, with at least two separate depositories accumulating sediment at different rates. A simplistic model where vertical thickness of sediment is considered useful as a guide to relative time of accumulation is misleading when applied to such a site.

Given this situation, it is entirely possible, for instance, that the normal-to-reversed transition above the hominid-rich gray breccia (Partridge's Member 3) could be the Gauss/Matuyama boundary at 2.43 m.y.a. The normal above this could be the "X" anomaly (Mankinen and Dalrymple, 1979), while the reversed-to-normal transition low in Partridge's Member 2 could be the Gilbert/Gauss boundary, with the Kaena and Mammoth events lost to unconformity or nondeposition. This alternative, no less plausible than the three given by McFadden *et al.* (1979), makes the minimum age of *A. africanus* at Makapansgat 2.4 instead of 2.9 million years (an estimate more consistent with faunal evidence discussed below).

The inapplicability of radiometric techniques, the unacceptable geomorphological estimates, and the ambiguity of paleomagnetic results suggest that the only reliable method of dating South African *Australopithecus* is biostratigraphy. Relative temporal placement of the fossils has been attempted almost from the beginning (Broom, 1930). Cooke and Maglio (1972, p. 328) recognized that ". . . the traditional faunal methods of relative age determination continue to be of primary importance." Further documentation of mammalian evolution within well-dated stratigraphic successions in eastern Africa has finally made possible accurate biostratigraphic placement of South African *Australopithecus*.

There is no temporally diagnostic fauna unequivocally associated with the Taung holotype of *A. africanus*. Even when association between the hominid and other fossils is assumed, relative placement of the Taung specimen ranges from the oldest (de Graff, 1960; Cooke, 1967, 1978*a*) to the youngest (Wells, 1967, 1969) South African *Australopithecus*. Despite claims of Butzer (1974), Partridge (1973), and Tobias (1973*b*, 1978*c*), the Taung specimen is best considered undated.

In contrast to Taung, the relevant Sterkfontein and Makapansgat nonhominid faunal elements are temporally diagnostic and apparently contemporary with remains of *A. africanus*. Faunas from these two sites have usually been considered equivalent in antiquity. Ewer (1957), Wells (1962), and Ewer and Cooke (1964) divided the "ape-man" fauna into two provisional stages, the earlier containing Sterkfontein and Makapansgat and the later Swartkrans. Cooke (1967) later introduced the provisional Sterkfontein and Swartkrans "faunal spans." Hendey (1974) places all of these sites into earlier and later parts of the Makapanian Mammal Age.

Further study has clarified the relationships between these faunas and their radiometrically dated counterparts in East Africa. Work on the carnivores (Collings *et al.*, 1975; Hendey, personal communication) and cercopithecoid primates (Delson, personal communication; Eck, personal communication) places the Makapansgat and Sterkfontein *A. africanus* strata between 2.0 and 3.0 m.y.a. Vrba's (1975, 1976, 1977) work on bovids suggests that the age for Makapansgat Members 2 and 3 ". . . may belong between 2–3 m.y. ago with the later part of that time range preferable to the earlier" (Vrba, 1977, p. 146). The Sterkfontein bovids probably slightly postdate the Makapansgat ones (Vrba, 1975).

Fossil suids have recently emerged as reliable biostratigraphic markers, frequently forecasting errors in radiometric and paleomagnetic age estimates

(White and Harris, 1977; Harris and White, 1979; Cooke and Maglio, 1972; Cooke, 1978*b*). Large collections of two suid genera are contemporary with *A. africanus* at Makapansgat. The taxa, *Metridiochoerus andrewsi* and *Notochoerus capensis,* occur in radiometrically controlled East African contexts. The stage of evolution featured by both Makapansgat pig lineages is the same as that seen in Omo Shungura Formation Member C (2.4–2.6 m.y.a.) in Ethiopia. The two Pliocene pig species were recently identified at the same evolutionary stage in the geographically intermediate Chiwondo Beds of Malawi in Central Africa (Kaufulu *et al.,* 1980). Sites in East Africa with suids clearly *predating* those in Omo Member C and the Makapansgat gray breccia include the De-nen Dora and Sidi Hakoma Members of the Hadar Formation, the Laetolil Beds, Shungura Member B and the Usno Formation of the Omo, and the circum-Tulu Bor units of Koobi Fora. All of these sites have also yielded *A. afarensis.*

Biostratigraphic evidence indicates an age of between 2.0 and 3.0 million years for known *A. africanus* in South Africa. Bovids and suids, which seem the most useful biostratigraphic indicators to date, point to an age of 2.6 million years for Makapansgat Member 3 fossils, including *A. africanus.* These data do not lend support to geomorphological or paleomagnetic "ages" of 3.0 million years for *A. africanus.* There are two possibilities—either the latter age estimates are excessive, or independent mammal lineages evolved faster in South Africa than they did in East Africa. We consider the first possibility more likely, but even if one accepts the *earliest* age estimates for the South African faunas, it is clear that the Transvaal's *A. africanus* postdated most East African *A. afarensis* by about 0.5 million years. It was therefore chrono-logically intermediate between *A. afarensis* and *A. robustus* + *A. boisei,* whose earliest representatives are first known from deposits about 2.2 million years old.

East African A. africanus: Mistaken identity?

Due to the historical precedence of the South African discoveries, there has been a tendency to view the more recent East African finds in a systematic framework developed for the South African fossils. The first attribution of East African fossils to *A. africanus* was Robinson's (1953, 1955) assessment of the Garusi maxilla. Remane (1954) and Şenyürek (1955) described dental fea-tures distinguishing this specimen from *A. africanus* (*sensu stricto*) and the fragment is now included in *A. afarensis* (Johanson *et al.,* 1978). The contro-versy of the late 1960s over allocation of Olduvai fossils to *A. africanus* was ended by recovery of hominid fossils from Koobi Fora that support the bio-logical validity of *Homo habilis.* Specimens of *A.* aff. *africanus* were reported from the Omo Beds [mostly isolated teeth from Shungura Formation Mem-bers B, C, D, E, F, and G and the Usno Formation (Howell and Coppens, 1976)], but Howell stated that, "The oldest specimens are generally small, with simple dental morphology and might ultimately prove, with additional, more complete material to represent a distinctive, though related lower taxonomic category" (Howell, 1978*b*, p. 123). We believe that these Omo fossils represent *A. afarensis,* a view now shared by Howell (personal communication).

The large lower Pleistocene hominid collections from Koobi Fora, Olduvai, and Omo include a number of small specimens distinct from the abundant *A. boisei.* These specimens were initially attributed to *H. habilis,* while the presence of a third hominid species in East Africa was rejected: ". . . there is no clear evidence for two contemporaneous species of *Australopithecus* in East Africa in the Lower Pleistocene" (R. Leakey, 1972, p. 268). Shortly thereafter, this evidence became "limited" (R. Leakey, 1973*d*, p. 172) and by 1974 the evidence had ". . . reopen[ed] the possibility of its [*A. africanus*] existence" (R. Leakey, 1974, p. 655). Finally, in 1976, the evidence for two species of *Australopithecus* contemporary with *Homo* could ". . . be established with some conviction" (R. Leakey, 1976*b*, p. 176).

Workers who are convinced of the existence of *A. africanus* in the lower Pleistocene of East Africa have provided the following definition for this taxon: ". . . gracile mandibles with small cheek teeth, cranial capacity values at 600cc or less, and sagittal crests rare or nonexistent" (R. Leakey, 1976*b*, p. 176). It has been pointed out that this definition is so broad as to encompass the common chimpanzee (Johanson and White, 1980). In fact, most specimens now attributed to East African *A. africanus* have been known since the early 1960s when they were included in the paratype series of *H. habilis* (e.g., O.H. 13). Other specimens from Koobi Fora now attributed by R. Leakey to *A. africanus* were originally placed in *Homo erectus* (e.g., KNM-ER 992) (Wood, 1974, 1976). In an attempt to distinguish *Australopithecus* from *Homo,* Walker (1976) applied two indices that placed the cranium KNM-ER 1470 within the *Australopithecus* range.* When these indices are applied to the KNM-ER 1813 cranium, allegedly the best evidence for *A. africanus* in East Africa, this specimen falls into Walker's *Homo* range even though he assigns the specimen to *Australopithecus* (R. Leakey and Walker, 1980).

The earliest purported *A. africanus* in East Africa dates from about 1.8 m.y.a., while the youngest comes from just below the Chari tuff at Koobi Fora (1.3 m.y.a.). It is evident that these specimens do not sample *A. boisei,* but it is not clear why these "*A. africanus*" specimens do not represent *Homo* individuals. These considerations, combined with problems in the definition, chronology, and composition of *A. africanus* in East Africa, have led Wood, an authority on Koobi Fora hominid crania, to comment: ". . . if there is taxonomic variation within East African 'gracile' hominids it is by no means certain that it is due to the presence of *Australopithecus africanus*" (Wood, 1978, p. 369). Bernard Campbell has echoed this hesitancy to recognize *A. africanus* in eastern Africa: "This suggestion, which is not theoretically impossible, would need rather careful demonstration, and it will take a lot of good evidence with large samples to demonstrate it" (Campbell, 1978, p. 574). For the

---

*The relatively large face on this specimen is not surprising, considering its early date and probable male status.

purpose of this study we shall consider *A. africanus* to be confined to the late Pliocene of Southern Africa.

Comparative Postcranial Morphology

It is evident that bipedal locomotion (and concomitant changes in the postcranial skeleton) was the primary hominid adaptation. By 3.7 m.y.a. the hominid postcranial skeleton was fully adapted to a striding bipedal form of

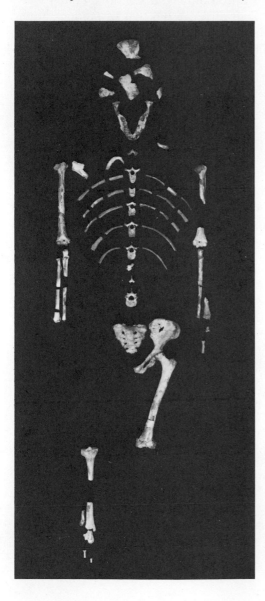

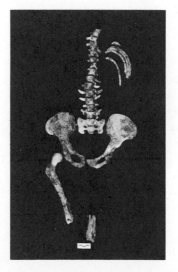

Fig. 7. Comparison of partial hominid skeletons: *A. afarensis* (A.L. 288–1, left) and *A. africanus* (STS 14, right). The specimens are very similar in postcranial anatomy.

locomotion (White, 1980*b;* Lovejoy, 1981). Hominids at this time, however, retained many primitive features in their skulls and dentitions. There is little reason to believe that the hominid postcranial skeleton underwent further major modification until the lower Pleistocene, when one clade embarked on a course of pronounced encephalization (Lovejoy, 1974). Theoretically, therefore, there is minimal reason to suspect that middle and upper Pliocene hominid species should be distinguishable on the basis of postcranial remains. The locomotor skeletons of *A. africanus, A. robustus + A. boisei,* and *H. habilis* are inadequately known. Even though *A. afarensis* is better represented by postcrania, comparisons with other taxa can only be made on a limited basis. For instance, the pisiform in *A. afarensis* differs from that of modern hominoids, but this bone is still unknown in other fossil hominid species.

For all these reasons, the diagnosis of *A. afarensis* did not include reference to postcranial features, despite recent statements to the contrary (Boaz, 1979; Anon, 1979*a,b*). The hominid species *A. robustus + A. boisei, A. africanus, H. habilis,* and *A. afarensis* are indistinguishable in much of their known postcranial anatomy (Fig. 7), but recovery of more fossils and further studies of the abundant *A. afarensis* hand and foot bones may yet reveal significant differences. In the meantime, systematic analyses must rely on cranial, dental, and mandibular morphology.

Comparative Cranial Morphology*

The Facial Skeleton

Despite their fragility, facial bones are common in Plio-Pleistocene hominid collections. Transvaal collections of *A. africanus* and *A. robustus* include numerous adult specimens preserving the facial skeleton. East African *A. boisei* faces are less common, but are represented by specimens from four localities. The facial skeleton of *H. habilis* is known in partial crania from Olduvai Gorge, Koobi Fora, and the Omo Shungura Formation. Sterkfontein (Member 5) specimen STW 53 and the SK 847 composite cranium from Swartkrans (Member 1) are probable South African representatives of this taxon. Four Hadar adult specimens of *A. afarensis* preserve facial parts exhibiting a range of size variation similar to that observed in these other hominid taxa.†

The *A. afarensis* facial skeleton is characterized by an anatomical complex distinguishing this taxon from other known hominid species (Fig. 8). Diagnos-

---

*We refer to specimens listed in parentheses only as *examples.* We do not mean to imply that our comparative conclusions are based on the tiny samples listed as exemplary. We considered all known specimens (in August, 1980) in our analysis.

†Compare TM 1512 vs. STS 71 (*A. africanus*), SK 48 vs. SK 83 (*A. robustus*), KNM-ER 732 vs. KNM-ER 406 (*A. boisei*), KNM-ER 1813 vs. KNM-ER 1470 (*H. habilis*), A.L. 199-1 vs. A.L. 333-1 (*A. afarensis*).

tic morphology here is under the predominating influence of relatively large anterior teeth with robust, curved roots. Expanded premaxillary portions are curved in sagittal and transverse planes. Lateral incisor roots are situated partly lateral to the nasal aperture's margins. Nasoalveolar clivuses extend well into the nasal cavity, intervening between the vomer and the anterior nasal spine. Prominent canine/P^3 juga form laterally placed pillars that do not participate in forming the nasal aperture's margins. The anterior roots of the zygomatic processes arise above M^1 or P^4/M^1 and are separated from the canine pillars by laterally directed canine fossae extending onto the anterior surface of the processes. The canine fossae are very deep, lending a "pinched" appearance to the maxillae of large individuals (A.L. 58-22 and 333-1). Zygomatics are not robustly constructed, but exhibit expanded and flared inferior margins (*m. masseter* origin) and tightly arched *incisurae malaris*. The A.L. 333-1 specimen exhibits a conspicuous transverse ridge extending above the canine fossa from the superior canine jugum to the zygomatic. Rak (1981) calls this feature, often seen in gorillas, the "transverse buttress."

Hadar palates are narrow, flat, and uniformly shallow, with little or no shelving of premaxillary portions. Low alveolar processes produce weakly arched coronal sections across the premolar region. Maxillary tooth rows converge posteriorly and are frequently interrupted by I^2/C diastemata. This constellation of characters is shared with extinct and extant apes. It distinguishes *A. afarensis* from other hominid species (Fig. 9), where palates tend to be deeper, with premaxillary shelving in most specimens and tooth rows that diverge posteriorly.

Due to their antiquity, it is not surprising that *A. afarensis* and *A. africanus* share several primitive facial characteristics, such as pronounced facial and subnasal prognathism and strong canine juga. However, superimposed on this symplesiomorphic "*Grundform,*" *A. africanus* faces display a derived morphological composite that makes them excellent structural and phylogenetic intermediates between *A. afarensis* and *A. robustus* + *A. boisei*. Although prognathic, the *A. africanus* nasoalveolar clivus is flatter and straighter than in *A. afarensis* (TM 1511; STS 5, 17, 71; STW 13, 73). This condition is like that seen in many *A. robustus* + *A. boisei* specimens, but in the latter taxa, the premaxilla projects only slightly, if at all, anterior to the bicanine axis. In nearly all *A. africanus* maxillae (MLD 9 is the single exception), the inferior nasal margin is expressed as a short sill and the nasoalveolar clivus fails to extend significantly behind this level. The anterior tip of the vomer inserts just below the nasal spine in these specimens. In *A. robustus* + *A. boisei* the clivus grades imperceptibly into the nasal cavity floor, the inferior nasal margin is deeply recessed, and the vomer inserts directly into the nasal spine. Canine juga in *A. africanus* vary, but in nearly all cases buttress the nasal aperture and impart a distinctive rounding to inferior corners of the aperture also seen in *A. robustus* + *A. boisei*. Canine juga in *A. robustus* + *A. boisei* also vary, but nearly always lie near the coronal plane of the forwardly thrust zygomatics, thereby eliminating the "canine/malar step" of Clarke (1977*a*).

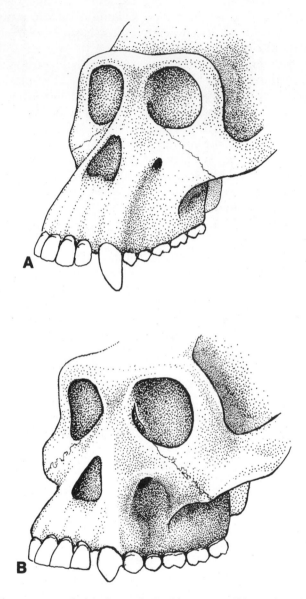

Fig. 8. Left three-quarter facial views of (A) chimpanzee; (B) *A. afarensis* (A.L. 333-1, -45 reconstruction); (C) *A. africanus* (based on STS 5 and STS 71); (D) *A. robustus* (based on SK 46 and SK 48); (E) *H. habilis* (based on KNM-ER 1813). See text for explanation [Figure format developed by Yoel Rak and used with permission.]

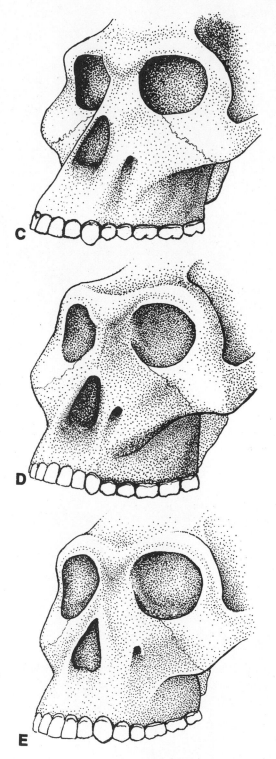

C

D

E

A

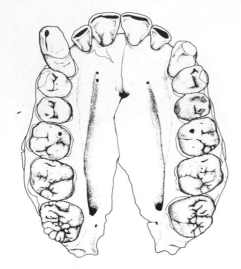

B

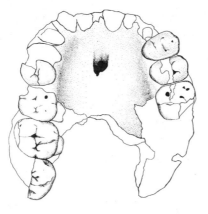

C

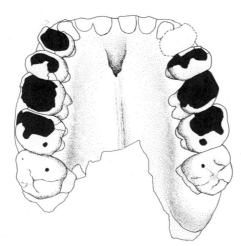

Fig. 9. Comparison of palates of (A) *A. afarensis* (A.L. 200-1a); (B) *A. africanus* (STS 17); (C) *A. robustus* (SK 46). Note the posteriorly convergent tooth rows, shallow palate, and I²/C diastemata in A.L. 200-1a (RC displaced *postmortem*).

Australopithecus africanus lateral incisor roots are usually placed more directly beneath the nasal aperture than in *A. afarensis,* due to anterior dental arch reduction (TM 1512; STS 5, 17, 71; MLD 9). This feature reaches extreme proportions in *A. robustus* + *A. boisei,* where the reduced incisor/canine row forms a straight line across the front of the jaw.

Zygomatics in *A. africanus* are robustly constructed, their anterior roots frequently arising in a more anterior position (usually above P^4/M^1 or P^4) than in *A. afarensis.* The malar surface of *A. africanus* zygomatics is inflated and angles strongly forward (TM 1511; STS 5, 52a, 71). The "transverse buttress" exhibited by A.L. 333-1 disappears into the facial surface in other *Australopithecus* species. These features combine with canine reduction to diminish the canine fossa to little more than a narrow groove running inferiorly from the infraorbital foramen in at least five *A. africanus* specimens (TM 1512; STS 5, 17; MLD 6, 9). Specimens STS 52a and TM 1514 exhibit larger canine fossae resembling the condition in *A. afarensis,* but in TM 1511 and STS 71 there is neither a fossa nor a groove. In *A. robustus* + *A. boisei* the canine fossa is virtually absent. The *incisura malaris* in *A. africanus* and *A. robustus* + *A. boisei* is always high and weakly arched, resulting in *m. masseter* origins higher above the occlusal plane than in A.L. 333-1. These features of the *A. africanus* facial skeleton foreshadow the extreme conditions seen in *A. robustus* + *A. boisei* and thus signal an early phase in the morphological and functional trend discussed below.

Known *H. habilis* facial skeletons evince a trend toward masticatory apparatus reduction. Thus, canine juga and subnasal prognathism are substantially reduced and nasal aperture margins become sharp and superiorly everted. Several characters not directly associated with dental arch reduction are probably retentions of the primitive state seen in *A. afarensis.* These include fairly slender, unexpanded zygomatics arising over M^1 and low, strongly arched *incisurae malaris.* The most robustly constructed *H. habilis* face, KNM-ER 1470 (probably male), has an anterior zygomatic process root originating above P^4, but it also exhibits a malar incisura of low radius, very weak canine juga, and an almost total lack of subnasal prognathism. The lack of appreciable subnasal prognathism also characterizes the SK 847 and STW 53 crania, while KNM-ER 1813 and O.H. 24 display a moderately projecting nasoalveolar clivus.

Juvenile crania with facial skeletons are rare in the Plio-Pleistocene hominid record. As a result, there is widepread reluctance to incorporate these specimens in systematic analyses (Le Gros Clark, 1955). We share Clarke's (1977a) view, however, that detailed morphological analysis can prove useful in the assignation of juveniles at the species level. Moreover, it is clear that many diagnostic features seen in juveniles presage homologous adult conditions.

All three juvenile specimens assigned to *A. afarensis* (L.H.-21; A.L. 333-86, -105) show laterally placed canine juga that do not contribute to nasal aperture formation. This results in a maxillary frontal process that is fairly flat mediolaterally and, in A.L. 333-105, sharp lateral aperture margins.* In addition, lateral deciduous incisor roots in both A.L. 333-86 and -105 lie mostly lateral to the nasal aperture. The Taung holotype of *A. africanus* (about the same dental age as the *A. afarensis* specimens) exhibits canine juga that buttress the nasal aperture margins and inflate inferior portions of the maxillary frontal processes. Lateral deciduous incisor roots lie completely beneath the nasal aperture in this specimen. Specimen A.L. 333-105 possesses an incipient "transverse buttress," a feature absent in the Taung child. Both the Hadar juveniles exhibit fairly low, tightly arched *incisurae malaris,* while that of the Taung specimen is higher and broader.

We find no substantive anatomical basis for placement of the Taung juvenile in *A. robustus.* This specimen agrees in detail with the adult *A. africanus* pattern and helps to establish a morphological link between *A. afarensis* and *A. robustus* + *A. boisei.* It does not exhibit typical *A. robustus* morphology seen in the informative Swartkrans juvenile maxilla SK 66. The same conclusion about Taung was also reached by Clarke (1977*a*) and Rak (1981) and is fully confirmed by our comparative dental analysis of the relevant specimens.

The Calvaria

Adult *A. afarensis* calvarial morphology reflects the combined influence of small brain size, large masticatory musculature, and extensive pneumatization. All three reasonably intact Hadar adult crania (A.L. 162-28, 288-1, 333-45) reflect relatively expanded posterior temporalis muscles. Compound temporal/nuchal (T/N) crests are present in two of these, a probable male (A.L. 333-45) and female (A.L. 162-28). All specimens exhibit closely approximated temporal lines that parallel midline for nearly the entire bregmalambda arc. In A.L. 333-45 and 288-1, the temporal lines do not diverge until well below the level of lambda. Anterior vault portions are not preserved in A.L. 162-28 or 333-45, but the temporal lines diverge gradually across the frontal squama in A.L. 288-1. In *A. afarensis* crania the occipital squama is divided into a short, broad occipital plane and a long, steeply inclined nuchal plane. Nuchal plane steepness is particularly evident in A.L. 162-28 and 288-1, where the nuchal portion of the occipital also exhibits distinct transverse curvature. Extensive pneumatization of lateral cranial base structures (A.L. 166-9, 333-45, 333-84) produces strongly flared parietal mastoid angles that are extensively overlapped by temporal squamae. The parietal mastoid angle, together with the occipitomastoid border of the occipital, forms a distinctive triangular notch (called here the "asterionic notch") for the mastoid portion of the temporal (A.L. 162-28, 333-45).

The inferior temporal surface (A.L. 58-22, 166-9, 333-45, 333-84) presents several distinctive features. Mediolaterally broad mandibular fossae are shallow, bounded anteriorly by only a hint of articular eminence. Conse-

---

*Neither L.H.-21 nor A.L. 333-86 preserves relevant nasal aperture portions.

quently, there is a smooth, gradual transition to the flat preglenoid plane. Tympanics lie completely posterior to the postglenoid processes, are tubular in basal view, and are usually directed inferiorly. The mastoids have extensive, flattened posterolateral faces that are very strongly inflected beneath the cranial base. There are few accurate cranial capacity values for *A. afarensis,* but brain size in this taxon seems to roughly approximate that in *A. africanus.**

Calvar 'l morphology in *A. africanus* demonstrates an emphasis on the anterior fibers of *m. temporalis.* No specimens from Sterkfontein or Makapansgat possess compound T/N crests (*contra* Wolpoff, 1974*b*). Specimen MLD 1, probably the largest known *A. africanus* cranium, exhibits temporal lines that closely approach the lateral extremities of the nuchal torus. This contrasts with the pattern in *A. afarensis* and extant pongids, where the closest approximation or meeting of temporal and nuchal musculature occurs in both large and small crania. Temporal lines in *A. africanus* specimens usually approach most closely at or a short distance behind bregma and then gradually diverge posteriorly (MLD 10, 37/38; STS 5, 71; this also appears to have been the case in MLD 1). In contrast, temporal lines of *A. afarensis* often remain parallel across the lambdoidal suture (Fig. 10). In *A. africanus* crania the occipital plane metrically dominates the nuchal plane (MLD 1,† 37/38; STS 5); the converse is usually true in *A. afarensis.* The nuchal plane is transversely flat and in most cases does not slope to the superior nuchal line as steeply as in *A. africanus* specimens. Although there is some overlap in this regard (STS 5 vs. A.L. 333-45), no *A. africanus* cranium displays the nuchal plane steepness or transverse curvature seen in A.L. 162-28 or 288-1. Pneumatization of *A. africanus* crania is reduced relative to the extensive lateral inflation characteristic of *A. afarensis.* Thus, lateral flare of the parietal mastoid angles is minimal and there is no "asterionic notch" (MLD 1, 37/38; STS 5, 25, 71).

Crania of *A. africanus* are well represented in the Transvaal collections, but complete cranial bases are rare (MLD 37/38; STS 5, 19). The *A. africanus* cranial base does, however, present several key features distinguishing it from *A. afarensis* (Fig. 11). Mandibular fossae, although broad, are deeper than in *A. afarensis* due to moderate or well-developed articular eminences (MLD 37/38; TM 1511; STS 5, 19, 25, 71). Tympanics are more vertically oriented, with strongly curved anterior faces in all specimens except STS 25. The auditory meatus is therefore cone-shaped in sagittal cross section, contrasting with its tubular shape in *A. afarensis.* The *A. africanus* mastoid process inflection is reduced relative to the *A. afarensis* condition, although mastoid tips lie in approximately the same sagittal plane as the external auditory apertures in both taxa.

---

*Holloway obtains an approximate cranial capacity estimate of 500 cm³ for A.L. 333-45. An estimate for A.L. 162-28 is not yet available, but this cranium is significantly smaller than any known *A. africanus* specimen.

†There are two possible ways to determine the position of lambda in this specimen, due to the presence of numerous lambdoid ossicles. One method yields a relatively short occipital plane, while the other, preferred by Tobias (1967), gives a relatively long occipital plane.

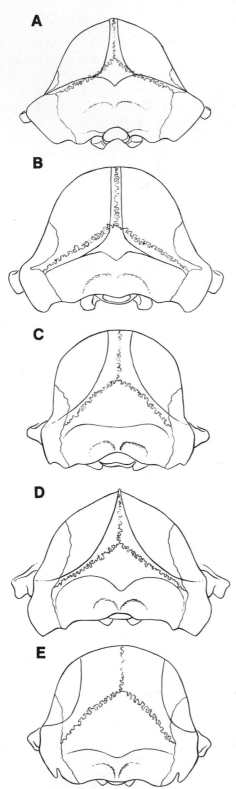

Fig. 10. Occipital views of (A) chimpanzee; (B) *A. afarensis* (A.L. 333-45); (C) *A. africanus* (STS 5); (D) *A. boisei* (KNM-ER 406); (E) *H. habilis* (KNM-ER 1813). In *A. afarensis* note the sutural arrangement at asterion ("asterionic notches"), posteriorly convergent temporal lines, and compound temporal/nuchal crests. These features characterize both large and small crania of *A. afarensis*. See text for additional details.

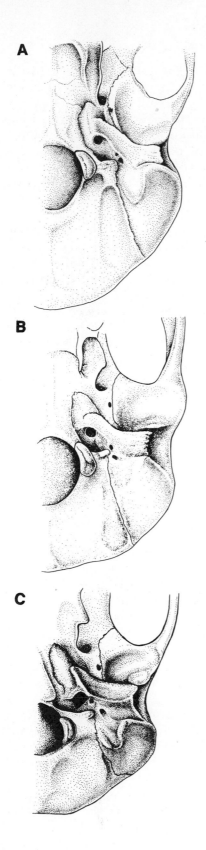

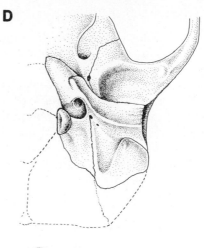

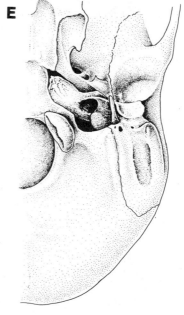

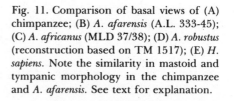

Fig. 11. Comparison of basal views of (A) chimpanzee; (B) *A. afarensis* (A.L. 333-45); (C) *A. africanus* (MLD 37/38); (D) *A. robustus* (reconstruction based on TM 1517); (E) *H. sapiens*. Note the similarity in mastoid and tympanic morphology in the chimpanzee and *A. afarensis*. See text for explanation.

In calvarial features distinguishing *A. afarensis* and *A. africanus* the former species more closely resembles extant pongids. Several of these similarities probably reflect, in part, convergence in aspects of masticatory apparatus form and function (e.g., hypertrophied anterior dentition). Others, such as tympanic morphology, may be associated with primitive lack of pronounced basicranial kyphosis. Although this cannot be precisely assessed in known *A. afarensis* crania, preserved portions indicate a weakly flexed cranial base.

The crania of *A. robustus* + *A. boisei* demonstrate further anterior development of the temporalis muscles. Anteriorly placed sagittal crests are common in both South and East African forms (SK 46, 48, 49, 83; KNM-ER 406; O.H. 5; Omo 323-1976-896). The crests usually bifurcate into bilateral temporal crests well above lambda, but compound T/N crests are present in two large, male East African specimens (O.H. 5 and Omo 323-1976-896). Occipital planes are normally relatively short and nuchal planes are flat and range from steep (KNM-ER 406) to nearly horizontal (O.H. 5). Despite marked lateral inflation of the mastoid region, *A. robustus* + *A. boisei* crania do not exhibit the strong mastoid process inflection typical of *A. afarensis*. Transversely long and concave tympanics result in mastoid tips that lie medial to external auditory apertures (Clarke, 1977b) (TM 1517; SK 46, 47, 83; KNM-ER 407; O.H. 5; Omo 323-1976-896). In addition, parietal mastoid angles are weakly flared and there is concomitant absence of a strong "asterionic notch." Overlap of the temporal squamae on the parietals is, however, extensive (Rak, 1978).

Calvariae of *A. robustus* + *A. boisei* exhibit many specializations concordant with the divergent facial anatomy discussed above. Interestingly, several of these specializations, particularly in the cranial base, also characterize the genus *Homo* (including *Homo habilis*). These include: a more centrally placed foramen magnum that makes a slight angle with the Frankfurt Horizontal, strongly concave and nearly vertical tympanics that merge with postglenoid processes superiorly, and deep mandibular fossae with strong, steep articular eminences from which the preglenoid plane rises abruptly and quite steeply. Some of these shared characters led Louis Leakey to declare that the *Zinjanthropus* holotype cranium, O.H. 5, ". . . represents a stage of evolution nearer to man as we know him today than to the near-man of South Africa" (L. Leakey, 1960b, p. 434). However, the possibility that parallelism could account for these similarities has not been adequately explored. Species in both lineages exhibit strongly flexed cranial bases and relatively orthognathous faces, but the functional and biomechanical consequences and interrelations of these characters should be ascertained before phylogenetic assessment is made.

Known *H. habilis* calvariae exhibit numerous morphological features related to encephalization. Postorbital constriction is moderate, both absolutely and relative to superior facial breadth and maximum biparietal breadth (KNM-ER 1590, 1470, 1805,* 1813; O.H. 16, 24). Frontal bones exhibit a

---

*The Koobi Fora skull KNM-ER 1805 deserves special mention. This specimen is attributed to the ill-defined "small" East African *Australopithecus* by Walker and Leakey (1978) and to *H.*

modest glabellar prominence and a distinct supraorbital sulcus characteristic of African and Asian *H. erectus* (KNM-ER 1813; O.H. 16, 24; STW 53; SK 847). Known *Australopithecus* frontals exhibit glabellar inflation and flat or evenly concave squamae. Parietal bossing is evident in most specimens (KNM-ER 1590, 1470, 1805, 1813; O.H. 7, 13, 24), while known *Australopithecus* specimens lack this feature.† Further testimony to brain enlargement in *H. habilis* lies in its occipital expansion. The nuchal plane in this species is long and fairly horizontal, yet the occipital plane is relatively longer (KNM-ER 1470, 1813; O.H. 13, 24). In addition, the occipital sagittal contour is more rounded. Pneumatization and ectocranial rugosity are markedly diminished in *H. habilis*. Mastoid processes are relatively uninflated, weakly inflected, and more laterally oriented compared to *Australopithecus* (KNM-ER 1470, 1813; O.H. 24; SK 847; STW 53). Temporal lines do not course far medially before arching posteriorly and do not normally form sagittal or compound T/N crests.

Comparative Mandibular Morphology

Hadar and Laetoli hominids bear more primitive mandibles than other early hominid species.†† Archaic features distinguishing *A. afarensis* mandibles are shared with extinct and extant pongids. The lateral surface of the *A. afarensis* mandible bears a hollow defined posteriorly by the lateral prominence and anteriorly by bulges over strong canine and P_3 roots. The entire area above and behind the mental foramen is consequently hollowed. The foramen itself lies below midcorpus height, opening anterosuperiorly. The ramus arises high on the corpus to form a narrow extramolar sulcus. The anterior corpus has a rounded, bulbous anterior profile. The posterior surface of the symphyseal region has a weak or moderate superior transverse torus and a rounded, basally set inferior transverse torus. Tooth rows are typically straight, but range from slight lateral convexity to slight lateral concavity (the latter condition shared only with pongids). In all of these features

---

erectus by Howell (1978a) and Wolpoff (1980). These widely divergent assignations are in no small part due to the specimen's bizarre morphological configuration. The face of KNM-ER 1805 is large, its brain case very long relative to breadth, and its mandibular corpus relatively small but robust. Features of the face, mastoid, and glenoid regions and brain case depart from known *Australopithecus* conditions, but posteriorly placed sagittal and compound T/N crests and a very steeply inclined nuchal plane are reminiscent of the *Australopithecus* pattern. The strong size incongruity between major constituents of the skull plus its persistent metopic suture (Day *et al.*, 1976) suggest the possibility of some growth abnormality. Our taxonomic assessment of this specimen is therefore provisional.

†The subadult status of KNM-ER 1590, O.H. 7, and 13 does not bias this observation, because the subadult *A. boisei* calvaria from the Omo Shungura Formation (L338y-6) exhibits minimal parietal bossing.

††Better preserved adult specimens of the taxa discussed in this section include: L.H.-4, A.L. 288-li, 400-la, 277-1, 145-35, 333w-1, -12, and -60 (*A. afarensis*); MLD 18, 40, STS 7, 36, 52b (*A. africanus*); TM 1517b, SK 12, 23, 34, 1586, Natron, Omo L7A, KNM-ER 729, 810, 818 (*A. robustus* + *A. boisei*); KNM-ER 1483, 1802, O.H. 7, 13 (*H. habilis*).

A. afarensis resembles apes, particularly the Miocene genera *Sivapithecus, Ramapithecus,* and *Ouranopithecus* (Pilbeam *et al.,* 1977, de Bonis and Melentis, 1977*b*).

Homo habilis mandibles are not as abundant or as well preserved as are those of *A. afarensis,* but differences between the two taxa are evident. Size and robusticity do not increase in *H. habilis.* The corpus retains a hollowed lateral contour but develops a low torus that divides the hollow into superior and inferior parts. The anterior face of the anterior corpus becomes more vertical, with the formation of an incipient mental trigon. The posterior surface of this region lacks well-defined superior and inferior transverse tori.

The anatomy of *A. robustus* + *A. boisei* mandibles common at South and East African lower Pleistocene sites is very distinct from that of the contemporary *Homo habilis.* Biomechanical demands of heavy mastication practiced by this specialized *Australopithecus* species have resulted in a distinctive mandibular form [morphological and biomechanical consequences of heavy mastication for the mandible are discussed by Ward (1974), Ward and Molnar (1980), Hylander (1979), and White (1977*a*)]. Lateral corpus contours in *A. robustus* + *A. boisei* are inflated. The mandibular base is deepened, normally placing the mental foramen above midcorpus. The foramen itself opens directly laterally. A tall, broad ramus arises more anteriorly, inferiorly, and laterally than in *A. afarensis* to define a wide extramolar sulcus. The anterior surface of the anterior corpus is vertical, usually developing a weak or moderate rounded basal prominence. The posterior symphyseal region shows a strong superior transverse torus overhanging the deep genioglossal fossa. The inferior transverse torus is substantially elevated above the base compared to the *A. afarensis* condition. Mean cross-sectional corpus area is greater than for either *A. afarensis* or *H. habilis.*

The mandible of *A. africanus* is not as well known as for the other *Australopithecus* species. In size and morphology it is derived in the direction of *A. robustus* + *A. boisei* relative to the primitive *A. afarensis* condition. Lateral corpus hollowing is entirely lost in some specimens (STS 7; MLD 40) (Fig. 12, 13) and is reduced to minor alveolar depression in others (STS 36; MLD 18) as the oblique line sweeps anteriorly to join a low eminence at the mental foramen. Swollen lateral corpus contours typically extend to the base and medial surface of *A. africanus* mandibles (MLD 18, 40). Even the largest Hadar mandible (A.L. 333w-60) lacks the total corpus inflation of specimens like MLD 40. Differences between *A. afarensis* and *A. africanus* are evident even at an early ontogenetic stage—the A.L. 333-43 specimen shows lateral corpus hollowing associated with a strong deciduous canine jugum. These features are absent in *A. africanus* subadults like Taung and are also lacking in the extensive *A. robustus* mandibular growth series. The mandible ramus in *A. africanus* is relatively broad and tall, arising more laterally and inferiorly than *A. afarensis* to demarcate a broad extramolar sulcus in some specimens, such as MLD 40.* In other specimens, such as STS 36 or MLD 18, the sulcus more

---

*Many of the better *A. africanus* mandibles have artificially low rami: STS 7, 36, and 52b are all crushed and MLD 40 lacks superior and posterior ramus margins.

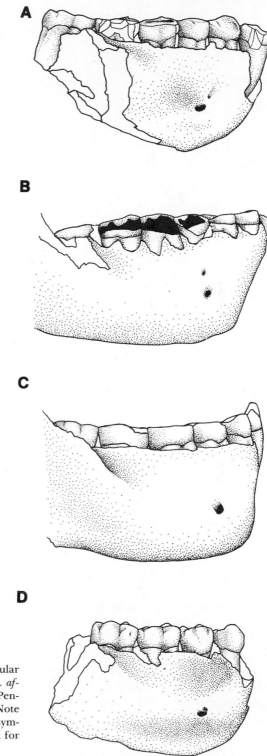

Fig. 12. Lateral views of the mandibular corpus: (A) *A. afarensis* (L.H. 4); (B) *A. africanus* (STS 7, reversed); (C) *A. boisei* (Peninj); (D) *H. habilis* (KNM-ER 1802). Note the corpus inflation and straight symphyseal profile in *A. africanus*. See text for explanation.

A. *africanus* A. *robustus* + A. *boisei*

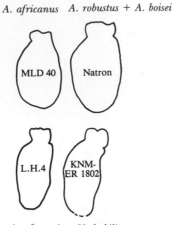

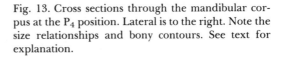

A. *afarensis*   H. *habilis*

Fig. 13. Cross sections through the mandibular corpus at the P_4 position. Lateral is to the right. Note the size relationships and bony contours. See text for explanation.

closely approximates the A. *afarensis* condition. Size of the lateral prominence overlaps between the taxa but tends to be more pronounced in A. *africanus* mandibles (no A. *afarensis* specimen is as swollen as MLD 40 and no A. *africanus* specimen is as flat as A.L. 198-1 or L.H.-4). The mental foramen in A. *africanus* normally opens directly outward (MLD 22, 40; STS 7), but occasionally opens slightly anteriorly (STS 36). It usually exits above midcorpus level (MLD 18, 40; STS 7), but occasionally is lower on the corpus.

The anterior corpus profile in A. *africanus* is not as bulbous or receding as in A. *afarensis*. It is normally a more vertical, planar region whose posterior inclination is intermediate between A. *afarensis* and A. *robustus* + A. *boisei* (MLD 18, 40; STS 7, 52b). A slight sulcus delimits a weak, rounded basal eminence in A. *africanus*. Comparison of mandibles in inferior aspect highlights these differences (contrast L.H.-4, STS 7, and Natron) and reveals a decreasing amount of anterior corpus anterior to the basal contour in the series A. *afarensis*, A. *africanus*, A. *robustus* + A. *boisei*. Total basal eversion in A. *africanus* and A. *robustus* + A. *boisei* is less than that in A. *afarensis*. Australopithecus africanus mandibles preserving the posterior symphyseal region show relatively stronger superior transverse tori than do A. *afarensis*. This torus often overhangs the genioglossal fossa and combines with a larger, higher inferior transverse torus than A. *afarensis* to buttress the symphysis (MLD 18). Known A. *africanus* tooth rows are convex laterally (MLD 18, 40; STS 7).

In all of these characters, A. *africanus* forms a good morphological intermediate between the primitive A. *afarensis* and the specialized A. *robustus* + A. *boisei*. Individual characters in some specimens overlap between the three fossil samples, but the morphological differences are functionally based (see below) and appear to be evolutionarily significant.

Comparative Dental Morphology

African Plio-Pleistocene hominids are represented by abundant dental remains. Teeth were analyzed by Broom (1929) and Dart (1934) in their arguments for the hominid status of the Taung specimen. Robinson's dietary hypothesis was substantially based on dental comparison. The following analysis is based on larger samples available in 1980 (see Table 1).

Pooling the Hadar and Laetoli hominids is recommended by close morphological resemblances in teeth from the two sites. The samples completely overlap in most details and we have found no significant anatomical differences between teeth from Hadar and Laetoli. Differences may ultimately be demonstrated, but small samples for many tooth categories often hinder comparison.

Le Gros Clark's (1950) work on the *Australopithecus* dentition reviewed characters differentiating modern pongids and hominids. Parabolic dental arcades, spatulate and nonprojecting canines, absent diastemata, and nonsectorial P_3's unambiguously allied *Australopithecus* with the Hominidae. When dental remains of *A. afarensis* are subjected to similar analysis the outcome is not so clear-cut—this earliest hominid species bears the unmistakeable stamp of ape ancestry on its teeth.

Our analysis of the dentition will focus on two sets of features: those that distinguish *A. afarensis* from other species and those shared between *A. africanus* and *A. robustus* + *A. boisei*.

Table 1. The Numbers of Fossil Hominid Individuals Represented by at Least One Crown Retaining Some Useful Morphology at Each Tooth Position[a]

	Maxillary			Mandibular		
	A. afar.	*A. afric.*	*A. rob.*	*A. afar.*	*A. afric.*	*A. rob.*
di1	0	2	1	2	1	1
di2	1	1	0	2	2	3
dc	4	1	0	4	2	6
dm1	4	3	3	4	5	9
dm2	5	4	5	2	7	13
I1	6	7	13	6	5	7
I2	10	7	9	6	3	6
C	10	7	15	13	16	11
P3	9	17	25	18	11	19
P4	10	24	28	15	11	20
M1	11	27	33	18	13	25
M2	6	30	26	22	18	22
M3	8	32	24	13	17	22

[a]This table provides a rough guide to sample sizes available in the comparative dental study of *Australopithecus*.

Dental Arcade

Primitive dental arcade shape distinguishes *A. afarensis* from other early hominid taxa (see above). Dental arcades preserving circumcanine regions are rare, but the wide $I^2/\underline{C}$ diastema of the A.L. 200-1a *A. afarensis* palate is unmatched in other known African early hominids. When maxillary ($I^2/\underline{C}$) and mandibular ($\bar{C}/P_3$) diastema frequencies are determined by presence or absence of interproximal facets the following results obtain: two of six adult *A. afarensis* individuals had maxillary diastemata, compared to three of eight *A. africanus* individuals and four of 19 *A. robustus* individuals. In the lower jaw, nine of 20 *A. afarensis* individuals retained a $\bar{C}/P_3$ diastema, while only one of 12 *A. africanus* and none of ten *A. robustus* individuals displayed this feature. *Homo habilis* and *H. erectus* occasionally retain small diastemata.

Deciduous Dentition

The deciduous dentition is poorly documented for early hominids, particularly *A. africanus* and *H. habilis* (see Table 1). The tiny known samples of deciduous incisors reveal no differences among Plio-Pleistocene hominid taxa. Deciduous canines are not much more abundant, but both upper and lower *A. afarensis* dc crowns (A.L. 333-35, -66; L.H.-2, L.H.-3/6) are commonly taller, narrower, and more pointed than are those of *A. africanus* (Taung, STS 24) or *A. robustus* (SK 62, 63).

The *A. afarensis* dm^1 (A.L. 333-86; L.H.-21) is sometimes distinguished from that of *A. africanus* (STS 2, 24a) and *A. robustus* (SK 91) by a large paracone relative to the metacone and always by a thinner distal marginal ridge on known specimens. The *A. afarensis* dm^2's (A.L. 333-86; L.H.-21) are distinguished from these taxa (STS 2; Taung, SK 90, 838a) by occlusally angled instead of bulging lingual protocone faces. Worn *A. afarensis* dm^2's show prominent buccal cusps even when the lingual cusps betray advanced wear. This contrasts with the *A. robustus* condition (SK 90), where the crown wears very flat early in tooth life. Specimens of *A. africanus* are intermediate in this character, with dentine exposed on the buccal dm^2 cusps slightly earlier (STS 24a; Taung) than in *A. afarensis* but substantially later than in *A. robustus*.

The dm$_1$ figures prominently in discussions concerning the hominid status of *Australopithecus* and in the definition of species within this genus. Broom's (1941) discovery of dm$_1$ hypermolarization in *A. robustus* still provides a basis for sorting advanced members of this hominid clade. The *A. africanus* dm$_1$ is poorly known but is most similar to that of *A. afarensis*. Characters distinguishing the two taxa include the dm$_1$ distal crown profile. It is tapering, narrow, and asymmetric in *A. afarensis* (A.L. 333-43; L.H.-2), square in *A. africanus* (STS 24b; Taung), and square, bulbous, and expanded in *A. robustus* + *A. boisei* (SK 61, 62.) Samples reflect a similar morphological series in talonid height. The basin is low in *A. afarensis,* higher in *A. africanus,* and very elevated relative to the trigonid in *A. robustus*. Relative trigonid/talonid wear reflects these changes, being relatively disparate in the *A. afarensis* dm$_1$ (L.H.-3), intermediate in *A. africanus* (Taung), and equal in *A. robustus* + *A. boisei* (SK 61, 62).

The *A. afarensis* dm$_2$ (A.L. 333-43; L.H.-2) is distinguished by shallow central foveae, strongly bilobate lateral crown profiles, and protoconids set strongly mesiad of metaconids. In contrast, *A. robustus* (SK 3978) and *A. africanus* (MLD 5) dm$_2$'s have deep central foveae. The dm$_2$ protoconid and metaconid are set at equal anteroposterior levels in *A. robustus* (SK 61; TM 1536) and the lateral crown profile is rounded. Specimens of *A. africanus* (STW 67; Taung) are intermediate in the last two characters.

Permanent Incisors

The permanent dentition is fairly well known in *Australopithecus* species, but sample size for *H. habilis* remains limited in most categories.

Canines and incisors are rare relative to both maxillary and mandibular postcanine teeth. Maxillary central incisors are relatively tall and narrow in *A. africanus* (STS 52b, MLD 43) and *A. robustus* (SK 70, 839), while the *A. afarensis* counterparts bear a more flared appearance in labial profile and a relatively longer incisal edge (A.L. 333x-20, 200-la; L.H.-3). The basal eminence on *A. afarensis* I^1's is prominent (A.L. 200-la). It is more subdued in *A. africanus* (MLD 43) and varies from pronounced (SK 2) to absent (SK 55) in *A. robustus*.

The I^2's of *A. afarensis* often display strong basal lingual tubercles that project incisally to give a bilobed base to the lingual fossa (A.L. 200-la; L.H.-3). Wear is confined to the occlusal edge. Only one of seven *A. africanus* individuals shows the lingual basal tubercle (MLD 11 + 30), the normal lingual I^2 morphology involving convergent crown shoulders bounding a narrow, V-shaped pit near the lingual crown base (STS 24a, 52a). This pattern is repeated in *A. robustus* (SK 52, 71). Both *A. africanus* and *A. robustus* display substantial polishing wear on the lingual I^2 surface.

Mandibular incisors are relatively rare. Samples display few significant morphological differences between *A. africanus* and *A. afarensis*. The lone unworn *A. afarensis* I$_1$ (L.H.-2) has seven mammelons, the single *A. africanus* counterpart (STS 24b) has five, and five unworn *A. robustus* I$_1$'s have only three mammelons each.

Permanent Canines

Upper and lower canines figure prominently in the *A. afarensis* diagnosis, even though *A. africanus* canines are morphologically similar. Upper canines in *A. africanus* (STS 2) and *A. robustus* (SK 55, 92) tend to show more symmetric labial crown profiles, even though asymmetric representatives (MLD 11) do occur. Wear is apical in both taxa. In contrast, *A. afarensis* permanent upper canines (A.L. 333x-3, 200-la, 400-lb; L.H.-5) are more often asym-

Fig. 14. Comparison of upper right canines, lingual view, slightly rotated. Left: SK 65a, *A. robustus;* center: TM 1527, *A. africanus;* right: A.L. 200-1a, *A. afarensis.* Note the distinctive wear on the Hadar specimen. See text for details.

metric and projecting. They do not usually wear exclusively from the apex, but instead resemble the pongid condition, where a distinct C̄ contact facet forms along the mesial occlusal edge and a P_3 contact occupies the longer distal occlusal edge. A thin dentine strip is often exposed along the second contact (Fig. 14).

All three *Australopithecus* species have asymmetric lower canines with lingual ridges. The lingual C̄ face in *A. afarensis* (A.L. 128-23, 198-1; L.H.-14) is, however, dominated by the prominent lingual ridge and is not "hollowed" along the vertical crown axis as in *A. africanus* (STS 51; STW 21; MLD 42) and *A. robustus* (SK 29, 87, 96). The mesial occlusal edges in *A. afarensis* canines are steep, straight, and elongate compared to the more horizontal, curved, shorter edges seen in *A. africanus* and *A. robustus.* The distal cingulum in *A. afarensis* is set basally, contributing to a relatively elongate distal occlusal edge. The cingulum reaches a higher relative crown position to limit the edge length in *A. robustus.* The *A. africanus* C̄ is intermediate. Wear in some *A. afarensis* (A.L. 198-1, BMNH M 18773) is pongid-like, since contact with the upper canine often flattens the distal occlusal edge of the lower. Thus, only three of nine *A. afarensis* individuals show exclusively apical C̄ wear, while ten of 12 *A. africanus* and five of six *A. robustus* canines are worn only from the apex. Apical wear in *A. africanus* and *A robustus* commonly has a strong lingual inclincation absent in *A. afarensis* lower canines. The O.H. 16 *H. habilis* C̄ has a wear pattern similar to that seen in *A. afarensis.*

Permanent Premolars

The P_3 is more primitive in *A. afarensis* (Fig. 15) than in other known hominid species. Occlusal crown outline is a long, narrow oval in *A. afarensis* (A.L. 288-1, 400-la; L.H.-3, 4, and 24) and the pongids. The distolingual crown corner is normally projecting and tightly curved in Hadar and Laetoli P_3's. Both *A. africanus* (STS 24, 52; MLD 40) and *A. robustus* (SK 74a, 831; TM 1517) have rounder occlusal outlines with more abbreviated distolingual corners. *Australopithecus afarensis* is the only known Plio-Pleistocene hominid with both female (A.L. 288-1) and male (A.L. 277-1) P_3's that are unicuspid. Of 15 individuals preserving relevant occlusal relief, six lack metaconids and bear

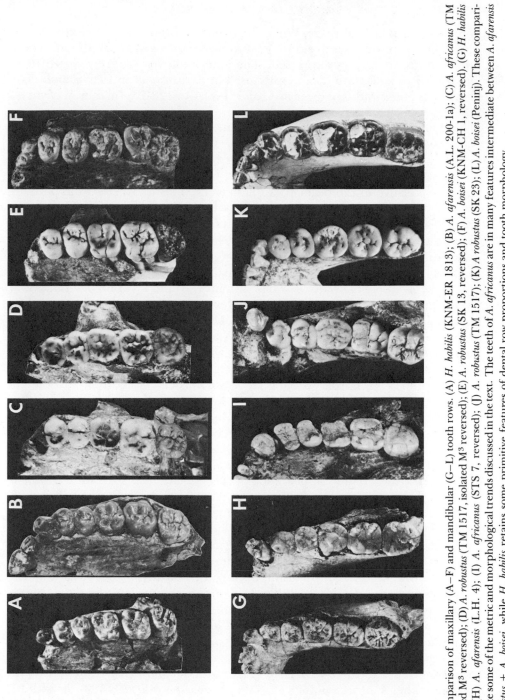

Fig. 15. Comparison of maxillary (A–F) and mandibular (G–L) tooth rows. (A) *H. habilis* (KNM-ER 1813); (B) *A. afarensis* (A.L. 200-1a); (C) *A. africanus* (TM 1511, isolated M³ reversed); (D) *A. robustus* (TM 1517, isolated M³ reversed); (E) *A. robustus* (SK 13, reversed); (F) *A. boisei* (KNM-CH 1, reversed). (G) *H. habilis* (O.H. 13); (H) *A. afarensis* (L.H. 4); (I) *A. africanus* (STS 7, reversed); (J) *A. robustus* (TM 1517); (K) *A robustus* (SK 23); (L) *A. boisei* (Peninj). These comparisons illustrate some of the metric and morphological trends discussed in the text. The teeth of *A. africanus* are in many features intermediate between *A. afarensis* and *A. robustus* + *A. boisei*, while *H. habilis* retains some primitive features of dental row proportions and tooth morphology.

only a swollen lingual ridge. When the metaconid is distinct its relative size is small compared to the average *A. africanus* or *A. robustus* condition, where metaconids are always expressed. The anterior fovea is rarely closed in *A. afarensis* P_3's and more commonly closed in *A. africanus* and *A. robustus*. The talonid is lower and narrower in *A. afarensis* (A.L. 128-23) than in *A. africanus* (STS 51) or *A. robustus* (SK 6).

The distal marginal ridge becomes increasingly cuspidate in the series *A. afarensis* (A.L. 128-23), *A. africanus* (STS 51, *A. robustus* (SK 857) and the buccal enamel line more frequently shows a strong mesiobuccal extension in *A. afarensis* (A.L. 198-1, 333w-58; L.H.-4). Occlusal P_3 wear is limited in *A. afarensis* relative to the other postcanine teeth. This may reflect the partially interlocking C/P_3 and the generalized masticatory complex of the species. In *A. africanus* and *A. robustus* the P_3 wears flat, with eventual total dentine exposure (MLD 18; SK 23). Postcanine teeth in *A. afarensis* show stronger wear differentials. For example, the P_4, M_1, and M_2 crowns of A.L. 311-1 are entirely worn away, but the P_3 crown remains little worn and very prominent.

The *A. afarensis* P^3 also exhibits primitive overall morphology relative to other hominids, a point recognized by Remane (1954) and Şenyürek (1955). Occlusal crown outline is asymmetric (A.L. 200-la; L.H.-6), with buccal cusp length exaggerated. This combines with an abbreviated distolingual crown corner to produce the asymmetry and give the occlusal fovea its triangular shape. In *A. robustus* the lingual and buccal cusps are about equal in length, crown shape is oval, the distolingual crown corner is filled out, and the occlusal fovea is more square (SK 13, 24). Specimens of *A. africanus* are intermediate in all these characters (STS 1, 24a, 52a; MLD 23). Mesiobuccal projection of the enamel line is found in all *Australopithecus* species, with complete overlap between *A. afarensis* and *A. africanus* (Garusi I; TM 1511), but only occasional projection in *A. robustus*. Specimens of *A. afarensis* (A.L. 200-la, 333-1) lack the inflated appearance of the lingual cusp seen in *A. africanus* (STW 73; MLD 23) and *A. robustus* (SK 48, 823). This inflation is based, in part, on swollen internal faces of the lingual cusps in the latter taxa (compare L.H.-6 with STS 1, 24 and SK 13, 33). When worn, the lingual cusp is evenly rounded, in contrast to the angular worn cusps of *A. afarensis* (A.L. 199-1, 200-la, 333-1) with their distinct mesial and distal wear planes.

The only salient feature distinguishing the *A. afarensis* P_4 is its asymmetric occlusal outline (A.L. 288-1, 400-la; L.H.-3). This shape is based on a projecting distolingual corner and an abbreviated distobuccal corner. In *A. africanus* (TM 1523; STW 14) and *A. robustus* (TM 1517; SK 826) the P_4 crown is rounder and more symmetric. Upper P^4's are more diagnostic, but there are few available unworn specimens. The mesial and distal marginal ridges are smaller relative to the buccal cusp length in *A. afarensis* (A.L. 200-la; L.H.-3) than in either *A. africanus* (STS 12, 55) or *A. robustus* (SK 28, 1589). As with P^3, cusps are often more inflated in *A. africanus* (STS 61; MLD 6) and *A. robustus* (SK 48, 49) than in *A. afarensis* (L.H.-3, 6). Worn lingual cusps in the latter taxon develop clear mesial and distal occlusal slopes even late in wear (A.L. 199-1, 200-la). In *A. africanus,* the tendency is for a more evenly rounded wear across the surface (TM 1511) and sometimes the development of large, flat occlusal platforms (STS 35, 42, 47). Such platforms are common in *A. robustus* (SK 65, 845) and unworn P^4's of this species show closely approximated cusp apices and fairly closed, square occlusal foveae (SK 13, 28, 825). *Australopithecus afarensis* has more open foveae (L.H.-3) and laterally placed cusp apices. Some *A. africanus* specimens resemble the *A. robustus* condition, while others have more open occlusal foveae.

Permanent Molars

The lingual cingulum is expressed strongly in only one *A. afarensis* upper molar (L.H.-17), but large, shelflike projections mark the lingual surfaces of several *A. africanus* M^2 (STS 12, 37) and M^3 (STS 28; STW 2) crowns. When present in *A. robustus*, the cingulum is reduced to weak Carabelli features (SK 13, 49). Instead of a primitive retention, the *A. africanus* cingular effects may represent the early phases of buccolingual crown expansion. First and second molars have the basic four cusps in all taxa under consideration, but *A. africanus* (STS 37, 52, TM 1561) and *A. robustus* (SK 31, 52, 3977) M^3's show a tendency for the metacone (and to a lesser extent, the hypocone) to divide into subequal cusps forming the distal occlusal rim. Compared to *A. afarensis* (A.L. 333x-1; L.H.-12), occlusal foveae tend to be wider and squarer in *A. africanus* M^3's. Crown area behind the major lingual and buccal grooves is often relatively large in *A. africanus* (TM 1511) compared to *A. afarensis* (A.L. 161-40).

The buccal cusps (paracone and metacone) of *A. afarensis* upper molars remain high, sharp, and relatively unscathed by wear compared to the lingual cusps (M^1, M^2, M^3: L.H.-17, A.L. 200-la, A.L. 161-40). Cusps in *A. robustus* upper molars are often blunt when unworn and all four cusps plane off quickly to produce a broad, flat occlusal platform early in tooth life (M^1, M^2, M^3: TM 1517, SK 877, SK 21a). Upper molar wear morphology is intermediate in *A. africanus*. Its buccal/lingual cusp wear disparity is usually reduced relative to the *A. afarensis* condition (M^1, M^2, M^3: STS 8, STS 35, TM 1514).

Lower molar occlusal wear is complementary to that described for the upper molars. These teeth in *A. afarensis* exhibit prominent lingual cusps (especially the metaconid) that project occlusally to form a sharp, fairly continuous lingual rim even late in tooth wear (M$_1$, M$_2$, M$_3$: A.L. 128-23, L.H.-23, A.L. 333-74). Lower molars in *A. robustus* wear to form a planar occlusal surface fairly early in tooth wear (M$_1$, M$_2$, M$_3$: SK 23, SK 1586, SK 885). Specimens of *A. africanus* are intermediate, their lingual cusps tending to reveal dentine earlier and wear flatter than those of *A. afarensis* (M$_1$, M$_2$, M$_3$: STS 52b, MLD 40, STS 41). Dentine exposure is relatively deep in *A. afarensis* and the buccal cusps are often perforated in crown wear earlier than in *A. robustus* (contrast A.L. 333w-60 and TM 1517) (Fig. 16).

Lower first and second molars in *A. afarensis* bear basally swollen hypoconids donating a bilobate appearance to the buccal crown profile (A.L.

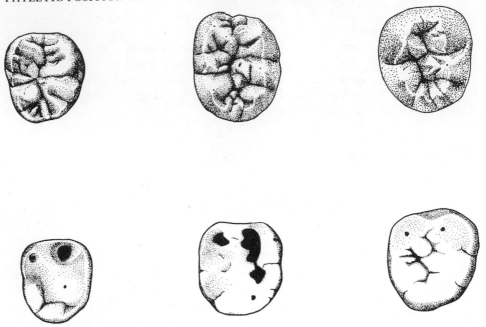

Fig. 16. Comparison of lower third molars, occlusal view. Top row, unworn teeth, left to right: *A. afarensis*, A.L. 400-1a; *A. africanus*, STW 14; *A. robustus*, SK 6. Bottom row, worn teeth, left to right: *A. afarensis*, A.L. 198-1, reversed; *A. africanus*, TM 1519; *A. robustus*, SK 12, reversed. See the dental descriptions in the text for explanation of morphological differences.

145-35, 266-1, 333w-12). This lobation is weak or absent in *A. africanus* (STS 52b; MLD 18) and *A. robustus* (SK 23; STW 5). Lower molar cusps in *A. afarensis* lack the swollen, inflated appearance often seen in *A. africanus* (TM 1518; STW 3) and typical of *A. robustus* (SK 843, 6). The *A. afarensis* M_3 bears a narrow talonid that produces a triangular crown outline (A.L. 333-74; L.H.-4). Talonids in *A. africanus* (TM 1519, STS 41) are squarer, but this feature is highly variable in *A. robustus*. The hypoconulid is appressed mesially in *A. afarensis* M_1's and M_2's to give the distal crown profile a square appearance (A.L. 128-23, 333w-60), in contrast to the rounded profile characteristic of *A. africanus* (STS 52b; STW 14) and *A. robustus* (SK 6). In *A. afarensis* M_2's the trigonid breadth often clearly exceeds the talonid breadth (A.L. 400-1a, 333w-60). These breadths more closely approximate each other in *A. africanus* (STS 9, 18) and *A. robustus* (SK 6, 23). Occlusal relief of *A. afarensis* M_3's is characteristically crenulate (A.L. 288-1, 400-1a), but lacks the clear major cusp differentiation seen in *A. africanus* (STS 52b; STW 5, 14) or *A. robustus* (TM 1517; SK 75).

Subocclusal Anatomy

Preliminary investigations of subocclusal morphology in the teeth and jaws of *A. afarensis* have revealed strong parallels with middle Miocene hominoids (Ward, 1979, 1980). Ward has recently completed comparative work on the South African hominid collections and his results already show several characters distinguishing *A. afarensis* and *A. africanus*.

Comparative Dental Metrics

The literature on fossil hominoids reflects a well-entrenched anthropological obsession with dental metrics. Tooth crown measures often receive primary and sometimes exclusive consultation when the systematic assessment of a fossil collection comes under scrutiny. This is true because fossil teeth commonly exist in numbers adequate for statistical treatment and in states of preservation amenable to measurement. It is also true because teeth are easily measured and only altered by wear once erupted. Unfortunately, many investigators have missed the morphology between the measuring points in their eagerness to quantify the hominid fossil record. The reduction of a tooth crown to a length, breadth, area, or index is bound to result in tremendous information loss, as length and breadth measurements are nothing more than crude indicators of tooth crown size. Used judiciously, these data can sometimes supplement morphological information bearing on the systematic position of fossil organisms.

We have summarized our dental metric data in Tables 2–4 and Fig. 17. Measurements used to construct these tables and figures were taken from original specimens by the authors. The measurement technique is described elsewhere (White, 1977*b*). The buccolingual (BL) dimension is a maximum crown breadth. The mesiodistal (MD) dimension is a maximum axial length corrected for interproximal wear only on postcanine teeth. It is impossible to accurately correct for interproximal wear in incisors and canines, so these values represent both worn and unworn teeth. Where specimens were only slightly damaged, estimates were made and incorporated. Sample sizes refer to the number of fossil hominid *individuals* measured: when both right and left representatives of a tooth category were preserved for one individual, only the right side tooth was included.* Obviously anomalous teeth (like the SK 6 RP$_4$) were excluded. Pongid-like C and P$_3$ crown profiles in *A. afarensis* render these measurements somewhat incomparable to other taxa. Similar discrepancy was noted by Schuman and Brace (1954) and Mahler (1973) for apes, but these two teeth were measured according to traditional anthropological practice only for the purposes of this paper.

Sample composition for dental metrics is as follows: Laetoli and Hadar are site samples—all fossil hominids are from the Hadar Formation and Laetolil Beds. The *A. afarensis* sample is a simple combination of Laetoli and Hadar data. The *A. africanus* sample is composed of all fossil hominid specimens from Taung, Makapansgat Member 3, and Sterkfontein Member 4. No significant differences were found among the three *A. africanus* site samples. The *A. robustus* sample comprises all representatives of this taxon from Kromdraai

*This is the same procedure used to construct Table 1 in Johanson and White (1979).

Table 2a. Maxillary Dentition: MD Length/BL Breadth Index

		Laetoli	Hadar	A. afarensis	A. africanus	A. robustus	A. boisei	Homo sp.
P^3	N	4	4	8	12	17	5	10
	R	66–73	66–73	66–73	67–78	66–84	64–84	66–78
	$\bar{X}$	70.3	70.5	70.4	72.5	71.35	71.8	73.5
	S	2.98	3.10	2.80	3.98	4.10	7.40	3.95
P^4	N	1	5	6	16	20	3	11
	R	—	68–79	68–79	66–81	65–77	68–74	69–84
	$\bar{X}$	73.0	72.6	72.66	72.6	71.45	70.66	77.0
	S	—	4.70	4.20	4.18	3.00	—	3.70
M^1	N	4	3	7	17	20	5	16
	R	85–104	90–97	85–104	80–100	82–99	86–97	91–104
	$\bar{X}$	94.0	92.7	93.4	91.88	90.75	92.8	99.19
	S	8.98	—	6.60	5.17	4.63	4.43	3.29
M^2	N	3	2	5	18	22	3	12
	R	84–86	90–91	84–91	84–99	82–99	82–92	85–101
	$\bar{X}$	85.0	90.05	87.2	91.61	91.0	88.33	90.67
	S	—	—	3.11	4.59	4.77	—	4.16
M^3	N	3	4	7	13	18	3	9
	R	81–84	87–95	81–95	84–97	77–97	79–98	81–98
	$\bar{X}$	82.7	89.3	86.4	88.77	89.17	90.67	88.33
	S	—	3.86	4.54	4.04	5.01	—	6.20

Table 2b. Mandibular Dentition: MD Length/BL Breadth Index

		Laetoli	Hadar	A. afarensis	A. africanus	A. robustus	A. boisei	Homo sp.
P_3	N	5	13	18	7	18	4	7
	R	87–118	71–99	71–118	73–94	77–113	79–97	78–104
	$\bar{X}$	100.2	87.23	90.83	82.57	87.33	86.5	91.0
	S	12.05	8.02	10.74	6.88	8.96	9.00	9.22
P_4	N	3	10	13	10	17	8	7
	R	92–96	74–99	74–99	79–93	82–95	86–107	78–102
	$\bar{X}$	94.3	86.8	88.54	88.2	88.35	96.86	90.29
	S	—	8.04	7.75	4.16	4.40`	6.92	8.44
M_1	N	4	11	15	7	19	6	13
	R	101–105	97–113	97–113	101–115	101–117	103–115	106–126
	$\bar{X}$	102.3	104.1	103.6	108.3	109.5	108.3	114.8
	S	1.89	4.55	4.03	4.92	4.33	5.05	4.87
M_2	N	2	15	17	14	20	6	12
	R	109–111	93–118	93–118	103–118	102–117	104–118	106–120
	$\bar{X}$	110.0	103.5	104.2	109.6	110.7	109.8	112.7
	S	—	6.21	6.21	3.65	4.37	5.42	4.21
M_3	N	2	8	10	13	21	8	13
	R	109–115	101–121	101–121	101–121	110–133	113–126	107–129
	$\bar{X}$	112.0	111.0	111.0	111.5	118.4	119.4	117.5
	S	—	7.10	6.40	5.24	6.70	4.66	7.20

and Swartkrans. Specimens SK 27, 45, 847, and 14252 are placed in *Homo* sp. following Clarke (1977*a,b*). The *A. boisei* sample includes known specimens from Chesowanja, Natron, Koobi Fora, Ileret, and Olduvai Gorge in East Africa. The *Homo* sp. sample includes specimens from Swartkrans Member 1, Sterkfontein Member 5, Olduvai Beds I and II, Koobi Fora, and Ileret. The latter samples represent *H. habilis* but are supplemented by a few early *H. erectus* specimens from Koobi Fora. The metric data incorporate all specimens available as of July 1980.

Table 3a. Maxillary Dentition: Mesiodistal Diameter (mm)[a]

		Laetoli	Hadar	*A. afarensis*	*A. africanus*	*A. robustus*	*A. boisei*	*Homo* sp.
I¹	N	1	3	4	6	10	1	5
	R	—	9.0vw–10.9sw	9.0vw–11.8	8.1–10.7	7.1–9.4	—	9.2–12.5
	$\bar{X}$	11.8	10.23	10.63	9.77	8.36	10.0w	10.86
	S	—	—	1.17	0.942	0.841	—	1.33
I²	N	2	4	6	6	6	2	9
	R	7.8	6.7vw–8.2w	6.7vw–8.2w	5.9sw–7.0	5.8–6.7	6.9–7.0	5.9w–8.0
	$\bar{X}$	7.8	7.6	7.67	6.68	6.32	6.95	7.16
	S	—	0.698	0.550	0.426	0.371	—	0.783
C̲	N	3	7	10	5	13	8	7
	R	9.6–11.6	8.9–10.4	8.9–11.6	8.8sw–10.0	7.6w–9.2sw	6.5w–10.8	8.3w–11.5
	$\bar{X}$	10.43	9.7	9.92	9.56	8.49	8.58	9.43
	S	—	0.529	0.744	0.483	0.433	1.18	1.35
P³	N	3	4	7	14	21	6	11
	R	8.9–9.3	7.5–8.9	7.5–9.3	8.7–9.6	9.2–10.7	9.5–11.8	8.2–10.3
	$\bar{X}$	9.03	8.5	8.73	9.04	9.92	10.62	8.9
	S	—	0.668	0.568	0.276	0.516	0.884	0.648
P⁴	N	5	5	10	21	26	4	12
	R	9.0–9.7	7.6–9.5	7.6–9.7	8.7–10.8	9.5–11.9	11.7–12.4	8.3–9.8
	$\bar{X}$	9.20	8.76	8.98	9.43	10.65	12.08	9.12
	S	0.283	0.792	0.607	0.564	0.648	0.299	0.702
M¹	N	5	3	8	28	25	5	15
	R	11.0–13.8	10.8–12.1	10.8–13.8	11.1–13.8	11.4–15.6	13.4–15.6	11.5–14.1
	$\bar{X}$	12.58	11.6	12.21	12.69	13.25	14.96	12.94
	S	1.13	—	1.06	0.686	0.761	0.888	0.960
M²	N	3	2	5	24	24	3	12
	R	12.7–12.9	12.1–13.5	12.1–13.5	12.6–16.4	12.8–15.8	15.6–17.2	11.4–14.7
	$\bar{X}$	12.77	12.80	12.78	14.04	14.48	16.5	12.85
	S	—	—	0.502	1.12	0.884	—	0.9596
M³	N	3	4	7	21	19	4	11
	R	10.9–11.6	11.4–14.3	10.9–14.3	11.1–15.3	13.2–17.1	16.3–17.3	11.2–13.9
	$\bar{X}$	11.13	12.73	12.04	13.4	14.99	16.93	12.87
	S	—	1.45	1.35	1.09	1.01	0.435	0.899

[a]vw, very worn; w, worn; sw, slightly worn.

Table 3b. Maxillary Dentition: Buccolingual Diameter (mm)

		Laetoli	Hadar	A. afarensis	A. africanus	A. robustus	A. boisei	Homo sp.
I¹	N	1	4	5	4	6	2	7
	R	—	7.1–8.6	7.1–8.6	8.3–8.9	6.8–7.7	7.4–8.0	6.0–8.5
	X̄	8.4	8.1	8.16	8.6	7.35	7.7	7.2
	S	—	0.678	0.602	0.245	0.339	—	1.02
I²	N	2	5	7	5	6	2	8
	R	7.4–8.1	6.2–7.9	6.2–8.1	5.6–7.0	5.6–6.7	6.4–7.5	5.5–8.1
	X̄	7.75	6.94	7.17	6.4	6.35	6.95	6.94
	S	—	0.650	0.692	0.543	0.446	—	0.923
C̲	N	3	7	10	6	14	4	7
	R	9.8–12.5+	9.3–12.4	9.3–12.5+	8.7–10.3	7.9–11.2	7.5–9.9	8.6–12.3
	X̄	10.77	11.01	10.94	9.7	9.13	8.85	10.66
	S	—	1.04	1.11	0.540	0.924	1.11	1.43
P³	N	2	4	6	12	19	5	10
	R	13.0–13.4	11.3–12.4	11.3–13.4	11.7–13.2	11.6–15.2	13.8–17.0	11.1–13.5
	X̄	13.2	12.0	12.4	12.46	13.78	15.18	12.07
	S	—	0.497	0.740	0.533	0.887	1.35	0.846
P⁴	N	1	5	6	16	23	4	11
	R	—	11.1–12.6	11.1–12.6	10.7–14.2	12.3–16.3	14.2–17.6	10.9–13.7
	X̄	12.5	12.0	12.08	12.97	14.89	16.23	11.94
	S	—	0.604	0.578	0.872	0.987	1.56	0.807
M¹	N	5	3	8	19	19	5	14
	R	12.8–14.6	12.0–13.3	12.0–14.6	12.9–15.7	13.0–16.8	14.9–17.7	11.9–14.2
	X̄	13.62	12.5	13.2	13.74	14.63	16.18	13.03
	S	0.694	—	0.867	0.700	0.854	1.03	0.710
M²	N	4	2	6	24	21	3	11
	R	14.6–15.1	13.4–14.8	13.4–15.1	13.7–18.3	14.1–16.9	17.1–21.0	13.1–16.8
	X̄	14.93	14.1	14.65	15.66	15.82	18.77	14.33
	S	0.222	—	0.638	1.23	0.951	—	1.15
M³	N	3	4	7	14	22	3	9
	R	13.0–14.0	13.1–15.5	13.0–15.5	13.1–18.3	15.8–18.2	17.4–20.5	12.7–16.7
	X̄	13.47	14.28	13.93	15.09	16.84	18.57	14.7
	S	—	1.20	0.996	1.12	0.710	—	1.37

Conversion of tooth crown dimensions to crown shape indices yields Table 2. These data firmly place the Hadar and Laetoli samples with *Australopithecus*. *Homo* has significantly narrower upper teeth only at the P⁴ and M¹ positions. In the lower dentition, only the *Homo* M_1's are significantly narrower than teeth of any *Australopithecus* species. General statements to the effect that all postcanine teeth are relatively narrow in *Homo habilis* (L. Leakey *et al.*, 1964). are effectively negated by these data and were perhaps based on inadequate samples. *Australopithecus afarensis* has relatively broad postcanine teeth at all positions, but relative to *A. africanus* this species has significantly broader M_1's and M_2's.

Absolute tooth crown dimensions are also informative. Whether lengths or breadths are considered, similar variation is seen in all early hominid taxa—the variation in *A. afarensis* is not excessive.

Sample sizes for anterior teeth are usually inadequate to demonstrate significant differences between taxa. A doubling of sample sizes here is critical

Table 4a. Mandibular Dentition: Mesiodistal Diameter (mm)[a]

		Laetoli	Hadar	*A. afarensis*	*A. africanus*	*A. robustus*	*A. boisei*	*Homo* sp.
I_1	N	1	2	3	4	8	5	5
	R	—	5.6–6.3w	5.6–8.0	5.3vw–6.2	4.8–5.6w	4.2w–5.9	4.5vw–7.1
	$\bar{X}$	8.0	5.95	6.63	5.68	5.25	5.36	6.50
	S	—	—	—	0.411	0.307	0.680	0.474
I_2	N	1	4	5	3	5	4	4
	R	—	6.1–7.2w	5.7w–7.2w	5.5vw–7.3	5.5–6.7w	6.0–6.6	5.8vw–7.6w
	$\bar{X}$	5.7w	6.65	6.46	6.60	6.10	6.3	6.95
	S	—	0.580	0.658	—	0.534	0.258	0.794
$\bar{C}$	N	2	4	6	11	10	5	6
	R	9.3w–11.7	7.5–9.5	7.5–11.7	8.8–11.0	7.0–8.7	7.2–8.5w	7.6–9.1w
	$\bar{X}$	10.5	8.48	9.15	9.48	7.84	7.78	8.52
	S	—	0.896	1.47	0.624	0.595	0.471	0.538
P_3	N	5	13	18	10	19	5	7
	R	10.2–12.6	8.2–9.8	8.2–12.6	8.4–11.2	9.1–11.4	8.9–13.0	8.7–10.6
	$\bar{X}$	10.74	9.17	9.61	9.57	10.21	11.1	9.71
	S	1.06	0.547	1.00	0.741	0.554	1.66	0.713
P_4	N	3	11	14	10	19	9	9
	R	10.3–10.9	7.7–11.1	7.7–11.1	9.5–11.0	10.6–12.6	10.1–15.6	9.0–11.6
	$\bar{X}$	10.70	9.40	9.68	10.24	11.41	13.52	9.81
	S	—	0.934	0.998	0.513	0.564	1.76	0.891
M_1	N	4	14	18	10	25	10	14
	R	13.1–14.0	10.1–14.6	10.1–14.6	12.1–15.1	13.2–16.6	15.4–18.6	12.5–14.6
	$\bar{X}$	13.5	12.73	12.9	13.66	14.69	16.53	13.55
	S	0.392	1.05	0.994	0.938	0.813	0.973	0.752
M_2	N	3	16	19	16	20	8	12
	R	14.4–14.8	12.1–15.4	12.1–15.4	14.3–17.8	14.8–18.2	16.4–20.0	13.5–16.8
	$\bar{X}$	14.67	13.95	14.06	15.53	16.33	18.16	14.67
	S	—	1.10	1.05	0.890	0.982	1.33	0.991
M_3	N	2	9	11	15	22	12	15
	R	13.7–16.3	14.0–15.4	14.0–16.3	13.7–17.2	15.1–20.5	17.6–22.4	12.6–18.0
	$\bar{X}$	15.0	14.53	14.62	15.95	17.15	19.98	14.97
	S	—	0.497	0.756	0.930	1.40	1.71	1.23

[a]vw, very worn; w, worn.

Table 4b.   Mandibular Dentition: Buccolingual Diameter (mm)[a]

		Laetoli	Hadar	A. afarensis	A. africanus	A. robustus	A. boisei	Homo sp.
I_1	N	1	3	4	4	6	4	5
	R	—	7.3–7.6	7.3–7.7	5.8–6.8	5.2–6.5	5.9–7.3	6.2–7.0
	$\bar{X}$	7.7	7.47	7.53	6.28	5.92	6.30	6.64
	S	—	—	0.171	0.427	0.478	0.673	0.321
I_2	N	1	3	4	3	3	4	5
	R	—	6.7w–8.2	6.7w–8.2	7.0–8.0	6.7–7.4	6.4–8.2	5.8–7.4
	$\bar{X}$	7.6	7.47	7.50	7.57	7.0	7.25	6.78
	S	—	—	0.616	—	—	0.885	0.665
$\bar{C}$	N	3	8	11	12	9	2	5
	R	10.1–10.4	8.8–12.4	8.8–12.4	8.8–12.0	7.5–10.5	8.3–9.1	7.6–10.9
	$\bar{X}$	10.27	10.4	10.36	9.84	8.49	8.7	9.32
	S	—	1.47	1.23	0.948	0.956	—	1.23
P_3	N	5	13	18	8	18	5	6
	R	10.0–11.5	9.5–12.6	9.5–12.6	11.0–13.4	9.0–13.7	11.4–13.7	9.0–12.0
	$\bar{X}$	10.74	10.57	10.62	11.76	11.66	13.04	10.68
	S	0.541	1.00	0.887	0.776	1.10	0.934	1.23
P_4	N	3	10	13	10	17	9	7
	R	10.7–11.9	9.8–12.8	9.8–12.8	10.3–12.4	11.5–14.7	12.3–16.5	9.7–11.9
	$\bar{X}$	11.37	10.83	10.95	11.62	12.92	14.28	10.90
	S	—	0.944	0.887	0.713	0.980	1.28	0.737
M_1	N	4	11	15	8	19	6	15
	R	12.6–13.9	11.0–13.5	11.0–13.9	11.4–14.0	11.8–14.7	14.4–17.6	10.4–13.2
	$\bar{X}$	13.18	12.52	12.69	12.9	13.57	15.48	11.91
	S	0.562	0.896	0.856	0.930	0.854	1.17	0.881
M_2	N	3	15	18	16	20	6	13
	R	13.3–14.0	12.1–15.2	12.1–15.2	12.7–16.2	12.8–16.2	15.8–18.6	11.5–14.6
	$\bar{X}$	13.63	13.49	13.51	14.24	14.75	16.92	13.18
	S	—	1.10	1.01	0.909	0.802	1.11	0.981
M_3	N	2	9	11	13	21	9	13
	R	12.6–14.2	12.1–14.9	12.1–14.9	12.5–16.0	12.6–17.0	14.7–19.2	11.4–14.5
	$\bar{X}$	13.4	13.38	13.38	14.22	14.55	16.43	12.92
	S	—	0.961	0.931	0.914	1.17	1.56	1.01

[a]w, worn.

to confirming the hypothesis that *A. africanus* has absolutely and relatively smaller anterior teeth than *A. afarensis* or *H. habilis*. It is clear from the tooth size profiles in Fig. 17 that *A. africanus* occupies an intermediate position in postcanine tooth size between *A. afarensis* and *H. habilis* on the one hand and *A. robustus* and *A. boisei* on the other. Summarizing tooth crown areas would magnify these results but not alter the basic relationship. In several cases, average *A. africanus* postcanine dimensions are closer to *A. robustus* than to *A.*

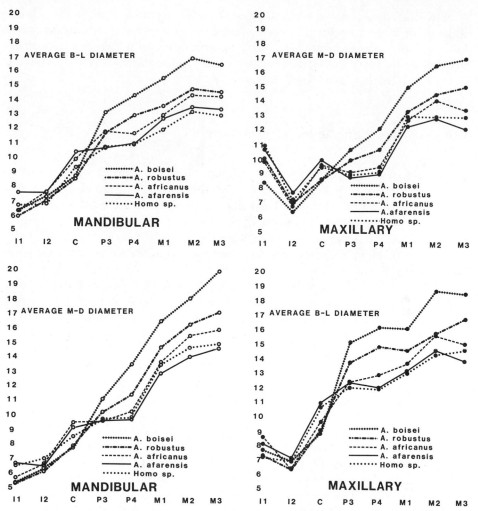

Fig. 17. Average tooth size profiles for early hominid taxa. See text for explanation.

afarensis. The slope of the line segments between buccolingual P³ and P⁴ diameters may reflect molarization in *A. africanus* relative to *A. afarensis* and *H. habilis.* It is evident from Miocene hominoids and *A. afarensis* that a relatively large P³ constitutes the primitive condition (*contra* Sperber, 1973).

Another striking feature of the plots is the large and consistent size difference between East African (*A. boisei*) and South African (*A. robustus*) "robust" *Australopithecus* samples. These differences could reflect real species level differences. Alternately, they could either indicate larger individual body sizes in subtropical East African populations of a single polytypic species or reflect variables of death assemblage composition. Rak's (1981) analysis of the facial skeleton suggests that species level distinction be retained for *A. boisei.*

In sum, dental metric data suggest that the primitive hominid condition was one of large anterior teeth and small but relatively broad premolars and molars. They also suggest that *A. africanus* occupies an intermediate position in postcanine tooth size between these early generalized hominids and the later, specialized *A. robustus* + *A. boisei*.

Comparative Functional Morphology

The foregoing sections provide a guide to the diagnostic anatomy of *A. afarensis*. We have described features of the face, calvaria, mandible, and dentition bearing on the hypothesis that *A. africanus* is the most appropriate candidate for the phyletic ancestry of *A. robustus* + *A. boisei* given the perspective afforded by *A. afarensis*. We will review some functional aspects of the morphological intermediacy seen in *A. africanus* before turning to phylogenetic considerations.

Robinson employed a dietary hypothesis to explain adaptive differentiation in early hominids as early as 1954 (Robinson, 1954*b*). Recent observation, experiment, and theory have generated new data bearing on the problem of progressive functional differentiation in the early hominid masticatory apparatus.

Cranial morphology in *A. robustus* + *A. boisei* is clearly tailored to generate and withstand large amounts of vertical occlusal force. Ward and Molnar (1980) have confirmed, in a series of masticatory simulation experiments, that variation in mandible ramus height and zygomatic process root position (i.e., position of anterior *m. masseter*) substantially alters force distribution and magnitude along the postcanine tooth row. A relatively tall ramus and anteriorly situated zygomatic process maximize the vertical occlusal force and minimize the force gradient. These experimental results conform to observations of early hominid morphology. The sequence *A. afarensis*, *A. africanus*, *A. robustus* + *A. boisei* shows increasing emphasis on anterior temporalis, a forward shift in the zygomatic process root, increasing height and verticality of the mandibular ramus, and progressive expansion and flattening of the postcanine occlusal platform.

Hylander (1979) and White (1977*a*) discuss morphological responses to large masticatory forces in *A. robustus* + *A. boisei* mandibles. Hylander interprets these deep and transversely thick mandibles as adaptations to counter increased sagittal bending about the balancing side mandible corpus and increased torsion about the long axis of the working side corpus. In addition to the resulting dorsoventral shear stress through the symphysis (Hylander, 1979), action of the hypertrophied, medially angulated anterior temporalis and strong medial pterygoid muscles would produce compression through the posterior part of the anterior corpus during mastication. Thus, the prominent double internal buttressing of the *A. robustus* + *A. boisei* symphyseal region can be related, in part, to the geometry of muscular action.

The skull and dentition of *A. afarensis* present a primitive morphological composite that parallels extant and middle Miocene apes. Posteriorly set zygomatic roots, emphasis on posterior temporalis, and low mandibular rami result in steep force gradients along the postcanine dentition relative to other hominids. Thus, *A. afarensis* P_3's retain occlusal topography, and canines continue to project even when molars show extreme wear. The relatively weak superior transverse torus across the posteriorly angled mandibular symphysis and the hollowed corpus contours in *A. afarensis* bespeak the absence of masticatory specialization seen in *A. robustus* + *A. boisei*. The large incisors and canines and the strongly prognathic maxillae and premaxillae supporting their massive roots imply significant utilization of the anterior dental arch in food preparation and other manipulative functions.

From a functional perspective, *A. africanus* crania, mandibles, and teeth foreshadow the *A. robustus* + *A. boisei* character state. Specimens of *A. africanus* exhibit robust zygomatics, relatively expanded and anteriorly situated zygomatic process roots, tall, vertical mandibular rami, and inflated mandible corpus contours. They have vertical and well-buttressed posterior symphyseal regions and enlarged postcanine teeth that wear to flat occlusal platforms. This morphological complex represents an adaptation to generating and withstanding increased amounts of vertical occlusal force, albeit to a less extreme degree than in *A. robustus* + *A. boisei*.

Although *A. afarensis* and *A. africanus* share many common primitive features,* the latter taxon exhibits a morphological composite of the skull and dentition derived toward the *A. robustus* + *A. boisei* character state. This pattern and all its components can be related to a functional trend toward maximizing and spreading vertical occlusal forces along the postcanine tooth row in response to dietary specialization. Members of the *Homo* clade do not show this specialization, but exhibit a pattern indicative of encephalization and masticatory apparatus reduction.

A. africanus Reconsidered

Johanson and White (1979) proposed that the Hadar and Laetoli hominids be subsumed under the name *Australopithecus afarensis*. We further suggested that this taxon was a common ancestor of later hominids and that *A. africanus* was a suitable and exclusive ancestor of *A. robustus* + *A. boisei*.

The attribution of South African *A. africanus* to the "robust" *Australopithecus* clade reflected an attempt to link fossils into a parsimonious interpretive framework. A fossil species must have been ancestral to *A. robustus* + *A. boisei* unless a complex, specialized musculoskeletal adaptation sprung *de novo* from a generalized *Australopithecus* ancestor. Our recognition of "robust"

---

*A simple matching of *these* characters is possible, but will always fail to provide any clues to the phylogenetic positions of these two species.

elements in *A. africanus* was hardly novel—Tobias (1967, 1973*c*), Wallace (1972, 1978), Aguirre (1970), Brace (1973), Wolpoff (1973), and others had reported similar findings, but we enjoyed the perspective afforded by the new, more primitive fossils from East Africa. The recognition that *A. africanus* might be the exclusive structural and phyletic precursor of *A. robustus* + *A. boisei* enhances its importance to evolutionary biologists.

Since 1979, we have undertaken further study of Miocene hominoids from Europe, Asia, and Africa. Our team has made exhaustive analyses of the South and East African fossil hominid collections. Many of our observations and results are presented here. They confirm the conspecific nature and generic attribution of the Laetoli and Hadar remains. Beyond this, our results allow us to assess in greater detail the relative merit of the three alternative evolutionary hypotheses presented in Fig. 18.

The first hypothetical phylogeny (Fig. 18A) would subsume the Laetoli and Hadar hominids in *Australopithecus africanus*. We have described and assessed the primitive morphology diagnostic of the combined middle Pliocene sample from eastern Africa. The morphological differences between these fossils and South African *A. africanus* are equivalent in magnitude to those employed by vertebrate paleontologists in the definition of species. A subspecific designation within *A. africanus* for the Hadar and Laetoli hominids is rejected by these differences. Facial, calvarial, mandibular, and dental differences between *A. africanus* and *A. afarensis* (see above) are of a greater magnitude than the minor, typically unfossilizable variations in pelage, distribution, or behavior that systematists employ in the recognition of most vertebrate subspecies. Moreover, these differences can be functionally interpreted and indicate the process of adaptive differentiation in early hominids. Mayr (1969) has warned against attributing evolutionary significance to

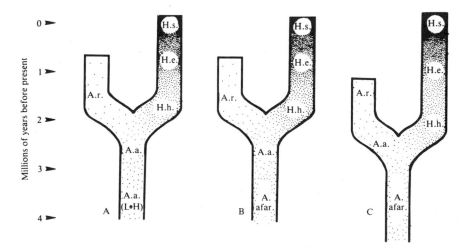

Fig. 18. Alternative phylogenetic hypotheses of Pliocene–Pleistocene hominid evolution. The authors favor hypothesis C, for reasons discussed in the text.

the subspecies category and has stressed, instead, its value merely as a classificatory device. The known hominid most similar to *A. africanus* in time and morphology is indeed *A. afarensis*. To combine these two fossil collections under the same species name would, however, obscure the real and evolutionarily significant differences between them. For these reasons we reject the taxonomic scheme proposed in Fig. 18A.

The second alternative (Fig. 18B) recognizes the specific distinctiveness of *A. afarensis* and retains *A. africanus* as common ancestor for both major hominid clades. This scheme is plausible, but we consider it to be less parsimonious than the third alternative (Fig. 18C). The latter phylogeny, essentially that presented by Johanson and White (1979), interprets *A. africanus* as exhibiting shared, derived characters indicative of a special phyletic relationship with later "robust" *Australopithecus*. For South African *A. africanus* to be considered a common ancestor for both later hominid clades it is necessary to postulate an evolutionary reversal involving many functionally related characters of an established morphological and adaptive complex. The derived characters of the mandible, face, and dentition in *A. africanus* would necessarily have been evolved from the *A. afarensis* condition and then been lost again in a relatively late transition to *Homo habilis*. It is possible that this was the case. It seems to us, however, more parsimonious to use these characters to phyletically link *A. africanus* exclusively with *A. robustus* + *A. boisei*. The link would persist even if *A. afarensis* and *A. africanus* were ultimately shown to overlap in time. It is morphology, not time, that is the key variable in our phylogenetic assessment. In its known morphology *and* chronological placement, however, *A. africanus* provides the most suitable phyletic ancestor for *A. robustus* + *A. boisei*. Indeed, lacking this intermediate taxon, paleontologists would have to postulate an ancestor for *A. robustus* + *A. boisei* very similar to *A. africanus*.

What was the nature of the cladogenesis that ultimately gave rise to *A. robustus* + *A. boisei*? Was it allopatric speciation, involving peripheral South African populations of *A. afarensis*? Was *A. africanus* a purely local, transitory, South African phenomenon or did it range more widely in time and space? When did the earliest, still unknown members of the *Homo* clade arise and how did they differ from *A. africanus* and *A. afarensis*? Our hypothesis predicts that structural and temporal intermediates between *A. afarensis* and *H. habilis* will not exhibit the *A. africanus* morphological pattern. If reliable evidence to the contrary is discovered, we may find the phylogenetic alternative in Fig. 18B more acceptable.

These predictions and questions involve evidence in the time period between about 2 and 3 m.y.a. The only adequate evidence comes from the geographically and temporally restricted samples of South African *A. africanus*. Hominids were almost certainly more widespread in Africa during the terminal Pliocene, but only one hominid-bearing deposit, the Omo Shungura Formation, is known for this critical time span in East Africa. The record from this site consists of very fragmentary hominid remains. Other sites, such as Olduvai Gorge, Koobi Fora, and Ileret, have not yielded fossils of equivalent antiquity. The absence of this crucial evidence hinders our ability to test our predictions and formuate answers to our questions.

Relatively more abundant hominid fossils are known from other time periods. Pleistocene taxa indicated in Fig. 18 are fairly well-established. In the absence of clear-cut evidence to the contrary, we have directly linked successive taxa in each of the two major hominid clades. Gould (1979) has commented that such linking implies more gradualism for later human evolution than he is inclined to accept. It is entirely possible that early hominid evolution was rectangular (Stanley, 1975) in nature—populations of *H. habilis* may have coexisted alongside the daughter species *H. erectus* for hundreds of thousands of years, even though we now lack any convincing evidence for this. It is probable that our preferred phylogeny has too much trunk and not enough branches—for example, evidence for a specific distinction between contemporary *A. robustus* and *A. boisei* is mounting. Unfortunately, the fossil record of early hominid evolution is not yet dense enough to test these hypotheses. We have only a scatter of data points across the vast void of the past 4 million years. Paleontologists are presently in a position to characterize hominid evolution by sketching lines between the data points, but the points are not yet numerous enough to decide whether the evolution was characterized by gradualism or a series of punctuated equilibria. The observation of morphological intermediacy of *A. africanus* is a case in point—such intermediacy does not imply a gradualistic mode of evolution between *A. afarensis* and *A. robustus* + *A. boisei*. We are in no position to determine the tempo or mode of that evolution without additional fossil data. The resolution of these and similar problems of evolutionary biology among early hominids requires greatly intensified fieldwork in Africa.

Conclusion

In 1968, five years before the first hominid discoveries at Hadar, Phillip Tobias wrote, ". . . our picture of the ancestral australopithecine is virtually indistinguishable from that of *A. africanus*." He went on to caution, "Of course, if we had an adequate Pliocene fossil record, we should not need to extrapolate from the Pleistocene hominids to their presumed ancestor" (Tobias, 1968, p. 307). We now possess an adequate fossil record for this once presumed Pliocene ancestor. It is a species distinguishable from, indeed, far more primitive than, *A. africanus*.

Australopithecus afarensis provides a crucial clue for elucidating hominid cladogenesis in the Pliocene. Until this species was identified, the most parsimonious interpretation of early hominid phylogeny was one where *A. africanus* represented a "perfect ancestor" (Tattersall and Eldredge, 1977, p. 208), and stood below the major fork in the hominid evolutionary tree. We now suggest that *A. afarensis* should replace *A. africanus* in this role. By virtue of its primitive morphology and its antiquity, *A. afarensis* appears to be the most suitable known ancestor for both later *Homo* and *Australopithecus*. As a hypothesis, our model is predictive and testable. It identifies specific regions and times where paleontologists might profitably focus their attention in the hope of obtaining crucial fossils. These fossils will judge the predictive strengths and weaknesses of our hypothesis.

ACKNOWLEDGMENTS

Our research was supported by The National Science Foundation, The National Geographic Society, The L. S. B. Leakey Foundation, and The Foundation for Research into the Origin of Man. We thank the following for access to collections and assistance in our study: L. de Bonis, A. R. Hughes, J. Kitching, M. D. Leakey, R. E. Leakey, B. Maguire, J. Maguire, A. A. Mturi, and M. Raath. In addition, the following people contributed to discussions helpful in the formulation of this paper: C. K. Brain, E. Delson, F. Grine, B. Hendey, T. C. Partridge, Y. Rak, E. Vrba, and S. Ward. Finally, we extend our gratitude to our friend and colleague, Prof. Phillip V. Tobias, who welcomed us into his laboratory and arranged for the presentation (by DCJ, and TDW) of our research results at a meeting of the Royal Society of South Africa in September 1980. Our manuscript was inspired by that invitation.

Luba Dmytryk-Gudz drew Figs. 8–12, 14, and 16. Steve Misencik drew Fig. 1. We thank John Aicher, Anson Laufer, Cindy Luchetti, and Bruce Frumker for photographic assistance.

NOTE ADDED

After this paper was submitted, Prof. Tobias kindly sent us reprints of his new article, "*Australopithecus afarensis*" and *A. africanus:* Critique and alternative hypothesis, [*Paleontol. Afr.* **23**:1–17 (1980)]. Tobias considers our phylogenetic findings part of what he dubs "the assault on *Australopithecus africanus*" (p. 3). He makes three major critical points. First, Tobias doubts whether our diagnosis of *A. afarensis* distinguishes this taxon from *A. africanus*. Second, he questions the validity of pooling the Hadar and Laetoli hominid samples. Finally, he considers the Johanson and White phylogenetic hypothesis to be based on a "misinterpretation" of South African evidence. Tobias has expressed these doubts in other publications (Tobias, 1978*b*, 1979, 1980). We discuss these and other criticisms in detail above. Tobias' alternative hypothesis is that the Hadar and Laetoli hominids represent subspecies of *A. africanus*, the common ancestor of all later hominids. He suggests that only a few dental metric traits separate the Laetoli sample from *A. africanus*. Moreover, on the basis of "originals and casts of *some* of the fossils" (p. 14; our emphasis) he *provisionally* proposes subspecific distinction for the Hadar hominids.

Even though Prof. Tobias has not studied the full collections of fossils attributed to *A. afarensis,* this most recent analysis is his most comprehensive treatment of the Pliocene hominids. We hope that Tobias' new article will be read carefully and examined critically in the face of the *entire* body of relevant evidence. The relative merits of our hypothesis should then be apparent.

References

Aguirre, E. 1970. Identification de "Paranthropus" en Makapansgat, in: *Crónica del XI Congreso Nacional de Arquelogia, Merida 1969*, pp. 98–124.

Anon. 1979a. Difficulties in the definition of new hominid species. *Nature (Lond.)* **278**:400–401.

Anon. 1979b. Problems in hominid taxonomy. *Nature (Lond.)* **281**:258.

Anon. 1979c. The Leakey footprints: An uncertain path. *Sci. News* **115**:196–197.

Aronson, J. L., Schmitt, T. J., Walter, R. C., Taieb, M., Tiercelin, J. J., Johanson, D. C., Naeser, C. W., and Nairn, A. E. M. 1977. New geochronologic and paleomagnetic data for the hominid-bearing Hadar Formation of Ethiopia. *Nature (Lond.)* **267**:323–327.

Bishop, W. W. 1978. Geochronological framework for African Plio-Pleistocene hominids: As Cerberus sees it, in: *Early Hominids of Africa* (C. J. Jolly, ed.), pp. 255–265, Duckworth, London.

Boaz, N. T. 1979. Hominid evolution in eastern Africa during the Pliocene and early Pleistocene. *Annu. Rev. Anthropol.* **8**:71–85.

Boaz, N. T., and Howell, F. C. 1977. A gracile hominid cranium from upper Member G of the Shungura Formation, Ethiopia. *Am. J. Phys. Anthropol.* **46**:93–108.

Brace, C. L. 1967. *The Stages of Human Evolution*, Prentice Hall, Englewood Cliffs, New Jersey. 116 pp.

Brace, C. L. 1973. Sexual dimorphism in human evolution. *Yearb. Phys. Anthropol.* **16**:31–49.

Brace, C. L., Mahler, P. E., and Rosen, R. B. 1973. Tooth measurements and the rejection of the taxon "Homo habilis." *Yearb. Phys. Anthropol.* **16**:50–68.

Brain, C. K. 1967. Procedures and some results in the study of Quaternary cave fillings, in: *Background to Evolution in Africa* (W. W. Bishop and J. D. Clark, eds.), pp. 285–301, University of Chicago Press, Chicago.

Brain, C. K. 1969. New evidence for climatic change during middle and late stone age times in Rhodesia. *S. Afr. Archeol. Bull.* **24**:127–143.

Brain, C. K. 1970. New finds at the Swartkrans australopithecine site. *Nature (Lond.)* **225**:1112–1119.

Brain, C. K. 1976a. A re-interpretation of the Swartkrans site and its remains. *S. Afr. J. Sci.* **72**:141–146.

Brain, C. K. 1976b. Some principles in the interpretation of bone accumulations associated with man, in: *Human Origins: Louis Leakey and the East African Evidence* (G. L. Isaac and E. R. McCown, eds.), pp. 97–116, Benjamin, Menlo Park, California.

Brain, C. K., Vrba, E. S., and Robinson, J. T. 1974. A new hominid innominate bone from Swartkrans. *Ann. Transvaal Mus.* **29**:55–63.

Brock, A., McFadden, P. L., and Partridge, T. C. 1977. Preliminary paleomagnetic results from Makapansgat and Swartkrans. *Nature (Lond.)* **266**:249–250.

Broom, R. 1929. Note on the milk dentition of *Australopithecus. Proc. Zool. Soc. Lond.* **1928**:85–88.

Broom, R. 1930. The age of *Australopithecus. Nature (Lond.)* **125**:814.

Broom, R. 1939. The dentition of the Transvaal Pleistocene anthropoids, *Plesianthropus* and *Paranthropus. Ann. Transvaal Mus.* **19**:302–314.

Broom, R. 1941. Mandible of a young *Paranthropus* child. *Nature (Lond.)* **147**:607–608.

Broom, R. 1950. The genera and species of the South African fossil ape-men. *Am. J. Phys. Anthropol. N.S.* **8**:1–13.

Broom, R., and Robinson, J. T. 1949. A new type of fossil man. *Nature (Lond.)* **164**:322–323.

Broom, R., and Robinson, J. T. 1950. Man contemporaneous with the Swartkrans ape-man. *Am. J. Phys. Anthropol. N.S.* **8**:151–155.

Broom, R., and Robinson, J. T. 1952. Swartkrans ape-man, *Paranthropus crassidens. Transvaal Mus. Mem.* **6**:1–123.

Broom, R., and Schepers, G. W. H. 1946. The South African fossil ape-men, the Australopithecinae. *Transvaal Mus. Mem.* **2**:1–272.

Broom, R., Robinson, J. T., and Schepers, G. W. H. 1950. Sterkfontein ape-man, *Plesianthropus. Transvaal Mus. Mem.* **4**:1–117.

Brown, F. H., Howell, F. C., and Eck, G. G. 1978. Observations on problems of correlation of late Cenozoic hominid-bearing formations in the North Lake Turkana basin, in: *Geological Background to Fossil Man* (W. W. Bishop, ed.), pp. 473–498, Scottish Academic Press, Edinburgh.

Butzer, K. W. 1974. Paleoecology of South African australopithecines: Taung revisited. *Curr. Anthropol.* **15**:367–382.

Butzer, K. W., 1976. Lithostratigraphy of the Swartkrans Formation. *S. Afr. J. Sci.* **72**:136–141.

Campbell, B. G., 1978. Some problems in hominid classification and nomenclature, in: *Early Hominids of Africa* (C. Jolly, ed.), pp. 565–581, Duckworth, London.

Cerling, T. E., Brown, F. H., Cerling, B. W., Curtis, G. H., and Drake, R. E. 1979. Preliminary correlations between Koobi Fora and Shungura Formations, East Africa. *Nature (Lond.)* **279**:118–121.

Clarke, R. J. 1977*a. The Cranium of the Swartkrans Hominid SK 847 and its Relevance to Human Origins,* Ph.D. Dissertation, University of the Witwatersrand, Johannesburg. 325 pp.

Clarke, R. J. 1977*b.* A juvenile cranium and some adult teeth of early *Homo* from Swartkrans. *S. Afr. J. Sci.* **73**:46–49.

Collings, G. E., Cruikshank, A. R. I., Maguire, J. M., and Randall, R. M. 1975. Recent faunal studies at Makapansgat Limeworks, Transvaal, South Africa. *Ann. S. Afr. Mus.* **71**:153–165.

Cooke, H. B. S. 1967. The Pleistocene sequence in South Africa and problems of correlation, in: *Background to Evolution in Africa* (W. W. Bishop and J. D. Clark, eds.), pp. 175–184, University of Chicago Press, Chicago.

Cooke, H. B. S. 1978*a.* Faunal evidence for the biotic setting of early African hominids, in: *Early Hominids of Africa* (C. J. Jolly, ed.), pp. 267–281, Duckworth, London.

Cooke, H. B. S. 1978*b.* Pliocene–Pleistocene Suidae from Hadar, Ethiopia. *Kirtlandia* **29**:1–63.

Cooke, H. B. S., and Maglio, V. J. 1972. Plio-Pleistocene stratigraphy in East Africa in relation to proboscidean and suid evolution, in: *Calibration of Hominoid Evolution* (W. W. Bishop and J. A. Miller, eds.), pp. 303–329, Scottish Academic Press, Edinburgh.

Curtis, G. H., Drake, R. E., Cerling, T. E. and Hampel, J. 1975. Age of KBS Tuff in Koobi Fora Formation, East Rudolf, Kenya. *Nature (Lond.)* **258**:395–398.

Dart, R. A. 1925. *Australopithecus africanus:* The man-ape of South Africa. *Nature (Lond.)* **115**:195–199.

Dart, R. A. 1934. The dentition of *Australopithecus africanus. Folia Anat. Jpn.* **12**:207–221.

Day, M. H., and Wickens, E. H. (1980). Laetoli Pliocene hominid footprints and bipedalism. *Nature (Lond.)* **286**:385–387.

Day, M. H., Leakey, R. E., Walker, A. C., and Wood, B. A. 1976. New hominids from East Turkana, Kenya. *Am. J. Phys. Anthropol.* **45**:369–436.

Day, M. H., Leakey, M. D., and Olson, T. R. 1980. On the status of *Australopithecus afarensis. Science* **207**:1102–1103.

De Bonis, L., and Melentis, J. 1977*a.* Les primates hominoides du Vallésien de Macédoine (Grèce). Étude de la machoire inférieure. *Géobios* **10**:849–885.

De Bonis, L., and Melentis, J. 1977*b.* Un nouveau genere de Primate hominide dans le Vallésien (Miocène supérieur) de Macédoine. *C. R. Acad. Sci. Paris D* **284**:1393–1396.

De Bonis, L., Johanson, D. C., Melentis, J., and White, T. D., 1981. Variations métriques dans la denture chez les Hominidés primitifs: Comparaison entre *Australopithecus afarensis* et *Ouranopithecus macedoniensis. C. R. Acad. Sci. Paris D* **292**:373–376.

De Graff, G. 1960. A preliminary investigation of the mammalian microfauna in Pleistocene deposits of caves in the Transvaal system. *Paleontol. Afr.* **7**:59–118.

De Swardt, A. M. J. 1974. Geomorphological dating of cave openings in South Africa. *Nature (Lond.)* **250:**683.

Drake, R. E., Curtis, G. H., Cerling, T. E., Cerling, B. W., and Hampel, J. 1980. KBS Tuff dating and geochronology of tuffaceous sediments in the Koobi Fora and Shungura Formations, East Africa. *Nature (Lond.)* **283:**368–372.

Ewer, R. F. 1957. Faunal evidence on the dating of the Australopithecinae, in: *Proceedings of the Third Pan-African Congress of Prehistory, Livingston, 1955* (J. D. Clark and S. Cole, eds.) pp. 135–142, Chatto and Windus, London.

Ewer, R. F., and Cooke, H. B. S. 1964. The Pleistocene mammals of southern Africa, in: *Ecological Studies in Southern Africa* (D. H. S. Davis, ed.), pp. 35–48, W. Junk, The Hague.

Fitch, F. J., and Miller, J. A. 1970. Radioisotopic age determinations of Lake Rudolf artefact site. *Nature (Lond.)* **226:**226–228.

Gleadow, A. J. W. 1980. Fission track age of the KBS Tuff and associated hominid remains in northern Kenya. *Nature (Lond.)* **284:**225–230.

Gould, S. J., 1979. Our greatest evolutionary step. *Nat. Hist.* **88**(6):40–44.

Greenfield, L. O. 1980. A late divergence hypothesis. *Am. J. Phys. Anthropol.* **52:**351–365.

Gregory, W. K., and Hellman, M. 1939. The dentition of the extinct South African man-ape *Australopithecus (Plesianthropus) transvaalensis* Broom. A comparative and phylogenetic study. *Ann. Transvaal Mus.* **19:**339–373.

Harris, J. M., and White, T. D. 1979. Evolution of the Plio-Pleistocene African Suidae. *Trans. Am. Phil. Soc.* **69:**1–128.

Hendey, Q. B. 1974. The late Cenozoic Carnivora of the southwestern Cape Province. *Ann. S. Afr. Mus.* **63:**1–369.

Hillhouse, J. W., Ndombi, J. W. M., Cox, A., and Brock, A. 1977. Additional results on paleomagnetic stratigraphy of the Koobi Fora Formation, east of Lake Turkana (Lake Rudolf), Kenya. *Nature (Lond.)* **265:**411–415.

Howell, F. C. 1978*a*. Hominidae, in: *Evolution of African Mammals* (V. J. Maglio and H. B. S. Cooke, eds.), pp. 154–248, Harvard University Press, Cambridge, Massachusetts.

Howell, F. C. 1978*b*. Overview of the Pliocene and earlier Pleistocene of the lower Omo basin, southern Ethiopia, in: *Early Hominids of Africa* (C. J. Jolly, ed.), pp. 85–130, Duckworth, London.

Howell, F. C., and Coppens, Y. 1976. An overview of Hominidae from the Omo succession, Ethiopia, in: *Earliest Man and Environments in the Lake Rudolf Basin* (Y. Coppens, F. C. Howell, G. L. Isaac, and R. E. F. Leakey, eds.), pp. 522–532, University of Chicago Press, Chicago.

Hughes, A. R., and Tobias, P. V. 1977. A fossil skull probably of the genus *Homo* from Sterkfontein, Transvaal. *Nature (Lond.)* **265:**310–312.

Hylander, W. 1979. The functional significance of primate mandibular form. *J. Morphol.* **160:**223–240.

Johanson, D. C., and Taieb, M. 1976. Plio-Pleistocene hominid discoveries in Hadar, Ethiopia. *Nature (Lond.)* **260:**293–297.

Johanson, D. C., and White, T. D. 1979. A systematic assessment of early African hominids. *Science* **203:**321–330.

Johanson, D. C., and White, T. D. 1980. On the status of *Australopithecus afarensis*. *Science* **207:**1104–1105.

Johanson, D. C., White, T. D., and Coppens, Y. 1978. A new species of the genus *Australopithecus* (Primates: Hominidae) from the Pliocene of eastern Africa. *Kirtlandia* **28:**1–14.

Kaufulu, Z., Vrba, E. S., and White, T. D. 1980. Age of the Chiwondo Beds, northern Malawi. *Ann. Transvaal Mus.* **33:**1–8.

Keith, A. 1925. The fossil anthropoid ape from Taungs. *Nature (Lond.)* **115:**234–235.

Keith, A. 1931. *New Discoveries Relating to the Antiquity of Man*, Williams and Northgate, London.

Kennedy, G. E. 1980. *Paleoanthropology*, McGraw-Hill, New York. 479 pp.

Kohl-Larsen, L. 1943. *Auf den Spuren des Vormenschen*, Strecker and Schroder Verlag, Stuttgart.

Leakey, L. S. B. 1959. A new fossil skull from Olduvai. *Nature (Lond.)* **184:**491–493.

Leakey, L. S. B. 1960*a*. Recent discoveries at Olduvai Gorge. *Nature (Lond.)* **188:**1050–1052.

Leakey, L. S. B. 1960*b*. Finding the world's earliest man. *Nat. Geogr. Mag.* **118:**420–435.

Leakey, L. S. B. 1961. New finds at Olduvai Gorge. *Nature (Lond.)* **189**:649–650.

Leakey, L. S. B. 1966. *Homo habilis, Homo erectus* and the australopithecines. *Nature (Lond.)* **209**:1279–1281.

Leakey, L. S. B., and Leakey, M. D. 1964. Recent discoveries of fossil hominids in Tanganyika: At Olduvai and near Lake Natron. *Nature (Lond.)* **202**:5–7.

Leakey, L. S. B., Tobias, P. V., and Napier, J. 1964. A new species of the genus *Homo* from Olduvai Gorge. *Nature (Lond.)* **202**:7–9.

Leakey, M. D. 1979. Footprints in the ashes of time. *Nat. Geogr. Mag.* **155**:446–457.

Leakey, M. D., and Hay, R. L. 1979. Pliocene footprints in the Laetolil Beds at Laetoli, northern Tanzania. *Nature (Lond.)* **278**:317–323.

Leakey, M. G., and Leakey, R. E. F., eds. 1978. *Koobi Fora Research Project, Volume 1: The Fossil Hominids and an Introduction to their Context, 1968–1974,* Clarendon Press, Oxford.

Leakey, M. D., Hay, R. L., Curtis, G. H., Drake, R. E., Jackes, M. K., and White, T. D. 1976. Fossil hominids from the Laetolil Beds, Tanzania. *Nature (Lond.)* **262**:460–466.

Leakey, R. E. F. 1970. In search of man's past at Lake Rudolf. *Nat. Geogr. Mag.* **137**:712–732.

Leakey, R. E. F. 1972. Further evidence of Lower Pleistocene hominids from East Rudolf, north Kenya. *Nature (Lond.)* **237**:264–269.

Leakey, R. E. F. 1973*a*. Skull 1470. *Nat. Geogr. Mag.* **143**:818–829.

Leakey, R. E. F. 1973*b*. Evidence for an advanced Plio-Pleistocene hominid from East Rudolf, Kenya. *Nature (Lond.)* **242**:447–450.

Leakey, R. E. F. 1973*c*. Australopithecines and hominines: A summary on the evidence from the early Pleistocene of eastern Africa. *Symp. Zool. Soc. Lond.* **33**:53–69.

Leakey, R. E. F. 1973*d*. Further evidence of lower Pleistocene hominids from East Rudolf, north Kenya, 1972. *Nature (Lond.)* **242**:170–173.

Leakey, R. E. F. 1974. Further evidence of lower Pleistocene hominids from East Rudolf, north Kenya, 1973. *Nature (Lond.)* **248**:653–656.

Leakey, R. E. F. 1976*a*. New hominid fossils from the Koobi Fora Formation in northern Kenya. *Nature (Lond.)* **261**:574–576.

Leakey, R. E. F. 1976*b*. Hominids in Africa. *Am. Sci.* **64**:174–178.

Leakey, R. E., and Lewin, R. 1977. *Origins.* Dutton, New York.

Leakey, R. E. F., and Walker, A. C. 1980. On the status of *Australopithecus afarensis. Science* **207**:1103.

Le Gros Clark, W. E. 1947. Observations on the anatomy of the fossil Australopithecinae. *J. Anat.* **81**:300–333.

Le Gros Clark, W. E. 1950. Hominid characters of the australopithecine dentition. *J. Roy. Anthropol. Inst.* **80**:37–54.

Le Gros Clark, W. E. 1955. *The Fossil Evidence for Human Evolution,* University of Chicago Press, Chicago.

Le Gros Clark, W. E. 1967. *Man-Apes or Ape-Men?,* Holt, Rinehart, Winston, New York.

Lovejoy, C. O. 1974. The gait of australopithecines. *Yearb. Phys. Anthropol.* **17**:147–161.

Lovejoy, C. O. 1981. The origin of man. *Science* **211**:341–350.

MacDougall, D., and Price, P. B. 1974. Attempt to date early South African hominids by using fission tracks in calcite. *Science* **185**:943–944.

Mahler, P. E. 1973. *Metric Variation in the Pongid Dentition,* University Microfilms, Ann Arbor.

Mankinen, E. A., and Dalrymple, G. B. 1979. Revised geomagnetic polarity timescale for the interval 0–5 mybp. *J. Geophys. Res.* **84**:615–626.

Mayr, E. 1969. *Principles of Systematic Zoology,* McGraw-Hill, New York. 428 pp.

McDougall, I., Maier, R., Sutherland-Hawkes, P., and Gleadow, A. J. W. 1980. K–Ar estimate for the KBS Tuff, East Turkana, Kenya. *Nature (Lond.)* **284**:230–234.

McFadden, P. L., Brock, A., and Partridge, T. C. 1979. Paleomagnetism and the age of the Makapansgat hominid site. *Earth Planet. Sci. Lett.* **44**:373–382.

Olson, T. R. 1981. Basicranial morphology of the extant hominoids and Pliocene hominids: The new material from the Hadar Formation, Ethiopia and its significance in early human evolution and taxonomy, in: *Aspects of Human Evolution* (C. B. Strınger, ed.), pp. 99–128, Taylor and Francis, London.

Oxnard, C. E. 1975. The place of the australopithecines in human evolution: Grounds for doubt? *Nature (Lond.)* **258**:389–395.

Oxnard, C. E. 1979. Relationship of *Australopithecus* and *Homo:* Another view. *J. Hum. Evol.* **8**:427–432.

Pilbeam, D. R., Meyer, G. E., Badgley, C. Rose, M. D., Pickford, M. H. L., Behrensmeyer, A. K., and Shah, S. M. I. 1977. New hominoid primates from the Siwaliks of Pakistan and their bearing on hominoid evolution. *Nature (Lond.)* **270**:689–694.

Partridge, T. C. 1973. Geomorphological dating of cave opening at Makapansgat, Sterkfontein, Swartkrans and Taung. *Nature (Lond.)* **246**:75–79.

Partridge, T. C. 1978. Re-appraisal of lithostratigraphy of Sterkfontein hominid site. *Nature (Lond.)* **275**:282–287.

Partridge, T. C. 1979. Re-appraisal of lithostratigraphy of Makapansgat Limeworks hominid site. *Nature (Lond.)* **279**:484–488.

Rak, Y. 1978. The functional significance of the squamosal suture in *Australopithecus boisei. Am. J. Phys. Anthropol.* **49**:71–78.

Rak, Y. 1981. *The Morphology and Architecture of the Australopithecine Face.* Ph.D. Dissertation, University of California, Berkeley. 349 pp.

Remane, A. 1954. Structure and relationships of *Meganthropus africanus. Am. J. Phys. Anthropol. N.S.* **12**:123–126.

Robinson, J. T. 1953. *Meganthropus,* australopithecines and hominids. *Am. J. Phys. Anthropol. N.S.* **11**:1–38.

Robinson, J. T. 1954a. The genera and species of the Australopithecinae. *Am. J. Phys. Anthropol. N.S.* **12**:181–200.

Robinson, J. T. 1954b. Prehominid dentition and hominid evolution. *Evolution* **8**:324–334.

Robinson, J. T. 1955. Further remarks on the relationship between "Meganthropus" and australopithecines. *Am. J. Phys. Anthropol. N.S.* **13**:429–445.

Robinson, J. T. 1956. The dentition of the Australopithecinae. *Transvaal Mus. Mem.* **9**:1–179.

Robinson, J. T. 1961. The australopithecines and their bearing on the origin of man and stone tool-making. *S. Afr. J. Sci.* **57**:3–13.

Robinson, J. T. 1963. Adaptive radiation in the australopithecines and the origin of man, in: *African Ecology and Human Evolution* (F. C. Howell and F. Bourliére, eds.), pp. 385–416., Aldine, Chicago.

Robinson, J. T. 1965. *Homo "habilis"* and the australopithecines. *Nature (Lond.)* **205**:121–124.

Robinson, J. T. 1966. The distinctiveness of *Homo habilis. Nature (Lond.)* **209**:957–960.

Robinson, J. T. 1967. Variation and the taxonomy of the early hominids, in: *Evolutionary Biology,* Volume 1 (T. Dobzhansky, M. K. Hecht, and W. Steere, eds.), pp. 69–99, Appleton-Century-Crofts, New York.

Robinson, J. T. 1972. *Early Hominid Posture and Locomotion,* University of Chicago Press, Chicago.

Schuman, E. L., and Brace, C. L. 1954. Metric and morphologic variations in the dentition of the Liberian chimpanzee: Comparisons with anthropoid and human dentitions. *Hum. Biol.* **26**:239–268.

Şenyurek, M. 1955. A note on the teeth of *Meganthropus africanus* Weinert from Tanganyika Territory. *Belleten (Ankara)* **19**:1–54.

Shuey, R. T., Brown, F. H., Eck, G. G., and Howell, F. C. 1978. A statistical approach to temporal biostratigraphy, in: *Geological Background to Fossil Man* (W. W. Bishop, ed.), pp. 103–124, Scottish Academic Press, Edinburgh.

Smith, G. E. 1925. The fossil anthropoid ape from Taungs. *Nature (Lond.)* **115**:235.

Smith-Woodward, A. 1925. The fossil anthropoid ape from Taungs. *Nature (Lond.)* **115**:235–236.

Sperber, G. H. 1973. *The Morphology of the Cheek Teeth of Early South African Hominids,* University Microfilms, Ann Arbor.

Stanley, S. M. 1975. A theory of evolution above the species level. *Proc. Natl. Acad. Sci. USA* **72**:646–650.

Swedlund, A. C. 1974. The use of ecological hypotheses in australopithecine taxonomy. *Am. Anthropol.* **76**:515–529.

Tattersall, I., and Eldredge, N. 1977. Fact, theory, and fantasy in human paleontology. *Am. Sci.* **65**:204–211.

Tobias, P. V. 1965. Early man in East Africa. *Science* **199**:22–33.

Tobias, P. V. 1967. *Olduvai Gorge*, Volume 2. *The Cranium and Maxillary Dentition of Australopithecus (Zinjanthropus) boisei*, Cambridge University Press, London. 264 pp.

Tobias, P. V. 1968. The taxonomy and phylogeny of the australopithecines, in: *Taxonomy and Phylogeny of Old World Primates with References to the Origin of Man* (B. Chiarelli, ed.), pp. 277–315, Rosenberg and Sellier, Torino.

Tobias, P. V. 1973a. A new chapter in the history of the Sterkfontein early hominid site. *J. S. Afr. Biol. Soc.* **14**:30–44.

Tobias, P. V. 1973b. Implications of the new age estimates of the early South African hominids. *Nature (Lond.)* **246**:79–83.

Tobias, P. V. 1973c. New developments in hominid paleontology in South and East Africa. *Annu. Rev. Anthropol.* **2**:311–334.

Tobias, P. V. 1978a. South African australopithecines in time and hominid phylogeny, with special reference to dating and affinities of the Taung skull, in: *Early Hominids of Africa* (C. J. Jolly, ed.), pp. 45–84, Duckworth, London.

Tobias, P. V. 1978b. Position et rôle des australopithécinés dans la phylogenèse humaine, avec étude particulière de *Homo habilis* et des théories controversées avancées à propos des premiers hominidés fossiles de Hadar et de Laetolil, in: *Les Origines Humaines et les Époques de l'Intelligence*, pp. 38–75, Foundation Singer-Polignac/Masson, Paris.

Tobias, P. V. 1978c. The place of *Australopithecus africanus* in hominid evolution, in: *Recent Advances in Primatology*, Volume 3, *Evolution* (D. J. Chivers and K. A. Joysey, eds.), pp. 373–394, Academic, London.

Tobias, P. V. 1979. The Silberberg Grotto, Sterkfontein, Transvaal, and its importance in paleoanthropological researches. *S. Afr. J. Sci.* **75**:161–164.

Tobias, P. V. 1980. A survey and synthesis of the African hominids of the late Tertiary and early Quaternary periods, in: *Current Argument on Early Man* (L. K. Königsson, ed.), pp. 86–113, Pergamon, Oxford.

Tobias, P. V., and Hughes, A. R. 1969. The new Witwatersrand University excavation at Sterkfontein. *S. Afr. Archeol. Bull.* **24**:158–169.

Vrba, E. S. 1975. Some evidence of chronology and palaeoecology of Sterkfontein, Swartkrans and Kromdraai from the fossil Bovidae. *Nature (Lond.)* **254**:301–304.

Vrba, E. S. 1976. The fossil Bovidae of Sterkfontein, Swartkrans and Kromdraai. *Transvaal Mus. Mem.* **21**:1–166.

Vrba, E. S. 1977. New species of *Parmularius* Hopwood and *Damaliscus* Sclater and Thomas (Alcelaphini, Bovidae, Mammalia) from Makapansgat and comments on faunal chronological correlation. *Paleontol. Afr.* **20**:137–151.

Vrba, E. S. 1979. The significance of bovid remains as indicators of environment and predation patterns, in: *Fossils in The Making* (A. K. Behrensmeyer and A. P. Hill, eds.), pp. 247–271, University of Chicago Press, Chicago.

Walker, A. C. 1976. Remains attributable to *Australopithecus* in the East Rudolf succession, in: *Earliest Man and Environments in the Lake Rudolf Basin* (Y. Coppens, F. C. Howell, G. L. Isaac, and R. E F. Leakey, eds.), pp. 490–506, University of Chicago Press, Chicago

Walker, A. C., and Leakey, R. E. F. 1978. The hominids of East Turkana. *Sci. Am.* **239**(2): 54–66.

Wallace, J. A. 1972. *The Dentition of the South African Early Hominids: A Study of Form and Function*, Ph.D. Dissertation, University of the Witwatersrand, Johannesburg.

Wallace, J. 1978. Evolutionary trends in the early hominid dentition, in: *Early Hominids of Africa* (C. J. Jolly, ed.), pp. 285–310, Duckworth, London.

Walter, R. C. 1980. *The Volcanic History of the Hadar Early Man Site and the Surrounding Afar Region of Ethiopia*, Ph.D. Dissertation, Case Western Reserve University, Cleveland, Ohio.

Walter, R. C., and Aronson, J. L. 1982. Revision of K/Ar ages for the Hadar hominid site. *Nature (Lond.)* **296**:122–127.

Ward, S. C., 1974. *Form and Function in Primate Jaw Mechanics*, University Microfilms, Ann Arbor.

Ward, S. C. 1979. Subocclusal dental morphology of Pliocene hominid mandibles from the Hadar Formation. *Am. J. Phys. Anthropol.* **50**:490 (Abstract).

Ward, S. C. 1980. Maxillary subocclusal morphology of Pliocene hominids from the Hadar

Formation and Miocene hominoids from the Siwalikds and Potwar Plateau. *Am. J. Phys. Anthropol.* **52:**290–291 (Abstract).

Ward, S., and Molnar, S. 1980. Experimental stress analysis of topographic diversity in early hominid gnathic morphology. *Am. J. Phys. Anthropol.* **53:**383–395.

Weidenreich, F. 1936. The mandibles of *Sinanthropus pekinensis,* A comparative study. *Paleontol. Sin. D* **7:**1–162.

Weiner, J. S., Oakley, K. P., and Le Gros Clark, W. E. 1955. The solution of the Piltdown problem. *Bull. Br. Mus. (Nat. Hist.) Geol.* **2:**141–146.

Wells, L. H. 1962. Pleistocene faunas and the distribution of mammals in southern Africa. *Ann. Cape Prov. Mus.* **2:**37–40.

Wells, L. H. 1967. Antelopes in the Pleistocene of southern Africa, in: *Background to Evolution in Africa* (W. W. Bishop and J. D. Clark, eds.), pp. 175–184, University of Chicago Press, Chicago.

Wells, L. H. 1969. Faunal subdivision of the Quaternary in southern Africa. *S. Afr. Archeol. Bull.* **24:**93–95.

White, T. D. 1977a. *The Anterior Mandibular Corpus of Early African Hominidae: Functional Significance of Shape and Size,* University Microfilms, Ann Arbor, Michigan. 385 pp.

White, T. D. 1977b. New fossil hominids from Laetolil, Tanzania. *Am. J. Phys. Anthropol.* **46:**197–230.

White, T. D. 1980a. Additional fossil hominids from Laetoli, Tanzania: 1976–1979 specimens. *Am. J. Phys. Anthropol.* **53:**487–504.

White, T. D. 1980b. Evolutionary implications of Pliocene hominid footprints. *Science* **208:**175–176.

White, T. D. 1981. Primitive hominid canine from Tanzania. *Science* **213:**348–349.

White, T. D., and Harris, J. M. 1977. Suid evolution and correlation of African hominid localities. *Science* **198:**13–21.

Wolpoff, M. H. 1968. "Telanthropus" and the single species hypothesis. *Am. Anthropol.* **70:**477–493.

Wolpoff, M. H. 1970. The evidence for multiple hominid taxa at Swartkrans. *Am. Anthropol.* **72:**576–607.

Wolpoff, M. H. 1973. Posterior tooth size, body size, and diet in South African gracile australopithecines. *Am. J. Phys. Anthropol.* **39:**375–393.

Wolpoff, M. H. 1974a. The evidence for two australopithecine lineages in South Africa. *Yearb. Phys. Anthropol.* **17:**113–139.

Wolpoff, M. H. 1974b. Sagittal cresting in the South African australopithecines. *Am. J. Phys. Anthropol.* **40:**397–408.

Wolpoff, M. H. 1980. *Paleoanthropology,* Knopf, New York. 379 pp.

Wood, B. A. 1974. Morphology of a fossil hominid mandible from East Rudolf, Kenya. *J. Anat.* **117:**652–653 (Abstract).

Wood, B. A. 1976. Remains attributable to *Homo* in the East Rudolf succession, in: *Earliest Man and Environments in the Lake Rudolf Basin* (Y. Coppens, F. C. Howell, G. L. Isaac, and R. E. F. Leakey, eds.), pp. 490–506, University of Chicago Press, Chicago.

Wood, B. A. 1978. The classification and phylogeny of East African hominids, in: *Recent Advances in Primatology,* Volume 3, *Evolution* (D. J. Chivers and K. A. Joysey, eds.), pp. 351–372, Academic, London.

Zihlman, A. L. 1979. Pygmy chimpanzee morphology and the interpretation of early hominids. *S. Afr. J. Sci.* **75:**165–168.

Concluding Remarks and Summary Comments VIII

Hominoid Cladistics and the Ancestry of Modern Apes and Humans

A Summary Statement

30

R. L. CIOCHON

Introduction

The major purpose for assembling this volume was to provide a forum for the presentation of alternative viewpoints on the subject of ape and human ancestry. In this chapter I will reduce those viewpoints to a summary statement reflecting the contributors' intent through an evaluation of the volume's major themes and by the presentation of a series of cladistic models and character analyses. Naturally it will be impossible to cover all of the points dealt with in the preceding 29 chapters. Therefore, only the salient points of the major themes of hominoid phylogeny will be considered. In particular I will consider (1) the branching order (cladogenesis) of fossil and recent hominoid primates, (2) the structural components (morphotype) of the last common ancestor of humans and the living apes as well as the morphotypes of earlier hypothetical ancestors in the diversification of the Hominoidea, (3) the timing and geographical placement of the ape–human divergence and the origin of the extant ape and human lineages, and (4) the adaptive nature and probable scenario of the Miocene hominoid cladogenesis with specific focus on the initial differentiation of hominids from their ape forebears.

R. L. CIOCHON • Department of Paleontology, University of California, Berkeley, California 94720.
Preparation of this chapter was supported by a grant from the L. S. B. Leakey Foundation.

In my view, the chapters of this volume represent a major step toward reconceptualization of the phyletic position and relationships of the Miocene hominoids *vis à vis* modern apes and humans. As was earlier suggested in a report of the Florence conference on which this volume is based (see Ciochon and Corruccini, 1982), this reconceptualization could foster new evolutionary paradigms of ape and human ancestry, melding together a wide variety of neontological, paleontological, and biomolecular data. This final chapter should help to document this reconceptualization.

Evaluation of Major Themes

Taxonomic Nomenclature

A persistent problem complicating the phyletic evaluation of Miocene to Recent hominoids involves the usage of taxonomic nomenclature. Though the naming of a group of organisms, whether at the specific or familial level, appears straightforward, the complexities arise when phylogenetic implications are attached to the taxonomic terms. In this volume where so many different phylogenetic schemes have been proposed it is obvious that there has been little agreement on taxonomic nomenclature. With this in mind I plan to briefly review some of the nomenclature that has been proposed and select a basic scheme that I will employ throughout the rest of this chapter.

Traditionally the hominoids of the Miocene and Pliocene have been grouped together in the single pongid subfamily Dryopithecinae (Simons and Pilbeam, 1965). Though some contributors to the volume continue to use the term "dryopithecine" in a general descriptive sense (e.g., Prasad, 1983) or employ the subfamily rank in a more limited sense (e.g., Gantt, 1983), there is general agreement that the diversity of Miocene hominoids is greater than can be subsumed under a single subfamily heading. For example, de Bonis (1983) employs two subfamily groupings to contain the diversity, whereas Gantt (1983) proposes two families and four subfamilies for nearly the same taxa. The basic theme that seems to run through the writings of most contributors is that there are at least two "kinds" of Miocene hominoids: (1) the early to middle Miocene primitive, thinner enameled species of East Africa, Europe, and possibly Asia, and (2) the middle to late Miocene derived thicker enameled species of Asia, Europe, and possibly Africa.

In this volume the first group of Miocene hominoids has been variously referred to as the Dryopithecidae, Dryopithecinae, Dryopithecini, dryopithecines, dryomorphs, Proconsulinae, and Proconsulini, depending on the chosen taxonomic rank or descriptive term and the taxa under consideration. The second group has been called the Sivapithecinae, Sivapithecini, "sivapithecine complex," sivapithecines, Ramapithecidae, Ramapithecinae, Ramapithecini, ramapithecines, and ramamorphs. The two newest additions to this parade of names are the dryomorph and ramamorph designations introduced by Ward and Pilbeam (1983). They consider these terms "convenient" linguistic devices that summarize the geographic, temporal, and morphological limits of the two groups, while avoiding uncritical and informal use of subfamily and family designations which connote biological relationships that are at the moment unclear" (Ward and Pilbeam, 1983, p. 215). I agree with this assessment and their use of informal terms, but since these designations were set up to characterize differences in maxillofacial morphology (see cranial/dental evidence), I propose to use an even more basic set of terms. These are Dryopiths and Ramapiths, which do have some systematic connotation but are clearly not formal taxonomic terms.

A second taxonomic issue raised by several contributors (e.g., Wolpoff, 1983; Greenfield, 1983; Gantt, 1983) is the usage of the family nomen Pongidae. Traditionally this family has included the Asian and African Great Apes and various extinct fossil groups. Recently much new biomolecular and morphological evidence has come to light which indicates that the African Great Apes are more closely related to hominids (*Australopithecus* and *Homo*) than they are to the Asian Great Ape, the orangutan [but see also Kluge (1983) and Kay and Simons (1983) for dissenting opinions]. In cladistic terms the African apes and hominids are sister groups, which means the traditional usage of Pongidae as the family designation for all the Great Apes is no longer valid. Gantt (1983) favors the usage of Paninae for the African Great Apes as a subfamily division of the Hominidae. Pongidae is then retained for the orangutan and related extinct forms. A similar proposal was also recently made by Andrews and Cronin (1982), although they used the subfamily designation Gorillinae for the African apes. Related taxonomic schema have been proposed by other researchers, all with the express aim of dealing with the nonvalidity of the Pongidae. Unfortunately, almost all of the proposals result in the expansion of the Hominidae to include at least some (or all) of the Great Apes. Since stability and usability are important elements of any taxonomic system, I do not favor these proposals. The term "hominid" should be retained to include "those taxa on the lineage leading to *Homo sapiens,* and any collateral sidebranches on this lineage, after the divergence of this lineage with the one leading to the African apes," as Wolpoff (1983, p. 651) has precisely defined it. If the African and Asian Great Apes can no longer be classified in the same family, I propose that the subfamily Paninae suggested by Gantt (1983) be raised to familial rank, resulting in a three-family division of the extant Great Apes and humans. A similar suggestion was made by de Bonis in 1979 at the Florence conference (see Ciochon and Corruccini, 1982, p. 184). This would retain the classical usage of "hominid" while not distorting the true cladistic relationship of the Great Apes.

The final taxonomic issue of concern here is the introduction and usage of new taxa at the generic and specific level. Chopra (1983) is the only contributor to formally describe a new Siwalik hominoid taxon in this volume, although both von Koenigswald (1983) and Dehm (1983) describe a number of important new Siwalik specimens. Chopra's new taxon, named *Sivasimia chinjiensis*, is based on an upper isolated first molar from the middle Miocene locality of Ramnagar in North India. It is doubtful whether any taxon based only on an isolated tooth can be sustained for long, especially when this tooth may actually be a germ (R. N. Vasishat, personal communication). Less doubtful are the affinities of this specimen; it is possibly related to the orangutan clade as Chopra (1983) suggests (see also evidence for orangutan lineage). Two other Siwalik taxa from the locality of Haritalyangar in Himachal Pradesh, North India have recently undergone changes in nomenclature. *Pliopithecus krishnaii* (Chopra and Kaul, 1979) has become *Krishnapithecus krishnaii* (Ginsburg and Mein, 1980), a change which I endorse since the specimen is "gibbon-like" and clearly is not *Pliopithecus* (see also evidence for gibbon lineage). *Gigantopithecus bilaspurensis* (Simons and Chopra, 1969) has become *Gigantopithecus giganteus* (Pilgrim, 1915) by a new combination proposed by Szalay and Delson (1979) and supported here by Bernor (1983), von Koenigswald (1983), Prasad (1983), and Kay and Simons (1983). Among contributors, there is almost universal recognition that the nomen *Proconsul* deserves separate generic rank, replacing its grouping as a subgenus of *Dryopithecus*. For the nomen *Sivapithecus* a similar recognition is made (e.g., Chopra, 1983, p. 546). Finally, the newly proposed hominid taxon *Australopithecus afarensis* (Johanson *et al.*, 1978) is sustained by all contributors discussing the issue, with only Boaz (1983) dissenting (see also section on differentiation of hominid lineage).

Biomolecular Evidence

The cladistic branch order of the extant Hominoidea (see Fig. 5) is now fully established and very well documented through analysis of blood groups and histocompatibility antigens, chromosome banding patterns, protein structure and antigenicity, amino acid sequences of proteins, and DNA endonuclease restriction mapping, sequencing, and reassociation kinetics (Cronin, 1983; Sarich, in Cronin, 1983; Goodman *et al.*, 1983; Zihlman and Lowenstein, 1983; Yunis and Prakash, 1982; Ferris *et al.*, 1981; Socha and Moor-Jankowski, 1979; Bruce and Ayala, 1979; King and Wilson, 1975). It is worth noting that Mai (1983) in his presentation of a Cartesian model of chromosome evolution does not specifically support the conventional biomolecular branch order of the Hominoidea, but rather considers it one of several likely alternatives (see Fig. 13, p. 108). Mai (1983) also suggests the proto-*Pan*, proto-*Gorilla*, proto-*Homo* trifurcation still supported by Cronin (1983) is untenable if not falsifiable. From another point of view, Kluge (1983) challenges the "conventional wisdom" of a *Gorilla–Pan–Homo* lineage in a series of well-reasoned morphological and biomolecular arguments. Specifically, he admonishes researchers in the fields of karyology and macromolecular studies to employ more cladistically oriented methodologies in analysis of their data. These reservations aside, the biomolecular data base is now stronger than ever and can no longer be ignored as an important source of data by paleoanthropologists, as it has all too often been in the past.

Some authors embracing the biomolecular evidence (Cronin, 1983; Sarich, in Cronin, 1983; Zihlman and Lowenstein, 1983) as a primary data source for timing the hominoid cladogenesis object that many primate paleontologists are only now beginning to interpret the fossil record along the lines of their previously presented biomolecular phylogenies. I find these objections justifiable; the full impact of the biomolecular data set for the interpretation of paleontological findings has yet to be realized. However, it should be noted that several authors in this volume who deal primarily with paleontological evidence (Ward and Pilbeam, 1983; Greenfield, 1983; Boaz, 1983) are now fully acknowledging the usefulness of the biomolecular data with regard to their phylogenetic assessments of the fossil evidence. Therefore, Zihlman and Lowenstein's (1983, p. 681) comment that "increasing numbers of morphologists are riding the molecular bandwagon without paying their fare—namely, the acknowledgment that their interpretations have been influenced, at least in part, by the increasing weight of the biochemical evidence" is not a wholly accurate representation.

Support of the biomolecular evidence by primate paleontologists does not always produce phylogenies with congruent sets of divergence dates. For example, de Bonis (1983) accepts the biomolecular evidence for hominoid cladistics. In this regard he sees *Pongo* as the sister group of the *Pan–Gorilla--Homo* clade. However, when de Bonis (1983) interprets the dental and postcranial evidence of the fossil record he derives a set of divergence dates for hominoid lineages that are totally at odds with the views of mainstream molecular biologists (Cronin, 1983). Specifically, de Bonis, by linking the origin of *Gorilla* and *Pan* to the Dryopiths and *Homo* and *Australopithecus* to the Ramapiths (as in Fig. 3 of this chapter), comes up with a divergence date of about 20 m.y.a. for the African ape–human split. He then reasons that since the biomolecular evidence indicates that the orangutan is the sister group of the African ape–human clade, its separation would have to be prior to 20 m.y.a. Problems with de Bonis' cladistic analysis produce these results, specifically with regard to the evolution of hominoid postcranial anatomy and locomotor systems (see postcranial characters in Table 3).

The concept of a "molecular clock" to accurately gauge the timing of the divergence sequence of the hominoid cladogenesis is discussed by a number of contributors. Attitudes ranged from very negative (Wolpoff, 1983) to totally positive (Sarich, in Cronin, 1983). Wolpoff states (1983, p. 661) "the best way to summarize the very disparate points raised [in his chapter] is that the 'clock' simply *should not* work. This conclusion supports the paleontological analyses that claim the 'clock' *does not* work when applied to divergence times over broad timespans." This reflects Sarich's evaluation that a "*prima facie* conviction that the molecular clock must be faulty in conception, or execution, or both, has tended to relieve the critics of seeing the implications of their specific criticisms; that is, as hypotheses to be tested in the world of real data" (Sarich, in Cronin, 1983, p. 142). I feel it is most productive to interpret the molecular clock as does Cronin (1983, p. 116) when he states, "the clock is one of an approximate, not metronomically perfect nature."

The concept of an *approximate* molecular clock is supported by a recent statistical analysis of the entire macromolecular data set (Corruccini *et al.,* 1980). The concept of an approximate clock also brings the Berkeley camp of biochemists and the Wayne State camp a little closer to a long-awaited "meeting of the minds." Goodman *et al.* (1983) still maintain that selection occurs at the molecular level and that evolutionary rates have decelerated toward the present. However, they have recently moved closer to accommodation with the Berkeley camp. Goodman *et al.* (1983) now believe that the "slowdown hypothesis" may be more applicable to their own serum protein sequencing data than to the data sources of Sarich and Cronin. However, they still hold out for some modification of the clock dates to bring them more in line with the paleontological record.

With regard to reconciling biomolecular vs. paleontological views of human origins it is relevant to mention a recent conference held in May 1982 at the Vatican in Rome under the auspices of the Pontifical Academy of Sciences. Participants at the conference, four of whom are contributors to this volume (Greenfield, Lowenstein, Pilbeam, and Simons), reached agreement regarding the nonhominid status of "*Ramapithecus*" and concluded that the majority of evidence now favors a relatively late divergence of hominids and African apes (Lowenstein, personal communication).

With the aim in mind of understanding the potential for reconciliation, one of the originators of the "molecular clock," Vincent Sarich, presents a detailed historical retrospective on hominoid macromolecular systematics and the clock (Sarich, in Cronin, 1983). He ponders the issue of why the clock divergence dates, which have withstood more than a decade of attempts to falsify them, have only recently begun to gain some acceptance. Sarich (in Cronin, 1983, p. 148) then philosophically summarizes what can be learned from the whole affair:

> We will benefit only if we learn to discriminate productive from nonproductive or counterproductive ways of asking questions; learn how to judge the quality of an argument in evolutionary reconstruction; learn how not to be blinded by prevailing orthodoxy; and, above all, to recognize that the fossil record and its interpreters represent but one line of evidence contributing to our understanding of the evolutionary process in general and specific evolutionary events in particular.

I agree with Sarich's sentiments and suggest that the present volume should go a long way toward vindicating his basic scientific premises and approach. However, it is only fair to point out that Sarich's data also constitute "but one line of evidence."

Cranial/Dental Evidence

The skull and dentition of the Miocene Hominoidea have provided the primary data source for drawing phyletic interpretations since the first fossil hominoids were recovered more than a century ago. All of the classic studies from Lydekker, Pilgrim, Gregory, and Lewis to the more current research efforts in the 1960s and 1970s by Simons, Pilbeam, and others have concentrated almost entirely on the cranial/dental data base. It is interesting to note that most of the phyletic interpretations put forward by these researchers on the basis of skull fragments and teeth have been substantially modified through the application of alternative data sources (e.g., postcranial, biomolecular) and by the realization that many "diagnostic" hominoid cranial and dental characteristics were actually primitive retentions from a common catarrhine ancestor (Szalay and Delson, 1979; Kay, 1982*a;* Fleagle and Kay, 1983; see also Table 3, this chapter). This reevaluation of the hominoid skull and dentition has resulted in the introduction of new, much more rigorous methods of analysis (Ward and Pilbeam, 1983; Gantt, 1983) and the application of cladistic analyses (Andrews and Cronin, 1982; Kay, 1982*a;* Kay and Simons, 1983; Falk, 1983). Thus, the papers on craniodental morphology in this volume break new ground in our understanding of Miocene hominoid interrelationships and often produce results concordant with a variety of other data sources.

Ward and Pilbeam (1983) outline the existence of a major dichotomy in the topographic relationships of the maxillofacial region of Miocene Dryopiths and Ramapiths. They demonstrate that this dichotomy is mirrored by the differences in the maxillofacial region of living African apes and humans on the one hand and the orangutan on the other. The Dryopith pattern (see morphotype 2 in Table 3) is interpreted as primitive and the Ramapith pattern (see morphotype 6 in Table 3) is judged as derived (Ward and Pilbeam, 1983). From this evidence they conclude that Ramapiths and orangutans are specially related (i.e., sister groups in a cladistic sense; see Table 2 and Fig. 4). Ward and Pilbeam (1983, p. 220) also declare that "the last common ancestor of the Asian and African hominoids must predate the oldest ramamorph hominoid," which is a logical outgrowth of the derived character states they attribute to Ramapiths. Ward and Pilbeam further suggest that the subnasal region of *Australopithecus afarensis* is most like that of living chimpanzees; in other words, both forms share the primitive Dryopith subnasal pattern. Furthermore, other aspects of the subnasal region appear uniquely derived in *Pan* and *A. afarensis* (Ward and Pilbeam, 1983; Ward *et al.,* 1983), which lends additional support to the view that *A. afarensis* is a descendant of an "African ape-like" ancestor.

cial region grew out of the discovery and description of the most complete Ramapith skull yet recovered (Pilbeam, 1982; see also Ward and Pilbeam, 1983, p. 212, Fig. 1). It is interesting to note that Andrews and Cronin (1982), independent of Ward and Pilbeam, developed a similar set of maxillofacial characters uniting Ramapiths and *Pongo* and indicating that Dryopiths are primitive and most like the African apes and humans (Pilbeam, personal communication; Andrews, personal communication). The basis for Andrews and Cronin's determinations was the analysis and description of another Ramapith skull (*S. meteai*) by Andrews and Tekkaya (1980). I have also listed the characters from Andrews and Cronin's (1982) study in Tables 2 and 3 (morphotypes 2 and 6), including a dual listing of those characters that overlap between the two independent studies.

An interesting application of the Ward and Pilbeam (1983) study is its usefulness in assessing the affinities of other fossil specimens. The site of Rudabánya in Hungary has produced a series of fossil hominoids whose phyletic relationships are not fully established. Ward and Pilbeam (1983, p. 220) suggest that Rud 12, previously attributed to *Rudapithecus* (Kretzoi, 1975) or *Ramapithecus* (Simons, 1976), "is more closely related to the African dryomorphs than it is to the late Miocene hominoids of Asia." Thus, Dryopiths may be represented at Rudabánya (see also section on postcranial evidence). This accords well with the view expressed by Martin and Andrews (1982), who consider the possibility that nearly all the Rudabánya hominoids may be Dryopiths. Kay and Simons (1983) also offer an opinion based on their analysis of the maxillofacial region. They conclude that Rud 44, previously attributed to *Bodvapithecus* (Kretzoi, 1975) or to *Sivapithecus indicus* (Kay, 1982a), actually has a very broad interorbital region, unlike Siwalik *Sivapithecus*. Kay and Simons (1983) attempt to use this discrepancy to invalidate the *Sivapithecus–Pongo* relationship on the basis that the interorbital distance character is variable, since both of these taxa are viewed as *sharing* a derived narrow interorbital region (see Table 2). To the contrary, I suggest that Rud 44 may actually be better attributed to the Dryopiths *on the basis* of its primitive broad interorbital region, a view that as was also considered by Martin and Andrews (1982). Thus, Rudabánya provides an excellent example of how our enhanced understanding of hominoid craniofacial morphology might result in more accurate interpretations of phyletic relationships.

In another cranial analysis Falk (1983) reinvestigates the cortical sulcal patterns of the endocast of the only known skull of *Proconsul africanus*. She undertakes a detailed cladistic analysis comparing the endocast of *Proconsul* with a wide spectrum of anthropoids and lower primates. Falk (1983) concludes from this study that the cortical sulcal patterns of *P. africanus* can only be characterized as primitive anthropoid. *Proconsul* exhibits not one shared derived cortical character which could be used to link it with extant hominoids. She states, "it appears, then, that during pongid as well as hominid evolution, evolution of the external morphology of the cerebral cortex lagged strikingly behind dental/skeletal evolution" (p. 246). Falk (1983) labels this phenomenon "neurological lag" and suggests that it removes *P. africanus* from anything but a very distant relationship with the extant Hominoidea.

Dental features of Miocene to Recent hominoids, such as metrical assessments of tooth variability, analyses of enamel prism patterns and microstructure, and determinations of molar enamel cap thickness, have been discussed by many contributors. Kay and Simons (1983) present an in-depth analysis of dental size variability in Recent and fossil hominoids. Their central premise is that new discoveries of fossil hominoids during the last decade have increased the sample sizes of extinct taxa to the point where they can now be compared with the kind and degree of dental variability seen in modern ape and monkey species. Kay and Simons (1983) establish the limits of sexual dimorphism and coefficients of variation in dental dimensions for living species and apply this to the fossil hominoid record to assess the viability of extinct taxa (see also Kay, 1982*a,b*). They present a major revision of the Ramapiths based on an impressive locality-by-locality analysis of dental variability. The number of Ramapith genera is decreased to include only two (*Sivapithecus* and *Gigantopithecus*) with the notable elimination of "*Ramapithecus*" (see Kay, 1982*a*). Species diversity within *Sivapithecus* is redefined with the naming of one new species, *S. simonsi* Kay 1982*a*, and potentially two other species (Kay and Simons, 1983, pp. 604–605). Kay and Simons' revision of the Ramapiths represents a major step forward in the definition of extinct taxa that have a semblance of biological reality *vis à vis* modern taxa. Though all researchers will not agree with their choice of names or the affinities they ascribe to the Ramapiths, few can deny the impact of their revision on the field of paleoanthropology and its solid foundation in modern biological thought.

Gantt (1983) presents a major review of the enamel prism patterns and microstructure of Neogene hominoids based on his extensive experience with the scanning electron microscope. He documents that all hominoids uniquely share developmental Pattern 3 in the makeup of their enamel and can be divided into two subgroups based on consistent differences in the structure of their enamel prism patterns. The first group, Pattern 3A, is described as tadpole-shaped and is found in *Proconsul, Ouranopithecus, Sivapithecus,* "*Ramapithecus,*" *Gigantopithecus blacki, Hylobates, Pongo, Gorilla,* and *Pan* (Gantt, 1983). The second group, Pattern 3B, is described as keyhole-shaped and characterizes *Australopithecus afarensis, A. africanus, A. robustus, Homo erectus,* and *H. sapiens.* Thus, the Pattern 3A/3B dichotomy reflects the taxonomic distinctiveness of the hominid family and is possibly related to a functional difference in enamel ultrastructure underlying this dichotomy (Gantt, 1983). One interesting facet of this study was Gantt's apparent inability to resolve significant enamel prism pattern differences reflecting the Dryopith–Ramapith dichotomy in the Hominoidea. Since this was an important focus of his investigation (Gantt, personal communication), its absence is particularly noteworthy and may signify that dental adaptations of these two groups are less different than some would believe.

The degree of thickness of enamel caps on the molars of Miocene to Recent hominoids has recently become a "key character" for deciphering hominoid evolution. This feature was probably discussed by more contributors than any other single issue. Simons and Pilbeam (1972, pp. 58–59) were apparently the first to document and clearly diagram the adaptive significance of thin vs. thick enamel in Miocene hominoids. However, it was not until much more recently that the thin vs. thick enamel differences became associated with the Dryopith vs. Ramapith dichotomy as expressions of primitive vs. derived character states (e.g., Kay, 1981). The thin vs. thick enamel dichotomy of hominoids is beginning to show signs of weakness as more researchers investigate the actual data base. For example, Ward and Pilbeam (1983) suggest that at least three categories of enamel thickness should be recognized: thin, intermediate, and thick. They characterize the African apes as thin-enameled, humans and orangutans as moderately thick-enameled, and Ramapiths as thick-enameled. They further argue (Ward and Pilbeam, 1983, p. 235, Fig. 10) that the common ancestor of Asian and African hominoids most probably had thick or moderately thick enamel, implying that the primitive ancestral Dryopith was something *other than* thin-enameled.

Gantt's (1983) interpretation of enamel thickness in Miocene to Recent hominoids differs somewhat from that of Ward and Pilbeam's. Gantt measured the thickness of enamel caps directly from actual thin sections of molars taken along the buccal–lingual axis (see p. 267, Fig. 9) and regressed these values against body weight. Gantt (1983, pp. 285–286) observes, "There is a close correlation between the thickness of enamel and body size among extant primates, the exception being *Homo sapiens* (see Fig. 23). Humans have much thicker enamel than any extant primate. . . ." Gantt further notes that among living apes the orangutan most closely approaches the thickness of enamel seen in humans, and the gorilla has the thinnest enamel. Gantt (1983) also included several fossil hominoids in this thin-section analysis. Gantt (1983, p. 279) states, "Comparison of the thickness of enamel in similar sized hominoids (e.g., *Ramapithecus* and *Proconsul africanus*) reveals that the thickness of enamel is identical, contrary to Kay (1981). Evidence does not support the division of Miocene hominoids into 'thin' (ancestral early Miocene forms) and 'thick' (descendant, middle to late Miocene forms)." These results are presented in Fig. 26 of Gantt (1983, p. 291). Note that the molars of "*Ramapithecus*" and *Proconsul africanus* used here have *identical* dental dimensions and *identical* enamel thicknesses. Thus, it is becoming increasingly difficult to argue that thin enamel caps on cheek teeth are the primitive hominoid (Dryopith) morph, as Kay and Simons (1983, p. 587) do when they state, "Gorilla and chimpanzee still retain thin enamel as do their early Miocene dryopithecine forebears and the extant hylobatids." Other contributors (Wolpoff, 1983; Greenfield, 1983; Zihlman and Lowenstein, 1983) suggest that enamel thickness has little validity at all for sorting hominoid groups or establishing phyletic relationships, since its expression as a *continuous* variable within groups can be demonstrated. Obviously, there is great need for the collection of statistically significant samples of data on enamel thickness to resolve these issues. As a preliminary statement, my view of the enamel thickness changes in hominoid evolution can be found in Table 3.

Postcranial/Locomotor Evidence

The postcranial skeleton and locomotor behavior of the Miocene Hominoidea have been the subject of lively debate over the last two decades. Much of this debate has centered around the partial skeleton of *Proconsul africanus* from the early Miocene of East Africa first described by Napier and Davis (1959). The proposed phylogenetic affinities of *P. africanus*, also known from a complete skull and other cranial remains, have ranged from a knuckle-walking chimpanzee analog to a suspensory-adapted ancestor of the apes to a generalized primitive hominoid. In each phylogenetic scheme the postcranial skeleton was used in some fashion to bolster the favored position. Given this historical perspective, it is easy to see why the recovery of much of the rest of the original skeleton of *P. africanus* (Walker and Pickford, 1983) would generate much new debate. In the first part of this section I will review some current controversies surrounding *P. africanus* and follow this with a discussion of evolutionary and locomotor trends in the Miocene hominoids.

In their description of the new limb material Walker and Pickford (1983, p. 350) comment, "It has been the consensus that the postcranial skeleton of *Proconsul* was, in many features, reminiscent of those of certain New and Old World monkeys. It now seems that in other respects there are many features that are also found in living apes, particularly *Pan*." They further suggest that the new postcranial elements indicate that in the forelimb there appears to be a gradient along the limb, with the more hominoid-like features occurring more proximally; the gradient appears to reverse in the hindlimb, with the hominoid-like features being found more distally. Had the new skeletal elements of *P. africanus* and *P. nyanzae* been available to Napier and Davis for study, it is quite possible that their now classic interpretations of the phylogenetic position and locomotor adaptations of *Proconsul* would have been different.

Reviewing aspects of the elbow joint of fossil hominoids, Fleagle (1983*a*, p. 320) observes that *P. africanus* "looks more like a living ape than does any other early Miocene or Oligocene hominoid. It shows, incipiently, the same elbow features which first appear fully developed in the late Miocene apes of Europe and Asia" as documented in this volume by Rose (1983) and Morbeck (1983). However, *P. africanus* also lacks a number of uniquely derived features of the dentition, cranium, and wrist shared by extant apes (see also Fleagle and Kay, 1983; McHenry and Corruccini, 1983). It may be that *P. africanus* was more than just a "dental ape" (Corruccini *et al.*, 1976). Postcranially it might be called a "formative ape" (Feldesman, 1982), but apparently it was not a "neural ape" (Falk, 1983). From a locomotor perspective it was most likely a slow-moving arboreal quadruped with a wide range of generalized postural capabilities (Walker and Pickford, 1983).

McHenry and Corruccini (1983) dispute the views of *P. africanus* shared by Fleagle (1983*a*) and Walker and Pickford (1983). They state, "The postcranium of *Proconsul africanus* is not in any way intermediate between modern apes and monkeys despite the fact that its dentition is so clearly hominoid" (McHenry and Corruccini, 1983, p. 365). There is also disagreement regarding what constitutes the ancestral morphotype of all catarrhines (see also Table 3). Both Fleagle (1983*a*) and Rose (1983) conclude that cercopithecoids are derived cranially *and postcranially* from the ancestral morphotype. McHenry and Corruccini (1983) argue that the cercopithecoid postcranial morph is primitive (plesiomorphic) for the Catarrhini. Fleagle (1983*a*) demonstrates that several early fossil catarrhines from widely separated geographic areas and spanning some 10 million years actually share a uniform platyrrhine-like limb morphology. If suggestions of Rose and Fleagle concerning the derived nature of the cercopithecoid postcranium are sustained, then the primitive forelimb morphology for early hominoids can only be viewed as platyrrhine-like. Taking this one step further, it could follow that many of the "derived" forelimb features of the extant Hominoidea are actually catarrhine symplesiomorphies (Fleagle, 1983a). These are often the *same* features that earlier workers used to demonstrate potential phyletic links between fossil hominoids and extant species. Similar arguments could also apply to the evolution of suspensory hindlimbs. It is presently not possible to say if these are wholly derived or alternately represent more or less the ancestral condition. It may prove "that many of the suspensory, climbing features of extant hominoids are just further developments of the primitive early catarrhine adaptation" (Fleagle, 1983a, 321).

On one issue there is general agreement by all investigators of the postcranial skeleton. All agree that by the middle Miocene the hominoid forelimb had begun to show the anatomical pattern characteristic of the extant Great Apes and humans (see also Table 3). It is possible to sort this variation out into a primitive group and a derived group which basically corresponds to the Dryopith and Ramapith morphs. However, the correspondence of taxa and morphology is not always as straightforward as it seems. For example, Morbeck (1983) cites the existence of two structural groups (morphs) in the forelimb elements of the middle to late Miocene hominoids she examined. One group (Potwar humeri, Rudabánya humerus and ulna, St. Gaudens humerus) appear derived in the same direction as extant hominoids, while the other (Maboko Island humerus, Fort Ternan humerus, and Klein Hadersdorf humerus and ulna) is more generalized, representing the primitive or ancestral condition. Where associations between postcranial and cranial remains exist the derived morph is apparently linked with Ramapiths and the generalized morph with Dryopiths. The existence of both morphs at one locality is thought to occur in several places. The site of Rudabánya in Hungary provides an interesting example.

Original descriptions of the two Rudabánya hominoids, *Rudapithecus* and *Bodvapithecus,* seemed to indicate affinities with the Ramapiths (Kretzoi, 1975; Simons, 1976). However, recent studies of the same specimens have led a number of researchers to conclude that one or both of these taxa should actually be included in the Dryopiths (Ward and Pilbeam, 1983; Kay and Simons, 1983; Martin and Andrews, 1982) (see also cranial/dental section of this chapter). If both Ramapiths and Dryopiths are present at Rudabánya, then Morbeck's (1983) conclusions regarding the existence of derived "extant hominoid-like" (=Ramapith) forelimb elements is fully in line with this evidence (she also cites the existence of a proximal radius that appears to represent the Dryopith morph). However, if the view expressed by Martin and Andrews (1982) that nearly all the Rudabánya specimens are referable to a single Dryopith taxon, *Dryopithecus carinthiacus* Mottl, should prevail, then derived hominoid postcranial elements would be associated with the seemingly primitive Dryopiths. This might necessitate regrouping Eurasian Dryopiths more closely with the Ramapiths (in part on the basis of derived postcranial features) than with the African Dryopiths. I think it is important to bring these issues forward here because the linking of Eurasian and African Dryopiths as suggested by many authorities in this volume (see also Table 3) is based almost entirely on shared primitive characters. This arrangement will probably not stand for long. Studies at European sites such as Rudabánya, where associated cranial, dental, and postcranial elements exist, could very well help to resolve the phyletic position of Eurasian Dryopiths.

Not all researchers who studied the Miocene postcranial evidence wish to approach the issue by imposing polarities or dichotomies on the material. Rose (1983) deals with this subject from a more fluid and behaviorally oriented perspective. He observes that the "differences between earlier and later Miocene hominoids are mostly of degree rather than of kind. The basic plan upon which these variations were made was evidently a very successful one. This success was based on generalized morphology underlying generalized capabilities" (p. 415). Rose (1983) later discusses the pertinent question posed in the title of his chapter: Were the Miocene hominoids more "monkey-like" or "ape-like" in their postcranial anatomy? In answering this question he first characterizes Old World monkeys as "specialized generalists" rather than "generalized generalists" and further observes that they resemble Miocene hominoids in only those parts of the limb skeleton that are the most generalized. Such resemblances are conservative retentions and relate only to the most common aspects of quadrupedal progression. Similarly, the ways in which later Miocene hominoids resemble living hominoids in the shoulder, elbow, and hip according to Rose (1983) also relate to very generalized abilities. Rose (1983, p. 415) then states, "To the extent that some cebids are morphologically and behaviorally quite generalized, the early Miocene hominoids are more 'New World monkey-like' than 'Old World monkey-like.'" Rose concludes that our understanding of this highly successful primate group might be better advanced not by drawing comparisons with contemporary groups but instead by characterizing them as "Miocene hominoid-like."

Paleoenvironmental Evidence

Studies of Miocene paleoenvironments of the Old World have provided a detailed framework for investigating the paleoecology, zoogeography, and evolutionary relationships of the Miocene Hominoidea. Indeed, many a morphologically-based investigation of the Miocene hominoids would not be considered complete without reference to some facet of paleoenvironmental research that either supports or denies an issue in question. With this sort of demand for accurate paleoenvironmental interpretations it is easy to see why controversy and divergence of opinion have often surrounded these investigations (for example, compare Bernor, 1983, and Kortlandt, 1983).

Perhaps the most universal theme in Miocene paleoenvironmental interpretation has been the documentation of climatic and vegetational shifts that ultimately bring about dramatic turnovers in fauna. Pickford (1983) provides such an example through analysis of paleoenvironmental parameters from the Miocene and Plio-Pleistocene of Kenya. In the lower Miocene of western Kenya he documents the existence of numerous fossil localities sampling forested to woodland environments each characterized by a diverse array of hominoids exhibiting a wide spectrum of body sizes. By the middle Miocene, fossil localities in this same region of western Kenya show evidence of an environmental shift to open woodland and bushland environments with a reduction in the diversity, body size range, and population density of hominoid primates (Pickford, 1983). The Plio-Pleistocene in Kenya is characterized by a further environmental shift to savanna-oriented assemblages with a very low diversity of hominoids. In this example Pickford (1983, p. 431–32) is testing the hypothesis that "the primitive condition among hominoids was an adaptation to life in forested environments, while the more advanced or derived condition is represented by an adaptation to life in open woodland and grassland." His results verify a progression through time from primitive forest-adapted hominoids toward more advanced open-country-adapted populations.

Kortlandt (1983) disputes the conclusion that the paleoenvironments of the early Miocene localities of East Africa were heavily forested. In a rather discordant presentation he marshalls evidence that such classic early Miocene localities as Rusinga, Napak, and Songhor had relatively dry climates with short seasonal rainfalls [*sahélien* or dry-*soudanien* type of climate according to Kortlandt, 1983, p. 470]. Such climates would be associated with mosaic habitats of woodlands and savanna that had little or no closed-forest components. Thus, according to Kortlandt (1983), Pickford's (1983) view that the primitive condition among hominoids was an adaptation to forested environments would have to be rejected.

Kortlandt (1983) takes this argument one step further when he discusses the adaptations of the Oligocene hominoids of Egypt's Fayum Province. He states, "the earliest known apes must apparently have been semiterrestrial woodland-, bushland-, and savanna-dwellers, and were adapted to survive prolonged annual droughts" (p. 483). Here, Kortlandt's conception of the primitive habitat for hominoids is taken to its logical extreme (see also Kortlandt, 1980), but this view has recently been seriously challenged by Bown *et al.* (1982). In a well documented study this interdisciplinary group, representing the fields of geology, sedimentology, paleobotany, vertebrate paleontology, and primate anatomy, demonstrate beyond any reasonable doubt that Fayum paleohabitats were heavily forested, with dense undergrowths, and that the earliest apes were demonstrably arboreal. It remains to be seen whether Kortlandt's (1983) views regarding the early Miocene paleoenvironments of East Africa will be similarly challenged. With this in mind, it should be pointed out that Walker and Pickford (1983) found that the R114 locality at Rusinga, which yielded the *Proconsul africanus* skeleton, was characteristic of a dry-forest to open-woodland habitat. This single locality appears to provide a small measure of support for Kortlandt's paleoenvironmental interpretation of the early Miocene [but also see Andrews (1981) and Bernor (1983) for alternative views].

One of the paleoenvironmental themes voiced by several contributors is the apparent relationship between Miocene hominoid group, habitat preference, and molar enamel thickness. For example, de Bonis (1983) concludes that Dryopiths inhabitated closed, heavily forested environments and had jugal teeth with thin enamel related to a soft diet of fruits, leaves, and young plants. Bernor (1983) observes that Dryopith habitats in Europe are associated with subtropical forest and closed woodland communities, while in Africa they are associated with tropical forest communities. Pickford (1983) also points out the relationship between Dryopiths, forest habitats, and "thin-enameled" cheek teeth. Ramapiths, on the other hand, were open-environment-dwelling primates with thick enamel on their jugal teeth related to a more abrasive diet according to de Bonis (1983). Bernor (1983) concludes that the origin, dispersal, and extinction of the Ramapiths were closely tied to the appearance, expansion, and replacement of an evergreen woodland vegetation type consisting of trees, chaparral, and herbaceous undergrowths. Andrews (1983) observes that in the Siwaliks, Ramapiths are associated with a "woodland–bushland pattern," at the Paşalar locality in Greece with "open woodland conditions," and at Lufeng in China with a "warm, temperate deciduous woodland" (Andrews, 1983). Finally, Pickford (1983) suggests that the adaptive radiation of hominoids with thick enamel (the Ramapiths) began at approximately the same time in Kenya as the development of open-country habitats. Pickford reasons that the Ramapiths could have evolved in response to this change in environment or their arrival in western Kenya could have been coincident with the development of appropriate niches for them.

In a survey of the natural history of *Sivapithecus,* Andrews (1983) presents probably the most detailed paleoecological study of a fossil hominoid yet attempted. Many of his comments about *Sivapithecus* can probably be applied to most other Ramapiths as well. With regard to paleohabitat, Andrews (1983, p. 445–46) concludes, "Tropical forest setting is never indicated, and tropical to subtropical woodland (adapted for a moderately seasonal climate and distinguished on the basis of a single tree canopy dominated by relatively few species and with grasses often common in the ground vegetation) may be the preferred habitat. . . ." From a series of consistent paleoenvironmental associations coupled with studies of the *Sivapithecus* dentition and projected body size (and based on analogies with living primates), Andrews expertly argues that *Sivapithecus* (and other Ramapiths?) probably exhibited some degree of terrestriality, developed aspects of polygynous social group behavior, and had a generalized omnivorous diet that was better adapted to herbivorous food stuffs than to frugivorous or folivorous foods. Regarding a potential paleoecological origin scenario for Ramapiths, Andrews (1983) speculates that these newly emergent thick-enameled hominoids could have expanded into less favorable woodland habitats containing geophytic plants with underground storage parts, as a result of competition with the expanding cercopithecoid monkey radiation. Once in seasonal woodland habitats these thick-enameled hominoids could have exploited underground food sources from geophytes as well as a wide variety of other nonfrugivorous food sources.

Before closing this section, mention should be made of Bernor's (1983) model of Miocene hominoid zoogeography, which is derived from his analysis of Miocene geochronology and paleoenvironments. Bernor (1983) concludes that the Hominoidea originated in Africa. In the late Oligocene and early Miocene the hominoids underwent a major adaptive radiation. At the beginning of the middle Miocene (about 16 m.y.a.) the first hominoids dispersed from Africa into Europe and Asia via a land corridor that had been established across the Arabian Peninsula when the African Plate docked with Eurasia (Bernor, 1983, p. 37, Fig. 3). During the late Miocene, hominoids became isolated in the tropical areas of southeast Asia and sub-Saharan Africa and by the latest Miocene most had become extinct.

Status of "Ramapithecus" and Affinities of Ramapiths

The position of "Ramapithecus" in hominoid phylogeny has been a controversial issue in paleoanthropology for the last two decades. Simons (1961, p. 5) sparked the rebirth of "Ramapithecus" when he claimed that this species "occurs in the proper time and place to represent a forerunner of Pleistocene Hominidae." Throughout the 1960s and well into the 1970s the hominid status of "Ramapithecus" was not seriously challenged from a morphological perspective. By the latter 1970s, however, a number of new fossil discoveries from the Potwar Plateau in Pakistan and the Afar Depression in Ethiopia together with the ever-expanding body of biomolecular data began to seriously undercut the position of "Ramapithecus" as a basal hominid. In July 1980 the Pre-Congress Symposium on which this volume is based was held in Florence, Italy [see report by Ciochon and Corruccini (1982)]. At that symposium both the generic status and hominid affinities of "Ramapithecus" were challenged by a small but vocal group of participants. That initially small base of support at the symposium has expanded to a significant majority in this volume. In the minds of many contributors "Ramapithecus" no longer exists as a separate genus and its phyletic relationship to the Plio-Pleistocene Hominidae is also seriously questioned. Though this volume will certainly contribute to the demise of the "Ramapithecus" taxon, the "Ramapithecus" concept in human evolution will live on, for, as Wolpoff (1983, p. 652) points out, "the importance of Ramapithecus has historically been in the claim that it represents the earliest hominid," *not* in the importance of the taxon itself. As our concepts of hominid origins continue to change, it will be interesting to observe what new fossil taxon assumes the position of earliest hominid.

The reasoning and mechanics that various contributors employ in the sinking of "Ramapithecus" are worth noting for the insights they yield regarding the nature of paleobiological inference. For example, Ward and Pilbeam (1983) conclude that the only significant difference between "Ramapithecus" and *Sivapithecus* other than size "involves the extent of maxillary sinus invasion into the alveolar process" (p. 233). From this they cautiously suggest that "Ramapithecus" is more likely congeneric with *Sivapithecus* but less likely conspecific. Greenfield (1983, p. 699) states that "Ramapithecus" is "a smaller version of *Sivapithecus* showing no *special* similarities to the earliest undoubted member of the human lineage, *A. afarensis*." Greenfield (1983) draws a parallel between the size differences of "*Ramapithecus*" and *Sivapithecus*, separate at the genus level, and the different sized species of *Proconsul*, arguing that no generic separation is warranted for "*Ramapithecus*." In his major paper on the adaptive pattern of "*Ramapithecus*" Greenfield (1979) assigns both Asian species ("*R.*" *punjabicus* and "*R.*" *brevirostris*) to *Sivapithecus brevirostris*.

Kay and Simons (1983) take a bolder approach in their sinking of "*Ramapithecus*" by placing the hypodigm of "*R.*" *punjabicus* (which includes "*R. brevirostris*") into *Sivapithecus sivalensis*. Using coefficients of variation derived from analyses of dental metrics of modern populations, they conclude that "Variability within *S. sivalensis* is sufficient to include all the specimens formerly attributed by Simons and Pilbeam (1965) to *Ramapithecus punjabicus*" (p. 582). The actual sinking of "*R.*" *punjabicus* is documented in Kay (1982a). Kay and Simons (1983) go on to point out that recognition that "*R.*" *punjabicus* and *S. sivalensis* are actually the same taxon is a suggestion that had never been made previously. Gantt (1983) presents a new analysis of "*Ramapithecus*" specimens (based on "*R.*" *punjabicus*, AMNH 19565-B, and "*R.*" *brevirostris*, YPM 13799), suggesting that on the basis of prism patterns and enamel thickness "*Ramapithecus*" cannot be distinguished from *S. sivalensis*, contrary to his earlier interpretations (Gantt *et al.*, 1977). Gantt (1983) questions the generic taxonomic status of "*Ramapithecus*" and observes that this taxon and *S. sivalensis* "are exceedingly similar forms and could very well represent sexual dimorphs of the same species, or alternately, two closely related species" (p. 287). Finally, de Bonis (1983), Andrews and Cronin (1982), and Boaz (1983) support the views of Ward, Pilbeam, Greenfield, Kay, Simons, and Gantt by stating their agreement on the issue of synonymizing "*Ramapithecus*" with *Sivapithecus*.

Boaz (1983) introduces the issue of "*Ramapithecus*"—*Kenyapithecus* relationships by observing that by the ways in which Asian "*Ramapithecus*" differs from East African *Kenyapithecus*, the former more closely approximates *Pongo*. Such a statement would argue for a separation, at least at the generic level, of *Kenyapithecus*. Indeed, both Boaz (1983) and Prasad (1983) suggest that the inclusion of *Kenyapithecus wickeri* in "*Ramapithecus*" (Simons and Pilbeam, 1965) is no longer warranted and that the former nomen should now be resurrected. Prasad (1983) bases his judgment on morphological, temporal, and ecological grounds. Boaz (1983, p. 707) observes that "'*Ramapithecus*' *sensu stricto* (now included in *Sivapithecus*) was likely already on a divergent lineage leading to *Pongo*, whereas the earlier *Kenyapithecus* may, in fact, represent a close approach to the common ancestor of the Great Apes and humans, or at least the African Hominoidea." Pickford (1982) also suggests that *Kenyapithecus* should be accorded separate generic rank. However, he considers this taxon to be the most primitive Ramapith and suggests it may have given rise to "*Ramapithecus*," *Sivapithecus*, and *Ouranopithecus*. Kay and Simons (1983) take the opposite viewpoint from Boaz, Prasad, and Pickford by indicating that a separate generic distinction is not warranted for the *Kenyapithecus* specimens. They interpret this taxon as congeneric with *Sivapithecus* (including

"*Ramapithecus*") and suggest *S. africanus* as the species name which has priority. Martin and Andrews (1982) conclude that the *Kenyapithecus* specimens, along with others from Czechoslovakia and Turkey, deserve separate generic rank because they all differ from *Sivapithecus* by preserving cingula and lacking a specialized upper canine that is set buccolingually. However, they do not suggest which generic nomen they favor. The position of *Kenyapithecus vis à vis* African Dryopiths, Eurasian Dryopiths, and Asian Ramapiths needs to be precisely established. It could prove that both Pickford (1982) and Boaz (1983) are correct; *Kenyapithecus* may represent the ancestral stock from which Asian Ramapiths *and* the African Hominoidea arose. In this case the *Kenyapithecus* group would be "pre-ramamorph" in the sense of Ward and Pilbeam (1983).

The phyletic affinities of "*Ramapithecus*" either as a separate genus or combined with *Sivapithecus* as part of the Ramapith group were discussed by most contributors. The position that Ramapiths (or specifically "*Ramapithecus*") are the earliest hominids or in some way are exclusively connected with the origin of the Hominidae is supported by Kay and Simons (1983), de Bonis (1983), and Chopra (1983). Partial support for the position can also be found in the writings of Pickford (1982, and 1983) and Prasad (1983), although not often explicitly stated, and in Kluge (1983) by inference from his cladistic analysis. The alternative view that Ramapiths (or specifically "*Ramapithecus*") are Miocene hominoids bearing no exclusive relationship to the origin of the Hominidae is supported by Boaz (1983), Cronin (1983), Andrews (1982, 1983), Greenfield (1983), Wolpoff (1983), Ward and Pilbeam (1983), Zihlman and Lowenstein (1983), Sarich, in Cronin (1983), Bernor (1983), and Gantt (1983).

The first position is most eloquently presented by Kay and Simons (1983). Based on an in-depth analysis of cranial and dental characters of Miocene to Recent hominoids (see Table 1), they conclude that "ramapithecines are ideally suited to be the ancestors of *Australopithecus* and *Homo*" (p. 619). They see the hominid lineage becoming separated in the middle Miocene from a Great Ape-like forebear with hominid-like dental adaptations that emphasized nut-cracking and the processing of other tough food stuffs (see also Kay, 1981). In a cladistic sense, Kay and Simons (1983) view the Ramapiths as the sister group of Pliocene to Recent hominids (see Fig. 3 of the present chapter). Regarding their classification, Kay and Simons (1983) ponder whether or not Ramapiths should be placed in the hominid family due to the fact they were quite apelike in some respects. They opt for defining the Hominidae cladistically "as being all creatures specially related to *Homo* and *Australopithecus* going back to the split between these forms and the ancestry of any of the extant apes" (p. 620). In this sense, then, Kay and Simons (1983) favor the placement of Ramapiths in the Hominidae. The derived characters they use to make this judgment (see characters 8, 10, 12, 14, 16, and 20 in Table 1) have been questioned by other contributors and therefore provide a basis for future testing and debate of the Kay and Simons position.

The alternative position, that Ramapiths are a Miocene hominoid group

Table 1. Some Characters Common to Ramapiths and Hominids (*Australopithecus* and *Homo*) with Interpretations of Their Polarity[a]

Character		Interpretations of polarity
1. Zygomatic process set well forward	P	Kay and Simons (1983), Kay (1982a)
2. Deep and flaring zygomatic process	P	Kay and Simons (1983), Kay (1982a)
3. Canine fossa	V	Kay (personal communication)
4. Parabolic (not U-shaped) upper dental arcade	V	Ciochon and Corruccini (1982), Corruccini and Ciochon (1983), de Bonis (1983), Greenfield (1978)
5. Arched palate	P	Kay and Simons (1983), Kay (1982a)
6. Low degree of premaxillary prognathism	P	Kay and Simons (1983), Kay (1982a)
7. Smaller upper and lower incisors relative to cheek tooth size	P	Kay and Simons (1983), Kay (1982a)
	C	McHenry and Corruccini (1980)
8. Small upper and lower canines relative to cheek tooth size	D	Kay and Simons (1983), Kay (1982a)
	V	Andrews and Cronin (1982)
9. Vertically oriented canine roots	V	Kay (personal communication)
10. Long axis of upper canine cross section set more or less buccolingually (externally rotated canine roots)	D	Kay and Simons (1983), Kay (1982a)
	C	Ward and Pilbeam (1983), Andrews and Cronin (1982)
11. V-shaped lower dental arcade	P	Kay and Simons (1983), Kay (1982a)
12. Shallow, broad mandibular corpora	D	Kay and Simons (1983), Kay (1982a)
	C	Andrews and Cronin (1982)
13. Megadont molars	C	Ward and Pilbeam (1983)
14. P_3 with well-developed metaconid	D	Kay and Simons (1983), Kay (1982a)
	V	Wolpoff (1983), Andrews and Cronin (1982)
15. Cheek teeth with low relief	P	Kay and Simons (1983), Kay (1982a)
16. Thick enamel caps on molars	D	Kay and Simons (1983), de Bonis (1983)
	C	Ward and Pilbeam (1983)
	V	Andrews and Cronin (1982)
	P	Gantt (1983)
17. Lower molar buccal cingula poorly developed or absent	P	Kay and Simons (1983), Kay (1982a)
18. Widely spread molar cusps with correspondingly wide fovea	V	Corruccini (1977), Greenfield (1979)
19. Extreme molar wear gradient, with much interstitial wear	V	Wolpoff (1971)
20. Low canine sexual dimorphism	D	Kay (1982b), Kay and Simons (1983)
	V	Andrews and Cronin (1982)

[a]The characters listed in the left column have been suggested, at one time or another, as anatomical features linking Ramapiths and Hominids. An attempt has been made to assess character correlation in order to avoid excessive duplication when compiling this tabulation. Some character overlap, however, still is evident. The polarity or status of the characters appears in the right-hand column according to the following conventions: D, derived unique character by comparison to sister taxon; P, primitive, ancestral character shared by sister taxon; C, convergent (or parallel) nonhomologous character indicating phenetic similarity only; V, variable character of doubtful phyletic significance due to its random expression in sister taxa under comparison or in other out-groups. Following the polarity or status designation is the source or sources which provided the interpretation. Note that some characters have more than one interpretation of polarity.

that bears no exclusive relationship to the Hominidae, has two subpositions: (1) Ramapiths as a group are specially related to the ancestry of the Asian Great Apes, the orangutans, and not to any other extant group, and (2) Ramapiths as a group are specially related to the ancestry of the Asian and African Great Apes and to hominids and thus provide the basal stock from which they are all derived. The first subposition, where Ramapiths and orangutans are viewed as sister groups (see Fig. 4), is supported by Andrews (1982, 1983), Ward and Pilbeam (1983), Boaz (1983), Cronin (1983), Gantt (1983), and Bernor (1983, and personal communication) among contributors to this volume. Further support can be found in Andrews and Cronin (1982), Lipson and Pilbeam (1982), and Martin and Andrews (1982). This view is premised on a series of shared derived characters (see Table 2) specially linking Ramapiths (especially *Sivapithecus*) with the ancestry of the Asian Great Apes (for morphological overview see section on cranial/dental evidence). Among contributors to this volume, Ward and Pilbeam (1983) most persuasively argue this view through the presentation of their dryomorph–ramamorph maxillofacial dichotomy. At the other end of the spectrum, Kay and Simons (1983) mount a rebuttal by challenging a number of the characters in Table 2. In my opinion, a major unresolved problem with this view is the relationship of Asian Ramapiths to African Ramapiths (e.g., *Kenyapithecus?*) with regard to the shared derived character complexes that indicate affinities with the Asian Great Apes.

The second subposition, where Ramapiths are viewed as the sister group of all extant Great Apes and the Hominidae (see Fig. 5), is supported by Wolpoff (1983), Zihlman and Lowenstein (1983), and Greenfield (1983). Further basic support for this view can be found among strict advocates of the biomolecular clock evidence (e.g., Sarich, in Cronin, 1983), since no direct linkages are drawn between extant hominoid genera and specific Miocene hominoid subgroups. Wolpoff (1983) provides the most well-rounded presentation of this view. Specifically, he sees the derived characters in Tables 1 and 2 supporting both views at the same time, indicating that the extant large hominoid cladogenesis took place *within* the Ramapith radiation. From the point of view of a cladist this situation is difficult to visualize. However, since the distributions of characters in Tables 1 and 2 are not known for each and every Ramapith taxon (especially with regard to potential African Ramapiths), this second subposition cannot be falsified on the basis of current knowledge.

Before leaving this section a few observations can be made about the origin, geographical range, and adaptations of Ramapiths. Assuming that African Ramapiths [in the sense of Pickford (1982)] and Eurasian Ramapiths are genealogically related, the earliest documented occurrence of a Ramapith is from western Kenya about 16.5 m.y.a. (Pickford, 1983). Ramapith origins can probably be traced to a *Proconsul*-like African Dryopith (Gantt, 1983; Pickford, 1982). In Asia, the earliest evidence of Ramapiths is from the Potwar Plateau of Pakistan about 12.5 m.y.a. with the suggestion that future collecting may produce finds on the order of 14 m.y.a. (Ward and Pilbeam, 1983). With the exception of *Sivapithecus*-related species from Greece and ?Czecho-

Table 2. Some Characters Common to Ramapiths and Orangutans with Interpretations of Their Polarity[a]

Character		Interpretations of polarity
1. Narrow interorbital septum	D	Andrews (1983)
	V	Kay and Simons (1983)
2. Deep zygomatic process with strong lateral flare	D	Andrews (1983)
	P	Kay and Simons (1983)
3. Zygomatic bone flattened and facing anteriorly	D	Andrews and Cronin (1982), Preuss (1982)
	V	Kay and Simons (1983)
4. Short, broad nasal aperture	D	Andrews (1983)
	P	Kay and Simons (1983)
5. High alveolar or premaxillary prognathism	D	Andrews and Tekkaya (1980), Wolpoff (1983)
	P	Kay and Simons (1983)
6. Broad I^1 that is much larger than I^2	D	Andrews (1983), Wolpoff (1983)
	C	Kay and Simons (1983)
	P	Preuss (1982)
7. Thick enamel caps on molars	D	Andrews (1983), Wolpoff (1983)
	P	Gantt (1983)
	V	Ward and Pilbeam (1983)
8. Deeply incised wrinkles that persist after crown morphology has been worn away	D	Wolpoff (1983), Gantt (1983)
	V	Andrews and Cronin (1982)
9. Central maxillary incisors that change angulation during life	D	Wolpoff (1983)
10. External rotation of upper canine roots	D	Ward and Pilbeam (1983)
11. Strong superior–medial angulation of canine roots	D	Wolpoff (1983)
12. Canine roots quadrangular in transverse section	D	Ward and Pilbeam (1983)
13. Well-developed canine fossa	D	Ward and Pilbeam (1983)
14. Orbits higher than broad, rather ovoid in shape	D	Andrews and Cronin (1982), Wolpoff (1983), Kay and Simons (1983), Preuss (1982)
15. Poor development of supra-orbital torus	P	Andrews (1982), Kay and Simons (1983)
16. No glabellar thickening	D	Andrews and Cronin (1982)
17. Zygomatic foramina relatively large	D	Andrews and Cronin (1982)
	V	Kay and Simons (1983)
18. Extremely small incisive foramina	D	Andrews and Cronin (1982), McHenry *et al.* (1980)
	V	Kay and Simons (1983)
19. Slit-like greater palatine foramina	D	Andrews and Cronin (1982)
	V	Kay and Simons (1983)

(continued)

[a]The characters listed in the left-hand column have recently been suggested as anatomical features linking Ramapiths and orangutans. An attempt has been made to assess character correlation in order to avoid excessive duplication when compiling this tabulation. Some character overlap, however, still is evident. The polarity or status of the characters appears in the right-hand column according to the following conventions: D, derived unique character by comparison to sister taxon; P, primitive, ancestral character shared by sister taxon; C, convergent (or parallel) nonhomologous character indicating phenetic similarity only; V, variable character of doubtful phyletic significance due to its random expression in sister taxa under comparison or in other out-groups. Following the polarity or status designation is the source or sources which provided the interpretation. Note that some characters have more than one interpretation of polarity.

Table 2 (*Continued*)

Character		Interpretations of polarity
20. Nasoalveolar clivus extends posteriorly into nasal cavity (subnasal plane smooth, not truncated)	D	Ward and Pilbeam (1983), Andrews and Cronin (1982), Lipson and Pilbeam (1982), Preuss (1982)
21. Subnasal alveolar process that appears elliptical in sagittal section	D	Ward and Pilbeam (1983)
22. Incisive fossa forms narrow depression just behind nasospinale (subnasal plane continuous with floor of nasal cavity, not stepped)	D	Ward and Pilbeam (1983), Andrews and Cronin (1982)
	V	Kay and Simons (1983)
23. Wide maxillary sinus that rapidly diminishes anteriorly	D	Ward and Pilbeam (1983)

slovakia at about 10–12 m.y.a., Ramapiths may have been largely absent from Europe (Martin and Andrews, 1982; Delson, personal communication). Bernor (1983, p. 57) suggests that the appearance of Ramapiths coincides with the development of a "seasonally adapted scherophyllous evergreen woodland biome" along with the adaptive radiation of woodland ruminants such as bovids and giraffids. The last record of Eurasian Ramapiths (with the exception of the *Gigantopithecus* and *Pongo* (?) lineages) occurs about 7 m.y.a. In Africa no fossil evidence exists to record their apparent demise. Bernor (1983) links the extinction of Ramapiths to the development of an increasingly seasonal environment, resulting in more woodlands and the evolution of a true open-country fauna.

Regarding the dietary and habitat preferences of Ramapiths, Kay (1981) notes that the thickly enameled cheek teeth characteristic of this group are not necessarily associated with ground feeding in primates but instead with specialized diets consisting of tough fruits, seeds, and nuts which are indicative of terrestrial *and* arboreal extant primates. Thick enamel in this sense can be viewed as an adaptation providing resistance against the occlusal stresses created by powerful vertical compression resulting from a "nut-cracking diet" (Kay, 1981). Gantt (1983) also notes that the thick enamel of Ramapiths is characteristic of living primates with frugivorous/graminivorous diets. Thus the inferred diet of Ramapiths cannot shed light on their habitat preference, although arguments based on body size and faunal associations (Andrews, 1983) indicate they were probably at least partially exploiting a terrestrial niche.

Cladistic Models of Hominoid Evolution

In the following sections a series of cladograms is presented that address the issue of relationships of modern hominoid groups to fossil groups from which they are possibly derived. Several of these cladograms deal with issues touched on in preceding sections of this chapter but are presented here to provide a more sequential view of the major events in hominoid evolution. It is naturally not possible to present every cladistic model considered by contributors to this volume. However, the models selected acknowledge all of the major issues and in the process provide a series of testable cases that can be modified as future evidence dictates.

Evidence for Initial Differentiation of Hominoids

The origins of the Hominoidea can be traced back to the Fayum catarrhine radiation (Fleagle and Kay, 1983; Fleagle, 1983a). In a cladistic sense, however, this is difficult to justify, since, as Fleagle and Kay (1983) demonstrate, the Fayum "hominoid" Propliopiths (*Aegyptopithecus* and *Propliopithecus*) share no derived features that link them with any Miocene taxa or with any extant catarrhine group. Falk's (1983) comments on the only known endocast of a Propliopith (*Aegyptopithecus*) also support this view. Thus, the Propliopiths occupy the position of primitive basal catarrhines and are suitable phyletic ancestors of the Hominoidea *sensu stricto* and Cercopithecoidea (Fig. 1).

Because of their uniformly primitive nature, the classification of Propliopiths within the Catarrhini presents a problem. Fleagle and Kay (1983)

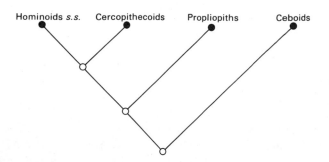

Fig. 1. Cladogram depicting relationship of the Hominoidea to other major groups in the Anthropoidea. Hominoids are here defined in the strict sense to include the six extant genera (*Hylobates, Symphalangus, Pongo, Gorilla, Pan, Homo*) and their hypothetical ancestors and collaterals, all of which share at least some of the suite of derived characters definable in the living apes and humans (see Table 3). The sister group of the Hominoidea *sensu stricto* is the Cercopithecoidea (Old World Monkeys), as diagrammed here. The Oligocene Fayum Propliopiths are, in turn, the sister group of the hominoid–cercopithecoid clade and may very well prove to be the ancestral stock from which that clade is derived (see text). Finally, the Ceboidea (New World Monkeys) join the others as the sister group of the Catarrhini. Note that the nonhominoid, noncercopithecoid Parapithecidae from the Fayum have not been diagrammed here since they have little bearing on the higher category relationships discussed herein. The solid circles in this cladogram represent existing phyletic groups, whereas the open circles define hypothetical ancestors.

argue for retaining the group within the Hominoidea in a separate subfamily of the family Pliopithecidae. This implies a broader definition of the Hominoidea and from a strict cladistic perspective is not wholly defendable since the Fayum Propliopiths are viewed as giving rise to *both* extant catarrhine superfamilies yet are ranked within the Hominoidea. One cladistically correct alternative would be to set up a separate superfamily for the Propliopiths; another would be to rank the Old World monkeys as a family within the Hominoidea. Since neither of these is an adequate arrangement in my opinion, I will support Fleagle and Kay's choice as a temporary albeit, "incorrect" solution.

Following Fleagle and Kay's (1983) lead, the realization that Old World monkeys are derived from a primitive hominoid ancestor might force reconsideration of the nature of the cercopithecoid adaptation. For instance, Andrews (1983) points out that the ancestral cercopithecoid morphotype represents the derived condition *vis à vis* hominoids with respect to their specialized bilophodont dentition, diet, habitat preference, and posture. Fleagle (1983a) and Rose (1983) add that the postcranial skeleton of cercopithecoids also represents the derived condition. Therefore, living cercopithecoids can be regarded, in many ways, as a wholly derived group, whereas living hominoids may approximate the primitive ancestral catarrhine condition to a much greater extent than has been previously considered.

Finally, the consideration that Propliopiths are ancestral to both cercopithecoids and hominoids is a delight to advocates of the biomolecular clock evidence (Cronin, 1983; Zihlman and Lowenstein, 1983), who have insisted for many years that the cercopithecoid–hominoid split postdated the Fayum catarrhine radiation. Cronin (1983) dates this divergence at 20 ± 2 m.y.a., while Boaz (1983) offers a paleontological–biomolecular averaged estimate of 26 m.y.a.

Evidence for Gibbon Lineage

The ancestry of the Hylobatidae is perhaps the most unresolved issue in the evolution of the Hominoidea. It is not so much a lack of fossil evidence that has caused this situation but rather the way that evidence is viewed. Fleagle (1983b) in a recent review of gibbon ancestry observes there is a notable lack of consensus of what a primitive gibbon should look like. Gibbons are phyletically the most primitive hominoids. Their clade represents the first branch of the hominoid cladogenesis that survived to the present. Yet, primitive as they are in some features, they are extraordinarily specialized in others (see morphotype 4 in Table 3). It is the muscular and skeletal specializations that make gibbons so unique but at the same time these specializations almost certainly mask much of the morphological evidence of gibbon heritage (Fleagle, 1983b).

Gibbon evolution was not widely discussed by contributors to this volume.

Ward and Pilbeam (1983) do point out that extant hylobatids evince the primitive "dryomorph" subnasal pattern. Gantt (1983) notes that *Hylobates* exhibits enamel prism Pattern 3A, like the great majority of the Hominoidea, and also has relatively thin molar enamel caps. Kluge (1983) observes that gibbons display a number of primitive hominoid soft anatomical features (see also morphotype 3 in Table 3). Fleagle and Kay (1983), Prasad (1983), and Chopra (1983) provide background discussion of some potential fossil gibbons. Finally, Cronin (1983) suggests the gibbon lineage diverged from the hominoid stem at 12 ± 3 m.y.a. based on molecular clock evidence and Boaz (1983) estimates a divergence date of 16.5 m.y.a. on the basis of averaged data sets.

Figure 2 illustrates three cladistic relationships of gibbons to other hominoid groups that have been proposed at various times. In a similar diagram Fleagle (1983*b*) cites the existence of four different relationships. These are both particularly apt representations of the unresolved nature of hylobatid ancestry. I will now briefly review some fossil evidence of the Lesser Apes in the hope of delimiting the choices available in Fig. 2.

Gibbons are today restricted to the rainforests of Southeast Asia. In the Pleistocene their range was considerably larger, since numerous sites in China, Indonesia, and Southeast Asia have yielded primate teeth indistinguishable from those of living hylobatids (Szalay and Delson, 1979). To my knowledge there is no corroborated evidence of Pliocene gibbons. From the late Miocene of North India Chopra and Kaul (1979) and Chopra (1983) describe the existence of a potential hylobatid left M^3 from the Nagri level beds of Quarry D at Haritalyangar. The Quarry D site has recently been dated by paleomagnetism at an age of 7.6 m.y.a. (G. Johnson, personal communication). In size and morphology, the left M^3 is remarkably similar to that of extant gibbons (Chopra, 1983, p. 543, Fig. 1). A significant derived character uniting this specimen with modern gibbons, according to Chopra (1983), is the relatively weak development of its lingual cingulum. Just as important in my estimation are both its phenetic similarity to living gibbons *and* its occurrence in late Miocene exposures of southern Asia, the correct temporal and zoogeographical provenience for an early gibbon. The specimen was named *Pliopithecus krishnaii* by Chopra and Kaul (1979) and more recently has been assigned to a new genus, *Krishnapithecus* Ginsburg and Mein 1980. I endorse this opinion since it is unlikely that the Quarry D M^3 was related to the primitive European catarrhine, *Pliopithecus.*

Prasad (1983) notes the occurrence of a mandibular fragment containing a partial M_3 from a Nagri level horizon near Haritalyangar. This specimen was originally described as *Hylopithecus* by Pilgrim (1927) and judged by him as a potential early hylobatid, a view favored by Simons and Pilbeam (1965). Prasad (1983) hints that the isolated M^3 from Quarry D at Haritalyangar now assigned to *Krishnapithecus krishnaii* may represent the same primate taxon as *Hylopithecus*. Recently, I had the opportunity to examine the type specimens of both of these species in India in 1982 and can confirm that they are quite

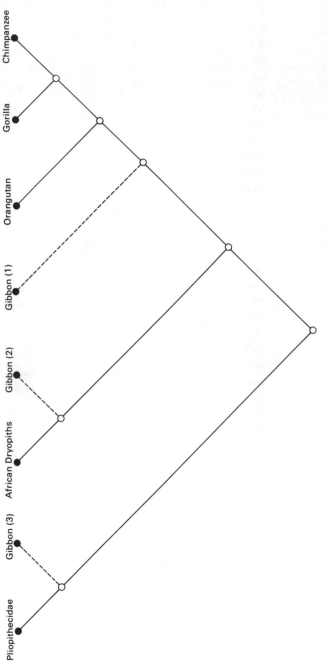

Fig. 2. Cladogram depicting three potential cladistic relationships of gibbons (Hylobatidae) to other members of the Hominoidea. In the gibbon (3) clade, hylobatids are represented as the sister group of the Pliopithecidae based on their long-noted resemblance to *Propliopithecus* and *Pliopithecus*. In the gibbon (2) clade, hylobatids are viewed as the sister group of the African Dryopiths because of their resemblance to *Dendropithecus* and *Micropithecus*. In the gibbon (1) clade hylobatids are presented as the sister group of the extant Great Apes without further identifying their relationship to any particular fossil group. This implies an ancestry more closely linked with the cladogenesis of the Great Apes. The solid circles in this cladogram represent existing phyletic groups, whereas the open circles define hypothetical ancestors. Note that in succeeding cladograms the position of the hylobatid clade is always indicated by a dashed line to indicate that its position is conjectural.

different in size and morphology. The *Hylopithecus* type is now best viewed as further evidence of a middle-sized and thin-enameled Dryopith in Asia, while the *K. krishnaii* specimen is very possibly an early gibbon.

The scant but tantalizing evidence of a fossil gibbon from Quarry D at Haritalyangar* in North India takes on additional significance in light of new discoveries at the Lufeng hominoid locality in southern China. This locality, recently dated to 8 m.y.a. (Flynn and Qi, 1982), has yielded a partial face, about a dozen jaws, and many isolated teeth of a small hominoid with distinct gibbon affinities (E. Delson, personal communication). An even earlier find in China, *Dionysopithecus* (Li, 1978), from the late middle Miocene (?12 m.y.a.), also has distinct gibbon affinities. Though only known from a partial maxilla, it is morphologically, temporally, and paleozoogeographically the most likely candidate for the title of earliest true fossil gibbon (see also Fleagle, 1983*b*). However, its uncanny resemblance to *Micropithecus* (Fleagle, 1978) from the early Miocene of East Africa forces consideration of that taxon as a possible precursor of *Dionysopithecus*. *Micropithecus* is known from a larger sample of specimens and is decidedly part of the early Miocene African Dryopith radiation. Yet its maxilla differs in virtually no significant way from the type of *Dionysopithecus* (Fleagle, 1983*b*). Both taxa were described within a few months of each other, but unfortunately neither had associated postcrania. Fleagle (1983*b*) points out that should future finds of *Dionysopithecus* establish its identity with *Micropithecus*, then the former nomen would assume priority. This would then extend the temporal and paleozoogeographic range of the ancestral gibbon lineage to between 17–19 m.y.a. in East Africa. This would conflict with the biomolecular data and would also usher in the return of an "African fossil gibbon," a concept with which I have never been comfortable. This concept seems especially unlikely when one considers the newly documented Asian *Sivapithecus*–orangutan link (see following section).

Reviewing the evidence for a still more distant ancestry of the Lesser Apes necessitates comparisons with the Pliopithecidae. *Propliopithecus* is a particularly unlikely candidate for the title of earliest gibbon, due to its complete lack of extant catarrhine synapomorphies (see also comments in preceding section). *Pliopithecus* also has neither the derived cranial features (Fleagle and Kay, 1983) nor postcranial specializations (Fleagle 1983*a*) that would link it with the ancestry of gibbons. As concluded by earlier researchers (e.g., Remane, 1965; Ciochon and Corruccini, 1977*a*; Szalay and Delson, 1979) and supported in this volume by Fleagle and Kay (1983), *Pliopithecus* is best regarded as a phyletically primitive catarrhine related most closely to the Fayum Propliopiths. Though slightly more derived in a few features, many of these same arguments can also be applied to the early Miocene African *Dendropithecus* in rejecting its ancestral gibbon status (Ciochon and Corruccini, 1977*a;* Fleagle, 1983*b*).

---

*New excavations at Haritalyangar under the direction of faculty at Panjab University and with the proposed participation of this volume's editors may soon supplement the currently meager fossil record of *K. krishnaii*.

Fleagle (1983*b*) summarizes and reflects on the way past interpretations have been drawn from the fossil record of gibbon ancestry. He states:

> . . . fossil gibbons have in virtually all cases been identified on the basis of either primitive hominoid features or 'trends,' rarely on unique 'shared-derived' features linking the fossils with the extant taxa. Therefore, at present, our understanding of extant hominoid phylogeny contributes more to our interpretation of the fossil record of gibbon evolution than *vice versa*.

From Fleagle's comments and my discussion of the fossil evidence for gibbon ancestry I feel it is now possible to reject the relationships (see Fig. 2) of the gibbon (3) clade and the gibbon (2) clade without further discussion. The gibbon (1) clade relationship has no fossil branches indicated, yet I do feel there is now evidence in the 7–12 m.y.a. time range in southern Asia for early gibbons. Perhaps the cladogram should be redrawn to include an Asian Dryopith (?) branch to represent such forms as *Krishnapithecus*, Lufeng fossil "gibbons," and possibly *Dionysopithecus* (see Fig. 8 in Appendix). Even if future discoveries prove that some or all of these taxa are not early gibbons, I suggest that the search for evidence of fossil hylobatids should be concentrated in the later Miocene and Pliocene of southern Asia among the small-bodied Asian Dryopiths.

Evidence for Orangutan Lineage

Until quite recently the fossil record of the orangutan lineage was limited to a few isolated teeth of questionable affinity from the Siwaliks, such as "*Paleosimia*" Pilgrim 1915 (see also Prasad, 1983), and to hundreds of isolated teeth from the early Pleistocene to Recent deposits of southern China, Thailand, Malaysia, Sumatra, Java, and Borneo (Szalay and Delson, 1979). With the announcement and interpretation of a new partial skull of *Sivapithecus* (Pilbeam, 1982; Andrews, 1982; Ward and Pilbeam, 1983) from the U level (~8 m.y.a.) of the Potwar Plateau of Pakistan our understanding of the fossil record of orangutans has increased dramatically. This unique discovery not only documented the earliest evidence of the orangutan lineage, it also altered many researchers' perceptions of the phyletic relationships of Ramapiths as a group (see previous discussions in the sections on cranial/dental evidence and on the status of "*Ramapithecus*").

The change in viewpoints regarding the affinities of Ramapiths in general and *Sivapithecus* in particular can be related directly to the identification of a suite of derived craniofacial features in several new fossil skulls (see Table 2). From this perspective it is useful to review the opinions expressed over the last decade regarding the position of *Pongo* in hominoid evolution (compare Figs. 3–5). In my opinion the current evidence from craniofacial anatomy makes Fig. 4 the most likely choice. This view is also supported by a plurality of contributors. Specifically, Boaz (1983, p. 708) suggests that *Sivapithecus* "is ancestral at least at a generic level to *Pongo*." Ward and Pilbeam (1983, p. 220) observe that the subnasal pattern of *Sivapithecus* and *Pongo* "indicates a close phylogenetic relationship." Finally, Andrews (1983, p. 457) concludes that *Sivapithecus* "may be considered to be broadly ancestral to the orangutan." Wolpoff (1983) takes a slightly different view by suggesting that *Pongo* as well as the African ape–human clade are derived from within the Ramapith radiation, indicating his support of a Fig. 5 arrangement. Kay and Simons (1983, p. 616), on the other hand, support the alternative in Fig. 3 by cautiously stating, "African Great Apes and *Pongo* derive from stocks of thin-enameled apes represented perhaps collaterally by western and northern European later Miocene *Dryopithecus*." They suggest that *Dryopithecus* in Europe may represent a similar grade of development in postcranial, dental, and gnathic structure relative to that seen in Ramapiths and the living Great Apes. Kay and Simons (1983) also strongly question the validity of the *Sivapithecus–Pongo* link (see disputed characters in Table 2), although their own scenario for the origin of *Pongo* is equally open to challenge.

Putting aside the question of later Miocene *Sivapithecus*, it is worth mentioning that Chopra (1983) describes and names a new genus and species of "orang-like" hominoid based on the occurrence of an isolated left M^1 from the upper middle Miocene (Chinji level) deposits of Ramnagar in North India. According to Chopra (1983, p. 544), "*Sivasimia chinjiensis* is significantly different from all other Chinji level hominoids in possession of strongly wrinkled enamel on its occlusal surface." Chopra feels this character and other aspects of the cusp morphology indicate potential phyletic ties to the orangutan. Von Koenigswald (1981) has also recently described a new genus and species of hominoid based on an isolated molar (right M_2) from Chinji level beds near Chinji, Pakistan (von Koenigswald, 1983, p. 520, Fig. 1). This M_2 exhibits wrinkling of its occlusal enamel surface like the Ramnagar specimen. However, von Koenigswald (1981, 1983) does not posit phyletic ties to *Pongo*, but rather, he proposes that *Chinjipithecus atavus* von Koenigswald was ancestral to the late Miocene *Gigantopithecus giganteus*. Both of these isolated molars appear to be unerupted tooth germs. This may account, in part, for the high degree of occlusal enamel wrinkling that characterizes each specimen. On the other hand, it is possible that one or both of these specimens could indicate earlier evidence of the orangutan lineage. However, observations made on the molars of the Potwar *Sivapithecus* skull (GSP 15000) show no significant degree of occlusal enamel wrinkling (Pilbeam, 1982, and personal communication). Therefore it is unlikely that these Chinji level molars would be exhibiting derived occlusal features shared with extant orangutans when the more recent Nagri level GSP 15000 does not appear to show these features. To complicate matters, the Chinese Lufeng *Sivapithecus* material (of Nagri age equivalence) is characterized by strong enamel wrinkling (Pilbeam, personal communication; Wolpoff, personal communication).

It should be noted here that there is one paleoecological obstacle which does not lend support to the derivation of *Pongo* from a *Sivapithecus*-like ancestor. Andrews (1983) acknowledges that if the temperate woodland-forest-adapted omnivore *Sivapithecus* is the closest known relative of the tropical-forest-adapted frugivorus orangutan, then (1) the ancestral orangutan could

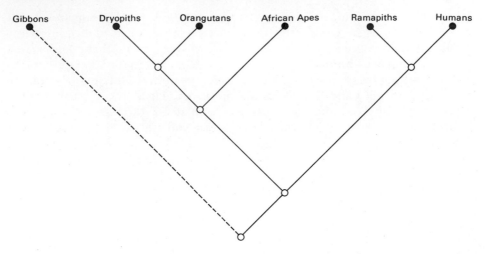

Fig. 3. Cladogram depicting the traditional view of hominoid relationships supported by many researchers during the 1960s and early 1970s. Ramapiths are viewed as the sister group of humans and were considered by many to be the earliest representatives of the Hominidae. Orangutans and the two African apes were thought to be independently derived from different groups of Dryopiths and are here represented as sister groups of the Dryopiths. Gibbons were considered a distant outgroup not closely related to the Great Ape–human radiation. This view of hominoid relationships has maintained a group of adherents and recently resurfaced in a much more sophisticated formulation (see text and Table 1). Solid circles and open circles defined as in Figs. 1 and 2.

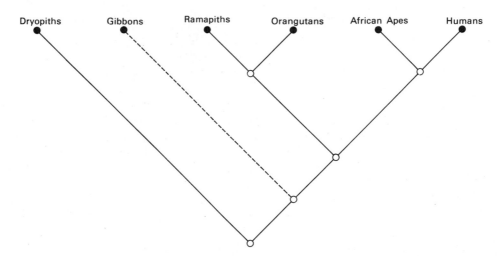

Fig. 4. Cladogram depicting a recent view of hominoid relationships supported by a growing number of researchers. The African apes are viewed as the sister group of humans, reflecting a close phyletic relationship between the two groups. Orangutans and Ramapiths are considered sister groups based on a suite of newly recognised derived characters (see text and Table 2). Dryopiths are considered a distant out-group defined mainly on the basis of primitive characters (see Table 3). The gibbon clade is placed in a position more closely related to the ancestry of the Great Apes but its exact position *vis à vis* Ramapiths and Dryopiths is still viewed as conjectural. Solid circles and open circles defined as in Figs. 1 and 2.

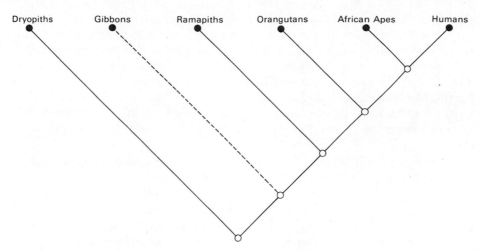

Fig. 5. Cladogram depicting relationships within the Hominoidea supported by morphologists and molecular biologists who favor a temporally recent cladogenesis of the extant apes and humans. The Ramapiths are here considered the sister group of the extant Great Apes and humans, which indicates that the derivation of those groups would postdate the earliest known occurrence of Ramapiths. Note also that humans and the African apes share a sister group relationship which is next joined by the orangutan clade. The position of the gibbon clade is once again indicated as conjectural; some researchers who favor the biomolecular evidence have even suggested its positioning as the sister group of the Great Ape–human clade. Following this view, researchers would have to look no further than the Ramapiths to derive all the extant Hominoidea. Dryopiths occupy the extreme out-group position based on their suite of shared primitive characters (see Table 3). Solid circles and open circles defined as in Figs. 1 and 2.

have been more terrestrial and omnivorous in the recent past and/or (2) a major adaptive shift could have taken place along the *Sivapithecus–Pongo* lineage, resulting in the development of an arboreal frugivore from partly terrestrial omnivore. Smith and Pilbeam (1980) and Ward and Pilbeam (1983) provide elements of support for both of these views. However, this is one issue where parsimony is not apparent and adequate resolution may have to await the recovery of a middle Pliocene orangutan with associated floral remains.

Finally, to close on a more positive note, it is worthwhile once again emphasizing that the newly proposed "*Sivapithecus* as orangutan ancestor model" is basically consistent with the molecular clock estimate of 10 ± 3 m.y.a. for the divergence of *Pongo* (Cronin, 1983; Andrews and Cronin, 1982). Boaz (1983) arrives at a similar "averaged age" estimate of 11.0 m.y.a. for the orangutan split with the African ape–hominid clade. Zihlman and Lowenstein (1983) also offer an estimate of 9–10 m.y.a. for this same event.

Evidence for African Great Ape Lineage

Fossil evidence for the initial differentiation and later evolution of the African Great Apes is lacking. Except for the brief mention of an incisor from the lower Pleistocene of Uganda described by Von Bartheld *et al.* (1970) as an early gorilla,* not one single piece of fossil evidence is available to document the lineage leading to the African apes. Therefore many contributors to this volume have chosen to combine their discussions of the early evolution of the African apes with the early evolution of the hominid line. This view has been adopted by those supporting African apes and humans as sister groups (as in Figs. 4 and 5). Especially prominent among this group are contributors suggesting a relatively recent divergence of the African Great Ape and hominid clades.

Wolpoff (1983) believes that a suitable Ramapith ancestor of the African ape–human clade will be found in later Miocene exposures of the African continent. This hypothetical Ramapith, according to Wolpoff, would have a chimpanzee-like cranial vault with a Ramapith-like face and dentition equipped for handling anterior dental loading stresses. Wolpoff (1983) sees this hypothetical ancestor fairly closely resembling the known cranial parts of *Australopithecus afarensis* [see figures in White *et al.* (1983) for examples]. Zihlman and Lowenstein (1983) approach the African ape–human common ancestor concept from a different perspective based on analogs to living forms. They argue that "pygmy chimpanzees (*Pan paniscus*) are more generalized than the common species and make a reasonable beginning point for the evolution of three very different adaptations: the slender, bipedal hominids, the large dimorphic gorillas, and the moderately dimorphic fruit-eating common chimpanzees" (p. 685). The term "generalized" to Zihlman and Lowenstein (1983) implies "approaching the ancestral state" and thus they consider *P. paniscus* has undergone the least amount of morphological change *vis à vis* the last common ancestor of the African ape–hominid clade, which makes it a suitable candidate to represent that ancestor.

Wolpoff (1983) comments further on the common ancestor of chimpanzees, gorillas, and humans, suggesting this ancestor will *not* resemble a pygmy chimpanzee to a significantly greater degree than *A. afarensis* already does. Now, since Zihlman and Lowenstein (1983, p. 687–688, Fig. 1 and Table 1) draw attention to a large number of features shared by both *Pan paniscus* and *Australopithecus*, it appears on the surface that Wolpoff's views and their own are not so different. However, Wolpoff (1983, p. 666) also suggests that "the last common ancestor of the African apes and humans was probably as different from living apes as it was from living humans. No living ape . . . could form an adequate model for this ancestral form, any more than a living human group could." This interchange leads me to suggest that Wolpoff and Zihlman–Lowenstein are arguing at cross purposes. In my estimation they actually have both defined elements of the same African ape–human ancestral morphotype (see morphotype 9 in Table 3) but are in

---

*Morphologically this specimen is somewhat different from both modern gorillas and chimpanzees, but in terms of size it is more like a chimpanzee. The structure of its enamel will soon be examined with a scanning electron microscope to establish its hominoid primate status (L. Martin, personal communication).

disagreement over its semantic expression. The ancestor probably was a generalized chimplike creature with a primitive hominoid "dryomorph" subnasal morphology (Ward and Pilbeam, 1983). It probably was a Ramapith, as Wolpoff (1983) suggests, but a uniquely African variety, which might better be termed a "dryomorph Ramapith."* When fossil evidence of the African ape/human clade is one day recovered, these observations will be ultimately falsifiable.

Those contributors supporting a more ancient origin of the African apes do not engage in the same sort of scenario-building as that presented in the discussion above. Rather, they emphasize the Dryopith–Ramapith dichotomy linking the origin of the African apes (and usually *Pongo*) to the Dryopiths and the origin of the Hominidae to the Ramapiths (see Fig. 3). As outlined in the preceding section on the orangutan lineage, Kay and Simons (1983) suggest deriving the African apes from later Miocene thin-enameled Dryopiths, perhaps represented, in a collateral sense, by European *Dryopithecus*. Pickford (1982, 1983) suggests a similar scheme, focusing on the East African late lower Miocene Dryopiths, from which he derives the "Pongidae." Pickford (1982) specifically excludes *Proconsul* from the "Pongidae" since it does not possess the apelike facial anatomy that he suggests later Dryopiths share. De Bonis (1983) also looks to the Dryopiths for the ancestral stock of the African apes. Based on dental and postcranial features, he suggests that European *Dryopithecus* and *Pan* are sister groups, with *Gorilla* joining the clade next, followed by *Proconsul*. From this arrangement he dates the *Pan–Gorilla* divergence to between 14 and 22 m.y.a.

One interesting outgrowth of de Bonis' (1983) cladistic analysis is that he regards knuckle-walking as a shared derived behavioral mode specific to the African apes. It has long been known that the African apes and humans are very similar in their upper arm anatomy but quite divergent in most aspects of lower limb anatomy. These differences naturally relate to human bipedality vs. African ape knuckle-walking quadrupedality. For many years it was assumed that knuckle-walking was part of the ancestral "pongid" morphotype

---

*Ward and Pilbeam (1983) and Ward *et al.* (1983) suggest that the ancestor of the African ape–human lineage was characterized by the primitive "dryomorph" subnasal pattern. Yet, in many other respects this ancestor appears derived in the direction of the Ramapiths. I would like to suggest that the dryomorph–ramamorph maxillofacial dichotomy may not totally reflect the morphological variation in the Dryopith–Ramapith categories I have proposed. Suppose that the dryomorph–ramamorph differences are analogous to the sciuromorph–hystricomorph maxillofacial differences seen in living and fossil rodents (Wood, 1980; Lavocat, 1980). If so, the terms might be applied in a more descriptive sense. For instance, African Dryopiths might be termed "dryomorph Dryopiths," whereas Asian Ramapiths could be termed "ramamorph Ramapiths," while African Ramapiths would be termed "dryomorph Ramapiths." This scheme would preserve the nature of the shared derived morphology that unites Asian Ramapiths with the orangutan clade but leave open the possibility that the African ape–human clade was derived from a African Ramapith such as *Kenyapithecus*. The split between African and Asian Ramapiths could be as old as 15 m.y.a. (Bernor, 1983). Therefore this proposal would not alter any of Ward and Pilbeam's conclusions. This proposal also would be clearly testable with the recovery of an African Ramapith that has the maxillofacial region intact.

and therefore that the earliest hominids had passed through a knuckle-walking stage in human ancestry. Once it became possible to isolate the anatomical correlates of knuckle-walking in the wrists and hands of the African Great Apes (Tuttle, 1969, 1975; Corruccini, 1978), this information could be applied directly to the hominid fossil record. To date, none of the African ape knuckle-walking features have been found in the wrist or hand bones of any fossil hominid, including the recently described *A. afarensis* (Tuttle, 1981; Bush *et al.*, 1982). It may therefore be assumed that knuckle-walking is a shared derived character complex that arose in the African Great Apes (see morphotype 10 in Table 3) after the separation of the hominid clade. Regarding the uniqueness of knuckle-walking in the African apes, it should be noted that the existence of a parallel behavior has been documented in a captive orangutan (Tuttle and Beck, 1972; Susman, 1974). This observation prompted Tuttle (1974, p. 397) to remark:

> . . . if a highly advanced arboreal climber and arm-swinger like the orangutan can develop ontogenetically into a knuckle-walker (albeit in special circumstances of captivity), ancestors of the African apes may have been somewhat similarly predisposed to knuckle-walking by their own arboreal heritage. It also seems reasonable to argue that if the common ancestor of chimpanzee and gorilla possessed somewhat advanced features of the hand for suspensory behavior and climbing, it could have developed into a knuckle-walker quite rapidly after adopting terrestrial habits.

Thus Tuttle (1974) offers a scenario for the unique development of knuckle-walking in the African apes based on "preadaptive" features of the common hominoid arboreal heritage.

Evidence for Differentiation of the Hominid Lineage

The origin of the Hominidae has been the central theme of paleo-anthropological studies for more than a century. It is therefore not surprising, then, that it is also the central theme of many chapters in this volume. The issues raised here by contributors focus specifically on the early differentiation of the hominid clade with regard to its geographical and temporal placement and its adaptive nature. Later aspects of the hominid cladogenesis were not the primary focus of this volume on Miocene Hominoidea, but nevertheless have also received a well-documented coverage (e.g., White *et al.*, 1983; Boaz, 1983).

As long ago as 1871 Charles Darwin observed that Africa was the probable continent of origin of our early progenitors, based on human resemblances to the African Apes. The first archiac human discoveries in Java and China appeared to shift the balance of opinion on continental origins to Asia. Thus the Africa/Asia issue in hominid origins became a major point of contention as paleoanthropological discoveries began to document the human fossil record. The issue of continent of origin of the hominid clade has been discussed by a variety of contributors in this volume. Basically, those who support Ramapiths as exclusive hominid ancestors (as in Fig. 3) favor the Asian continent, where most of the fossil evidence has been found. Those contributors viewing Ramapiths as specially related to the orangutan (as in Fig. 4) or who suggest Ramapiths as the common ancestor of all the Great Ape and human line (as in Fig. 5) tend to favor Africa. It should be noted that my use of the word "favor" in reference to a contributors' position is a rather reductionist view. The basic issue is that few contributors actually discuss the paleozoogeographic implications of their viewpoints, with the notable exception of Bernor (1983).

Bernor presents an in-depth overview of the zoogeographic relationships of the Miocene Hominoidea. In the later Miocene he documents the spread of truly open-country habitats (Bernor, 1983, p. 51, Fig. 6) that developed in parts of the Old World in the Turolian (10–5 m.y.a.). Bernor argues that though Ramapiths were the most open-habitat-adapted of the Miocene hominoids, they could not make the shift to these new habitats. Thus the presence of a seasonally-adapted Turolian open-country biotope interposed between the Siwalik Province and the East African Province provided an effective ecological barrier to hominoid migrations between the two regions by the earliest Turolian (~10 m.y.a.). Bernor (1983) also notes that throughout the Miocene the predominant mammalian dispersal pattern appears to have been into, not out of, the Siwalik Province, making it a sort of provincial biogeographic *cul-de-sac*. Combining these points with the earliest occurrence of bipedal hominids in Africa at the 3–4 m.y.a. time range, Bernor (1983, p. 59) concludes that "the evolution of *Australopithecus* would most likely have been an African event," probably confined to sub-Saharan Africa. Thus Bernor presents specific zoogeographic evidence excluding Siwalik Ramapiths (and other Eurasian hominoids?) from participation in the evolution of the earliest documented hominids. This does not prove an African origin of the Hominidae, but it strongly supports one.

The temporal placement (dating) of the origin of the Hominidae has always been a thorny issue. This probably can be traced back to the Victorian Age, when Thomas Henry Huxley first documented our close affinity with the African apes and Charles Darwin expounded on the "missing link" concept in human evolution. Simply stated, if the common link between ape and human could be demonstrated as ancient, then it was surmised that humans would be less closely related to the living apes. Naturally, the converse would apply if the "link" was more recent. Among contributors to this volume, the actual dating of the hominid lineage was closely dependent upon one's cladistic views of hominoid relationships (compared in Figs. 3–5). If it can be assumed that views favoring the African apes and humans as sister groups among living hominoids are the more correct interpretations [*contra* Kluge (1983), Kay and Simons (1983)], then the biomolecular data set can be brought to bear directly on this issue. Zihlman and Lowenstein (1983) estimate the human–African ape divergence at 5–6 m.y.a. based on their interpretation of the molecular clock. Cronin (1983) also discusses a similar age estimate, but later states "I am increasingly confident that the hominid–pongid divergence is in the range of 4.0–8.0 m.y.a." (p. 134). I support this more liberal phrasing of the dating issue and also submit that Goodman *et al.* (1983) and others will probably tone down their criticisms of clock dates based on this interpretation.

If the origin of the Hominidae is in the 4.0–8.0 m.y.a. time range, how early can we reasonably expect to document this lineage in the fossil record (based on shared derived characters like those in morphotype 11 of Table 3)? Greenfield (1983, p. 700) indirectly answers this question by concluding that divergences "may or may not have immediately given rise to clearly distinguishable members of independent lineages leading to humans and the extant Great Apes." Therefore, I feel that isolating the exact point of divergence of earliest Hominidae in the fossil record may be an unresolvable issue. This is precisely why students of cladistic analysis emphasize the need to document sister group relationships rather than search for perhaps unknowable ancestors. This is also why the biomolecular evidence may prove more useful than the fossil evidence for defining the precise divergence date of two extant lineages. However, the biomolecular data cannot at present resolve the relationship of fossil groups to one another or to modern groups (but see Lowenstein, 1982) so both sources of evidence *must* be used in tandem to develop an "alpha phylogeny."

Cronin (1983) presents a genetically based, time-oriented scenario for the origin of the hominid lineage. He suggests that the actual morphological shift from an apelike forebear to a distinguishable early hominid occurred relatively rapidly, perhaps in only 1 or 2 million years (approximately 50,000 to 100,000 generations). Cronin (1983, p. 131) states:

> A rapid shift in locomotor adaptations was the first major derived change, followed by unique changes in the brain and dentition, etc. One need not posit millions of years for the initial hominid adaptations to evolve from the ape ancestral conditions. Rates of evolution vary dramatically and traits may be responding to either intense selection or random fixation. Small population size, coupled with inbreeding, may facilitate rapid change during the process of speciation. This type of small population bottleneck, or small population mode of allopatric speciation, may have predominated in ancestral hominid populations.

Zihlman and Lowenstein (1983) also observe that origin of the Hominidae from an apelike (pygmy chimp-like) precursor may have taken place in 1 million years or perhaps less time. They speculate on whether 1 million years of transformation can be viewed as a gradual or punctuated event in the context of evolutionary processes. I think it probably comes as no surprise that these biomolecular-related views of hominid origins are characterized by a temporally rapid shift for the origin of the hominid lineage. Since these contributors both favor a late ape–human divergence, the transitions they propose are concordantly rapid. Those who favored an earlier divergence, such as in the now-dated "*Ramapithecus*" as early hominid model, usually supported a more gradual transition in morphology. Because I consider that the origin of higher categories occurs primarily by rapid, steplike, punctuated events, I find myself in full agreement with the views offered by Cronin and by Zihlman and Lowenstein.

The adaptive nature of the earliest hominids, or rather the delineation of the behavioral and morphological feedback systems that resulted in the acquisition of our distinctly human characteristics, has been the subject of much recent discussion (e.g., Lovejoy, 1981). Wolpoff (1983) among contributors to this volume touches on this issue more than any other contributor. He suggests that the common ancestor of hominids and the African apes was probably a special form of Ramapith. Based on shared similarities of African apes and humans, together with evidence from the hominid fossil record, Wolpoff (1983, p. 667–668) characterizes this proto-African ape, proto-human as:

> . . . behaviorally complex, a tool user, and a rudimentary tool maker, an omnivore utilizing a wide variety of dietary resources ranging from foods difficult to masticate to protein obtained through organized hunting and systematically shared by at least part of the social group, an incipient biped (the behavior was possible but not morphologically efficient), and just possibly a more complex communicator than generally thought, utilizing an albeit limited but symbol-based open communication system.

Wolpoff observes that a shift in importance of these features was as much responsible for the origin of the hominid clade as the acquisition of additional hominid apomorphous features. Therefore, it may not have been development of new characteristics that marked the adaptive shift of the earliest hominids, but rather an emphasis on specialization of an already existing suite of features. This view of the hominization process differs from the conventional one in that many of the previously considered uniquely hominid adaptations would have been at least incipiently present in the earliest African ape. As the African ape lineage developed, these features would have been subject to an alternate set of selective forces and modified still further after the separation of the *Pan* and *Gorilla* lineages. If Wolpoff (1983) is correct in this view, then the early members of the African ape clade would have been more "human-like" in their behavior and morphology than either of the living African Great Apes. Of course, the need then arises to explain how many of the traditional, once sacred, hominid features arose in the common proto-African ape proto-human ancestor. So the question of adaptive significance has simply been shifted to a different level.*

---

*In this discussion, one additional question should be posed: What factors can be *causally related* to the divergence of the African ape and hominid lineages? Wolpoff (1983) along with many other paleoanthropologists, considers the exploitation of open woodland/parkland and savanna habitats by the earliest hominids an important factor in the origin of the Homindae. What event in the 4–8 m.y.a. time range may have triggered the spread of these particular habitats in Africa? I suggest that the desiccation of the waters of the Mediterranean Basin and Black Sea together with the development of a cooling trend in the adjacent oceans (Hsü, 1972, 1978; Bensen, 1976, Hsü *et al.*, 1978) is causally related to dramatic changes in climate on the African subcontinent. This drying out of the Mediterranean region, the "Messinian Event" as it is often termed, is dated to 6.5 m.y.a. Perhaps as huge deposits of salt accumulated in the Mediterranean Basin and desertic conditions prevailed in the region, this climatic shift could have triggered the development of more seasonal environments in parts of Sub-Saharan Africa resulting in the appearance of dry open-woodland and savanna habitats. Thus the Messinian Event may have been a determining factor in the divergence of the African ape and hominid lineages about 6.5 m.y.a.

One aspect of Wolpoff's (1983) hominid origins model that deserves further discussion is his characterization of the common African ape–human ancestor as an "incipient biped." I take this to mean that the common ancestor was more predisposed to bipedalism than are the extant knuckle-walking chimpanzee and gorilla. If so, then Wolpoff's model supports aspects of Tuttle's (1974, 1975, 1981) hylobatian model. Specifically, Tuttle (1981) argues that arboreal bipedalism on horizontal boughs and vertical climbing on trunks and vines were aspects of a locomotor repertoire that evolved *before* the emergence of the Hominidae. Tuttle's pre-hominids utilized bipedal running and hindlimb-propelled leaps to forage in the arboreal milieu. In this sense their arboreal behavior was preadaptive for terrestrial bipedalism. Stern and Susman (1981) also relate the arboreal activity of vertical climbing to the emergence of terrestrial bipedalism, emphasizing its preadaptive basis. They conclude that the earliest hominids were at least partially arboreal. These points all focus attention, in my opinion, on the adaptive mosaic that must have characterized the very early Hominidae.

This early phase of hominid evolution is not well documented in the fossil record. However, by the 3–4 m.y.a. time range in East Africa the record becomes considerably more complete. It is from this time period that the earliest evidence of bipedalism is documented in taxon *Australopithecus afarensis* Johanson *et al.* 1978. With regard to *A. afarensis*, the critical issue lies not in the nature of its bipedal adaptation, but rather in its phylogenetic affinities. Boaz (1983, p. 712) observes that "The primary question in regard to *A. afarensis* is not whether it represents a separate (though perhaps earlier and closely related) species distinct from *A. africanus,* but is the assessment of its phylogenetic position." Once raising this issue, Boaz (1983) addresses two questions: (1) Whether the *A. afarensis–Homo* relationship is valid and (2) whether *A. africanus* is the ancestor of robust australopithecines. Boaz (1983) argues the evidence for a link between the hominid populations of Laetoli and Hadar (*A. afarensis vide* Johanson *et al.* 1978) and the first species of *Homo* is tenuous. He observes there is a temporal and morphological gap spanning this phylogenetic connection that is better bridged by the gracile australopithecines from South Africa. In this regard he would place the southern African (Sterkfontein and Makapansgat) and eastern African (Laetoli and Hadar) gracile australopithecines in the same species, *A. africanus.* Boaz (1983, p. 713) concludes that a more parsimonious hypothesis is one where the "South African *A. africanus* represents the population descendent from earlier, slightly more primitive *A. africanus* and ancestral to early *Homo.*"

Regarding the second question concerning Johanson and White's (1979) views on the origin of the *A. robustus/boisei* lineage, Boaz (1983) critiques the evidence for *in situ* evolution of robust australopithecines in South Africa. Since they appear rather suddenly in the fossil record of East and South Africa (about 2.0–2.1 m.y.a.), Boaz suggests the robust forms may have evolved elsewhere on the African continent (by allopatric or parapatric speciation) and dispersed into East and South Africa in the latest Pliocene. He then briefly discusses the South African australopithecine record, commenting on the lack of appropriate "robust-like" ancestors.

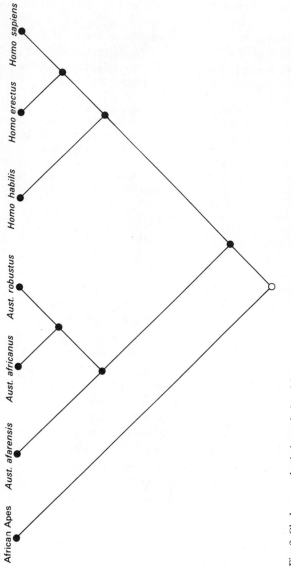

Fig. 6. Cladogram depicting relationships among currently recognized taxa of the Hominidae based on the most recent synthesis of the data (see Chapter 29 and text). Note that the genus *Australopithecus* shares a sister group relationship with the genus *Homo*. Also note that the African apes are viewed as the sister group of Hominidae, which would indicate that these groups shared a common ancestry subsequent to the Ramapith radiation. Within the Hominidae, *A. afarensis* occupies a unique position as the sister group of all later hominids. *Homo habilis* occupies a similar position within the genus *Homo*. Solid circles in this cladogram represent existing phyletic groups; the single open circle represents the hypothetical common ancestor of the African apes and the earliest known hominid, *Australopithecus*.

Boaz (1983) does focus on two valid points of contention concerning the phylogeny and scenario proposed by Johanson and White (1979). However, a more recent contribution on the subject (White *et al.*, 1983) in my opinion firmly establishes *A. africanus* as the sister group of the robust australopithecines (see Fig. 6) and thus very probably the ancestral stock from which that group is derived. Furthermore, White *et al.* (1983) demonstrate beyond any reasonable doubt that *A. afarensis* is a species distinct from *A. africanus* and in those characters in which they differ, the former represents a better ancestral morphotype for *Homo* than the latter. This was all accomplished by carefully circumscribing and then minutely defining the variation present in the taxon *A. africanus*. White *et al.* (1983) observe:

> Although *A. afarensis* and *A. africanus* share many common primitive features, the latter taxon exhibits a morphological composite of the skull and dentition derived toward the *A. robustus* + *A. boisei* character state. This pattern and all its components can be related to a functional trend toward maximizing and spreading vertical occlusal forces along the postcanine tooth row in response to dietary specializations. Members of the *Homo* clade do not show this specialization but exhibit a pattern indicative of encephalization and masticatory apparatus reduction (p. 769). Our hypothesis predicts that structural and temporal intermediates between *A. afarensis* and *H. habilis* will not exhibit the *A. africanus* morphological pattern (p. 771).

I view Fig. 6 as defined by White *et al.* (1983, p. 729, Fig. 4) as the most parsimonious arrangement of the known paleontological evidence for Plio-Pleistocene hominid evolution. As is evident from the cladogram in Fig. 6, I feel there is reasonable evidence that all the ancestors for the various clades presented here have been identified, with the single exception of the first node at the bottom, which links the African apes with the Plio-Pleistocene Hominidae. This in no way signifies that all of Plio-Pleistocene hominid evolution is now understood. What it does show is that, insofar as the diagramming of relationships is a rough approximation of past events, we now have enough evidence down to the species level to draw an "alpha" cladogram of hominid relationships that is fasifiable. In my opinion if we possessed as much knowledge concerning Miocene hominoid evolution as we do concerning Plio-Pleistocene hominid evolution this volume would be considerably shorter and much more succinct! But knowing the relationships of taxa in a cladogram does not always signify a concomitant understanding of the evolutionary process and factors that brought about the neatly diagrammed speciation events. More evidence from the fossil record and studies based on living representatives will always be necessary to further reconstruct the scenario of events responsible for the origin of each clade.

Summary Cladistic Model

In order to provide a final overview of the many facts and opinions expressed throughout this volume I have decided to put forth a summary cladistic model and combine this with a tabular presentation of hominoid

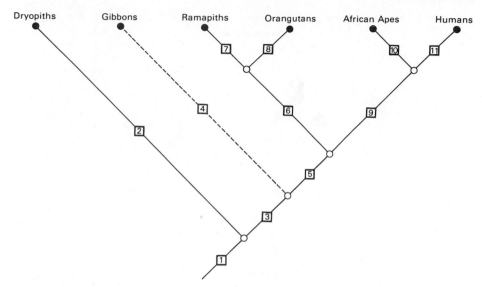

Fig. 7. Summary cladogram of hominoid relationships supported by a plurality of contributors to this volume and reflecting the views of the editors and a growing number of researchers in the field of paleoanthropology. To this cladogram, which was first presented in Fig. 4, I have added a series of numbered boxes that correspond to the 11 hominoid morphotypes described in Table 3. These boxes have been centered between nodes in this cladogram so that the morphotypes might not be confused with ancestral stages in hominoid evolution but rather used to define the character states of features likely to be diagnostic of those ancestors. In reality, no hypothetical or actual ancestor would be likely to possess all of the characters of any one morphotype described in Table 3. In this sense the morphotypes might best be viewed as a summary of mosaic evolutionary events along a clade (see text for further explanation). It is not possible to label this cladogram the "consensus" view of the volume, since contributor support was split between Figs. 3, 4, and 5 (see text). However, a plurality (including the editors) did specifically favor this cladogram, along with a number of other contributors who described the anatomical characters listed in Table 3 that support this cladogram. The 11 morphotypes are labeled as follows: (1) Proto-Hominoid, (2) Dryopith, (3) Stem extant Hominoid, (4) Proto-gibbon, (5) Pre-Great Ape/human, (6) Stem Ramapith, (7) Ramapith, (8) Proto-orangutan, (9) Pre-African ape/human, (10) Proto-African ape, (11), Proto-human.

characters that support this model. Figure 7 represents the views of a plurality of contributors, including those of the editors. I have added a series of numbered boxes between the nodes of this cladogram, which represent the morphotypes of each clade. The characters of each morphotype appear in Table 3 under the appropriate numerical heading. The analysis of character states for each morphotype was conducted following the principles of the cladistic method as outlined in this volume by Kluge (1983) and de Bonis (1983) and elsewhere. Each morphotype can be viewed as a list of features likely to be diagnostic of the ancestor of one or more taxa in its clade. It is unlikely that any single fossil specimen will possess all of the characters of any one morphotype. On the other hand, if it can be shown that a fossil possesses none of the derived characters of a morphotype, then it can be rejected as a member of that clade. Four different categories of characters are defined in the foot-

Table 3. Selected Characters of the Eleven Hominoid Morphotypes[a] Diagrammed in Fig. 7

Polarity	Character	Source
(1) Proto-Hominoid morphotype		
D	Reduction of snout together with development of short, nonprojecting nasals	Fleagle (1983b), Fleagle and Kay (1983)
D	Choanal shape with widening toward its base	Delson and Andrews (1975)
D	Alveolar recess of maxillary sinus invades the alveolar process	Cave and Haines (1940)
D	Presence of tubular ectotympanic bone	Fleagle and Kay (1983)
D	Dorsal reduction of premaxillae	Fleagle and Kay (1983)
D	Reduction of postorbital constriction	Fleagle and Kay (1983)
D	Appearance of enamel prism Pattern 3A	Gantt (1983)
D	Loss of entepicondylar foramen of humerus	Fleagle and Kay (1983)
P	Humerus with large projecting medial epicondyle and with spherical capitulum separated from trochlea by shallow groove	Fleagle (1983a)
P	Ontogenetically early appearance of ischial callosities	Delson and Andrews (1975)
P	Diploid ($2n$) number of chromosomes equal to 44	Chiarelli (1975), Kluge (1983)
P	Moderate sexual dimorphism	Fleagle et al. (1980)
(2) Dryopith morphotype		
P	Broad interorbital septum	Andrews and Cronin (1982), Kay (1982a)
P	Nasal aperture higher than broad, oval-shaped	Andrews and Cronin (1982), Kay (1982a)
P	No fronto-ethmoidal sinus	Martin and Andrews (1982)
P	Subnasal alveolar process appears as a flattened oval in sagittal section	Ward and Pilbeam (1983)
P	Incisive fossa forms transversely broad basin that opens directly into oral cavity (subnasal plane stepped down to floor of nasal cavity)	Ward and Pilbeam (1983), Andrews and Cronin (1982)

(continued)

[a]The polarity or status of each character appears in the left-hand column preceding the character description according to the following conventions: D, derived unique character by comparison to sister taxon; P, primitive, ancestral character shared by sister taxon; C, convergent (or parallel) nonhomologous character indicating phenetic similarity only; V, variable character of doubtful phyletic significance due to its random expression in sister taxa under comparison or in other out-groups. The source or sources which provided the character and in most cases the interpretation of polarity appears in the right-hand column. In some cases where the source made no explicit statement of polarity, I made the determination. I did not always choose the author who first mentioned the existence of a feature, but rather chose as the source the most cladistically phrased view of that character's representation. In many cases, the characters and sources are derived from this volume. When a difference of opinion existed over a character the alternative polarity and author can be found listed together in the source column. An attempt has been made to assess character correlation throughout this tabulation to avoid duplication of characters. Some repetition of anatomical features, however, was unavoidable. The exact positioning of a few characters within particular morphotypes is open to question due to the lack of appropriate fossil evidence. Therefore, even though a character's primitive or derived state may be firmly established, its position within a particular morphotype is sometimes conjectural. Data from biomolecular studies, such as amino acid sequencing of proteins and DNA endonuclease restriction mapping of base sequences, can easily be added to this tabulation in the future to further augment the designated morphotypes.

Table 3 (*Continued*)

Polarity	Character	Source
P	Nasoalveolar clivis abbreviated, terminating in incisive fossa (subnasal plane truncated)	Ward and Pilbeam (1983), Andrews and Cronin (1982)
P	Narrow troughlike maxillary sinus that tends to the dental alveoli	Ward and Pilbeam (1983)
P	Occasional thickening of glabellar region	Andrews and Cronin (1982)
P	Flaring zygomata with strong posterior slope	Kay (1982a), Andrews and Cronin (1982)
P	Zygomatic foramina situated at or below the lower rim of the orbits	Andrews and Cronin (1982)
P	Zygomatic foramina small	Andrews and Cronin (1982)
P	Large, oval-shaped greater palatine foramina	Andrews and Cronin (1982)
P	Infra-orbital foramina well removed from zygomaticomaxillary suture	Andrews and Cronin (1982)
P	Uniformly small lower incisors	Kay (1982a)
P	Upper incisors lacking large size discrepancy	Andrews and Cronin (1982), Kay (1982a)
P	Single cusped and sectorial P_3	Martin and Andrews (1982), Kay (1982a)
P	Slender canines	Martin and Andrews (1982)
P	High degree of canine sexual dimorphism	Kay (1982a, 1982b)
P	Canine root aligned along tooth row axis or rotated internally	Ward and Pilbeam (1983)
P	Canine roots elliptical in transverse section	Ward and Pilbeam (1983)
P	Molars with moderate to large cingula	Von Koenigswald (1973), Martin and Andrews (1982), Kay (1982a)
P	M_2^2 larger than M_1^1	Corruccini and Henderson (1978)
P	Intermediate thickness of enamel caps on molars	Gantt (1983), Ward and Pilbeam (1983)
P	Gracile mandibles with symphyses buttressed by large inferior and superior transverse tori	Martin and Andrews (1982), Kay (1982a)
P	Presence of superior parallel sulcus (a^1) and anterior occipital sulcus (a^3) on frontal lobe of cortex	Falk (1983)
P	Prominent to moderate expression of deltopectoral crest of humerus	Morbeck (1983)
D?	Presence of relatively broad, spool-shaped trochlea on humerus	Rose (1983)
D?	Reduction of olecranon beak of ulna	Rose (1983)
D?	Presence of spherical femoral head with articular surface extending onto neck anteriorly	Rose (1983)

(3) Stem extant Hominoid morphotype

D	Acquisition of fronto-orbital sulcus (*fo*), opercular sulcus (*io*), and frontalis superior sulcus (*fs*) on frontal lobe of cortex	Falk (1983)
D	Lengthening and torsioning of clavicle	Napier and Napier (1967), Oxnard (1968)

(*continued*)

Table 3 (*Continued*)

Polarity	Character	Source
D	Broadened sternum with progressive fusion of the sternal elements in adults	Tuttle (1974)
D	Dorsal positioning of scapula with glenoid fossa directed more cranially	Le Gros Clark (1959), Washburn (1963a)
D	Glenoid fossa widened, rounded, and flat with nonprojecting supraglenoid tubercle	Corruccini and Ciochon (1976)
D	Spinoglenoid notch of scapula deepened	Corruccini and Ciochon (1976)
D	Presence of coraco-acromial ligament spanning the laterally projecting coracoid process and acromion	Ciochon and Corruccini (1977b)
D	Lengthening of humerus with humeral head becoming large and globular	Le Gros Clark (1959), Corruccini and Ciochon (1976)
D	Intertubercular sulcus of humerus narrow and deep with lateral extent of head articular surface restricted from extending into the sulcus	Corruccini and Ciochon (1976)
D	Distal humerus with broad trochlea relative to capitulum with midportion constricted; capitulum is bulbous, coranoid fossa is deep, and radial fossa is shallow	Morbeck (1983)
D	Lateral edge of trochlea that separates trochlea and capitulum anteriorly distinctly wraps around distally to meet with olecranon fossa	Morbeck (1983)
D	Proximal ulna with segmented trochlear notch and U-shaped deep radial notch	Morbeck (1983)
D	Some reduction of olecranon process of ulna	Tuttle (1975)
D	Moderate reduction of ulnar styloid process with development of interarticular meniscus and occasional bony lunula but maintaining some direct ulna–triquetral contact	Lewis (1972), Corruccini (1978)
D	Radial head approaches circularity and is not tilted	Rose (1983)
D	Loss of sesamoid bones from the tendons of the hand	Washburn (1963a)
D	Shortening of lumbar region of vertebral column by 1–2 vertebrae, yielding a mode of five	Schultz (1961, 1963)
D	Reduction of the tail not by atrophy but by transformation into the shelflike coccyx	Andrews and Groves (1976)
D	Development of suspensory foraging and feeding adaptations including some underbranch leaping and hauling actions	Tuttle (1975), Washburn (1968a)
D	Development of pelvic diaphragm	Tuttle (1975)
D	Presence of vermiform appendix	Andrews and Cronin (1982)
D	Fetal membranes with interstitial implantation, decidua capsularis but no uterine symplasma	Luckett (1975)

Table 3 (*Continued*)

Polarity	Character	Source
P	Urethrovaginal septum distinct with labia majora forming relatively defined narrow, cutaneous folds	Kluge (1983)
P	Well-developed glans penis with prominent corona glandis	Kluge (1983)
P	Flexor pollicus longus muscle and tendon well developed	Kluge (1983)
P	Tendons of flexor hallucis longus and flexor digitorum longus are fused and evenly distributed to five digits	Kluge (1983)

(4) Proto-gibbon morphotype

Polarity	Character	Source
D	Substantial elongation of all forelimb elements with emphasis on ricochetal arm-swinging and vertical climbing	Tuttle (1975)
D	Interarticular meniscus in wrist joint strengthened by full development of bony lunala, the os Daubentonii	Lewis (1969), Delson and Andrews (1975)
D	Proximal capitate and hamate form distinctive narrow, knoblike process	Jenkins and Fleagle (1975)
D	Carpal and pedal digits II–V elongated	Biegert (1963), Tuttle (1972*b*)
P	Long, fully opposable thumb	Fleagle (1983*b*)
D	Carpal digit I forming ball-and-socket joint at the trapezium and metacarpal base	Van Horn (1972), Andrews and Groves (1976)
D	Ventral curvature (bowing) of digital rays II–V of hand notable in proximal and middle phalanges	Tuttle (1972*b*), Andrews and Groves (1976)
D	Phalanges of pedal digits II–V curved ventrally and act together with the widely divergent, heavily muscled hallux to aid in the development of above-branch bipedal running	Tuttle (1972*b*, 1975)
D	Lateral flexion and rotation of spine with lowered center of gravity	Tuttle (1981)
D	Reduction of sexual dimorphism	Delson and Andrews (1975)
D	Loss of superior parallel sulcus (a^1) on frontal lobe of cortex	Falk (1983)
D	Thin enamel cap on molars	Gantt (1983), Frisch (1973)

(5) Pre-Great Ape/human morphotype

Polarity	Character	Source
D	Choanal shape, narrow and high	Delson and Andrews (1975)
P or D	Intermediate or thick enamel caps on molars	Ward and Pilbeam (1983)
D	Cartilaginous meniscus fully interposed between greatly reduced ulnar styloid process and pisiform, resulting in total exclusion of ulnar–carpal articulations	Lewis (1969, 1972), McHenry and Corruccini (1983)
D	Reduction of pisiform size and loss of its articular facet for the ulnar styloid process	McHenry and Corruccini (1983), Morbeck (1972)

(*continued*)

Table 3 (*Continued*)

Polarity	Character	Source
D	Presence of fossa on distal ulna for attachment of interarticular meniscus	McHenry and Corruccini (1983)
D	Development of a pronounced hook on the hamate	O'Conner (1975), Corruccini *et al.* (1975), McHenry and Corruccini (1983)
D	Dorsal tubercle (Lister's) of radius shifts laterally to overlie the scaphoid facet	O'Conner (1975), Corruccini *et al.* (1975), McHenry and Corruccini (1983)
D	Sickle-shaped development of scaphoid bone	O'Conner (1975), Corruccini *et al.* (1975)
D	Lack of extension of articular cartilage onto dorsum of metacarpal V	O'Conner (1975), Corruccini *et al.* (1975)
D	Progressive increase in mobility of wrist, elbow, and shoulder joints until full range of forelimb movement capabilities is achieved, resulting in various suspensory postures and hanging/feeding adaptations	Rose (1983), Morbeck (1983), Corruccini *et al.* (1976), Fleagle (1983*a*)
D	Ontogenetically late appearance of ischial callosities	Delson and Andrews (1975)
D	Acquisition of arcurate (*arc*) sulcus on frontal lobe of cortex	Falk (1983)
D	Diploid (2*n*) number of chromosomes equal to 48	Yunis and Prakash (1982), Mai (1983), Kluge (1983)

(6) Stem-Ramapith morphotype

Polarity	Character	Source
D	Narrow interorbital septum	Andrews (1983), Kay and Simons (1983, ?V)
D	Short, broad nasal aperture	Andrews (1983), Kay and Simons (1983, ?P)
D	Subnasal alveolar process that appears elliptical in sagittal section	Ward and Pilbeam (1983)
D	Incisive fossa forms narrow depression just behind nasospinale (subnasal plane continuous with floor of nasal cavity, not stepped)	Ward and Pilbeam (1983), Andrews and Cronin (1982), Lipson and Pilbeam (1982), Kay and Simons (1983, ?V)
D	Nasoalveolar clivus extends posteriorly into nasal cavity (subnasal plane smooth, not truncated)	Ward and Pilbeam (1983), Andrews and Cronin (1982), Lipson and Pilbeam (1982), Preuss (1982)
D	High alveolar or premaxillary prognathism	Andrews and Tekkaya (1980), Wolpoff (1983), Kay and Simons (1983, ?P)
D	Wide maxillary sinus that rapidly diminishes anteriorly	Ward and Pilbeam (1983)
D	Orbits higher than broad, rather ovoid in shape	Andrews and Cronin (1982), Wolpoff (1983), Kay and Simons (1983), Preuss (1982)
D	Temporal edge of orbit marked by a well-defined marginal process	Preuss (1982)

Table 3 (*Continued*)

Polarity	Character	Source
D	Face is concave in lateral view, which can be termed "simognathic"	Preuss (1982), Napier and Napier (1967)
D	No glabellar thickening	Andrews and Cronin (1982)
D	Deep zygomatic process with strong lateral flare	Andrews (1983), Kay and Simons (1983, ?P)
D	Zygomatic bone flattened and facing anteriorly	Andrews and Cronin (1982), Preuss (1982), Kay and Simons (1983, ?V)
D	Presence of pronounced malar notch (or incisura malaris) on inferolateral surface of zygomatic body	Preuss (1982)
D	Zygomatic foramina relatively large	Andrews and Cronin (1982), Kay and Simons (1983, ?V)
D	Slitlike greater palatine foramina	Andrews and Cronin (1982), Kay and Simons (1983, ?V)
D	Extremely small incisive foramina	Andrews and Cronin (1982), Kay and Simons (1983, ?V)
D	Broad I^1 that is much larger than I^2	Andrews (1983), Wolpoff (1983), Preuss (1982), Kay and Simons (1983, ?C)
D	Central maxillary incisors that change angulation during life	Wolpoff (1983)
D	Some external rotation of upper canine roots	Ward and Pilbeam (1983)
D	Canine roots quadrangular in transverse section	Ward and Pilbeam (1983)
D	Well-developed canine fossa	Ward and Pilbeam (1983)
D	Strong superior–medial angulation of canine roots	Wolpoff (1983)
P or D	Intermediate thickness of enamel caps on molars	Ward and Pilbeam (1983), Gantt (1983)
D	Deeply incised wrinkles that persist after crown morphology has been worn away	Wolpoff (1983), Gantt (1983), Andrews and Cronin (1982, ?V)

(7) Ramapith morphotype

P	Poor development of supra-orbital torus	Andrews (1982), Kay and Simons (1983)
P	Arched palate	Kay and Simons (1983), Kay (1982a)
P	Zygomatic process set well forward	Kay and Simons (1983), Kay (1982a)
P	Infra-orbital foramina few in number	Andrews and Cronin (1982)
D	Shallow, broad mandibular corpora	Kay and Simons (1983), Kay (1982a), Andrews and Cronin (1982, ?C)
P	V-shaped lower dental arcade	Kay and Simons (1983), Kay (1982a)
P	Small upper and lower incisors relative to cheek tooth size	Kay and Simons (1983), Kay (1982a)

(*continued*)

Table 3 (*Continued*)

Polarity	Character	Source
D	Small upper and lower canines relative to cheek tooth size	Kay and Simons (1983), Kay (1982a), Andrews and Cronin (1982, ?V)
D	Long axis of upper canine cross section set more or less buccolingually	Kay and Simons (1983), Kay (1982a), Ward and Pilbeam (1983, ?C), Andrews and Cronin (1982, ?C)
D?	P_3 with well-developed metaconid	Kay and Simons (1983), Kay (1982a), Wolpoff (1983, ?V), Andrews and Cronin (1982, ?V)
P	Lower molar buccal cingula poorly developed or absent	Kay and Simons (1983), Kay (1982a)
D	Thick enamel caps on molars	Kay and Simons (1983), de Bonis (1983), Wolpoff (1983), Ward and Pilbeam (1983), Andrews (1983), Preuss (1982)
D	Low canine sexual dimorphism	Kay (1982b), Kay and Simons (1983), Andrews and Cronin (1982, ?V)

(8) Proto-orangutan morphotype

D	Zygomatic foramen situated above the level of the lower rim of the orbit	Andrews and Cronin (1982)
P	Poor development of supra-orbital torus	Andrews (1982), Kay and Simons (1983)
D	Increase in size of upper premolars relative to molars	Preuss (1982)
P	Intermediate thickness of enamel on molar caps with prominent wrinkling	Ward and Pilbeam (1983)
D?	Shortening of lumbar region of vertebral column by one vertebra, yielding a mode and mean of four	Schultz (1961, 1963), Kluge (1983)
V	Occasional ossification of the os centrale and scaphoid of the wrist occurring only in aged or arthritic individuals	Schultz (1936, 1963), Corruccini (1978)
C	Proximal capitate and hamate form distinctive narrow, knoblike process	Jenkins and Fleagle (1975)
C?	Further lengthening of forelimb elements with emphasis on vertical climbing, hanging, and terminal branch foraging and feeding	Tuttle (1975)
D	Hallux reduced dramatically, often resulting in absence of distal phalanx	Tuttle and Rogers (1966)
D	Pedal digits II–V elongate and the extrinsic pedal digital flexor musculature is well developed	Tuttle (1975)
D	Metatarsal and proximal and middle phalangeal bones of digits II–V possess marked degree of plantar curvature important in powerful grasping	Tuttle (1970)

Table 3 (*Continued*)

Polarity	Character	Source
D	Tensor fasciae lata muscle and iliotibial tract are lost	Tuttle (1975)
D	Refinement of suspensory adaptations to specialized canopy dwelling takes place	Tuttle (1975)
D	Urethrovaginal septum indistinct together with absence of well-defined labia majora	Kluge (1983)
D	Glans penis either not differentiated or only moderately developed with little or no corona glandis	Kluge (1983)
D	Flexor pollicus longus muscle and tendon greatly reduced	Kluge (1983)
D	Tendons of flexor hallucis longus and flexor digitorum longus are markedly discrete and unevenly distributed to five digits	Kluge (1983)

(9) Pre-African ape/human morphotype

P	Broad interorbital septum	Andrews and Cronin (1982), Kay (1982*a*)
D?	Nasoalveolar clivis projects well back into nasal cavity and drops sharply into incisive fossa, which is transversely broad	Ward and Pilbeam (1983)
D	Incisive fossa is divided into two chambers by the vomeronasal contact with the hard palate being deflected beneath nasospinale, resulting in formation of true incisive canel	Ward and Pilbeam (1983)
D	Large sphenopalatine fossae	Andrews and Cronin (1982)
D	Presence of fronto-ethmoidal sinus	Cave and Haines (1940), Andrews and Cronin (1982)
P	P_3 sectorial (C^1 honing) and bilaterally compressed	Delson and Andrews (1975), Kay (1982*a*)
D	Marked reduction of trigonid in lower molars	Delson *et al.* (1977)
P	Canine roots aligned along tooth row axis or rotated internally and are elliptical in transverse section	Ward and Pilbeam (1983)
P or D	Intermediate thickness of enamel caps on molars	Ward and Pilbeam (1983)
D	Nearly universal and complete fusion of the os centrale to the scaphoid at a very early age (usually prenatal)	Schultz (1936), Corruccini (1978), Kluge (1983)
D?	Foot is fully plantigrade with prominent plantar flexing mechanism	Tuttle (1975)
D	Iliac blades exhibit some shortening, widening, and lateral projection	Tuttle (1981)
D?	Vertical climbing (hoisting, hauling, and transferring) emphasized together with suspensory posturing and incipient bipedal behavior	Tuttle (1975)

(continued)

Table 3 (*Continued*)

Polarity	Character	Source
D	Incipient bipedalism is performed with knee joints flexed and femora abducted, flexed, and laterally rotated about the hip joint	Tuttle (1975), Wolpoff (1983)
P	Diploid (2*n*) number of chromosomes equal to 48	Yunis and Prakash (1982), Mai (1983), Kluge (1983)
D	Reduction or loss of dorsal hair from middle segments of fingers and toes	Washburn (1968*b*)

(10) Proto-African ape morphotype

Polarity	Character	Source
D	Thin enamel caps on molars	Ward and Pilbeam (1983), Gantt (1983), Kay and Simons (1983, ?P)
D	Volar and ulnar inclination of concave articular surface of the distal radius	Jenkins and Fleagle (1975), Tuttle (1974, 1975)
D	Prominent bony ridge on dorsodistal aspect of radial articular surface and on distal surface of scaphoid	Tuttle (1975)
D?	Concavoconvex facet on the capitate for the os centrale	Jenkins and Fleagle (1975)
D	Prominent transverse ridge at base of dorsal articular surface of metacarpal heads	Tuttle (1967, 1969)
D	Pronounced extension of articular surface onto the dorsal aspect of metacarpal heads II–V	Tuttle (1967)
D	Presence of friction skin pads (knuckle pads) over the dorsal aspects of the middle phalanges and their associated osteoligamentous support mechanisms	Schultz (1936), Tuttle (1969)
D	Remarkably strong development of the flexor digitorum superficialis which guards against excessive stress on the metacarpophalangeal weight-bearing joints	Tuttle (1975)
D	Development of steepness on the lateral aspect of the olecranon fossa forming a ridgelike structure superior to fossa together with a deepening of the fossa to accommodate overextension, which stabilizes the humeroulnar joint	McHenry (1975), Tuttle (1975)
C	Shortening of lumbar region of vertebral column by 1–2 vertebrae to yield a mode of four and mean of 3.5	Schultz (1961, 1963), Kluge (1983, ?D)
D	Occurrence of knuckle-walking as a distinctive mode of locomotion	De Bonis (1983)
C	Urethovaginal septum indistinct together with absence of well-defined labia majora	Kluge (1983, ?D)
C	Glans penis either not differentiated or only moderately developed with little or no corona glandis	Kluge (1983, ?D)
C	Flexor pollicus longus muscle and tendon greatly reduced	Kluge (1983, ?D)

Table 3 (*Continued*)

Polarity	Character	Source
C	Tendons of flexor hallucis longus and flexor digitorum longus are markedly discrete and unevenly distributed to the five digits	Kluge (1983, ?D)
D	Set V, Y, Y′, Y″, and Z events involving unique changes in chromosomes 1, 2, 4, 5, 7, 8, 9, 10, 12, 13, 14, 16, 17, and 22	Mai (1983)

(11) Proto-human morphotype

Polarity	Character	Source
D	Progressive increase in cranial capacity	Tobias (1971)
D	Lack of external evidence of the fronto-orbital sulcus (*fo*) on the cortical surface of the frontal lobe	Connolly (1950), Falk (personal communication)
D	Progressive anterior shift in position of foramen magnum and occipital condyles	Le Gros Clark (1964), Tobias (1971)
D	Development of a less flattened and straight nasoalveolar clivis	Boaz (1983)
D	Reduction in overall size of canine	Le Gros Clark (1964), Washburn and Ciochon (1974), White *et al.* (1983)
D	Reduction in degree of canine sexual dimorphism	Kay (1982b), Kay and Simons (1983)
D	Progressive loss of canine–premolar honing with incipient development of metaconid on P_3 and enlargement of protocone or paracone on P_3, P_4, resulting in bicuspid upper premolars	Delson *et al.* (1977), de Bonis (1983), Kay and Simons (1983)
D	Progressive reduction in size and frequency of I^2–$\bar{C}$ and $\underline{C}$–P_3 diastemata	Boaz (1983)
C	Long axis of upper canine cross section set more or less buccolingually (externally rotated canine roots)	Ward and Pilbeam (1983), Andrews and Cronin (1982), Kay and Simons (1983, ?D)
D	Decrease in molar length relative to molar width	Corruccini and McHenry (1980), Boaz (1983)
D	Increase in molar crown height	Corruccini (1977), Corruccini and McHenry (1980)
D	Reduction in frequency of Carabelli cusp complexes in M^1	Boaz (1983)
P	Intermediate thickness of enamel caps on molars	Ward and Pilbeam (1983)
D	Appearance of enamel prism Pattern 3B	Gantt (1983)
C	Shallow and broad mandibular corpora	Andrews and Cronin (1982), Kay and Simons (1983, ?D)
D	Progressive reduction in forelimb length relative to trunk height	Jungers (1982)
D	Shortening of ischium and splaying and broadening of illia	Washburn (1963b), Zihlman and Lowenstein (1983)
D	Pronounced curvature in lumbosacral region, where the broadened sacrum is abruptly bent back (dorsoflexed) so that it forms a striking promontory with the lumbar region	Schultz (1968), Washburn (1963b), Zihlman and Lowenstein (1983)

(*continued*)

Table 3 (*Continued*)

Polarity	Character	Source
D	Well-developed anterior inferior iliac spine	Lovejoy (1975, 1978)
D	Clear definition of ilio-psoas groove	Lovejoy (1978)
D	Increase in bicondylar angle of femur	Lovejoy (1978)
D	Lengthening of femoral neck	Lovejoy (1978)
D	Deepening of patellar groove of femur	Lovejoy (1975, 1978)
D	Development of rotational astragalocalcaneal joint	Delson and Andrews (1975)
D	Permanent convergence of the hallux	Biegert (1963), Tuttle (1975)
D	Cranial retreat of the lower portion of gluteus maximus	Tuttle *et al.* (1975)
D	Occurrence of bipedalism as a distinctive locomotor pattern	White *et al.* (1983), de Bonis (1983)
D	Complete loss of ischial callosities	Delson and Andrews (1975)
D	Diploid ($2n$) number of chromosomes reduced to 46 via event 2b, which changed the number of acrocentrics in the diploid karyotype from 14 to ten, the number of metacentrics by two, and the diploid number by two, and also set X events involving unique changes in chromosomes 1, 2, 4, 9, and 18	Mai (1983)

note to Table 3. Naturally, only the derived (apomorphic) characters are of use in establishing the sister group relationships diagrammed in Figure 7. However, some morphotypes are defined primarily on the basis of primitive characters, in many cases due to the lack of appropriate fossil material. Figure 7 therefore is probably best viewed at this stage as a heuristic device capable of enhancing our understanding of hominoid evolution rather than providing any definitive answers.

In the reconstruction of phylogeny no evolutionary scenario can be completely parsimonious. The fact that reversals and parallelisms are common features in mammalian evolution is now considered a truism. Therefore a certain number of characters used to establish any cladogram or phylogeny can be expected to show homoplasy. In some cases the occurrence of reversals, parallelisms, or convergence can involve as many as half of all characters (Cartmill, 1982). Parsimony at this point becomes a relative term. The cladogram presented in Figure 7 and documented in Table 3 is, in my opinion, the most parsimonious arrangement of the data assembled in this volume and gleaned from other recent sources. Some instances of parallelism or convergence have been noted; others almost certainly exist but have not been recognized. With the acknowledgment of this caveat, I feel it is possible to view Figure 7 and Table 3 as a series of falsifiable steps or stages that quite possibly occurred in the evolution of the Hominoidea. Many of these steps are testable through the recovery of additional fossil remains. New fossils also allow more precise determination of homoplasy. Though no ancestor is expected to completely resemble the reconstructed morphotypes presented in Table 3, the demonstration of one or more of the derived character states in newly discovered fossil forms provides an independent assessment of the morphotype's validity, and thus the sequence of events that occurred in hominoid evolution. Conversely, if enough new fossil discoveries constantly lacked one set of derived character states and yet preserved others, then one might question the validity of the morphotype.

It should be noted that some characters that cannot be assessed in the fossil record have also been included in Table 3. These characters, such as soft anatomical features or chromosome number, can be significant additions since they often provide an independent check on the "hard" osteological data and may one day also be used to corroborate evolutionary scenarios based on new neontological analyses. Finally, I want to emphasize, once again, the preliminary nature of the relationships and character determinations presented in Figure 7 and Table 3, respectively. A more complete set of data is now being assembled and will be presented in a future publication (Ciochon and Myers, in preparation). That publication will also contain further elaboration of the hypothetical nodal ancestors identified in Figure 7.

Conclusion

It is my hope that a volume of this sort will demonstrate that the field of paleoanthropology is progressing toward the development of new, fully testable paradigms of ape and human ancestry. The summary model of hominoid cladogenesis outlined above certainly represents progress toward the attainment of an enduring consensus. Concrete evidence that progress has occurred during the last two decades can be found by comparing this volume's contents with what is now regarded as a classic in paleoanthropology, Washburn's (1963*b*) *Classification and Human Evolution.* Both volumes deal with a remarkably similar range of topics (comparative anatomy, biochemistry, hominid phylogeny), although Washburn's volume emphasized the new field of primate behavior to a much greater extent than can be found here.

Classification and Human Evolution achieved a major milestone when it provided the first legitimate forum for the presentation of the biomolecular viewpoint of human origins. I suggest that that viewpoint has assumed a preeminent position in the minds of many contributors to *New Interpretations of Ape and Human Ancestry,* as the preceding chapters illustrate. This can be legitimately labeled as progress. In further comparisons of the two volumes it is possible to see that progress has also occurred in our understanding of the fossil record. In several of the chapters on hominid phylogeny in Washburn's volume, much discussion and debate took place over the position of the enigmatic primate from Italy, *Oreopithecus,* in human phylogeny. Twenty years later in this volume on ape and human ancestry *Oreopithecus* receives hardly a mention throughout the 29 preceding chapters, and rightly so, since this "swamp ape" has little bearing on current theories of ape and human origins. Perhaps a decade from now *"Ramapithecus"* will be relegated to a similar position within human phylogeny. This illustrates that progress in paleo-anthropology is often slow to come, but once a major shift occurs, the implications can be suddenly far-reaching.

Appendix

As stated in the Introduction, the purpose of this chapter was to provide a summary statement reflecting the various points of view concerning hominoid phylogeny expressed by contributors to this volume. While compiling the data for presentation in Tables 1–3 and Figs. 1–7 it became apparent that no cladogram presented or implied in the pages of this volume accurately reflected my own developing viewpoint. Hewing to the principle that this chapter should primarily summarize, rather than engage in speculative clade-building, an attempt was made to limit my own opinions to a minimum. However, this volume would not be complete unless I had an opportunity to air my viewpoint like so many have done in the preceding chapters. Figure 8 therefore represents my view of hominoid relationships.

As is apparent, this cladogram is based on the summary cladogram of Fig. 7. To this a number of new clades have been added which are based only on fossil groups. The groups are defined as follows: Prop, Propliopiths (*Aegyptopithecus, Propliopithecus, Pliopithecus*); Proc, Proconsulmorphs (*Proconsul, Rangawapithecus, Limnopithecus, Micropithecus*); L Dry, larger Dryopiths (*Dryopithecus,* "*Hispanopithecus*" "*Rudapithecus,*" *Hylopithecus,* Domeli jaw, some iso-

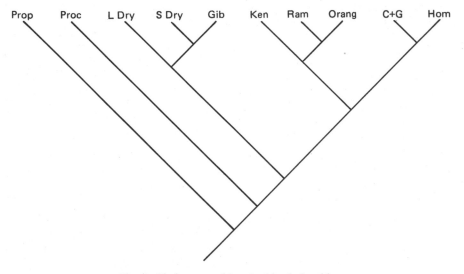

Fig. 8. Cladogram of hominoid relationships.

lated teeth from Ramnagar); S Dry, small Dryopiths (*Dionysopithecus, Krishnapithecus,* smaller new specimens from Lufeng); Gib, Gibbons (*Hylobates, Symphalangus*); Ken, Kenyapiths [= dryomorph Ramapiths (*Kenyapithecus*)]; Ram, Ramapiths [= ramamorph Ramapiths (*Sivapithecus,* "*Ramapithecus,*" *Gigantopithecus,* "*Ouranopithecus*")]; Orang, orangutan (*Pongo*); C+G, chimpanzee and gorilla (*Pan* and *Gorilla*); Hom, Hominids (*Australopithecus, Homo*). It is easy to see why I did not present this cladogram in the summary cladistic model section. The addition of all these new groups would have necessitated the expansion of Table 3 to include several new morphotypes for which there is not adequate fossil evidence. Hence, I consider this a speculative view which in some instances is not supported by much evidence. My reasoning for these steps is nevertheless presented throughout the text (see also footnote on page 815).

An abbreviated classification of the Hominoidea that is at least partially consistent with this cladogram is as follows: Hominoidea—Pliopithecidae: Propliopithecinae (*Aegyptopithecus*), Pliopithecinae (*Pliopithecus*)—Dryopithecidae: Proconsulinae (*Proconsul*), Dryopithecinae (*Dryopithecus*)—Hylobatidae: Subfamily Indet. (*Krishnapithecus*), Hylobatinae (*Hylobates*)—Pongidae: Ramapithecinae (*Sivapithecus*), Ponginae (*Pongo*)—Panidae (*Pan* and *Gorilla*)—Hominidae (*Australopithecus* and *Homo*). Only one or two of the genera belonging to each group have been listed in this abbreviated classification.

ACKNOWLEDGMENTS

I would like to thank Rob Corruccini for assistance and advice throughout the preparation of this chapter. For helpful background discussions on the topic of hominoid cladistics I thank Peter Andrews, P. K. Basu, Eric Delson, John Fleagle, Dave Gantt, and Lawrence Martin. I also acknowledge Joy Myers for providing background material and suggestions during the early development of this chapter. I am grateful to the UNCC Cartographic Laboratory Staff for drafting the figures, to Frankie Baucom for typing the text and tables, and to Wenda Trevathan for carefully proofreading the manuscript.

References

Andrews, P. 1981. Hominoid habitats of the Miocene. *Nature (Lond.)* **289:**749.

Andrews, P. J. 1982. Hominoid evolution. *Nature (Lond.)* **295:**185–186.

Andrews, P. J. 1983. The natural history of *Sivapithecus,* in: *New Interpretations of Ape and Human Ancestry* (R. L. Ciochon and R. S. Corruccini, eds.), pp. 441–463, Plenum, New York.

Andrews, P., and Cronin, J. E. 1982. The relationship of *Sivapithecus* and *Ramapithecus* and the evolution of the orang-utan. *Nature (Lond.)* **297:**541–546.

Andrews, P., and Groves, C. P. 1976. Gibbons and brachiation. *Gibbon and Siamang* **4:**167–218.

Andrews, P. J., and Tekkaya, I. 1980. A revision of the Turkish Miocene hominoid *Sivapithecus meteai*. *Palaeontology* **23**(1):85–95.

Benson, R. H. 1976. Testing the Messinian salinity crisis biodynamically: An introduction. *Palaeogeogr. Palaeoclimatol. Palaeoecol.* **20**:3–11.

Bernor, R. L. 1983. Geochronology and zoogeographic relationships of Miocene Hominoidea, in: *New Interpretations of Ape and Human Ancestry* (R. L. Ciochon and R. S. Corruccini, eds.), pp. 21–64, Plenum, New York.

Biegert, J. 1963. The evaluation of characteristics of the skull, hands and feet for primate taxonomy, in: *Classification and Human Evolution* (S. L. Washburn, ed.), pp. 116–145, Aldine, Chicago.

Boaz, N. T. 1983. Morphological trends and phylogenetic relationships from middle Miocene hominoids to late Pliocene hominids, in: *New Interpretations of Ape and Human Ancestry* (R. L. Ciochon and R. S. Corruccini, eds.), pp. 705–720, Plenum, New York.

Bown, T. M., Kraus, M. J., Wing, S. L., Fleagle, J. G., Tiffany, B. H., Simons, E. L., and Vondra, C. F. 1982. The Fayum primate forest revisited. *J. Hum. Evol.* **11**(7):603–632.

Bruce, E. J., and Ayala, F. J. 1979. Phylogenetic relationships between man and the apes: Electrophoretic evidence. *Evolution* **33**:1040–1056.

Bush, M. E., Lovejoy, C. O., Johanson, D. C., and Coppens, Y. 1982. Hominid carpal, metacarpal and phalangeal bones recovered from the Hadar Formation: 1974–1977 collections. *Am. J. Phys. Anthropol.* **57**:651–677.

Cartmill, M. 1982. Basic primatology and prosimian evolution, in: *Fifty Years of Physical Anthropology in North America* (F. Spencer, ed.), pp. 147–186, Academic, New York.

Cave, A. J. E., and Haines, R. W. 1940. The paranasal sinuses of the anthropoid apes. *J. Anat.* **72**:493–523.

Chiarelli, B. 1975. The study of primate chromosomes, in: *Primate Functional and Evolutionary Biology* (R. H. Tuttle, ed.), pp. 103–127, Mouton, The Hague.

Chopra, S. R. K. 1983. Significance of recent hominoid discoveries from the Siwalik Hills of India, in: *New Interpretations of Ape and Human Ancestry* (R. L. Ciochon and R. S. Corruccini, eds.), pp. 539–557, Plenum, New York.

Chopra, S. R. K., and Kaul, S. 1979. A new species of *Pliopithecus* from the Indian Sivaliks. *J. Hum. Evol.* **8**:475–477.

Ciochon, R. L., and Corruccini, R. S. 1977a. The phenetic position of *Pliopithecus* and its phylogenetic relationship to the Hominoidea. *Syst. Zool.* **26**:290–299.

Ciochon, R. L., and Corruccini, R. S. 1977b. The coraco-acromial ligament and projection index in man and other anthropoid primates. *J. Anat.* **124**:627–632.

Ciochon, R. L., and Corruccini, R. S. 1982. Miocene hominoids and new interpretations of ape and human ancestry, in: *Advanced Views in Primate Biology* (A. B. Chiarelli and R. S. Corruccini, eds.), pp. 149–159, Springer-Verlag, Berlin.

Connolly, C. J. 1950. *External Morphology of the Primate Brain*, Thomas, Springfield.

Corruccini, R. S. 1977. Crown component variation in hominoid lower third molars. *Z. Morphol. Anthropol.* **68**:14–25.

Corruccini, R. S. 1978. Comparative osteometrics of the hominoid wrist joint, with special reference to knuckle-walking. *J. Hum. Evol.* **7**:307–321.

Corruccini, R. S., and Ciochon, R. L. 1976. Morphometric affinities of the human shoulder. *Am. J. Phys. Anthropol.* **45**:19–38.

Corruccini, R. S., and Ciochon, R. L. 1983. Overview of ape and human ancestry: Phyletic relationships of Miocene and later Hominoidea, in: *New Interpretations of Ape and Human Ancestry* (R. L. Ciochon and R. S. Corruccini, eds.), pp. 1–19, Plenum, New York.

Corruccini, R. S., and Henderson, A. M. 1978. Palatofacial comparison of *Dryopithecus* (*Proconsul*) with extant catarrhines. *Primates* **19**:35–44.

Corruccini, R. S., and McHenry, H. M. 1980. Cladometric analysis of Pliocene hominoids. *J. Hum. Evol.* **9**:209–221.

Corruccini, R. S., Ciochon, R. L., and McHenry, H. M. 1975. Osteometric shape relationships in the wrist joint of some anthropoids. *Folia Primatol.* **24**:250–274.

Corruccini, R. S., Ciochon, R. L., and McHenry, H. M. 1976. The postcranium of Miocene hominoids: Were dryopithecines merely dental apes? *Primates* **17**:205–223.

Corruccini, R. L., Baba, M., Goodman, M., Ciochon, R. L., and Cronin, J. E. 1980. Non-linear macro-molecular evolution and the molecular clock. *Evolution* **34**:1216–1219.

Cronin, J. E. 1983. Apes, humans and molecular clocks: A reappraisal, in: *New Interpretations of Ape and Human Ancestry* (R. L. Ciochon and R. S. Corruccini, eds.), pp. 115–149, Plenum, New York.

De Bonis, L. 1983. Phyletic relationships of Miocene hominoids and higher primate classification, in: *New Interpretations of Ape and Human Ancestry* (R. L. Ciochon and R. S. Corruccini, eds.), pp. 625–649, Plenum, New York.

Dehm, R. 1983. Miocene hominoid primate dental remains from the Siwaliks of Pakistan, in: *New Interpretations of Ape and Human Ancestry* (R. L. Ciochon and R. S. Corruccini, eds.), p. 527–537, Plenum, New York.

Delson, E., and Andrews, P. 1975. Evolution and interrelationships of the catarrhine primates, in: *Phylogeny of the Primates* (W. P. Luckett and F. S. Szalay, eds.), pp. 405–446, Plenum, New York.

Delson, E., Eldredge, N., and Tattersall, I. 1977. Reconstruction of hominoid phylogeny: A testable framework based on cladistic analysis. *J. Hum. Evol.* **6**:263–278.

Falk, D. 1983. A reconsideration of the endocast of *Proconsul africanus:* Implications for primate brain evolution, in: *New Interpretations of Ape and Human Ancestry* (R. L. Ciochon and R. S. Corruccini, eds.), pp. 239–248, Plenum, New York.

Feldesman, M. R. 1982. Morphometric analysis of the distal humerus of some Cenozoic catarrhines. The late divergence hypothesis revisited. *Am. J. Phys. Anthropol.* **59**:73–95.

Ferris, S. D., Wilson, A. C., and Brown, W. M. 1981. Evolutionary trees of apes and humans based on cleavage maps of mitochondrial DNA. *Proc. Natl. Acad. Sci USA* **78**:2432–2436.

Fleagle, J. G. 1978. *Micropithecus clarki*, a small ape from the Miocene of Uganda. *Am. J. Phys. Anthropol.* **49**:427–440.

Fleagle, J. G. 1983*a*. Locomotor adaptations of Oligocene and Miocene hominoids and their phyletic implications, in: *New Interpretations of Ape and Human Ancestry* (R. L. Ciochon and R. S. Corruccini, eds.), pp. 301–324, Plenum, New York.

Fleagle, J. G. 1983*b*. Are there any fossil gibbons?, in: *The Lesser Apes: Evolutionary and Behavioral Biology* (D. J. Chivers, H. Preuschoft, N. Creel, and W. Brockelman, eds.), Edinburgh University Press, Edinburgh.

Fleagle, J. G., and Kay, R. F. 1983. New interpretations of the phyletic position of Oligocene hominoids, in: *New Interpretations of Ape and Human Ancestry* (R. L. Ciochon and R. S. Corruccini, eds.), pp. 181–210, Plenum, New York.

Fleagle, J. G., Kay, R. F., and Simons, E. L. 1980. Sexual dimorphism in early anthropoids, *Nature (Lond.)* **287**:328–330.

Flynn, L. J., and Qi, G. 1982. Age of the Lufeng, China, hominoid locality. *Nature (Lond.)* **298**:746–747.

Frisch, J. E. 1973. The hylobatid dentition. *Gibbon and Siamang* **2**:56–95.

Gantt, D. G. 1983. The enamel of Neogene hominoids: Structural and phyletic implications, in: *New Interpretations of Ape and Human Ancestry* (R. L. Ciochon and R. S. Corruccini, eds.), pp. 249–298, Plenum, New York.

Gantt, D. G., Pilbeam, D., and Steward, G. 1977. Hominoid enamel prism patterns. *Science* **189**:135–137.

Ginsburg, L., and Mein, P. 1980. *Crouzelia rhodanica*, nouvelle espèce de Primate catarhinien, et essai sur la position systématique des Pliopithecidae. *Bull. Mus. Nat. Hist. Nat. Paris, Ser. 4* **2**(C):57–85.

Goodman, M., Baba, M. L., and Darga, L. L. 1983. The bearing of molecular data on the cladogenesis and times of divergence of hominoid lineages, in: *New Interpretations of Ape and Human Ancestry* (R. L. Ciochon and R. S. Corruccini, eds.), pp. 67–86, Plenum, New York.

Greenfield, L. 1978. On the dental arcade reconstructions of *Ramapithecus J. Hum. Evol.* **7**:345–359.

Greenfield, L. 1979. On the adaptive pattern of *"Ramapithecus." Am. J. Phys. Anthropol.* **50:**527–548.

Greenfield, L. O. 1983. Toward the resolution of discrepancies between phenetic and paleontological data bearing on the question of human origins, in: *New Interpretations of Ape and Human Ancestry* (R. L. Ciochon and R. S. Corruccini, eds.), pp. 695–703, Plenum, New York.

Hsü, K. J. 1972. When the Mediterranean dried up. *Sci. Am.* **227**(6):26–36.

Hsü, K. J. 1978. When the Black Sea was drained. *Sci. Am.* **228**(5):53–63.

Hsü, K. J., Mantadert, L., Bernoulli, D., Cita, M. B., Erickson, A., Garrison, R. E., Kidd, R. B., Melieres, F., Muller, C., and Wright, R. 1977. History of the Mediterranean Salinity Crisis. *Nature (Lond.)* **267:**399–403.

Jenkins, F. A., Jr. and Fleagle, J. G. 1975. Knuckle-walking and the functional anatomy of the wrists in living apes, in: *Primate Functional Morphology and Evolution* (R. H. Tuttle, ed.), pp. 213–227, Mouton, The Hague.

Johanson, D. C., and White, T. D. 1979. A systematic assessment of early African hominids. *Science* **203:**321–330.

Johanson, D. C., White, T. D., and Coppens, Y. 1978. A new species of the genus *Australopithecus* (Primates: Hominidae) from the Pliocene of Eastern Africa. *Kirtlandia* **28:**1–14.

Jungers, W. L. 1982. Lucy's limbs: Skeletal allometry and locomotion in *Australopithecus afarensis. Nature (Lond.)* **297:**676–678.

Kay, R. F. 1981. The Nut-Crackers—A new theory of the adaptations of the Ramapithecinae. *Am. J. Phys. Anthropol.* **55:**141–151.

Kay, R. F. 1982a. *Sivapithecus simonsi,* a new species of Miocene hominoid with comments on the phylogenetic status of the Ramapithecinae. *Int. J. Primatol.* **3:**113–174.

Kay, R. F. 1982b. Sexual dimorphism in Ramapithecinae. *Proc. Natl. Acad. Sci. USA* **79:**209–212.

Kay, R. F., and Simons, E. L. 1983. A reassessment of the relationship between later Miocene and subsequent Hominoidea, in: *New Interpretations of Ape and Human Ancestry* (R. L. Ciochon and R. S. Corruccini, eds.), pp. 577–624, Plenum, New York.

King, M. C., and Wilson, A. C. 1975. Evolution at two levels in humans and chimpanzees. *Science* **188:**107–116.

Kluge, A. G. 1983. Cladistics and the classification of the Great Apes, in: *New Interpretations of Ape and Human Ancestry* (R. L. Ciochon and R. S. Corruccini, eds.), pp. 151–177, Plenum, New York.

Kortlandt, A. L. 1980. The Fayum primate forest. Did it exist? *J. Hum. Evol.* **9:**277–297.

Kortlandt, A. L. 1983. Facts and fallacies concerning Miocene ape habitats, in: *New Interpretations of Ape and Human Ancestry* (R. L. Ciochon and R. S. Corruccini, eds.), pp. 465–515, Plenum, New York.

Kretzoi, M. 1975. New ramapithecines and *Pliopithecus* from the lower Pliocene of Rudabánya in northeastern Hungary. *Nature (Lond.)* **257:**578–581.

Lavocat, R., 1980. The implication of rodent paleontology and biogeography to the geographical sources and origin of platyrrhine primates, in: *Evolutionary Biology of the New World Monkeys and Continental Drift* (R. L. Ciochon and A. B. Chiarelli, eds.), pp. 93–102, Plenum, New York.

Le Gros Clark, W. E. 1959. *The Antecedents of Man,* Edinburgh University Press, Edinburgh.

Le Gros Clark, W. E. 1964. *The Fossil Evidence for Human Evolution,* 2nd ed., University of Chicago Press, Chicago.

Lewis, O. J. 1969. The hominoid wrist joint. *Am. J. Phys. Anthropol.* **30:**251–268.

Lewis, O. J. 1972. Osteological features characterizing the wrists of monkeys and apes with a reconsideration of this region in *Dryopithecus (Proconsul) africanus. Am. J. Phys. Anthropol.* **36:**45–58.

Li, C. 1978. A Miocene gibbon-like primate from Shihhung, Kiangsu Province. *Vertebr. Palasiat.* **16:**187–192.

Lipson, S., and Pilbeam, D. 1982. *Ramapithecus* and hominoid evolution. *J. Hum. Evol.* **11:**545–548.

Lovejoy, C. O. 1975. Biomechanical perspectives on the lower limb of early hominids, in: *Primate Functional Morphology and Evolution* (R. H. Tuttle, ed.), pp. 291–326, Mouton, The Hague.

Lovejoy, C. O. 1978. A biomechanical view of the locomotor diversity of early hominids, in: *Early Hominids of Africa* (C. Jolly, ed.), pp. 403–429, St. Martin's, New York.

Lovejoy, C. O. 1981. The origin of man. *Science* **211:**341–350.

Lowenstein, J. M. 1982. Fossil proteins and evolutionary time, in: *Proceedings of the Pontifical Academy of Sciences,* The Vatican, Rome.

Luckett, W. P. 1975. Ontogeny of the fetal membranes and placenta: Their bearing on primate phylogeny, in: *Phylogeny of the Primates* (W. P. Luckett and F. S. Szalay, eds.), pp. 157–182, Plenum, New York.

Mai, L. L. 1983. A model of chromosome evolution in Primates and its bearing on cladogenesis in the Hominoidea, in: *New Interpretations of Ape and Human Ancestry* (R. L. Ciochon and R. S. Corruccini, eds.), pp. 87–114, Plenum, New York.

Martin, L., and Andrews, P. J. 1982. New ideas on the relationships of the Miocene hominoids. *Primate Eye* **18:**4–7.

Matthew, W. D. 1915. Climate and evolution. *Ann. N.Y. Acad. Sci* **24:**171–318.

McHenry, H. M. 1975. Multivariate analysis of early hominid humeri, in: *Measures of Man* (E. Giles and J. S. Friedlander, eds.), pp. 338–371, Shenkman, Cambridge, Massachussets.

McHenry, H. M., and Corruccini, R. S. 1980. On the status of *Australopithecus afarensis. Science* **207:**1103–1104.

McHenry, H. M., and Corruccini, R. S. 1983. The wrist of *Proconsul africanus* and the origin of hominoid postcranial adaptations, in: *New Interpretations of Ape and Human Ancestry* (R. L. Ciochon and R. S. Corruccini, eds.), pp. 353–367, Plenum, New York.

McHenry, H. M., Andrews, P., and Corruccini, R. S. 1980. Miocene hominoid palatofacial morphology. *Folia Primatol.* **33:**241–252.

Morbeck, M. E. 1972. *A Re-Examination of the Forelimb of the Miocene Hominoidea,* Ph.D. Dissertation, University of California, Berkeley.

Morbeck, M. E. 1983. Miocene hominoid discoveries from Rudabanya: Implications from the postcranial skeleton, in: *New Interpretations of Ape and Human Ancestry* (R. L. Ciochon and R. S. Corruccini, eds.), pp. 369–404, Plenum, New York.

Napier, J. R. and Davis, P. R. 1959. The forelimb skeleton and associated remains of *Proconsul africanus. Fossil Mammals of Africa* (Br. Mus. Nat. Hist.) **16:**1–70.

Napier, J. R., and Napier, P. H. 1967. *A Handbook of Living Primates,* Academic, London. 456 pp.

O'Conner, B. L. 1975. The functional morphology of the cercopthecoid wrist and inferior radioulnar joints and their bearing on some problems in the evolution of the Hominoidea. *Am. J. Phys. Anthropol.* **43:**113–122.

Oxnard, C. E. 1968. The architecture of the shoulder in some mammals. *J. Morphol.* **126:**249–290.

Pickford, M. 1982. New higher primate fossils from the middle Miocene deposits at Majiwa and Kaloma, western Kenya. *Am. J. Phys. Anthropol.* **58:**1–19.

Pickford, M. 1983. Sequence and environments of the lower and middle Miocene hominoids of western Kenya, in: *New Interpretations of Ape and Human Ancestry* (R. L. Ciochon and R. S. Corruccini, eds.), pp. 421–439, Plenum, New York.

Pilbeam, D. 1969. Newly recognized mandible of *Ramapithecus. Nature* (*Lond.*) **222:**1093–1094.

Pilbeam, D. 1982. New hominoid skull material from the Miocene of Pakistan. *Nature* **295:**232–234.

Pilgrim, G. 1915. New Siwalik primates and their bearing on the question of evolution of man and the Anthropoidea. *Rec. Geol. Surv. India* **45:**1–74.

Pilgrim, G. 1927. A new *Sivapithecus* palate and other primate fossils from India. *Mem. Geol. Surv. India (Palaeontol. Ind.)* **14:**1–26.

Prasad, K. N. 1983. Historical notes on the geology, dating and systematics of the Miocene hominoids of India, in: *New Interpretations of Ape and Human Ancestry* (R. L. Ciochon and R. S. Corruccini, eds.), pp. 559–574, Plenum, New York.

Preuss, T. M. 1982. The face of *Sivapithecus indicus:* Description of a new, relatively complete specimen from the Siwaliks of Pakistan. *Folia Primatol.* **38:**141–157.

Remane, A. 1965. Die geschichte der menschenaffen, in: *Menschliche Abstammungslehre* (G. Heberer, ed.), pp. 249–309, Fischer, Stuttgart.

Rose, M. D. 1983. Miocene hominoid postcranial morphology: Monkey-like, ape-like, neither, or both?, in: *New Interpretations of Ape and Human Ancestry* (R. L. Ciochon and R. S. Corruccini, eds.), pp. 405–417, Plenum, New York.

Schultz, A. H. 1936. Characters common to higher primates and characters specific to man. *Q. Rev. Biol.* **11**:259–283, 425–455.

Schultz, A. H. 1961. Vertebral column and thorax. *Primatologia* **1**:887–964.

Schultz, A. H. 1963. Age changes, sex differences, and variability as factors in the classification of primates, in: *Classification and Human Evolution* (S. L. Washburn, ed.), pp. 85–115, Aldine, Chicago.

Schultz, A. H. 1968. The recent hominoid primates, in: *Perspectives on Human Evolution*, Volume 1 (S. L. Washburn and P. C. Jay, eds.), pp. 122–195, Holt, Rinehart and Winston, New York.

Schwartz, J. H., Tattersall, I., and Eldredge, N. 1978. Phylogeny and classification of the Primates revisited. *Yearb. Phys. Anthropol.* **21**:95–133.

Simons, E. L. 1961. The phyletic position of *Ramapithecus*. *Postilla* (Peabody Mus. Nat. Hist., Yale Univ.) **57**:1–9.

Simons, E. L. 1972. *Primate Evolution*, Macmillan, New York. 322 pp.

Simons, E. L. 1976. Relationship between *Dryopithecus, Sivapithecus* and *Ramapithecus* and their bearing on hominid origins, in: *Les Plus Anciens Hominides* (Colloque VI, IX Union Internationale des Sciences Prehistorique et Protohistorique, Nice) (P. V. Tobias and Y. Coppens, eds.), pp. 60–67, CNRS, Paris.

Simons, E. L., and Chopra, S. R. K. 1969. *Gigantopithecus* (Pongidae, Hominoidea) a new species from North India. *Postilla* (Peabody Mus. Nat. Hist., Yale Univ.) **138**:1–18.

Simons, E. L., and Pilbeam, D. R. 1965. Preliminary revision of the Dryopithecinae (Pongidae, Anthropoidea). *Folia Primatol.* **3**:81–152.

Simons, E. L., and Pilbeam, D. R. 1972. Hominoid paleoprimatology, in: *The Functional and Evolutionary Biology of Primates* (R. Tuttle, ed.), pp. 36–62, Aldine Atherton, Chicago.

Smith, R. J., and Pilbeam, D.R. 1980. Evolution of the orang-utan. *Nature (Lond.)* **284**:447–448.

Socha, W. W., and Moor-Jankowski, J. 1979. Blood groups of Anthropoid apes and their relationship to human blood groups. *J. Hum. Evol.* **8**:453–465.

Stern, J. T., Jr., and Susman, R. L. 1981. Electromyography of the gluteal muscles in *Hylobates, Pongo* and *Pan:* Implications for the evolution of hominid bipedality. *Am. J. Phys. Anthropol.* **55**:153–166.

Susman, R. 1974. Facultative terrestrial hand posture in an orangutan (*Pongo pygmaeus*) and pongid evolution. *Am. J. Phys. Anthropol.* **40**:27–37.

Szalay, F. S., and Delson, E. 1979. *Evolutionary History of the Primates*, Academic, New York. 580 pp.

Tobias, P. V. 1971. *The Brain in Hominid Evolution*, Columbia University Press, New York.

Tuttle, R. H. 1967. Knuckle-walking and the evolution of hominoid hands. *Am. J. Phys. Anthropol.* **26**:171–206.

Tuttle, R. H. 1969. Knuckle-walking and the problem of human origins. *Science* **166**:953–961.

Tuttle, R. H. 1970. Postural, propulsive, and prehensile capabilities in the cheirida of chimpanzees and other Great Apes, in: *The Chimpanzee*, Volume 2 (G. H. Bourne, ed.), pp. 167–253, Karger, Basel.

Tuttle, R. H. 1972a. Knuckle-walking hand postures in an orangutan (*Pongo pygmaeus*). *Nature (Lond.)* **236**:33–34.

Tuttle, R. 1972b. Functional and evolutionary biology of hylobatid hands and feet. *Gibbon and Siamang* **1**:136–206.

Tuttle, R. H. 1974. Darwin's apes, dental apes, and the descent of man: Normal science in evolutionary anthropology. *Curr. Anthropol.* **15**:389–398.

Tuttle, R. H. 1975. Parallelism, brachiation and hominoid phylogeny, in: *Phylogeny of the Primates* (W. P. Luckett and F. S. Szalay, eds.), pp. 447–480, Plenum, New York.

Tuttle, R. H. 1981. Evolution of hominid bipedalism and prehensile capabilities. *Phil. Trans. R. Soc. Lond. B* **292**:89–94.

Tuttle, R. H., and Beck, B. B. 1972. Knuckle-walking hand postures in an orangutan (*Pongo pygmaeus*). *Nature (Lond.)* **236**:33–34.

Tuttle, R. H., and Rogers, C. M. 1966. Genetic and selective factors in reduction of the hallux in *Pongo pygmaeus. Am. J. Phys. Anthropol.* **24:**191–198.

Tuttle, R. H., Basmajian, J. V., and Ishida, H. 1975. Electromyography of the gluteus maximus muscle in *Gorilla* and the evolution of hominid bipedalism, in: *Primate Functional and Evolutionary Biology* (R. H. Tuttle, ed.), pp. 251–269, Mouton, The Hague.

Van Horn, R. N. 1972. Structural adaptations to climbing in the gibbon hand. *Am. Anthropol.* **74:**326–333.

Von Bartheld, F., Erdbrink, D. P., and Krommenhoek, W. 1970. A fossil incisor from Uganda and a method for its determination. *Proc. K. Ned. Akad. Wet. Amsterdam B* **73:**426–431.

Von Koenigswald, G. H. R. 1973. The position of *Proconsul* among the Pongidae, in: *Symposia of the 4th International Congress of Primatology,* Vol. 3, *Craniofacial Biology of Primates,* pp. 148–153 Karger, Basel.

Von Koenigswald, G. H. R. 1981. A possible ancestral form of *Gigantopithecus* (Mammalia, Hominoidea) from the Chinji layers of Pakistan. *J. Hum. Evol.* **10:**511–515.

Von Koenigswald, G. H. R. 1983. The significance of hitherto undescribed Miocene hominoids from the Siwaliks of Pakistan in the Senckenberg Museum, Frankfurt, in: *New Interpretations of Ape and Human Ancestry* (R. L. Ciochon and R. S. Corruccini, eds.), pp. 517–526, Plenum, New York.

Walker, A. C., and Pickford, M. 1983. New postcranial fossils of *Proconsul africanus* and *Proconsul nyanzae,* in: *New Interpretations of Ape and Human Ancestry* (R. L. Ciochon and R. S. Corruccini, eds.), pp. 325–351, Plenum, New York.

Ward, S. C., and Pilbeam, D. R. 1983. Maxillofacial morphology of Miocene hominoids from Africa and Indo-Pakistan, in: *New Interpretations of Ape and Human Ancestry* (R. L. Ciochon and R. S. Corruccini, eds.), pp. 211–238, Plenum, New York.

Ward, S. C., Kimbel, W. H., and Pilbeam, D. 1983. Subnasal alveolar morphology and the systematic position of *Sivapithecus. Am. J. Phys. Anthropol.* **61**(1).

Washburn, S. L. 1963a. Behavior and human evolution, in: *Classification and Human Evolution* (S. L. Washburn, ed.), pp. 190–203, Aldine, Chicago.

Washburn, S. L. (ed.). 1963b. *Classification and Human Evolution,* Aldine, Chicago, 371 pp.

Washburn. S. L. 1968a. Speculation on the problem of man's coming to the ground, in: *Changing Perspectives on Man* (B. Rothblatt, ed.), pp. 191–206, University of Chicago Press, Chicago.

Washburn, S. L. 1968b. *The Study of Human Evolution.* Condon Lecture Series, Oregon State System of Higher Education, Eugene.

Washburn, S. L., and Ciochon, R. L. 1974. Canine teeth: Notes on controversies in the study of human evolution. *Am. Anthropol.* **76:**765–784.

White, T. D., Johanson, D. C., and Kimbel, W. H. 1983. *Australopithecus africanus:* Its phyletic position reconsidered, in: *New Interpretations of Ape and Human Ancestry* (R. L. Ciochon and R. S. Corruccini, eds.), pp. 721–780, Plenum, New York.

Wolpoff, M. H. 1971. Interstitial wear. *Am. J. Phys. Anthropol.* **34:**205–228.

Wolpoff, M. H. 1983. *Ramapithecus* and human origins: An anthropologist's perspective of changing interpretations, in: *New Interpretations of Ape and Human Ancestry* (R. L. Ciochon and R. S. Corruccini, eds.), pp. 651–676, Plenum, New York.

Wood, A. E. 1980. The origin of the caviomorph rodents from a source in Middle America: A clue to the area of origin of the platyrrhine primates, in: *Evolutionary Biology of the New World Monkeys and Continental Drift* (R. L. Ciochon and A. B. Chiarelli, eds.), pp. 79–91, Plenum, New York.

Yunis, J. J., and Prakash, O. 1982. The origin of man: A chromosomal pictorial legacy. *Science* **215:**1525–1529.

Zihlman, A. L., and Lowenstein, J. M. 1983. *Ramapithecus* and *Pan paniscus:* Significance for human origins, in: *New Interpretations of Ape and Human Ancestry* (R. L. Ciochon and R. S. Corruccini, eds.), pp. 677–694, Plenum, New York.

Author Index

Page numbers in italics refer to citations in the references or, if followed by an *f* or *t*, indicate information contained in figures or tables, respectively, whereas page numbers in roman type refer to citations in the text.

Harper, C.W., 110, *113*

Harrington, G.N., 471, *510* (*see also* Ross, I.C. *et al.*, 1976)

Harris, J., 437, *438*

Harris, J.M., 713, *719*, 723, 726, 730, 734, *776*, *780*

Harris, L.D., 471, *510*

Harrison, C.J.O., 477, *510*

Harrison, G.A., 102, *113*

Harrison, T., 5, *17*, 290t, 407, 412, 414, *416*, 434, 436, *438* (*see also* Andrews, P.J. *et al.*, 1981*a*)

Hartenberger, J.-L., *see* Heintz, E. *et al.*, 1978

Hatley, T., 459, 460, *462*

Hay, R., 428, *438*, 710, *719*, 728, 777 (*see also* Leakey, M.D. *et al.*, 1976)

Hazael-Massieux, P., *see* Boué, J. *et al.*, 1975

Heck, H., 684, *693*

Heintz, E., 41, 62

Heintz, E. *et al.*, 1978, 52, *62*

Heiple, K.G., *see* Johanson, D.C., *et al.*, 1976; Lovejoy, C.O. *et al.*, 1972, 1973

Heissig, K., *see* Sickenberg, O. *et al.*, 1975

Hellman, M., 536, *537*, 540, 555, 645, *646*, 662, *672*, 696, 700, *703*, 706, *718*, 723, *776* (*see also* Brown, B. *et al.*, 1924; Gregory, W.K. *et al.*, 1938)

Helmcke, J.G., 256, 257, *296*

Henderson, A.M., *825t*, *838*

Hendey, Q.B., 734, *776*

Hennig, W., 103, *113*, 152, *175*, 263, 292, *296*, 627, *647*

Herring, S.W., 451, 452, 458, *462*

Hewett-Emmett, D., *see* Tashian, R.E. *et al.*, 1980

Hiiemae, K.M., 261, 262, *295*, *296*

Hill, A., 214

Hill, A.P., *see* Bishop, W.W. *et al.*, 1971

Hill, R., *see* Buckton, K.E. *et al.*, 1976

Hill, W.C.O., 164, 166, *175*, 200, *209*, 542, *555*

Hillhouse, J.W. *et al.*, 1977, 733, *776*

Hofmann, R.R., 502, *510*

Holloway, R.L., 655, 667, *672*, *673*, 745

Holmes, A., 467, *510*

Holmquist, R., 102, *113*, 661, *673*

Hooijer, D.A., 545, 552, *555*, 567, *573*, 708, *719*

Hooker, P., *see* Shipman, P.L. *et al.*, 1981

Hopwood, A.T., 422, *438*, 645, *647*

Horn, A.D., 684, *692*

Houtz, R.E., *see* Kennett, J.P. *et al.*, 1974

Howell, F.C., 50, 51, 54, *62*, 126, 130, *149*, 189, *209*, 710, 713, 714, *718*, *719*, 724, 726, 733, 735, 736, 749, *774*, *776* (*see also*

Howell, F.C. (*cont.*)
Brown, F.H. *et al.*, 1978; Shuey, R.T. *et al.*, 1978)

Howell, F.C. *et al.*, 1978, 13, *17*

Howells, W.W., 710, *719*

Hoyer, B.H. *et al.*, 1972, 124, 142, *149*

Hrdlička, A., 232, *237*, 370, *401*, 654, *673*

Hsu, Chun-hua *et al.*, 1975, 662, *673*

Hsü, K.J., 819, *840*

Hsü, K.J. *et al.*, 1973, 29, 62

Hsü, K.J. *et al.*, 1977, 29, 63

Hsü, K.J. *et al.*, 1978, 819, *840*

Hsu, T.C., 90, 92, *112*, *113*

Huag, J., 587, *622*

Hughes, A.R., 722, *776*, *779*

Hull, D.L., 155, *175*

Hünermann, K.A., 499, *510* (*see also* Sickenberg, O. *et al.*, 1975)

Hürzeler, J., 52, *63*, 634, 637, *646*, *647*

Hutchinson, J., 460, *462*, 473, *510*

Huxley, T.H., 115, 134, *149*, 151, 173, 316, *323*, 578, 616, *622*, 660, *673*, 683, *692*

Hylander, W.L., 261, 262, *296*, 451, 459, *462*, 503, *511*, 588, *622*, 656, *673*, 750, 768, *776*

Imai, H.T., 162, *175*

Institut des Parcs Nationaux du Congo Belge, *510*

International Congress on Mediterranean Neogene Stratigraphy, 31

International Symposiums on Tooth Enamel, 282

Ishida, H., *see* Tuttle, R.H. *et al.*, 1975

Izawa, K., 481, 482, *511*

Jackes, M.K., *see* Leakey, M.D. *et al.*, 1976

Jackson, J.K., 471, *511*

Jacobs, L.L., 58, *63*, 525, *526*, 661, *673*, 717, *719* (*see also* Pilbeam, D. *et al.*, 1977*a*)

Jacobs, P.A., 103, *112* (*see also* Buckton, K.E. *et al.*, 1976)

Jaeger, J.-J., 41, 49, *63* (*see also* Ameur, R. *et al.*, 1979; Thomas, H. *et al.*, 1981)

Jaeger, J.-J. *et al.*, 1977, 49, 51, *63*

Jarman, P.J., 452, 453, 455, *462*

Jenkins, D.G., *see* Kennett, J.P. *et al.*, 1974

Jenkins, F.A., Jr., 166, *175*, 308, 313, 319, *323*, 354, 357, 358, 359, 360, 361, 363, 364, 365, *366*, 407, 410, 411, *417*, *827t*, *830t*, *832t*, *840*

Jevons, W.S., 151, *175*

Jia, Lan-Po, 7, *17*, 369, *401*

Johanson, D.C., 3, 12, 13, *17*, 67, 73, *86*, 126, 130, 132, *149*, 170, *175*, 211, 215, *237*,

Taxonomic Index

Page numbers in italics followed by an *f* or *t* indicate taxa contained in figures or tables respectively whereas page numbers in roman type refer to taxa in the text.

Specimen Index

Page numbers in italics followed by an *f* or *t* indicate specimens occurring in figures or tables respectively whereas page numbers in roman type refer to specimens in the text. Key to specimen abbreviations appears in frontpiece, pp. xxiii–xxiv.

Subject Index

Page numbers in italics followed by an *f* or *t* indicate information occurring in figures or tables respectively, whereas page numbers in roman type refer to information in the text.